Lecture Notes in Computer Science 16413

The series Lecture Notes in Computer Science (LNCS), including its subseries Lecture Notes in Artificial Intelligence (LNAI) and Lecture Notes in Bioinformatics (LNBI), has established itself as a medium for the publication of new developments in computer science and information technology research, teaching, and education.

LNCS enjoys close cooperation with the computer science R & D community, the series counts many renowned academics among its volume editors and paper authors, and collaborates with prestigious societies. Its mission is to serve this international community by providing an invaluable service, mainly focused on the publication of conference and workshop proceedings and postproceedings. LNCS commenced publication in 1973.

Jakub Lokoč · Ladislav Peška · Jan Zahálka ·
Stevan Rudinac · Marc Kastner · Jingjing Chen ·
Min-Chun Hu · Jiaxin Wu · Ujjwal Sharma
Editors

MultiMedia Modeling

32nd International Conference on Multimedia Modeling, MMM 2026
Prague, Czech Republic, January 29–31, 2026
Proceedings, Part II

Springer

Editors
Jakub Lokoč
Charles University
Prague, Czech Republic

Ladislav Peška
Charles University
Prague, Czech Republic

Jan Zahálka
Czech Technical University in Prague
Prague, Czech Republic

Stevan Rudinac
University of Amsterdam
Amsterdam, The Netherlands

Marc Kastner
Hiroshima City University
Hiroshima, Japan

Jingjing Chen
Fudan University
Shanghai, China

Min-Chun Hu
National Tsing Hua University
Hsinchu, Taiwan

Jiaxin Wu
Shenzhen University
Shenzhen, China

Ujjwal Sharma
University of Amsterdam
Amsterdam, The Netherlands

ISSN 0302-9743 ISSN 1611-3349 (electronic)
Lecture Notes in Computer Science
ISBN 978-981-95-6956-4 ISBN 978-981-95-6957-1 (eBook)
https://doi.org/10.1007/978-981-95-6957-1

This Springer imprint is published by the registered company Springer Nature Singapore Pte Ltd.
The registered company address is: 152 Beach Road, #21-01/04 Gateway East, Singapore 189721, Singapore

Preface

These four proceedings volumes contain the papers presented at the 32nd International Conference on Multimedia Modeling (MMM 2026), held in Prague, Czech Republic, from January 29 to 31, 2026. Traditionally, MMM brings together experts from various fields related to multimedia modeling and machine learning models dealing with multimedia data. The conference venue in Prague provided a unique combination of medieval experience with modern city attractions. All these aspects offered great opportunities for researchers and industry practitioners to meet, present new ideas, share original research results, and discuss practical experiences. As in previous years, the topics included exciting challenges in multimedia content analysis, multimedia signal processing and communications, multimedia applications and services, as well as the ethical, legal, and societal aspects of multimedia. Part of the conference consisted of special sessions focusing on specific cutting-edge areas. MMM also featured a practical demonstration session that enabled interactive experiences with scientific applications. The Video Browser Showdown (VBS) competition was another unique aspect of MMM, a live demonstration event where teams competed in various challenging video retrieval tasks using advanced video search tools. The program further included three inspiring keynote talks by Jiri Matas from the Czech Technical University, Marcel Worring from the University of Amsterdam, and Giuseppe Amato from CNR-ISTI.

This year, MMM 2026 received a total of 251 valid regular paper submissions. Of these, 137 high-quality papers were accepted following a rigorous double-blind peer-review process. Each submission was evaluated by at least three program committee members or reviewers. In addition, 8 demonstration papers and 12 VBS systems were selected for presentation at the conference. The program also included 11 special session papers, each addressing one of the following topics:

- Human-Centric Multimodal Behavior Analysis
- Modelling Robustness and Security for Multimedia AI (MARS)
- Multi-Object Multi-Sensor Tracking (MOMST)

Behind the success of a conference is always a huge amount of work by the people who participated in all phases of the organization. Therefore, we would like to express our sincere thanks to the members of the organizing committee, special session, demonstration and VBS organizers, the steering and technical program committee members, reviewers, keynote speakers, authors, and Springer for making MMM 2026 a great event.

December 2025

Jakub Lokoč
Ladislav Peška
Jan Zahálka
Stevan Rudinac
Marc Kastner
Jingjing Chen
Min-Chun Hu
Jiaxin Wu
Ujjwal Sharma

Organization

General Chairs

Jakub Lokoč	Charles University, Czech Republic
Ladislav Peška	Charles University, Czech Republic

Program Committee Chairs

Jingjing Chen	Fudan University, China
Min-Chun Hu	National Tsing Hua University, Taiwan
Marc Kastner	Hiroshima City University, Japan
Stevan Rudinac	University of Amsterdam, The Netherlands
Jan Zahálka	Czech Technical University in Prague, Czech Republic

Special Session Chairs

Rahel Arnold	University of Basel, Switzerland
Werner Bailer	Joanneum Research, Austria
Luca Rossetto	Dublin City University, Ireland

Demonstration Chairs

Kai Uwe Barthel	HTW Berlin, Germany
Lucia Vadicamo	CNR-ISTI, Italy

Video Browser Showdown Chairs

Werner Bailer	Joanneum Research, Austria
Cathal Gurrin	Dublin City University, Ireland
Jakub Lokoč	Charles University, Czech Republic
Klaus Schöffmann	Klagenfurt University, Austria

Publicity Chairs

Cathal Gurrin	Dublin City University, Ireland
Richang Hong	Hefei University of Technology, China

Proceedings Chairs

Ujjwal Sharma	University of Amsterdam, The Netherlands
Jiaxin Wu	Shenzhen University, China

Steering Committee

Kiyoharu Aizawa	University of Tokyo, Japan
Phoebe Chen	La Trobe University, Australia
Wen-Huang Cheng	National Taiwan University, Taiwan
Tat-Seng Chua	National University of Singapore, Singapore
Peng Cui	Tsinghua University, China
Cathal Gurrin	Dublin City University, Ireland
Richang Hong	Hefei University of Technology, China
Benoit Huet	Median Technologies Inc., France
Björn Þór Jónsson	Reykjavik University, Iceland
Guo-Jun Qi	University of Central Florida, USA
Klaus Schöffmann	Klagenfurt University, Austria

Special Session Organizers

Human-Centric Multimodal Behavior Analysis

Yu-Gang Jiang	Fudan University, China
Hong Liu	Xiamen University, China
Wenjing Li	Hefei University of Technology, China
Nan Pu	University of Trento, Italy
Zhun Zhong	Hefei University of Technology, China

Modelling Robustness and Security for Multimedia AI (MARS)

Nandor Knust	UIT Arctic University of Norway, Norway
Michael A. Riegler	Simula Research Laboratory, Norway

Multi-object Multi-sensor Tracking (MOMST)

Mario Döller	University of Applied Sciences Kufstein, Austria
Ruben Tous	Universitat Politècnica de Catalunya, Spain

Program Committee

Avinash Anand	Singapore Institute of Technology, Singapore
Rahel Arnold	University of Basel, Switzerland
Qian Cao	Renmin University of China, China
Wei-Ta Chu	National Cheng Kung University, Taiwan
Keisuke Doman	Chukyo University, Japan
Zhicheng Du	Tsinghua University, China
Peter Eisert	Humboldt Universität zu Berlin, Germany
Alex Falcon	University of Udine, Italy
Gylfi Þór Guðmundsson	Reykjavik University, Iceland
Jaime Roblero	Dublin City University, Ireland
Shintami Hidayati	Institut Teknologi Sepuluh Nopember, Indonesia
Jian Hou	Dongguan University of Technology, China
Bastian Jäckl	University of Konstanz, Germany
Naye Ji	Communication University of Zhejiang, China
Vijay John	Lawrence Technological University, USA
Björn Þór Jónsson	Reykjavik University, Iceland
Klaus Jung	HTW Berlin, Germany
Yasutomo Kawanishi	RIKEN, Japan
Chutisant Kerdvibulvech	National Institute of Development Administration, Thailand
Tran Khoa	National Institute of Information and Communications Technology, Japan
Takahiro Komamizu	Nagoya University, Japan
Yongqiang Kong	Beijing Polytechnic College, China
Satoshi Kosugi	Tokyo Institute of Technology, Japan
Haruya Kyutoku	Aichi University of Technology, Japan
Bao Le Hoang	Dublin City University, Ireland
Mario Leopold	Klagenfurt University, Austria
Feng Li	Anhui University of Finance and Economics, China
Chih-Wei Lin	Fujian Agriculture and Forestry University, China
Yu Liu	Tongji University, China
Vasileios Mezaris	Centre for Research and Technology Hellas, Greece

Atsuyuki Miyai	University of Tokyo, Japan
Hisashi Miyamori	Kyoto Sangyo University, Japan
Eisei Nakahara	Nihon University, Japan
Takayuki Nakatsuka	National Institute of Advanced Industrial Science and Technology, Japan
Thu Nguyen	SimulaMet, Norway
Thang-Long Nguyen-Ho	Dublin City University, Ireland
Naoko Nitta	Kansai University, Japan
Tse-Yu Pan	National Taiwan University of Science and Technology, Taiwan
Nick Pantelidis	CERTH, Greece
Maria Pegia	Centre for Research & Technology Hellas, Greece
Halvorsen Pål	SimulaMet, Norway
Na Qi	Beijing University of Technology, China
Joseph Robinson	Meta, USA
Albert Ali Salah	Utrecht University, Netherlands
Shin'ichi Satoh	National Institute of Informatics, Japan
Konstantin Schall	vviinn, Germany
Udo Schlegel	LMU München, Germany
Francesca Scotti	CNR-ISTI, Italy
Palaiahnakote Shivakumara	University of Salford, UK
Thitirat Siriborvornratanakul	National Institute of Development Administration, Thailand
Ivana Sixtova	Charles University, Czech Republic
Shih-Wei Sun	Taipei National University of the Arts, Taiwan
Cheng-Kang Tan	National Cheng Kung University, Taiwan
Marion Taschwer	Klagenfurt University, Austria
Masahiro Toyoura	University of Yamanashi, Japan
Quang-Linh Tran	Dublin City University, Ireland
Kosetsu Tsukuda	National Institute of Advanced Industrial Science and Technology, Japan
Huy Quang Ung	KDDI Research, Inc., Japan
Stefanos Vrochidis	Centre for Research and Technology Hellas, Greece
Haodian Wang	CHN Energy Digital Intelligence Technology Development (Beijing) Co., Ltd., China
Ling Xiao	Hokkaido University, Japan
Yoko Yamakata	University of Tokyo, Japan
Toshihiko Yamasaki	University of Tokyo, Japan
Yuan Zeng	Shenzhen Technology University, China
Fan Zhang	Communication University of Zhejiang, China
Pengcheng Zhao	Nanjing Audit University, China

Additional Reviewers

Ali Abdari
Abir Baâzaoui
Yannan Bai
Werner Bailer
Amit Banerjee
Walid Barhoumi
Tianyu Cai
Changxin Chen
Feiyu Chen
Hongyou Chen
Lucy Chen
Zhiting Chen
Ziyu Chen
Doru Constantin
Yan Deng
Zhixuan Deng
Thomas Derrien
Haizhou Du
Qing Duan
Shijie Fan
Ziyang Fan
Xiaojing Feng
Cheng Fu
Tian Gao
Ziyuan Gao
Hajer Ghodhbani
Chao Gong
Timothy Greer
Yuanyuan Guan
Lili Guo
YuNong Guo
Hongze Han
Shijie Hao
Xu Hao
Haoran Hu
Yuka Hashizume
Steven Hicks
Luong Ho
Jiayi Hu
Min-Chun Hu
Weiyi Hu
Yizheng Hu
Yunteng Hu
Yunming Hui
Adam Jakobsen
Caiyan Jia
Heng Jia
Meng kun Jia
Dongming Jiang
Xuerong Jiang
Pengkun Jiao
Xin Jin
Zhe Jin
Marc A. Kastner
Seon Ho Kim
Onanong Kongmeesub
Haishan Lai
Hoang Bao Le
Lehan Zhang
Hai Li
Hao Li
Hongyang Li
Jiafeng Li
Jiashuo Li
Jiayi Li
Junhao Li
Leixin Li
Li Li
Qingfeng Li
Qingguan Li
Yiqian Li
Yizhi Li
Yuxuan Li
Fang Liang
Lanbin Liang
Peirou Liang
Yun Liao
Hailan Lin
Yuzhen Lin
Anqi Liu
Chengji Liu
Dongqin Liu
Guoshan Liu
Guozhen Liu
Jing Liu
Na Liu

Nana Liu
Wenlong Liu
Yanfei Liu
Yuming Liu
Jakub Lokoč
Nengbo Lu
Weihai Lu
Na Lyu
Zhiyong Ma
Zhiyang Mai
Fuyou Mao
Iván Martín-Fernández
Lyu Meiyi
Xiangwu Meng
Zixiang Meng
Ryota Mibayashi
Nicolas Müller
Phuong-Anh Nguyen
Thanh Binh Nguyen
Thu Nguyen
Trong-Thuan Nguyen
Trung Thanh Nguyen
Wenlong Niu
Stefanie Onsori-Wechtitsc
Yangjun Ou
Hongting Pan
Bo Peng
Ladislav Peška
Itthisak Phueaksri
Wenzhi Ping
Mengyu Qiao
Wanxiang Qin
Weibin Qiu
Ruichao Ren
Xuena Ren
Loris Sauter
Ujjwal Sharma
Yutong Shen
Qifeng Shi
Tong Shi
Mantek Singh
Yingying Song
Florian Spiess
Dimitrios Stefanopoulos
Jianmei Su
Zijin Su
Hao Sun
Zhicong Sun
Junliang Tan
Zhenjun Tang
Adane Tarekegn
Shangzhi Teng
Vajira Thambawita
Hao Tian
Allie Tran
Quoc Khanh Tran
Teng Tu
Kazuya Ueki
Haihui Wan
Chaoyue Wang
Dongsheng Wang
Haoyang Wang
Jiwei Wang
Lin Wang
Quan Wang
Shibin Wang
Shuai Wang
Yongxiang Wang
Yue Wang
Zhangyi Wang
Wentong Wang
Kento Watanabe
Jobin Idiculla Wattasseril
Guangcun Wei
Lijing Wei
Xinyue Wei
Jie Wen
Junchao Wen
Bo Wu
Chunzhuang Wu
Jiaxin Wu
Xinlan Wu
Wen Xie
Zhihua Xie
Zhonghao Xie
Jiahang Xin
Jian Xing
Feifei Xu
Hang Xu
Man Xu

Xu Yan
Gang Yang
Jiale Yang
Jiarui Yang
Lili Yang
Tianchi Yang
Xuan Yang
Yuanzhe Yang
Zhenguo Yang
Yuheng Yao
Haiyang Ye
Xiong Yi
JiaYu Yu
Qiqun Yu
Ming Yuan
Xin Yuan
Jan Zahálka
Long Zeng
Hanling Zhan
Congxuan Zhang
Lehan Zhang
Tianyun Zhang
Wenli Zhang
Xuan Zhang
Youxuan Zhang
Yuanping Zhang
Baoquan Zhao
Faqi Zhao
Liurui Zhao
Ruixiang Zhao
Yawei Zhao
Dequan Zheng
Cankun Zhong
Aoxiang Zhou
Shuaishuai Zhou
Xinying Zhou
Yan Zhou
Zhangfeng Zhou
Ziyu Zhou
Fengbin Zhu
Hongyi Zhu
Jianggang Zhu

Contents

A Comparative Study of Decoding Strategies in Medical Text Generation

Oriana Presacan[1(✉)], Alireza Nik[2,3], Vajira Thambawita[2], Bogdan Ionescu[1], and Michael Riegler[4]

[1] AI Multimedia Lab, CAMPUS Research Institute, National University of Science and Technology Politehnica Bucharest, 060042 Bucharest, Romania
oriana.presacan@stud.etti.upb.ro
[2] Department of Holistic Systems, SimulaMet, 0170 Oslo, Norway
[3] Oslo Metropolitan University, 0176 Oslo, Norway
[4] Cyber Security, Simula Research Laboratory, 0164 Oslo, Norway

Abstract. Large Language Models (LLMs) rely on various decoding strategies to generate text, and these choices can significantly affect output quality. In healthcare, where accuracy is critical, the impact of decoding strategies remains underexplored. We investigate this effect in five open-ended medical tasks, including translation, summarization, question answering, dialogue, and image captioning, evaluating 11 decoding strategies with medically specialized and general-purpose LLMs of different sizes. Our results show that deterministic strategies generally outperform stochastic ones: beam search achieves the highest scores, while η and top-k sampling perform worst. Slower decoding methods tend to yield better quality. Larger models achieve higher scores overall but have longer inference times and are no more robust to decoding. Surprisingly, while medical LLMs outperform general ones in two of the five tasks, statistical analysis shows no overall performance advantage and reveals greater sensitivity to decoding choice. We further compare multiple evaluation metrics and find that correlations vary by task, with MAUVE showing weak agreement with BERTScore and ROUGE, as well as greater sensitivity to the decoding strategy. These results highlight the need for careful selection of decoding methods in medical applications, as their influence can sometimes exceed that of model choice.

Keywords: decoding strategies · LLM · medical benchmarks

1 Introduction

In recent years, we have witnessed a growing interest in adopting generative Artificial Intelligence (AI), including Large Language Models (LLMs), in healthcare applications [9,16,22]. Medical tasks ranging from decision making and question answering to translation can now be carried out using state-of-the-art models, such as OpenAI's GPT-4 [17] and Google's Gemini [8]. These models have shown impressive potential in the analysis, interpretation, and generation

J. Lokoč et al. (Eds.): MMM 2026, LNCS 16413, pp. 1–15, 2026.
https://doi.org/10.1007/978-981-95-6957-1_1

of high-quality responses in medical exams [16,20]. Considering the sensitive nature of the medical domain, ensuring that these AI models generate medically accurate responses in real-world situations is a key challenge.

An important component affecting the quality, diversity, and fluency of the generated LLM responses is the decoding strategy used during inference. LLMs generate text in an autoregressive manner, sequentially predicting the next token conditioned on all previously generated tokens in the sequence [23]. At each step, the model computes a probability distribution over the vocabulary and selects the next token accordingly. The simplest approach, greedy decoding [26], always chooses the token with the highest probability. While this deterministic method can produce coherent and consistent text, it often leads to "flat" and repetitive outputs, limiting diversity. In contrast, stochastic decoding methods, such as top-k [6], top-p [11], etc., introduce controlled randomness, which can yield more dynamic, engaging, and varied responses. Decoding strategies, therefore, involve a trade-off between the familiar and the unexpected, balancing predictability with creativity [29]. In domains such as healthcare, this balance is particularly important: some deterministic methods may ensure precision but risk generating generic or overly rigid content, whereas overly stochastic methods may enhance contextual richness but at the cost of factual reliability. Selecting an appropriate decoding strategy is thus a trade-off between generating medically precise, consistent information and producing diverse, contextually rich text, which are essential for tasks like medical dialogue and summarization [3,27].

The current literature shows that decoding techniques in general text generation are rapidly evolving [25]. Research has progressed from deterministic [7,24] and stochastic methods [6,10,11] to more advanced strategies aimed at improving the factuality and coherence of the output text [4,21], or even reducing decoding latency [13]. Although the effect of decoding methods has recently been studied on general open-ended benchmarks [2,19], their impact remains under-explored in the medical domain. As new medical benchmarks are introduced [28], most research has focused on model architecture, overlooking decoding strategies despite their influence on generative quality. Furthermore, the role of decoding methods in medical-domain LLMs, such as models fine-tuned or pre-trained for clinical or biomedical tasks, has received even less attention, leaving a gap in understanding how these techniques affect performance in specialized models.

The main contribution of this research is to investigate the impact of various decoding strategies on LLM performance in the medical domain, including:

- **Comprehensive analysis of various decoding strategies in healthcare:** we systematically study how different decoding strategies influence the generative performance of general-purpose and medical LLMs on medical datasets.
- **Task-specific evaluation of decoding methods:** we investigate which decoding strategies are most effective and efficient for different types of medical tasks.
- **Model performance and sensitivity to decoding strategies:** we analyze the performance and sensitivity of various LLMs, ranging from small to large, general-purpose to medical models, to changes in decoding configurations.

- **Evaluation metrics sensitivity and agreement:** we assess the stability and sensitivity of the evaluation metrics (BLEU, ROUGE, BERTScore, and MAUVE) in different tasks, and their agreement with each other.
- **Computational efficiency analysis:** We report the inference time across decoding strategies and models, offering practical insights into their runtime trade-offs, which are beneficial for resource-constrained healthcare applications.

The remainder of the paper is organized as follows: Sect. 2 provides background information on the decoding strategies evaluated in this study. Section 3 describes the implementation details, including the logic behind LLM and metric selection, hyperparameters, benchmarks, and hardware. Section 4 presents the results and statistical analyses, followed by a discussion of the findings in Sect. 5. Finally, Sect. 6 concludes the paper. The code for text generation and statistical analysis, along with the complete set of result tables, is available at our GitHub repository.

2 Background

2.1 Decoding Strategies

Deterministic Strategies. ***Greedy*** [26] method is the simplest strategy as the model always chooses the token with the highest probability. However, it limits the diversity and can lead to repetitive phrasing.

Beam Search (BS) [7] method keeps the top-k most probable sequences, where k is a user-defined hyperparameter. At each decoding step, it expands the sequence by all possible next tokens, calculates their cumulative probabilities, and retains the top-k sequences. Unlike greedy decoding, BS aims to find a globally optimal sequence rather than making the best choice at each step.

Diverse Beam Search (DBS) [24] decoding strategy enhances BS by introducing more diversity among candidate outputs. It does this by dividing the beams into a user-defined number of groups and adding a diversity penalty to reduce overlap between groups.

Contrastive Search (CS) [21] is an extension of BS that aims to improve text quality by reducing repetition. It selects the next token based on both model confidence and a semantic similarity penalty. It uses two hyperparameters: k (top-k candidates) and α (penalty weight), to balance fluency and diversity.

DoLa [4] takes a different approach by using outputs from both the final layer and an earlier layer of the model. It compares these outputs, giving more weight to the final one to guide generation. This helps highlight tokens that are consistently supported across layers.

Stochastic Strategies. ***Temperature Sampling*** [1] strategy rescales the logits before applying the softmax function. A lower temperature sharpens the distribution, favoring high-probability tokens, while a higher temperature flattens it, increasing the likelihood of selecting lower-probability tokens and thereby enhancing output diversity. Temperature sampling can be combined with any of the following stochastic strategies; in our experiments, we fix $T = 1$ when evaluating the other strategies.

Top-k Sampling [6] limits the sampling process to the top-k most probable tokens. The next token is then sampled only from this restricted subset, where k is a predefined hyperparameter.

Top-p (nucleus) Sampling [11] strategy selects the smallest set of top-ranked tokens whose cumulative probability reaches a threshold p, allowing the number of candidates to vary with the model's confidence rather than being fixed as in top-k sampling.

η-Sampling [10] strategy keeps only tokens above a probability threshold that adapts based on the entropy of the token distribution. When the model is confident (low entropy), only top tokens are kept. When it's uncertain (high entropy), more tokens are included for flexibility.

Min-p Sampling [15] sampling relies solely on the most probable token, while η-sampling adapts its threshold based on the overall entropy of the distribution. It sets the cutoff relative to this top token's probability, scaling it by a hyperparameter to determine the sampling pool. By adapting directly to the model's confidence in its best prediction, min-p remains stable even at high temperatures, where other sampling methods often produce more chaotic outputs.

Typical Sampling [14] method selects tokens closest to the average entropy of the distribution. A threshold hyperparameter controls how much deviation from this average is allowed, filtering out tokens that are either too predictable or too unlikely, and focusing on those that are most typical for the given context.

3 Implementation Details

3.1 LLM Selection

We adopted a systematic approach to the selection of LLMs, focusing exclusively on open-source models and imposing a maximum size limit of 14 billion parameters due to computational constraints. The selection covered three categories: general-purpose, medical, and multimodal models. For general-purpose LLMs, we grouped candidates by model size (small, medium, and large) and selected representative models within each group. All models were evaluated consistently across tasks using greedy decoding, and for each task, we retained the best-performing model within each category and size group. Medical-specific models were selected based on their prominence in recent biomedical NLP literature and availability in open-source form. For multimodal tasks, we evaluated several vision-language models, including both general-purpose and medical-specialized variants. Figure 1 presents an overview of the selected models, grouped by type and size, as well as the final model choices per task.

(A)

General-Purpose Models

Small (1–2B):

- Qwen3-1.7B
- gemma-3-1b-it
- Llama-3.2-1B-Instruct

Medium (7–8B):

- Qwen3-8B
- Mistral-7B-Instruct-v0.3
- Llama-3.1-8B-Instruct

Large (12–14B):

- Qwen3-14B
- gemma-3-12b-it
- Llama-2-13b-chat-hf

Medical-Specific Models

- BioMistral-7B
- meditron-7B
- medalpaca-7B
- medgemma-4B-it

Multimodal Models

Medical:

- medgemma-4b-it
- llava-med-v1.5-mistral-7b

General:

- gemma-3-12b-it
- llava-1.5-7b-hf

(B)

Type	Size	Trans	Summ	QA	Dialogue	Images
General	Small	Qwen	Qwen	Llama	Qwen	–
	Medium	Qwen	Llama	Mistral	Qwen	–
	Large	Qwen	Qwen	Llama	Qwen	–
Medical	–	medalpaca	medgemma	BioMistral	medgemma	–
Multimodal	General	–	–	–	–	llava
	Medical	–	–	–	–	llava-med

Fig. 1. Panel A: All evaluated LLMs, categorized by type and size, with their Hugging Face model names. Panel B: Final selected models per task.

3.2 Hyperparameters

We explored a range of hyperparameters for each method and selected those that yielded the best performance for our final analysis. The hyperparameter values were informed by prior literature [19] and are summarized in Table 1.

Table 1. Ranges of hyperparameter values tested for the decoding strategies.

Strategies	Hyperparameter	Values
BS	beam size	{3, 5, 10}
DBS	(beam size, diversity penalty)	{(4,2), (4,4), (8,2), (8,4)}
CS	penalty α, top-k candidate pool size	{α: {0.1, 0.2, 0.3, 0.4, 0.5, 0.6}, k= 6}
Dola	dola layer	{low, high}
temperature	temperature	{0.3, 0.5, 0.7, 0.9}
top p	nucleus sampling probability threshold	{0.8, 0.85, 0.9, 0.95}
top k	candidate pool size	{5, 25, 50, 100}
min p	minimum token probability threshold	{0.05, 0.1, 0.3, 0.5}
eta	η-sampling cutoff threshold	{0.0003, 0.0006, 0.0009, 0.002, 0.004}
typical	typicality threshold	{0.2, 0.5, 0.9, 0.95}

3.3 Medical Benchmarks

We selected five different open-ended medical tasks, widely used in healthcare: translation, summarization, question answering, dialogue, and image captioning. For each, we used only 100 samples, due to the experimental nature of this study and constraints on available resources.

For the medical translation task, we used the UFAL Medical Corpus 1.0[1], which contains medical text translations in multiple languages. We focused on the German–English pairs and applied preprocessing to retain only medical-domain entries, remove duplicates, and filter out short or low-quality sentences based on length, numeric content, and symbolic artifacts. The final dataset was uniformly sampled across sources.

Pubmed-summarization dataset [5], which consists of full-text biomedical articles from PubMed paired with their corresponding abstracts, was used for the summarization task.

The medical QA dataset[2] we selected contains open-ended medical question–answer pairs covering topics like anatomy, physiology, and pharmacology. It was derived from flashcards created by students and the content was cleaned for use in training and evaluating medical QA models.

For open-ended healthcare dialogues, we used Healthbench[3], a benchmark by OpenAI to assess LLMs in a range of real-world medical tasks. The benchmark includes multi-turn dialogues between LLMs and individuals, evaluated using a rubric designed by medical professionals. However, in our study, we only considered the physicians' completions of the dialogue as the ground truth and did not apply the rubric-based evaluation. We randomly selected single-turn dialogues with the global health theme.

[1] https://ufal.mff.cuni.cz/ufal_medical_corpus.
[2] https://huggingface.co/datasets/medalpaca/medical_meadow_medical_flashcards.
[3] https://openai.com/index/healthbench/.

Lastly, for image captioning, we used ROCOv2 (Radiology Objects in Context Version 2) [18], a multimodal dataset consisting of medical images and corresponding figure captions. It was derived from the PubMed Open Access subset and includes radiology images across various anatomical regions and seven different imaging modalities.

3.4 Metric Selection

To enable consistent comparisons across tasks, we applied two common metrics to all benchmarks: ROUGE, a traditional metric, and BERTScore, a newer embedding-based metric. Additionally, we used BLEU for translation and MAUVE for dialogue and QA. This approach allowed us to evaluate individual task performance while also identifying trends across models and metrics.

3.5 Hardware and Software Setup

We conducted all experiments on a machine equipped with NVIDIA A100-PCIE-40GB GPUs and a dual 64-core AMD EPYC 7763 CPU @ 2.45GHz. Our implementation used Python 3.10.12 and PyTorch 2.6.0+cu126. Model loading and inference were handled using the Hugging Face *Transformers*[4] library, and evaluation metrics were computed using the *Evaluate*[5] library from Hugging Face.

4 Evaluation Results

In our analysis, we evaluated 11 decoding strategies across 5 medical tasks, using 4 LLMs for each benchmark (and 2 for the multimodal task). Each configuration was assessed with 2-3 evaluation metrics, in addition to inference time, resulting in a total of 718 observations (with CS results missing for MedGemma due to a technical incompatibility encountered during implementation). Given the non-normal distribution of our data, we applied non-parametric statistical methods. The following subsections present our findings, organized into three categories: decoding strategies, models, and evaluation metrics.

4.1 Decoding Strategies

Table 2 reports the results for the top-performing LLMs across benchmarks; the complete set of results is provided in our repository. The impact of decoding strategies differs by task: largest in dialogue and QA and moderate in translation, summarization, and captioning. Overall, BS had the best mean rank, followed by CS and DBS; top-p, eta, and top-k performed the worst.

To assess whether there are statistically significant differences between the decoding strategies, we used ROUGE as a consistent evaluation metric across

[4] https://huggingface.co/docs/transformers/en/index.
[5] https://huggingface.co/docs/evaluate/en/index.

Table 2. Results across decoding strategies and tasks for the best-performing LLM; bold values indicate the highest result for that model.

	Translation Qwen 14B BLEU	Summarization Llama 8B ROUGE	QA MedAlpaca 7B ROUGE	Dialogue Qwen 14B MAUVE	Captioning Llava Med ROUGE
greedy	0.4422	0.2252	0.3839	0.9141	0.1725
BS	0.4379	**0.2374**	**0.4197**	0.8870	0.1718
DBS	0.4388	0.2329	0.3578	0.8196	0.1715
CS	**0.4428**	0.2305	0.3845	**0.9379**	0.1723
DoLa	0.4285	0.2263	0.3798	0.8967	0.1632
temperature	0.4379	0.2255	0.3617	0.9322	0.1645
top p	0.4415	0.2217	0.3354	0.9309	0.1590
top k	0.4392	0.2162	0.3137	0.8776	0.1543
min p	0.4419	0.2272	0.3649	0.9252	**0.1756**
eta	0.4392	0.2181	0.2972	0.9158	0.1547
typical	0.4415	0.2223	0.3377	0.9080	0.1577

tasks. Scores were normalized within each task to account for scale differences. We then applied the *Friedman* test, a non-parametric statistical test suitable for comparing multiple treatments (decoding strategies) across repeated measures (task×model blocks). The test was significant ($\chi^2 = 44.25$, $df = 10$, $p = 2.97 \times 10^{-6}$; $n = 16$ blocks), indicating that at least one strategy performs differently from the others and that further tests are needed to identify where the differences lie. Therefore, we ran pairwise *Wilcoxon* signed-rank tests across task×model blocks. Unlike the *Friedman* test, which only detects overall differences, *Wilcoxon* enables direct comparisons between strategy pairs. We applied the *Holm* correction to control the family-wise error rate, which is important when making multiple pairwise comparisons. This analysis revealed significant differences between:

- BS > eta ($W = 7.5$, $p_{\text{Holm}} = 0.011$, $\Delta\text{rank} = 6.50$, $n = 18$);
- BS > top-p ($W = 10$, $p_{\text{Holm}} = 0.017$, $\Delta\text{rank} = 4.75$, $n = 18$);
- BS > top-k ($W = 11$, $p_{\text{Holm}} = 0.022$, $\Delta\text{rank} = 7.00$, $n = 18$);
- DBS > eta ($W = 7.5$, $p_{\text{Holm}} = 0.011$, $\Delta\text{rank} = 4.75$, $n = 18$);
- DBS > top-k ($W = 11$, $p_{\text{Holm}} = 0.022$, $\Delta\text{rank} = 5.25$, $n = 18$);

where W is the *Wilcoxon* test statistic, p_{Holm} is the *Holm*-adjusted p-value, Δ is the median rank difference between strategies, n is the number of paired samples (task×model blocks).

While the above analysis was conducted across all tasks, we now examine each task individually to determine whether certain decoding strategies are better suited for specific tasks. Based on average ranks across models, DBS performed best for translation, min p sampling for summarization, CS for image captioning, and BS for dialogue and QA. When applying the *Friedman* test per task, we found statistically significant differences between strategies only in QA and dialogue. However, paired *Wilcoxon* signed-rank tests found no statistically significant differences between individual strategy pairs, suggesting that, although some strategies appear to perform better on average, the differences are not strong enough within individual tasks.

Next, we grouped strategies as deterministic vs. stochastic and compared the groups. A paired *Wilcoxon* signed-rank test on mean differences showed a small but statistically significant advantage for deterministic decoding (median paired difference ΔROUGE $= 0.0029$, $p = 0.0003$, $n = 18$, $W = 10$, $n = 18$). At the per-task level ($n = 4$ blocks each), tests were not significant, but the medians consistently favored deterministic decoding.

Lastly, we also analyzed how decoding strategies impact inference time (measured in seconds per token) and found significant differences across methods (*Friedman test*: $\chi^2 = 125.33$, $p < 0.001$). Top-p, top-k, typical were the fastest, while BS, CS, DBS were the slowest. Additionally, we found a statistically significant positive correlation between the normalized inference time and ROUGE performance (Kendall's $\tau = 0.4909$, $p = 0.0405$), suggesting that, on average, slower decoding strategies tend to yield better output quality.

4.2 Models

We evaluated the models from three perspectives: performance, measured by the mean ROUGE score; sensitivity, quantified using the coefficient of variation ($CV = std/mean$) to assess consistency across decoding strategies; and inference time, measured as the number of seconds per generated token. Figure 2 presents bar charts showing these for all models divided by tasks.

Among general models (excluding multimodal ones), we found a strong and statistically significant positive correlation between model size and task-normalized performance, with larger models achieving higher scores (Kendall's $\tau = 0.7807, p = 0.0151$). A similar trend was found between model size and inference time: larger models decoded more slowly (Kendall's $\tau = 0.6141, p = 0.0081$). In contrast, model size did not show a significant correlation with sensitivity.

While general models occasionally outperformed medical domain-specific LLMs on individual tasks, this advantage was inconsistent, and no statistically significant difference in average performance was found between the two groups. Inference speed likewise showed no meaningful difference between the two groups. Sensitivity to decoding, however, did differ. We applied *Levene's* test, which assesses whether two or more groups have equal variances. Because our distributions are not normal, we also report the Brown–Forsythe modification, which substitutes the group median for the mean when computing deviations, making the test more robust. The results were consistent in both tests (Levene: $F_{(1,16)} = 20.10$, $p = 0.0004$; Brown–Forsythe: $F_{(1,16)} = 7.74$, $p = 0.013$; $n = 18$), showing that medical LLMs are more sensitive to the decoding strategy.

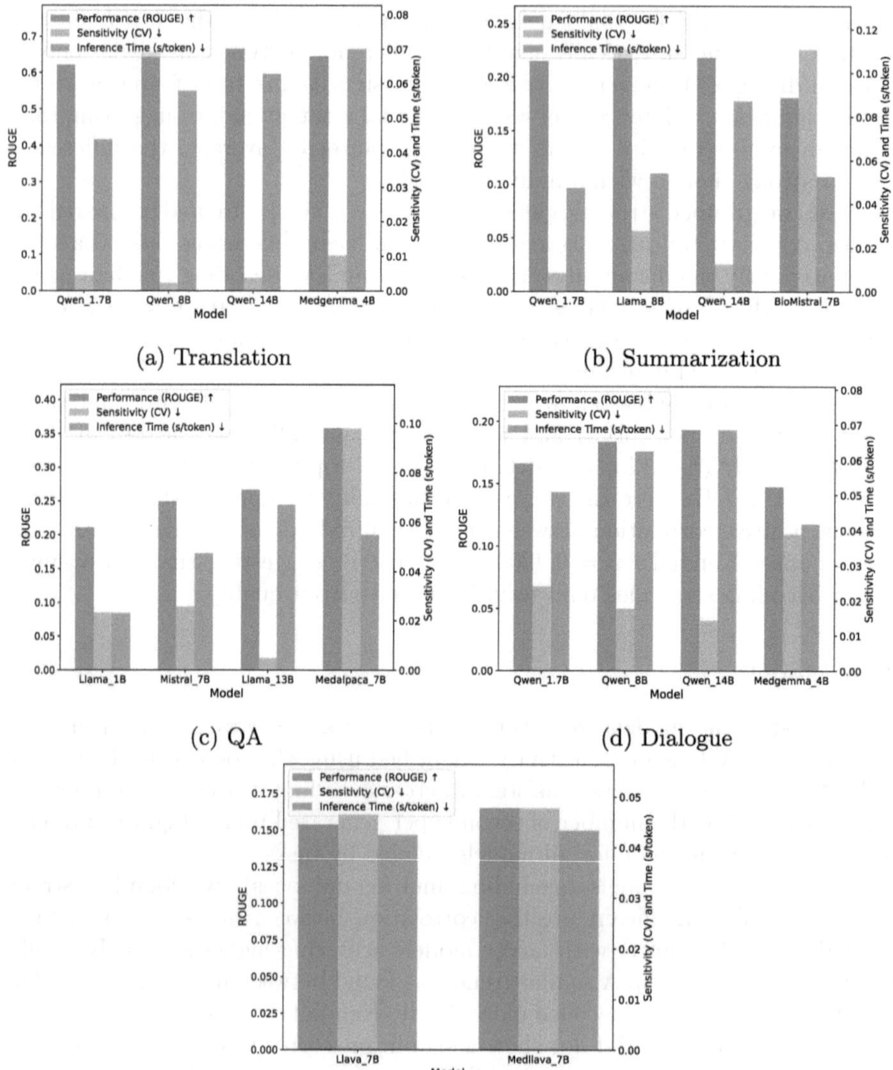

(a) Translation

(b) Summarization

(c) QA

(d) Dialogue

(e) Image Captioning

Fig. 2. Per–model performance, measured as mean ROUGE across decoding strategies, and sensitivity, measured as the coefficient of variation (CV), for the five tasks. Model names are followed by their number of parameters and are ordered from smallest to largest; the final model is the medical LLM.

4.3 Metrics

We measured rank agreement between metrics using Kendall's τ. Figure 3a shows a forest plot of ROUGE–BERTScore correlations, reported per task and overall. Error bars indicate 95% confidence intervals across task–model blocks. The agreement is highest for summarization, followed by captioning, QA, and translation, and lowest for dialogue. Figure 3b shows correlations for BLEU with ROUGE and BERTScore (translation), and MAUVE with ROUGE and BERTScore (Dialogue+QA). BLEU aligns strongly with both metrics, whereas MAUVE shows weak, sometimes negative correlation with them.

Furthermore, we quantified the metric sensitivity to decoding using the CV, computed within each task×model block across strategies. BERTScore was the most stable (mean CV ≈ 0.002), followed by BLEU (≈ 0.019) and ROUGE (≈ 0.029), whereas MAUVE was by far the most variable (≈ 0.162).

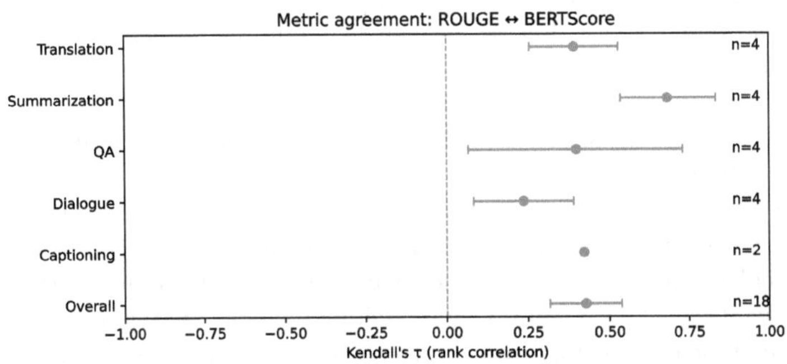

(a)

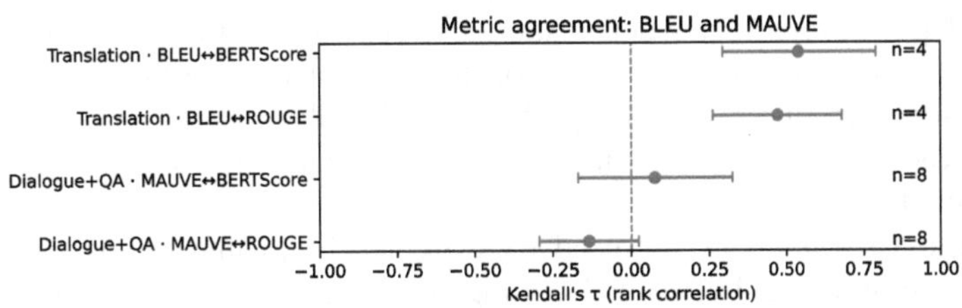

(b)

Fig. 3. Metric agreement (Kendall's τ): (a) ROUGE–BERTScore overall and by task; (b) BLEU–BERTScore and BLEU–ROUGE (translation), and MAUVE–BERTScore and MAUVE–ROUGE (dialogue+QA).

5 Discussion

Deterministic decoding strategies, especially search-based ones like BS, CS, and DBS, achieved higher ROUGE scores than stochastic methods such as top-p, top-k, and eta sampling, at the cost of increased inference time. The positive correlation between decoding latency and performance highlights a trade-off in medical applications, where precision and reliability are essential. Our results indicate that the choice of decoding strategy should be regarded as an important design decision in the deployment of LLMs for medical use, as it can affect both patient safety and the quality of clinical decision support; for instance, by producing unsafe dosage recommendations or, conversely, overly generic guidance.

Interestingly, in summarization, the min-p strategy achieved the highest performance, suggesting that its adaptive balance between coherence and diversity is particularly effective in the context of medical summarization. By dynamically adjusting the probability threshold, min-p can preserve key factual content while allowing sufficient linguistic variety to produce fluent summaries. This property can be advantageous in healthcare contexts, where summaries must be both accurate and readable for diverse audiences, including clinicians and patients.

Beyond decoding, our model analysis shows that bigger models are more performant but slower, as expected; however, they are not more robust with respect to decoding sensitivity. Medical LLMs outperformed general models in only two of the five tasks, indicating that domain-specific pre-training does not guarantee performance gains. They were also slower, although size-matched comparisons are too limited for firm conclusions. Another interesting observation is that medical models are far more sensitive than general ones to the choice of decoding strategy. This means that a medical model that performs well with greedy decoding can perform much worse if the strategy is changed, whereas general models shift only marginally. Medical models are fine-tuned on a smaller, domain-specific dataset, and fine-tuning has shown to make language models less calibrated [12]. Moreover, when a model is specialized in a certain domain, it tends to assign high probabilities to tokens relevant to in-domain questions, but for out-of-domain questions, the probability distribution may be flatter, which increases the influence of decoding methods and can lead to larger variations in output quality.

For statistical analysis, we used ROUGE as the primary metric to ensure a consistent comparison between decoding methods. We also included BLEU, a traditional n-gram–based measure for translation; BERTScore, which uses contextual embeddings to capture semantic similarity; and MAUVE, which measures distributional similarity and is often applied to dialogue.

Metric agreement patterns varied: BERTScore and ROUGE aligned most closely for summarization and least for dialogue, the latter expected given its open-ended nature. Translation might be expected to show high agreement due to its rigid structure, but ROUGE's reliance on exact token matches may penalize valid paraphrases that BERTScore can capture. As anticipated, BLEU and ROUGE correlated strongly, and both correlated moderately with BERTScore. MAUVE, in contrast, showed weak or negative correlations with the others, likely because it emphasizes diversity and naturalness over reference overlap. It was also very sensitive to decoding strategy, while BERTScore was the most stable.

In the medical domain, where accuracy is essential, MAUVE alone may be insufficient; combining it with a precision-oriented metric could yield a more balanced evaluation, a direction future work should explore. A limitation of our study is the lack of analysis on repetition and frequency penalties, despite their growing use in practice. While automatic metrics are valuable for scalability and rapid experimentation, they cannot fully capture aspects such as factual correctness or user satisfaction. Thus, human evaluation continues to play a vital role alongside metric-based assessment.

6 Conclusion

Our findings show in summary that while deterministic, search-based decoding strategies like beam search generally produce higher quality results at the cost of increased inference time, the optimal choice is task-dependent. Notably, the impact of the decoding strategy can be as significant as the choice of the LLM itself. We also found that larger models, while more performant, are not more robust to decoding sensitivity. A surprising result is that medical-specific LLMs do not show a consistent performance advantage over general models and are significantly more sensitive to the selected decoding method. This highlights a need for careful, task-specific tuning and evaluation, especially as metrics like MAUVE show weak agreement with traditional measures and greater sensitivity to decoding choices. Future work should examine whether these findings generalize to larger datasets, as our evaluation was limited in sample size.

Acknowledgments. The research presented in this paper has benefited from the Experimental Infrastructure for Exploration of Exascale Computing (eX3), which is financially supported by the Research Council of Norway under contract 270053.

Disclosure of Interests. The authors have no competing interests to declare that are relevant to the content of this article.

References

1. Ackley, D.H., Hinton, G.E., Sejnowski, T.J.: A learning algorithm for boltzmann machines. Cogn. Sci. **9**(1), 147–169 (1985)
2. Arias, E.G., Li, M., Heumann, C., Aßenmacher, M.: Decoding decoded: understanding hyperparameter effects in open-ended text generation. arXiv preprint arXiv:2410.06097 (2024)
3. Calle, P., et al.: Towards ai-driven healthcare: systematic optimization, linguistic analysis, and clinicians' evaluation of large language models for smoking cessation interventions. In: Proceedings of the 2024 CHI Conference on Human Factors in Computing Systems, pp. 1–16 (2024)
4. Chuang, Y.S., et al.: Dola: decoding by contrasting layers improves factuality in large language models (2024). https://arxiv.org/abs/2309.03883
5. Cohan, A., et al.: A discourse-aware attention model for abstractive summarization of long documents. In: Proceedings of the 2018 Conference of the North American Chapter of the Association for Computational Linguistics: Human Language Technologies, Volume 2 (Short Papers), pp. 615–621. Association for Computational Linguistics (2018). https://doi.org/10.18653/v1/N18-2097

6. Fan, A., Lewis, M., Dauphin, Y.: Hierarchical neural story generation (2018). https://arxiv.org/abs/1805.04833
7. Freitag, M., Al-Onaizan, Y.: Beam search strategies for neural machine translation. In: Proceedings of the First Workshop on Neural Machine Translation, pp. 56–60. Association for Computational Linguistics, Copenhagen, Denmark (2017). https://doi.org/10.18653/v1/W17-3207
8. Google: introducing gemini: our largest and most capable ai model (2023). https://blog.google/technology/ai/google-gemini-ai/
9. He, K., et al.: A survey of large language models for healthcare: from data, technology, and applications to accountability and ethics. Inf. Fusion **118**, 102963 (2025)
10. Hewitt, J., Manning, C., Liang, P.: Truncation sampling as language model desmoothing. In: Goldberg, Y., Kozareva, Z., Zhang, Y. (eds.) Findings of the Association for Computational Linguistics: EMNLP 2022, pp. 3414–3427. Association for Computational Linguistics, Abu Dhabi, United Arab Emirates (2022). https://doi.org/10.18653/v1/2022.findings-emnlp.249, https://aclanthology.org/2022.findings-emnlp.249/
11. Holtzman, A., Buys, J., Du, L., Forbes, M., Choi, Y.: The curious case of neural text degeneration (2020). https://arxiv.org/abs/1904.09751
12. Kong, L., et al.: Calibrated language model fine-tuning for in- and out-of-distribution data. In: Webber, B., Cohn, T., He, Y., Liu, Y. (eds.) Proceedings of the 2020 Conference on Empirical Methods in Natural Language Processing (EMNLP), pp. 1326–1340. Association for Computational Linguistics, Online (2020). https://doi.org/10.18653/v1/2020.emnlp-main.102
13. Leviathan, Y., Kalman, M., Matias, Y.: Fast inference from transformers via speculative decoding. In: International Conference on Machine Learning, pp. 19274–19286. PMLR (2023)
14. Meister, C., Pimentel, T., Wiher, G., Cotterell, R.: Locally typical sampling. Trans. Assoc. Comput. Linguistics **11**, 102–121 (2023). https://doi.org/10.1162/tacl_a_00536, https://aclanthology.org/2023.tacl-1.7/
15. Nguyen, M.N., et al.: Turning up the heat: min-p sampling for creative and coherent LLM outputs (2025). https://arxiv.org/abs/2407.01082
16. Nori, H., King, N., McKinney, S.M., Carignan, D., Horvitz, E.: Capabilities of GPT-4 on medical challenge problems. arXiv preprint arXiv:2303.13375 (2023)
17. OpenAI: introducing GPT-4o (2024). https://openai.com/index/gpt-4o-and-more-tools-to-chatgpt-free/
18. Rückert, J., et al.: ROCOv2: radiology objects in context version 2, an updated multimodal image dataset (2024). https://doi.org/10.5281/zenodo.8333645
19. Shi, C., et al.: A thorough examination of decoding methods in the era of LLMS. arXiv preprint arXiv:2402.06925 (2024)
20. Stribling, D., et al.: The model student: GPT-4 performance on graduate biomedical science exams. Sci. Rep. **14**(1), 5670 (2024)
21. Su, Y., et al.: A contrastive framework for neural text generation. In: Proceedings of the 36th International Conference on Neural Information Processing Systems. NIPS '22, Curran Associates Inc. (2022)
22. Tian, S., et al.: Opportunities and challenges for ChatGPT and large language models in biomedicine and health. Briefings Bioinf. **25**(1) (2023)
23. Vaswani, A., et al.: Attention is all you need (2023). https://arxiv.org/abs/1706.03762

24. Vijayakumar, A.K., et al.: Diverse beam search for improved description of complex scenes. In: Proceedings of the Thirty-Second AAAI Conference on Artificial Intelligence (AAAI-18), pp. 7371–7379. AAAI Press, New Orleans, Louisiana, USA (2018)
25. Wang, H., Xu, X., Philip, S.Y., Shu, K.: Beyond tokens: a survey on decoding methods for large language models and large vision-language models. Authorea Preprints (2025)
26. Wiher, G., Meister, C., Cotterell, R.: On decoding strategies for neural text generators. Trans. Assoc. Comput. Linguistics **10**, 997–1012 (2022)
27. Xie, Q., et al.: Faithful ai in medicine: a systematic review with large language models and beyond. MedRxiv (2023)
28. Yan, L.K., et al.: Large language model benchmarks in medical tasks. arXiv preprint arXiv:2410.21348 (2024)
29. Zhang, H., Duckworth, D., Ippolito, D., Neelakantan, A.: Trading off diversity and quality in natural language generation. arXiv preprint arXiv:2004.10450 (2020)

DiSCo: Disrupting Semantic Consistency for Transferable Cross-Modal Adversarial Attacks

Peirou Liang[1], Meng Yang[1], Zhiqian Wu[1], Peng Yuan Zhou[2], and Yong Liao[1](✉)

[1] University of Science and Technology of China, Hefei, China
prliang@mail.ustc.edu.cn, yliao@ustc.edu.cn
[2] Aarhus University, Aarhus, Denmark
pengyuan.zhou@ece.au.dk

Abstract. Current vision-language pretraining (VLP) models exhibit significant vulnerability to adversarial examples. Compared to white-box attacks, transfer attacks on unseen models better reflect real-world scenarios and offer greater research value. However, existing methods generally fail to effectively disrupt vision-language semantic alignment mechanism, limiting the transferability of adversarial examples. To address this issue, we propose DiSCo, a unified adversarial example generation framework. Specifically, we propose a perturbation generation strategy that disrupts semantic topological structures and image–text feature similarity, introducing cross-modal inconsistency that degrades downstream task performance. To mitigate the influence of redundant feature dimensions, we project image-text features into a semantic subspace constructed via clustering and hierarchical sampling before optimizing the adversarial trajectory. Furthermore, we design a cross-modal attention-based perturbation modulation mechanism that identifies text-relevant image regions and adjusts perturbations accordingly. Extensive experiments across various downstream tasks and VLP architectures demonstrate that our method significantly outperforms state-of-the-art approaches, with superior transferability and generalization.

Keywords: Cross-modal Adversarial Attacks · Adversarial Transferability · Vision-language Pre-training Models

1 Introduction

Vision-language pretraining (VLP) models trained on large-scale data enhance multi-modal understanding by aligning semantics across visual and textual modalities [11,15,22]. However, recent studies show that adversarial attacks can severely degrade the performance of VLP models, where small input perturbations can cause large prediction errors [3,6]. Most previous work [9,28] considers white-box settings that allow access to model parameters and gradients, which is often impractical. This motivates studying adversarial transferability, where adversarial examples fool unseen models with different architectures.

J. Lokoč et al. (Eds.): MMM 2026, LNCS 16413, pp. 16–30, 2026.
https://doi.org/10.1007/978-981-95-6957-1_2

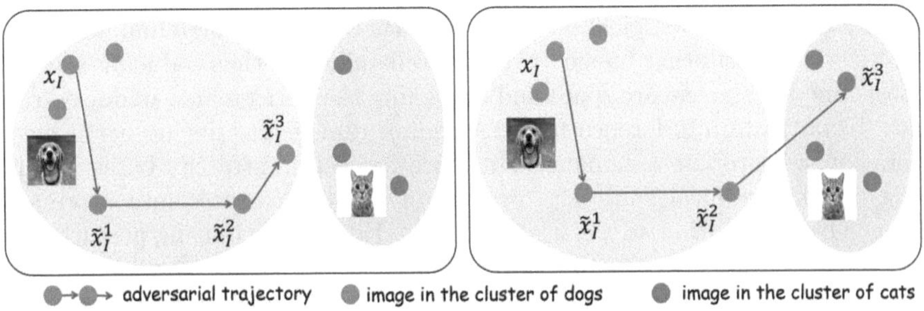

Fig. 1. Left: In traditional adversarial trajectory optimization methods, perturbed features remain within the original semantic cluster. Right: Our method moves features into a different semantic cluster, effectively altering the semantics.

Adversarial trajectory optimization has become a key strategy for improving the cross-model transferability of adversarial examples in vision-language models. It crafts instance-specific perturbations by modeling perturbation trajectories, which mitigates overfitting to the source model and improves generalization to unseen targets. However, most of these methods, including Co-Attack [27] and DRA [5], fail to effectively disrupt vision-language semantic alignment mechanism. First, they focus on controlling perturbation directions to increase diversity and reduce overfitting, without leveraging topology disruption to shift features across semantic clusters. As shown in Fig. 1, the adversarial example often remains in its original semantic cluster, even when it is far from its benign counterpart in the feature space, limiting decision boundary shifts and cross-model transferability. In contrast, AdvCLIP [29] employs a generative adversarial network to synthesize a universal adversarial patch, which induces topological changes by altering the feature distributions across diverse inputs. Although effective in shifting the local neighborhood structure of samples in the feature space, this approach is limited by its universal design and cannot adapt to instance-specific features. As a result, its adversarial transferability across diverse inputs and models is limited compared to methods that explicitly model individualized perturbation trajectories. In addition, previous studies [14] show that image features exhibit substantial dimensional redundancy, as many dimensions are not related to textual semantics, meaning that the impact of carefully crafted perturbations aimed at influencing similarity and topology in the original feature space may be diluted by redundant dimensions. Furthermore, computing perturbations solely in the feature space often leads to overfitting to the source model, since different architectures encode high-level features in different ways. Consequently, adversarial examples generated solely in feature space show limited transferability [7,24].

To address the above issues, we propose an adversarial example generation framework named **DiSCo**(**Di**srupting **S**emantic **Co**nsistency). We propose a perturbation generation strategy that optimizes for topological deviation and image-text feature misalignment. In particular, our objective maximizes the discrepancy in global topological structural change between benign and adversarial samples while disrupting image–text feature similarity, thus inducing semantic inconsistency in the feature space and degrading the performance of downstream tasks. To mitigate interference from redundant dimensions during perturbation generation, we propose a semantic subspace projection strategy based on clustering and hierarchical sampling, projecting original features into a semantic subspace before crafting adversarial examples. Beyond optimizing perturbations solely in the feature space, we further leverage the pixel space to improve transferability. Recent studies [10,21] show that visual patterns in pixel space, such as salient objects, are consistently captured across different VLP models. These architecture-invariant cues allow us to anchor perturbations computed in feature space to shared regions, enhancing cross-model transferability. Consequently, We leverage the cross-modal attention mechanisms inherent in VLP models to identify these cues, and adaptively refine perturbations to maximize their impact on disrupting modality-consistent representations.

We conduct extensive experiments on Flickr30K [19] and MSCOCO [16] datasets, evaluating multiple mainstream vision-language models across various multimodal tasks. The results demonstrate that the proposed framework consistently outperforms existing methods in transferability.

Our main contributions are summarized as follows:

- We propose DiSCo, a framework that jointly perturbs cross-modal feature and pixel spaces to disrupt cross-modal semantic consistency and generate transferable adversarial examples.
- We propose a strategy that projects image-text features into a clustered semantic subspace, where a joint objective is optimized to maximize topological deviation and feature misalignment. Additionally, we incorporate a cross-modal attention-guided mechanism to amplify perturbations in image regions that are most relevant to the text, further enhancing attack transferability.
- Extensive experiments on the MSCOCO and Flickr30K datasets show that DiSCo improves attack success rate of Text-to-Image Retrieval (TR R@1) and Image-to-Text Retrieval (IR R@1) on average by 12.17% and 9.59% on MSCOCO, and by 7.25% and 5.32% on Flickr30K, respectively, proving significant improvement in adversarial transferability and generalization across unseen models and tasks.

2 Related Work

2.1 Vision-Language Pre-training Models

VLP models advance Vision-and-Language tasks by learning multi-modal representations from large-scale image-text pairs. Early approaches predominantly relied on pre-trained object detectors to extract visual features [1,23]. Recent advancements leverage end-to-end image encoders like Vision Transformers

(ViT) [4], which offer faster inference speeds and computational efficiency compared to traditional detectors, prompting their adoption in modern VLP frameworks. VLP architectures generally follow two paradigms: fused and aligned models. Fused architectures, such as ALBEF [12] and TCL [26], employ separate unimodal encoders for text and images, followed by a multi-modal encoder to merge their features. In contrast, aligned models like CLIP [20] emphasize cross-modal feature space alignment through contrastive learning, significantly boosting downstream task performance. While fused models enable deep modality interaction, aligned architectures excel at efficient cross-modal retrieval and zero-shot generalization.

2.2 Transferability of Adversarial Examples

Improving the transferability of multi-modal adversarial examples is challenging due to limited access to model architectures and parameters. This work focuses on perturbation-based approaches, which introduce imperceptible noise and are widely adopted in uni- and multi-modal attacks. Early methods like Sep-Attack [27] treat each modality independently, neglecting cross-modal dependencies. MDA [25] leverages diffusion models to generate perturbations but suffers from high computational costs. To enhance cross-model transferability, trajectory-based methods have emerged, modeling the evolution of perturbations to promote input diversity and mitigate overfitting. Co-Attack [27] enforces a modality-consistency constraint during joint optimization. SGA [17] employs set-level augmentation, while DRA [5] encourages divergence in shared adversarial trajectories. SA-AET [8] introduces adversarial evolution triangles to guide sampling and optimization. However, these approaches often struggle to disrupt vision-language alignment, as perturbed samples typically remain within original semantic clusters, limiting prediction shifts. Furthermore, their emphasis on embedding space while overlooking pixel-level cross-modal cues constrains adversarial transferability.

3 Problem Formulation

In this work, we focus on cross-modal non-targeted attack for discriminative vision-language pretraining models, such as CLIP [4]. Formally, given a VLP model $f : (\mathbf{x}_I, \mathbf{x}_T) \mapsto y$, the attack maximizes the discrepancy between outputs on benign and adversarial inputs:

$$\max_{\substack{\tilde{\mathbf{x}}_I \in B[\mathbf{x}_I, \epsilon_I] \\ \tilde{\mathbf{x}}_T \in B[\mathbf{x}_T, \epsilon_T]}} \mathcal{L}\left(f(\mathbf{x}_I, \mathbf{x}_T),\ f(\tilde{\mathbf{x}}_I, \tilde{\mathbf{x}}_T)\right), \tag{1}$$

where $\mathcal{L}$ measures prediction discrepancy. Since predictions are largely determined by intermediate representations [17,27], this objective can be approximated by maximizing feature discrepancies. The final loss is given in Sect. 4.1.

4 Method

In this section, we propose an adversarial example generation framework named **DiSCo**. Section 4.1 details a perturbation generation strategy that projects image–text features into a semantic subspace and generates adversarial perturbations by jointly optimizing topological structure disruption and feature similarity constraints. Section 4.2 presents a cross-modal attention–guided perturbation modulation mechanism. The overall framework is shown in Fig. 2.

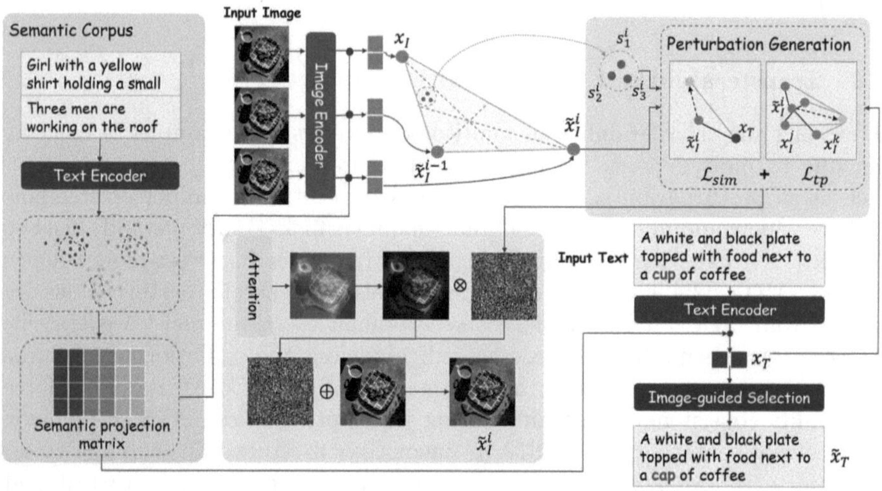

Fig. 2. We construct a semantic subspace via corpus clustering, disrupt topology and feature alignment jointly, and apply attention-guided perturbation modulation.

4.1 Topology-Aware Perturbation with Semantic-Aware Projection

We formulate a joint optimization objective that disrupts both the cross-modal feature similarity and the global topological structure in the feature space of vision-language models.

First, we introduce a semantic similarity loss $\mathcal{L}_{\text{sim}}$ that encourages misalignment between adversarial image features and their corresponding text features in the semantic subspace. Specifically, we define the similarity loss as follows:

$$\mathcal{L}_{\text{sim}} = \frac{\tilde{\mathbf{x}}_I \cdot \tilde{\mathbf{x}}_T}{d}, \tag{2}$$

where $\tilde{\mathbf{x}}_I$ and $\tilde{\mathbf{x}}_T$ denote perturbed image and text features respectively, and d is the dimensionality of the semantic feature space.

We introduce a topological deviation loss $\mathcal{L}_{\text{tp}}$ to disrupt semantic consistency by measuring the global topological structural change between the original and perturbed feature spaces. To quantify the neighborhood relations among samples, we construct an adjacency probability matrix $\mathbf{D}$, where $\mathbf{D}_{ij}$ denotes the probability that sample j belongs to the neighborhood of sample i. Specifically, it is defined as:

$$\mathbf{D}_{ij} = \frac{2 - (v_{ij} - r_j)}{\sum\limits_{k=1,\, k\neq j}^{N} [2 - (v_{jk} - r_j)]}, \quad 0 < i, j \leq N, \tag{3}$$

where v_{ij} represents the distance between samples i and j, and r_j denotes the distance from sample j to its nearest neighbor.

The structural discrepancy before and after perturbation is measured by the cross-entropy between the corresponding adjacency probability distributions:

$$\mathcal{L}_{\mathrm{T}}(\mathbf{x}, \tilde{\mathbf{x}}) = \mathrm{CrossEntropy}(\mathbf{D}_{\mathbf{x}}, \mathbf{D}_{\tilde{\mathbf{x}}}), \tag{4}$$

leading to the definition of the topological deviation loss:

$$\mathcal{L}_{\mathrm{tp}} = -\left(\mathcal{L}_{\mathrm{T}}(\mathbf{x}_I, \tilde{\mathbf{x}}_I) + \lambda\, \mathcal{L}_{\mathrm{T}}(\mathbf{x}_T, \tilde{\mathbf{x}}_I)\right), \tag{5}$$

where λ balances the structural contributions from image and text modalities. $\mathcal{L}_{\mathrm{tp}}$ encourages global topological structural change by penalizing the structural similarity between the benign and perturbed similarity graphs.

The overall joint loss is as follows:

$$\mathcal{L} = \mathcal{L}_{\mathrm{sim}} + \beta \mathcal{L}_{\mathrm{tp}}, \tag{6}$$

with β controlling the influence of the topological perturbation.

Since most of the image feature dimensions are misaligned with text and contain redundancy, perturbations on similarity and topology may be weakened. To address this, we construct a semantic subspace. We apply k-means clustering to the training text features to obtain K groups, and perform stratified sampling by uniformly selecting a proportion $\sigma \in (0, 1]$ of features from each group to form a representative subset. Compared to random sampling, this strategy ensures broad coverage across diverse feature groups while reducing intra-group redundancy, which forms a more discriminative semantic subspace and improves adversarial transferability (see Sect. 5.4). These features are stacked into a matrix $\mathbf{Z} \in \mathbb{R}^{d \times N}$, where N is the total number of sampled features. We then apply Singular Value Decomposition (SVD) on $\mathbf{Z}$ to extract the dominant semantic directions:

$$\mathbf{U},\ \mathbf{S},\ \mathbf{V}^{\top} = \mathrm{SVD}(\mathbf{Z}). \tag{7}$$

We retain the top-k components of $\mathbf{U}$ to form a low-rank projection matrix $\mathbf{P} = \mathbf{U}_k \mathbf{U}_k^{\top}$, which maps input features into a k-dimensional semantic subspace embedded within the original space. Let $\mathbf{F}_I(\cdot)$ and $\mathbf{F}_T(\cdot)$ denote the image and text encoders, and let x_I and x_T be the corresponding inputs. The projected features are given by $\mathbf{x}_I = \mathbf{P}\mathbf{F}_I(x_I)$ and $\mathbf{x}_T = \mathbf{P}\mathbf{F}_T(x_T)$.

Inspired by [5,8], we adopt a gradient-based triangular sampling strategy to guide adversarial trajectory optimization. As shown in Fig. 2, we form $\triangle(\mathbf{x}_I, \tilde{\mathbf{x}}_I^{i-1}, \tilde{\mathbf{x}}_I^i)$ from the benign, previous, and current adversarial images in the i-th iteration and partition it into six sub-triangles based on the relative distances to the three vertices. We sample intermediate points $\{\mathbf{s}_k^i\}_{k=1}^K$ from the sub-triangle near the benign and previous adversarial examples to compute gradient directions.

$$\boldsymbol{\epsilon}_k^i = \alpha \cdot \text{sign}\left(\frac{\nabla_{\mathbf{s}_k^i}\mathcal{L}(\mathbf{s}_k^i, \mathbf{x}_T)}{\left\|\nabla_{\mathbf{s}_k^i}\mathcal{L}(\mathbf{s}_k^i, \mathbf{x}_T)\right\|}\right), \tag{8}$$

where α is a step-size hyperparameter. We then select the perturbation direction that maximizes the loss:

$$o = \arg\max_k \mathcal{L}(\tilde{\mathbf{x}}_I^i + \boldsymbol{\epsilon}_k^i, \mathbf{x}_T). \tag{9}$$

To further enhance the selected optimal point s_o^i, we apply multi-scale augmentation, generating an enhanced set $\{s_o^{i(1)}, \ldots, s_o^{i(n)}\}$, and aggregate their gradients to update the perturbation:

$$\boldsymbol{\epsilon}_o^i = \alpha \cdot \text{sign}\left(\frac{\nabla_{\mathbf{s}_o^i}\sum_{j=1}^n \mathcal{L}(\mathbf{s}_o^{i(j)}, \mathbf{x}_T)}{\left\|\nabla_{\mathbf{s}_o^i}\sum_{j=1}^n \mathcal{L}(\mathbf{s}_o^{i(j)}, \mathbf{x}_T)\right\|}\right), \tag{10}$$

where $\boldsymbol{\epsilon}_o^i$ is used to update the adversarial image as $\tilde{\mathbf{x}}_I^{i+1} = \tilde{\mathbf{x}}_I^i + \boldsymbol{\epsilon}_o^i$.

For adversarial text generation, we consider the benign image $\mathbf{x}_I$, the penultimate adversarial image $\tilde{\mathbf{x}}_I^{N-1}$, and the final adversarial image $\tilde{\mathbf{x}}_I^N$:

$$\tilde{\mathbf{x}}_T = \max_{\mathbf{x}'_T \in B[\mathbf{x}_T, \epsilon_T]} \kappa \cdot \mathcal{L}(\mathbf{x}_I, \mathbf{x}'_T) + \eta \cdot \mathcal{L}(\tilde{\mathbf{x}}_I^{N-1}, \mathbf{x}'_T) + \nu \cdot \mathcal{L}(\tilde{\mathbf{x}}_I^N, \mathbf{x}'_T), \tag{11}$$

where $\kappa + \eta + \nu = 1$ and $B[\mathbf{x}_T, \epsilon_T]$ denotes the valid substitution set.

4.2 Perturbation Modulation via Cross-Modal Attention

To further disrupt cross-modal semantic alignment, we design a cross-modal attention-guided perturbation modulation method based on [2]. Given a single image-text pair, the model computes the logit score representing their semantic similarity. To extract relevant features, we compute the gradient of the logit with respect to the self-attention maps of Vision Transformer. Let ℓ denote the computed logit score, and let $\mathbf{A}^{(l,h)}$ represent the attention weights of head h in layer l. The corresponding gradient is denoted by $\mathbf{Q}^{(l,h)}$:

$$\mathbf{Q}^{(l,h)} = \frac{\partial \ell}{\partial \mathbf{A}^{(l,h)}}. \tag{12}$$

The gradient-weighted attention map for each head is computed as:

$$\mathbf{C}^{(l,h)} = \left(\mathbf{Q}^{(l,h)} \odot \mathbf{A}^{(l,h)}\right)^{+}, \tag{13}$$

where $\odot$ denotes the Hadamard product, and $(\cdot)^{+}$ retains only the positive values. We then average the results over all attention heads H:

$$\bar{\mathbf{A}}^{(l)} = \frac{1}{H}\sum_{h=1}^{H} \mathbf{C}^{(l,h)}. \tag{14}$$

Let $\mathbf{R}$ denote the relevance matrix, which is initialized as the identity matrix $\mathbb{I}^{N\times N}$. Starting from the i_{th} layer, we iteratively update $\mathbf{R}$ as follows:

$$\mathbf{R} = \mathbb{I}^{N\times N}, \quad \mathbf{R} \leftarrow \mathbf{R} + \bar{\mathbf{A}}^{(l)} \odot \mathbf{R}, \quad \text{for } l = i, i+1, \ldots, L. \tag{15}$$

We interpret $\mathbf{R}_{0,1:N}$ as the final relevance map, capturing the influence of the [CLS] token on the image patch tokens. In the remainder of this paper, we refer to it simply as $\mathbf{R}$.

The perturbation ϵ_o^i in Equation (10) is first scaled by a factor μ and then modulated element-wise using the attention map $\mathbf{R}$. The weighted perturbation, denoted as $\epsilon_o^{i'}$, is computed as:

$$\epsilon_o^{i'} = \frac{(\mu \cdot \epsilon_o^i) \odot [\exp(\gamma(\mathbf{R} - \delta))]}{\max(|(\mu \cdot \epsilon_o^i) \odot [\exp(\gamma(\mathbf{R} - \delta))]|)}, \tag{16}$$

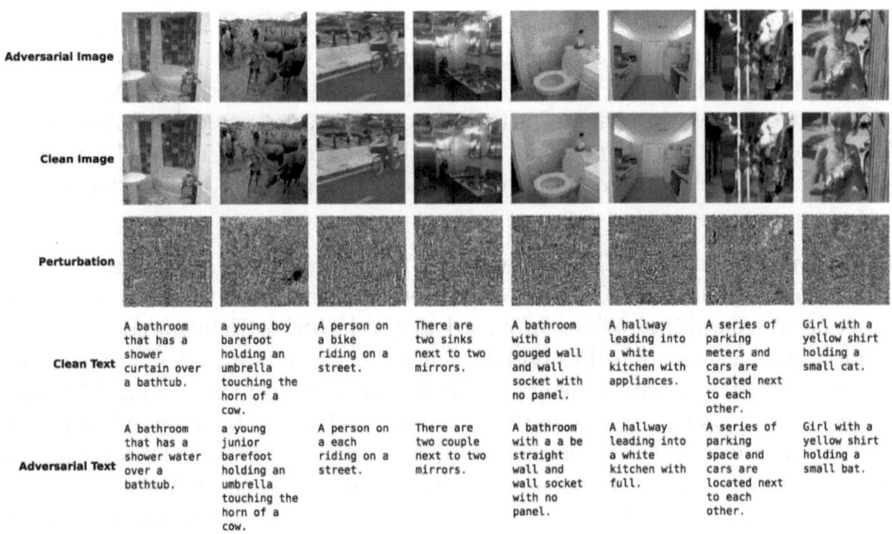

Fig. 3. Adversarial perturbation visualization on MSCOCO dataset.

where δ is a bias term and γ is a scaling factor controlling the strength of the modulation. The perturbation $\epsilon_o^{i'}$ replaces the original ϵ_o^i in Equation (10).

5 Experiments

5.1 Experiment Settings

We evaluated three ViT-based VLP models: CLIP with ViT-B/16(CLIP$_{\text{ViT}}$ [4]), ALBEF [12] and TCL [26]. Benchmarks include image-text retrieval on MSCOCO [16] and Flickr30K [19], visual grounding on RefCOCO+, and image captioning on MSCOCO. Our method is compared with seven baselines: PGD [18], BERT-Attack [13], Sep-Attack [27], Co-Attack [27], SGA [17], DRA [5], and SA-SET [8].

We set the text perturbation bound to $\epsilon_T = 1$, with a candidate word list size of $W = 10$, and the image perturbation bound to $\epsilon_I = \frac{8}{255}$ under the L_∞ norm. The attack is conducted over $T = 15$ iterations with a step size of $\alpha = \frac{2}{255}$. Sampling is performed from the sub-triangle near both the clean and previous adversarial examples, using $m = 5$ samples. Other hyper-parameters are configured as follows: the number of clusters $K = 10$, semantic corpus sampling ratio $\sigma = 0.4$, and the coefficients for adversarial text generation are set to $\kappa = 0.6$, $\eta = 0.2$, and $\nu = 0.2$. The attention scaling factor is $\gamma = 2$, the perturbation magnitude is $\mu = 2$, and the attention bias term is $\delta = 0.5$. Additionally, the topology deviation coefficient is set to $\lambda = 5$, and the loss weighting factor is $\beta = 1$.

5.2 Cross-Model Transferability

We generate multimodal adversarial samples using source models and assess their transferability on target models to evaluate the transferability of adversarial examples across three VLP models in black-box image-text retrieval. We evaluate attack effectiveness in terms of attack success rate, measured by Top-1 Recall (R@1) for two retrieval directions: Text-to-Image Retrieval (TR) and Image-to-Text Retrieval (IR), where higher values indicate stronger attack performance.

Performance on MSCOCO Dataset. Table 1 compares various adversarial attack methods on MSCOCO for cross-modal retrieval, clearly demonstrating the superior transferability of our proposed method DiSCo. Figure 3 shows adversarial perturbations applied to both image and text modalities. Under black-box settings, DiSCo achieves significantly better performance: when using CLIP$_{VIT}$ as the source model, it attains 90.98% and 91.27% TR R@1 on ALBEF and TCL respectively, far surpassing the 57.64% and 88.78% of the second-best method (SA-AET). Across all source-target combinations, DiSCo consistently outperforms other methods. Notably, the largest performance gain reaches up to 33.34% on TR R@1 (CLIP$_{VIT}$ to ALBEF), underscoring its strong cross-model and cross-modal adversarial transferability.

Table 1. Cross-Modal Adversarial Attack Evaluation on **MSCOCO**: **Bold** and underlined results denote the best and second-best methods, respectively. The △ row shows our method's performance gap from the best baseline.

Source	Method	ALBEF		TCL		$CLIP_{VIT}$	
		TR R@1	IR R@1	TR R@1	IR R@1	TR R@1	IR R@1
ALBEF	PGD	94.35	93.26	34.15	36.86	21.71	27.06
	BERT-Attack	24.39	36.13	24.34	33.39	44.94	52.28
	Sep-Attack	97.01	96.59	61.06	66.13	57.19	65.82
	Co-Attack	65.22	72.41	94.95	97.87	55.28	62.33
	SGA	99.95	99.94	87.46	88.17	63.72	69.71
	DRA	99.90	99.93	88.81	90.06	69.25	75.31
	SA-AET	100.0	99.99	97.28	96.88	76.57	80.24
	DiSCo(Ours)	**100.0**	**99.99**	**98.99**	**99.64**	**93.48**	**95.02**
	△	0.00	0.00	+1.71	+2.76	+16.91	+14.78
TCL	PGD	40.81	44.09	98.54	98.20	21.79	26.92
	BERT-Attack	35.32	45.92	38.54	48.48	51.09	58.80
	Sep-Attack	66.38	72.19	99.15	98.98	59.94	65.95
	Co-Attack	49.84	60.36	91.68	95.48	32.64	42.69
	SGA	92.70	92.99	100.0	100.0	59.79	65.31
	DRA	94.72	95.89	100.0	**100.0**	70.51	74.95
	SA-AET	97.78	98.08	100.0	99.99	76.12	79.74
	DiSCo(Ours)	**99.54**	**99.36**	**100.0**	99.99	**92.94**	**95.17**
	△	+1.76	+1.28	0.00	-0.01	+16.82	+15.43
$CLIP_{VIT}$	PGD	10.26	13.69	12.72	15.81	82.91	90.51
	BERT-Attack	20.34	29.74	21.08	29.61	45.06	51.68
	Sep-Attack	25.91	36.84	28.20	38.47	88.36	97.09
	Co-Attack	26.35	36.69	88.78	**96.72**	28.23	38.42
	SGA [42]	43.75	51.08	44.05	51.02	100.0	100.0
	DRA	52.69	61.50	51.88	61.06	100.0	100.0
	SA-AET	57.64	66.88	57.30	65.16	100.0	100.0
	DiSCo(Ours)	**90.98**	**93.28**	**91.27**	93.62	**100.0**	**100.0**
	△	+33.34	+26.40	+2.49	-3.1	0.00	0.00

Performance on Flickr30K Dataset. Table 2 reports cross-modal retrieval results on Flickr30K under adversarial attacks. DiSCo consistently outperforms existing methods, demonstrating superior adversarial transferability. In white-box settings, DiSCo achieves an attack success rate of 99.9% on both ALBEF and TCL. In black-box settings, it significantly boosts the attack success rate. For example, when using ALBEF as the source model, DiSCo achieves 96.84% on TR R@1 and 96.76% on IR R@1 against TCL, outperforming the best baseline (SA-AET) by 0.42% and 0.74%, respectively. Similarly, when attacking ALBEF from $CLIP_{ViT}$, DiSCo improves the TR R@1 by 17.00% and IR R@1 by 12.19% over the strongest baseline. Across all source-target combinations, DiSCo shows substantial performance gains (up to 17.00%), highlighting its effectiveness and robustness in cross-model adversarial attacks.

5.3 Cross-Task Transferability

Table 3 reports the cross-task transferability of adversarial examples generated on the Image-Text Retrieval (ITR) task and evaluated on Visual Grounding

Table 2. Cross-Modal Adversarial Attack Evaluation on **Flickr30K**: **Bold** and underlined results denote the best and second-best methods, respectively. The △ row shows our method's performance gap from the best baseline.

Source	Method	ALBEF		TCL		CLIP$_{VIT}$	
		TR R@1	IR R@1	TR R@1	IR R@1	TR R@1	IR R@1
ALBEF	PGD	93.74	94.43	24.03	27.90	10.67	15.82
	BERT-Attack	11.57	27.46	12.64	28.07	29.33	43.17
	Sep-Attack	95.72	96.14	39.30	51.79	34.11	45.72
	Co-Attack	97.08	98.36	39.52	51.24	29.82	38.92
	SGA	99.90	99.98	87.88	88.05	36.69	46.78
	DRA	99.90	99.93	91.57	91.17	46.26	56.80
	SA-AET	<u>99.90</u>	**99.98**	<u>96.42</u>	<u>96.02</u>	<u>55.58</u>	<u>63.89</u>
	DiSCo(Ours)	**99.90**	<u>99.95</u>	**96.84**	**96.76**	**63.56**	**67.91**
	△	0.00	-0.03	+0.42	+0.74	+7.98	+4.02
TCL	PGD	35.77	41.67	99.37	99.33	10.18	16.3
	BERT-Attack	11.89	26.82	14.54	29.17	29.69	44.49
	Sep-Attack	52.45	61.44	99.58	99.45	37.06	45.81
	Co-Attack	49.84	60.36	91.68	95.48	32.64	42.69
	SGA	92.50	92.77	100.0	100.0	36.81	46.97
	DRA	95.20	95.58	100.0	99.98	47.24	57.28
	SA-AET	<u>98.85</u>	<u>98.50</u>	<u>100.0</u>	<u>100.0</u>	<u>56.20</u>	<u>63.47</u>
	DiSCo(Ours)	**99.48**	**98.95**	**100.0**	**100.0**	**58.28**	**65.85**
	△	+0.63	+0.45	0.00	0.00	+2.08	+2.38
CLIP$_{VIT}$	PGD	3.13	6.48	4.43	8.83	69.33	84.79
	BERT-Attack	9.59	22.64	11.80	25.07	28.34	39.08
	Sep-Attack	7.61	20.58	10.12	20.74	76.93	87.44
	Co-Attack	8.55	20.18	10.01	21.29	78.53	87.50
	SGA	22.42	34.59	25.08	36.45	100.0	100.0
	DRA	27.84	42.84	27.82	44.60	100.0	100.0
	SA-AET	<u>36.60</u>	<u>50.44</u>	<u>39.20</u>	<u>51.10</u>	<u>100.0</u>	<u>100.0</u>
	DiSCo(Ours)	**53.60**	**62.63**	**54.58**	**63.21**	**100.0**	**100.0**
	△	+17.00	+12.19	+15.38	+12.11	0.00	0.00

(VG) and Image Captioning (IC). The "clean" row denotes baseline performance without attack, while rows for SGA [17], DRA [5], SA-AET [8], and our DiSCo show results under attack. On VG (columns "Val", "TestA", "TestB"), DiSCo yields the largest drops (40.62, 46.73, and 32.91), compared to clean scores of 58.46, 65.89, and 46.25. Similarly, for IC (columns "B@4", "METEOR", "ROUGE-L", "CIDEr", "SPICE"), DiSCo reduces CIDEr from 133.3 to 64.1, with consistent declines across all metrics. These results demonstrate the superior ability of DiSCo to generate transferable adversarial examples.

5.4 Ablation Study

The ablation results in Fig. 4 show that DiSCo consistently achieves the highest attack success rates across the six transfer settings in Flickr30K dataset. To evaluate the contribution of each module, we conduct ablations by removing one component at a time. Removing Semantic-Aware Projection (SAP), the semantic projection matrix is constructed by randomly sampling features from

Table 3. Cross-Task Transferability: Adversarial examples are generated using ALBEF for Visual Grounding (RefCOCO+) and Image Captioning (MSCOCO). The "clean" row shows baseline performance; lower values indicate stronger attacks. Bold denotes the most effective results, and <u>underline</u> marks the second-best. The final row (△) shows the gap between our method and the strongest baseline.

Method	ITR→VG			ITR→IC				
	ValâĘŞ	TestAâĘŞ	TestBâĘŞ	B@4↓	METEORâĘŞ	ROUGE-L↓	CIDEr↓	SPICE↓
Clean	58.46	65.89	46.25	39.7	31.0	60.0	133.3	23.8
SGA	50.56	57.42	40.66	28.0	24.6	51.2	91.4	17.7
DRA	49.70	56.32	40.54	27.2	24.2	50.7	88.3	17.2
SA-AET	<u>47.44</u>	<u>53.27</u>	<u>38.58</u>	<u>21.0</u>	<u>20.5</u>	<u>45.2</u>	<u>65.7</u>	<u>13.6</u>
DiSCo	**40.62**	**46.73**	**32.91**	**20.7**	**20.4**	**45.1**	**64.1**	**13.4**
△	-6.82	-6.54	-5.67	-0.30	-0.10	-0.10	-1.60	-0.20

the corpus. Removing TAP (Topology-Aware Perturbation) discards the topological objective while retaining the similarity loss. Removing PM (Perturbation Modulation) applies perturbations directly from the feature space without attention-based modulation. In the ALBEF to CLIP_{ViT} setting, the attack success rate of TR R@1 drops from 63.56% to 55.58%, and IR R@1 decreases from 67.91% to 63.89% when all components are removed. Similarly, in the CLIP_{ViT} to ALBEF setting, TR R@1 falls from 53.60% to 36.60%, while IR R@1 declines from 62.63% to 50.44% when all components are removed. These results confirm the necessity of each component and highlight their synergistic effect in enhancing adversarial transferability.

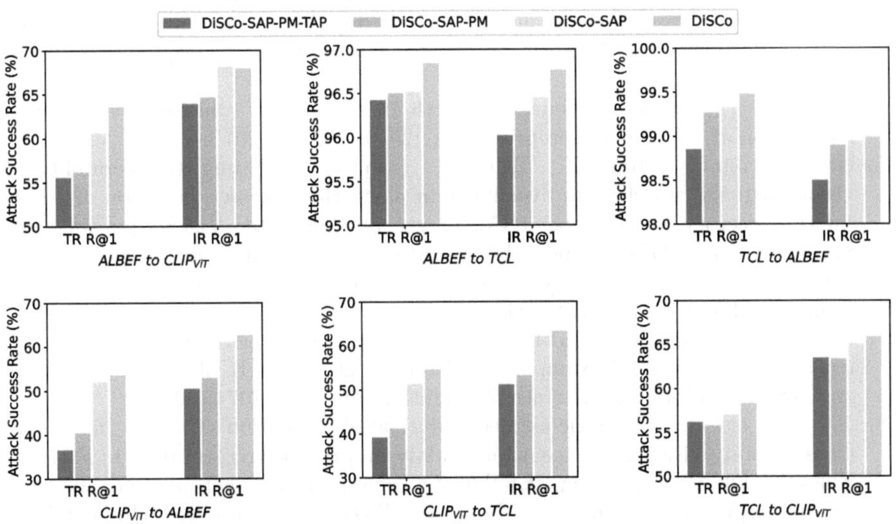

Fig. 4. Ablation study on Flickr30K: attack success rate (%) of DiSCo and its variants obtained by successively removing the SAP, PM, and TAP modules.

6 Conclusion

In this paper, we proposed DiSCo, a framework that jointly optimizes perturbations in the feature and pixel spaces to disrupt the semantic consistency between image and text in vision-language pre-training models. By generating perturbations through the joint optimization of topological deviation and semantic misalignment, and further modulating them with attention guidance in the pixel space, DiSCo effectively enhances the transferability of adversarial attacks across different VLP models. Our experiments show that this approach outperforms existing methods, achieving higher attack success rates and better cross-model transferability. DiSCo represents a significant advancement in generating transferable adversarial examples and offers valuable insights for improving the robustness of multimodal models against adversarial threats.

Acknowledgment. This work was supported by Anhui Province Science and Technology Innovation Breakthrough Plan (202423l10050033). Corresponding author: Yong Liao.

References

1. Alayrac, J.B., et al.: Flamingo: a visual language model for few-shot learning. Adv. Neural. Inf. Process. Syst. **35**, 23716–23736 (2022)
2. Chefer, H., Gur, S., Wolf, L.: Generic attention-model explainability for interpreting bi-modal and encoder-decoder transformers. In: Proceedings of the IEEE/CVF International Conference on Computer Vision, pp. 397–406 (2021)
3. Cheng, H., et al.: Unveiling typographic deceptions: insights of the typographic vulnerability in large vision-language models. In: European Conference on Computer Vision, pp. 179–196. Springer, Cham (2024). https://doi.org/10.1007/978-3-031-73202-7_11
4. Dosovitskiy, A., et al.: An image is worth 16 × 16 words: transformers for image recognition at scale. arXiv preprint arXiv:2010.11929 (2020)
5. Gao, S., Jia, X., Ren, X., Tsang, I., Guo, Q.: Boosting transferability in vision-language attacks via diversification along the intersection region of adversarial trajectory. In: European Conference on Computer Vision, pp. 442–460. Springer,Cham. (2024). https://doi.org/10.1007/978-3-031-72998-0_25
6. Huang, Y., et al.: Texture re-scalable universal adversarial perturbation. IEEE Trans. Inf. Forensics Secur. (2024)
7. Inkawhich, N., Wen, W., Li, H.H., Chen, Y.: Feature space perturbations yield more transferable adversarial examples. In: Proceedings of the IEEE/CVF Conference on Computer Vision and Pattern Recognition, pp. 7066–7074 (2019)
8. Jia, X., et al.: Semantic-aligned adversarial evolution triangle for high-transferability vision-language attack. arXiv preprint arXiv:2411.02669 (2024)
9. Khan, Z., Fu, Y.: Exploiting BERT for multimodal target sentiment classification through input space translation. In: Proceedings of the 29th ACM International Conference on Multimedia, pp. 3034–3042 (2021)
10. Kim, G., Kim, J., Lee, J.S.: Exploring adversarial robustness of vision transformers in the spectral perspective. In: Proceedings of the IEEE/CVF Winter Conference on Applications of Computer Vision, pp. 3976–3985 (2024)

11. Li, J., Li, D., Savarese, S., Hoi, S.: BLIP-2: bootstrapping language-image pre-training with frozen image encoders and large language models. In: International Conference on Machine Learning, pp. 19730–19742. PMLR (2023)
12. Li, J., Selvaraju, R., Gotmare, A., Joty, S., Xiong, C., Hoi, S.C.H.: Align before fuse: vision and language representation learning with momentum distillation. Adv. Neural. Inf. Process. Syst. **34**, 9694–9705 (2021)
13. Li, L., Ma, R., Guo, Q., Xue, X., Qiu, X.: BERT-ATTACK: adversarial attack against BERT using BERT. In: Proceedings of the 2020 Conference on Empirical Methods in Natural Language Processing (EMNLP), pp. 6193–6202 (2020)
14. Liang, V.W., Zhang, Y., Kwon, Y., Yeung, S., Zou, J.Y.: Mind the gap: understanding the modality gap in multi-modal contrastive representation learning. Adv. Neural. Inf. Process. Syst. **35**, 17612–17625 (2022)
15. Lin, J., Yin, H., Ping, W., Molchanov, P., Shoeybi, M., Han, S.: VILA: on pre-training for visual language models. In: Proceedings of the IEEE/CVF Conference on Computer Vision and Pattern Recognition, pp. 26689–26699 (2024)
16. Lin, T.Y., et al.: Microsoft COCO: common objects in context. In: Fleet, D., Pajdla, T., Schiele, B., Tuytelaars, T. (eds.) ECCV 2014. LNCS, vol. 8693, pp. 740–755. Springer, Cham (2014). https://doi.org/10.1007/978-3-319-10602-1_48
17. Lu, D., Wang, Z., Wang, T., Guan, W., Gao, H., Zheng, F.: Set-level guidance attack: boosting adversarial transferability of vision-language pre-training models. In: Proceedings of the IEEE/CVF International Conference on Computer Vision, pp. 102–111 (2023)
18. Madry, A., Makelov, A., Schmidt, L., Tsipras, D., Vladu, A.: Towards deep learning models resistant to adversarial attacks. arXiv preprint arXiv:1706.06083 (2017)
19. Plummer, B.A., Wang, L., Cervantes, C.M., Caicedo, J.C., Hockenmaier, J., Lazebnik, S.: Flickr30k entities: collecting region-to-phrase correspondences for richer image-to-sentence models. In: Proceedings of the IEEE International Conference on Computer Vision, pp. 2641–2649 (2015)
20. Radford, A., Kim, J.W., Hallacy, C., Ramesh, A., Goh, G., Agarwal, S., Sastry, G., Askell, A., Mishkin, P., Clark, J., et al.: Learning transferable visual models from natural language supervision. In: International Conference on Machine Learning, pp. 8748–8763. PMLR (2021)
21. Wang, H., et al.: Transferable multimodal attack on vision-language pre-training models. In: 2024 IEEE Symposium on Security and Privacy (SP), pp. 1722–1740. IEEE (2024)
22. Wang, W., et al.: CogVLM: visual expert for pretrained language models. Adv. Neural. Inf. Process. Syst. **37**, 121475–121499 (2024)
23. Wang, W., et al.: VisionLLM: large language model is also an open-ended decoder for vision-centric tasks. Adv. Neural. Inf. Process. Syst. **36**, 61501–61513 (2023)
24. Wu, W.,et al.: Boosting the transferability of adversarial samples via attention. In: Proceedings of the IEEE/CVF Conference on Computer Vision and Pattern Recognition, pp. 1161–1170 (2020)
25. Xu, W., Chen, K., Gao, Z., Wei, Z., Chen, J., Jiang, Y.G.: Highly transferable diffusion-based unrestricted adversarial attack on pre-trained vision-language models. In: Proceedings of the 32nd ACM International Conference on Multimedia, pp. 748–757 (2024)
26. Yang, J.,et al.: Vision-language pre-training with triple contrastive learning. In: Proceedings of the IEEE/CVF Conference on Computer Vision and Pattern Recognition, pp. 15671–15680 (2022)

27. Zhang, J., Yi, Q., Sang, J.: Towards adversarial attack on vision-language pre-training models. In: Proceedings of the 30th ACM International Conference on Multimedia, pp. 5005–5013 (2022)
28. Zhou, Y., Wu, J., Wang, H., He, J.: Adversarial robustness through bias variance decomposition: a new perspective for federated learning. In: Proceedings of the 31st ACM International Conference on Information & Knowledge Management, pp. 2753–2762 (2022)
29. Zhou, Z., Hu, S., Li, M., Zhang, H., Zhang, Y., Jin, H.: AdvCLIP: downstream-agnostic adversarial examples in multimodal contrastive learning. In: Proceedings of the 31st ACM International Conference on Multimedia, pp. 6311–6320 (2023)

Auditing M-LLMs for Privacy Risks: A Synthetic Benchmark and Evaluation Framework

Junhao Li[1(✉)], Jiahao Chen[2], Zhou Feng[2], and Chunyi Zhou[2]

[1] Guangzhou University, Guangzhou, People's Republic of China
lijh@e.gzhu.edu.cn

[2] Zhejiang University, Hangzhou, People's Republic of China
{xaddwell,zhou.feng,zhouchunyi}@zju.edu.cn

Abstract. Recent advances in multi-modal Large Language Models (M-LLMs) have demonstrated a powerful ability to synthesize implicit information from disparate sources, including images and text. These resourceful data from social media also introduce a significant and under-explored privacy risk: the inference of sensitive personal attributes from seemingly daily media content. However, the lack of benchmarks and comprehensive evaluations of SOTA (state-of-the-art) M-LLM capabilities hinders the research of private attribute profiling on social media. Accordingly, we propose (1) **PRISM**, the first multi-modal, multi-dimensional and fine-grained synthesized dataset incorporating a comprehensive privacy landscape and dynamic user history; (2) an Efficient evaluation framework that measures the cross-modal privacy inference capabilities of advanced M-LLM. Specifically, PRISM is a large-scale synthetic benchmark designed to evaluate cross-modal privacy risks. Its key feature is 12 sensitive attribute labels across a diverse set of multi-modal profiles, which enables targeted privacy analysis. These profiles are generated via a sophisticated LLM agentic workflow, governed by a prior distribution to ensure they realistically mimic social media users. Additionally, we propose a Multi-Agent Inference Framework that leverages a pipeline of specialized LLMs to enhance evaluation capabilities. We evaluate the inference capabilities of six leading M-LLMs (Qwen, Gemini, GPT-4o, GLM, Doubao, and Grok) on PRISM. The comparison with human performance reveals that these MLLMs significantly outperform in accuracy and efficiency, highlighting the threat of potential privacy risks and the urgent need for robust defenses.

Keywords: multi-modal Large Language Models (M-LLMs) · Privacy · Benchmark · Synthetic Data Generation

1 Introduction

Since the emergence of powerful multi-modal Large Language Models (M-LLMs), such as OpenAI's GPT-4o [18] and Google's Gemini [22], models are no longer

J. Lokoč et al. (Eds.): MMM 2026, LNCS 16413, pp. 31–44, 2026.
https://doi.org/10.1007/978-981-95-6957-1_3

limited to analyzing text or images in isolation; they now exhibit sophisticated reasoning capabilities that allow them to synthesize clues across modalities [24]. This leap in capability, particularly when applied to social media platforms, introduces a new privacy threat: M-LLMs could infer sensitive personal attributes by connecting the dots between a user's benign photos and textual posts.

To validate the real-world user perception of the privacy threat on social media, we conducted a large-scale **User Study** with 569 participants. As detailed in Sect. 3.1 and illustrated in Fig. 2, our key findings highlight the problem's urgency:

- **User Concern:** 82.% reported feeling "very concerned" or "somewhat concerned" about their information being used for privacy inference.
- **Platform Trust:** Over 80% believe that current privacy protections are "insufficient" or "completely insufficient".

Such findings provide strong evidence that multi-modal privacy risks constitute not merely a conceptual vulnerability but rather an imminent security challenge causing widespread user apprehension.

As illustrated in Fig. 1, a single social media post can become a rich source for inference. While prior research has acknowledged attribute inference from single modalities like text [21] or images [13, 15], these unimodal analyses fail to capture the integrated threat posed by M-LLMs. The privacy [10,20] risks arising from such cross-modal synthesis, where models can exploit a far richer context, remain largely unexplored.

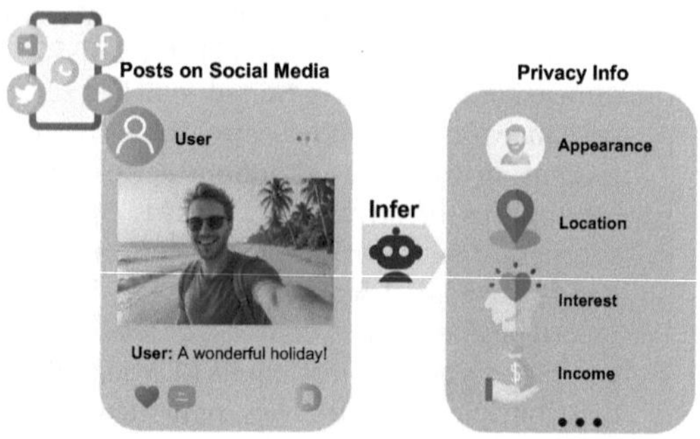

Fig. 1. Scenario Overview. A photograph of an overseas vacation not only directly exposes the user's appearance and location, but may also be inferred as personal attributes such as hobbies and income.

Despite this well-founded user concern, a significant research gap hinders the systematic evaluation of this threat. **The academic community currently lacks a benchmark specifically designed for multi-modal privacy inference.** Specifically, existing datasets are often limited to single image-text pairs for general vision-language tasks (e.g., MS-COCO [12]) and do not contain the fine-grained, privacy-centric attribute annotations necessary for this analysis.

Furthermore, beyond the technical limitations of existing datasets, the direct use of real-world social media data for such research presents profound **ethical and legal challenges**. Essentially, this line of inquiry involves attempting to

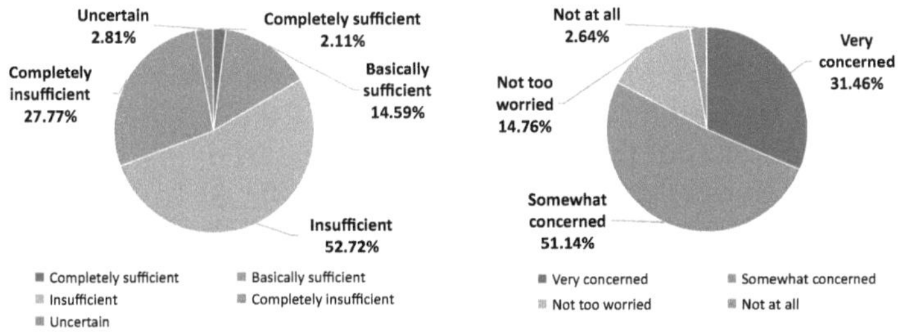

Fig. 2. Results from user study with 569 participants, highlighting public perception of multi-modal privacy risks. **Left:** User concern levels regarding the leakage of their personal information. **Right:** User perception of the adequacy of current platform privacy protections.

infer sensitive attributes that users have not explicitly consented to be analyzed for this purpose. Obtaining informed consent for a study designed to test the limits of privacy violation is a paradoxical and often insurmountable ethical barrier. Therefore, to conduct a systematic and reproducible investigation without compromising individuals' privacy, the creation of a synthetic benchmark is not merely a practical alternative, but a scientific and ethical imperative.

Besides, there are no established methodologies to probe how M-LLMs can be leveraged for multi-modal privacy inference, to quantify the privacy risks in the wild. This leaves a critical question unanswered: *To what extent can M-LLMs systematically integrate disparate, multi-modal information from social media to construct complete and accurate user profiles?*

To fill this critical technical and ethical gap, this paper introduces **PRISM**, a novel framework and benchmark designed to systematically quantify the risks of cross-modal privacy inference on social media. Our framework leverages generative AI (e.g., Qwen, DeepSeek, Doubao) to create realistic synthetic profiles that mimic social media users, where ground-truth private attributes are linked to a series of posts containing subtle textual and visual clues. A comprehensive user study testified to the high quality of the generated dataset, with participants expressing strong satisfaction with both images and text. We then evaluate the capabilities of six leading M-LLMs (Qwen [2], Gemini [22], GPT-4o [18], GLM [30], Doubao [3], and Grok [25]) on the PRISM benchmark for privacy inference. The privacy inference task is performed by our multi-agent framework, where specialized models first analyze clues from the text and image data, respectively. These unimodal findings are then channeled into a final multi-modal model that performs a holistic synthesis to produce the comprehensive inference. The results are assessed using a robust scoring system [23] to ensure a fair and comprehensive analysis. Our major contributions are as follows:

1. To our knowledge, we are the first to systematically investigate the multi-modal privacy risk of M-LLMs profiling sensitive user attributes by synthesizing clues from disparate multi-modal social media content.

2. We develop an integrated framework for privacy risk assessment, consisting of PRISM, the first large-scale synthetic and multi-modal benchmark purpose-built for this task, and a Multi-Agent Inference Architecture to automate the evaluation and measure the potential risks in the wild.
3. We conduct a comprehensive evaluation of leading M-LLMs using PRISM. Supported by robust evaluation metrics and our novel architecture, our experiments systematically reveal the models' alarming proficiency and quantify the critical impact of multi-modality on privacy inference.
4. Through a direct comparison with human investigators, we provide the first quantitative evidence that M-LLMs can outperform them in both accuracy and efficiency for this task, highlighting the scalability of the threat.

2 Related Work

2.1 M-LLMs and Emergent Risks for Privacy Inference

The emergence of M-LLMs [22] has dramatically enhanced AI's ability to interpret diverse data by integrating vision with language understanding [5,22]. This introduces new privacy risks, as these models can infer sensitive information like geolocation [15] from seemingly benign content on social media. Unlike traditional attacks that require specialized skills, modern LLMs lower the barrier for non-experts and can achieve superhuman performance. For instance, a recent study showed that M-LLMs outperform humans in image geolocation with up to 21 times lower error distances, demonstrating their potential for large-scale, low-effort privacy attacks [9]. Concurrently, existing defense mechanisms like image perturbation [17] and prompt-based refusal often prove insufficient, highlighting an urgent need for privacy-aligned defense strategies.

2.2 Attribute Inference from Multi-modal Social Media Content

The shift to M-LLMs has transformed social media analysis, enabling sophisticated reasoning from subtle multi-modal clues [1] and introducing new privacy risks [16]. These models can infer a wide range of sensitive attributes from routine posts [26], effectively democratizing the threat by lowering the barrier for non-expert adversaries [15]. The risk is particularly acute in the visual domain, where models like PIGEON [9] demonstrate superhuman capabilities in tasks such as image geolocation, sometimes achieving up to 21$\times$ lower error distances than humans [15]. This capability to turn subtle visual details into precise, sensitive data constitutes a tangible and scalable privacy attack vector.

Compounding this threat is the inadequacy of current defense mechanisms. Data-level strategies like image perturbation [6,11] often face a critical trade-off between effectiveness and content utility. Meanwhile, model-level safeguards [14], such as refusing malicious prompts [8], are proving brittle and can often be circumvented by carefully engineered jailbreak prompts [28]. This dangerous imbalance between rapidly advancing inference capabilities and lagging defense strategies highlights the critical need for standardized benchmarks to systematically evaluate and understand these privacy vulnerabilities.

3 A Methodological Framework for Privacy Risk Evaluation

Our methodology is structured to first establish the practical significance of multi-modal privacy threats through a user study, and then to detail the technical framework we developed to systematically evaluate these threats.

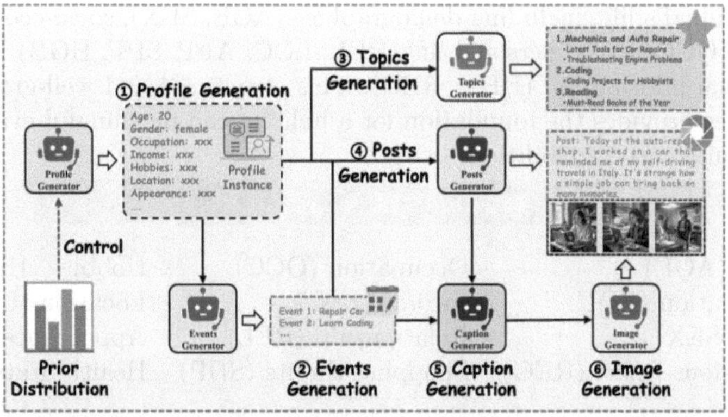

Fig. 3. The data generation workflow of PRISM begins by creating a realistic user Profile from a controllable Prior Distribution. Specialized generators then synthesize corresponding Events, Topics, Posts, and a Caption. Finally, an Image Generator renders a high-fidelity image from the caption, creating a complete and contextually consistent multi-modal post.

3.1 Motivation Study

To validate the real-world significance of this threat, we first conducted a comprehensive user study. The survey was designed to create a holistic picture of the current privacy landscape, covering user demographics, social media [27] sharing habits, and awareness of AI-driven data analysis. We collected responses from 569 participants, representing a diverse cross-section of the online population. The respondents spanned a wide range of age groups and professional backgrounds, including students, IT professionals, educators, and individuals from various other industries, ensuring our findings reflect a broad user perspective.

The results, partially summarized in Fig. 2, reveal widespread public anxiety. A significant majority of respondents, 82.6%, reported feeling "very concerned" or "somewhat concerned" about their publicly shared personal information being used for privacy inference or profit. Concurrently, over 80% of users believe that the privacy protection measures currently offered by online platforms are "somewhat insufficient" or "completely insufficient". These findings strongly validate that privacy leakage stemming from multi-modal content is not merely a potential technical threat, but a pressing, real-world issue causing broad public apprehension.

Based on our analysis of multi-modal social media data, we define a comprehensive set of twelve sensitive attributes for evaluation: AGE, EDU, SEX, OCC, INC, REL, LOC, APP, SHP, HOB, HEA, and REG. The selection of these attributes was a deliberate process, informed by a comprehensive review of existing literature on privacy inference [23,29] and the findings from our own user study. By synthesizing these sources, we consolidated the list to cover key facets of an individual's life, including demographics (AGE, SEX), socio-economic status (EDU, OCC, INC), personal life (REL, LOC, APP, SHP, HOB), and sensitive personal information (HEA, REG). This structured and well-grounded set of attributes provides the foundation for a holistic and meaningful evaluation of M-LLMs' inference capabilities.

Private Attributes of Adversaries' Interests		
– Age (AGE)	– Occupation (OCC)	– Hobbies (HOB)
– Education (EDU)	– Income (INC)	– Location (LOC)
– Sex (SEX)	– Relationship (REL)	– Appearance (APP)
– Religious-Belief (REG)	– Shopping-Habits (SHP)	– Health-Status (HEA)

Our preliminary experiments on existing large-scale, open-source datasets, such as Reddit image-text pairs, showed that while M-LLMs could infer some private attributes, the results were highly inconsistent and difficult to evaluate systematically. We identified the critical bottleneck as the lack of a predefined, structured set of target privacy attributes, which caused the models' inferences to be unfocused, non-specific, and speculative. For instance, a model might output five attributes for one image but a completely different set of thirteen for another, including speculative categories. This absence of a consistent attribute schema made a fair and scalable evaluation impossible. This underscored the necessity for a purpose-built benchmark with a fixed set of privacy attributes, and therefore, to enable a rigorous and reproducible scientific investigation, we developed our own synthetic dataset and evaluation framework.

3.2 A Synthetic Benchmark for Auditing Cross-Modal Privacy Risks

Our evaluation is based on PRISM, a high-quality synthetic dataset generated through an automated agentic workflow that mirrors the complexity of real-world social media content. As given in Fig. 3, the generation process begins with a Profile Generator, an LLM that creates a ground-truth "Profile Instance" for each virtual user containing 12 sensitive attributes, which are sampled from a controlled prior distribution to ensure realistic diversity and representation.

Prior Distribution. To ensure the generated dataset reflects the complexity and diversity of real-world user populations, the Profile Generator does not create attributes randomly. Instead, it samples from a carefully defined prior distribution grounded in empirical data to enforce realistic demographic constraints and inter-attribute correlations.

For instance, the location attribute is sampled according to the regional breakdown of global internet users, with higher probabilities assigned to regions with larger online populations, such as Asia, based on data from global digital reports [4]. The sex and age distributions are calibrated to align with demographic surveys of social media users. Crucially, we enforce logical correlations between attributes to prevent unrealistic profiles. The income level is conditioned on the user's occupation, reflecting official wage statistics, and the selection of hobbies is conditioned on their age group, consistent with patterns found in sociological research. This principled, data-grounded approach is vital for creating a benchmark that poses a realistic challenge and yields meaningful insights.

Our workflow automatically generates the multi-modal dataset by synthesizing a rich user footprint through a sequence of five specialized generators, as illustrated in Fig. 3. The roles within this pipeline are assigned as follows:

Events Generator: Synthesizes a series of plausible life events based on the attributes in the ground-truth Profile Instance to provide narrative context.

Topics Generator: Enriches the user's semantic profile by expanding on their hobbies and events to create a list of associated interests.

Posts Generator: Creates the textual content of a social media post, taking the user's persona and a specific event as input to craft a natural language post.

Caption Generator: A critical intermediate step in our visual synthesis process. We observed in preliminary experiments that directly tasking a text-to-image model with interpreting abstract event and profile data yielded inconsistent and low-quality images. To solve this, the Caption Generator acts as a "translator", converting the event, topics, and post text into a detailed, descriptive prompt optimized for a text-to-image model. This ensures a high degree of control and relevance in the final generated image.

Image Generator: Takes the detailed caption from the previous step as input and renders a high-fidelity, contextually appropriate image to be paired with the text post, completing the multi-modal data point.

The final output of this pipeline is a complete, multi-modal dataset where each user profile is linked to a chronological series of posts. This structured approach ensures each data point is contextually consistent, providing a solid foundation for our subsequent evaluation experiments.

3.3 The Multi-agent Inference Architecture

To perform the privacy inference simulation, we design a structured inference pipeline composed of three distinct nodes, each leveraging a specialized LLM.

Textual Analysis Agent: Acting as the textual specialist, this model is first responsible for processing all written content from a user's profile, such as posts and topics. It utilizes a foundational LLM to perform a deep analysis of this text, extracting a preliminary set of potential privacy-related clues in a structured format.

Image Analysis Agent: Operating in parallel, a visual specialist processes all the image data. For each of the user's images, this model prompts a vision-capable LLM with both the image and its corresponding text description. This method enables a context-aware analysis, allowing it to accurately identify objects, scenes, and other visual elements that could serve as privacy clues.

Multi-modal Synthesis Agent: Finally, the textual and visual clues extracted in the previous stages are channeled into the synthesis model, along with all the original data (including original posts, images, etc.). This model utilizes a powerful, vision-capable LLM to perform the final synthesis and inference. It integrates all the disparate pieces of information to generate a comprehensive and structured final user profile.

4 Experiment

This section presents a comprehensive evaluation of our proposed framework. We conducted two main experiments on PRISM dataset. First, we evaluated six SOTA LLM on their privacy inference capabilities. Second, Human Evaluation demonstrates that these models significantly surpass human performance in both inference accuracy and efficiency, underscoring the scale of the automated threat. By systematically benchmarking multiple SOTA LLMs on PRISM and contrasting them with human baselines, we validate the robustness and representativeness of our benchmark and framework: the results reveal the universality of models' inference capabilities across diverse attributes and highlight the severity of privacy risks in real-world contexts.

4.1 Experimental Setup

All experiments were conducted on our PRISM dataset, comprising over 500 unique user profiles. Each profile is richly detailed, associated with 2-3 sets of images and corresponding posts, as well as more than 3 topics of interest with multiple associated articles. The ground-truth for each profile consists of 12 distinct privacy attributes as defined in Sect. 3.

We evaluated six prominent LLMs: Qwen, GPT-4o, Gemini, Doubao, Grok, and GLM. The models were tasked with inferring user profiles using our Multi-Agent Inference Architecture. This architecture is composed of three distinct nodes: a text LLM that processes textual information (posts, topics, articles); an image LLM that analyzes visual data and its associated text; and a multi-modal LLM that aggregates the findings from the previous two nodes to generate the final, comprehensive inference result.

To ensure a fair and robust assessment, we employed an LLM-as-a-Judge protocol. The inference results from each target model were evaluated by two independent, powerful judge models: Qwen (qwen3-coder-480b-a35b-instruct) and

GPT-4o. The judges scored the accuracy of each inferred attribute against the ground-truth on a scale of 0 to 100. The final score reported is the average of the scores from these two judges. We report results in the format of mean. To answer our research questions, we first conduct an automatic evaluation focused on the impact of multi-modality, followed by a human evaluation.

4.2 Automatic Evaluation of Private Attribute Inference in M-LLMs

To answer our research questions, we conducted an ablation study to quantify the impact of modality.

Ablation Study. The inference task was run under two conditions: (1) In the Text-Only setting, the inference pipeline was simplified to only use the Text LLM before final aggregation. (2) In the Multi-Modal setting, the full three-node architecture was employed, providing the models with both text and their corresponding images.

The comprehensive results are presented in Table 1. Our findings reveal that the inclusion of visual information can substantially enhance inference performance, though the degree of improvement varies across different models and attributes. The enhancement is particularly dramatic for attributes with strong visual correlations, such as sex and appearance. For instance, with the Qwen model, the inference accuracy for appearance surged from a mere 1.28 to 33.93 when introducing visual data. Similarly, for the sex attribute, several models achieved accuracy rates exceeding 95% in the multi-modal setting, a stark contrast to the less reliable text-only results. This provides strong empirical support for our central hypothesis: visual data constitutes a rich, often essential, channel for privacy leakage that is effectively exploited by M-LLMs.

This ablation study confirms that the visual modality is not merely supplementary but often the primary channel for privacy inference in these systems. The experiment aims to answer two key research questions:

RQ1: How proficient are SOTA models at inferring private attributes? Our findings indicate that current SOTA models possess a strong capability in profiling users from multi-modal data. As shown in the Multi-Modal results in Table 1, models like Gemini consistently lead in performance, achieving the highest scores across 7 of the 12 attributes. Certain attributes are exceptionally vulnerable. For instance, SEX is inferred with near-perfect accuracy by multiple models, with Gemini reaching a score of 99.82. Attributes that are often explicitly mentioned or strongly implied in social media discourse, such as OCC (Occupation) and HOB (Hobbies), also exhibit high inference scores, with top models like Gemini and Grok scoring 78.39 and 64.50, respectively. Conversely, attributes that are typically more private and less frequently disclosed, such as INC, prove to be the most difficult to infer, with all models scoring below 6. This demonstrates that while models are highly capable, their proficiency varies significantly depending on the nature of the private attribute.

RQ2: What is the quantitative impact of visual information on this inference task? By comparing the Multi-Modal and Text-Only results within Table 1, we can quantify the dramatic impact of visual data. The inclusion of images serves as a powerful catalyst for privacy inference, particularly for attributes that are inherently visual. The most striking example is APP. In the text-only setting, models struggled immensely, with an average score of just 4.37 across all models. However, when images were introduced, the average score surged to 28.53, an increase of over 550%. For some models like Doubao, the score leaped from a mere 0.28 to 32.21. Similarly, the SEX attribute saw its average inference score skyrocket from 18.53 (Text-Only) to 92.92 (Multi-Modal), trans-

Table 1. Comprehensive performance comparison of six MLLMs on the PRISM benchmark. The table details the inference accuracy, scored on a scale of 0 to 100, for each of the 12 privacy attributes under two distinct settings: Text-Only and Multi-Modal.

Attribute	Modality	Qwen	GPT-4o	Gemini	Doubao	Grok	GLM
AGE	Text	5.81	13.07	4.65	2.62	5.98	10.99
	Multi-Modal	**58.83**	**15.56**	**65.98**	**23.14**	**57.64**	**42.84**
EDU	Text	10.91	3.16	3.96	3.13	7.04	11.42
	Multi-Modal	**38.33**	**3.87**	**7.20**	**5.43**	**2.67**	**24.14**
SEX	Text	21.73	42.55	15.20	4.15	14.64	12.89
	Multi-Modal	**98.90**	**77.26**	**99.82**	**88.58**	**96.39**	**96.55**
OCC	Text	65.70	55.09	71.07	49.17	61.56	52.88
	Multi-Modal	**71.84**	**67.25**	**78.39**	**67.66**	**71.78**	**76.50**
INC	Text	2.35	0.28	0.42	0.12	0.19	0.67
	Multi-Modal	**5.36**	**0.12**	**1.20**	**0.14**	**0.58**	**1.73**
REL	Text	21.64	22.82	25.05	21.80	23.62	19.48
	Multi-Modal	**28.71**	**24.70**	**27.29**	**22.43**	**26.71**	**10.78**
LOC	Text	59.69	50.59	50.55	43.68	52.18	42.32
	Multi-Modal	**48.98**	**47.15**	**56.36**	**47.42**	**50.15**	**37.78**
APP	Text	1.28	12.96	9.54	0.28	0.67	1.50
	Multi-Modal	**33.93**	**18.72**	**33.85**	**32.21**	**20.78**	**31.69**
SHP	Text	35.11	27.98	41.64	25.73	35.35	20.24
	Multi-Modal	**25.14**	**27.61**	**50.84**	**32.46**	**35.86**	**20.26**
HOB	Text	70.90	72.23	68.80	68.69	78.98	61.63
	Multi-Modal	**58.27**	**65.56**	**59.49**	**61.08**	**64.50**	**60.42**
HEA	Text	40.52	37.55	38.75	28.02	40.63	24.68
	Multi-Modal	**47.12**	**38.21**	**50.67**	**38.77**	**47.89**	**34.49**
REG	Text	54.26	52.10	54.74	43.99	52.65	36.56
	Multi-Modal	**52.48**	**54.26**	**59.09**	**56.91**	**52.53**	**44.83**

forming the task from a difficult guess to a near-certainty. This is because images provide undeniable visual evidence of gender and appearance that text can only vaguely describe. The AGE attribute also saw a massive average performance gain, jumping from 7.19 to 43.99.

Interestingly, for a few attributes, the impact of visual data is less pronounced or even negative. For HOB, the average score slightly decreased from 70.22 in the text setting to 61.55 in the multi-modal setting. This suggests that for topics often explicitly stated in text, accompanying images may add little new information or even introduce ambiguity. A similar trend was observed for LOC with some models. This nuanced result underscores that the value of visual data is context-dependent, but for visually-grounded attributes, it represents the most significant vector for privacy leakage.

4.3 Human Evaluation

Setup. To further evaluate the effectiveness and efficiency of employing MLLMs to profile sensitive attributes from multi-modal content, we conducted a human evaluation. For this study, we recruited 20 graduate students, all majoring in computer science. Participants were familiar with and accustomed to using various search engines and LLM services. In this study, we randomly selected a subset of user profiles from our PRISM dataset. Participants were instructed to view the multi-modal posts (both images and text) associated with these individuals and infer the 12 privacy-related attributes. They were allowed to view the content multiple times as needed and permitted to use search engines or LLMs to retrieve relevant information, enabling more accurate inferences.

Table 2. Performance comparison between MLLMs and human participants. The table shows inference accuracy (%) for each of the 12 attributes, the overall average (Avg), and the total time spent on the task in minutes.

Models	Spent Time	AGE	EDU	SEX	OCC	INC	REL	LOC	APP	SHP	HOB	HEA	REG	Avg
Human	485.75 min	21.64	**35.00**	100.00	55.00	0.00	**30.00**	56.25	38.12	12.50	16.83	50.00	30.00	37.11
Doubao	75.6 min	49.35	2.00	90.00	78.50	0.00	14.50	41.50	35.42	33.50	49.63	45.50	53.50	49.46
Qwen	64.8 min	46.83	18.50	80.00	79.00	**2.00**	23.00	47.00	23.01	27.50	54.54	**53.50**	44.50	49.95
GLM	56.5 min	50.42	14.00	98.00	74.50	0.50	11.00	39.42	**38.88**	15.00	48.70	37.50	41.50	47.45
Gemini	196.3 min	**60.55**	16.50	**100.00**	**92.00**	0.00	20.50	**56.57**	33.00	**50.00**	**56.99**	50.00	**54.50**	**54.25**

Comparison between Humans and MLLMs. The results of our Human Evaluation study are detailed in Table 2. The findings reveal a significant performance gap between MLLM agents [7] and human participants across both accuracy and efficiency. In terms of average inference accuracy, all evaluated models surpassed the human average of 36.28. Notably, our top-performing model, Gemini, achieved an average accuracy of 54.25, representing a substantial absolute improvement of nearly 18 points over human performance. This superiority was particularly evident in attributes requiring deep contextual understanding or broad knowledge, such as OCC, where Gemini scored 92.00 compared to the human score of 55.00.

Furthermore, the M-LLM agents demonstrated a staggering advantage in time efficiency. Human participants required a total of 485.75 min to complete the task, whereas the fastest model, GLM, finished in just 56.5 min—over 8.5 times faster. This highlights the most critical aspect of this threat: its scalability. M-LLMs can perform automated, large-scale privacy reconnaissance at a speed and scale that is simply unattainable for human actors, posing a severe and scalable risk to user privacy. We found a positive correlation between human and LLM inference capabilities, suggesting that attributes easier for humans to infer are generally also easier for the models to infer.

5 Discussion and Potential Defenses

Our findings underscore the urgent need for effective defenses against cross-modal privacy inference. Current research has explored two main categories of defense, each with a significant trade-off. The first, data-level defense, involves perturbing inputs with methods like Gaussian noise for images [15] or signal jamming for audio [23]. While this approach can directly disrupt clues, it faces a critical dilemma: strong perturbations severely degrade content quality and user experience, whereas weaker ones are often insufficient to fool sophisticated model reasoning. The second category, model-level defense, uses prompt engineering to compel models to refuse sensitive requests [15]. The advantage is that it preserves the original data, but the drawback is that models often become over-defensive, rejecting benign queries and thus harming their overall utility.

Given these limitations, we argue that future defense strategies must be more intelligent and multi-layered. At the user level, social media platforms could integrate proactive warning systems that analyze content before a user posts, alerting them to potential privacy risks and empowering them with informed control. At the model level, more advanced techniques like In-context Unlearning [19] offer a promising direction. This method can dynamically suppress a model's ability to associate specific features with sensitive attributes on a per-query basis, offering a targeted defense without permanently degrading the model's capabilities. We believe that combining proactive user-side warnings with intelligent model-side suppression is key to building a robust defense ecosystem for the future.

6 Conclusion

In this paper, we systematically investigated the emergent threat of cross-modal privacy inference on social media. Our work makes two primary contributions. First, we introduce PRISM, a novel Dataset Benchmark that is multi-modal, multi-dimensional, fine-grained, and designed for quantifiable privacy risk analysis. Second, using this benchmark, we established the first comprehensive Model Performance Benchmark for this task by evaluating six leading M-LLMs. Our findings reveal their alarming proficiency, with the inclusion of visual data boosting inference accuracy by over 550% for key attributes like appearance. Furthermore, the models significantly surpassed human investigators in both accuracy and efficiency. They completed the task over 8.5 times faster, confirming the scalability of the threat. This work validates the feasibility of automated privacy attacks and provides a foundational framework to facilitate future research into robust defenses.

Acknowledgments. This work was partly supported by the National Key Research and Development Program of China (No. 2022YFB3102100), the NSFC–Yeqisun Science Foundation (No. U244120033), the National Natural Science Foundation of China (Nos. U24A20336, 62172243, 62502432, 62402425, 62402418), the China Postdoctoral Science Foundation (No. 2024M762829), the Zhejiang Provincial Natural Science Foundation (No. LD24F020002), and the Zhejiang Provincial Priority-Funded Postdoctoral Research Project (No. ZJ2024001).

References

1. AlSaad, R., et al.: Multimodal large language models in health care: applications, challenges, and future outlook. J. Med. Internet Res. **26**, e59505 (2024). https://doi.org/10.2196/59505
2. Bai, J., et al.: Qwen2: a family of large-scale language models. arXiv preprint arXiv:2406.04832 (2024)
3. ByteDance: Doubao-llm (2024). https://www.volcengine.com/product/doubao-llm
4. DataReportal: Digital 2024: Global overview report (2024). https://datareportal.com/reports/digital-2024-global-overview-report. Accessed 16 Aug 2025
5. Fang, J., et al.: SafeMLRM: Demystifying Safety in Multi-modal Large Reasoning Models. arXiv preprint (2025)
6. Feng, Z., et al.: Enkidu: universal frequential perturbation for real-time audio privacy protection against voice deepfakes. In: Proceedings of the 33rd ACM International Conference on Multimedia (MM 2025), pp. 11638–11647. Association for Computing Machinery, New York (2025). https://doi.org/10.1145/3746027.3755629
7. Gan, Y., et al.: Navigating the risks: a survey of security, privacy, and ethics threats in LLM-based agents (2024). https://arxiv.org/abs/2411.09523
8. Ge, Y., Kirtane, N., Peng, H., Hakkani-Tür, D.: LLMs are Vulnerable to Malicious Prompts Disguised as Scientific Language. arXiv preprint (2025)
9. Haas, L., Skreta, M., Alberti, S., Finn, C.: PIGEON: Predicting Image Geolocations. arXiv preprint (2024)
10. He, X., et al.: Artificial intelligence security and privacy: a survey. SCIENCE CHINA Inf. Sci. **68**(8), 181101 (2025). https://doi.org/10.1007/s11432-025-4388-5
11. Li, J., Ji, R., Liu, H., Hong, X., Gao, Y., Tian, Q.: Universal perturbation attack against image retrieval. In: Proceedings of the IEEE/CVF International Conference on Computer Vision (ICCV) (2019)
12. Lin, T., et al.: Microsoft COCO: common objects in context. CoRR (2014)

13. Liu, F., et al.: The Eye of Sherlock Holmes: Uncovering User Private Attribute Profiling via Vision-Language Model Agentic Framework. arXiv preprint (2025)
14. Luo, W., et al.: Dynamic guided and domain applicable safeguards for enhanced security in large language models. In: Findings of the Association for Computational Linguistics: NAACL 2025, pp. 6599–6620. Association for Computational Linguistics (2025). https://doi.org/10.18653/V1/2025.FINDINGS-NAACL.368
15. Luo, W., et al.: Doxing via the Lens: Revealing Location-related Privacy Leakage on Multi-modal Large Reasoning Models. arXiv preprint (2025)
16. Novikova, E., Doynikova, E., Kotenko, I.V.: What are your privacy risks? Privacy risk assessment based on privacy policies analysis. Expert Syst. Appl. **280**, 127270 (2025). https://doi.org/10.1016/J.ESWA.2025.127270
17. Oh, S.J., Fritz, M., Schiele, B.: Adversarial image perturbation for privacy protection: a game theory perspective. In: 2017 IEEE International Conference on Computer Vision (ICCV), pp. 1491–1500 (2017). https://doi.org/10.1109/ICCV.2017.165
18. OpenAI: Gpt-4o technical report. arXiv preprint arXiv:2405.20712 (2024)
19. Pawelczyk, M., Neel, S., Lakkaraju, H.: In-context unlearning: language models as few shot unlearners (2024)
20. Pu, Y., Chen, J., Pan, J., Yan, D., Zhang, X., Ji, S.: Facial data minimization: shallow model as your privacy filter. IEEE Trans. Dependable Secure Comput. **22**(5), 5186–5202 (2025). https://doi.org/10.1109/TDSC.2025.3563477
21. Staab, R., Vero, M., Balunović, M., Vechev, M.: Beyond Memorization: Violating Privacy Via Inference with Large Language Models. arXiv preprint (2024)
22. Team, G., Google: Gemini: A family of highly capable multimodal models. arXiv preprint arXiv:2312.11805 (2023)
23. Wang, L., et al.: The man behind the sound: demystifying audio private attribute profiling via multimodal large language model agents (2025). https://arxiv.org/abs/2507.10016
24. Wang, S., Hu, M., Li, Q., Safari, M., Yang, X.: Capabilities of GPT-5 on Multimodal Medical Reasoning. arXiv preprint (2025)
25. xAI: Announcing grok (2023). https://x.ai/blog/grok
26. Xu, W., Fan, W., Lu, S., Li, T., Wang, B.: Polarized Patterns of Language Toxicity and Sentiment of Debunking Posts on Social Media. arXiv preprint (2025)
27. Yang, Y., et al.: Invisible-face: rethinking facial attribute privacy in social media photo sharing. IEEE Trans. Inf. Forensics Secur. **20**, 6101–6116 (2025). https://doi.org/10.1109/TIFS.2025.3579592
28. Yu, Z., Liu, X., Liang, S., Cameron, Z., Xiao, C., Zhang, N.: Don't listen to me: understanding and exploring jailbreak prompts of large language models. In: 33rd USENIX Security Symposium (USENIX Security 2024), pp. 4675–4692. USENIX Association, Philadelphia, PA (2024)
29. Zhang, S., Yuan, W., Yin, H.: Comprehensive privacy analysis on federated recommender system against attribute inference attacks. IEEE Trans. Knowl. Data Eng. **36**(3), 987–999 (2024). https://doi.org/10.1109/TKDE.2023.3295601
30. Zheng Xiao Du, Yujie Qian, X.L., et al.: Glm-4: An open bilingual multimodal large model. arXiv preprint arXiv:2405.19793 (2024)

FGR: Frequency Aware and Geometric Structure-Guided Multi-modality Image Registration Framework

Qihao Ye and Zhuowei Wang(✉)

Guangdong University of Technology, Guangzhou, China
wangzhuowei0710@163.com

Abstract. Current unsupervised learning approaches employ image-to-image translation networks to reduce inter-modal discrepancy before applying mono-modal registration techniques. However, significant inherent differences across imaging modalities limit GAN-based spatial domain translation methods in bridging the modality gap and result in insufficient prediction accuracy for texture-dependent registration networks. To address this, we propose a Frequency-aware and Geometric structure-guided multi-modality image Registration framework (FGR). The framework utilizes both spatial and frequency domain information for modality translation, and incorporates the proposed Texture-Structure Aware Module (T-SAM) to guide the deep registration network in extracting geometric structural features and integrating them with texture information for precise deformation prediction, thereby enabling it to excavate geometric structural features with stronger modality correlations. This design avoids purely mono-modal registration paradigms. Experimental results demonstrate that our method achieves state-of-the-art performance in multi-modality registration task.

Keywords: Multi-modality image registration · Modality translation · Image fusion · Frequency-aware · Geometric structure-guided

1 Introduction

Numerous complex image processing tasks such as object detection, semantic segmentation, and pedestrian detection utilize infrared and visible (IR&VI) sensors to enhance accuracy in challenging environments. Current approaches typically fuse IR&VI images (IVIF) to create comprehensive scene representations before downstream tasks. However, most deep learning-based IVIF algorithms are designed for pre-registered image pairs and cannot handle misalignment. Directly fusing misaligned images causes ghosting artifacts, impairing the performance of downstream tasks. Therefore, automated IR&VI registration is the essential prerequisite, highlighting its critical significance.

Deep learning has advanced end-to-end IR&VI registration. Early methods integrated deep learning into traditional pipelines, while unsupervised techniques

J. Lokoč et al. (Eds.): MMM 2026, LNCS 16413, pp. 45–58, 2026.
https://doi.org/10.1007/978-981-95-6957-1_4

enhanced Spatial Transformer Networks (STN) [6] using similarity metrics [5,8]. To bridge modality gaps, methods like NEMAR [1], UMF [20], and IMF [21] implemented a 'translate-then-align' paradigm: image-to-image (I2I) translation networks transform VI to pseudo-IR before mono-modal registration.

Despite high precision, existing methods exhibit inherent limitations. Significant radiation and texture differences stemming from distinct imaging mechanisms cause spatial-domain I2I translation networks to struggle to bridge the modality gap. Consequently, subsequent mono-modal registration networks based on texture features fail to predict precise deformation fields, hindering registration accuracy. To address these limitations, we propose a Frequency-aware and Geometric structure-guided multi-modality image Registration framework (FGR). First, the Multi-modality Frequency-aware dual-Attention Translation Network (MFATN) enhances frequency features, optimized with frequency-guided loss for superior spatial fidelity pseudo-IR images. The Geometry-guided Progressive Multiscale Registration Network (GPMRN) then predicts fields using the Texture-Structure Aware Module (T-SAM) to identify structural correspondences via a features fusion strategy.

Overall, our main contributions are as follows:

- We propose FGR, a novel multi-modality framework for image registration, aimed at mitigating ghosting artifacts from misaligned pairs, addressing geometric deformations.
- We introduce the MFATN, a frequency-aware dual-attention network incorporating frequency-enhanced modules to extract cross-modal details, optimized via both frequency-domain and spatial loss to generate high-quality pseudo-infrared images.
- To achieve precise deformation field prediction, we introduce the GPMRN, a geometry-guided multi-scale registration network using hierarchical texture and edge features extraction. Its T-SAM leverages edge-texture disparities in corresponding regions to focus on structural edges for robust alignment.

2 Related Work

2.1 Multi-modality Image Translation

Multi-modality image translation involves converting images from one modality to another, such as transforming visible light images to thermal or infrared representations. This technology has significant applications in computer vision tasks including surveillance, autonomous navigation, and person re-identification, where different imaging modalities provide complementary information.

ThermalGAN [9] built a modality transfer network using a BiCycleGAN-derived architecture, which synthesizes long-wave infrared images from visible images for person re-identification. Leveraging the outstanding capabilities of classic GANs like CycleGAN [27] in image translation, some studies [10,15,19] further adopted their core principles for modality transfer task.

2.2 Multi-modality Image Registration

Traditional multi-modality registration methods can be categorized into feature-based and region similarity-based approaches. Feature-based methods [2,12] demonstrate robustness against rotation and scale variations but struggle with radiometric discrepancies. Region similarity-based approaches [5,8] utilize statistical metrics to perform regional template matching to find relevant regions, which tolerate grayscale differences but fail under severe radiometric mismatches.

With the development of deep learning, some studies have begun employing deep neural networks to predict deformation field parameters, wrapping them onto the moving images to accomplish the registration task. MU-Net [24] improved generalization through multi-scale architecture and feature-driven loss constructed by the high-level features. To further reduce modality discrepancies in image pairs, recent studies have employed image-to-image (I2I) translation networks to perform modality transfer, followed by mono-modal deep registration networks for deformation field prediction. Nemar [1] adopted a GAN-based I2I translation network to mitigate modality differences, while jointly training the modality translation network and registration network to preserve geometric consistency. DFMIR [4] proposed a discriminator-free framework using contrastive learning to avoid structural artifacts. Closest to our work, UMF [20] integrated CPSTN, which generates pseudo-infrared images through perceptual style transfer and cross-regularized two-cycle learning, with MRRN for iterative features refinement, while its enhanced version IMF [21] advanced the framework by incorporating multi-scale progressive alignment.

3 Methods

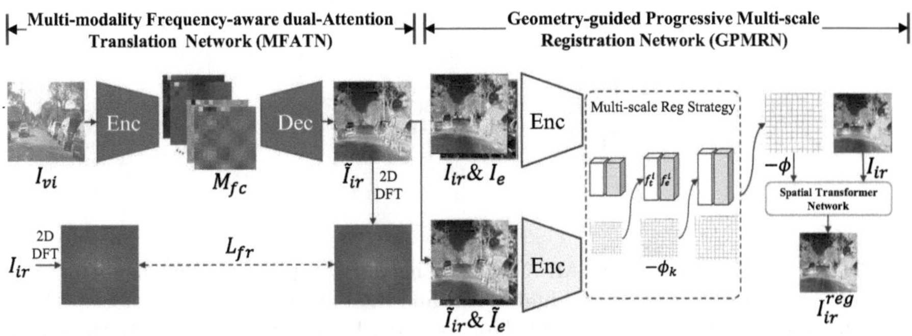

Fig. 1. The overall workflow of the proposed FGR. MFATN leverages spatial and frequency domain information for modality translation, generating the pseudo-infrared image $\tilde{I}_{ir}$. The GPMRN registers I_{ir} against $\tilde{I}_{ir}$, extracting and learning texture features f_t^i and structural features f_e^i. A multi-scale progressive strategy refines the distorted image, yielding the registered IR image I_{ir}^{reg}.

3.1 Overall Workflow

Figure 1 presents the workflow of our FGR. The visible image I_{vis} is initially transformed into the pseudo-infrared image $\tilde{I}_{ir}$ through the proposed MFATN, which leverages frequency domain information to reduce intra-modal discrepancies and enhance subsequent registration performance. Subsequently, the proposed GPMRN aligns $\tilde{I}_{ir}$ as the fixed image with I_{ir} as the moving image, utilizing their corresponding edge structure and texture features. This network iteratively refines spatial deformations through a multi-scale strategy, ultimately producing the registered infrared image I_{ir}^{reg}.

3.2 Multi-modality Frequency-Aware Image Translation

To bridge the modality gap between visible and infrared images and facilitate precise deformation field estimation in subsequent geometric registration, the Multi-modality Frequency-aware dual-Attention Translation Network (MFATN) is proposed for modality translation. The translation process is defined as:

$$\tilde{I}_{ir} = T\left(I_{vi}, \theta_T\right), \tag{1}$$

where $T\left(\cdot\right)$ refers to the MFATN with parameter θ_T.

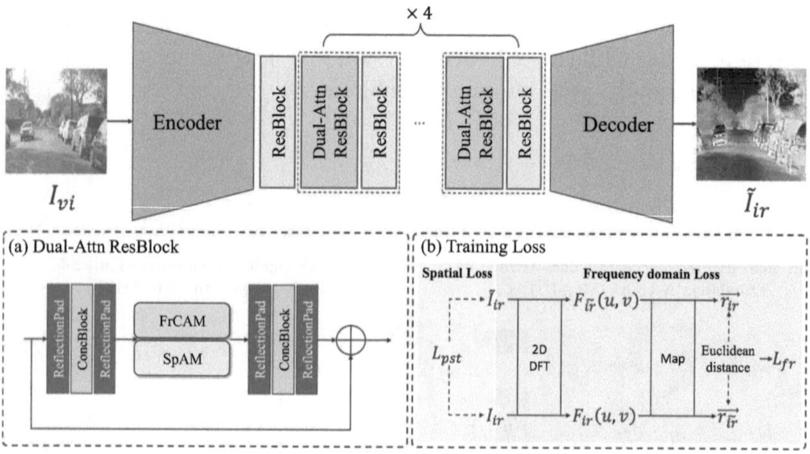

Fig. 2. The architecture of our MFATN. The network employs a UNet-like generator that integrates both standard residual blocks and dual-attention residual blocks (Dual-Attn ResBlock). (a) The structure of the Dual-Attn ResBlock. We integrate the multi-spectral channel attention module (FrCAM) [16] and spatial attention module (SpAM). (b) We use dual-domain loss to train MFATN. L_{pst} is the spatial loss proposed by CPSTN [20] and L_{fr} is the Focal Frequency Loss [7].

As illustrated in Fig. 2, MFATN employs a UNet-like generator integrating standard residual blocks and dual-attention residual blocks (Dual-Attn ResBlock). Given the intermediate feature map of the visible image $F_v \in \mathbb{R}^{C\times H\times W}$, the proposed dual-attention mechanism sequentially computes the multispectral channel attention map $M_{fc}(F_v) \in \mathbb{R}^{C\times1\times1}$ and the spatial attention map $M_{sp}(F_v{}') \in \mathbb{R}^{1\times H\times W}$. This process can be formulated as:

$$F_v{}' = M_{fc}\left(F_v\right) \otimes F_v, \tag{2}$$

$$F_v{}'' = M_{sp}\left(F_v{}'\right) \otimes F_v{}', \tag{3}$$

where $\otimes$ denotes element-wise multiplication and $F_v{}''$ denotes the feature map of the visible image after refinement by the dual-attention mechanism.

We first adopt multi-spectral channel attention mechanism (FrCAM) of FcaNet [16] to generate attention maps. Compared to conventional global average pooling (GAP), FrCAM better captures diverse feature map information. Given input feature map $F_v \in \mathbb{R}^{C\times H\times W}$, it is partitioned into n channel-wise sub-feature maps $\left[F_v^0, F_v^1, \cdots, F_v^{n-1}\right]$ where $F_v^i \in \mathbb{R}^{C'\times H\times W}$. Each sub-feature map undergoes 2D Discrete Cosine Transform (2D DCT), which can be formulated as:

$$\begin{aligned} Freq^i &= \mathrm{2DDCT}^{u_i,v_i}(F_v^i) \\ s.t.\ i &\in \{0, 1, \cdots, n-1\}, \end{aligned} \tag{4}$$

where $[u_i, v_i]$ denote the 2D frequency-domain coordinates, $Freq^i \in \mathbb{R}^{C'}$ represents the 2D DCT output of F_v^i, producing a C'-dimensional compressed vector. These vectors are then aggregated to yield the final frequency-compressed representation $M_{fc}\left(F_v\right)$, which can be described as follows:

$$M_{fc}\left(F_v\right) = Sigmoid\left(fc\left(Cat([Freq^0, Freq^1, \cdots, Freq^{n-1}])\right)\right), \tag{5}$$

where fc denotes the fully connector layer.

For spatial attention SpAM, we employ the Spatial Attention Module proposed by CBAM [22].

3.3 Geometry-Guided Progressive Multi-scale Registration

We propose a Geometry-guided Progressive Multi-scale Registration Network (GPMRN) to predict the deformation field between image pairs. The predicted deformation field is subsequently applied to I_{ir} to generate the registered infrared image I_{ir}^{reg}. The registration process is defined as:

$$-\phi = R\left(\tilde{I}_{ir}, I_{ir} \mid \theta_R\right), I_{ir}^{reg} = ST\left(I_{ir}, -\phi\right), \tag{6}$$

where $R\left(\cdot\right)$ denotes our proposed GPMRN, θ_R represents the trainable network parameters of GPMRN, and $ST\left(\cdot\right)$ indicates the spatial deformation operation.

As shown in Fig. 3, input images $\tilde{I}_{ir}$ and I_{ir} undergo grayscale conversion and Gaussian smoothing for noise reduction, followed by Laplace operator application to generate edge structure maps $\tilde{I}_e$ and I_e. A multi-scale extractor then constructs pyramids of edge-texture features, yielding paired multi-scale features $\left[\left(\tilde{f}_t^1, \tilde{f}_e^1\right), \cdots, \left(\tilde{f}_t^K, \tilde{f}_e^K\right)\right]$ and $\left[\left(f_t^1, f_e^1\right), \cdots, \left(f_t^K, f_e^K\right)\right]$. The complementary representation enhances deformation field prediction accuracy through multi-modality feature fusion.

To achieve progressive refinement in registration, we propose Texture-Structure Aware Module (T-SAM) to fully leverage both feature types for accurately predicting k-scale deformation fields. These fields will transform the unregistered infrared feature maps at the k-th scale. This process can be written as:

$$\hat{f}_e^k = ST\left(f_e^k, -\phi_{k-1}\right), \hat{f}_t^k = ST\left(f_t^k, -\phi_{k-1}\right), \quad s.t.\ k \in \{2, 3\}. \tag{7}$$

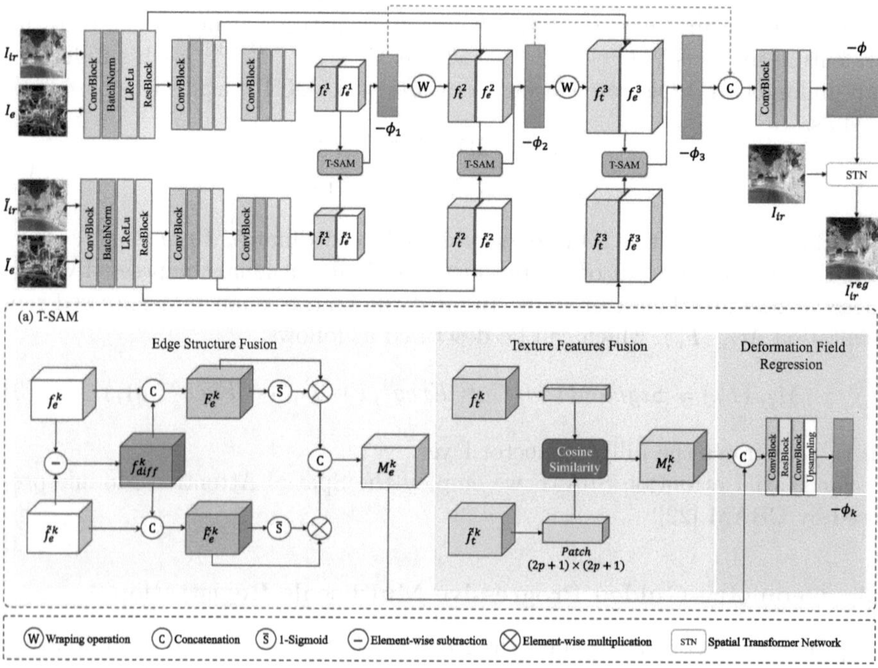

Fig. 3. The architecture of the Geometry-guided Progressive Multi-scale Registration Network (GPMRN). (a) The structure of the Texture-Structure Aware Module (T-SAM). T-SAM integrates edge structure and texture features to predict a deformation field.

As shown in Fig. 3(a), T-SAM first performs perception fusion on edge structure features and texture features separately, and finally regresses the intermediate deformation field results based on the fusion results. The edge structure fusion part first utilizes subtraction to suppress redundant features while enhancing discrepancies, producing f_{diff}^k. This is concatenated with $\tilde{f}_e^k$ and f_e^k to form refined features $\widetilde{F}_e^k$ and F_e^k. A reverse Sigmoid function subsequently transforms these features into probability maps, which are multiplied element-wise with original features to detect latent matches, generating $\widetilde{E}_e^k$ and E_e^k. These maps are concatenated to create the edge-aware correlation map M_e^k. The process is defined as:

$$\begin{gathered} f_{diff}^k = f_e^k - \tilde{f}_e^k, \\ \widetilde{F}_e^k = Cat\left(\tilde{f}_e^k, f_{diff}^k\right), F_e^k = Cat\left(f_e^k, f_{diff}^k\right), \\ \widetilde{E}_e^k = \widetilde{F}_e^k \otimes \left(1 - \sigma\left(\widetilde{F}_e^k\right)\right), E_e^k = F_e^k \otimes \left(1 - \sigma\left(F_e^k\right)\right), \\ M_e^k = Cat\left(\widetilde{E}_e^k, E_e^k\right). \end{gathered} \tag{8}$$

In texture features fusion part, to compute texture-aware correlation maps, we utilize localized similarity. As depicted in Fig. 3(a), for each feature point in the distorted texture feature map f_t^k, we calculate its similarity to a $(2p+1) \times (2p+1)$ local neighborhood centered at the corresponding position in the target texture feature map $\tilde{f}_t^k$ using cosine similarity, which can be represented as:

$$c^i = \frac{f_t^k(u,v) \cdot \tilde{f}_t^k(u-m, v-n)}{\|f_t^k(u,v)\| \|\tilde{f}_t^k(u-m, v-n)\|}, \tag{9}$$

where $m, n \in [-p, \ldots, p]$. We set $P = 2p+1$ to 3 (i.e., $p = 1$), resulting in a 3×3 local region. Here, $c^i \in \mathbb{R}^{1 \times 1 \times P^2}$ denotes a local similarity vector. The final texture-aware correlation map M_t^k has dimensions $H_k \times W_k \times P^2$, where H_k and W_k are the height and width of the input feature map, respectively.

After acquiring the correlation maps M_e^k and M_t^k, we first apply a 3×3 convolutional layer with LReLU activation to compress channel dimensions. The processed features are then passed to the last part for deformation field prediction. This part employs a decoder with deformable convolutions to estimate the deformation field $-\phi_k$.

3.4 Loss Functions

Loss Function for Image Translation. As shown in Fig. 2(b), to synthesize pseudo-infrared images with enhanced geometric fidelity, we employ a hybrid loss function comprising a spatial domain term L_{sp} and a frequency domain term L_{fr}, which collectively optimize MFATN by enforcing consistency in both pixel and spectral spaces.

For spatial domain loss, we adopt the Perceptual style transfer loss proposed by CPSTN [20] to form our spatial domain loss, i.e. $L_{sp} = L_{pst}$.

To guide the image translation network to emphasize underrepresented infrared frequency components, we introduce the Focal Frequency Loss (FFL) [7], which augments spatial loss to enhance pseudo-infrared image reconstruction fidelity. As shown in Fig. 2(b), FFL employs two-dimensional Discrete Fourier Transform (2D DFT) to derive the frequency representation $F(u,v)$ of an image. Through decomposition into imaginary $I(u,v)$ and real $R(u,v)$ components via Euler's formula, these components are mapped to the Cartesian coordinate system, yielding mapping vectors $\boldsymbol{r}_{ir}$ and $\boldsymbol{r}_{\widetilde{ir}}$. The Euclidean distance between these vectors quantifies frequency domain discrepancy by incorporating both amplitude $|F(u,v)|$ and phase $\angle F(u,v)$ information, enabling accurate modeling of frequency domain divergence between real and pseudo-infrared images.

The overall loss for image translation is defined as:

$$L_{trans} = L_{sp} + L_{fr} + L_{GAN}, \tag{10}$$

where L_{GAN} is an adversarial loss inherited from CycleGAN for optimizing the discriminator parameters.

Loss Function for Image Registration. We design a bidirectional pixel similarity loss, which measures consistency between the warped and target images across both directions. This loss is expressed as:

$$L_{sim}^{bi} = \| \psi_j (I_{ir}^{reg}) - \psi_j \left(\tilde{I}_{ir}\right) \|_1 + \lambda_{inv} \| \psi_j \left(\phi \circ \tilde{I}_{ir}\right) - \psi_j (I_{ir}) \|_1, \tag{11}$$

where $\lambda_{inv} = 0.3$. Furthermore, a smoothness loss L_{smooth} is introduced to ensure the generation of smooth deformation fields, formulated as:

$$L_{smooth} = \|\nabla\phi\|_1. \tag{12}$$

Finally, the overall registration loss is written as:

$$L_{reg} = L_{sim}^{bi} + \lambda_{sm} L_{smooth}, \tag{13}$$

where we set the smoothing term λ_{sm} to 10.

4 Experiments

4.1 Datasets and Evaluation Metrics

We evaluate our method on three multi-spectral datasets: TNO [18], RoadScene [23] and M3FD [11]. As these datasets provide pre-aligned image pairs, we introduced unregistered infrared images by applying random affine and elastic transformations to their infrared modality.

For multi-modality image registration evaluation, we employe three metrics: mean square error (MSE), normalized cross-correlation (NCC) [3] and mutual information (MI) [14]. These metrics are calculated between manually registered and algorithm-registered infrared images. MSE quantifies pixelwise intensity differences, whereas NCC and MI assess similarity via correlation windows and image entropy, respectively. Lower MSE and higher NCC and MI indicate superior registration performance.

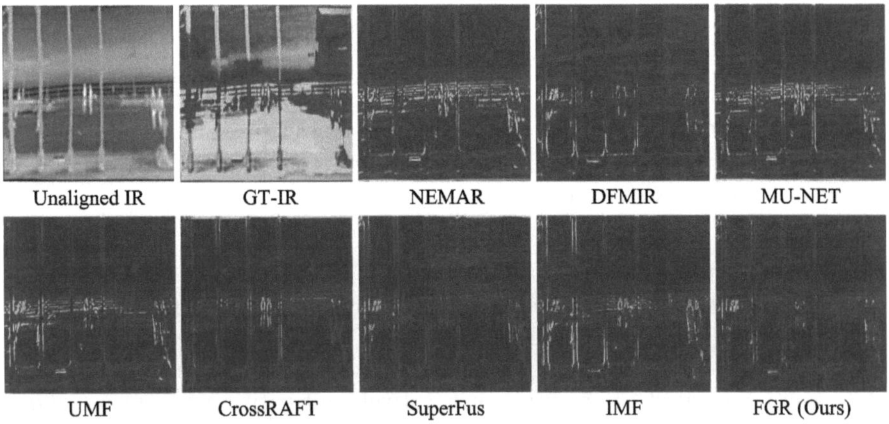

Fig. 4. Error maps visualization of the registration results on the RoadScene dataset.

4.2 Implementation Details

The proposed framework is implemented in PyTorch. During the training phase, we train on all images from the RoadScene and M3FD datasets, and randomly selected 4 image patches of size 256×256 to form a batch data. In the testing phase, we selected 107 images from the RoadScene dataset and all images from the TNO and M3FD datasets to construct the test dataset. We use a single NVIDIA 4090 GPU for training and testing. Model parameters are optimized using the Adam optimizer ($\beta_1 = 0.9$, $\beta_2 = 0.999$) with a fixed initial learning rate of 0.001 throughout the 400 training epochs.

Table 1. Quantitative comparison of multi-modality image registration methods on the TNO [18], RoadScene [23] and M3FD [11] datasets. The top results are highlighted in **bold**.

Methods	TNO			RoadScene			M3FD		
	MSE↓	NCC↑	MI↑	MSE↓	NCC↑	MI↑	MSE↓	NCC↑	MI↑
Unalign	0.009	0.818	1.397	0.018	0.834	1.370	0.008	0.888	2.048
NEMAR [1]	0.008	0.824	1.374	0.017	0.847	1.377	0.007	0.867	2.031
DFMIR [4]	0.034	0.447	0.916	0.004	0.957	1.766	0.004	0.941	2.099
MU-Net [24]	0.013	0.827	1.451	0.010	0.907	1.723	0.008	0.907	2.110
UMF [20]	0.006	0.874	1.443	0.010	0.909	1.514	0.005	0.929	2.128
CrossRAFT [26]	0.009	0.828	1.578	0.011	0.900	1.840	0.005	0.920	2.260
SuperFus [17]	0.007	0.862	1.485	0.005	0.952	**1.955**	0.004	0.941	2.189
IMF [21]	0.004	0.916	1.546	0.006	0.939	1.638	0.004	0.934	2.167
FGR (ours)	**0.003**	**0.935**	**1.613**	**0.004**	**0.959**	1.794	**0.003**	**0.962**	**2.358**

4.3 Comparison with State-of-the-Art Methods

Quantitative Comparison. We evaluate registration accuracy against state-of-the-art approaches: NEMAR [1], DFMIR [4], MU-NET [24], UMF [20], CrossRAFT [26], SuperFus [17] and IMF [21]. As showed in Table 1, our method achieves the highest scores across all metrics, indicating statistically significant improvements over existing methods.

Qualitative Comparison. As illustrated in Fig. 4, we visualize registration errors between algorithm-registered infrared images and ground-truth manually aligned targets on the synthetically misaligned RoadScene dataset, where brighter regions denote larger misalignment magnitudes. The proposed method exhibits superior registration accuracy, effectively mitigating artifacts and ghosting in subsequent fusion steps, thereby facilitating the generation of high-quality fused images.

4.4 Downstream Multi-modality Image Fusion Task

Setup and Evaluation Metrics. We perform the image fusion task on the unregistered M3FD dataset in conjunction with a SOTA multi-modality registration network, and the selected SOTA fusion network is CDDFuse [25]. The training epoch, batch size, optimizer, and initial learning rate were set to 400, 8, SGD optimizer, and 0.001, respectively. For fusion evaluation, we employ six metrics: information entropy (EN), mutual information (MI), cross-correlation (CC), sum of correlation differences (SCD), visual information fidelity (VIF), Q_{abf} and structural similarity index (SSIM). Detailed definitions of these metrics are provided in [13].

Table 2. Quantitative comparison results of the downstream multi-modality image fusion task. The best results are highlighted in **bold**.

Methods	EN↑	MI↑	CC↑	SCD↑	VIF↑	Qabf↑	SSIM↑
NEMAR [1]	6.966	3.801	0.399	1.477	0.592	0.532	0.823
DFMIR [4]	6.984	**3.987**	0.412	1.544	0.633	0.544	0.854
MU-NET [24]	6.968	3.781	0.409	1.516	0.560	0.521	0.824
UMF [20]	6.960	3.926	0.408	1.517	0.608	0.539	0.849
CrossRAFT [26]	6.934	3.862	0.404	1.498	0.599	0.535	0.859
SuperFus [17]	6.938	3.922	0.412	1.534	0.611	0.539	0.862
IMF [21]	6.951	3.923	0.410	1.529	0.609	0.540	0.855
FGR(Ours)	**7.041**	3.961	**0.419**	**1.562**	**0.636**	**0.545**	**0.878**

Comparison with SOTA Methods. As shown in Table 2, compared with existing SOTA multi-modality registration networks, the multi-modality fusion network incorporating the proposed FGR achieves superior performance on nearly all fusion metrics. This indicates that the proposed FGR can significantly enhance the performance of existing SOTA fusion networks in fusing unregistered multi-modality image pairs.

4.5 Ablation Studies

Effectiveness of MFATN. To validate the effectiveness of MFATN for multi-modal registration tasks, we disabled it and employed only GPMRN to accomplish the entire multimodal registration. As indicated in Table 3, disabling MFATN leads to a decline in registration accuracy. This underscores the critical role of MFATN in the registration task.

Table 3. Quantitative ablation studies of the MFATN on the RoadScene dataset. The best results are highlighted in **bold**.

MFATN	GPMRN	MSE↓	NCC↑	MI↑
×	✓	0.007	0.923	1.547
✓	✓	**0.004**	**0.959**	**1.794**

Effectiveness of Frequency-Domain Guidance in MFATN. To evaluate the effectiveness of frequency-domain Guidance in visible-to-infrared translation, we disabled key frequency-domain components in MFATN, including the FrCAM module and the frequency-domain loss term L_{fr}.

As shown in Table 4, by comparing the MSE, PNSR, SSIM and MI metrics between synthetic pseudo-infrared images and reference infrared images, the quantitative results reveal that removing FrCAM or L_{fr} individually reduces modality translation fidelity, while ablating both causes a severe performance drop. This confirms that frequency-domain constraints guide cross-modal feature alignment, enabling high-fidelity pseudo-infrared synthesis.

Effectiveness of T-SAM in GPMRN. To evaluate the effectiveness of T-SAM, we deactivated its two key parts.

The corresponding results are provided in the Exp. I, II and III of Table 5. Evidently, deactivating either key part of T-SAM degrades the registration performance of GPMRN. The results demonstrate that the proposed T-SAM significantly contributes to predicting precise registration deformation fields.

Table 4. Quantitative ablation studies of the FrCAM and frequency domain loss term L_{fr} of MFATN on the RoadScene dataset. The best results are highlighted in **bold**.

FrCAM	L_{fr}	MSE↓	PNSR↑	SSIM↑	MI↑
×	✓	0.017	18.005	0.405	1.149
✓	×	0.018	18.014	0.423	1.161
×	×	0.046	14.052	0.368	0.958
✓	✓	**0.012**	**19.320**	**0.446**	**1.223**

Effectiveness of Progressive Refinement Strategy in GPMRN. In the Exp. IV of Table 5, we validate the effectiveness of the progressive refinement strategy in GPMRN. The results demonstrate that removal of the progressive refinement strategy leads to degraded registration performance of FGR. This not only indicates the superiority of this strategy over single-step registration, but also confirms its capability to comprehensively rectify spatial distortions across more locations in infrared images.

Table 5. Ablation studies of the modules and registration strategy in GPMRN on the RoadScene dataset. The best results are highlighted in **bold**.

Exp.	Modules or Strategy	MSE↓	NCC↑	MI↑
I	w/o Edge Structure Fusion	0.005	0.956	1.765
II	w/o Texture Features Fusion	0.005	0.952	1.735
III	w/o T-SAM	0.006	0.947	1.685
IV	w/o progressive refinement	0.005	0.954	1.761
V	**FGR (Ours)**	**0.004**	**0.959**	**1.794**

5 Conclusion

In this work, we propose FGR, a novel multi-modality registration framework addressing misalignment between heterogeneous sensors. Our solution employs frequency-aware modality translation to mitigate the modality gap through cross-domain information, reducing registration complexity and strengthening inter-modal correlations. The framework incorporates a Texture-Structure Aware Module that guides the deep registration network to extract discriminative geometric edge structures and integrate them with texture features across progressive registration stages. This dual-pathway design enables precise alignment by identifying geometrically consistent, modality-invariant features. Experiments on three benchmark datasets show FGR achieves superior registration accuracy and significantly improves downstream fusion models' performance on unregistered image pairs.

Acknowledgments. This work was supported by the National Natural Science Foundation of China (62472106), Guangdong Provincial Key Laboratory of Cyber-Physical System (2020B1212060069).

References

1. Arar, M., Ginger, Y., Danon, D., Bermano, A.H., Cohen-Or, D.: Unsupervised multi-modal image registration via geometry preserving image-to-image translation. In: Proceedings of the IEEE/CVF Conference on Computer Vision and Pattern Recognition, pp. 13410–13419 (2020)
2. Bay, H., Ess, A., Tuytelaars, T., Van Gool, L.: Speeded-up robust features (surf). Comput. Vis. Image Underst. **110**(3), 346–359 (2008)
3. Cao, X., Yang, J., Wang, L., Xue, Z., Wang, Q., Shen, D.: Deep learning based inter-modality image registration supervised by intra-modality similarity. In: Machine Learning in Medical Imaging: 9th International Workshop, MLMI 2018, Held in Conjunction with MICCAI 2018, Granada, Spain, September 16, 2018, Proceedings 9, pp. 55–63. Springer (2018)
4. Chen, Z., Wei, J., Li, R.: Unsupervised Multi-modal Medical Image Registration Via Discriminator-Free Image-to-image Translation. arXiv preprint arXiv:2204.13656 (2022)
5. Hisham, M., Yaakob, S.N., Raof, R., Nazren, A.A., Wafi, N.: Template matching using sum of squared difference and normalized cross correlation. In: 2015 IEEE Student Conference On Research and Development (SCOReD), pp. 100–104. IEEE (2015)
6. Jaderberg, M., Simonyan, K., Zisserman, A., et al.: Spatial transformer networks. Adv. Neural Inf. Proc. Syst. **28** (2015)
7. Jiang, L., Dai, B., Wu, W., Loy, C.C.: Focal frequency loss for image reconstruction and synthesis. In: Proceedings of the IEEE/CVF International Conference on Computer Vision, pp. 13919–13929 (2021)
8. Kern, J.P., Pattichis, M.S.: Robust multispectral image registration using mutual-information models. IEEE Trans. Geosci. Remote Sens. **45**(5), 1494–1505 (2007)
9. Kniaz, V.V., Knyaz, V.A., Hladuvka, J., Kropatsch, W.G., Mizginov, V.: ThermalGAN: Multimodal color-to-thermal image translation for person re-identification in multispectral dataset. In: Proceedings of the European Conference on Computer Vision (ECCV) workshops, pp. 0–0 (2018)
10. Li, B., Xian, Y., Su, J., Zhang, D.Q., Guo, W.L.: I-GANs for infrared image generation. Complexity **2021**(1), 6635242 (2021)
11. Liu, J., et al.: Target-aware dual adversarial learning and a multi-scenario multi-modality benchmark to fuse infrared and visible for object detection. In: Proceedings of the IEEE/CVF Conference on Computer Vision and Pattern Recognition, pp. 5802–5811 (2022)
12. Lowe, D.G.: Distinctive image features from scale-invariant keypoints. Int. J. Comput. Vision **60**, 91–110 (2004)
13. Ma, J., Ma, Y., Li, C.: Infrared and visible image fusion methods and applications: a survey. Inf. Fusion **45**, 153–178 (2019)
14. Mahapatra, D., Ge, Z., Sedai, S., Chakravorty, R.: Joint registration and segmentation of xray images using generative adversarial networks. In: Machine Learning in Medical Imaging: 9th International Workshop, MLMI 2018, Held in Conjunction with MICCAI 2018, Granada, Spain, September 16, 2018, Proceedings 9, pp. 73–80. Springer (2018)

15. Qian, X., Zhang, M., Zhang, F.: Sparse GANs for thermal infrared image generation from optical image. IEEE Access **8**, 180124–180132 (2020)
16. Qin, Z., Zhang, P., Wu, F., Li, X.: FcaNet: frequency channel attention networks. In: Proceedings of the IEEE/CVF International Conference on Computer Vision, pp. 783–792 (2021)
17. Tang, L., Deng, Y., Ma, Y., Huang, J., Ma, J.: Superfusion: a versatile image registration and fusion network with semantic awareness. IEEE/CAA J. Automatica Sin. **9**(12), 2121–2137 (2022)
18. Toet, A., Hogervorst, M.A.: Progress in color night vision. Opt. Eng. **51**(1), 010901–010901 (2012)
19. Uddin, M.S., Hoque, R., Islam, K.A., Kwan, C., Gribben, D., Li, J.: Converting optical videos to infrared videos using attention GAN and its impact on target detection and classification performance. Remote Sens. **13**(16), 3257 (2021)
20. Wang, D., Liu, J., Fan, X., Liu, R.: Unsupervised misaligned infrared and visible image fusion via cross-modality image generation and registration. arXiv preprint arXiv:2205.11876 (2022)
21. Wang, D., Liu, J., Ma, L., Liu, R., Fan, X.: Improving misaligned multi-modality image fusion with one-stage progressive dense registration. IEEE Trans. Circuits Syst. Video Technol. (2024)
22. Woo, S., Park, J., Lee, J.Y., Kweon, I.S.: CBAM: convolutional block attention module. In: Proceedings of the European conference on computer vision (ECCV), pp. 3–19 (2018)
23. Xu, H., Ma, J., Le, Z., Jiang, J., Guo, X.: FusionDN: a unified densely connected network for image fusion. In: Proceedings of the AAAI conference on artificial intelligence, vol. 34, pp. 12484–12491 (2020)
24. Ye, Y., Tang, T., Zhu, B., Yang, C., Li, B., Hao, S.: A multiscale framework with unsupervised learning for remote sensing image registration. IEEE Trans. Geosci. Remote Sens. **60**, 1–15 (2022)
25. Zhao, Z., et al.: CDDFuse: correlation-driven dual-branch feature decomposition for multi-modality image fusion. In: Proceedings of the IEEE/CVF Conference on Computer Vision and Pattern Recognition, pp. 5906–5916 (2023)
26. Zhou, S., Tan, W., Yan, B.: Promoting single-modal optical flow network for diverse cross-modal flow estimation. In: Proceedings of the AAAI Conference on Artificial Intelligence, vol. 36, pp. 3562–3570 (2022)
27. Zhu, J.Y., Park, T., Isola, P., Efros, A.A.: Unpaired image-to-image translation using cycle-consistent adversarial networks. In: Proceedings of the IEEE International Conference on Computer Vision, pp. 2223–2232 (2017)

MedFuse-GRM: Multi-scale Feature Extraction and Medically-Guided Graph Relation Modeling for Multimodal Skin Lesion Classification

Zhangyi Wang and Zongze Li(✉)

Nanyang Technological University, Singapore, Singapore
{zhangyi001,li0005ze}@e.ntu.edu.sg

Abstract. Multimodal skin lesion classification methods typically suffer from three limitations: relying on single-scale features, using generic ImageNet pre-training, and treating classification tasks independently without considering medical relationships. We propose MedFuse-GRM, a medically-driven framework with three key components: (1) a multi-scale feature extraction network with cross-attention guided fusion for comprehensive visual representations, (2) contrastive learning pre-training to enhance cross-modal feature consistency between dermoscopy and clinical images, and (3) a medically-guided graph relation module that leverages domain knowledge to model structured connections between diagnostic categories and pathological characteristics. Experiments on the Seven-Point Checklist dataset show MedFuse-GRM achieves 79.7% average accuracy, outperforming state-of-the-art methods by 2.1% points, with ablation studies confirming each component's substantial contribution.

Keywords: Deep Learning · Medical and Biological Imaging · Graph-based Machine Learning · Multi-modal Vision · Skin Lesion Diagnosis · Contrastive learning

1 Introduction

Skin cancer is one of the most prevalent malignancies worldwide, with incidence rates continuing to rise globally [1,30]. Early and accurate diagnosis is critical, as timely detection can elevate 10-year survival rates to over 93% [3]. However, the severe global shortage of dermatology specialists, particularly in resource-limited regions, presents a significant barrier to effective screening and diagnosis [35]. As a result, automated skin cancer diagnosis has become a key focus in medical artificial intelligence research, aiming to enhance diagnostic efficiency and accuracy.

In clinical practice, dermatologists routinely rely on two imaging modalities: **clinical images** (Fig. 1(a)), captured with standard digital cameras,

J. Lokoč et al. (Eds.): MMM 2026, LNCS 16413, pp. 59–73, 2026.
https://doi.org/10.1007/978-981-95-6957-1_5

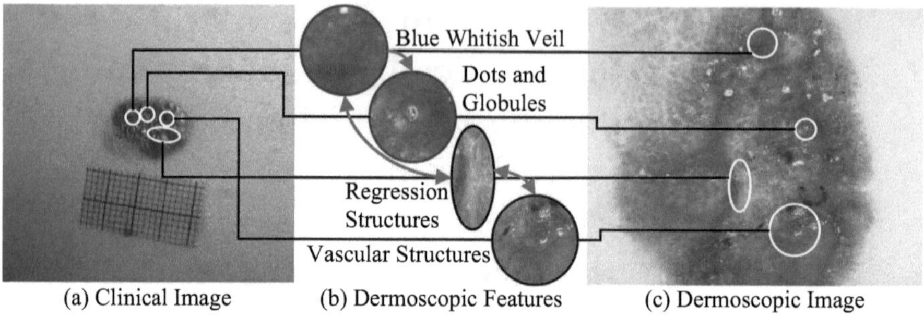

Fig. 1. Example of a biopsy-confirmed melanoma case from the Seven-Point Checklist dataset, shown in both clinical and dermoscopic views. Red arrows highlight medically-guided relationships between dermoscopic features (e.g., blue-whitish veil and regression structures) that are strongly indicative of melanoma. (Color figure online)

which reveal the geometric morphology and color of lesions while preserving macroscopic shape and boundary information [13]; and **dermoscopic images** (Fig. 1(c)), acquired through optical magnification and specialized illumination, which expose fine subsurface structures and texture patterns [6]. These two modalities offer macroscopic and microscopic perspectives of skin lesions, respectively, and their integration has shown to improve diagnostic performance by providing a more comprehensive view of lesion characteristics [7]. In addition to visual assessment, clinicians also follow established diagnostic criteria such as the Seven-Point Checklist [18], which evaluates the presence of dermoscopic features (Fig. 1(b)), such as dots and globules or blue-whitish veils, to guide diagnosis and assess malignancy risk.

With the advent of deep learning, convolutional neural networks (CNNs) have significantly advanced medical image analysis by enabling automatic, end-to-end feature extraction [13,18]. Despite their strong performance, most CNN-based studies focus on a single modality (either dermoscopic or clinical images), thereby failing to leverage the complementary information offered by multimodal inputs. In recent years, multimodal fusion methods [5,14,34] have gained prominence in skin lesion classification research, demonstrating superior performance compared to unimodal approaches. However, current existing multimodal frameworks encounter several critical limitations that hamper their effectiveness in accurately diagnosing skin lesions.

First, many existing multimodal approaches [18,33,38] rely exclusively on single-scale features from the final layers of backbone networks like ResNet-50. This limits their ability to capture the multi-scale nature of skin lesions and often leads to the loss of fine-grained details, such as pigment network irregularities, dot patterns, and vascular structures, that are critical for accurate diagnosis. As a result, these models produce incomplete representations that fail to fully characterize the complex features of skin lesions.

Second, multimodal skin lesion classification faces significant challenges from the substantial distribution mismatch between dermoscopy and clinical

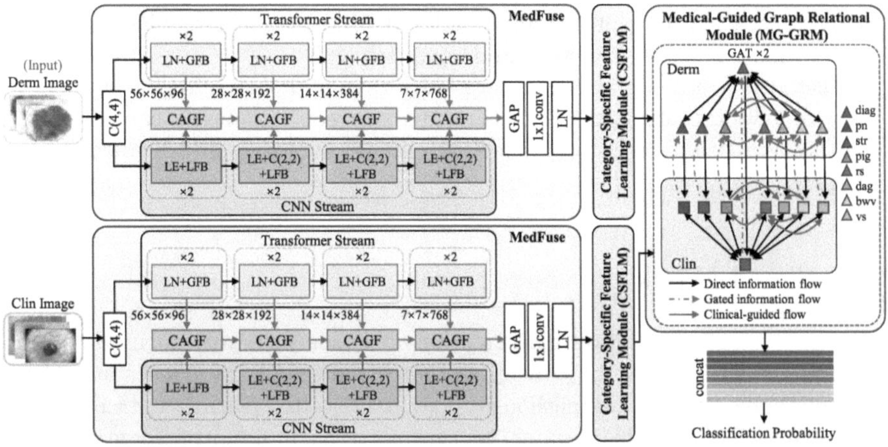

Fig. 2. The MedFuse-GRM framework performs multi-modal skin lesion classification by first extracting multi-scale features from dermoscopy and clinical images (MedFuse), then refining these features for specific diagnostic classes (CSFLM), and finally modeling medical relationships between the diagnostic category and seven pathological characteristics using a graph network (MG-GRM).

images. Contemporary methods [5,14,40] typically use ImageNet-pretrained feature extractors, which introduce fundamental incompatibilities between natural image characteristics and specialized medical imaging properties. Moreover, these approaches extract features from each modality independently, without optimizing for cross-modal consistency. This results in poorly aligned feature spaces that hinder effective integration of complementary information and lead to suboptimal performance in clinical applications.

Third, existing approaches [29,36,41] treat the eight classification tasks (seven dermoscopic features and a diagnostic category) in the Seven-Point Checklist [18] as independent and unrelated tasks. However, extensive clinical studies have demonstrated that certain dermoscopic features exhibit strong medically-guided relationships. For example as illustrated in Fig. 1(b), the presence of a blue-whitish veil (BWV) is often associated with dots and globules (DAG) [2,23], while vascular structures (VS) are known to co-occur with regression structures (RS) [23,39]. Additionally, BWV and RS are frequently observed together in melanoma cases [28]. By ignoring these medically established interdependencies, current methods fail to leverage valuable medical knowledge that could substantially enhance diagnostic accuracy.

To address these critical limitations, we propose MedFuse-GRM, a novel medically-driven multimodal framework that incorporates three core components, as illustrated in Fig. 2:

- Multi-scale feature extraction with cross-attention guided fusion that integrates CNN and Transformer capabilities to capture detailed and contextual lesion features;

- Contrastive learning pre-training tailored for skin images, which enhances cross-modal feature alignment and consistency;
- A medically-guided graph relation modeling module (MG-GRM) that explicitly models the structured relationships between diagnostic categories and pathological characteristics, reflecting clinical reasoning and medical knowledge.

2 Related Work

Early automated skin lesion classification methods primarily focused on dermoscopic images and can be categorized into traditional approaches and CNN-based methods. Traditional methods rely on handcrafted features extracted from images, including color histograms, texture patterns, and morphological characteristics, which are then classified using support vector machines, random forests, or AdaBoost classifiers [4,27]. However, these approaches exhibit strong dependencies on image preprocessing and expert knowledge, and demonstrate limited robustness when dealing with dermoscopic images captured under varying imaging devices and lighting conditions.

Deep learning approaches for skin lesion classification have evolved significantly, yet existing methods still suffer from three critical limitations that our MedFuse-GRM framework is designed to address:

Feature Extraction. CNN-based architectures [15,31] have gained widespread adoption, with methods like [19,24] showing performance improvements through ImageNet pretraining and fine-tuning. More recently, Transformer-based approaches [16] have been explored for their ability to capture long-range dependencies and contextual information through hierarchical attention structures. Despite these advances, both CNN- and Transformer-based methods are often single-modal and predominantly rely on single-scale features, typically extracted from the final layers of their backbone networks. To address modality complementarity, recent multimodal fusion methods have been proposed. For instance, MSMA [29] employs cross-modal feature interaction modules, TFormer [40] uses hierarchical multimodal Transformer blocks, MBF-HA [36] incorporates hybrid attention mechanisms, and HcCNN [33] applies multi-scale attention blocks. While these models have introduced more sophisticated fusion strategies, they still largely rely on limited feature scales, which constrains their ability to fully capture and integrate fine-grained pathological cues. In contrast, our MedFuse network introduces a cross-attention guided fusion mechanism that combines the strengths of CNNs and Transformers across multiple feature levels. By jointly capturing both local details and global semantic context, MedFuse produces comprehensive multi-scale representations of skin lesions, which is essential for accurate and robust diagnosis.

Representation Learning. Existing approaches including Deep Multi-Modal Learning (DMML) methods [26] and their applications across various medical imaging tasks [10,11,17,20–22] rely heavily on ImageNet pre-training while neglecting specialized representation learning for medical images. This introduces substantial domain gaps between natural and skin lesion images, which in turn degrade cross-modal feature consistency. Our framework addresses this limitation through specialized contrastive learning pre-training, which optimizes cross-modal feature alignment by pulling together dermoscopic and clinical representations from the same case in feature space, while pushing apart representations from different cases. This approach significantly reduces domain-specific representation gaps and enhances the model's ability to integrate multimodal information effectively.

Medical Knowledge Integration. Current methods treat the eight classification tasks in the Seven-Point Checklist as independent problems [17], overlooking the inherent medical relationships between diagnostic categories and pathological characteristics that form the foundation of clinical diagnostic reasoning. Our medically-guided graph relation modeling module establishes structured associations between diagnostic categories (DIAG) and pathological characteristic tasks (7PC) based on domain knowledge. This design enables mutual task influence in a way that aligns with clinical diagnostic logic and leverages the interdependencies that dermatologists naturally consider during diagnosis.

3 Methods

3.1 Overview

Figure 2 illustrates the comprehensive architecture of MedFuse-GRM, comprising three core modules: (1) MedFuse (Multi-scale Feature Extraction Network); (2) CSFLM (Class-Specific Feature Learning Module); and (3) MG-GRM (Medically-Guided Graph Relation Modeling Module). The model integrates features from these modules to achieve precise multi-label skin disease classification.

3.2 MedFuse

Stem Structure. The Stem structure serves to reduce dimensionality and generate initial feature representations. Given an input image $X \in \mathbb{R}^{224\times224\times3}$, a 4×4 convolution with stride 4 is applied for downsampling, producing feature maps $F_0 \in \mathbb{R}^{56\times56\times96}$ that serve as input for subsequent feature extraction modules. The Stem structure then bifurcates into two parallel pathways: the Global Path feeding into the GFB and the Local Path feeding into the LFB.

GFB and LFB. We get inspiration from [17] to design our feature extraction blocks. The Global Feature Block implements windowed multi-head self-attention (W-MSA) to capture long-range dependencies and global pathological structures, while the Local Feature Block employs depthwise convolution to efficiently extract fine-grained local details such as pigment networks and vascular structures.

CAGF CAGF establishes bidirectional information exchange between global features (GFB output F_g^{out}) and local features (LFB output F_l^{out}) to optimize multi-scale feature fusion. It incorporates a Cross-Channel Spatial Attention (CCSA) mechanism to enhance feature compatibility.

The fusion process consists of:

$$\hat{F}_g = \sigma\Big(f_{7\times7}\big(\text{AvgPool}(F_l^{\text{out}}) + \text{MaxPool}(F_l^{\text{out}})\big)\Big) \otimes F_g^{\text{out}} \tag{1}$$

$$\hat{F}_l = \sigma\Big(\text{MLP}\big(\text{AvgPool}(F_g^{\text{out}})\big) + \text{MLP}\big(\text{MaxPool}(F_g^{\text{out}})\big)\Big) \otimes F_l^{\text{out}} \tag{2}$$

$$\hat{F} = f_{1\times1}(\text{Concat}[\hat{F}_g, \hat{F}_l]) \tag{3}$$

$$F^{\text{final}} = \text{IRMLP}(\text{LN}(\hat{F})) + \hat{F} \tag{4}$$

where σ represents the Sigmoid activation function; $\otimes$ denotes Hadamard product (element-wise multiplication); AvgPool and MaxPool represent average pooling and max pooling operations, respectively; $f_{7\times7}$ denotes 7×7 convolution; MLP refers to multi-layer perceptron; Concat represents feature concatenation operation; and IRMLP denotes inverted residual multi-layer perceptron with skip connection.

The resulting fused features F^{final} are then forwarded to subsequent modules. For clarity, throughout the remainder of this paper, we denote these MedFuse output features as F_{MedFuse}^{D} for dermoscopy images and F_{MedFuse}^{C} for clinical images.

Through the integration of global-local feature interaction and CAGF, MedFuse effectively addresses the limitations of single-scale feature extraction approaches. The following section will detail the contrastive learning pre-training strategy employed to enhance cross-modal feature consistency.

3.3 Contrastive Learning Pre-Training for MedFuse

To address cross-modal feature inconsistency and overfitting on small datasets, we introduce contrastive learning during MedFuse pre-training. Our approach aims to: (1) enhance cross-modal feature consistency between dermoscopy and clinical images from the same case; (2) improve discrimination between different cases; and (3) provide higher-quality feature representations for downstream tasks.

We implement an in-batch contrastive learning scheme with:

- **Positive Pairs**: (D_i, C_i) - dermoscopy and clinical images from the same case
- **Negative Pairs**: $(D_i, C_j), j \neq i$ - images from different cases within the mini-batch

Two separate MedFuse branches extract features from each modality, where F^D_{MedFuse} is the feature extracted by the MedFuse dermoscopy branch, F^C_{MedFuse} is the feature extracted by the MedFuse clinical branch, and $F^{C_j}_{\text{MedFuse}}$ is the feature of a randomly selected negative sample from the batch. We optimize using Normalized Temperature-scaled Cross-Entropy Loss (NT-Xent) [9,32]:

$$L = -\log \frac{\exp(s_{DC}/\tau)}{\exp(s_{DC}/\tau) + \sum_{j \neq i} \exp(s_{DC_j}/\tau)} \tag{5}$$

where $s_{DC} = \text{sim}(F^D_{\text{MedFuse}}, F^C_{\text{MedFuse}})$ represents cosine similarity between positive pairs, and $s_{DC_j} = \text{sim}(F^D_{\text{MedFuse}}, F^{C_j}_{\text{MedFuse}})$ for negative pairs. The temperature parameter τ is set to the same 0.07 as in [37] to control the distribution smoothness.

This approach enhances modality alignment while improving the model's cross-case discrimination ability without requiring additional negative sample banks.

After training, the MedFuse model pre-trained through contrastive learning provides robust feature representations for subsequent tasks. These features feed into CSFLM, which further refines them with category-specific supervision signals.

3.4 CSFLM

The Category-Specific Feature Learning Module optimizes features through auxiliary classification signals before they enter the Graph Relational Module. CSFLM contains eight independent units corresponding to diagnostic category and seven pathological characteristics.

For each modality, CSFLM transforms features and generates auxiliary predictions through:

$$h_j^{\text{CSFLM},D/C} = W_0^{D/C,j} \cdot F^{D/C}_{\text{MedFuse}} + b_0^{D/C,j} \tag{6}$$

$$p_j^{\text{CSFLM},D/C} = \sigma\left(W_1^{D/C,j} \cdot \text{ReLU}(h_j^{\text{CSFLM},D/C}) + b_1^{D/C,j}\right) \tag{7}$$

where $h_j^{\text{CSFLM}} \in \mathbb{R}^{512\times 1}$ represents intermediate feature vectors, and $p_j^{\text{CSFLM}} \in \mathbb{R}^{1\times k}$ represents auxiliary predictions. The D/C notation indicates identical mathematical structures applied separately to dermoscopy and clinical modalities, with four trainable parameter: weight matrices $W_0^{D/C,j} \in \mathbb{R}^{512\times 512}$ for feature transformation, $W_1^{D/C,j} \in \mathbb{R}^{512\times k}$ for auxiliary classification, and corresponding bias vectors $b_0^{D/C,j} \in \mathbb{R}^{512\times 1}$ and $b_1^{D/C,j} \in \mathbb{R}^{1\times k}$. These auxiliary predictions generate auxiliary cross-entropy losses $L_{\text{CSFLM},D}$ and $L_{\text{CSFLM},C}$ (detailed in Section Final Category Classification and Model Training) that contribute to training but are discarded during inference.

3.5 MG-GRM

The Medically-Guided Graph Relational Module establishes structured relationships between diagnostic categories and pathological characteristics to enhance the model's interpretability and classification performance. As illustrated in Fig. 2, MG-GRM consists of a two-layer Graph Convolutional Network (GCN) where both layers share the same topology structure. The network models relationships between categories based on medical knowledge, with the diagnostic category (DIAG) serving as a central node connected bidirectionally to all 7PC categories: Pigment Network (PN), Streaks (STR), Pigmentation (PIG), Regression Structures (RS), Dots and Globules (DAG), Blue Whitish Veil (BWV), and Vascular Structures (VS), forming a star-like structure. Each GCN layer maintains two identical topological structures corresponding to the dermoscopy and clinical modalities, with information flow facilitated through direct connections from dermoscopy to clinical modality and gated connections from clinical to dermoscopy, while additional medically-informed connections between pathological characteristics are established based on clinical evidence and co-occurrence patterns.

Medically-Informed Graph Topology. We optimized the basic star topology by incorporating additional intra-Skin Pathological Characteristics (intra-SPC) connections supported by clinical evidence. These connections reflect co-occurrence and hierarchical relationships among skin pathological characteristics as documented in dermatological literature, and are represented as Clinical-guided flow in Fig. 2. Key medical connections include:

- **BWV → PIG**: Studies indicate that blue-whitish veil often co-occurs with pigment irregularities in melanoma lesions, typically appearing at later stages accompanied by significant pigment abnormalities [2,8].
- **BWV → DaG**: Literature documents the frequent co-occurrence of blue-whitish veil and irregular dots and globular structures in progressive melanoma. Blue-whitish veil often marks lesions that have undergone irregular pigmentation phases, supporting this unidirectional connection [2,23].
- **PIG → DaG**: Research shows that irregular pigment network boundaries commonly accompany irregular dots or globular structures, reflecting typical progression sequence in the radial growth phase of melanoma [2].
- **VS ↔ RS**: Multiple clinical studies have found strong bidirectional association between regression phenomena and abnormal vascular patterns in melanoma [23,39].
- **PIG ↔ RS**: Early melanoma studies demonstrate frequent simultaneous appearance of pigment network irregularities and regression structures, serving as important combined indicators for early melanoma diagnosis [25].
- **BWV ↔ RS**: Clinical observations reveal that blue-whitish veil and regression structures often coexist in different regions of the same lesion, suggesting simultaneous tumor processes (invasion and regression), warranting this bidirectional connection [28].

These medically-supported connections create a more realistic representation of feature relationships in actual clinical diagnosis, enhancing the model's ability to capture complex pathological patterns and combination phenomena that are crucial for accurate diagnosis.

Cross-Modal Information Flow. We implement direct connections from dermoscopy to clinical modality using standard Graph Attention Network mechanics to ensure effective transfer of fine-grained structural features, acknowledging dermoscopy's superiority in providing detailed diagnostic information. For the clinical to dermoscopy direction, we employ a Gated Modulation mechanism to regulate information propagation because clinical modality typically contains more noise that could potentially degrade high-quality dermoscopy features, ensuring only high-confidence contextual information influences dermoscopy features while preserving their intrinsic quality.

Graph Attention Mechanism. The MG-GRM employs Graph Attention Networks (GAN) with multi-head attention mechanisms to propagate and aggregate information among connected nodes, adaptively emphasizing important features while downplaying irrelevant inputs through learned attention weights. Through this two-layer GAN design with medically-informed topology and cross-modal information flow mechanisms [12], MG-GRM effectively models the inherent relationships among diagnostic categories and pathological characteristics, capturing both direct connections and indirect associations through hierarchical propagation.

3.6 Final Category Classification and Model Training

After processing through the two-layer MG-GRM, the category feature vectors integrate complementary information from both dermoscopy and clinical modalities. The output node vectors are concatenated to form eight task-specific feature vectors (1024-dimensional), which serve as inputs to the Final Category Prediction (FCP) units corresponding to seven pathological characteristics and one diagnostic category. Each FCP unit employs a two-layer fully connected network with ReLU activation and Softmax output to generate final prediction probabilities.

The model training employs a joint optimization strategy with a composite loss function consisting of three components: the primary FCP cross-entropy loss for final predictions, and auxiliary CSFLM losses from dermoscopy and clinical branches (averaged with equal weights of 0.5). This multi-task learning framework enables information sharing across classification tasks and creates deeper supervision through intermediate auxiliary losses, enhancing model robustness and generalization capability while guiding the learning process more effectively in early training phases. The joint optimization strategy leverages category-specific features from CSFLM for additional supervision, improving both accuracy and interpretability of skin lesion classification.

Table 1. Comparison of classification accuracy (%) on the Seven-Point Checklist (SPC) dataset for different methods across all eight classification tasks

Method	BWV	DAG	PIG	PN	RS	STR	VS	DIAG	avg
Inception-unbalanced [18]	87.6	56.7	65.6	68.1	78.2	75.9	81.3	68.4	72.7
Inception-balanced [18]	87.3	60.3	64.8	68.9	78.2	75.7	81.5	70.8	73.4
Inception-combine [18]	87.1	60.0	66.1	70.9	77.2	74.2	79.7	74.2	73.7
EmbeddingNet [38]	84.3	57.5	64.3	65.1	78.0	73.4	82.5	68.6	71.7
TripleNet [13]	87.9	61.3	67.3	63.3	76.0	74.4	83.0	68.6	72.7
HcCNN [5]	87.1	65.6	68.6	70.6	80.8	71.6	84.8	69.9	74.9
CAFNet [14]	87.8	61.5	73.4	70.1	81.8	77.0	84.3	78.2	76.8
AMFAM [34]	88.1	63.8	70.9	70.6	80.8	74.7	83.3	75.4	76.0
MBF-HA [36]	87.9	63.3	72.9	70.4	81.0	75.4	83.5	76.0	76.3
FusionM4Net-SS [33]	$88.5_{\pm 0.4}$	$64.4_{\pm 1.3}$	$71.3_{\pm 1.3}$	$69.2_{\pm 1.5}$	$81.4_{\pm 0.8}$	$76.1_{\pm 1.1}$	$81.6_{\pm 0.5}$	$77.6_{\pm 1.5}$	$76.3_{\pm 0.7}$
TFormer [40]	$86.1_{\pm 0.8}$	$65.2_{\pm 1.4}$	$69.3_{\pm 3.4}$	$71.4_{\pm 1.1}$	$79.8_{\pm 1.5}$	$75.0_{\pm 0.8}$	$81.8_{\pm 3.2}$	$74.6_{\pm 1.3}$	$75.4_{\pm 0.8}$
MSMA [29]	$88.2_{\pm 0.9}$	$66.0_{\pm 1.4}$	$71.5_{\pm 1.0}$	$74.0_{\pm 1.0}$	$81.8_{\pm 1.2}$	$76.7_{\pm 1.0}$	$83.7_{\pm 1.1}$	$79.0_{\pm 1.5}$	$77.6_{\pm 0.2}$
SkinM2Former [41]	$88.8_{\pm 0.1}$	$65.8_{\pm 0.1}$	$71.9_{\pm 0.1}$	$73.6_{\pm 0.2}$	$81.2_{\pm 0.1}$	$76.1_{\pm 0.2}$	$82.7_{\pm 0.1}$	$77.8_{\pm 0.1}$	$77.2_{\pm 0.5}$
MedFuse-GRM (Ours)	$\mathbf{90.1}_{\pm 0.7}$	$\mathbf{68.9}_{\pm 1.2}$	$\mathbf{75.8}_{\pm 1.1}$	$\mathbf{75.6}_{\pm 0.9}$	$\mathbf{83.5}_{\pm 0.8}$	$\mathbf{78.4}_{\pm 1.0}$	$\mathbf{85.3}_{\pm 0.9}$	$\mathbf{80.2}_{\pm 1.3}$	$\mathbf{79.7}_{\pm 0.4}$

4 Experimental Setup

We evaluated MedFuse-GRM on the Seven-Point Checklist dataset [18] (1011 cases with paired dermoscopy/clinical images; 413 training, 203 validation, 395 test samples) for 8 classification tasks, using average accuracy (primary), sensitivity, specificity, precision, and AUC as metrics. Implemented in PyTorch on an NVIDIA RTX 3090 GPU, the framework used 224×224×3 resized images, trained end-to-end for 150 epochs (batch size 32) with Adam (1×10^{-5}), data augmentation, and retained weights with highest validation accuracy. Total training time: 2 h, suitable for timely clinical use.

5 Results and Discussion

5.1 Overall Performance

Table 1 demonstrates that our MedFuse-GRM framework consistently outperforms state-of-the-art multimodal classification methods across all eight classification tasks. The model achieves 79.7% average accuracy, surpassing the previous best method MSMA (77.6%) by 2.1% points, with particularly notable improvements in Blue-Whitish Veil detection (90.1%), Pigmentation classification (75.8%), Pigment Network analysis (75.6%), and diagnostic categorization (80.2%). These results validate the effectiveness of our medically-guided graph relational approach in capturing complex relationships between pathological characteristics and diagnostic categories.

The results including sensitivity, specificity, precision, and AUROC metrics which are consistent with those used in previous references [5,18] are provided in the Table 2.

Table 2. Results of the state-of-the-art multimodal classification methods on the seven-point checklist dataset in terms of sensitivity, specificity, precision, and AUROC. In the first row of this table, the selected category in each classification task (e.g., ATP in PN, PRS in BWV, and IR in VS) is positively associated with the diagnosis of melanoma.

		PN ATP	BWV PRS	VS IR	PIG IR	STR IR	DaG IR	RS PRS	DIAG MEL	avg
Inception-unbalanced	Sensitivity	35.5	49.3	0.0	57.3	34.0	67.8	31.1	46.6	40.0
[18]	Specificity	93.7	96.6	100.0	80.1	94.0	67.4	95.5	92.2	89.9
	Precision	63.5	77.1	0.0	56.8	64.0	62.8	71.7	66.2	57.8
	AUROC	78.6	87.0	73.4	78.1	78.3	76.4	79.9	83.2	79.4
Inception-balanced	Sensitivity	41.9	65.3	10.0	67.7	50.0	66.1	62.3	55.4	52.3
[18]	Specificity	92.1	92.5	99.5	73.4	87.7	73.4	84.1	88.4	86.4
	Precision	61.9	67.1	60.0	53.8	56.0	66.9	58.9	62.2	60.9
	AUROC	78.9	87.5	76.1	79.4	78.7	78.0	83.5	84.2	80.8
Inception-combined	Sensitivity	48.4	77.3	13.3	59.7	51.1	62.1	66.0	61.4	54.9
[18]	Specificity	90.7	89.4	97.5	80.1	85.7	78.9	81.3	88.8	86.6
	Precision	61.6	63.0	30.8	57.8	52.7	70.5	56.5	65.3	57.3
	AUROC	79.9	89.2	76.1	79.0	78.9	79.9	82.9	86.3	81.5
HcCNN	Sensitivity	40.9	92.2	20.0	55.7	35.1	80.2	95.2	58.4	59.7
[5]	Specificity	92.4	65.3	98.4	86.3	90.0	71.6	41.5	88.1	79.2
	Precision	62.3	91.9	50.0	65.1	52.4	69.6	81.6	62.8	67.0
	AUROC	78.3	89.8	82.7	81.3	77.6	82.6	81.9	85.6	82.5
CAFNet	Sensitivity	65.9	68.8	45.0	60.3	67.1	74.1	42.7	75.3	62.4
[14]	Specificity	90.2	95.1	98.7	89.1	91.2	74.0	96.5	93.6	91.1
	Precision	50.8	75.4	58.3	70.1	54.8	67.8	76.7	77.9	66.5
	AUROC	75.3	94.7	83.4	85.0	85.4	78.7	85.4	87.2	85.0
MedFuse-GRM	Sensitivity	69.5	74.6	53.6	65.3	71.2	69.4	72.8	71.5	68.5
(Ours)	Specificity	93.7	94.1	87.9	89.2	91.7	75.8	90.7	91.8	89.4
	Precision	70.6	75.3	24.7	72.1	68.4	66.2	75.2	76.5	66.1
	AUROC	86.4	89.7	83.0	84.6	84.9	79.6	84.8	88.7	85.2

5.2 Ablation Study

To demonstrate the effectiveness of our multi-modal approach and each component of the MedFuse-GRM framework, we conducted comprehensive ablation studies. As shown in Table 3, our full MedFuse-GRM framework achieves 79.7% average accuracy, significantly outperforming single-modality models and a simple fusion approach. Table 4 validates the contribution of each key component, with the full framework showing an 8.9% improvement over the baseline ResNet-50 model. We further analyzed different configurations of our graph relational approach in Table 5, confirming that the full MG-GRM achieves the best performance at 79.7% (+0.4%). Additionally, Table 6 compares different feature extraction backbones, demonstrating that our specialized MedFuse backbone with contrastive learning achieves superior performance (79.7%) compared to general-purpose vision backbones like ResNet50, ConvNeXt-Tiny, and Swin-Tiny. These ablation studies confirm that each component of our proposed MedFuse-GRM framework contributes significantly to the overall performance improvement.

Table 3. Comparison of single-modality and multi-modality approaches.

Method	avg
Dermoscopy only	74.1
Clinical only	71.4
Simple Fusion	74.6
MedFuse-GRM (Ours)	79.7

Table 4. Contribution of different components.

Model	avg
Baseline (ResNet-50)	70.8
Only MedFuse	74.6
MedFuse + CL	76.3
MedFuse + CL + CSFLM	78.2
MedFuse + CL + CSFLM + MG-GRM	79.7

Table 5. Different variants of Graph Relational Module.

Method	avg
without GRM	78.2
GRM	78.9
GRM (with Gated Modulation)	79.3
MG-GRM	79.7

Table 6. Comparison of different backbone networks.

Backbone	avg
ResNet50	76.8
ConvNeXt-Tiny	77.8
Swin-Tiny	78.5
MedFuse (w/o CL)	79.0
MedFuse (full)	79.7

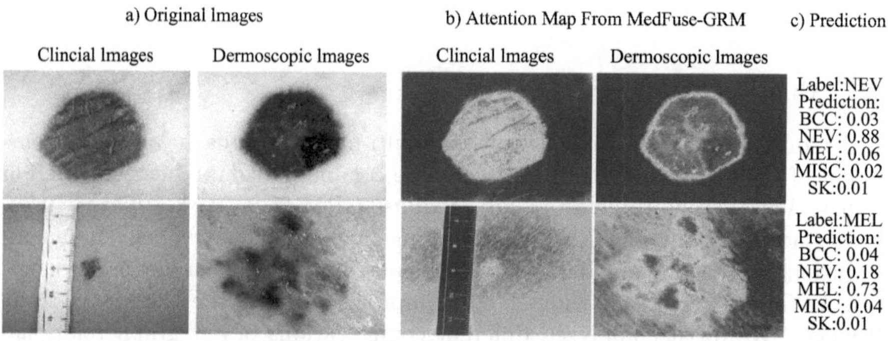

Fig. 3. Visualization of MedFuse-GRM performance on representative cases. a) Original clinical and dermoscopic images; b) Attention heatmaps for both modalities; c) Ground truth labels and prediction probabilities (highlighted in red). The heatmaps reveal model's focus on pathologically significant regions across diverse lesion types. (Color figure online)

5.3 Attention Map

As visualized in Fig. 3, the attention heatmaps reveal modality-specific focus areas: dermoscopic features emphasize fine-grained structures (e.g., dendritic vessels in BCC, irregular pigment patterns in MEL), while clinical features capture macroscopic morphology and contextual information. This complementary attention distribution confirms the model's ability to leverage cross-modal information synergistically.

6 Conclusion

In this paper, we proposed MedFuse-GRM, a novel multi-modal framework for skin lesion classification that addresses critical challenges through three key components: MedFuse for multi-scale feature extraction with cross-attention guided fusion, contrastive learning pre-training to enhance cross-modal feature consistency, and MG-GRM for medically-guided graph relation modeling. Experimental results on the Seven-Point Checklist dataset demonstrate that MedFuse-GRM achieves 79.7% average accuracy, outperforming state-of-the-art methods by 2.1% points. Future work will focus on systematic analysis of graph topologies, advanced contrastive learning strategies for imbalanced datasets, framework extension to other multi-modal medical diagnosis tasks, and selective incorporation of high-confidence metadata when available.

References

1. Apalla, Z., et al.: Skin cancer: epidemiology, disease burden, pathophysiology, diagnosis, and therapeutic approaches. Dermatol. Ther. **7**, 5–19 (2017)
2. Argenziano, G., et al.: Epiluminescence microscopy for the diagnosis of doubtful melanocytic skin lesions: comparison of the ABCD rule of dermatoscopy and a new 7-point checklist based on pattern analysis. Arch. Dermatol. **134**(12), 1563–1570 (1998)
3. Balch, C.M., et al.: Final version of 2009 AJCC melanoma staging and classification. J. Clin. Oncol. **27**(36), 6199–6206 (2009)
4. Barata, C., et al.: Improving dermoscopy image classification using color constancy. IEEE J. Biomed. Health Inform. **19**(3), 1146–1152 (2014)
5. Bi, L., et al.: Multi-label classification of multi-modality skin lesion via hyper-connected convolutional neural network. Pattern Recognit. **107**, 107502 (2020)
6. Bi, L., et al.: Step-wise integration of deep class-specific learning for dermoscopic image segmentation. Pattern Recogn. **85**, 78–89 (2019)
7. Borsari, S., et al.: Clinical indications for use of reflectance confocal microscopy for skin cancer diagnosis **152**(10), 1093–1098 (2016)
8. Carrera, C., et al.: Validity and reliability of dermoscopic criteria used to differentiate nevi from melanoma: a web-based international dermoscopy society study. JAMA Dermatol. **152**(7), 798–806 (2016)
9. Chen, T., et al.: A simple framework for contrastive learning of visual representations. In: International Conference on Machine Learning, pp. 1597–1607. PmLR (2020)

10. Das, B., et al.: A novel hybrid model combining vision transformers and graph convolutional networks for monkeypox disease effective diagnosis. Inf. Fus. **117**, 102858 (2025)
11. Dou, Q., et al.: Unpaired multi-modal segmentation via knowledge distillation **39**(7), 2415–2425 (2020)
12. Fu, X., et al.: Graph-based intercategory and intermodality network for multilabel classification and melanoma diagnosis of skin lesions in dermoscopy and clinical images. IEEE Trans. Med. Imaging **41**(11), 3266–3277 (2022)
13. Ge, Z., Demyanov, S., Chakravorty, R., Bowling, A., Garnavi, R.: Skin disease recognition using deep saliency features and multimodal learning of dermoscopy and clinical images. In: Descoteaux, M., Maier-Hein, L., Franz, A., Jannin, P., Collins, D.L., Duchesne, S. (eds.) MICCAI 2017. LNCS, vol. 10435, pp. 250–258. Springer, Cham (2017). https://doi.org/10.1007/978-3-319-66179-7_29
14. He, X., et al.: Co-attention fusion network for multimodal skin cancer diagnosis. Pattern Recognit. **133**, 108990 (2023)
15. He, K., et al.: Deep residual learning for image recognition. In: Proceedings of the IEEE Conference on Computer Vision and Pattern Recognition, pp. 770–778 (2016)
16. He, X., et al.: Fully transformer network for skin lesion analysis. Med. Image Anal. **77**, 102357 (2022)
17. Huo, X., et al.: HiFuse: hierarchical multi-scale feature fusion network for medical image classification. Biomed. Signal Process. Control **87**, 105534 (2024)
18. Kawahara, et al.: Seven-point checklist and skin lesion classification using multitask multimodal neural nets. IEEE J. Biomed. Health Inf. **23**(2), 538–546 (2018)
19. Kawahara, J., et al.: Deep features to classify skin lesions. In: 2016 IEEE 13th International Symposium on Biomedical Imaging (ISBI), pp. 1397–1400. IEEE (2016)
20. Liang, M., et al.: Integrative data analysis of multi-platform cancer data with a multimodal deep learning approach. IEEE/ACM Trans. Comput. Biol. Bioinf. **12**(4), 928–937 (2014)
21. Liao, X., et al.: MMTLNet: multi-modality transfer learning network with adversarial training for 3D whole heart segmentation. Comput. Med. Imaging Graph. **85**, 101785 (2020)
22. Liu, Y., et al.: A deep learning system for differential diagnosis of skin diseases. Nat. Med. **26**(6), 900–908 (2020)
23. Massi, D., et al.: Diagnostic significance of the blue hue in dermoscopy of melanocytic lesions: a dermoscopic–pathologic study. Am. J. Dermatopathol. **23**(5), 463–469 (2001)
24. Menegola, A., et al.: Knowledge transfer for melanoma screening with deep learning. In: 2017 IEEE 14th International Symposium on Biomedical Imaging (ISBI 2017), pp. 297–300. IEEE (2017)
25. Pizzichetta, M.A., et al.: Dermoscopic criteria for melanoma in situ are similar to those for early invasive melanoma. Cancer **91**(5), 992–997 (2001)
26. Ramachandram, D., Taylor, G.W.: Deep multimodal learning: a survey on recent advances and trends. IEEE Signal Process. Mag. **34**(6), 96–108 (2017)
27. Saez, A., et al.: Model-based classification methods of global patterns in dermoscopic images. IEEE Trans. Med. Imaging **33**(5), 1137–1147 (2014)
28. Scope, A., et al.: Correlation of dermoscopic structures of melanocytic lesions to reflectance confocal microscopy. Arch. Dermatol. **143**(2), 176–185 (2007)

29. Shu, C., et al.: MSMA: a multi-stage and multi-attention algorithm for the classification of multimodal skin lesions. Biomed. Sig. Process. Control **93**, 106180 (2024)
30. Simoes, M.C.F., et al.: Skin cancer and new treatment perspectives: a review. Cancer :Lett. **357**(1), 8–42 (2015)
31. Simonyan, K., Zisserman, A.: Very deep convolutional networks for large-scale image recognition. arXiv preprint arXiv:1409.1556 (2014)
32. Sohn, K.: Improved deep metric learning with multi-class n-pair loss objective. In: Advances in Neural Information Processing Systems, vol. 29 (2016)
33. Tang, P., et al.: Fusionm4Net: a multi-stage multi-modal learning algorithm for multi-label skin lesion classification. Med. Image Anal. **76**, 102307 (2022)
34. Wang, Y., et al.: Adversarial multimodal fusion with attention mechanism for skin lesion classification using clinical and dermoscopic images. Med. Image Anal. **81**, 102535 (2022)
35. Wang, H., Xu, T., Xu, J.: Factors contributing to high costs and inequality in China's health care system. JAMA **298**(16), 1928–1930 (2007)
36. Wei, Y., Ji, L.: Multi-modal bilinear fusion with hybrid attention mechanism for multi-label skin lesion classification. Multimedia Tools Appl. **83**(24), 65221–65247 (2024)
37. Wu, Z., et al.: Unsupervised feature learning via non-parametric instance discrimination. In: Proceedings of the IEEE Conference on Computer Vision and Pattern Recognition, pp. 3733–3742 (2018)
38. Yap, J., et al.: Multimodal skin lesion classification using deep learning. Exp. Dermatol. **27**(11), 1261–1267 (2018)
39. Zalaudek, I., et al.: How to diagnose nonpigmented skin tumors: a review of vascular structures seen with dermoscopy: part I. melanocytic skin tumors. J. Am. Acad. Dermatol. **63**(3), 361–374 (2010)
40. Zhang, Y., et al.: TFormer: a throughout fusion transformer for multi-modal skin lesion diagnosis. Comput. Biol. Med. **157**, 106712 (2023)
41. Zhang, Y., et al.: A novel perspective for multi-modal multi-label skin lesion classification. arXiv preprint arXiv:2409.12390 (2024)

AxialUNet: A Lightweight Network for Medical Image Segmentation with Axial Operators

Meiyi Lyu, Jiawei Mo, Xuewen Chen, and Chaoqun Wang(✉)

School of Artificial Intelligence, South China Normal University, Guangdong, China
2024025433@m.scnu.edu.cn, cqwang96@gmail.com

Abstract. Accurate medical image segmentation is essential for clinical diagnosis, disease monitoring, and surgical planning. However, convolutional neural networks (CNNs) are constrained by limited receptive fields, making it difficult to capture long-range dependencies, while Transformer-based methods incur quadratic computational costs and substantial parameter overhead as the input resolution grows. To address these challenges, we propose AxialUNet, a lightweight U-shaped architecture that efficiently models global context while preserving the native 2D spatial structure. At its core, the Multi-scale Spatial Mixing (MSSM) module integrates axial mechanisms with multi-scale convolutions to capture fine-grained local details and long-range dependencies in parallel. Additionally, a Dynamic Heterogeneous Expert Attention (DHEA) module is introduced in the skip connections, leveraging a Mixture-of-Experts system that also utilizes axial decomposition for precise boundary refinement. Extensive experiments on multiple benchmarks show that AxialUNet, with only 5.14M parameters and 11.83 GFLOPs, achieves performance comparable to state-of-the-art methods, establishing it as an effective and efficient solution for medical image segmentation.

Keywords: Medical Image Segmentation · Lightweight Architecture · Axial Operators

1 Introduction

Medical image segmentation is a fundamental task in computer vision, providing the precise anatomical delineations essential for clinical diagnosis and treatment planning [1–4]. In recent years, deep learning models, particularly convolutional neural networks (CNNs) [5] and Transformers [6], have achieved demonstrated exceptional performance in this domain. CNN architectures, exemplified by U-Net [7], rely on convolutional operations within localized receptive fields. This design is highly efficient for extracting local features such as edges and textures. Yet, this same operational locality fundamentally restricts the model's ability to capture long-range spatial dependencies. Transformer models leverage

J. Lokoč et al. (Eds.): MMM 2026, LNCS 16413, pp. 74–88, 2026.
https://doi.org/10.1007/978-981-95-6957-1_6

self-attention to establish global contextual relationships, addressing the limitations of CNNs [8,9]. This approach, however, incurs a substantial computational burden. The self-attention mechanism scales quadratically with image size, rendering its application to high-resolution medical imaging computationally prohibitive [10]. Consequently, existing methods present a fundamental trade-off. CNNs achieve computational efficiency by sacrificing global context, whereas Transformers attain comprehensive global modeling at the expense of computational efficiency.

This paper aims to resolve the trade-off between the limited receptive fields of CNNs and the quadratic computational complexity of Transformers. To this end, we propose AxialUNet, an efficient and lightweight architecture designed specifically for medical image segmentation. The core of AxialUNet is an axial decomposition computational strategy, which factorizes 2D global context modeling into two sequential 1D computational processes along the height and width axes. This strategy differs fundamentally from traditional Transformers. Specifically, the practice of flattening a 2D feature map into a 1D sequence in Transformers disrupting the image's inherent grid structure and spatial adjacency. In contrast, our axial decomposition strategy consistently operates on the original grid structure, thereby avoiding structural distortion. It achieves this through a two-step information aggregation process, first along the height and then the width, ultimately building a complete global receptive field with linear computational complexity.

To effectively apply this strategy to different parts of the network, we designed two complementary core innovative modules. In the backbone, we introduce the Multi-scale Spatial Mixing (MSSM) module to efficiently encode long-range spatial context. MSSM applies attention sequentially along each axis, and its two-step mechanism ensures that any two pixels on the feature map can exchange information, establishing a global receptive field enables dynamic re-weighting of multi-scale features. Meanwhile, in the skip connections, we designed the Dynamic Heterogeneous Expert Attention (DHEA) module to addresses the semantic gap between deep and shallow features and to precisely capture anisotropic features such as directional boundaries. DHEA innovatively employs a Mixture-of-Experts (MoE) framework, which contains a gating network and multiple expert sub-networks with varied architectures. Based on the input features, this gating network learns to selectively activate the most appropriate experts for feature refinement. This adaptive mechanism significantly enhances the model's capability for modeling complex structures, transforming the traditional skip connection into an adaptive feature fusion unit. Our main contributions are as follows:

- We propose AxialUNet, an efficient and lightweight framework based on axial decomposition. Its design facilitates deployment in resource-constrained clinical environments.
- We introduce two novel modules: the Multi-scale Spatial Mixing (MSSM) module to efficiently encode long-range spatial context in the backbone, and

the Dynamic Heterogeneous Expert Attention (DHEA) module to adaptively refine and fuse features in the skip connections.
- We demonstrate through extensive experiments on challenging public datasets, including ISIC and Synapse, that AxialUNet achieves state-of-the-art segmentation performance while maintaining high computational efficiency.

2 Related Work

2.1 Medical Image Segmentation

Medical image segmentation is a fundamental visual task that delineates anatomical structures or lesions at the pixel level, providing quantitative spatial information critical for clinical diagnosis, surgical planning, and disease monitoring. Nevertheless, blur, noise, and low contrast inherent to medical images continue to pose major challenges to accurate and robust segmentation.

Convolutional Neural Networks (CNNs) [5], exemplified by U-Net [7] and its variants, are the dominant approach in medical image segmentation. The classic U-shaped architecture and its skip-connection mechanism have inspired numerous refinements. These include the incorporation of dense skip connections [11], residual learning [12], and attention mechanisms [13] to enhance feature fusion, as well as specialized edge-supervision modules and structural re-parameterization [10] to improve boundary precision and create lightweight designs. Nevertheless, the intrinsic locality of CNNs constrains their ability to model long-range dependencies, leading to suboptimal delineation of fine boundaries [14].

To overcome these limitations, Vision Transformers (ViTs) [6] were introduced, leveraging self-attention to capture global context efficiently. This has led to two primary architectural approaches. One is a hybrid design, exemplified by TransUNet [14], which couples a CNN encoder for fine-grained local feature extraction with a Transformer for global context modeling. In contrast, the other is a pure Transformer architecture, represented by Swin-UNet [15], where the entire U-shaped network is built upon a hierarchical Transformer backbone. However, the computational demands of ViTs remain prohibitive; attempts to reduce complexity via approximations such as linear self-attention typically compromise model accuracy [16]. Consequently, the trade-off between the local efficiency of CNNs and the global expressiveness of computationally intensive ViTs continues to hinder the widespread clinical adoption of advanced segmentation methods.

2.2 Mixture of Experts

The Mixture of Experts (MoE) approach tackles complex tasks by dynamically routing input to a small subset of expert subnetworks, selected by a lightweight top-k gating function [17]. The core idea is to significantly increase model capacity without a proportional rise in computational cost, thus balancing performance and efficiency [18]. By activating only the most relevant experts for each

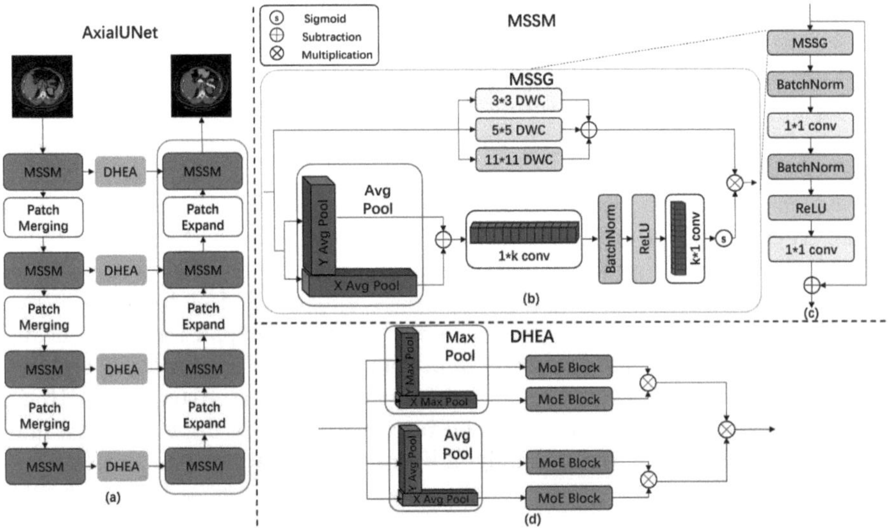

Fig. 1. Illustration of the AxialUNet architecture and its core components. (a) AxialUNet is a symmetric U-shaped network that uses MSSM blocks as feature extractors and enhances skip connections with the innovative DHEA module. (b) The MSSG unit performs spatial gating by element-wise modulation of multi-scale features using an axial attention map. (c) The MSSM block utilizes parallel depthwise convolutions and an MLP to efficiently capture multi-scale spatial features. (d) The DHEA module uses dynamic routing to assign features to experts, enhancing details and suppressing noise before fusion.

input, MoE models can learn richer, more specialized representations while keeping computational costs low. This offers a practical solution to the trade-off between performance and efficiency [19].

MoE has recently seen significant success in both Natural Language Processing (NLP) [20,21] and Computer Vision (CV) [22]. Research has progressed in two main directions: One involves the evolution of the gating mechanism from dense (activating all experts) to sparse (selective activation) [18], and more recently to fully differentiable soft gating [23]. The other focuses on diversifying the types of experts used.

While Feedforward Networks are the most common choice for experts [18], their high computational and memory costs have spurred alternatives. These include MoE-Infinity, which uses zero-computation experts [24], and MoELoRA for multi-task fine-tuning [25]. However, most current MoE frameworks use a single layer of identical, or homogeneous, experts. The effectiveness of these frameworks for tasks requiring fine-grained information, like image segmentation, especially when combined with skip connections, is still largely unknown.

3 Methodology

As illustrated in Fig. 1(a), our proposed AxialUNet is based on a symmetric encoder-decoder structure, featuring a four-stage encoder, a four-stage decoder, and skip connections enhanced by our Dynamic Heterogeneous Expert Attention (DHEA) module.

The encoder starts with a convolutional stem layer and comprises four stages. At each stage, features are extracted using the Multi-scale Spatial Mixing (MSSM) module. Subsequently, spatial resolution is halved via patch merging, and the channel dimension is doubled. The symmetrical decoder mirrors this process: at each stage, features are upsampled via patch expansion and concatenated with skip-connected features. Importantly, these skip features are first adaptively refined by the Dynamic Heterogeneous Expert Attention (DHEA) module to enhance key details, and then fused with the upsampled feature map using another MSSM module. Finally, a linear projection head generates pixel-level segmentation maps.

3.1 Multi-Scale Spatial Mixing Module

The Multi-scale Spatial Mixing (MSSM), shown in Fig. 1(c), uses an axial decomposition strategy within a single block to capture both global context and local details simultaneously.

First, global context is generated via axial pooling. Global average pooling is applied independently along the height and width axes of the input feature map X. This process decomposes the spatial features into two separate 1D context vectors. These vectors are then combined through addition and broadcasting to create the global context descriptor Z:

$$Z = \mathrm{Pool}h(X) + \mathrm{Pool}w(X), \tag{1}$$

where $\mathrm{Pool}_h(\cdot)$ and $\mathrm{Pool}_w(\cdot)$ denote average pooling along the height and width axes, respectively. This descriptor is then transformed into a 2D spatial attention map A by a lightweight convolutional branch:

$$A = \sigma(\mathrm{DWConv}K \times 1(\mathrm{ReLU}(\mathrm{BN}(\mathrm{DWConv}1 \times K(Z))))), \tag{2}$$

where BN is batch normalization, σ is the sigmoid activation, and DWConv is a depthwise convolution. The sequential $1 \times K$ and $K \times 1$ convolutions efficiently fuse information across the two axes.

Concurrently, a multi-scale branch extracts local features X_{ms} using three parallel depthwise convolutions with varying kernel sizes ($k_s \times k_s$, $k_m \times k_m$, and $k_l \times l_l$):

$$X_{\mathrm{ms}} = \mathrm{DWConv}k_s \times k_s(X) + \mathrm{DWConv}k_m \times k_m(X) + \mathrm{DWConv}_{k_l \times k_l}(X), \tag{3}$$

The attention map A, which encodes global context, then modulates the local features X_{ms} via element-wise multiplication to produce the output of the Multi-Scale Spatial Gating (MSSG) unit (Fig. 1(b)):

$$\mathrm{MSSG}(X) = X_{\mathrm{ms}} \odot A, \tag{4}$$

where $\odot$ denotes element-wise multiplication.

Finally, the MSSG unit is integrated into a residual block to form the complete MSSM module. The MSSG output is processed by a batch normalization layer and a channel-wise MLP. A residual connection, regularized by a stochastic depth (DropPath) layer, is added to the original input X. The final output Y is:

$$Y = X + \text{DropPath}(\text{MLP}(\text{BN}(\text{MSSG}(X)))). \tag{5}$$

3.2 Dynamic Heterogeneous Expert Attention

To further refine boundary details with global context, we introduce the Dynamic Heterogeneous Expert Attention (DHEA) module (Fig. 1(d)), which transforms axial decomposition from a simple feature extraction step into an adaptive, specialized process for axial information.

The two-dimensional skip-connection feature map S is first decomposed along the height and width axes. Recognizing that a single pooling strategy is insufficient for both smooth regions and intricate boundaries, we employ dual pooling (global average and global max pooling) to obtain four complementary one-dimensional axial descriptors: $D_{h,\text{avg}}$, $D_{h,\text{max}}$, $D_{w,\text{avg}}$, and $D_{w,\text{max}}$.

These descriptors are routed through a sparse Mixture of Experts (MoE). A shared gating network g computes routing weights for each descriptor D_{1D} as:

$$g(D_{\text{1D}}) = \text{Softmax}(D_{\text{1D}} \cdot W_g), \tag{6}$$

where W_g denotes a learnable weight matrix. The expert pool comprises spatial experts (3×1 axial convolutions) specialized in complex boundaries and lightweight experts for efficiently processing simpler patterns.

The activated experts' outputs are aggregated via a weighted sum to yield the refined one-dimensional axial attention vector:

$$A(D_{\text{1D}}) = \sum_{k \in \text{TopK}(g)} g_k \cdot \text{Expert}k(D1\text{D}), \tag{7}$$

The four refined axial attention vectors are broadcast back to their 2D shape and applied element-wise to re-weight the original skip-connection feature map S, producing two distinct attention-modulated pathways. Specifically, the pathway guided by average-pooled descriptors is:

$$S_{\text{avg}} = S \odot A_{h,\text{avg}} \odot A_{w,\text{avg}}, \tag{8}$$

The average- and max-guided pathways are then summed to generate the final module output, allowing the model to simultaneously suppress background noise and preserve fine details:

$$S' = S_{\text{avg}} + S_{\text{max}}, \tag{9}$$

where S' denotes the final output, S_{avg} and S_{max} are the feature maps modulated by the average- and max-pooled descriptors, respectively, and $\odot$ indicates element-wise multiplication.

Table 1. Model complexity and segmentation performance metrics. K(L) and K(R) stand for Left and Right Kidney, respectively. Gall. stands for Gallbladder.

Method	#Params(M)↓	FLOPs(G)↓	DSC↑	HD95↓	Aorta	Gall.	K(L)	K(R)	Liver	Pancreas	Spleen	Stomach
UNet [7]	31.04	83.87	76.85	39.70	<u>89.07</u>	69.72	77.77	68.60	93.43	53.98	86.67	75.58
Att-UNet [13]	34.90	101.90	77.77	36.02	**89.55**	68.88	77.98	71.11	93.57	58.04	87.30	75.75
TransUNet [14]	105.28	24.66	77.48	31.69	87.23	63.13	81.87	77.02	94.08	55.86	85.08	75.62
SAMed [30]	86.57	33.92	<u>81.88</u>	<u>20.64</u>	87.77	69.11	80.45	79.95	<u>94.80</u>	**72.17**	88.72	**82.06**
FocalUNet [31]	44.25	<u>17.44</u>	80.81	20.66	85.74	71.37	<u>85.23</u>	**82.99**	94.38	59.34	88.49	78.94
DA-TransUNet [32]	94.51	50.98	79.80	23.48	86.47	65.27	81.70	80.45	94.57	<u>61.62</u>	88.53	79.73
Swin-T+MCA [33]	<u>27.67</u>	26.93	80.29	22.15	86.71	70.03	84.09	80.47	94.11	60.39	<u>89.49</u>	77.00
AxialUNet(Ours)	**5.14**	**11.83**	**81.92**	**16.86**	88.47	**72.18**	**85.96**	<u>81.16</u>	**95.25**	59.76	**91.31**	<u>81.29</u>

Table 2. Results on ISIC 2017 and ISIC 2018 Datasets.

Method	ISIC 2017					ISIC 2018				
	mIoU(%) ↑	DSC(%) ↑	Acc(%) ↑	Spe(%) ↑	Sen(%) ↑	mIoU(%) ↑	DSC(%) ↑	Acc(%) ↑	Spe(%) ↑	Sen(%) ↑
UNet [7]	76.98	85.99	<u>94.64</u>	96.56	**86.82**	77.86	86.74	94.05	96.71	87.46
Att-UNet [13]	76.37	80.82	91.45	97.76	79.98	76.43	88.20	94.13	97.66	<u>90.75</u>
TransUNet [14]	77.69	81.23	92.07	95.77	82.63	77.05	84.99	94.56	96.53	88.23
SAMed [30]	77.81	90.04	94.36	97.54	79.55	79.72	91.36	94.48	94.48	89.88
DAE-Former [34]	<u>77.87</u>	90.81	93.39	95.78	85.64	79.42	91.47	95.41	<u>97.80</u>	**91.20**
DA-TransUNet [32]	77.61	89.74	92.20	97.27	<u>86.62</u>	**82.78**	88.88	95.27	97.13	90.36
Swin-T+MCA [33]	77.58	<u>91.02</u>	94.41	<u>97.86</u>	86.35	78.43	<u>91.72</u>	<u>96.12</u>	97.10	90.47
AxialUNet(Ours)	**77.89**	**92.25**	**95.99**	**98.10**	84.64	<u>81.04</u>	**92.68**	**96.32**	**97.83**	87.97

3.3 Hybrid Loss

Our total loss function, $\mathcal{L}_{\text{total}}$, is formulated to jointly optimize for segmentation accuracy and model stability. The primary component is a hybrid segmentation loss, which is a weighted sum of Cross-Entropy Loss ($\mathcal{L}_{\text{CE}}$) and Dice Loss ($\mathcal{L}_{\text{Dice}}$).

Additionally, we introduce an auxiliary load-balancing loss, $\mathcal{L}_{\text{MoE}}$, to encourage uniform expert utilization. This loss is formulated as:

$$\mathcal{L}_{\text{MoE}} = N \sum_{i=1}^{N} f_i P_i, \tag{10}$$

where f_i is the fraction of tokens routed to expert i, P_i is the average gate probability for that expert, and N is the total number of experts.

The complete loss function is then defined as a weighted sum of these components:

$$\mathcal{L}_{\text{total}} = \lambda \mathcal{L}_{\text{CE}} + (1 - \lambda)\mathcal{L}_{\text{Dice}} + \alpha \mathcal{L}_{\text{MoE}}, \tag{11}$$

where λ and α are hyperparameters that balance the contributions of the segmentation and load-balancing losses. This composite objective function guides the model to learn a precise segmentation policy while maintaining an efficient expert routing strategy.

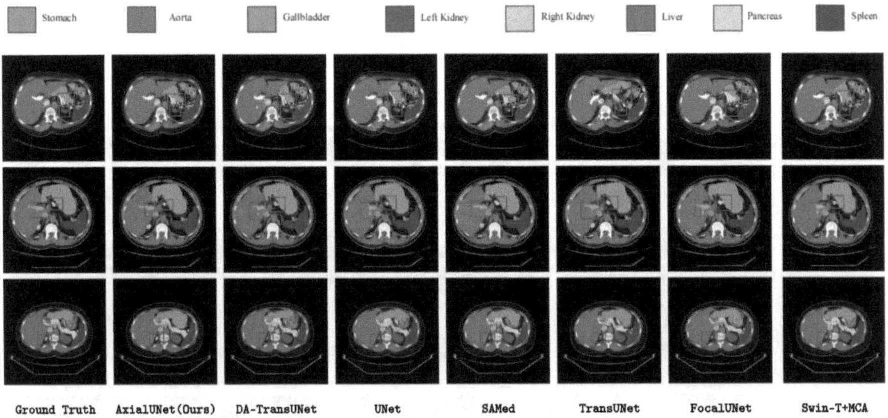

Fig. 2. Qualitative comparison of segmentation results on the Synapse dataset. Red rectangles highlight regions where our method shows significant advantages. (Color figure online)

4 Experiments

4.1 Experimental Settings

Datasets. We evaluate our model's performance on three public medical datasets covering different anatomical regions and imaging modalities: the Synapse multi-organ segmentation dataset [26], the ISIC-2017 Challenge dataset [27], and the ISIC-2018 Challenge dataset [28]. The Synapse dataset, from the MICCAI 2015 Multi-Atlas Abdominal Labeling Challenge, contains 30 abdominal computed tomography (CT) scans, totaling 3,779 axial CT images. The ISIC-2017 dataset includes 2,150 dermoscopic images with ground truth annotations for skin lesion segmentation. Similarly, the ISIC-2018 dataset provides an extended benchmark with 2,594 dermoscopic images. Following prior work [29], we split both ISIC datasets into training and testing sets using a 7:3 ratio.

Implementation Details. All experiments were conducted using the PyTorch framework on a single NVIDIA 4090 GPU. The model was trained for 250 epochs (approximately 30,000 iterations) with a batch size of 12. We used the AdamW optimizer with a cosine annealing learning rate schedule, an initial learning rate of 5×10^{-4}, and a weight decay of 0.001. Automatic Mixed Precision (AMP) was enabled to accelerate training. The loss function is a weighted sum of Dice loss ($\lambda_{\text{Dice}} = 0.2$) and Cross-Entropy loss ($\lambda_{\text{CE}} = 0.8$). All input images were resized to 224×224 pixels. The MSSM modules were configured with an MLP expansion ratio of 2.0, while the DHEA modules utilized 8 heterogeneous experts organized into 2 groups (4 experts per group), alternating between spatial and lightweight experts, with Top-2 dynamic gating and an auxiliary loss weight of $\alpha = 0.01$.

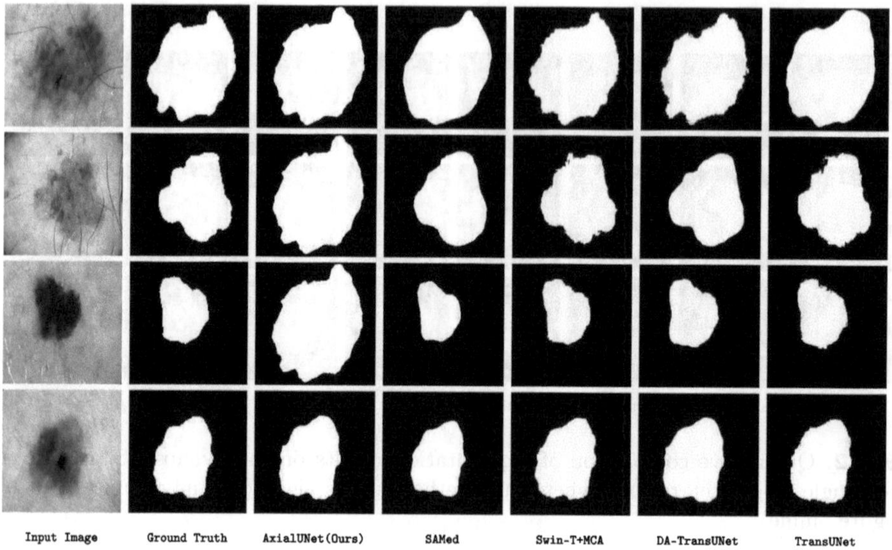

Fig. 3. Qualitative comparison of segmentation results on the ISIC dataset.

Evaluation Metrics. We assess model performance using standard, dataset-specific metrics. For the Synapse dataset, we report the Dice Score (DSC) and the 95th percentile Hausdorff Distance (HD95). For the ISIC-2017 and ISIC-2018 datasets, we report the mean Intersection over Union (mIoU), DSC, Accuracy, Specificity, and Sensitivity. Model efficiency is evaluated based on the number of parameters (Params) and GFLOPs.

4.2 Comparison with Other Methods

We compare our method against several state-of-the-art (SOTA) methods on the Synapse dataset, evaluating segmentation performance on eight abdominal organs. The results are presented in Table 1. Our AxialUNet demonstrates highly competitive performance, achieving a mean Dice score of 81.92% and surpassing all competing models on boundary metrics, attaining the best HD95 score of 16.86. Notably, competing methods with comparable performance rely on complex architectures with parameter counts nearly an order of magnitude larger and a more than four-fold increase in GFLOPs. In contrast, our significantly more efficient model not only matches this performance but demonstrates superior precision, particularly achieving the highest Dice scores for major organs such as the liver and right kidney (Table 5).

To verify the model's generalization capability, we conduct further evaluations on the ISIC 2017 and ISIC 2018 skin lesion segmentation benchmarks, with the results summarized in Table 2. Our method consistently outperforms existing models across multiple evaluation metrics. On the ISIC 2017 dataset, AxialUNet ranks first in four out of five reported metrics. For the ISIC 2018 dataset, it leads

Table 3. Ablation study on model components

	MSSM	DHEA	#Params(M)↓	FLOPs(G)↓	DSC(%)↑	HD95(mm)↓
Model A			4.37	9.77	74.03	29.23
Model B	✓		4.10	7.08	80.17	27.04
Model C		✓	3.91	12.10	78.31	29.60
Model D	✓	✓	5.14	11.83	**81.92**	**16.85**

Table 4. Ablation study on network depth

Depth	#Params(M)↓	FLOPs(G)↓	DSC(%)↑	HD95(mm)↓
3	1.37	9.51	78.91	31.05
4	5.14	11.83	**81.92**	**16.85**
5	19.99	14.09	77.81	17.24

in Dice, accuracy, and specificity. This consistent performance across different datasets underscores the strong generalization ability of our model, enabling it to robustly handle diverse and challenging segmentation tasks.

Qualitative segmentation results are illustrated in Fig. 2 and Fig. 3. We observe that AxialUNet accurately segments fine and complex boundary structures, producing more precise segmentation maps. This improvement is attributed to the core architectural innovations of our method: the axial decomposition within the MSSM module achieves a global receptive field at a low computational cost, while the DHEA module operates as an adaptive filter to selectively extract multi-scale features.

As detailed in Table 1, by leveraging a lightweight design, AxialUNet demonstrates a significant advantage in model complexity. It sets a new efficiency record with only 5.14 M parameters and 11.83 GFLOPs while effectively capturing long-range dependencies. This efficient architecture enables it to achieve an excellent trade-off among boundary accuracy, regional precision, and computational efficiency, making it highly suitable for resource-constrained clinical deployment.

4.3 Ablation Study

To systematically validate the contribution of each proposed component, we conducted a comprehensive ablation study. The baseline is a standard U-Net architecture that employs conventional convolutional blocks as its core operators and utilizes direct identity skip connections without any attention mechanism.

Effect of Multi-scale Spatial Mixing (MSSM). We designed the Multi-scale Spatial Mixing (MSSM) to capture long-range dependencies using an axial decomposition strategy. To evaluate its effectiveness, we replaced the standard convolutional blocks in the baseline's encoder and decoder with MSSM, creating

Table 5. Ablation study on multi-scale convolution kernel sizes

Multi-scale Convolution	DSC(%)↑	HD95(mm)↓
1, 3, 5	80.61	24.25
1, 3, 7	81.14	21.44
1, 3, 11	76.63	37.97
1, 5, 7	80.44	20.57
1, 5, 11	78.75	30.42
1, 7, 11	81.19	20.43
3, 5, 7	81.43	18.43
3, 5, 11	**81.92**	**16.85**
3, 7, 11	80.64	19.86

Model B. As shown in Table 3, Model B improves the Dice score by 6.14% over the baseline. These results show that MSSM overcomes the limited receptive fields of standard convolutions, improving the model's feature extraction.

Effect of Dynamic Heterogeneous Expert Attention (DHEA). We proposed the Dynamic Heterogeneous Expert Attention (DHEA) to adaptively refine encoder features in skip connections and bridge the semantic gap. We performed two experiments to evaluate its contribution. First, to isolate its effect, we integrated the DHEA module into the baseline's skip connections to create Model C. As shown in Table 3, Model C improves the Dice score by 4.28% over the baseline (from 74.03% to 78.31%), showing that DHEA alone significantly improves performance by filtering features. Second, when DHEA is integrated into Model B to form the final AxialUNet, its effectiveness is enhanced. The last row of Table 3 shows that AxialUNet improves the Dice score by an additional 1.75% and reduces the HD95 metric from 27.04 to 16.85. Qualitative results on the Synapse dataset (Fig. 4) show that organ boundaries generated with DHEA are sharper and more accurate. Together, these experiments demonstrate that DHEA is crucial for filtering features before fusion, which improves boundary precision.

Impact of Encoder-Decoder Depth. To evaluate the impact of network depth L on AxialUNet performance, we varied the number of encoderâĂŞdecoder stages. Table 4 indicates that a 4-stage configuration yields the optimal balance between segmentation accuracy and computational cost. A shallower network (L = 3 stages) suffers from limited feature abstraction and degrades performance despite its minimal parameter count (1.37 M). Conversely, a deeper network (L = 5 stages) increases the parameter count to 19.99 M yet reduces the Dice score to 77.81 %, presumably due to overfitting or optimization difficulties. Conse-

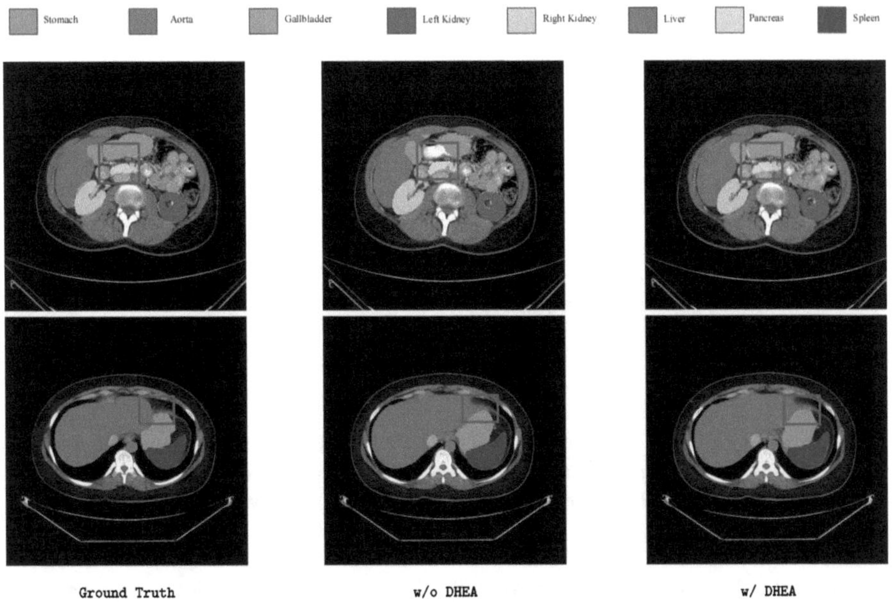

Fig. 4. Comparison of segmentation outputs from our model with ("w/ DHEA") and without ("w/o DHEA") the DHEA module. The red boxes highlight differences in boundary segmentation.

quently, we adopt the 4-stage design as the default and optimal configuration for AxialUNet.

Impact of Multi-scale Convolution Kernel Sizes. We conducted an ablation study on kernel size combinations within the Multi-scale Spatial Mixing (MSSM) to find the optimal configuration. As shown in Table 3, we tested nine combinations of small, medium, and large kernels. The results show the {3, 5, 11} combination achieves the best performance, with a DSC of 81.92% and an HD95 of 16.85 mm. Other combinations, like {3, 5, 7}, also perform well but are slightly inferior. Configurations without a large receptive field (e.g., {1, 3, 5}) or a medium-sized kernel (e.g., {1, 3, 11}) show a noticeable performance drop. These findings support using the {3, 5, 11} kernel combination, as it effectively captures multi-scale features and leads to more precise segmentation.

5 Conclusion

This paper introduces AxialUNet, an efficient architecture for medical image segmentation. Its backbone features the Multi-Scale Spatial Mixing (MSSM) module, which uses parallel convolutions with multiple kernel sizes to capture multi-scale contextual features. Furthermore, a Dynamic Heterogeneous Expert Attention (DHEA) mechanism is integrated into the skip connections to adaptively refine and fuse representations across different levels. Comprehensive evaluations on the Synapse and ISIC datasets demonstrate that AxialUNet outperforms recent CNN and Transformer-based baselines while maintaining a lightweight and computationally efficient design.

Acknowledgments. This work was supported by Guangdong Basic and Applied Basic Research Foundation under grants 2024A1515140109 and 2023A1515110695.

Disclosure of Interests. The authors have no competing interests to declare that are relevant to the content of this article.

References

1. Nguyen-Mau, T.H., Trinh, Q.H., Bui, N.T., et al.: PEFNet: positional embedding feature for polyp segmentation. In: El Yacoubi, M., et al. (eds.) International Conference on Multimedia Modeling (MMM 2023). LNCS, vol. 13812, pp. 240–251. (2023)
2. Elnakib, A., Gimel'farb, G., Suri, J.S., et al.: Medical image segmentation: a brief survey. In: Suri, J.S. (ed.) Multi Modality State-of-the-Art Medical Image Segmentation and Registration Methodologies, pp. 1–39. (2011)
3. Pham, D.L., Xu, C., Prince, J.L.: Current methods in medical image segmentation. Annu. Rev. Biomed. Eng. **2**(1), 315–337 (2000)
4. Sharma, N., Aggarwal, L.M.: Automated medical image segmentation techniques. J. Med. Phys. **35**(1), 3–14 (2010)
5. Kim, Y.: Convolutional neural networks for sentence classification. arXiv preprint arXiv:1408.5882 (2014)
6. Dosovitskiy, A., et al.: An image is worth 16x16 words: transformers for image recognition at scale. arXiv preprint arXiv:2010.11929 (2020)
7. Ronneberger, O., Fischer, P., Brox, T.: U-Net: convolutional networks for biomedical image segmentation. In: Navab, N., Hornegger, J., Wells, W.M., Frangi, A.F. (eds.) MICCAI 2015. LNCS, vol. 9351, pp. 234–241. Springer, Cham (2015). https://doi.org/10.1007/978-3-319-24574-4_28
8. Valanarasu, J.M.J., Oza, P., Hacihaliloglu, I., Patel, V.M.: Medical transformer: gated axial-attention for medical image segmentation. In: de Bruijne, M., Cattin, P.C., Cotin, S., Padoy, N., Speidel, S., Zheng, Y., Essert, C. (eds.) MICCAI 2021. LNCS, vol. 12901, pp. 36–46. Springer, Cham (2021). https://doi.org/10.1007/978-3-030-87193-2_4
9. Huang, X., Deng, Z., Li, D., et al.: MissFormer: an effective transformer for 2D medical image segmentation. IEEE Trans. Med. Imaging **42**(5), 1484–1494 (2022)
10. Xiao, H., Li, L., Liu, Q., et al.: Transformers in medical image segmentation: a review. Biomed. Signal Process. Control **84**, 104791 (2023)
11. Zhou, Z., Rahman Siddiquee, M.M., Tajbakhsh, N., et al.: UNet++: a nested U-Net architecture for medical image segmentation. In: International Workshop on Deep Learning in Medical Image Analysis, pp. 3–11 (2018)
12. Diakogiannis, F.I., Waldner, F., Caccetta, P., et al.: ResUNet-a: a deep learning framework for semantic segmentation of remotely sensed data. ISPRS J. Photogramm. Remote. Sens. **162**, 94–114 (2020)
13. Oktay, O., Schlemper, J., Folgoc, L.L., et al.: Attention U-Net: learning where to look for the pancreas. arXiv preprint arXiv:1804.03999 (2018)

14. Chen, J., Lu, Y., Yu, Q., et al.: TransUNet: transformers make strong encoders for medical image segmentation. arXiv preprint arXiv:2102.04306 (2021)
15. Cao, H., Wang, Y., Chen, J., et al.: Swin-UNet: UNet-like pure transformer for medical image segmentation. In: European Conference on Computer Vision, pp. 205–218. (2022)
16. Khan, S., Naseer, M., Hayat, M., et al.: Transformers in vision: a survey. ACM Comput. Surv. (CSUR) **54**(10s), 1–41 (2022)
17. Jacobs, R.A., Jordan, M.I., Nowlan, S.J., Hinton, G.E.: Adaptive mixtures of local experts. Neural Comput. **3**(1), 79–87 (1991)
18. Shazeer, N., Mirhoseini, A., Maziarz, K., et al.: Outrageously large neural networks: the sparsely-gated mixture-of-experts layer. arXiv preprint arXiv:1701.06538 (2017)
19. Lin, B., Tang, Z., Ye, Y., et al.: Moe-LLAVA: mixture of experts for large vision-language models. arXiv preprint arXiv:2401.15947 (2024)
20. Lepikhin, D., Lee, H.J., Xu, Y., et al.: GSHARD: scaling giant models with conditional computation and automatic sharding. arXiv preprint arXiv:2006.16668 (2020)
21. Fedus, W., Zoph, B., Shazeer, N.: Switch transformers: scaling to trillion parameter models with simple and efficient sparsity. J. Mach. Learn. Res. **23**(120), 1–39 (2022)
22. Riquelme, C., Puigcerver, J., Mustafa, B., et al.: Scaling vision with sparse mixture of experts. Adv. Neural. Inf. Process. Syst. **34**, 8583–8595 (2021)
23. Puigcerver, J., Riquelme, C., Mustafa, B., et al.: From sparse to soft mixtures of experts. arXiv preprint arXiv:2308.00951 (2023)
24. Zhou, Y., Lei, T., Liu, H., et al.: Mixture-of-experts with expert choice routing. Adv. Neural. Inf. Process. Syst. **35**, 7103–7114 (2022)
25. Liu, Q., Wu, X., Zhao, X., et al.: MOELORA: an MOE-based parameter efficient fine-tuning method for multi-task medical applications. arXiv preprint arXiv:2309.14101 (2023)
26. Landman, B., Xu, Z., Igelsias, J., Styner, M., Langerak, T., Klein, A.: MICCAI multi-atlas labeling beyond the cranial vault – workshop and challenge. In: Proceedings of the MICCAI Multi-Atlas Labeling Beyond Cranial Vault – Workshop Challenge, vol. 5, p. 12 (2015)
27. Codella, N.C., et al.: Skin lesion analysis toward melanoma detection: a challenge at the 2017 International Symposium on Biomedical Imaging (ISBI), hosted by the International Skin Imaging Collaboration (ISIC). In: 2018 IEEE 15th International Symposium on Biomedical Imaging (ISBI), pp. 168–172 (2018)
28. Codella, N., et al.: Skin lesion analysis toward melanoma detection 2018: a challenge hosted by the International Skin Imaging Collaboration (ISIC). arXiv preprint arXiv:1902.03368 (2019)
29. Ruan, J., Xiang, S.: VM-UNet: vision mamba UNet for medical image segmentation. arXiv preprint arXiv:2402.02491 (2024)
30. Zhang, K., Liu, D.: Customized segment anything model for medical image segmentation. arXiv preprint arXiv:2304.13785 (2023)
31. Naderi, M.R., Givkashi, M.H., Piri, F., et al.: Focal-UNet: UNet-like focal modulation for medical image segmentation. In: IEEE World AI IoT Congress (AIIoT), pp. 820–825. IEEE (2025)
32. Sun, G., et al.: DA-TransUNet: integrating spatial and channel dual attention with transformer U-Net for medical image segmentation. Front. Bioeng. Biotechnol. **12**, 1398237 (2024)

33. Shao, H., Zeng, Q., Hou, Q., Yang, J.: MCANet: Medical image segmentation with multi-scale cross-axis attention. In: Proceedings of the AAAI Conference on Artificial Intelligence, pp. 2034–2042 (2024)
34. Azad, R., Arimond, R., Khodapanah Aghdam, E., Kazerouni, A., Merhof, D.: DAE-former: dual attention-guided efficient transformer for medical image segmentation. In: International Workshop on Predictive Intelligence in Medicine, pp. 83–95 (2023)

Integrating Symbolic and Waveform Music Into Large Language Models

Teng Tu[1], Xiaohao Liu[1], Yunshan Ma[2(✉)], Ji Qi[3], and Tat-Seng Chua[1]

[1] National University of Singapore, Singapore, Singapore
{teng.tu,xiaohao.liu}@u.nus.edu, dcscts@nus.edu.sg
[2] Singapore Management University, Singapore, Singapore
ysma@smu.edu.sg
[3] Tsinghua University, Beijing, China
qj20@mails.tsinghua.edu.cn

Abstract. Music, as a unique and integral element of human life, is characterized by its complex structures, intricate details, and the fusion of multimodal information. Recent study advance music understanding by leveraging knowledge and reasoning capabilities derived from Large Language Models (LLMs). However, they often lack compatibility and fail to fully utilize the complementary strengths of diverse representations (e.g., ABC, MIDI, Waveform). To address these limitations, we propose a unified music-language model framework, named UniMuLM, transitioning from single-representation approaches to the integration of multiple music representations for LLM. Unifying different music representation formats poses challenges such as patch integrity and boundary ambiguity that arise from temporal discrepancies across these representations. To address these issues, UniMuLM employs a unified encoder that hierarchically aligns representations across multiple granularities, using contrastive learning and cross-reconstruction training to support coherent integration. Fine-tuned in multiple stages on open-source datasets, UniMuLM demonstrates the potential to handle dual-representation inputs. Notably, it achieves performance competitive with specialized waveform-only models on music understanding tasks, while surpassing open-source baselines in downstream applications such as music knowledge answering and ABC melody completion.

Keywords: Music Language Model · Music Understanding · Multimodal Language Model · Sound and Music Computing

1 Introduction

LLMs have achieved significant progress in linguistic tasks and also demonstrated potential in understanding other modalities (*e.g.*, vision), motivating the exploration of Multimodal Large Language Models [22]. Among various modalities, music stands out as a particularly challenging and underexplored domain due to its unique interplay of rhythm, melody, harmony, and lyrics, which collectively

J. Lokoč et al. (Eds.): MMM 2026, LNCS 16413, pp. 89–103, 2026.
https://doi.org/10.1007/978-981-95-6957-1_7

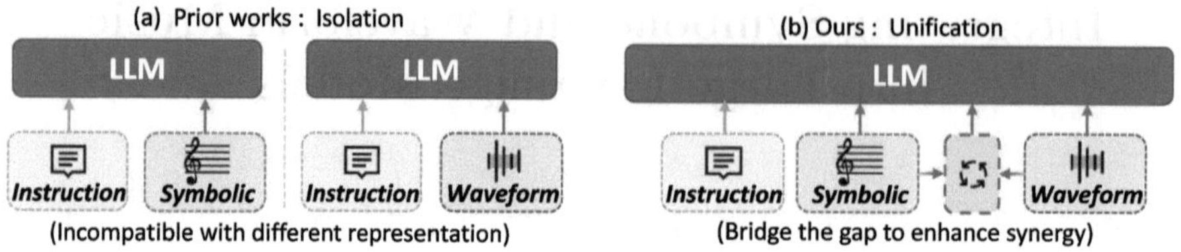

Fig. 1. Paradigm comparison between our method and prior works in music representation integration. (a) Prior works focus on utilizing either symbolic or waveform representations, typically in isolation. (b) In contrast, our approach introduces a unification mechanism bridging the gap and enhances their synergy.

evoke human emotions. This has sparked significant research interest in integrating musical knowledge into LLMs for music understanding tasks, *e.g.*, music captioning, question answering, and reasoning [5,10,23]. In this paper, we refer to such models as Music Understanding Language Models (MuLMs), following the terminology introduced by MU-LLaMA [23].

A pivotal obstacle preventing MuLMs from approaching human-level expertise in music lies in the treatment of music representations, which are typically either symbolic notations [4,35] or performance waveforms [11,13], with most models designed to handle only one of these forms. These two representations are inherently complementary: symbolic music captures structural and harmonic relationships, while waveforms preserve timbral characteristics and performance nuances. This synergy mirrors how human musicians perceive music and underscores the potential for more comprehensive music modeling.

Despite the salient research gap, unifying both representation formats is non-trivial due to the temporal scale inconsistency between the two formats. Specifically, symbolic music is organized into uneven time segments based on note durations and maintains semi-structured boundaries between notes, bars, phrases and movements. In contrast, waveform signals are sampled at much higher frequencies with overlapping note sounds, making clear segmentation boundaries difficult to establish. While various efforts have been made to encode waveforms [7,9,21,28,34], none have effectively bridged the representational gap to enable consistent unified modeling. This limitation has prevented LLMs from learning both representation formats simultaneously, motivating our development of a unified music encoder (Fig. 1).

Achieving this unification requires addressing two key technical challenges. The first challenge lies in the accurate integration of information during unification: conventional waveform patching methods rely on temporal segmentation [8,34], which complicates maintaining patch integrity while managing boundary ambiguities. Second, training stability when adapting language models for multiple modalities, particularly given the scarcity of training scenarios where symbolic music, waveform audio, and textual instructions co-occur.

To this end, we propose UniMuLM, a novel Music Understanding Language Model framework that employs a unified encoder compatible with both symbolic and waveform music representations (Fig. 2). First, we train the encoder to seamlessly pre-align music representations, ensuring synchronization between different representation formats while preserving token efficiency. Specifically, we utilize a hierarchical aligning training mechanism that enhances representational capabilities at both high-level temporal scales and fine-grained levels. Subsequently, to adapt LLMs for music tasks, we begin by applying LoRA tuning [17] to leveraging music knowledge and symbolic music datasets [31,33] to infuse music knowledge into the LLM. Afterward, we adapt the unified encoder to handle diverse music representations in shared embedding space across all downstream tasks using various datasets [2,23,26,27,31]. Our contributions are threefold:

- We emphasize the often-overlooked complementarity of music representations, proposing UniMuLM as the first framework to integrate symbolic and waveform music.
- We introduce and train a unified encoder that hierarchically aligns different music representations, merging the advantages of efficiency and effectiveness across different granularities.
- We benchmark UniMuLM using four music understanding datasets and explore its capability in various downstream tasks.

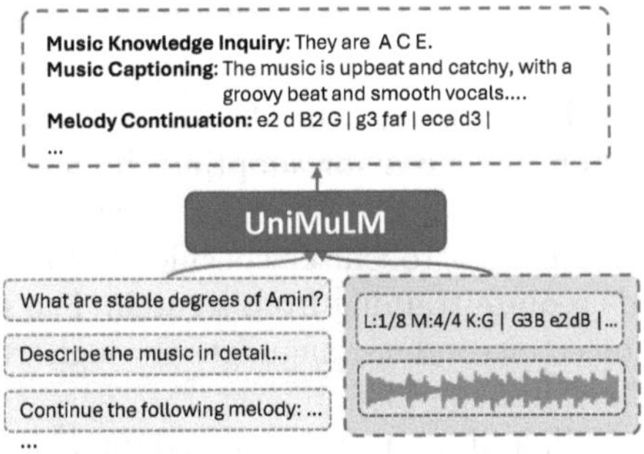

Fig. 2. A demonstration of UniMuLM in handling various types of music tasks.

2 Related Work

2.1 Music Encoding

Music representations in deep learning are broadly categorized into waveform and symbolic formats.

Waveform Representation consists of one-dimensional signals sampled at high frequencies. While some approaches process raw waveforms [28], most adopt spectrograms with the Fourier Transform [12]. Recent advances employ Variational Autoencoders to discretize waveforms into fixed-granularity tokens for conditioned generation tasks [6–8,25].

Symbolic Representation encodes discrete musical elements with precise, albeit non-uniform, temporal structures. MIDI, as the dominant protocol for real-time performance data, remains the industry standard among musicians and producers. Beyond direct MIDI processing [18], structured representations such as REMI [19], OctupleMIDI [35], and Compound Word [16] are designed to compress sequence length and enhance information density. Recently ABC notation [29,31–33] has gained traction in MuLM communities due to its easy-to-read structure and natural compatibility with natural language encoders.

2.2 Music Understanding Language Model

Recent advances in multimodal language models [22] have enabled new paradigms for music understanding. The key factor in MuLMs is how musical representations (waveform or symbolic), are integrated into LLMs.
Waveform-based MuLMs primarily follow two approaches. The first leverages vector-quantized variational autoencoders to discretize audio signals into tokens [3,7,8,34], achieving high-fidelity reconstruction but requires intensive training. The second employs acoustic feature adapters, mapping waveform features (e.g., MERT [21], CLAP [9]) into LLMs-either via direct projection [10] or cross-attention mechanisms [23].
Symbolic-based MuLMs employ two main tokenization strategies: creating a custom tokenizer, or adopting the pre-trained LM's text tokenizer. For the first approach, [35] use on-off or duration-based representations, while MuPT [29] customizes a tokenizer specifically for ABC notation. Nonetheless, these methods require training from scratch. The second approach processes musical notations as secondary languages through existing LLM vocabularies, exemplified by ChatMusician [33].

While discrete token or adapter approaches can technically combine waveform and symbolic encoders, the scarcity of co-occurring multi-representation training data induces *modality missing* scenarios. UniMuLM tackles this via a unified embedding space construction. Inspired by modality-binding [15,24], CLaMP3 [30], waveform-symbolic proximity enables latent space regularization, allowing single-modality training to enhance cross-modal understanding and support robust inference with partial inputs.

3 Problem Formulation

Our task is to generate a target output $\mathbf{y}$ given music inputs $\mathbf{m}^{\text{inputs}}$ and a task prompt $\mathbf{x}$ (e.g., *"Describe the music in detail"* for music captioning). In this work, $\mathbf{S}$ denotes symbolic notation (ABC), while $\mathbf{W}$ denotes waveform representation. We consider $\mathbf{m} = \{\mathbf{m}_\text{S}, \mathbf{m}_\text{W}\}$, where these representations are optional and may coexist, describing either identical or distinct music segments. Formally, we define a language token set $\mathcal{T}$. The prompt $\mathbf{x} \in \mathcal{T}^{l_x}$, output $\mathbf{y} \in \mathcal{T}^{l_y}$, and symbolic input $\mathbf{m}_\text{S} \in \mathcal{T}^{l_m}$ are all text token sequences. The waveform input $\mathbf{m}_\text{W} \in \mathbb{R}^{s \cdot r}$ is a sequence of audio samples with duration s and sampling rate r. Following standard language modeling, we frame this as an autoregressive estimation:$P(\mathbf{y}_i \mid \mathbf{x}, \mathbf{m}^{\text{inputs}}, \mathbf{y}_{1:i-1})$.

4 Methodology

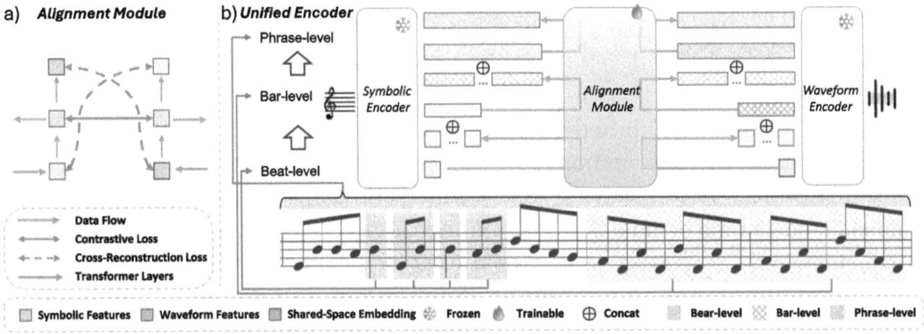

Fig. 3. The illustration of Unified Encoder. (a) The Alignment Module is trained using contrastive loss and cross-reconstruction loss, aligning symbolic feature and waveform features into a shared embedding space; (b) The hierarchical structure is presented, where features at different granularities are interconnected, aiming to enhance the model's information extraction capabilities.

As Fig. 4 shows, the UniMuLM framework employs a unified encoder architecture designed to process diverse music representations with a decoder-only transformer serving as the backbone, as illustrated. During training, we first train the encoder and then jointly optimize the adapter and LLM parameters.

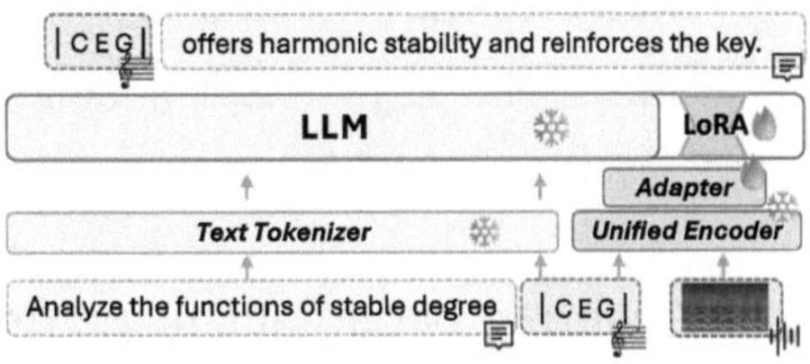

Fig. 4. The overall framework of UniMuLM.

4.1 Unified Encoder

To resolve inherent incompatibilities between music representations in existing systems, we propose a unified encoder architecture that establishes a shared embedding space while preserving modality-specific features, as Fig. 3 shown.

Overview. We address the structural gap between continuous audio signals, which involve continuous temporal sampling without explicit segmentation, and discrete symbolic representations, which organize into irregular and semantically defined units, as mentioned in Sect. 2, by using musically informed units as patch boundaries to maintain tempo alignment. Conventional fixed-duration audio windowing often misaligns with symbolic timing. To balance temporal resolution and structural consistency, we introduce a three-level hierarchical encoding at the beat, bar, and phrase levels. Larger temporal units improve computational efficiency, while finer segmentation preserves local detail. Our multi-scale design employs learnable attention to enable dynamic interaction between granularities, with task-specific effects evaluated in Sect. 5.

In this paper, we define $\Gamma = \{\text{beat}, \text{bar}, \text{phrase}\}$, where $\gamma \in \Gamma$ denotes an arbitrary granularity, and $\Psi = \{\text{symbolic}, \text{waveform}\}$, where $\psi \in \Psi$ denotes a music representation, with $\bar{\psi}$ representing the complementary one.

Implementation. The unified encoder processes symbolic and waveform inputs through distinct yet interconnected pathways. For waveform inputs ($\mathbf{m}_\text{W}$), we employ EnCodec [7] to generate latent representations $\mathbf{E}_\text{W}^\gamma \in \mathbb{R}^{l_W \times d}$, where r_c denotes the frame rate and d the latent dimension size. For symbolic inputs ($\mathbf{m}_\text{S}$), we tokenize ABC notation at the character level. Features across granularity levels are structured as $\mathbf{E}_\text{S}^\gamma \in \mathbb{R}^{l_S \times d}$, where l_S is the sequence length, with embedding dimension consistent with waveform representations.

The alignment module integrates dual encoding streams with cross-modal reconstruction. Enc_ψ^γ and Dec_ψ^γ are 3-layer Transformers that respectively map input $\mathbf{E'}_\psi^\gamma$ into shared-space $\mathbf{Z}_\phi^\gamma$, and decode them into reconstructed embeddings $\hat{\mathbf{E}}_{\bar{\psi}}^\gamma$ of the complementary modality. This design enables rich cross-modal interaction while preserving core information in the shared latent space.

The encoder input $\mathbf{E'}_\psi^\gamma$ differs from $\mathbf{E}_\psi^\gamma$ through hierarchical information propagation, where higher-level embeddings incorporate lower-level context:[1]

$$\text{Beat-level:} \quad \mathbf{E'}_\psi^{\text{beat}} = \mathbf{E}_\psi^{\text{beat}}, \tag{1}$$

$$\text{Bar-level:} \quad \mathbf{E'}_\psi^{\text{bar}} = \mathbf{E}_\psi^{\text{bar}} + \text{Concat}\big(\{\mathbf{Z}_\psi^{\text{beat}}\}_n\big), \qquad n \in \{2,3,4\}, \tag{2}$$

$$\text{Phrase-level:} \quad \mathbf{E'}_\psi^{\text{phrase}} = \mathbf{E}_\psi^{\text{phrase}} + \text{Concat}\big(\{\mathbf{Z}_\psi^{\text{bar}}\}_n\big), \qquad n = 4. \tag{3}$$

4.2 Adapting Large Language Model

Text Encoding. We utilize the pre-defined embedding table from the LM to encode instruction tokens (e.g., prompt $\mathbf{x}$) via the look-up function LM-Emb : $\mathcal{T} \to \mathbb{R}^{d_T}$, where d_T denotes the dimensionality of the textual embedding. Specifically, $\mathbf{Z}_\mathbf{x} = \text{LM-Emb}(\mathbf{x}) \in \mathbb{R}^{l_x \times d}$ represents the transformed embeddings, having the same length l_x as the input tokens. It is worth noting that ABC inputs will also be included as part of text input, following ChatMusician [33], which makes ABC have a dual representation.

Music Encoding. Both symbolic ABC $\mathbf{m}_\text{S}^\gamma$ and waveform inputs $\mathbf{m}_\text{W}^\gamma$ are encoded by the Unified Encoder into sequences of embeddings $\{\mathbf{Z}_\text{S}^\gamma\}$ and $\{\mathbf{Z}_\text{W}^\gamma\}$. For waveform processing under default settings, we adopt a 4/4 time signature assumption with BPM=120, resulting in beat-level segmentation every 0.5 s and bar-level segmentation every 2 s.

As we define in the problem formulation section, the LM backbone takes multi-modal embeddings to generate a sequence of textual tokens, expressed as:

[1] At the Bar-level, n represents the number of beats per bar, which depends on the time signature.

$$\mathrm{P}(\mathbf{y}_i \mid \mathbf{x}, \mathbf{m}, \mathbf{y}_{1:i-1}) = \mathrm{LM}\Big(\mathbf{Z}_{\mathbf{x},\mathbf{y}_{1:i-1}}, \mathrm{A}_{\mathrm{S}}^{\gamma}(\mathbf{Z}_{\mathrm{S}}^{\gamma}), \mathrm{A}_{\mathrm{W}}^{\gamma}(\mathbf{Z}_{\mathrm{W}}^{\gamma})\Big) \tag{4}$$

where the $\mathrm{A}_{\psi}^{\gamma}$ is a fully connected layer that adapts music embeddings to the embedding space and dimensionality of the LM.

4.3 Training Strategy

To mitigate data scarcity of multimodal co-occurrences, we propose a three-stage training strategy: 1) Music Representation Aligning (aligning symbolic and waveform music), 2) Knowledge Injecting (aligning symbolic music and text), 3) Multi-Task Fine-tuning (using waveform tasks to align all modalities).

Stage 1 (Music Representation Aligning). During training, paired training data $\mathcal{D}_{\mathsf{align}} = \{(\mathbf{m}_{\mathrm{S}}^{i,\gamma}, \mathbf{m}_{\mathrm{W}}^{i,\gamma})\}_{i=1}^{n}$ is synthesized by rendering symbolic sequences with randomized instrumental timbres. To align the symbolic and waveform intermediate embeddings within the shared latent space, we apply InfoNCE [14] as contrastive loss:

$$\mathcal{L}_{\mathrm{InfoNCE}}^{\gamma,\psi,(i)} = -\log \frac{\exp(\cos(\mathbf{Z}_{\psi,i}^{\gamma}, \mathbf{Z}_{\bar{\psi},i}^{\gamma})/\tau)}{\sum_{k=1}^{N} \exp(\cos(\mathbf{Z}_{\psi,i}^{\gamma}, \mathbf{Z}_{\bar{\psi},k}^{\gamma})/\tau)} \tag{5}$$

where $\cos(\cdot,\cdot)$ represents cosine similarity, τ is a temperature parameter, and N is the number of negative samples. To mitigate excessive information loss, we apply a cross-reconstruction loss, represented as: $\mathcal{L}_{\mathrm{rec}}^{\mathrm{S},\gamma} = \|\hat{\mathbf{E}}_{\mathrm{S}} - \mathbf{E}_{\mathrm{S}}\|_2^2$, $\mathcal{L}_{\mathrm{rec}}^{\mathrm{W},\gamma} = \|\hat{\mathbf{E}}_{\mathrm{W}}^{\gamma} - \mathbf{E}_{\mathrm{W}}^{\gamma}\|_2^2$. Thus, the loss for representation alignment, which combines both contrastive and reconstruction losses, is denoted as:

$$\underset{\Theta_{\mathrm{Dec,\ Enc}}}{\operatorname{argmin}}\ \mathcal{L} = \sum_{\gamma \in \Gamma} \left(\mathcal{L}_{\mathrm{InfoNCE}}^{\gamma} + \mathcal{L}_{\mathrm{rec}}^{\mathrm{S},\gamma} + \mathcal{L}_{\mathrm{rec}}^{\mathrm{W},\gamma}\right) \tag{6}$$

Stage 2 (Knowledge Injecting). We begin by using music knowledge and symbolic music datasets to warm up the pre-trained LLM. During this stage, the unified encoders remain inactive. The training dataset incorporates a mixture of symbolic music represented in ABC notation format, which is processed purely as textual data. Training is achieved through a negative log-likelihood (NLL) objective, where the model predicts the next token $\mathbf{y}_i$ in the sequence based on the previous tokens $\mathbf{y}_{1:i-1}$:

$$\underset{\Theta_{\mathrm{LoRA}}}{\operatorname{argmin}}\ \mathcal{L} = -\frac{1}{l_{\mathbf{y}}} \sum_{i=1}^{l_{\mathbf{y}}} \log \mathrm{P}(\mathbf{y}_i \mid \mathbf{x}, \mathbf{y}_{1:i-1}) \tag{7}$$

Stage 3 (Multi-task Fine-tuning). In the final stage, we freeze the Unified Encoder, LoRA-tune the LLM and train adapters to accommodate musical representations for all downstream tasks across different datasets that include symbolic music, waveform music, and textual instructions. We formally present the final stage training as follows:

$$\underset{\Theta_{\mathrm{LoRA}}, \mathrm{A_W}, \mathrm{A_S}}{\operatorname{argmin}} \quad \mathcal{L} = -\frac{1}{l_{\mathbf{y}}} \sum_{i=1}^{l_{\mathbf{y}}} \log \mathrm{P}(\mathbf{y}_i \mid \mathbf{x}, \mathbf{m}, \mathbf{y}_{1:i-1}) \tag{8}$$

5 Experiments

To comprehensively evaluate UniMuLM's performance and effectiveness, we conducted experiments on four music understanding benchmarks and explored its applicability to two downstream tasks: melody completion and music knowledge answering. We also performed detailed model studies and ablations to analyze the contributions of different components and configurations.

5.1 Experimental Settings

Hyperparameter Settings. We employ Llama3-8B as the LM backbone with a hidden dimension of 4096, a learning rate of 5e-6, and a total batch size of 16 across 4 devices, using a 64-rank LoRA with $\alpha = 16$. The Unified Encoder is followed by lightweight adapter modules consisting of a self-attention layer and an MLP, which encode each music feature into a 4096-dimensional embedding aligned with Llama3's hidden size.
Baselines. Waveform-based methods involve GPT-4o [1], LTU [13], Audio-Flamingo [20], LLark [11] and Mu-LLaMA [23]; Symbolic-based methods involve ChatMusician [33] and MuPT [29].[2]
Datasets. We categorize datasets into four core tasks, as summarized in Table 1. (1) Music Knowledge (MK) targets theoretical understanding, drawing on a corpus derived from MusicPile [33] and evaluated with MusicTheoryBench (MTB), a multiple-choice benchmark on composition theory and ABC notation. (2) Music Understanding (MU) aims to bridge waveform music and text, integrating multi-modal datasets: MusicCaps, SongDescriber [26], and MidiCaps [27] for music descriptions, together with MusicQA [23] for complex reasoning. SongDescriber is used only for evaluation, and MIDI sequences in MidiCaps are synthesized into waveforms. (3) Melody Completion (MC) focuses on symbolic generation, while (4) Melody Transcription (MT) involves converting waveform into symbolic notation. Both tasks are constructed from MelodyHub [31]'s ABC melodies at bar-, phrase-, and melody-level granularity, and support both multiple-choice and open-ended generation formats.

Table 1. Summary of training datasets.

Task	Datasets Source	Sampled
Music Knowledge (MK)	MusicPile	200K
Music Understanding (MU)	MusicCaps, MidiCaps, SongDescriber, MusicQA	25K
Melody Completion (MC)	MelodyHub	160K
Melody Transcription (MT)	MelodyHub	125K

[2] To keep the manuscript concise yet self-contained, we moved the detailed description of baselines to our online appendix https://tuteng0915.github.io/UniMuLM.

Table 2. Performance comparison on music understanding tasks. Best in **bold**, second best <u>underlined</u>.

Category	Model	MusicCaps			SongDescriber			MidiCaps			MusicQA		
		BLEU	R-L	Acc	BLEU	R-L	Acc	BLEU	R-L	Acc	BLEU	R-L	Acc
Baseline	GPT-4o	0.280	0.310	<u>0.875</u>	0.262	0.292	<u>0.830</u>	**0.305**	**0.312**	**0.895**	0.230	0.315	**0.955**
	LTU	0.216	0.248	0.805	0.222	0.237	0.805	0.201	0.223	0.780	0.242	0.328	0.905
	Audio-Flamingo	0.221	**0.320**	0.810	0.218	0.302	0.825	0.213	0.297	0.830	0.234	0.337	0.900
	LLark	<u>0.278</u>	0.250	0.770	0.243	0.237	0.730	0.248	0.268	0.810	0.201	0.194	<u>0.935</u>
	Mu-LLaMA	<u>0.281</u>	0.316	0.780	0.278	0.313	0.710	<u>0.271</u>	0.306	0.735	**0.306**	**0.466**	0.925
UniMuLM	Beat-level	0.217	0.205	0.830	0.204	0.234	0.700	0.190	0.224	0.810	0.212	0.333	0.880
	Bar-level	0.275	0.310	**0.890**	<u>0.290</u>	**0.341**	<u>0.875</u>	0.263	<u>0.311</u>	<u>0.875</u>	<u>0.295</u>	<u>0.409</u>	0.910
	Phrase-level	**0.281**	<u>0.317</u>	0.855	**0.287**	<u>0.329</u>	0.835	0.241	0.287	0.850	0.271	0.369	0.900

5.2 Performance Evaluation

In addition to utilizing BLEU and ROUGE-L (R-L) metrics to evaluate waveform music understanding in an open-ended manner across four datasets, we also created a 4-option closed-ended evaluation. This was achieved by pairing ground-truth answers with three randomly selected distractors from other music samples (200 test samples per dataset). Taken together, the combination of these perspectives provides a more holistic and reliable assessment of model capability, offering complementary insights into performance. The overall results are summarized in Table 2.

Among baseline models, Mu-LLaMA delivers competitive results on MusicCaps (BLEU: 0.281, R-L: 0.316) and achieves the highest BLEU (0.306) and R-L (0.466) on MusicQA. GPT-4o, in contrast, attains the top scores on MidiCaps (BLEU: 0.305, R-L: 0.312, Acc: 0.895) and also leads in MusicQA accuracy (0.955). Within the UniMuLM variants, the Bar-level encoding achieves the best results on SongDescriber (R-L: 0.341, Acc: 0.875) and maintains strong performance on MusicQA (R-L: 0.409), whereas the Phrase-level encoding slightly surpasses Bar-level on MusicCaps (BLEU: 0.281, R-L: 0.317) and on SongDescriber BLEU (0.287). Overall, Bar-level and Phrase-level encodings exhibit comparable and consistently superior performance over Beat-level, particularly in tasks requiring long-range musical context. By contrast, baseline models demonstrate dataset-specific peaks but display greater variability across BLEU and ROUGE-L metrics.

5.3 Downstream Tasks

The music understanding capabilities of UniMuLM enable it to perform well on various downstream tasks. In this study, we evaluate two categories of downstream tasks: Music knowledge answering and Melody Completion as shown in Table 3 and Table 4.

Music Knowledge Answering. On the MusicTheoryBench, in Table 3, UniMuLM with bar-level encoding achieves the highest accuracy (0.460), outperforming both general-purpose waveform-based models such as GPT-4o (0.441)

Table 3. Evaluation on MTB.

Category	Model	ACC
General	GPT-4o	0.441
	Llama3-8B	0.398
MuLM	ChatMusician	0.354
	Mu-LLaMA	0.256
	LTU	0.319
UniMuLM	Beat-level	0.401
	Bar-level	**0.460**
	Phrase-level	0.411

Table 4. Performance on melody continuation and inpainting tasks.

Category	Model	Continuation					Inpainting				
		Acc	Valid	RC	BLEU	R-L	Acc	Valid	RC	BLEU	R-L
General	GPT-4o	0.586	0.912	0.645	0.341	0.556	0.330	**0.963**	0.255	0.122	0.262
	Llama3-8B	0.502	0.756	0.457	0.205	0.213	0.312	0.799	0.120	0.114	0.121
MuLM	MuPT	-	0.798	0.385	0.272	0.295	-	-	-	-	-
	ChatMusician	0.553	0.852	0.630	0.487	0.532	0.454	0.885	0.121	0.069	0.082
UniMuLM	Beat-level	**0.688**	0.942	0.618	0.497	0.623	0.611	0.955	0.327	**0.133**	0.243
	Bar-level	0.673	**0.952**	0.644	**0.502**	**0.643**	**0.615**	0.948	**0.323**	0.128	**0.238**
	Phrase-level	0.668	0.947	**0.650**	0.495	0.638	0.611	0.965	0.322	0.128	0.222

and symbolic-focused models like ChatMusician (0.354). This suggests that UniMuLM effectively integrates structural and symbolic music features, enabling more accurate reasoning over musical knowledge. The comparison among different encoding granularities further reveals that bar-level encoding provides the best balance between local rhythmic detail and global structural coherence, that appears crucial for representing hierarchical concepts in music theory.

Melody Completion. We evaluate melody completion using two complementary subtasks: Continuation, which extends a given melodic fragment, and Inpainting, which reconstructs missing segments, in Table 4. Performance is assessed via both discrete-choice accuracy (Acc) and open-ended generation metrics, including BLEU, R-L, Rhythmic Consistency (RC) for pitch-agnostic temporal alignment, and Validity for syntactic correctness in ABC notation (e.g., consistent beat count). Across both subtasks, UniMuLM consistently outperforms baseline models. The bar-level variant demonstrates the strongest overall results, achieving, for example, 0.673 Acc in Continuation and 0.625 Acc in Inpainting. It also attains the highest BLEU and R-L scores while preserving high Validity and RC, indicating that it maintains rhythmic integrity without sacrificing melodic structure. Notably, the beat-level variant trails the other two granularities, further highlighting the importance of capturing longer-range musical dependencies.

5.4 Ablation and Model Analysis

We systematically evaluate our framework by (1) examining how the unified encoder design impacts Stage 1 effectiveness; (2) verifying if unified encoder outperforms existing encoders under identical experimental settings, and (3) assessing whether Stage 2 knowledge injection as warm-up training benefits Stage 3 LLM performance; (4) exploring the interaction synergies among tasks. Additional analyses and extended results are also provided in the online appendix.

Unified Encoder Design. To validate the Unified Encoder design, we conduct systematic ablation studies focusing on three critical components: (1) contrastive objective selection (InfoNCE, replaced in ablation with cosine similarity), (2) the cross-reconstruction object, and (3) inter-granularity concatenation. We evaluate cross-format retrieval ability, where S2W denotes retrieving a waveform representation from a symbolic query, and W2S denotes retrieving a symbolic representation from a waveform query. Retrieval accuracy is reported as hit@1 among 10 candidates, serving as a proxy for cross-modal alignment capability. As shown in Table 5, all three components are essential: removing any one results in a substantial drop in retrieval accuracy. Removing the InfoNCE objective or inter-granularity concatenation lead to moderate but consistent declines in retrieval—for instance, while eliminating the cross-reconstruction objective causes severe performance degradation with S2W accuracy plunging to as low as 0.27 for Phrase-level and W2S dropping to 0.33–0.44 across granularities, confirming its pivotal role in enabling thorough bidirectional information fusion.

Table 5. Ablation study on unified encoder designs.

Config	InfoNCE Loss ↓			Rec. Loss ↓			S2W Retrieval ↑			W2S Retrieval ↑		
	Beat	Bar	Phrase	Beat	Bar	Phrase	Beat	Bar	Phrase	Beat	Bar	Phrase
Proposed	0.14	0.15	0.18	0.11	0.24	0.35	0.94	0.93	0.97	0.89	0.88	0.93
w/o InfoNCE	-	-	-	0.13	0.19	0.42	0.90	0.89	0.82	0.83	0.85	0.88
w/o Recon.	0.25	0.23	0.21	-	-	-	0.43	0.38	0.27	0.38	0.33	0.44
w/o Concat	0.18	0.17	0.21	0.15	0.29	0.44	0.82	0.73	0.90	0.80	0.77	0.85

Comparison with Existing Encoders. We report ROUGE-L scores on four music understanding benchmarks and the accuracy on MusicTheoryBench, comparing our proposed encoder with MERT [21], CLAP [9], and CLaMP2 [32] (Table 6). Our model achieves the highest scores across all benchmarks, with especially large gains on knowledge-intensive datasets such as MusicQA (+0.087) and MusicTheoryBench (+0.078). These improvements indicate a stronger and more consistent integration of symbolic and audio representations.

Table 6. Comparison with existing encoders and Ablation of Stage 2

Encoder	MusicCaps	SongDescriber	MidiCaps	MusicQA	MTB
MERT	0.310	0.275	0.287	0.322	0.368
CLAP	0.305	0.316	0.259	0.288	0.354
CLaMP2	0.297	0.310	0.241	0.302	0.385
Proposed	0.310	0.341	0.311	0.409	0.463
- w/o Stage2	0.281	0.323	0.282	0.378	0.356

Effect of Stage 2 Knowledge Injection. Stage 2 pretraining serves to inject symbolic-music-specific knowledge before Stage 3 LLM adaptation. Skipping Stage 2 leads to modest degradation on SongDescriber (−0.018), but much sharper declines on knowledge-sensitive benchmarks like MusicQA (−0.031) and MusicTheoryBench (−0.107), as shown in Table 6. This indicates that Stage 2 is particularly beneficial for tasks requiring deeper reasoning over musical form and theory, whereas more descriptive tasks can partially rely on Stage 3 adaptation alone, and the knowledge injection phase plays a crucial role in bridging symbolic structure with downstream reasoning ability.

5.5 Multi-task Synergy

This experiment quantifies cross-task interactions by evaluating four core music-related tasks under different training configurations: Music Knowledge Modeling (MK, Accuracy on MusicTheoryBench), Music Understanding (MU, R-L across four benchmarks), Music Captioning (MC, multiple-choice accuracy), and Music Transcription (MT, open-ended generation F1), as summarized in Table 7. The results demonstrate several clear trends. First, adding MU to MK slightly decreases knowledge accuracy (-0.014), suggesting that shared representations introduce a trade-off between factual precision. Second, incorporating MC further improves both MK and MU, and additionally provides strong captioning accuracy (0.652), highlighting that tasks requiring semantic grounding benefit from multi-modal supervision. Finally, training on all four tasks achieves the highest overall scores across all metrics, indicating that a fully joint setup promotes complementary learning rather than interference.

Table 7. Multi-task synergy evaluation.

Tasks	Metrics			
	MK (Acc)	MU (R-L)	MC (Acc)	MT (F1)
MK	0.454	–	–	–
MK, MU	0.441	0.324	–	–
MK, MU, MC	0.457	0.338	0.652	–
All	**0.460**	**0.342**	**0.673**	**0.303**

6 Conclusion

We propose UniMuLM, a unified framework that bridges symbolic and waveform music within LLMs, using a novel encoder to coordinate multi-granularity features and maintain structural integrity while mitigating temporal inconsistencies across different music representation. To train UniMuLM efficiently and effectively, we implement a multi-stage optimization strategy, enabling the model to process diverse musical inputs and achieve competitive performance across various music-related tasks. This work shifts the paradigm from single-representation reliance to synergistic multi-representation integration, advancing music understanding and enriching the landscape of multimodal language models.

Limitations. UniMuLM currently has three main limitations. First, the alignment module was only trained on single-track, single-instrument synthesized music, primarily within Western tonal and metrical conventions; non-Western traditions and non-isochronous rhythms are out of scope. Second, due to computational constraints, our use of 4-bit quantization and LoRA-based fine-tuning limits the model's capacity. This is a trade-off for efficiency and may yield non-leading performance compared to full-parameter optimization. Third, UniMuLM's output is limited to text and ABC notation because it relies on a text-based LLM decoder. As a result, it does not directly generate MIDI or waveforms.

Future Work. Future research may focus on three directions: First, enhancing the alignment training by integrating multi-track symbolic parsing with acoustic scene analysis in combination with real recordings and synthesizer timbres. Second, developing a hybrid decoder that generates both symbolic and acoustic tokens. This will allow for direct MIDI generation using event-based tokens [16,19] and waveform synthesis using audio codecs [7,34].

References

1. Achiam, J., Adler, S., Agarwal, S., Ahmad, L., Akkaya, I., Aleman, F.L., et al.: GPT-4 technical report. arXiv preprint arXiv:2303.08774 (2023)
2. Agostinelli, A., Denk, T.I., Borsos, Z., Engel, J.H., Verzetti, M., Caillon, A., et al.: MusiCLM: generating music from text. CoRR abs/2301.11325 (2023)
3. Borsos, Z., Marinier, R., Vincent, D., Kharitonov, E., Pietquin, O., Sharifi, M., et al.: AudioLM: a language modeling approach to audio generation. IEEE ACM Trans. Audio Speech Lang. Process. **31**, 2523–2533 (2023)
4. Chou, Y., Chen, I., Chang, C., Ching, J., Yang, Y.: MidiBERT-piano: large-scale pre-training for symbolic music understanding. CoRR abs/2107.05223 (2021)
5. Chu, Y., Xu, J., Zhou, X., Yang, Q., Zhang, S., Yan, Z., et al.: QWEN-audio: advancing universal audio understanding via unified large-scale audio-language models. CoRR abs/2311.07919 (2023)
6. Copet, J., Kreuk, F., Gat, I., Remez, T., Kant, D., Synnaeve, G., et al.: Simple and controllable music generation. In: NeurIPS (2023)
7. Défossez, A., Copet, J., Synnaeve, G., Adi, Y.: High fidelity neural audio compression. Trans. Mach. Learn. Res. (2023)

8. Dhariwal, P., Jun, H., Payne, C., Kim, J.W., Radford, A., Sutskever, I.: Jukebox: a generative model for music. CoRR abs/2005.00341 (2020)
9. Elizalde, B., Deshmukh, S., Ismail, M.A., Wang, H.: CLAP learning audio concepts from natural language supervision. In: IEEE ICASSP, pp. 1–5. IEEE (2023)
10. Deng, Z., Ma, Y., Liu, Y., Guo, R., Zhang, G., Chen, W., et al.: Musilingo: bridging music and text with pre-trained language models for music captioning and query response. In: NAACL-HLT (Findings), pp. 3643–3655. Association for Computational Linguistics (2024)
11. Gardner, J., Durand, S., Stoller, D., Bittner, R.M.: LLARK: a multimodal foundation model for music. In: ICML (2024)
12. Gong, Y., Chung, Y., Glass, J.R.: AST: audio spectrogram transformer. In: Interspeech, pp. 571–575. ISCA (2021)
13. Gong, Y., Luo, H., Liu, A.H., Karlinsky, L., Glass, J.R.: Listen, think, and understand. In: ICLR (2024)
14. Gutmann, M., Hyvärinen, A.: Noise-contrastive estimation: a new estimation principle for unnormalized statistical models. In: AISTATS. JMLR Proceedings, vol. 9, pp. 297–304. JMLR.org (2010)
15. Girdhar, R., et al.: ImageBind: one embedding space to bind them all. In: CVPR (2023)
16. Hsiao, W., Liu, J., Yeh, Y., Yang, Y.: Compound word transformer: learning to compose full-song music over dynamic directed hypergraphs. In: AAAI, pp. 178–186. AAAI Press (2021)
17. Hu, E.J., Shen, Y., Wallis, P., Allen-Zhu, Z., Li, Y., Wang, S., et al.: LoRA: low-rank adaptation of large language models. In: ICLR (2022)
18. Huang, C.A., Vaswani, A., Uszkoreit, J., Simon, I., Hawthorne, C., Shazeer, N., et al.: Music transformer: generating music with long-term structure. In: ICLR (2019)
19. Huang, Y., Yang, Y.: Pop music transformer: beat-based modeling and generation of expressive pop piano compositions. In: ACM Multimedia, pp. 1180–1188. ACM (2020)
20. Kong, Z., Goel, A., Badlani, R., Ping, W., Valle, R., Catanzaro, B.: Audio flamingo: a novel audio language model with few-shot learning and dialogue abilities. In: ICML (2024)
21. Li, Y., Yuan, R., Zhang, G., Ma, Y., Chen, X., Yin, H., et al.: MERT: acoustic music understanding model with large-scale self-supervised training. In: ICLR (2024)
22. Liu, H., Li, C., Wu, Q., Lee, Y.J.: Visual instruction tuning. In: NeurIPS (2023)
23. Liu, S., Hussain, A.S., Sun, C., Shan, Y.: Music understanding llama: advancing text-to-music generation with question answering and captioning. In: ICASSP, pp. 286–290. IEEE (2024)
24. Liu, X., Xia, X., Huang, Z., Chua, T.-S.: Towards modality generalization: a benchmark and prospective analysis. CoRR abs/2412.18277 (2024)
25. Liu, X., Tu, T., Ma, Y., Chua, T.-S.: Extending visual dynamics for video-to-music generation. CoRR abs/2504.07594 (2025)
26. Manco, I., Weck, B., Doh, S., Won, M., Zhang, Y., Bogdanov, D., et al.: The song describer dataset: a corpus of audio captions for music-and-language evaluation. CoRR abs/2311.10057 (2023)
27. Melechovský, J., Roy, A., Herremans, D.: MidiCaps - a large-scale MIDI dataset with text captions. In: ISMIR, pp. 858–865. ISMIR (2024)
28. van den Oord, A., Dieleman, S., Zen, H., Simonyan, K., Vinyals, O., Graves, A., et al.: WaveNet: a generative model for raw audio. In: SSW, p. 125. ISCA (2016)

29. Qu, X., Bai, Y., Ma, Y., Zhou, Z., Lo, K.M., Liu, J., et al.: MUPT: a generative symbolic music pretrained transformer. In: ICLR (2025)
30. Wu, S., Guo, Z., Yuan, R., Jiang, J., Doh, S., Xia, G., et al.: Clamp 3: universal music information retrieval across unaligned modalities and unseen languages. In: ACL (Findings), pp. 2605–2625. Association for Computational Linguistics (2025)
31. Wu, S., Wang, Y., Li, X., Yu, F., Sun, M.: MeloDyt5: a unified score-to-score transformer for symbolic music processing. In: ISMIR, pp. 642–650. ISMIR (2024)
32. Wu, S., Wang, Y., Yuan, R., Guo, Z., Tan, X., Zhang, G., et al.: Clamp 2: multimodal music information retrieval across 101 languages using large language models. In: NAACL (Findings), pp. 435–451. Association for Computational Linguistics (2025)
33. Yuan, R., Lin, H., Wang, Y., Tian, Z., Wu, S., Shen, T., et al.: ChatMusician: understanding and generating music intrinsically with LLM. In: ACL, pp. 6252–6271. Association for Computational Linguistics (2024)
34. Zeghidour, N., Luebs, A., Omran, A., Skoglund, J., Tagliasacchi, M.: SoundStream: an end-to-end neural audio codec. IEEE ACM Trans. Audio Speech Lang. Process. **30**, 495–507 (2022)
35. Zeng, M., Tan, X., Wang, R., Ju, Z., Qin, T., Liu, T.: MusicBERT: symbolic music understanding with large-scale pre-training. In: ACL/IJCNLP (Findings), pp. 791–800. Association for Computational Linguistics (2021)

Video Hashing via a Mamba-Transformer Network for Retrieval

Likai Yang[1,2,3], Nianqiao Li[1,2,3], Xiaoping Liang[1,2,3], Lv Chen[1,2,3], and Zhenjun Tang[1,2,3(✉)]

[1] Key Lab of Education Blockchain and Intelligent Technology, Ministry of Education, Guangxi Normal University, Guilin 541004, China
tangzj230@163.com
[2] Guangxi Key Lab of Multi-Source Information Mining and Security, Guangxi Normal University, Guilin 541004, China
[3] University Engineering Research Center of Educational Intelligent Technology, Guangxi Normal University, Guilin 541004, China

Abstract. Self-Supervised Video Hashing (SSVH) has been widely applied to efficient video retrieval. Existing methods mostly use Siamese-like pairwise training to model inter-sample similarities, yet lack direct optimization for high-quality video content representation, limiting practical performance. To address this issue, we propose a novel SSVH method based on the Mamba-Transformer network (hereafter VHMT), which employs a teacher-student architecture. The teacher model extracts robust low-frequency semantic features using the Discrete Wavelet Transform (DWT) and captures temporal dependencies through a Vision Transformer (ViT)-based Temporal Attention Module (TAM). Concurrently, the introduced Temporal Reconstruction Enhancement (TRE) block further improves fine-grained reconstruction, thereby generating high-quality temporal modeling signals. The student model employs the Mamba architecture to model temporal relationships and achieves knowledge transfer and efficient single-path inference by approximating the teacher model's feature reconstruction and attention outputs. Furthermore, we introduce a joint loss function to enhance the student model's approximation of the teacher model's temporal modeling capability while optimizing the clustering structure of the generated hash codes. Experimental results demonstrate the VHMT outperforms several state-of-the-art methods in mean Average Precision.

Keywords: Video hashing · Video retrieval · Vision Transformer (ViT) · Mamba

1 Introduction

With the rise of video-sharing platforms, efficient video retrieval has become crucial. Video hashing addresses this by compressing high-dimensional data into

L. Yang and N. Li—Contributed equally to this work.

J. Lokoč et al. (Eds.): MMM 2026, LNCS 16413, pp. 104–118, 2026.
https://doi.org/10.1007/978-981-95-6957-1_8

compact binary codes, enabling fast similarity search with low storage cost [1–3]. Recent deep learning advances have boosted its performance, among which Self-Supervised Video Hashing (SSVH) stands out by generating pseudo-labels via pre-training, achieving high accuracy without costly manual annotations [4]. This makes SSVH more practical than supervised methods in real-world scenarios.

Most existing SSVH methods still rely on a pairwise training paradigm similar to Siamese networks [5], focusing primarily on modeling inter-sample similarity, but they lack direct optimization for high-quality video content representation, thereby limiting model performance in real-world tasks [6–8]. Moreover, these methods exhibit insufficient capability in temporal modeling and feature reconstruction [9–11]. To address these issues, we propose a novel SSVH method based on the Mamba-Transformer network (hereafter VHMT), with the following main contributions:

(1) A novel teacher-student network is proposed, in which the teacher model extracts robust low-frequency semantic features via Discrete Wavelet Transform (DWT) and captures temporal dependencies through a Temporal Attention Module (TAM) derived from the Vision Transformer (ViT). The student model employs a Mamba to model temporal relationships and achieves knowledge transfer and efficient single-path inference by approximating the teacher model's feature reconstruction and attention outputs.
(2) The proposed TAM employs a Multi-Head Attention (MHA) mechanism to model temporal dependencies across video segments, and introduces a novel Temporal Reconstruction Enhancement (TRE) module to improve the fine-grained reconstruction of temporal features.
(3) A novel loss function is proposed to jointly optimize temporal reconstruction, hash contrastive learning, and local structure preservation, thereby enhancing the student model's approximation of the teacher model's temporal modeling capabilities while simultaneously optimizing the clustering structure of the hash codes.

Experiments on three public datasets [12–14] show that the VHMT outperforms several state-of-the-art (SOTA) methods in video hashing retrieval under the mean Average Precision (mAP) metric. The rest of the paper is organized as follows: Sect. 2 reviews related work; Sect. 3 details the VHMT method; Sect. 4 reports results; and Sect. 5 concludes the paper.

2 Related Work

Recent advances in deep learning have propelled the application of SSVH in large-scale retrieval. Early methods, such as SSTH [15] and JTAE [16], primarily adopted Recurrent Neural Networks (RNNs) to extract temporal features from videos. Subsequently, several LSTM-based encoder-decoder framework methods were proposed [4], attempting to reconstruct original frame features from relaxed hash codes. Although the memory capacity of these models hinders modeling high-dimensional temporal information, these methods have contributed foundational training paradigms for early SSVH development.

Later, the introduction of Transformers has driven SSVH towards more sophisticated temporal modeling. BTH [9] was the first Transformer-based SSVH method, proposing a bidirectional encoder structure to fully exploit long-range bidirectional correlations among frames. DKPH [17] introduced a Gaussian-adaptive similarity graph into the Transformer model, decoupling temporal reconstruction from semantic retrieval, offering a new perspective for SSVH advancement. Following this, ConMH [6] designed a Siamese network architecture under a Masked Autoencoder (MAE)-inspired training paradigm, significantly improving retrieval accuracy through masked contrastive learning. Benefiting from this progress, subsequent Transformer-based methods built on Siamese networks have emerged, including TSVH [10], SPVH [11], and AutoSSVH [8].

Meanwhile, several methods based on the MLP-Mixer [18] have been developed, such as MCMSH [19] and EUVH [20], which leverage the simplicity of MLPs to build more efficient retrieval models. Additionally, MAGRH [21], based on Graph Neural Networks [22], offers a novel direction in relational reasoning. Recently, S5VH [7] introduced an encoder-decoder structure based on Mamba, which is both effective and efficient in capturing temporal relationships, providing new potential for SSVH.

Most existing SSVH methods still rely on a Siamese-like pairwise training paradigm that emphasizes inter-sample similarity, but they lack direct optimization for high-quality video content representation, thereby limiting their effectiveness in real-world tasks. Furthermore, these methods exhibit limited capability in temporal modeling and feature reconstruction. This study is dedicated to addressing these issues.

3 Proposed VHMT Method

3.1 Problem Definition and Overview

Consider an unlabeled video dataset containing N videos, represented as $V = \{V_i\}_{i=1}^{N}$, where each $V_i \in \mathbb{R}^{T \times D}$ denotes the frame-level features of the i-th video. These features are extracted using a pretrained CNN backbone (e.g., VGG or ResNet [23,24]), with T and D indicating the number of frames and the feature dimension, respectively. The proposed VHMT network, depicted in Fig. 1, takes V_i as input and outputs a compact binary hash code $b_i \in \{-1, +1\}^k$, ensuring that the Hamming distance between any two codes corresponds to the semantic similarity between the respective videos.

3.2 Video Hash Network

The VHMT adopts a teacher-student dual-path architecture [17] to jointly optimize temporal feature extraction and hash code learning. The teacher branch M^{T} employs a TAM to capture discriminative temporal patterns from full-frame

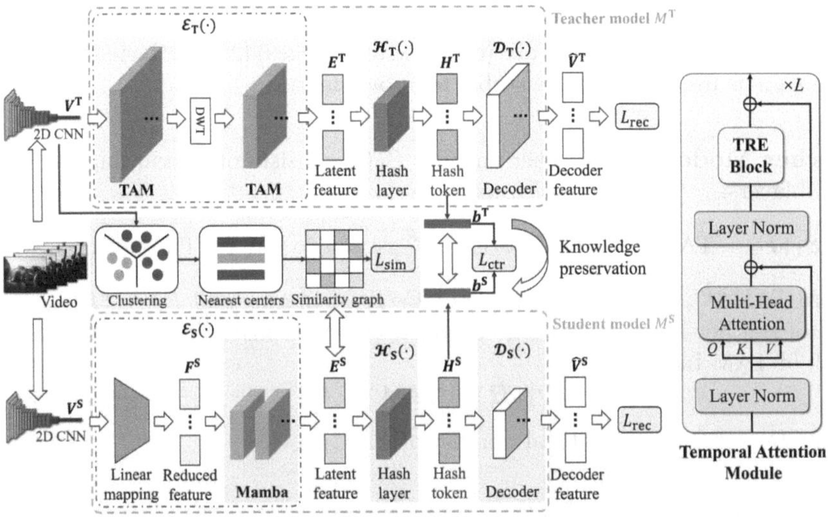

Fig. 1. Overview of the VHMT network.

features. The student branch M^{S} uses a lightweight Mamba structure [25] for efficient encoding. Between the teacher and student models, a parameter interaction mechanism is introduced to address the insufficient knowledge transfer commonly observed in conventional Siamese networks.

Preliminaries: State-Space Models and Mamba. State-space models (SSMs) process sequential data $\boldsymbol{x}(t) \in \mathbb{R}^d$ by maintaining a hidden state $\boldsymbol{h}(t) \in \mathbb{R}^u$, governed by

$$\dot{\boldsymbol{h}}(t) = \boldsymbol{A}\boldsymbol{h}(t) + \boldsymbol{B}\boldsymbol{x}(t), \quad \boldsymbol{y}(t) = \boldsymbol{C}\boldsymbol{h}(t), \tag{1}$$

where $\boldsymbol{A} \in \mathbb{R}^{u\times u}$, $\boldsymbol{B} \in \mathbb{R}^{u\times 1}$, and $\boldsymbol{C} \in \mathbb{R}^{1\times u}$ are system matrices. Discretizing the continuous model using zero-order hold with step size Δ yields

$$\bar{\boldsymbol{A}} = e^{\Delta \boldsymbol{A}}, \quad \bar{\boldsymbol{B}} = \boldsymbol{A}^{-1}(e^{\Delta \boldsymbol{A}} - \boldsymbol{I})\boldsymbol{B}, \tag{2}$$

$$\boldsymbol{h}_t = \bar{\boldsymbol{A}}\boldsymbol{h}_{t-1} + \bar{\boldsymbol{B}}\boldsymbol{x}_t, \quad \boldsymbol{y}_t = \boldsymbol{C}\boldsymbol{h}_t, \tag{3}$$

where $\boldsymbol{x}_t$, $\boldsymbol{h}_t$, and $\boldsymbol{y}_t$ are the input, hidden state, and output at discrete time step t. Mamba [25] enhances SSMs by making Δ, $\boldsymbol{B}$, and $\boldsymbol{C}$ input-adaptive, enabling selective feature propagation for better sequence modeling.

Preliminaries: 1D Discrete Wavelet Transform. The 1D DWT decomposes an input sequence $\boldsymbol{F}$ into low-frequency ($\boldsymbol{F}_{\mathrm{low}}$) and high-frequency ($\boldsymbol{F}_{\mathrm{high}}$) components, with Inverse Wavelet Transform (IWT) enabling full reconstruction:

$$[\boldsymbol{F}_{\mathrm{low}}, \boldsymbol{F}_{\mathrm{high}}] = \mathrm{DWT}(\boldsymbol{F}), \quad \boldsymbol{F} = \mathrm{IWT}(\boldsymbol{F}_{\mathrm{low}}, \boldsymbol{F}_{\mathrm{high}}), \tag{4}$$

where $\boldsymbol{F}_{\text{low}}$ captures stable global structural information of the video features, while $\boldsymbol{F}_{\text{high}}$ contains local details (e.g., textures, edges) [26]. In the teacher model, only $\boldsymbol{F}_{\text{low}}$ is used to promote stable temporal features.

Teacher Model. The teacher encoder $\mathcal{E}_{\text{T}}(\cdot)$ consists of alternating TAM and DWT layers:

$$\mathcal{E}_{\text{T}}(\boldsymbol{F}) = \text{TAM}^{\times E_n} \circ \text{DWT} \circ \text{TAM}^{\times E_{n-1}} \circ \cdots \circ \text{DWT} \circ \text{TAM}^{\times E_1}(\boldsymbol{F}), \tag{5}$$

where $\text{TAM}^{\times E_i}$ denotes a sequence of E_i TAM blocks, and $E_{\text{T}} = \{E_1, \ldots, E_n\}$ defines the per-layer depth. The decoder $\mathcal{D}_{\text{T}}(\cdot)$ reconstructs the input using a stack of TAM blocks:

$$\mathcal{D}_{\text{T}}(\boldsymbol{H}) = \text{TAM}^{\times D_{\text{T}}}(\boldsymbol{H}), \tag{6}$$

where D_{T} represents the total number of TAM blocks.

Student Model. The student model adopts a simplified architecture. Its encoder $\mathcal{E}_{\text{S}}(\cdot)$ and decoder $\mathcal{D}_{\text{S}}(\cdot)$ each consist of a single Mamba block stack:

$$\mathcal{E}_{\text{S}}(\boldsymbol{F}) = \text{Mamba}^{\times E_{\text{S}}}(\boldsymbol{F}), \tag{7}$$

$$\mathcal{D}_{\text{S}}(\boldsymbol{H}) = \text{Mamba}^{\times D_{\text{S}}}(\boldsymbol{H}), \tag{8}$$

where E_{S} and D_{S} denote the number of Mamba blocks in the encoder and decoder, respectively.

Hash Layer. To generate hash codes, the latent features $\boldsymbol{E}_i \in \mathbb{R}^{T\times d}$ are mapped to real-valued vectors $\tilde{\boldsymbol{H}}_i \in \mathbb{R}^{T\times k}$ via a fully connected layer $\text{FC}(\cdot)$, Layer Normalization $\text{LN}(\cdot)$, and the $\tanh(\cdot)$ function:

$$\tilde{\boldsymbol{H}}_i = \tanh(\text{LN}(\text{FC}(\boldsymbol{E}_i))) \in (-1, 1), \quad \boldsymbol{H}_i = \text{sign}(\tilde{\boldsymbol{H}}_i). \tag{9}$$

The final binary code $\boldsymbol{b}_i \in \{-1, 1\}^{1\times k}$ is obtained by average pooling over the temporal dimension T and binarization using $\text{sign}(\cdot)$:

$$\tilde{\boldsymbol{b}}_i = \frac{1}{T}\sum_{j=1}^{T} \boldsymbol{H}_{i,j}, \quad \boldsymbol{b}_i = \text{sign}(\tilde{\boldsymbol{b}}_i). \tag{10}$$

3.3 Temporal Attention Module

The teacher model uses the TAM module to integrate low-frequency features extracted via DWT, generating temporal modeling signals:

$$y = \text{MHA}(\text{LN}(x)) + \text{LN}(x), \quad z = \text{TRE}(\text{LN}(y)) + \text{LN}(y), \tag{11}$$

where x and z are the input and output features, respectively, $\text{MHA}(\cdot)$ is Multi-Head Attention from ViT, $\text{TRE}(\cdot)$ is the Temporal Reconstruction Enhancement block, and $\text{LN}(\cdot)$ is layer normalization.

Multi-head Attention. Following standard ViT blocks [27], we project input matrix $\boldsymbol{F}_i$ into query $\boldsymbol{Q}_i$, key $\boldsymbol{K}_i$, and value $\boldsymbol{V}_i$ matrices using learnable parameters $\boldsymbol{W}^Q$, $\boldsymbol{W}^K$, and $\boldsymbol{W}^V$. The scaled dot-product attention is defined as

$$\boldsymbol{F}_i^{\text{update}} = \text{softmax}\left(\frac{\boldsymbol{F}_i\boldsymbol{W}^Q(\boldsymbol{F}_i\boldsymbol{W}^K)^\top}{\sqrt{d^K}}\right)\boldsymbol{F}_i\boldsymbol{W}^V, \tag{12}$$

where d^K is a scaling factor. After E_T MHA layers, inputs are mapped to d-dimensional latent features $\boldsymbol{H}_i \in \mathbb{R}^{T\times d}$, encoding both frame-specific content and temporal dependencies.

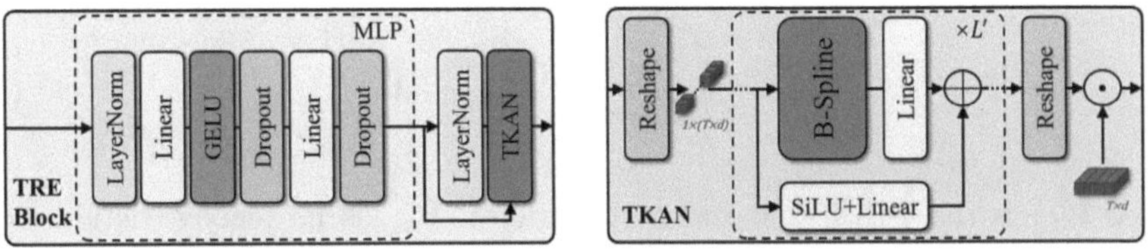

(a) TRE Block (b) TKAN

Fig. 2. Overview of the TRE block.

Temporal Reconstruction Enhancement Block. As shown in Fig. 2, the TRE block integrates a Temporal KAN (TKAN) [28] to overcome the fixed activation limitation in MLPs [19,20]. For input x_in, the transformation is

$$x_\text{MLP} = \text{MLP}(x_\text{in}) = \text{Drop}(\text{Linear}(\text{Drop}(\text{GELU}(\text{Linear}(\text{LN}(x_\text{in})))))), \tag{13}$$

$$x'_\text{MLP} = \text{Reshape}(\text{LN}(x_\text{MLP})), \tag{14}$$

$$x_\text{TKAN} = \text{Reshape}(\text{TKAN}(x'_\text{MLP})), \tag{15}$$

$$x_\text{out} = x_\text{TKAN} * x_\text{MLP}, \tag{16}$$

where x_MLP is the MLP$(\cdot)$ output, LN$(\cdot)$, GELU$(\cdot)$, Drop$(\cdot)$ denote Layer Normalization, GELU activation, and Dropout, Reshape$(\cdot)$ aligns dimensions for fusion, and $*$ is the Hadamard product. Let $\widetilde{x} = x'_\text{MLP}$, TKAN is defined as $\text{TKAN}(\widetilde{x}) = (\Phi_{L'} \circ \cdots \circ \Phi_1)\widetilde{x}$, with each Φ_i being a KANLinear layer:

$$\Phi(\widetilde{x}) = \omega \cdot \frac{\widetilde{x}}{1+e^{-\widetilde{x}}} + \omega' \cdot \sum_{p=1}^{N} c_p F_{p,q}(\widetilde{x}), \tag{17}$$

where ω, ω' are trainable weights, c are spline coefficients, and $F_{p,q}(\widetilde{x})$ are degree-q B-spline bases. The TRE block enables adaptive fusion for enhanced reconstruction.

3.4 Loss Function

To jointly optimize temporal reconstruction, hash contrastive learning, and local structure preservation, we propose a multi-component loss:

$$\mathcal{L}_{\text{VHMT}} = \alpha\mathcal{L}_{\text{rec}} + \beta\mathcal{L}_{\text{ctr}} + \gamma\mathcal{L}_{\text{sim}}, \tag{18}$$

where α, β, γ represent adjustable hyperparameters for task balancing.

Temporal Reconstruction. Inspired by [7,8], we perform temporal reconstruction via hash code decoding to maximize semantic capacity. For video sequences $\boldsymbol{V}_i \in \mathbb{R}^{T\times D}$, the reconstruction loss measures feature accuracy:

$$\mathcal{L}_{\text{rec}} = \frac{1}{NTD}\sum_{i=1}^{N}\sum_{j=1}^{T}\|\boldsymbol{V}_{i,j} - \hat{\boldsymbol{V}}_{i,j}\|_2^2, \tag{19}$$

where $\boldsymbol{V}_{i,j}$ is the j-th frame feature of video i, $\hat{\boldsymbol{V}}_{i,j}$ is its reconstruction, N is the batch size, T is the number of frames, and D is the feature dimension.

Hash Contrastive. To improve consistency between network views and mitigate sampling bias in teacher-student frameworks, we adopt a debiased InfoNCE loss [29]. For each video, correlated views $\boldsymbol{b}_i^{\text{S}}$ (student) and $\boldsymbol{b}_i^{\text{T}}$ (teacher) are treated as positives. The loss dynamically reweights negative samples:

$$\mathcal{L}_{\text{ctr}} = -\frac{1}{N}\sum_{i=1}^{N}\log\frac{\exp\left(\mathcal{S}(\boldsymbol{b}_i^{\text{S}}, \boldsymbol{b}_i^{\text{T}})/\tau\right)}{\exp\left(\mathcal{S}(\boldsymbol{b}_i^{\text{S}}, \boldsymbol{b}_i^{\text{T}})/\tau\right) + (2N-2)\bar{f}_i}, \tag{20}$$

where $\bar{f}_i$ is a dynamic baseline:

$$\bar{f}_i = \max\left(\frac{\theta - \rho\exp\left(\mathcal{S}(\boldsymbol{b}_i^{\text{S}}, \boldsymbol{b}_i^{\text{T}})/\tau\right)}{1-\rho}, \exp(-1/\tau)\right), \tag{21}$$

and θ estimates the average negative similarity:

$$\theta = \frac{1}{2N-2}\sum_{\substack{w=1\\ w\neq i}}^{N}\left[\exp\left(\mathcal{S}(\boldsymbol{b}_i^{\text{S}}, \boldsymbol{b}_w^{\text{S}})/\tau\right) + \exp\left(\mathcal{S}(\boldsymbol{b}_i^{\text{S}}, \boldsymbol{b}_w^{\text{T}})/\tau\right)\right], \tag{22}$$

where N is the batch size, $\mathcal{S}(\cdot,\cdot)$ is cosine similarity, $\tau > 0$ controls temperature, $\rho \in [0, 1]$ estimates intra-video similarity, and θ modulates negative contributions. This formulation enhances hash code discriminability by reducing label noise in contrastive learning.

Local Structure Preservation. To preserve semantic similarity and discrimination, we construct a similarity matrix $\boldsymbol{S} \in \{-1, 0, 1\}^{N \times N}$ using K-means and PCA hybrid clustering [9]. Video-level representations $\bar{\boldsymbol{v}}_i$ are obtained by aggregating frame-level features V^{T}. A sparse affinity matrix $\boldsymbol{P} \in \mathbb{R}^{N \times M}$ is then computed based on proximity to cluster centroids $\{\boldsymbol{c}_{i,1}, \ldots, \boldsymbol{c}_{i,M}\}$:

$$\boldsymbol{P}_{i,m} = \begin{cases} \frac{\exp(-\|\bar{\boldsymbol{v}}_i - \boldsymbol{c}_{i,m}\|^2/\sigma)}{\sum_{l=1}^{M} \exp(-\|\bar{\boldsymbol{v}}_i - \boldsymbol{c}_{i,l}\|^2/\sigma)}, & \text{if } \boldsymbol{c}_{i,m} \text{ is the nearest centroid,} \\ 0, & \text{otherwise,} \end{cases} \tag{23}$$

where $\sigma > 0$ controls the kernel bandwidth. An affinity matrix $\boldsymbol{A} \in \mathbb{R}^{N \times N}$ is then computed as

$$\boldsymbol{A} = \boldsymbol{P}\boldsymbol{\Lambda}^{-1}\boldsymbol{P}^{\top}, \tag{24}$$

where $\boldsymbol{\Lambda} = \mathrm{diag}(\boldsymbol{P}^{\top}\mathbf{1})$ is a diagonal matrix of cluster weights. $\boldsymbol{A}$ is thresholded at three levels $M_1 < M_2 < M_3$ to obtain binary matrices $\boldsymbol{B}^{(1)}, \boldsymbol{B}^{(2)}, \boldsymbol{B}^{(3)} \in \{1, -1\}^{N \times N}$. The refined similarity matrix $\boldsymbol{S}$ is defined as

$$\boldsymbol{S}_{i,m} = \begin{cases} 1, & \text{if } \boldsymbol{B}^{(1)}_{i,m} = 1, \\ -1, & \text{if } \boldsymbol{B}^{(2)}_{i,m} = -1 \text{ and } \boldsymbol{B}^{(3)}_{i,m} = 1, \\ 0, & \text{otherwise,} \end{cases} \tag{25}$$

encoding positive, negative, and neutral pairs. The local structure loss is

$$\mathcal{L}_{\mathrm{sim}} = \frac{1}{N} \sum_{i=1}^{N} \sum_{m=1}^{N} \left(\frac{1}{d} e_i^{\top} e_m - \boldsymbol{S}_{i,m} \right)^2, \tag{26}$$

where d is the latent feature E^{S} dimension, e_i, e_m are pooled latent vectors, and the loss preserves neighborhood structure while reducing compression distortion.

4 Experiments

4.1 Datasets

The VHMT is evaluated on three standard large-scale video datasets: FCVID [12], ActivityNet [13], and YFCC [14], with their statistics detailed in Table 1.

Table 1. Datasets used in the experiments.

Dataset	Videos	Train	Test	Classes
FCVID [12]	91,223	45,585	45,600	239
ActivityNet [13]	20,000	9,722	4,758	203
YFCC [14]	800,000	409,788	101,256	80

4.2 Evaluation Metrics

Following [7,8], we use mAP@k as the main metric:

$$\text{mAP@}k = \frac{1}{N}\sum_{q=1}^{N} \text{AP@}k(q), \tag{27}$$

where N is the number of queries and AP@$k(q)$ is

$$\text{AP@}k(q) = \frac{1}{g_q}\sum_{r=1}^{k} P(r|q)\cdot\delta_r(q), \quad g_q = \min(k, R_q), \tag{28}$$

with R_q the total relevant items for query q, $P(r|q)$ the precision at rank r, and $\delta_r(q) = 1$ if the r-th result is relevant, else 0. For overall evaluation, we compute GmAP as the root-aggregate score over key k values:

$$\text{GmAP} = \sqrt{\sum_{k\in\mathcal{K}}(\text{mAP@}k)^2}, \quad \mathcal{K} = \{5, 20, 40, 60, 80, 100\}. \tag{29}$$

4.3 Implementation Details

Preprocessing. We adopt the standard preprocessing pipeline used in SSVH [6–11]. For FCVID and YFCC, 25 frames per video are sampled, and 4,096-D VGG-16 [30] features are used; for ActivityNet, 30 frames and 2,048-D ResNet50 [24] features are extracted. All backbones are pre-trained on ImageNet [31]. Features for FCVID and YFCC are sourced from [4], and those for ActivityNet from [19].

Model Settings. Following [7], the student encoder of VHMT has a depth of 6 and a hidden dimension of 256, while the decoder has a depth of 1 and a hidden dimension of 192. The preset state dimension u in Mamba is set to 16 [25]. TKAN employs 2 KANLinear layers (L'), 5 spline coefficients (c), and a recursion degree of $q = 3$ [28]. The Haar wavelet is used as the DWT basis.

Loss Settings. The temperature parameter τ in the contrastive loss is set to 0.5, and the class prior ρ is set to 0.1 [29]. The number of clusters M is fixed at 2000, with $M_1 = 3$, $M_2 = 4$, and $M_3 = 5$ [19]. Loss weights are set to $\alpha = 1$, $\beta = 1$, and $\gamma = 0.1$.

Training. Experiments are run on NVIDIA A800 GPUs with PyTorch 2.1.1 and CUDA 12.3. We use a batch size of 512 and train for 800 (FCVID), 500 (ActivityNet), and 40 (YFCC) epochs. The learning rate is initialized at 1×10^{-4}, reduced by 90% every 20 epochs, and clamped to a minimum of 1×10^{-6}. Optimization is performed using Adam [32].

Testing. Retrieval performance is evaluated on each dataset using mAP@k [7, 8]. The test sets consist of 45,600 (FCVID), 4,758 (ActivityNet), and 101,256 (YFCC) samples. The best and second-best results are highlighted in **bold** and *italic*, respectively.

4.4 Comparison with State-of-the-Arts

Baselines. We compare 6 influential open-source SSVH methods from the past 5 years: BTH [9], DKPH [17], MCMSH [19], ConMH [6], S5VH [7], and AutoSSVH [8].

Performance Under Standard Protocols. As illustrated in Fig. 3(a)–(i), the VHMT demonstrates competitive performance across three datasets and various hash lengths, with consistent improvements in mAP@k over the baselines. This performance improvement highlights the effectiveness of our method: (1) the proposed teacher-student network; (2) the TAM module; and (3) the novel loss function, which jointly contribute to enhanced retrieval accuracy.

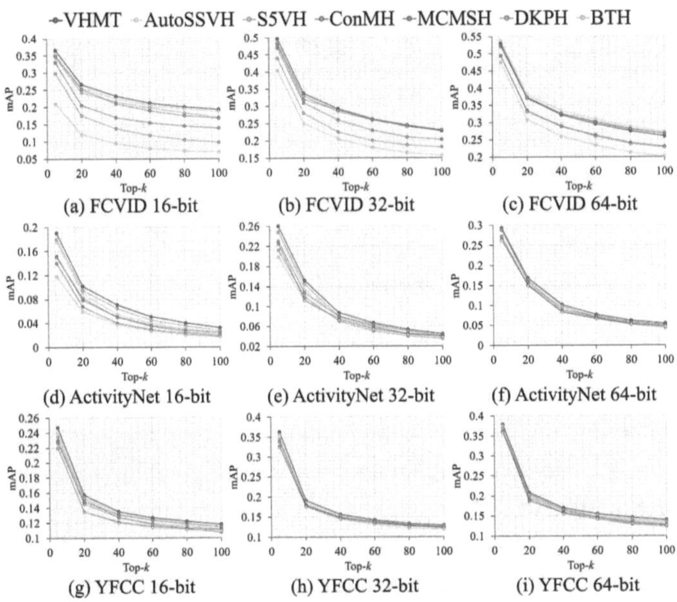

Fig. 3. VHMT vs. baselines: Performance (mAP@k) on three datasets with 16-bit, 32-bit, and 64-bit hash codes.

Cross-Dataset Transferability. As reported in Table 2, models are trained on FCVID and evaluated on YFCC. The VHMT shows a relatively smaller decline in mAP@20 compared to baselines, suggesting advantages in cross-dataset generalization.

Table 2. VHMT vs. baselines: Performance (mAP@20) on FCVID→YFCC with 64-bit hash codes and ↓drop (%).

BTH	DKPH	MCMSH	ConMH	S5VH	AutoSSVH	VHMT
0.191 ↓5.7	*0.199* ↓2.8	0.186 ↓3.2	0.188 ↓3.6	0.190 ↓3.6	0.194 ↓3.5	**0.201** ↓3.2

t-SNE Visualization. Figure 4(a)–(f) presents t-SNE visualizations [33] of hash codes on FCVID-small, constructed by sampling 10 classes with 80 videos each from FCVID following [6]. The VHMT achieves better class separation, more compact intra-class clustering, and a more uniform code distribution compared to baselines, thereby mitigating retrieval bias due to excessive false positives.

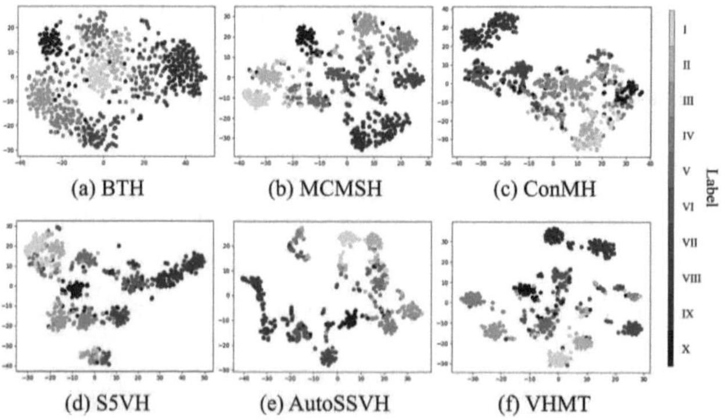

Fig. 4. VHMT vs. baselines: t-SNE visualization of feature distributions on FCVID with 64-bit hash codes.

4.5 Ablation Study

Effectiveness of the VHMT Network. As shown in Table 3, deeper TAM modules (Exps. (I)–(V)) yield moderate GmAP gains, with encoding time increasing as the encoder size grows. We adopt the configuration of Exp. (III) for optimal efficiency. Using smoother wavelet bases (db4 in Exp. (VI), sym8 in Exp. (VII)) improves GmAP by 0.11% and 0.20%, while max pooling (Exp. (VIII)) and average pooling (Exp. (IX)) degrade performance by 4.66% and 3.23%, respectively, demonstrating the superiority of wavelets in extracting robust low-frequency semantics. Under the same setting as Exp. (III), extending the student model to a Siamese structure (Exp. (X)) results in a 4.27% lower GmAP, confirming the effectiveness of the teacher-student network.

Table 3. Performance (GmAP) and Encoding Time (ms) of the VHMT network scale on FCVID with 64-bit hash codes, where $E_\mathrm{T}/D_\mathrm{T}$ and $E_\mathrm{S}/D_\mathrm{S}$ denote the encoder/decoder sizes of the teacher (M^T) and student (M^S) models, respectively.

Exp.	Network	Technique	E_T	D_T	E_S	D_S	GmAP	Time (ms)
(I)	M^T+M^S	Haar	{4,2}	1	3	1	0.8461	1.62
(II)	M^T+M^S	Haar	{4,2,2}	1	6	1	0.8592	3.13
(III)	M^T+M^S	Haar	{8,4,2}	1	6	1	0.8681	3.13
(IV)	M^T+M^S	Haar	{8,4,2}	2	12	2	0.8768	6.40
(V)	M^T+M^S	Haar	{16,8,4,2}	2	16	2	0.8839	8.64
(VI)	M^T+M^S	db4	{8,4,2}	1	6	1	0.8692	3.13
(VII)	M^T+M^S	sym8	{8,4,2}	1	6	1	0.8701	3.13
(VIII)	M^T+M^S	MaxPooling	{8,4,2}	1	6	1	0.8215	3.13
(IX)	M^T+M^S	AvePooling	{8,4,2}	1	6	1	0.8358	3.13
(X)	M^S Only	Haar	-	-	6	1	0.8254	3.13

Effectiveness of the TAM Module. As shown in Table 4, ablation studies evaluate the TAM module's components with other settings fixed. Combining MHA with MLP (Exp. (II)) improves performance over MHA alone (Exp. (I)), increasing mAP@5 from 0.504 to 0.519, while replacing MLP with KAN (Exp. (III)) degrades it, indicating KAN's limited fit for attention outputs. Substituting MHA with Mamba (Exp. (IV)) yields suboptimal results, suggesting the inductive bias of MHA is better aligned with the requirements of the teacher model. Integrating MHA with TRE (Exp. (V)) achieves the best mAP, validating their synergistic design.

Table 4. Performance (mAP@k) of TAM variants on FCVID with 64-bit hash codes.

Exp.	MHA	TRE	$k=5$	$k=20$	$k=40$	$k=60$	$k=80$	$k=100$
(I)	✓		0.504	0.353	0.306	0.281	0.266	0.247
(II)	✓	MLP	*0.519*	*0.366*	*0.318*	*0.292*	*0.274*	*0.258*
(III)	✓	KAN	0.500	0.350	0.303	0.279	0.262	0.245
(IV)	Mamba	✓	0.510	0.355	0.310	0.283	0.268	0.252
(V)	✓	✓	**0.530**	**0.372**	**0.322**	**0.296**	**0.275**	**0.260**

Effectiveness of the Loss Function. Table 5 shows that $\mathcal{L}_{\text{rec}}+\mathcal{L}_{\text{sim}}$ (Exp. (V)) improves mAP@5 from 0.497/0.381 to 0.507, $\mathcal{L}_{\text{rec}}+\mathcal{L}_{\text{ctr}}$ (Exp. (VI)) raises it to 0.528, and the full loss $\mathcal{L}_{\text{rec}}+\mathcal{L}_{\text{ctr}}+\mathcal{L}_{\text{sim}}$ (Exp. (VII)) achieves the best performance across all mAP@k metrics, with mAP@5 reaching 0.530, confirming the complementarity of the loss terms.

Table 5. Performance (mAP@k) of loss variants on FCVID with 64-bit hash codes.

Exp.	Variant	$k=5$	$k=20$	$k=40$	$k=60$	$k=80$	$k=100$
(I)	$\mathcal{L}_{\text{rec}}$ Only	0.497	0.302	0.249	0.209	0.191	0.171
(II)	$\mathcal{L}_{\text{ctr}}$ Only	0.388	0.276	0.247	0.230	0.217	0.208
(III)	$\mathcal{L}_{\text{sim}}$ Only	0.381	0.255	0.214	0.193	0.181	0.170
(IV)	w/o $\mathcal{L}_{\text{rec}}$	0.406	0.291	0.259	0.242	0.227	0.219
(V)	w/o $\mathcal{L}_{\text{ctr}}$	0.507	0.310	0.244	0.212	0.194	0.171
(VI)	w/o $\mathcal{L}_{\text{sim}}$	*0.528*	*0.368*	*0.318*	*0.292*	*0.271*	*0.256*
(VII)	$\mathcal{L}_{\text{VHMT}}$	**0.530**	**0.372**	**0.322**	**0.296**	**0.275**	**0.260**

5 Conclusions

This paper has proposed a novel SSVH method called the VHMT based on a teacher-student architecture. The teacher model extracts robust low-frequency semantic features via DWT and captures temporal dependencies using a TAM, enhanced by a TRE block to improve fine-grained reconstruction and generate high-quality temporal modeling signals. The student model employs a Mamba architecture to model temporal relationships, achieving knowledge transfer and efficient single-path inference by approximating the teacher's feature reconstructions and attention outputs. A joint loss function enables the student to closely mimic the teacher's temporal modeling while simultaneously optimizing the clustering structure of the hash codes. Experiments have shown that the VHMT outperforms several SOTA methods in retrieval performance under the mAP metric.

Acknowledgements. This work is partially supported by the Guangxi Natural Science Foundation (2025GXNSFBA069370, 2025GXNSFBA069304), and the National Natural Science Foundation of China (62272111, 62302108).

References

1. Tang, Z., Chen, L., Zhang, X., Zhang, S.: Robust image hashing with tensor decomposition. IEEE Trans. Knowl. Data Eng. **31**(3), 549–560 (2019)
2. Liang, X., Tang, Z., Zhang, X., Yu, M., Zhang, X.: Robust hashing with local tangent space alignment for image copy detection. IEEE Trans. Dependable Secure Comput. **21**(4), 2448–2460 (2024)
3. Liang, X., Tang, Z., Zhang, X., Zhang, X., Yang, C.N.: Robust image hashing with weighted saliency map and laplacian eigenmaps. IEEE Trans. Inf. Forensics Secur. **20**, 665–676 (2025)
4. Song, J., Zhang, H., Li, X., Gao, L., Wang, M., Hong, R.: Self-supervised video hashing with hierarchical binary auto-encoder. IEEE Trans. Image Process. **27**(7), 3210–3221 (2018)

5. Bromley, J., Guyon, I., LeCun, Y., Säckinger, E., Shah, R.: Signature verification using a "siamese" time delay neural network. Adv. Neural Inf. Process. Syst. **6** (1993)
6. Wang, Y., Wang, J., Chen, B., Zeng, Z., Xia, S.T.: Contrastive masked autoencoders for self-supervised video hashing. In: Proceedings of the AAAI Conference on Artificial Intelligence, vol. 37, no. 3, pp. 2733–2741 (2023)
7. Wang, J., et al.: Efficient self-supervised video hashing with selective state spaces. In: Proceedings of the AAAI Conference on Artificial Intelligence, vol. 39, no. 7, pp. 7753–7761 (2025)
8. Lian, N., et al.: Autossvh: exploring automated frame sampling for efficient self-supervised video hashing. In: Proceedings of the IEEE/CVF Conference on Computer Vision and Pattern Recognition, pp. 18881–18890 (2025)
9. Li, S., Li, X., Lu, J., Zhou, J.: Self-supervised video hashing via bidirectional transformers. In: Proceedings of the IEEE/CVF Conference on Computer Vision and Pattern Recognition, pp. 13544–13553 (2021)
10. Li, Q., Tian, X., Ng, W.W.Y.: Self-supervised temporal sensitive hashing for video retrieval. IEEE Trans. Multimedia **26**, 9021–9035 (2024)
11. Du, L., Liang, X., Yang, L., Tang, Z.: Structure-preserving video hashing via self-supervised transformer for retrieval. In: Proceedings of the IEEE International Conference on Acoustics, Speech and Signal Processing, pp. 1–5 (2025)
12. Jiang, Y.G., Wu, Z., Wang, J., Xue, X., Chang, S.F.: Exploiting feature and class relationships in video categorization with regularized deep neural networks. IEEE Trans. Pattern Anal. Mach. Intell. **40**(2), 352–364 (2018)
13. Heilbron, F.C., Escorcia, V., Ghanem, B., Niebles, J.C.: Activitynet: a large-scale video benchmark for human activity understanding. In: Proceedings of the IEEE Conference on Computer Vision and Pattern Recognition, pp. 961–970 (2015)
14. Thomee, B., et al.: YFCC100M: the new data in multimedia research. Commun. ACM **59**(2), 64–73 (2016)
15. Zhang, H., Wang, M., Hong, R., Chua, T.S.: Play and rewind: optimizing binary representations of videos by self-supervised temporal hashing. In: Proceedings of the ACM International Conference on Multimedia, pp. 781–790 (2016)
16. Li, C., Yang, Y., Cao, J., Huang, Z.: Jointly modeling static visual appearance and temporal pattern for unsupervised video hashing. In: Proceedings of the ACM Conference on Information and Knowledge Management, pp. 9–17 (2017)
17. Li, P., Xie, H., Ge, J., Zhang, L., Min, S., Zhang, Y.: Dual-stream knowledge-preserving hashing for unsupervised video retrieval. In: Proceedings of the European Conference on Computer Vision, pp. 181–197 (2022)
18. Tolstikhin, I.O., et al.: MLP-mixer: an all-MLP architecture for vision. Adv. Neural. Inf. Process. Syst. **34**, 24261–24272 (2021)
19. Hao, Y., Duan, J., Zhang, H., Zhu, B., Zhou, P., He, X.: Unsupervised video hashing with multi-granularity contextualization and multi-structure preservation. In: Proceedings of the ACM International Conference on Multimedia, pp. 3754–3763 (2022)
20. Duan, J., Hao, Y., Zhu, B., Cheng, L., Zhou, P., Wang, X.: Efficient unsupervised video hashing with contextual modeling and structural controlling. IEEE Trans. Multimedia **26**, 7438–7450 (2024)
21. Zeng, Z., Wang, J., Chen, B., Wang, Y., Xia, S.T., Intelligence, P.C.: Motion-aware graph reasoning hashing for self-supervised video retrieval. In: Proceedings of the British Machine Vision Conference, p. 82 (2022)

22. Veličković, P., Cucurull, G., Casanova, A., Romero, A., Lio, P., Bengio, Y.: Graph attention networks. In: International Conference on Learning Representations (2017)
23. Simonyan, K., Zisserman, A.: Very deep convolutional networks for large-scale image recognition. In: International Conference on Learning Representations (2015)
24. He, K., Zhang, X., Ren, S., Sun, J.: Deep residual learning for image recognition. In: Proceedings of the IEEE Conference on Computer Vision and Pattern Recognition, pp. 770–778 (2016)
25. Gu, A., Dao, T.: Mamba: linear-time sequence modeling with selective state spaces. arXiv preprint arXiv:2312.00752 (2023)
26. Zou, W., Gao, H., Yang, W., Liu, T.: Wave-mamba: wavelet state space model for ultra-high-definition low-light image enhancement. In: Proceedings of the ACM International Conference on Multimedia, pp. 1534–1543 (2024)
27. Vaswani, A., et al.: Attention is all you need. In: Advances in Neural Information Processing Systems, vol. 30 (2017)
28. Liu, Z., et al.: Kan: Kolmogorov-Arnold networks. In: International Conference on Learning Representations (2025)
29. Chuang, C.Y., Robinson, J., Lin, Y.C., Torralba, A., Jegelka, S.: Debiased contrastive learning. Adv. Neural. Inf. Process. Syst. **33**, 8765–8775 (2020)
30. Tammina, S.: Transfer learning using VGG-16 with deep convolutional neural network for classifying images. Int. J. Sci. Res. Publ. **9**(10), 143–150 (2019)
31. Russakovsky, O., et al.: Imagenet large scale visual recognition challenge. Int. J. Comput. Vision **115**, 211–252 (2015)
32. Kingma, D.P., Ba, J.: Adam: a method for stochastic optimization. In: International Conference on Learning Representations (2015)
33. Van der Maaten, L., Hinton, G.: Visualizing data using t-SNE. J. Mach. Learn. Res. **9**(11) (2008)

DFRF-MIAD: Multimodal Industrial Anomaly Detection via Feature Reconstruction and Fusion

Feng Wu, Zhaojing Wang, and Li Li(✉)

School of Computer Science and Artificial Intelligence, Wuhan Textile University, Wuhan 430200, China
lli@wtu.edu.cn

Abstract. Multimodal industrial anomaly detection is a challenging computer vision task. Recent unsupervised approaches rely on large-scale feature memory banks, which incur heavy storage and retrieval costs during inference and hinder their deployment in resource-constrained environments. Additionally, existing cross-modal interaction schemes lack an effective feature-fusion strategy, which makes it difficult to preserve geometric structure and texture details simultaneously, thereby leading to imprecise localization of anomalous regions. To address these issues, we propose an unsupervised multimodal industrial anomaly detection framework based on feature restoration, comprising feature extraction, feature alignment, feature reconstruction, cross-modal feature enhancement, and anomaly scoring modules. Within the framework, a dual-branch symmetric encoder–decoder network reconstructs RGB and point cloud features respectively, while self-attention and cross-attention modules further accomplish complementary cross-modal feature fusion. The entire framework is optimized with a joint reconstruction loss and a fusion consistency constraint. Extensive experiments on the MVTec 3D-AD and Eyecandies datasets show that our method outperforms previous state-of-the-art in 2D-3D multimodal industrial anomaly detection, thereby confirming its effectiveness and robustness in both intra-modal and cross-modal representation learning.

Keywords: Anomaly Detection · Multimodal Fusion · Feature Reconstruction

1 Introduction

Industrial anomaly detection [7], as an essential component of quality control, aims to identify abnormal regions of products during or after the production process to minimize the potential defects of industrial products. In practical application scenarios, the proportion of anomalous regions in industrial products is relatively low, anomaly patterns are unknown, and the cost of labeling is high, making it difficult to obtain high-quality annotated datasets. Traditionally,

J. Lokoč et al. (Eds.): MMM 2026, LNCS 16413, pp. 119–133, 2026.
https://doi.org/10.1007/978-981-95-6957-1_9

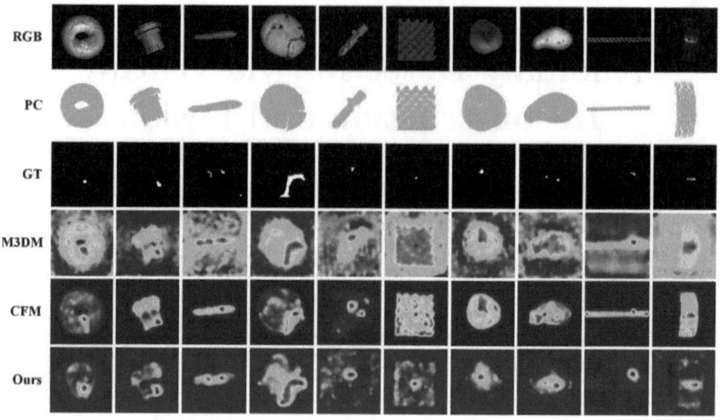

Fig. 1. Visual comparison with existing methods on the MVTec 3D-AD dataset [2]. Each column represents a category, with the first row showing the original RGB images and the second row displaying the aligned 3D point clouds. The third row shows the ground truth, followed by the anomaly detection results from M3DM [31], CFM [10], and our proposed method, all visualized as heatmaps.

anomaly detection relies on manual inspectors or image-based methods [16]. In this context, research and applications in industrial anomaly detection are gradually shifting from traditional 2D images to more comprehensive 3D data. Recently, with the introduction of 3D anomaly detection benchmark datasets, such as MVTec 3D-AD [2], Eyecandies [4] and Real3D-AD [17], especially the combination of point cloud and RGB image data, more comprehensive product surface information is provided, which is crucial for capturing product defects and has promoted the research on multimodal industrial anomaly detection.

Despite the great potential of 3D data in anomaly detection [11,18], effectively fusing and utilizing these multimodal data remains a challenge. For instance, while 3D data can provide unique spatial location information [6], it is not as sensitive to color perception as RGB image data [14]. Therefore, leveraging the redundancy and complementarity between modalities has become the key to improving the performance of industrial anomaly detection [16].

To address the aforementioned challenges, we propose a new multimodal industrial anomaly detection framework, DFRF-MIAD, which comprises feature extraction, feature alignment, feature reconstruction, cross-modal feature enhancement, and anomaly scoring. First, we employ pre-trained 2D and 3D feature extractors [1,9,10,31] to obtain high-level representations from RGB images and point clouds, respectively. We then apply bilinear upsampling [10,25,29] to the image features and spatially align the 3D features with their 2D counterparts to ensure one-to-one correspondence across modalities. We design a dual-branch feature reconstruction network, where each branch employs a symmetric encoder–decoder architecture to reconstruct its corresponding modality feature. In the cross-modal feature enhancement module, we incorporate self-attention and cross-attention mechanisms [19,26] to effectively integrate and complement the information between the two modalities. To promote both intra-modal and inter-modal representation learning, we further propose a joint reconstruction loss and a fusion consistency constraint, and optimize the overall training pipeline, thereby improving the accuracy and robustness of anomaly detection. Figure 1 illustrates the visual results of our method on the MVTec 3D-AD dataset, showing more accurate localization of anomalous regions compared to existing approaches. For clarity, our main contributions can be summarized as follows:

- We design symmetric encoder–decoder architectures for the RGB branch and the point cloud branch, integrating DSC_SE modules and skip connections to restore intra-modal feature structures and details.
- We propose a cross-modal feature enhancement method that leverages self-attention and cross-attention mechanisms to achieve efficient fusion and complementarity between the two modalities.
- We employ a joint reconstruction loss that combines cosine similarity with an L2 discrepancy, further regularized by a fusion consistency constraint, thereby enhancing the stability and robustness of unsupervised training.
- Extensive experiments are conducted on the MVTec 3D-AD and Eyecandies datasets, demonstrating the effectiveness of the proposed method and achieving a new state-of-the-art performance.

2 Related Work

In this section, we review the literature relevant to our proposed method. We begin by introducing anomaly detection approaches in the 3D single-modal domain. Subsequently, we discuss multimodal anomaly detection methods, which are closely related to our work.

2.1 3D Industrial Anomaly Detection

3D anomaly detection [15] remains in a relatively early stage, constrained by inherent challenges such as data sparsity, high dimensionality, and pervasive noise. 3D anomaly detection methods typically take point clouds or depth images as input [16] and leverage spatial structures, shape information, or depth features to identify and localize anomalies. Unlike RGB images, 3D data lack color information, leading to reduced accuracy when detecting anomalies, such as stains and rust, which exhibit indistinct structural characteristics. However, for defects with significant structural features, such as bends, contamination, cuts, cracks, and holes [2], 3D methods generally achieve higher detection precision.

Currently, purely 3D anomaly detection methods specifically designed for point clouds or depth images remain limited. Bergmann et al. [3] proposed the 3D-ST method, the first unsupervised anomaly detection framework directly on point clouds using a student-teacher network [3]. In this method, the teacher network extracts local geometric descriptors from normal 3D point cloud samples, and the student network [23,32] learns to match the teacher features [15], thereby localizing anomalies through regression errors. Liu et al. [17] constructed the high-precision point cloud dataset Real3D-AD [17] using a high-resolution laser scanner and proposed a baseline method for point cloud anomaly detection, Reg3D-AD [17], which is based on registration and feature memory banks.

2.2 Multimodal Industrial Anomaly Detection

Compared to single-modality anomaly detection based on 2D or 3D data, multimodal anomaly detection [9,10,31] requires processing data from different modalities, facing different pre-processing and feature extraction strategies. Moreover, effective information fusion, such as feature fusion or result aggregation [16], is crucial to improve interactions between different modalities [6].

Rudolph et al. [23] proposed an asymmetric teacher-student network architecture, AST [23], where the teacher network uses normalizing flows [8,13] for density estimation, and the student network employs a conventional feedforward network [28]. This method uses RGB and depth as inputs, and although it performs well in certain scenarios, its reliance on depth information to suppress the background may lead to the omission of subtle anomaly features. Horwitz et al. [14] proposed a fusion method combining classical 3D handcrafted features (FPFH) [24] with 2D color features (PatchCore) [22], significantly improving performance on the MVTec 3D-AD dataset [2] and demonstrating the complementary advantages of integrating 3D and color modalities. Wang et al. [31] pointed out that existing methods [5,7] directly concatenate multimodal features, which leads to strong interference between features and affects detection performance. Therefore, they proposed a hybrid fusion [31] scheme for handling multimodal data. M3DM [31] promotes the interaction of different modality features through block-level contrastive learning, uses decision-level fusion and multiple memory banks to avoid information loss [27], and employs an additional new classifier for final decision-making. However, this method incurs a memory cost of 6.52 GB during inference, with an FPS of only 0.51, limiting its feasibility in some practical applications. Costanzino et al. [10] proposed a cross-modal feature mapping method, CFM [10], which learns two cross-modal mapping functions. Given 2D features, it can predict the corresponding 3D features, and vice versa. CFM [10] outperforms existing multimodal anomaly detection methods [14,23,31] in both inference speed and memory usage, particularly by further optimizing memory and time efficiency through pruning techniques, making it more suitable for practical industrial applications.

3 Methodology

Our proposed anomaly detection method processes inputs from RGB images and corresponding 3D point clouds, relying on the learning of both intra-modal and inter-modal feature restoration. Figure 2 illustrates the multimodal anomaly

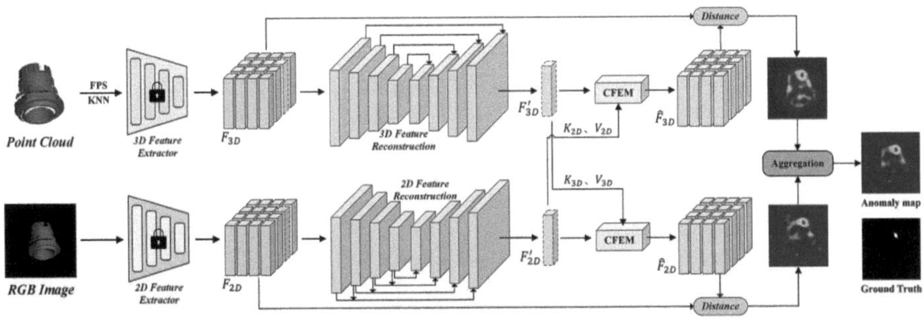

Fig. 2. The overall architecture of the proposed DFRF-MIAD framework. Given the point cloud data and the corresponding RGB image data, the two modalities go through feature extraction, feature alignment, feature reconstruction, cross-modal feature enhancement, and anomaly score computation, ultimately accomplishing anomaly detection and localization.

detection network architecture, which integrates 2D image features and 3D point cloud features through modules such as feature extraction, feature reconstruction, cross-modal feature enhancement, and anomaly score calculation, ultimately achieving anomaly detection and localization.

3.1 Multimodal Feature Extraction and Alignment

Given paired samples $\{x^{rgb} \in \mathbb{R}^{H \times W \times 3}, x^{pc} \in \mathbb{R}^{N \times 3}\}$, we employ pre-trained 2D and 3D feature extractors to obtain their feature representations. Specifically, the image branch uses ViT as the feature extractor, while the point cloud branch uses PointNet++ as the feature extractor. Both feature extractors are pre-trained on large-scale public datasets [15] and are kept frozen in this work (i.e., their weights are not updated). Due to the typically lower resolution of the features output by ViT compared to the original image's $H \times W$, to ensure pixel-level correspondence between the 2D and 3D branches in the feature space, we apply bilinear upsampling to the image features, restoring them to the input image resolution, denoted as $F_{2D} \in \mathbb{R}^{H \times W \times D_{img}}$, Let $\phi_{RGB}(\cdot)$ represent the 2D feature extractor, and $\mathrm{UP}_{2D}(\cdot)$ represent the bilinear upsampling operator, resulting in:

$$F_{2D} = \mathrm{UP}_{2D}(\phi_{RGB}(x^{rgb})) \tag{1}$$

To address the mismatch between image features and point cloud features in terms of spatial information, we refer to the point feature alignment strategy from M3DM [31]. We align the 3D features to the 2D image plane, thereby simplifying cross-modal interaction and improving detection performance. Specifically, we first interpolate the point cloud features output by PointNet++ [20] according to the original point cloud coordinates, then project the point cloud features onto the 2D image plane using camera parameters [16], obtaining $F_{3D} \in \mathbb{R}^{H \times W \times D_{pc}}$. During the projection process, for the "holes" caused by sparse point clouds, we

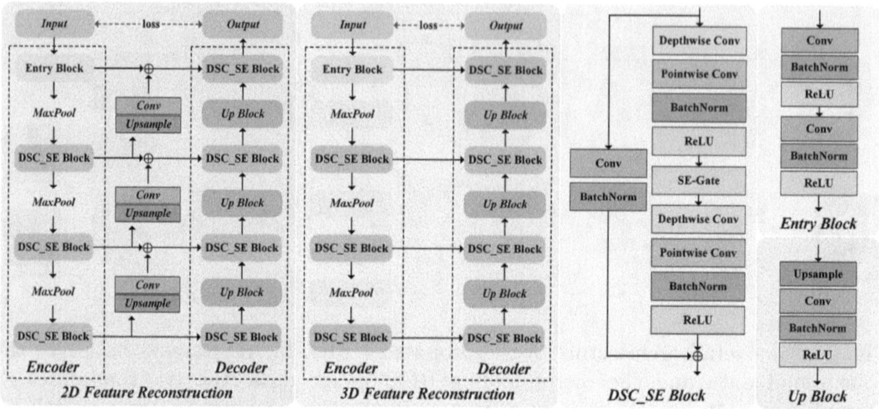

Fig. 3. Overview of the proposed dual-branch feature reconstruction network.

set the feature vector at the corresponding position to zero. Let $\phi_{PC}(\cdot)$ represent the 3D feature extractor, and the process is as follows:

$$F_{3D} = \mathrm{Align}_{3D \to 2D} \left(\mathrm{UP}_3 D \left(\phi_{PC}(x^{pc}) \right) \right) \tag{2}$$

3.2 Dual-Branch Feature Reconstruction Network

In order to accurately reconstruct 2D image features and 3D point cloud features for the unsupervised anomaly detection task, as shown in Fig. 3, we design symmetric encoder-decoder network architectures for each of these two modalities.

2D Feature Reconstruction. The input to the network is a 2D feature map $F_{2D} \in \mathbb{R}^{H \times W \times D_{img}}$, and the network structure consists of several modules, including the Entry block, the DSC_SE block, and the Up block. These modules progressively map the input features to a higher semantic space and finally recover the original feature map shape through the decoder.

In the input part of the 2D feature encoder, the features are first processed through the Entry block to extract initial spatial features from the input feature map F_{2D}. Specifically, the Entry block contains two convolution layers, each followed by batch normalization and ReLU, forming a standard convolution unit. Then, the features undergo dimension reduction through a MaxPool module to reduce the feature map's size. The DSC_SE block extracts local features through depthwise separable convolution and combines the SE (Squeeze-and-Excitation) module to automatically adjust channel weights, thereby enhancing the network's expressive power. We process the features layer by layer through multiple DSC_SE blocks, progressively reducing the feature map's size while extracting deeper semantic information.

In the 2D feature decoder, the restoration of spatial resolution relies on a series of alternating Up blocks and DSC_SE blocks. By combining these modules, the decoder gradually restores the low-resolution feature maps to the same

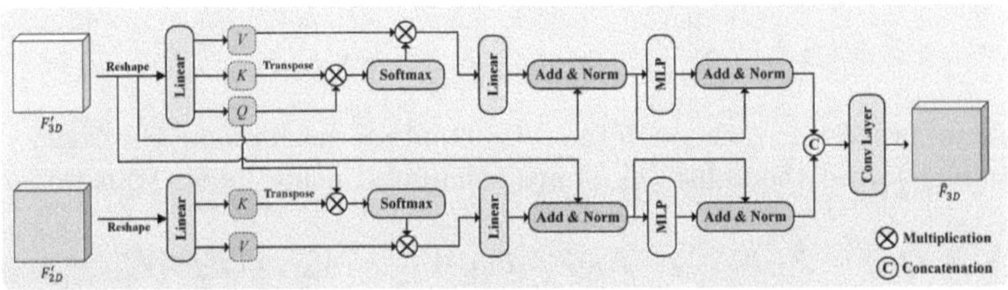

Fig. 4. Detailed structure of the proposed CFEM.

size as the input features. Unlike traditional U-Net [21], our network aggregates encoder output features at different scales from bottom to top, and effectively enhances feature diversity in the decoder through feature concatenation and fusion, reducing the loss of small-scale features during convolution. Finally, while preserving high-resolution details, the decoder fully leverages the contextual information captured by the encoder, outputting a reconstruction representation F'_{2D} of the same size as the input 2D features.

3D Feature Reconstruction. In the 3D feature reconstruction part, we design a symmetric encoder-decoder structure similar to the 2D feature reconstruction, with the network input being the 3D feature map $F_{3D} \in \mathbb{R}^{H \times W \times D_{pc}}$. The 3D feature encoder mainly consists of several DSC_SE modules, which extract features through depthwise separable convolutions and channel attention mechanisms, progressively reducing the feature map's size. Unlike the 2D feature decoder, the 3D feature decoder doesn't employ upsampling and convolution operations on the encoder layers, but rather directly utilizes skip connections and convolutions to effectively restore the features. At each stage, high-resolution features from the encoder are passed to the decoder via skip connections. In the decoder, features processed through convolutions are combined with features from the encoder, ensuring that detailed information is preserved while enhancing the final reconstruction result F'_{3D} by leveraging the contextual information captured by the encoder.

3.3 Cross-Modal Feature Enhancement Module

To achieve efficient fusion and complementarity of information from different modalities, we propose a cross-modal feature enhancement method (CFEM) based on self-attention and cross-attention [12,19,26] mechanisms. This module generates enhanced 3D features $\hat{F}_{3D}$ and 2D features $\hat{F}_{2D}$, which integrate 2D and 3D information, fully exploiting the complementary advantages between the two modalities, thereby improving the accuracy and robustness of multimodal anomaly detection [10,31] tasks. Figure 4 illustrates the complete 3D feature enhancement process.

First, we reshape the reconstructed 3D feature F'_{3D} and the 2D feature F'_{2D}, obtaining:

$$\dot{F}_{3D} \in \mathbb{R}^{N\times d}, \quad \dot{F}_{2D} \in \mathbb{R}^{N\times d} \tag{3}$$

Next, we process $\hat{F}_{3D}$ through the self-attention mechanism, the query (Q_{3D}), the key (K_{3D}), and the value (V_{3D}) are calculated using linear transformations:

$$Q_{3D} = \dot{F}_{3D} W_{3D}^{Q}, \quad K_{3D} = \dot{F}_{3D} W_{3D}^{K}, \quad V_{3D} = \dot{F}_{3D} W_{3D}^{V} \tag{4}$$

where $W_{3D}^{Q}, W_{3D}^{K}, W_{3D}^{V}, W_{3D}^{O} \in \mathbb{R}^{d\times d}$ are learnable linear transformation matrices. Then, the self-attention weights are computed through `Softmax(·)` and are multiplied by the values V_{3D}. We apply a residual connection and layer normalization to the result of the self-attention [19] and the feature $\dot{F}_{3D}$:

$$\tilde{F}_{3D} = \texttt{LayerNorm}\left(\left(\texttt{Softmax}(\frac{Q_{3D}{K_{3D}}^{T}}{\sqrt{d_k}}) \cdot V_{3D}\right) \cdot W_{3D}^{O} + \dot{F}_{3D}\right) \tag{5}$$

Next, we further enhance the feature representation through `MLP(·)` and apply a residual connection and layer normalization to the feature $\tilde{F}_{3D}$:

$$\tilde{F}_{3D}^{'} = \texttt{LayerNorm}\left(\texttt{MLP}(\tilde{F}_{3D}) + \tilde{F}_{3D}\right) \tag{6}$$

For the 2D feature, the key K_{2D} and the value V_{2D} are calculated through linear transformations of $\dot{F}_{2D}$:

$$K_{2D} = \dot{F}_{2D} W_{2D}^{K}, \quad V_{2D} = \dot{F}_{2D} W_{2D}^{V} \tag{7}$$

where $W_{2D}^{K}, W_{2D}^{V}, W_{2D}^{O} \in \mathbb{R}^{d\times d}$ are learnable linear transformation matrices. We use the query Q_{3D} from the 3D feature and the key K_{2D} from the 2D feature to compute the cross-attention weights, and these weights [12] are multiplied by the values V_{2D}. Then, we apply a residual connection and layer normalization to the result of the cross-attention [26] and the feature $\dot{F}_{3D}$:

$$\tilde{F}_{3D\to 2D} = \texttt{LayerNorm}\left(\left(\texttt{Softmax}(\frac{Q_{3D}{K_{2D}}^{T}}{\sqrt{d_k}}) \cdot V_{2D}\right) \cdot W_{2D}^{O} + \dot{F}_{3D}\right) \tag{8}$$

Similarly, we further process the result through `MLP(·)`, applying a nonlinear transformation to $\tilde{F}_{3D\to 2D}$ and combining it with a residual connection and layer normalization:

$$\tilde{F}_{3D\to 2D}^{'} = \texttt{LayerNorm}\left(\texttt{MLP}(\tilde{F}_{3D\to 2D}) + \tilde{F}_{3D\to 2D}\right) \tag{9}$$

Finally, the features processed through self-attention and cross-attention [12, 19], $\tilde{F}_{3D}^{'}$ and $\tilde{F}_{3D\to 2D}^{'}$, are concatenated and passed through a 1×1 convolution layer [26] to further fuse channel information, resulting in the final enhanced 3D feature $\hat{F}_{3D}$:

$$\hat{F}_{3D} = \texttt{Conv}\left(\texttt{Concat}(\tilde{F}'_{3D}, \tilde{F}'_{3D\rightarrow 2D})\right) \tag{10}$$

Through the above self-attention and cross-attention mechanisms, we are able to effectively fuse 2D image information into 3D point cloud features. The cross-attention mechanism not only captures the relationship between 2D and 3D modalities but also enhances feature representation through residual connections and MLP, thereby improving the model's performance in multimodal anomaly detection tasks. Similarly, using the CFEM method, we can also obtain the enhanced 2D feature $\hat{F}_{2D}$.

3.4 Loss Function

In order to simultaneously constrain the feature reconstruction of 2D and 3D modalities and ensure cross-modal feature fusion consistency in an unsupervised setting, we decompose the overall loss into three parts: 2D reconstruction loss, 3D reconstruction loss, and fusion consistency loss. The sum of these three terms forms the final optimization objective:

$$L_{total} = L_{2D} + L_{3D} + L_{fusion} \tag{11}$$

Single-Modal Feature Reconstruction Constraint. F_{2D} and F_{3D} represent the interpolated and aligned 2D image features and 3D point cloud features [29], respectively. F'_{2D} and F'_{3D} denote the reconstructed features. To simultaneously account for both directional consistency and numerical difference, we apply a combined loss function composed of "directional term + difference term," which is expressed as:

$$L_{2D} = \left(1 - \frac{F'_{2D} \cdot F_{2D}}{\|F'_{2D}\|\|F_{2D}\|}\right) + \alpha\|F'_{2D} - F_{2D}\|_2 \tag{12}$$

$$L_{3D} = \left(1 - \frac{F'_{3D} \cdot F_{3D}}{\|F'_{3D}\|\|F_{3D}\|}\right) + \beta\|F'_{3D} - F_{3D}\|_2 \tag{13}$$

The directional term applies a cosine similarity constraint on the angle between the reconstructed and original features, ensuring directional consistency. The difference term measures the Euclidean distance [13] between the two features using the L_2 norm. The parameters α and β are the weighting coefficients for the relative importance of the directional and difference terms. All operations are computed pixel-wise and averaged.

Fusion Consistency Constraint. During the cross-modal feature enhancement stage, in order to ensure the effective fusion between the 2D and 3D modalities, we designed a fusion consistency loss L_{fusion}. Specifically, the fusion consistency loss ensures the directional consistency and the accuracy of information transmission by measuring the cosine similarity between the final fused 2D and 3D features ($\hat{F}_{2D}$ and $\hat{F}_{3D}$) and their respective source features (F_{2D} and F_{3D}). The loss function is expressed as follows:

$$L_{fusion} = 2 - \frac{\hat{F}_{2D} \cdot F_{2D}}{\|\hat{F}_{2D}\| \|F_{2D}\|} - \frac{\hat{F}_{3D} \cdot F_{3D}}{\|\hat{F}_{3D}\| \|F_{3D}\|} \tag{14}$$

3.5 Anomaly Score Calculation

In the inference stage, we first normalize the feature vectors using L2 normalization and then use the Euclidean distance function to compute the differences between the feature maps at each pixel, obtaining the anomaly score maps for the 2D and 3D modalities respectively:

$$\mathcal{A}_{2D} = \mathcal{D}(F_{2D}, \hat{F}_{2D}) = \left\| \frac{F_{2D}}{\|F_{2D}\|_2} - \frac{\hat{F}_{2D}}{\|\hat{F}_{2D}\|_2} \right\|_2 \tag{15}$$

$$\mathcal{A}_{3D} = \mathcal{D}(F_{3D}, \hat{F}_{3D}) = \left\| \frac{F_{3D}}{\|F_{3D}\|_2} - \frac{\hat{F}_{3D}}{\|\hat{F}_{3D}\|_2} \right\|_2 \tag{16}$$

Subsequently, the anomaly score maps from the two modality branches are aggregated pixel-wise by summing them, resulting in the final anomaly score map:

$$\mathcal{A} = \mathcal{A}_{2D} + \mathcal{A}_{3D} \tag{17}$$

Finally, following the approach in CFM [10], we apply a Gaussian kernel with a standard deviation $\sigma = 4$ to smooth $\mathcal{A}$, and select the maximum value from the anomaly map $\mathcal{A}$ as the global anomaly score, which is further used for image-level anomaly detection and localization [29].

4 Experiments

Anomaly detection is a well-established topic that aims to distinguish between normal and anomalous samples [7]. In the experimental setup, normal samples are typically used for model training, while the capability of the trained model to distinguish anomalies is evaluated on the test dataset.

4.1 Dataset

In this study, we utilize two widely used benchmark datasets for 3D anomaly detection [15], MVTec 3D-AD [2] and Eyescandies [4]. The MVTec 3D-AD [2] dataset includes 10 categories of industrial objects. Eyescandies [4] is a synthetic dataset containing 10 object categories collected in an industrial conveyor belt setting. The RGB images and 3D point cloud data in both datasets have the same resolution and are pixel-wise aligned [30].

4.2 Evaluation Metrics

To ensure consistency with previous studies [1,10,31], we use AUROC to evaluate anomaly detection performance at both the image level (I-AUROC) and the pixel level (P-AUROC) [30]. For a more precise evaluation of the per-pixel prediction results, we use the per-region overlap metric (PRO) and compute the area under the PRO curve (AUPRO) as the evaluation metric for anomaly localization [9]. This metric considers both anomaly scores and their connectivity to ground truth components. Previous studies [7,14,23,31] have typically used an integration threshold of 0.3 to compute AUPRO. However, we argue that a threshold of 0.3 may be too lenient for real-world industrial applications. Therefore, we adopt a more stringent integration threshold of 0.01 when computing AUPRO [10,25,29]. Higher values of all the aforementioned metrics indicate better anomaly detection and localization performance [30].

Table 1. I-AUROC/AUPRO@$_{0.3}$ score for anomaly detection of MVTec 3D-AD.

Method	Bagel	Cable Gland	Carrot	Cookie	Dowel	Foam	Peach	Potato	Rope	Tire	Mean
Voxel GAN [2]	68.0/66.4	32.4/62.0	56.5/76.6	39.9/74.0	49.7/78.3	48.2/33.2	56.6/58.2	57.9/79.0	60.1/63.3	48.2/48.3	51.7/63.9
Voxel AE [2]	51.0/46.7	54.0/75.0	38.4/80.8	69.3/55.0	44.6/76.5	63.2/47.3	55.0/72.1	49.4/91.8	72.1/1.9	41.3/17.0	53.8/56.4
Voxel VM [2]	55.3/51.0	77.2/33.1	48.4/41.3	70.1/71.5	75.1/68.0	57.8/27.9	48.0/30.0	46.6/50.7	68.9/61.1	61.1/36.6	60.9/47.1
BTF [14]	93.8/97.6	76.5/96.7	97.2/97.9	88.8/**97.4**	96.0/**97.1**	66.4/88.4	90.4/97.6	92.9/98.1	98.2/95.9	72.6/97.1	87.3/96.4
EasyNet [7]	99.1/83.9	**99.8**/86.4	91.8/95.1	96.8/61.8	94.5/82.8	**94.5**/83.6	90.5/94.2	80.7/88.9	99.4/91.1	79.3/52.8	92.6/82.1
AST [23]	98.3/97.0	87.3/94.7	97.6/98.1	97.1/93.9	93.2/91.3	88.5/90.6	97.4/97.9	**98.1**/98.2	**100**/88.9	79.7/94.0	93.7/94.4
M3DM [31]	99.4/97.0	90.9/97.1	97.2/97.9	97.6/95.0	96.0/94.1	94.2/93.2	97.3/97.7	89.9/97.1	97.2/97.1	85.0/97.5	94.5/96.4
CFM [10]	99.4/**97.9**	88.8/97.2	**98.4**/**98.2**	99.3/94.5	98.0/95.0	88.8/96.8	94.1/**98.0**	94.3/98.2	98.0/97.5	**95.3**/98.1	95.4/97.1
Ours	**99.7**/97.2	93.2/**98.1**	98.2/97.5	**99.6**/96.2	**98.9**/96.9	90.5/**97.0**	**98.2**/97.8	95.7/**98.3**	96.9/**97.7**	94.1/**98.3**	**96.5**/**97.5**

Table 2. Mean performance metrics on the MVTec 3D-AD and Eyecandies datasets.

Method	MVTec 3D-AD				Eyecandies			
	I-AUROC	P-AUROC	AUPRO@$_{0.3}$	AUPRO@$_{0.01}$	I-AUROC	P-AUROC	AUPRO@$_{0.3}$	AUPRO@$_{0.01}$
BTF [14]	87.3	99.3	96.4	38.8	81.6	95.9	84.5	28.0
AST [23]	93.7	97.6	94.4	39.8	75.8	90.2	87.8	22.4
M3DM [31]	94.5	99.2	96.4	39.4	**89.7**	**97.7**	88.2	33.1
CFM [10]	95.4	99.3	97.1	45.5	88.1	97.4	88.7	33.5
Ours	**96.5**	**99.5**	**97.5**	**46.8**	88.9	97.4	**89.1**	**34.2**

4.3 Experimental Results

We conducted extensive experiments on the MVTec 3D-AD [2] and Eyecandies [4] datasets to evaluate the anomaly detection and localization performance of the proposed method. We compared our method with multiple state-of-the-art approaches, including BTF [14], AST [23], M3DM [31], and CFM [10]. As shown in Table 1, on the MVTec 3D-AD dataset [2], our method outperforms previous approaches in both multimodal anomaly detection and localization tasks, achieving the best performance with I-AUROC and AUPRO@$_{0.3}$ scores of 96.5% and 97.5%, respectively. Specifically, our method exhibits remarkable performance in multiple categories, particularly for cable gland, cookie, foam, and peach. As shown in Table 2, our method achieves a score of 46.8% on the more stringent AUPRO@$_{0.01}$ metric, surpassing CFM [10] by 1.3%, which indicates stronger localization capability under low false positive conditions. In addition, Table 2 shows that our method achieves the best anomaly localization performance on the Eyecandies [4] dataset, surpassing other approaches in terms of overall average metrics, with AUPRO@$_{0.3}$ and AUPRO@$_{0.01}$ scores of 89.1% and 34.2%, respectively. Meanwhile, our method ranks second in terms of the AUROC metric, with the M3DM [31] approach performing slightly better. Figure 5 further illustrates a visual comparison of our method with other approaches in the anomaly detection task. From the heatmap comparison, it is evident that our method captures anomalous regions more accurately, particularly excelling in detecting anomalies related to shape and texture.

4.4 Ablation Study

To comprehensively validate the effectiveness of the proposed cross-modal feature enhancement module and multimodal feature collaboration, we conducted systematic ablation experiments on the MVTec 3D-AD [2] dataset. As shown in Table 3, the results of the ablation study indicate that using multimodal (2D+3D) significantly improves anomaly detection and localization performance compared to using a single modality (2D or 3D). In addition, after incorporating the cross-modal feature enhancement module to the multimodal features (2D+3D), the performance of all evaluation metrics was further improved. Among them, I-AUROC improved by 2.4%, while AUPRO@$_{0.3}$ and AUPRO@$_{0.01}$ increased by 1.1% and 1.6%, respectively, indicating that CFEM efficiently integrates and complements information from the two modalities, thereby enhancing model's performance and accuracy.

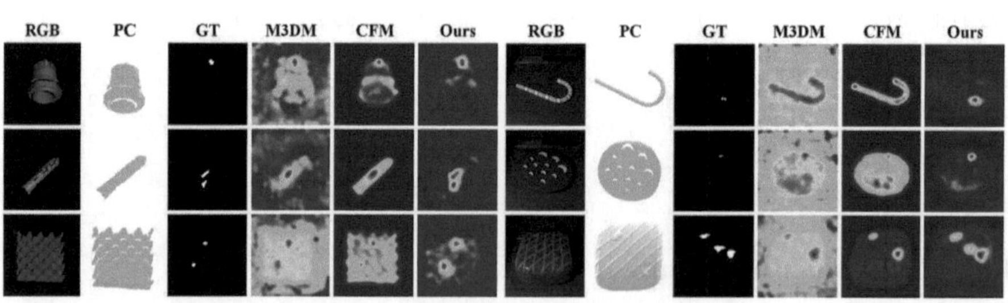

Fig. 5. Qualitative comparison of anomaly localization results on the MVTec 3D-AD [2] (left) and Eyecandies [4] (right) datasets.

Table 3. Ablation study on data modalities and CFEM.

Method	I-AUROC	P-AUROC	AUPRO@$_{0.3}$	AUPRO@$_{0.01}$
2D only	89.1	98.9	95.1	39.7
3D only	87.2	97.6	91.3	37.2
2D+3D	94.1	99.3	96.4	45.2
2D+3D+CFEM	**96.5**	**99.5**	**97.5**	**46.8**

5 Conclusion

In this work, our objective is to fully leverage the complementary advantages of multimodal information to enhance the accuracy of anomaly detection and localization. Accordingly, we propose an unsupervised multimodal industrial anomaly detection framework, DFRF-MIAD, based on feature restoration. Our proposed approach not only performs effective intra-modal feature learning within a single modality, but also achieves efficient cross-modal feature interaction learning between the point cloud and RGB modalities. Through extensive experiments on the MVTec 3D-AD [2] and Eyecandies [4] datasets, the results demonstrate that the proposed DFRF-MIAD framework achieves superior performance in anomaly detection and localization tasks, accomplishing high detection accuracy while effectively reducing the false positive rate. Particularly in identifying anomalies in geometric structures and texture details, the proposed method demonstrates stronger robustness and generalization capability, showing great potential for practical industrial applications. In future work, we will focus on efficiently and deeply integrating multimodal information to fully leverage the complementary advantages between modalities, and enhancing the model's capability for efficient deployment in low-resource environments.

References

1. Asad, M., Azeem, W., Jiang, H., Mustafa, H.T., Yang, J., Liu, W.: 2M3DF: advancing 3D industrial defect detection with multi perspective multimodal fusion network. IEEE Trans. Circ. Syst. Video Technol. (2025)
2. Bergmann, P., Jin, X., Sattlegger, D., Steger, C.: The MVTec 3D-AD dataset for unsupervised 3D anomaly detection and localization. arXiv preprint arXiv:2112.09045 (2021)
3. Bergmann, P., Sattlegger, D.: Anomaly detection in 3D point clouds using deep geometric descriptors. In: Proceedings of the IEEE/CVF Winter Conference on Applications of Computer Vision, pp. 2613–2623 (2023)
4. Bonfiglioli, L., Toschi, M., Silvestri, D., Fioraio, N., De Gregorio, D.: The eyecandies dataset for unsupervised multimodal anomaly detection and localization. In: Proceedings of the Asian Conference on Computer Vision, pp. 3586–3602 (2022)
5. Cao, Y., Xu, X., Shen, W.: Complementary pseudo multimodal feature for point cloud anomaly detection. Pattern Recogn. **156**, 110761 (2024)

6. Cao, Y., et al.: A survey on visual anomaly detection: challenge, approach, and prospect. arXiv preprint arXiv:2401.16402 (2024)
7. Chen, R., et al.: Easynet: an easy network for 3D industrial anomaly detection. In: Proceedings of the 31st ACM International Conference on Multimedia, pp. 7038–7046 (2023)
8. Chiu, L.L., Lai, S.H.: Self-supervised normalizing flows for image anomaly detection and localization. In: Proceedings of the IEEE/CVF Conference on Computer Vision and Pattern Recognition, pp. 2927–2936 (2023)
9. Chu, Y.M., Liu, C., Hsieh, T.I., Chen, H.T., Liu, T.L.: Shape-guided dual-memory learning for 3D anomaly detection. In: Proceedings of the 40th International Conference on Machine Learning, pp. 6185–6194 (2023)
10. Costanzino, A., Ramirez, P.Z., Lisanti, G., Di Stefano, L.: Multimodal industrial anomaly detection by crossmodal feature mapping. In: Proceedings of the IEEE/CVF Conference on Computer Vision and Pattern Recognition, pp. 17234–17243 (2024)
11. Du, J., Tao, C., Cao, X., Tsung, F.: 3D vision-based anomaly detection in manufacturing: a survey. Front. Eng. Manag. 1–18 (2025)
12. Gao, Y., et al.: Unsupervised automated track inspection based on symmetrical teacher-student network and cross-modal feature alignment attention. Available at SSRN 5172984
13. Gudovskiy, D., Ishizaka, S., Kozuka, K.: Cflow-ad: real-time unsupervised anomaly detection with localization via conditional normalizing flows. In: Proceedings of the IEEE/CVF Winter Conference on Applications of Computer Vision, pp. 98–107 (2022)
14. Horwitz, E., Hoshen, Y.: Back to the feature: classical 3D features are (almost) all you need for 3D anomaly detection. In: Proceedings of the IEEE/CVF Conference on Computer Vision and Pattern Recognition, pp. 2968–2977 (2023)
15. Li, K., Dai, B., Fu, J., Hou, X.: Das3d: dual-modality anomaly synthesis for 3D anomaly detection. In: European Conference on Computer Vision, pp. 148–165. Springer (2024)
16. Lin, Y., et al.: A survey on RGB, 3D, and multimodal approaches for unsupervised industrial image anomaly detection. Inf. Fusion 103139 (2025)
17. Liu, J., et al.: Real3d-ad: a dataset of point cloud anomaly detection. Adv. Neural. Inf. Process. Syst. **36**, 30402–30415 (2023)
18. Liu, Y., Zhang, C., Dong, X., Ning, J.: Point cloud-based deep learning in industrial production: a survey. ACM Comput. Surv. **57**(7), 1–36 (2025)
19. Petit, O., Thome, N., Rambour, C., Themyr, L., Collins, T., Soler, L.: U-net transformer: self and cross attention for medical image segmentation. In: International Workshop on Machine Learning in Medical Imaging, pp. 267–276. Springer (2021)
20. Qi, C.R., Yi, L., Su, H., Guibas, L.J.: Pointnet++: deep hierarchical feature learning on point sets in a metric space. In: Advances in Neural Information Processing Systems, vol. 30 (2017)
21. Ronneberger, O., Fischer, P., Brox, T.: U-net: convolutional networks for biomedical image segmentation. In: International Conference on Medical Image Computing and Computer-Assisted Intervention, pp. 234–241. Springer (2015)
22. Roth, K., Pemula, L., Zepeda, J., Schölkopf, B., Brox, T., Gehler, P.: Towards total recall in industrial anomaly detection. In: Proceedings of the IEEE/CVF Conference on Computer Vision and Pattern Recognition, pp. 14318–14328 (2022)
23. Rudolph, M., Wehrbein, T., Rosenhahn, B., Wandt, B.: Asymmetric student-teacher networks for industrial anomaly detection. In: Proceedings of the

IEEE/CVF Winter Conference on Applications of Computer Vision, pp. 2592–2602 (2023)
24. Rusu, R.B., Blodow, N., Beetz, M.: Fast point feature histograms (FPFH) for 3D registration. In: 2009 IEEE International Conference on Robotics and Automation, pp. 3212–3217. IEEE (2009)
25. Shangguan, W., et al.: CPIR: multimodal industrial anomaly detection via latent bridged cross-modal prediction and intra-modal reconstruction. Adv. Eng. Inform. **65**, 103240 (2025)
26. Shen, J., Chen, Y., Liu, Y., Zuo, X., Fan, H., Yang, W.: Icafusion: iterative cross-attention guided feature fusion for multispectral object detection. Pattern Recogn. **145**, 109913 (2024)
27. Sui, W., Lichau, D., Lefèvre, J., Phelippeau, H.: Cross-modal distillation in industrial anomaly detection: exploring efficient multi-modal IAD. arXiv e-prints pp. arXiv–2405 (2024)
28. Sun, Z., Li, X., Li, Y., Ma, Y.: Memoryless multimodal anomaly detection via student-teacher network and signed distance learning. Electronics **13**(19), 3914 (2024)
29. Wang, J., Niu, Y., Huang, B.: Fusion-restoration model for industrial multimodal anomaly detection. Neurocomputing **637**, 130073 (2025)
30. Wang, J., et al.: Incremental template neighborhood matching for 3D anomaly detection. Neurocomputing **581**, 127483 (2024)
31. Wang, Y., Peng, J., Zhang, J., Yi, R., Wang, Y., Wang, C.: Multimodal industrial anomaly detection via hybrid fusion. In: Proceedings of the IEEE/CVF Conference on Computer Vision and Pattern Recognition, pp. 8032–8041 (2023)
32. Zhu, J., Yan, P., Jiang, J., Cui, Y., Xu, X.: Asymmetric teacher–student feature pyramid matching for industrial anomaly detection. IEEE Trans. Instrum. Meas. **73**, 1–13 (2024)

PSR-Diff: Polarization-Guided Diffusion Model for Single Image Specular Highlight Removal

Guobin Zhang, Li Li(✉), Zhaojing Wang, Qihang Wang, Tao Peng, and Xinrong Hu

School of Computer Science and Artificial Intelligence, Wuhan Textile University, Wuhan, China
lli@wtu.edu.cn

Abstract. Specular highlights on smooth surfaces obscure image details and degrade the performance of vision tasks such as segmentation and detection. Traditional RGB-based highlight removal methods rely heavily on supervised data, suffering from high acquisition costs and limited generalization. Polarization imaging provides complementary cues by separating specular reflections from diffuse components, offering a physical strategy to address this challenge. Inspired by this, we propose the Polarization-guided Specular Removal Diffusion Model (PSR-Diff), which integrates diffusion models with polarization priors. PSR-Diff extracts polarization cues via Fresnel theory to construct a pseudo-diffuse estimate and fuses it with RGB features through a cross-modal attention mechanism. To mitigate data scarcity and modality differences, we introduce a dual-module fusion strategy that leverages a pre-trained RGB diffusion model while enabling polarization-guided learning. Experiments on polarization and RGB-only datasets show that PSR-Diff achieves superior highlight suppression, enhanced image quality, and robust performance under diverse lighting and materials.

Keywords: Specular highlights removal · Diffusion models · Polarization imaging · Multimodal fusion

1 Introduction

Specular highlights, resulting from the mirror-like reflection of light sources on object surfaces, are prominent and pervasive phenomena in natural images, as illustrated in Fig. 1. These intense reflections are closely associated with the shape, material, and texture of the object, and they pose significant challenges to visual understanding tasks such as image segmentation [1], object detection [2], and light source estimation [3]. Specular regions often occlude authentic surface details, making it difficult for models to accurately recover texture and structural information. Consequently, the effective removal of specular highlights and the restoration of occluded content is a critical problem in computer vision.

J. Lokoč et al. (Eds.): MMM 2026, LNCS 16413, pp. 134–148, 2026.
https://doi.org/10.1007/978-981-95-6957-1_10

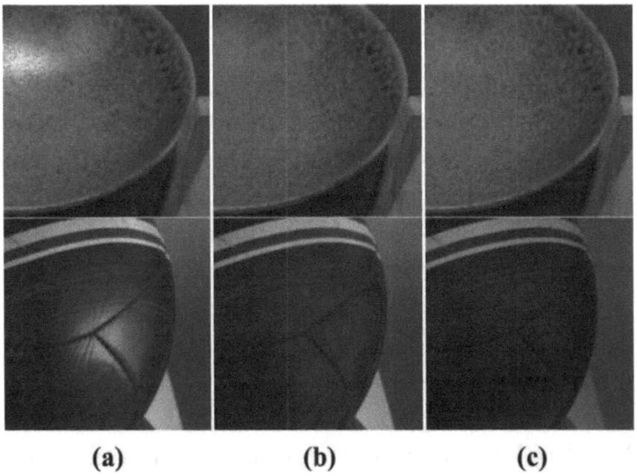

Fig. 1. The specular highlight removal results of our method: (a) input image with specular highlights, (b) removal results of our method, (c) ground truth.

Traditional specular highlight removal methods are typically based on physical models, treating highlights as sparse noise for filtering [4], or estimating highlight components through color variation ratios [5]. While effective under specific assumptions, these approaches rely heavily on prior knowledge and struggle in complex reflection scenarios. Recent advances in deep learning, particularly convolutional neural networks (CNNs) and generative adversarial networks (GANs), have demonstrated progress in suppressing highlights and restoring textures by learning the nonlinear relationships between specularities and underlying surfaces [6]. However, these methods still face limitations in accurately reconstructing details in areas with strong reflections.

Polarization imaging technology, as a physical imaging technique, has recently been introduced into the task of specular highlight removal. By capturing the polarization states of light, it provides additional physical information beyond standard RGB imagery. Polarization images taken at different angles enable the separation of specular and diffuse reflections [7], significantly enhancing highlight detection and removal in complex scenes. However, polarization images are highly sensitive to noise under low-light or strong-reflection conditions, and semantic gaps between polarization and RGB modalities hinder effective fusion. Furthermore, the limited availability of large-scale polarization datasets restricts model training.

To overcome these limitations, we propose PSR-Diff, a *Polarization-guided Specular Removal Diffusion Network*, as illustrated in Fig. 3. This method leverages a diffusion model pre-trained on large-scale RGB datasets to compensate for the scarcity of polarization data. Additionally, we construct pseudo-diffuse estimate from polarization images to guide the model in suppressing highlights and restoring structures. Finally, we design two *cross-modal feature refinement modules* with a spatial attention-like operation to enhance the fusion efficiency and expressive power between RGB and polarization features.

Our main contributions are summarized as follows:

- We propose a novel diffusion-based framework guided by polarization prompts for specular highlight removal. The model effectively integrates RGB and polarization modalities, enabling accurate highlight suppression and surface restoration in complex reflective scenes.
- We introduce a polarization-guided mechanism based on Fresnel theory to generate a pseudo-diffuse estimate. By extracting polarization angles and degrees from multi-angle images, we construct a reliable polarization prompt that effectively guides the denoising process, significantly improving the reconstruction of highlight regions.
- We propose a dual-module fusion architecture that enhances the interaction between RGB and polarization features. Through cross-modal attention, the network dynamically aligns semantics across modalities, preserves the dominant RGB structure, and effectively exploits the complementary cues from polarization inputs.

The remainder of this paper is organized as follows: Sect. 2 reviews related work in highlight removal. Section 3 presents the proposed PSR-Diff network in detail. Section 4 provides experimental setups and results. Section 5 concludes the paper.

2 Related Work

2.1 Specular Highlight Removal with RGB Images

Specular highlight removal has been extensively studied in computer vision. Classical methods typically exploit photometric constraints, color-space transformations, or statistical priors. Shafer's dichromatic reflection model [8] decomposes RGB images into diffuse and specular components, forming the basis for many color-space separation approaches. Subsequent works refine chromaticity-based separation through entropy maximization [9] or chromaticity-shift normalization [10], while sparse and low-rank decomposition methods [11] model diffuse layers as low-rank and specular layers as sparse. Although effective under constrained settings, these approaches rely on strong priors and generalize poorly to real-world scenes.

Learning-based methods alleviate these limitations by directly learning mappings from data. Funke *et al.* [6] employed CycleGAN for highlight detection in endoscopic imagery. Fu *et al.* [5] developed a multi-scale segmentation network and later extended it into a multitask framework for highlight modeling, detection, and removal. Hu *et al.* [12] proposed a mask-guided GAN with NMF-based masks to emphasize specular regions. Despite their advantages, these RGB-only methods often fail under strong, spatially dispersed, or large-area highlights.

2.2 Multimodal Highlight Removal

Polarization provides physically grounded cues for separating diffuse and specular reflections, as diffuse reflections show weak polarization while specular reflections exhibit stronger, oriented polarization. Early work [13] estimated polarization parameters (DoLP, AoLP) from multi-angle observations to detect highlights, while Lei *et al.* [14] leveraged polarization differences between natural and flash illumination. However, polarization imaging is sensitive to noise, and nonlinear computations amplify measurement errors.

To mitigate these limitations, recent approaches explore multimodal fusion of RGB and polarization cues. For example, SHMGAN [15] learns illumination variation from multi-angle polarization images via adversarial training, while other works [14] construct aligned RGB-polarization datasets for reflection removal. Nevertheless, large-scale polarization datasets [16] remain scarce, limiting generalization. This scarcity motivates us to adopt a pretraining-based transfer strategy, inspired by recent advances in prompt learning where Ikemura *et al.* [17] demonstrated multimodal prompt fusion for depth completion.

2.3 Diffusion-Based Image Enhancement

Diffusion probabilistic models [18] have recently become powerful generative frameworks for image synthesis and restoration. By progressively denoising Gaussian noise, they capture both global semantics and fine textures, achieving state-of-the-art results in super-resolution [19], inpainting [20], and deblurring [21]. Recent advances further extend them to low-level vision tasks such as image-to-image translation and reflection removal.

To address the underexplored use of polarization information in diffusion-based restoration, we integrate polarization-guided multimodal fusion into a diffusion framework for robust highlight suppression and texture reconstruction. While *PolarFree* [22] also leverages polarization cues with diffusion modeling, it targets reflection removal and adopts a two-stage design. In contrast, our PSR-Diff embeds polarization information directly into the diffusion process via the Polarization Prompt Learning Module (PPLM), enabling single-stage end-to-end optimization.

3 Methodology

As illustrated in Fig. 3, the proposed PSR-Diff framework integrates physical polarization priors with a pretrained diffusion model for specular highlight removal. The pipeline begins with a forward diffusion process, where RGB images are gradually perturbed into noisy representations. In parallel, polarization observations captured at multiple angles are utilized to compute a pseudo-diffuse estimate, which is incorporated as an auxiliary modality to provide physics-based guidance during reverse diffusion. During the denoising process, the pretrained RGB diffusion backbone is extended with a multimodal fusion design. At each

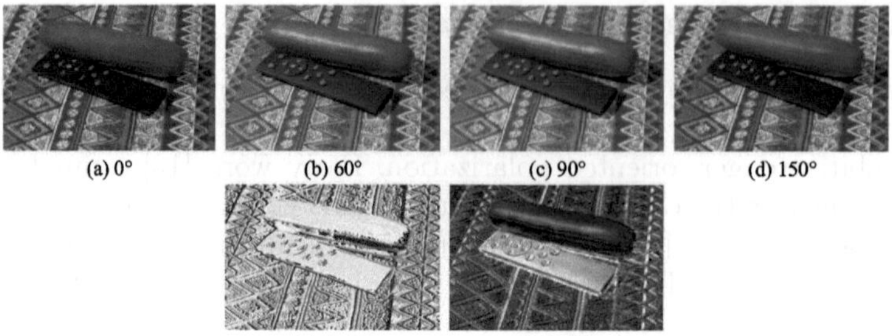

Fig. 2. Polarization images captured at four orthogonal polarization angles, along with the computed AoLP and DoLP.

stage, RGB features are enhanced by polarization cues, and polarization prompts are aligned with the RGB feature space to ensure consistent cross-modal representation. The fused features progressively guide the reverse diffusion to reconstruct highlight-free outputs with high fidelity and robustness.

In the following sections, we will introduce the extraction of polarization cues, the multimodal feature fusion strategy and the design of the loss function.

3.1 Polarization Cue Extraction

The proposed diffusion-based specular highlight removal network leverages polarization cues to exploit the inherent physical properties of polarized light and effectively fuse them with conventional RGB information. As illustrated in Fig. 2, the distinct polarization responses of specular and diffuse reflections, as described by Fresnel theory, are utilized to extract informative polarization features from multi-angle polarization images. These features serve as a prompt modality to guide the pretrained diffusion model in generating high-quality highlight-free images.

According to Fresnel theory, the observed intensity I can be decomposed into a diffuse component I_d and a specular component I_s:

$$I = I_d + I_s, \tag{1}$$

where I_s consists of a constant term I_{sc} and a cosine term I_{sv} modulated by the polarization angle ψ_{pol}:

$$I_s = I_{sc} + I_{sv} \cos 2(\psi_{pol} - \alpha), \tag{2}$$

with α denoting the angle of linear polarization (AoLP).

By capturing images at multiple ψ_{pol}, the pixel intensity I_p can be expressed as:

$$I_p(\psi_{pol}) = \frac{I_{total}}{2} \left[1 + \rho \cos\left(2\psi_{pol} - 2\alpha\right)\right], \tag{3}$$

where ρ is the degree of linear polarization (DoLP). Recovering α, ρ, and ψ_{pol} requires at least three polarization images; in practice, four orthogonal angles (e.g., 0°, 45°, 90°, 135°) are preferred to maximize sinusoidal intensity variation and improve estimation robustness.

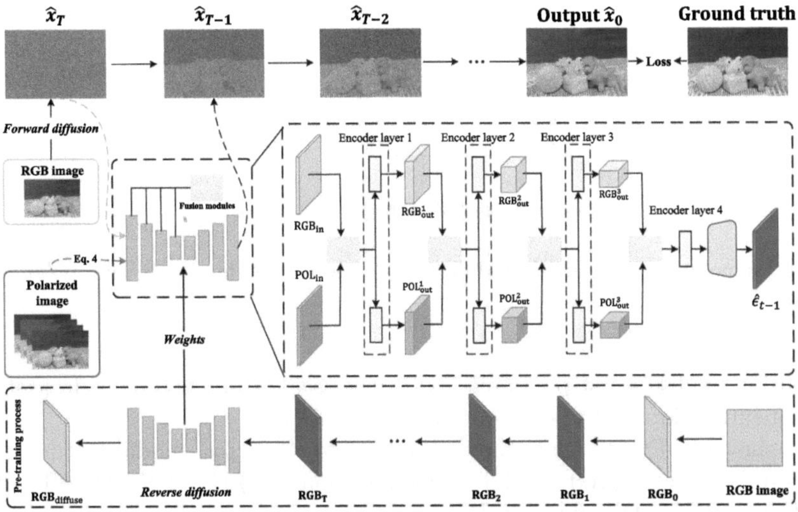

Fig. 3. The overall architecture of the proposed PSR-Diff.

In natural scenes, image intensity is a combination of diffuse reflection, polarized specular highlights, and unpolarized specular highlights. The polarized component varies with ψ_{pol}, while the diffuse and unpolarized components remain largely unchanged. Exploiting this property, following [15], a pseudo-diffuse estimate I_{DE} can be computed as:

$$I_{DE}(x, y) = \min_{\forall \psi_{pol}} I_{\psi_{pol}}(x, y), \tag{4}$$

which suppresses the polarized specular term while preserving diffuse information.

This I_{DE} serves as the polarization modality input to the network, providing an effective prior for the diffusion model. Unlike traditional polarization-based highlight removal methods that require strict imaging conditions (e.g., perfectly orthogonal polarizations, single light sources), this strategy offers greater applicability in complex real-world scenarios.

3.2 Multimodal Feature Fusion

Polarization-guided image processing faces two key challenges: (*i*) the scarcity of publicly available polarization datasets limits model generalization; and (*ii*) the substantial modality gap between polarization and RGB data complicates effective fusion. To address these issues, we adopt a transfer learning strategy:

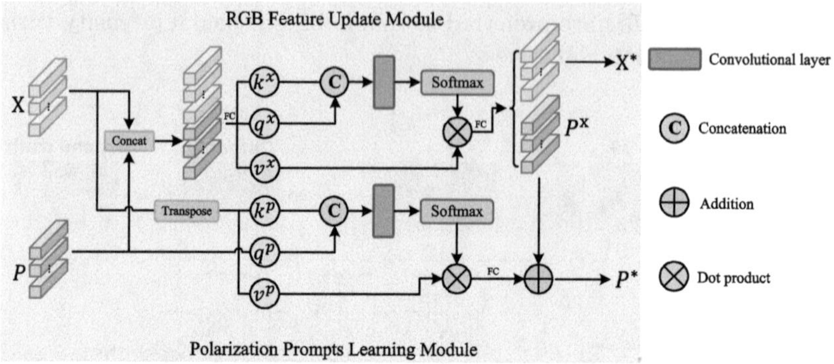

Fig. 4. Structure of the dual-module fusion framework for RGB and polarization modalities.

a diffusion model is first pretrained on a large-scale RGB highlight dataset and subsequently adapted to polarization-guided tasks.

As shown in Fig. 4, our framework integrates two cross-modal feature learning modules into the pretrained diffusion backbone: **(1) RGB Feature Update Module (RFUM)** enhances RGB semantics using polarization cues; **(2) Polarization Prompt Learning Module (PPLM)** aligns the semantic spaces of both modalities via cross-modal attention and adaptively adjusts their regional contribution. At each encoder stage, RGB and polarization features are updated and fused through these modules, ensuring progressive interaction and robust representation. The fused features are then decoded to predict the Gaussian noise $\hat{\epsilon}_t$ at timestep t, which is used in the diffusion process to estimate $\hat{\mathbf{x}}_{t-1}$ while polarization input remains fixed, providing consistent guidance for highlight removal.

Inspired by Ye *et al.* [23], we detail the implementation of the dual modules. In the RFUM, polarization information P and input features X are spatially interacted to update and produce fused features X^*. Specifically, convolutional layers extract feature maps from RGB and polarization modalities, which are then summed and passed through a multilayer perceptron (MLP) to generate key k^x, query q^x, and value v^x vectors with learnable parameters $W_k^x, W_q^x, W_v^x \in R^{2C\times C}$:

$$(k^x, q^x, v^x) = [P; X] \times (W_k^x, W_q^x, W_v^x) \tag{5}$$

$$[X^*, P^x] = FC_d\left(\lambda \cdot \text{softmax}\big(FC([q^x; k^x])\big) \odot v^x\right) \tag{6}$$

where $\odot$ denotes element-wise multiplication, and FC_d is a dropout-enabled fully connected layer. This module enhances the adaptability of RGB features to usable highlight cues from the polarization modality, improving the representational capacity of fused features.

In the PPLM, the RGB modality serves as auxiliary guidance to update the polarization prompt features P^* with the RFUM-enhanced representation P^x.

Here, RGB input X from RFUM is used to compute key k^p and value v^p vectors, while polarization prompt P computes the query q^p vector:

$$(k^p, v^p) = X \times (W_k^p, W_v^p), \quad q^p = P \times W_q^p \tag{7}$$

$$P^* = P^x + FC\left(\lambda \cdot \text{softmax}\big(FC([q^p; k^p])\big) \odot v^p\right) \tag{8}$$

While both modules share a cross-attention structure, their objectives differ: RFUM prioritizes RGB feature enhancement with polarization cues, whereas PPLM optimizes polarization feature expression for semantic alignment with RGB. This dual-module design maintains strong generative capacity and delivers effective highlight removal even under noisy polarization signals or limited data.

3.3 Loss Function

To simultaneously preserve structural consistency and restore fine details in the specular highlight removal task, we adopt a combined L_1+L_2 loss as the training objective, constraining the discrepancy between the generated image and the ground-truth highlight-free image. Let $\mathbf{x}^*$ denote the ground-truth highlight-free image and $\hat{\mathbf{x}}$ denote the generated result, the loss function is defined as

$$L(\mathbf{x}^*, \hat{\mathbf{x}}) = \frac{1}{|\mathcal{M}|} \sum_{p \in \mathcal{M}} \left(\|\mathbf{x}_p^* - \hat{\mathbf{x}}_p\|_1 + \|\mathbf{x}_p^* - \hat{\mathbf{x}}_p\|_2^2 \right), \tag{9}$$

where $\mathcal{M}$ represents the set of valid pixels (i.e., the region participating in loss computation), and p indexes individual pixels.

The L_1 term, $\|\mathbf{x}_p^* - \hat{\mathbf{x}}_p\|_1$, effectively preserves edges and texture details, mitigating the tendency toward over-smoothing. In contrast, the L_2 term, $\|\mathbf{x}_p^* - \hat{\mathbf{x}}_p\|_2^2$, is more sensitive to overall intensity errors, thereby enhancing the precision and stability of the reconstruction.

4 Experiments and Results

4.1 Experimental Settings

Our network is implemented in PyTorch and trained on an NVIDIA RTX 3080Ti for 300 epochs using the ADAM optimizer. Input images are resized to multiples of 16, with batch size 16. Data augmentation includes random rotations and flips. The learning rate is initialized at 2×10^{-5} and decays cyclically every 100 epochs.

4.2 Datasets and Evaluation Metrics

We evaluate our method on two benchmark datasets for specular highlight removal: PSD [24] and SHIQ [25]. PSD contains polarized images used for both training and evaluation, while SHIQ comprises RGB images only, assessing our model's generalization on RGB specular highlight removal.

SHIQ includes 43,300 images in four subsets: *highlight*, *highlight-free*, *highlight intensity*, and *highlight mask* (10,825 images each). We evaluate on the *highlight* subset and its *highlight-free* ground truth. Since the SHIQ dataset does not provide polarization images, we design a lightweight I_{DE}-Predictor network to enable PSR-Diff to operate under RGB-only conditions. The predictor estimates a pseudo-diffuse component from a single RGB image and is trained on the polarization-available PSD dataset, using the pseudo-diffuse estimate I_{DE} as the supervision target.

PSD has 13,380 high-resolution images from 2,210 real-world scenes, captured via 1,010 groups at fixed polarization angles (12 images per group, 30° increments) and 1,200 pairs at random angles. For training, we use four orthogonal angles (0°, 60°, 90°, 150°) from the fixed-angle subset with the RGB images.

For quantitative evaluation, we adopt two widely used metrics: Peak Signal-to-Noise Ratio (PSNR) and Structural Similarity Index Measure (SSIM). PSNR quantifies pixel-wise fidelity, while SSIM evaluates perceptual structural similarity considering luminance, contrast, and structure. Higher values of both metrics correspond to better reconstruction quality.

4.3 Comparison with State-of-the-Art Methods

To comprehensively assess the proposed method, we perform both qualitative and quantitative comparisons with representative traditional and deep learning-based approaches on the SHIQ and PSD datasets. All competing methods are trained and tested under identical experimental protocols to ensure fair and objective evaluation.

Results on the PSD Dataset. To evaluate the effectiveness of the proposed method in real-world scenarios, we conduct experiments on the PSD dataset. Baselines include classical physics-based methods (Akashi [26], Yamamoto [27]) and recent deep learning approaches such as CycleGAN-based networks.

Quantitative Results: Table 1 reports PSNR and SSIM. PSR-Diff achieves the best results (29.42 PSNR, 0.93 SSIM), surpassing existing methods by about 1.3 dB and 0.01. This highlights the advantage of our dual-module fusion strategy in preserving textures and structural details, especially in highlight regions.

Qualitative Results: Figure 5 compares visual results. Traditional methods (Fig. 5(c,d)) often show color distortion and blurring, while SpecularityNet [24] still leaves noticeable highlights (third and fourth rows of Fig. 5(e)). In contrast, PSR-Diff consistently removes highlights and preserves textures, benefiting from accurate highlight localization via polarization cues, and reliable cross-modal alignment enabled by dual-module fusion.

Table 1. Quantitative comparison on the PSD dataset. Best results are highlighted in **bold**.

Method	PSNR ↑	SSIM ↑
Akashi [26]	21.32	0.65
Yamamoto [27]	24.66	0.73
Wu [28]	27.89	0.90
SpecularityNet [24]	28.13	0.92
Ours	**29.42**	**0.93**

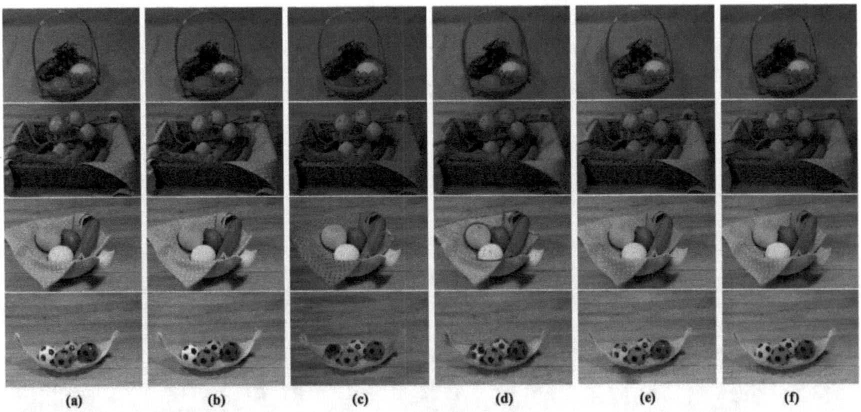

Fig. 5. Visual comparison of highlight removal methods on the PSD dataset: (a) input image, (b) GT, (c) Akashi et al. [26], (d) Yamamoto et al. [27], (e) SpecularityNet [24], (f) Ours.

Results on the SHIQ Dataset. The SHIQ dataset evaluates our method's generalization on highlight images under varying textures and material conditions, compared with four classical and four deep learning-based methods.

Quantitative Results: Table 2 presents the PSNR and SSIM scores on SHIQ. Our method outperforms all baselines, achieving 36.46 PSNR and 0.98 SSIM. These improvements demonstrate the effectiveness of the feature-update mechanism within the diffusion process in enhancing highlight suppression and structural fidelity.

Qualitative Results: We compare our method with several representative traditional and deep learning-based approaches on the SHIQ dataset, as illustrated in Fig. 6. Traditional methods generally struggle under complex illumination and material variations, often leaving residual highlights or introducing artifacts. For instance, Akashi [26] fails to remove highlights with high color contrast, while Spec-CGAN [33] produces severe color distortions. Although Yuan [32] achieves slightly better results, noticeable highlight remnants remain. Wu [28] and

Table 2. Quantitative comparison on the SHIQ dataset. Best results are shown in **bold**.

Method	PSNR ↑	SSIM ↑
Yang [29]	14.31	0.50
Akashi [26]	14.01	0.52
Guo [30]	17.18	0.58
Yamamoto [27]	19.54	0.63
Shi [31]	18.21	0.61
Fu [25]	34.13	0.86
Yuan [32]	33.85	0.95
Wu [28]	31.78	0.96
Ours	**36.46**	**0.98**

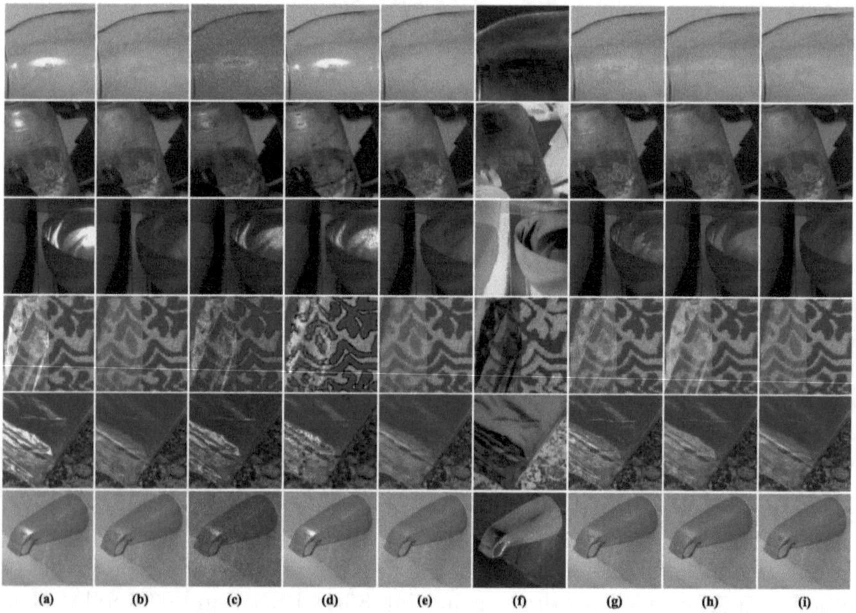

Fig. 6. Visual comparison of highlight removal methods on the SHIQ dataset: (a) input image, (b) GT, (c) Akashi et al. [26], (d) Yamamoto et al. [27], (e) Fu et al. [25], (f) SPEC-CGAN [33], (g) Yuan et al. [32], (h) Wu et al. [28], (i) Ours.

Fu [25] remove most highlights but exhibit unstable performance across different materials.

Through comparisons, our method consistently produces realistic, highlight-free images across diverse materials and scenes, with minimal distortion and artifacts, demonstrating strong generalization and yielding results perceptually closest to the ground truth.

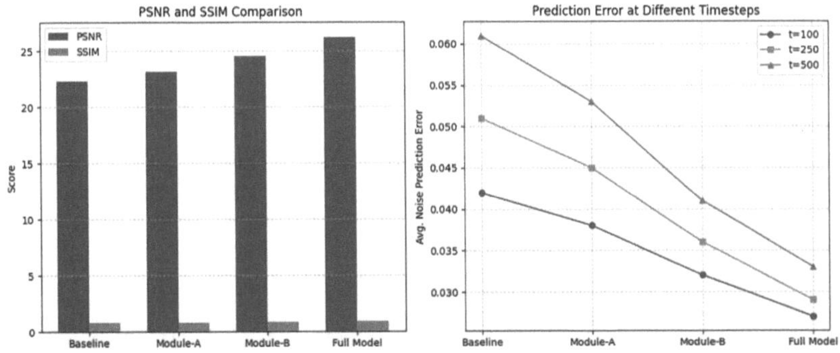

Fig. 7. Impact of the dual-module fusion on network performance. The left figure shows PSNR and SSIM under different configurations, while the right shows average noise prediction error at $t = 100, 250, 500$. The results indicate that dual-module fusion significantly improves highlight removal accuracy and diffusion stability.

4.4 Ablation Study

To assess the effectiveness of the proposed dual-module modality fusion, we compare four configurations: **(1) Baseline:** RGB-only diffusion model without fusion modules; **(2) Module-A only (RFUM):** using polarization information solely to update RGB features; **(3) Module-B only (PPLM):** using RGB features solely to guide polarization prompt learning; **(4) Full Model:** both modules applied at all encoder stages.

Figure 7 (left) shows PSNR/SSIM on SHIQ. The full model achieves the best performance (26.2/0.91), outperforming the baseline (22.3/0.78) by a large margin. Adding only Module-A or Module-B yields moderate gains, confirming their complementary effects. Figure 7 (right) reports average noise prediction error at timesteps $t = 100, 250, 500$, representing early, middle, and late diffusion stages. Errors increase over time for all models, but the full model consistently achieves the lowest values—especially at $t = 500$ (0.033 vs. 0.061 for baseline)—indicating greater stability in highlight reconstruction.

The gains stem from two factors: (*i*) polarization cues provide explicit structural priors early in the diffusion process, guiding convergence in the semantic space; and (*ii*) the dual-module design aligns and enhances cross-modal features via attention, preserving both complementarity and consistency across modalities.

5 Conclusion

In this paper, we propose a diffusion-based framework with dual modules—an RGB feature update module and a polarization information learning module—to bridge semantic and distributional gaps between modalities. Built on a pre-trained RGB diffusion model, polarization-derived guidance cues enhance structural modeling in highlight regions, with cross-modal attention enabling efficient feature coupling. Experiments show that our method surpasses classical and state-of-the-art approaches in highlight restoration and noise suppression, while ablation studies validate the role of each module in boosting generalization and robustness. Polarization imagery offers complementary, color-independent cues under complex lighting, effectively mitigating RGB limitations. Future work will investigate more advanced algorithms and improved training strategies.

References

1. Minaee, S., Boykov, Y., Porikli, F., Plaza, A., Kehtarnavaz, N., Terzopoulos, D.: Image segmentation using deep learning: a survey. IEEE Trans. Pattern Anal. Mach. Intell. **44**(7), 3523–3542 (2021)
2. Zhao, Z.Q., Zheng, P., Xu, S.T., Wu, X.: Object detection with deep learning: a review. IEEE Trans. Neural Netw. Learn. Syst. **30**(11), 3212–3232 (2019)
3. Jiddi, S., Robert, P., Marchand, E.: Detecting specular reflections and cast shadows to estimate reflectance and illumination of dynamic indoor scenes. IEEE Trans. Visual Comput. Graphics **28**(2), 1249–1260 (2020)
4. Yang, Q., Tang, J., Ahuja, N.: Efficient and robust specular highlight removal. IEEE Trans. Pattern Anal. Mach. Intell. **37**(6), 1304–1311 (2014)
5. Fu, G., Zhang, Q., Song, C., Lin, Q., Xiao, C.: Specular highlight removal for real-world images. In: Computer Graphics Forum, vol. 38, pp. 253–263. Wiley Online Library (2019)
6. Funke, I., Bodenstedt, S., Riediger, C., Weitz, J., Speidel, S.: Generative adversarial networks for specular highlight removal in endoscopic images. In: Medical Imaging 2018: Image-Guided Procedures, Robotic Interventions, and Modeling, vol. 10576, pp. 8–16. SPIE (2018)
7. Umeyama, S., Godin, G.: Separation of diffuse and specular components of surface reflection by use of polarization and statistical analysis of images. IEEE Trans. Pattern Anal. Mach. Intell. **26**(5), 639–647 (2004)
8. Shafer, S.A.: Using color to separate reflection components. Color. Res. Appl. **10**(4), 210–218 (1985)
9. Quan, L., Shum, H.Y., et al.: Highlight removal by illumination-constrained inpainting. In: Proceedings Ninth IEEE International Conference on Computer Vision, pp. 164–169. IEEE (2003)
10. Tan, R.T., Ikeuchi, K.: Separating reflection components of textured surfaces using a single image. IEEE Trans. Pattern Anal. Mach. Intell. **27**(2), 178–193 (2005)
11. Shen, H.L., Zheng, Z.H.: Real-time highlight removal using intensity ratio. Appl. Opt. **52**(19), 4483–4493 (2013)
12. Hu, G., Zheng, Y., Yan, H., Hua, G., Yan, Y.: Mask-guided cycle-GAN for specular highlight removal. Pattern Recogn. Lett. **161**, 108–114 (2022)
13. Dashpute, A., Anand, C., Sarkar, M.: Depth resolution enhancement in time-of-flight cameras using polarization state of the reflected light. IEEE Trans. Instrum. Meas. **68**(1), 160–168 (2018)
14. Lei, C., Huang, X., Zhang, M., Yan, Q., Sun, W., Chen, Q.: Polarized reflection removal with perfect alignment in the wild. In: Proceedings of the IEEE/CVF Conference on Computer Vision and Pattern Recognition, pp. 1750–1758 (2020)
15. Anwer, A., Ainouz, S., Saad, N.M., Ali, S.S.A., Meriaudeau, F.: Joint network for specular highlight detection and adversarial generation of specular-free images trained with polarimetric data. Neurocomputing **559**, 126769 (2023)

16. Verdié, Y., Song, J., Mas, B., Busam, B., Leonardis, A., McDonagh, S.: Cromo: cross-modal learning for monocular depth estimation. In: 2022 IEEE/CVF Conference on Computer Vision and Pattern Recognition (CVPR), pp. 3927–3937. IEEE (2022)
17. Ikemura, K., Huang, Y., Heide, F., Zhang, Z., Chen, Q., Lei, C.: Robust depth enhancement via polarization prompt fusion tuning. In: Proceedings of the IEEE/CVF Conference on Computer Vision and Pattern Recognition, pp. 20710–20720 (2024)
18. Ho, J., Jain, A., Abbeel, P.: Denoising diffusion probabilistic models. Adv. Neural. Inf. Process. Syst. **33**, 6840–6851 (2020)
19. Saharia, C., Ho, J., Chan, W., Salimans, T., Fleet, D.J., Norouzi, M.: Image super-resolution via iterative refinement. IEEE Trans. Pattern Anal. Mach. Intell. **45**(4), 4713–4726 (2022)
20. Lugmayr, A., Danelljan, M., Romero, A., Yu, F., Timofte, R., Van Gool, L.: Repaint: inpainting using denoising diffusion probabilistic models. In: Proceedings of the IEEE/CVF Conference on Computer Vision and Pattern Recognition, pp. 11461–11471 (2022)
21. Whang, J., Delbracio, M., Talebi, H., Saharia, C., Dimakis, A.G., Milanfar, P.: Deblurring via stochastic refinement. In: Proceedings of the IEEE/CVF Conference on Computer Vision and Pattern Recognition, pp. 16293–16303 (2022)
22. Yao, M., Wang, M., Tam, K.M., Li, L., Xue, T., Gu, J.: Polarfree: polarization-based reflection-free imaging. In: Proceedings of the Computer Vision and Pattern Recognition Conference, pp. 10890–10899 (2025)
23. Ye, H., Xu, D.T.: Spatial-channel multi-task prompting for dense scene understanding. In: Proceedings of the Eleventh International Conference on Learning Representations, Kigali, Rwanda, pp. 1–5 (2023)
24. Wu, Z., et al.: Single-image specular highlight removal via real-world dataset construction. IEEE Trans. Multimedia **24**, 3782–3793 (2021)
25. Fu, G., Zhang, Q., Zhu, L., Li, P., Xiao, C.: A multi-task network for joint specular highlight detection and removal. In: Proceedings of the IEEE/CVF Conference on Computer Vision and Pattern Recognition, pp. 7752–7761 (2021)
26. Akashi, Y., Okatani, T.: Separation of reflection components by sparse non-negative matrix factorization. In: Asian Conference on Computer Vision, pp. 611–625. Springer (2014)
27. Yamamoto, T., Nakazawa, A.: General improvement method of specular component separation using high-emphasis filter and similarity function. ITE Trans. Media Technol. Appl. **7**(2), 92–102 (2019)
28. Wu, Z., Guo, J., Zhuang, C., Xiao, J., Yan, D.M., Zhang, X.: Joint specular highlight detection and removal in single images via unet-transformer. Comput. Visual Media **9**(1), 141–154 (2023)
29. Yang, Q., Wang, S., Ahuja, N.: Real-time specular highlight removal using bilateral filtering. In: European Conference on Computer Vision, pp. 87–100. Springer (2010)
30. Guo, J., Zhou, Z., Wang, L.: Single image highlight removal with a sparse and low-rank reflection model. In: Proceedings of the European Conference on Computer Vision (ECCV), pp. 268–283 (2018)
31. Shi, J., Dong, Y., Su, H., Yu, S.X.: Learning non-lambertian object intrinsics across shapenet categories. In: Proceedings of the IEEE Conference on Computer Vision and Pattern Recognition, pp. 1685–1694 (2017)

32. Yuan, S., Li, L., Chen, H., Li, X.: Surface defect detection of highly reflective leather based on dual-mask-guided deep-learning model. IEEE Trans. Instrum. Meas. **72**, 1–13 (2023)
33. Muhammad, S., Dailey, M.N., Farooq, M., Majeed, M.F., Ekpanyapong, M.: Spec-net and spec-cgan: deep learning models for specularity removal from faces. Image Vis. Comput. **93**, 103823 (2020)

NPFML: Non-isotropic Potential Fields with Hierarchical Decay for Deep Metric Learning

Shuai Li[1], Xin Yuan[1,2(✉)], Minshi Chen[1], Yi Yin[1,2], and Xin Xu[1,2]

[1] School of Computer Science and Technology, Wuhan University of Science and Technology, Wuhan 430065, China

[2] Hubei Province Key Laboratory of Intelligent Information Processing and Real-Time Industrial System, Wuhan University of Science and Technology, Wuhan 430065, China

yuanxincherry@gmail.com

Abstract. Deep metric learning (DML) is designed to learn effective sample similarity metrics, which can improve the performance of various tasks, such as image retrieval and face recognition. In prior research, traditional DML methods predominantly rely on isotropic metrics, assuming uniform data distribution in all directions. However, in real-world data, this assumption often fails to hold, as data typically exhibit complex directional dependencies and manifold structures. To overcome these limitations, we propose NPFML: a **N**on-isotropic **P**otential **F**ield based deep **M**etric **L**earning framework with Hierarchical Decay. The core of the NPFML lies in constructing a direction-sensitive model of attractive and repulsive forces. Unlike traditional isotropic metrics, this potential field model can differentiate sample relationships in different directions based on directional information of the data, thereby guiding the neural network to learn an embedding space with directional selectivity. We have conducted comprehensive evaluations on multiple widely used benchmark datasets. The experimental results clearly demonstrate that the proposed NPFML effectively maintains intra-class compactness while significantly enhancing inter-class separability.

Keywords: Deep metric learning · Image retrieval · Loss function

1 Introduction

The core objective of Deep Metric Learning (DML) is to learn a nonlinear embedding space via Deep Neural Networks (DNN), so that semantically similar samples are closer together and semantically dissimilar samples are pushed further apart. The metric of semantic similarity plays a crucial role in fields such as image and video retrieval [2,9,25], and object re-identification [3,19,27]. By utilizing loss functions, DML can effectively uncover the intrinsic semantic associations of data, thereby significantly enhancing the core performance metrics of visual

J. Lokoč et al. (Eds.): MMM 2026, LNCS 16413, pp. 149–163, 2026.
https://doi.org/10.1007/978-981-95-6957-1_11

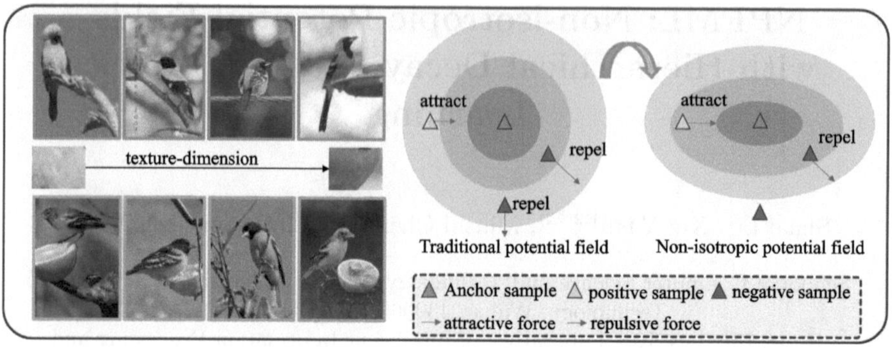

Fig. 1. Illustration of the limitations of the potential field framework in traditional DML. Compared to the second type of bird, the first type of bird has a broader distribution in the texture and color dimensions. In the potential field framework, the distance metric treats all feature dimensions equally, failing to highlight dimensions with discriminative power. Therefore, we assign adaptive weights to each dimension to construct a potential field framework with different attenuation structures for each dimension.

tasks. Existing loss functions can be broadly categorized into two paradigms: pair-based and proxy-based methods.

Pair-based methods optimize the structure of the embedding space by constructing tuples containing two or more category samples and exploring the relative distances between samples within the tuples. The core idea of the early tuple designs lies in directly optimizing distance relations between samples to achieve the learning objective of "pulling positive samples of the same class closer and pushing negative samples of different classes away." This idea gave rise to classic methods such as contrastive loss [5] and triplet loss [6,19]. However, it is difficult to obtain effective gradient update signals from these subsets during training, which leads to slow model convergence and makes it easy to fall into suboptimal solutions. To provide useful training signals, DML methods have emerged with the following strategies: 1) hard sample mining within batches, such as N-pair [20], Lifted loss [16], 2) synthesis of difficult samples [4,13,26,29], and 3) construction of cross-batch memory banks, such as XBM [24]. Furthermore, to compensate for the limited global information in the embedding space caused by mini-batch training, researchers assign one or more learnable proxies as class representatives for each category.

Proxy-based methods aim to learn class-discriminative feature representations by introducing a learnable proxy for each class. By constructing learnable proxies related to categories and comparing samples directly with these proxies rather than pairwise matching with other samples, these methods [8,15,22,30] effectively mitigate the issue of insufficient tuple information during sampling. However, since the positions of the proxies are learned from the samples, the proxies essentially lag behind the centers of the true sample clusters.

To address these issues, researchers have proposed a metric learning approach based on potential methods [1]. By simulating the effects of electric charges in physics for samples within a mini-batch, this method endows samples with attractive and repulsive fields. It establishes an optimal global potential field position to constrain the locations of the samples. However, traditional methods generally rely on isotropic metrics [18] such as Euclidean distance, which essentially assume that all dimensions in the feature space contribute equally to the potential field. In other words, the potential field method implicitly assumes the manifold structure [11] around each sample: the local neighborhood of each sample is uniform, which means that the repulsive or attractive potential field decays at an equal rate in every dimension. Nevertheless, this assumption is relatively idealized. As shown in Fig. 1, real-world data often exhibit complex manifold structures, with significant variations in the changes of features in different dimensions. For instance, in image data, some dimensions may correspond to texture features, while others may correspond to shape features. The magnitudes and significance of these feature changes can vary across dimensions. Therefore, it is unreasonable to assume that all dimensions contribute equally to the potential field decay.

To address limitations of existing methods, we propose a non-isotropic potential field-based deep metric learning framework with hierarchical decay (NPFML). This framework introduces a dual-branch module and a multistage hierarchical potential field, effectively overcoming the limitations of traditional methods.

- **Dual-branch Module**: A non-uniform Euclidean field is designed by assigning new weights to each dimension of the Euclidean distance. This method can better capture the variations of data in different dimensions, thereby enhancing the discriminative power of the embedding space. Specifically, the dual-branch module consists of a sample embedding branch and a structure embedding branch. The sample embedding branch is responsible for learning the semantic embeddings of samples, while the structure embedding branch focuses on learning the local neighborhood structure of samples.
- **Multi-stage Hierarchical Potential Field:** Building upon the existing potential field, a multi-stage hierarchical potential field is introduced to further enhance the discriminative power of the embedding space. Through the multi-stage hierarchical potential field, we can more precisely control the distances between samples, thereby better handling data with complex structures.

2 Proposed Method

2.1 Problem Statement

Given a training set $\mathbf{D}_{train} = \{(\mathbf{x}_i, y_i)\}_{i=1}^{N}$ consisting of N samples from class C, where $\mathbf{x}_i \in \mathcal{X}$ denotes the i-th input sample and $y_i \in \mathcal{Y}$ represents its corresponding label from the label space $\mathcal{Y} \in \{1, \ldots, C\}$. DML pipelines map the input $\mathbf{x}_i$ to a high-dimensional embedding through the function $\mathbf{v}_i = \mathbf{f}_\theta(\mathbf{x}_i)$, where $\mathbf{v}_i \in \mathbf{R}^d$ denotes the d-dimensional embedding vector and θ represents the parameters of the neural network. The distance between the embeddings $\mathbf{v}_i$ and $\mathbf{v}_j$ is defined as $\mathbf{d}(\mathbf{v}_i, \mathbf{v}_j)$, such as euclidean distance or cosine similarity. The goal is to ensure that $\mathbf{d}(\mathbf{v}_i, \mathbf{v}_j) < \mathbf{d}(\mathbf{v}_i, \mathbf{v}_k)$ when $\mathbf{y}_i = \mathbf{y}_j$ and $\mathbf{y}_i \neq \mathbf{y}_k$.

2.2 Review of Potential Field Metric Learning

To achieve this goal, researchers introduced a potential field [1] into DML, optimizing potential energy to simulate the attraction and repulsion interactions between samples, thereby guiding the embedding distribution. Specifically, for samples from each class, there are sets of similar and dissimilar points. These points are constrained by the entire set, rather than being constrained by sample selection as in the traditional pair-based method [6,16,20].

Formally, for a given embedding $\mathbf{v}_i$, the attractive potential field ϕ_{attr} and repulsive potential field ϕ_{repu} at a distance $\mathbf{r}$ from $\mathbf{v}_i$ are defined as:

$$\phi_{\text{attr}}(\mathbf{r}) = \begin{cases} -\frac{1}{\delta^\alpha} & \text{if } \mathbf{r} < \delta, \\ -\frac{1}{\mathbf{r}^\alpha} & \text{otherwise.} \end{cases} \tag{1}$$

$$\phi_{\text{repu}}(\mathbf{r}) = \begin{cases} \frac{1}{\mathbf{r}^\alpha} & \text{if } \mathbf{r} < \delta, \\ \frac{1}{\delta^\alpha} & \text{otherwise.} \end{cases} \tag{2}$$

Here, $\mathbf{r}$ represents the distance between $\mathbf{v}_i$ and another embedding $\mathbf{v}_j$, calculated as $\mathbf{r} = \|\mathbf{v}_i - \mathbf{v}_j\|_2$. The parameter δ is a predefined threshold distance, and α controls the decay rate of the potential field. Essentially, for a certain positive sample, gradient optimization is no longer provided for other samples within a certain distance range.

To represent the unsampled image embeddings in the current batch $\mathcal{B}$, proxies $\mathbf{p}_i^k$ are introduced to model the global potential field, where $i \in \{1, \ldots, C\}$ indexes the class and $k \in \{1, \ldots, M\}$ indexes the proxy number within each class. Given a training batch $\mathcal{B}$, the overall potential field is calculated to find its lowest landscape. The formula for the total potential field energy Φ for current batch $\mathcal{B}$ can be expressed as:

$$\Phi = \Phi_{\text{attr}}(\mathbf{r}) + \Phi_{\text{repu}}(\mathbf{r}) \tag{3}$$

where

$$\Phi_{\text{attr}} = \sum_{i=1}^{\mathcal{B}} \left(\sum_{\substack{j=1 \\ y_i = y_j}}^{\mathcal{B}} \phi_{\text{attr}}(r_{ij}) + \sum_{\substack{c=1 \\ y_i = y_{\mathbf{p}_c^k}}}^{C} \sum_{k=1}^{M} \phi_{\text{attr}}(r_{i\mathbf{p}_c^k}) \right), \tag{4}$$

$$\Phi_{\text{repu}} = \sum_{i=1}^{\mathcal{B}} \left(\sum_{\substack{j=1 \\ y_i \neq y_j}}^{\mathcal{B}} \phi_{\text{repu}}(r_{ij}) + \sum_{\substack{c=1 \\ y_i \neq y_{\mathbf{p}_c^k}}}^{C} \sum_{k=1}^{M} \phi_{\text{repu}}(r_{i\mathbf{p}_c^k}) \right), \tag{5}$$

Here, $\mathbf{r}_{ij} = \|\mathbf{v}_i - \mathbf{v}_j\|_2$ denotes the euclidean distance between embeddings $\mathbf{v}_i$ and $\mathbf{v}_j$, and $r_{i\mathbf{p}_c^k} = \|\mathbf{v}_i - \mathbf{p}_c^k\|_2$ denotes the euclidean distance between embedding $\mathbf{v}_i$ and proxy $\mathbf{p}_c^k$.

2.3 Anisotropic Potential Field Model Construction

The traditional potential field model often relies on isotropic metrics [18], such as Euclidean distance, which essentially assume that all dimensions in the feature space contribute equally to the distance. This assumption leads to a decay field structure of high-dimensional hyperspheres. However, this assumption often does not hold in real-world data [11]. Real-world data typically exhibits complex manifold structures, with significant variations in the changes of features across different dimensions. This contradiction between the isotropic assumption and the true structure of the data leads to the dilution of the repulsive and attractive boundaries of the traditional potential field in critical dimensions. To address this issue, we propose a dual-branch module design that incorporates Structure Embedding Learning [11]. This design enables the feature encoder to dynamically adjust the scale weights of each dimension based on the local data structure, ultimately generating an anisotropic potential field that conforms to the manifold

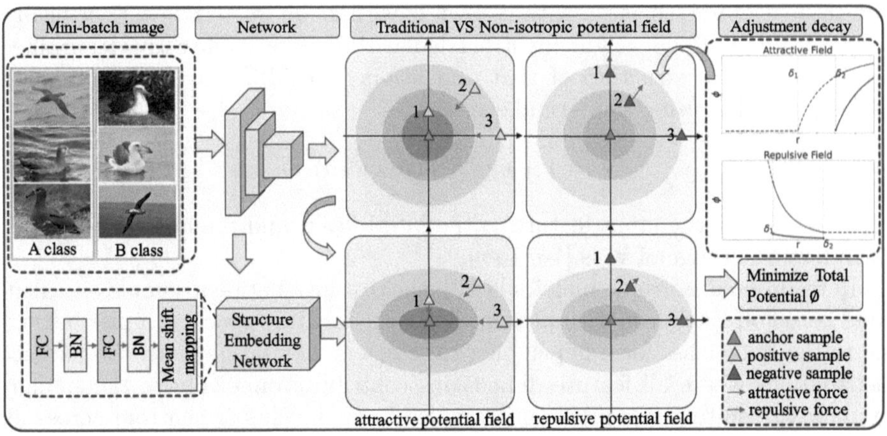

Fig. 2. The overview of the pipeline for non-isotropic potential field based deep metric learning with decay. The traditional and non-isotropic potential fields differ in optimization strength for the same samples. Staged adjustment of potential field decay controls optimization at different stages. Red: original potential field; Blue: new potential field; Green: common parts. (Color figure online)

characteristics of the data. Specifically, as shown in the framework in Fig. 2, the dual-branch module consists of two main components: the sample embedding branch and the structure embedding branch.

Sample Embedding Branch: This branch learns the semantic embeddings of the samples. Given an input sample $\mathbf{x}_i$, the sample embedding $\mathbf{v}_i$ is obtained by $\mathbf{x}_i$ passing through a backbone network such as ResNet or BNInception.

$$\mathbf{v}_i = \mathbf{f}_\theta(\mathbf{x}_i) \tag{6}$$

Structure Embedding Branch: The structure embedding branch is specifically designed to capture the local neighborhood structure of samples. By dynamically adjusting the scale weights of each dimension according to the local data distribution, this branch enables the model to generate a potential field that more accurately reflects the intrinsic structure of the data.

The structure embedding network, denoted as $\mathbf{g}_\theta$, consists of two fully connected layers, two batch normalization (BN) layers, and a Mean Shift Mapping (MSM) layer. Its core function is to learn the local structural patterns of the embedding $\mathbf{v}_i$, and output corresponding scale weights:

$$\mathbf{w}_i = \mathbf{g}_\Theta(\mathbf{v}_i) \tag{7}$$

where Θ represents the learnable parameters of the network, and the output $\mathbf{w}_i$ corresponds to the scale weights for each dimension of the embedding. In order to achieve adaptive weight learning for each dimension, we adopted the MSM layer, which limits the range of values for each weight.

The MSM layer is designed to enable the weight adjustment for each dimension to be deeply associated with its own local data characteristics. Specifically, the magnitude of the weight for a particular dimension is directly determined by the local data structure of that dimension. The MSM layer computes the adaptive weights using the formula:

$$\mathbf{w} = 1 + \eta \cdot (2 \cdot \text{Sigmoid(h)-1}) \tag{8}$$

where $\mathbf{h}$ denotes the input feature to the MSM layer, and the bias factor $\eta < 1$. Therefore, the range of $\mathbf{w}$ is $[1\text{-}\eta, 1+\eta]$.

In traditional isotropic metrics [8,22], the distance between samples is calculated based on Euclidean distance. In the learned feature space, samples belonging to the same class form hyperspherical clusters, exhibiting isotropic distance distributions where all feature dimensions contribute uniformly to the similarity metric, resulting in equidistant samples from the cluster centroid across all dimensions In this case, a fixed threshold $\mathbf{m}$ can clearly define the boundary. However, in weighted distance designs, when the weights of different dimensions vary significantly (some dimensions have weights much greater than 1, while others are much less than 1), the sample distribution can be distorted from a hypersphere to a hyperellipsoid. Dimensions with large weights amplify the distance contribution, "stretching" the boundary outward; dimensions with small weights compress the contribution, "contracting" the boundary inward. In this scenario, the discriminative logic relying solely on the threshold $\mathbf{m}$ becomes entirely ineffective. Thus, in this design, the weights $\mathbf{w}$ generated by the MSM layer represent projection transformations of the original Euclidean space along each dimension. The projected values in the transformed space are given by:

$$\nu_i = \mathbf{w}_i \odot \mathbf{v}_i \tag{9}$$

It is precisely through this weighted form of bounded scaling across dimensions that the boundaries of each dimension in the new embedding space can be effectively constrained by a unified threshold $\mathbf{m}$. Formally, for a given embedding ν_i, the transformed attractive potential field ϕ_{attr} at a transformed distance $\hat{r}$ is defined as:

$$\hat{\phi}_{\text{attr}}(\hat{\mathbf{r}}) = \begin{cases} -\frac{1}{\delta^\alpha} & \text{if } \hat{\mathbf{r}} < \delta, \\ -\frac{1}{\hat{\mathbf{r}}^\alpha} & \text{otherwise.} \end{cases} \tag{10}$$

Here, $\hat{\mathbf{r}}$ represents the transformed distance between ν_i and another embedding ν_j, calculated as $\hat{\mathbf{r}} = \|\nu_i - \nu_j\|_2$. The parameter δ is a predefined threshold distance, and α controls the decay rate of the potential field. The repulsive field is similar to the above structure. Then, our optimisation goals are as follows:

$$\hat{\Phi} = \hat{\Phi}_{\text{attr}}(\mathbf{r}) + \hat{\Phi}_{\text{repu}}(\mathbf{r}) \tag{11}$$

2.4 Multi-stage Hierarchical Potential Field

To further enhance the discriminative power of the embedding space, we introduce a multi-stage hierarchical potential field. Specifically, in the early training stages, samples have not yet formed compact intra-class clusters. Positive samples within the core range often exhibit weak constraints due to initialization differences or insufficient feature representation. Therefore, we expand the threshold range of the attractive potential field to enhance gradient signal constraints and facilitate clustering. Similarly, within the repulsive range, we apply a "weak repulsive force" to negative samples at medium and short distances to prevent inter-class embeddings from diffusing excessively outward. By refining the potential field in stages, we can more precisely control the attractive and repulsive forces between samples, thereby improving the discriminative power of the embedding space. Formally, for a given embedding ν_i, the transformed attractive potential field ϕ_{attr} at a transformed distance $\hat{r}$ is defined as:

$$\phi_{\text{attr}}(\mathbf{r}) = \begin{cases} -\frac{1}{\delta^\alpha} & \text{if } \mathbf{r} < \delta_{attr_1}, \\ -\frac{1}{\mathbf{r}^\alpha} & \text{else if } \delta_{attr_1} < \mathbf{r} < \delta_{attr_2}, \\ -\frac{1}{\mathbf{r}^{\alpha+0.5}} & \text{otherwise.} \end{cases} \tag{12}$$

Similarly, the repulsive potential field ϕ_{repu} at a distance $\mathbf{r}$ from $\mathbf{v}_i$ is defined as:

$$\phi_{\mathrm{repu}}(\mathbf{r}) = \begin{cases} \frac{1}{\mathbf{r}^{\alpha}} & \text{if } \mathbf{r} < \delta_{repu_1}, \\ \frac{1}{\mathbf{r}^{\alpha-0.5}} & \text{else if } \delta_{repu_1} < \mathbf{r} < \delta_{repu_2}, \\ \frac{1}{\delta^{\alpha}} & \text{otherwise.} \end{cases} \tag{13}$$

By refining the attractive and repulsive potential fields in stages, we can adjust the constraints between samples as needed. This method ensures that inter-class samples are moderately constrained, thereby avoiding excessive separation. Through optimized sample constraints, the discriminative power of the embedding space is ultimately enhanced.

Then, our optimization goals are as follows:

$$\Phi = \Phi_{\mathrm{attr}}(\mathbf{r}) + \Phi_{\mathrm{repu}}(\mathbf{r}) \tag{14}$$

Finally, we optimise the potential field energy function as the final loss function, which is defined as follows:

$$\mathcal{L} = (1 - \lambda)\Phi + \lambda\hat{\Phi} \tag{15}$$

3 Experiment

3.1 Dataset

Following established setting in [1], our comparative analysis employs three standard benchmarks: the CUB-200-2011 (CUB200) [23], CARS196 [10], and Stanford Online Products (SOP) [16].

- The CUB200 dataset used 5,864 images from the first 100 categories for training and 5,924 images from the last 100 categories for testing.
- The CARS196 dataset contains 16,185 images across 196 car categories. We used 8,054 images from the first 98 categories for training and 8,131 images from the last 98 categories for testing.
- We also tested our method on the large-scale dataset. The SOP dataset includes 120,053 images and 22,634 product categories.

For dataset division, we adopt the same standard settings as those employed in existing DML methods. To evaluate performance, our method utilizes recall@K and normalized mutual information (NMI) as evaluation metrics. Specifically, for CUB200 and CARS196, we evaluate recall@K at K=1, 2, 4, 8; for SOP, we evaluate it at K=1, 10, 100, 1000. Additionally, we use normalized mutual information (NMI) to evaluate clustering performance.

3.2 Implementation Details

To compare our method with existing DML methods, we conducted experiments on image retrieval and clustering tasks following the most commonly used experimental settings in the field. To ensure the fairness of comparisons, we first reproduced the performance of PMFL [1] using the parameter settings provided in their work, which serves as a critical baseline for subsequent comparative analyses. As illustrated in Fig. 2, the core architecture of our dual-branch model comprises two principal components. The sample embedding branch utilizes a ResNet-50 network as its feature extraction backbone, focusing on extracting discriminative features from input samples. On the other hand, the structure embedding branch is primarily responsible for weighting different dimensions, which helps in emphasizing the structural information within the data. The ResNet-50 backbone is initialized with weights pre-trained on the ImageNet-1K dataset. The output embedding dimension for our method is set to 512. All input image size of 224 $\times$ 224 is used for all experiments and comparisons. We train the backbone for 200 epochs on each dataset. We use the Adam optimizer with the learning rate chosen as 5e-4. Additionally, following the methods in [8,22], we scale up the learning rate of our proxies by 100 times. In our method, we set the number of proxies M=30 for the CUB200 and Cars196 datasets, while M=3 is used for the SOP dataset. The parameters η and α are set to 0.5 and 3, respectively. The segment boundaries for the attractive potential field are set to 0.2 and 0.3, while the segment boundaries for the repulsive potential field are set to 0.1 and 0.2. All experiments were performed on a single NVIDIA 2080Ti.

3.3 Analysis of Experimental Results

To validate the effectiveness of our proposed method, we conducted comparative experiments with state-of-the-art metric learning methods from recent work, including PFML [1], Anti-Collapse [7], and HIST [12]. The results are shown in

Table 1. Recall@K (%) Comparison on CUB200, Cars196, and SOP under Standard Settings. * indicates our reproduced results. BN for Inception with batch normalization, and R for ResNet50.

Method	Arch/Dim	CUB				Cars				SOP			
		R@1	R@2	R@4	R@8	R@1	R@2	R@4	R@8	R@1	R@10	R@10^2	R@10^3
Proxy-NCA++ [22]	R512	69.0	79.8	87.3	92.7	86.5	92.5	95.7	97.7	80.7	92.0	96.7	98.9
NSoftmax [28]	R2048	65.3	76.7	85.4	91.8	89.3	94.1	96.4	98.0	79.5	91.5	96.7	-
PA [8]	R512	69.7	80.0	87.0	92.2	89.0	93.5	96.2	97.9	79.7	91.1	96.2	98.7
MS+DAS [13]	R512	69.2	79.2	87.1	92.6	87.8	93.1	96.0	97.8	80.5	91.8	96.7	98.9
HIST [12]	R512	71.4	81.1	81.1	-	89.6	93.9	96.4	-	**81.4**	**92.0**	96.7	-
DIVA [14]	BN512	66.4	77.2	-	-	83.1	90.0	-	-	77.5	90.3	-	-
SoftTriplet [17]	BN512	65.4	76.4	84.5	90.4	84.5	90.7	94.5	96.9	78.3	90.3	95.9	-
CircleLoss [21]	R512	66.7	77.4	86.2	91.2	83.4	89.8	94.1	96.5	78.3	90.5	96.1	98.6
Anti-Collapse [7]	R512	71.7	81.2	-	-	90.5	94.6	-	-	81.0	91.6	-	-
PA+NIR [18]	R512	70.5	80.6	-	-	89.1	93.4	-	-	80.4	91.4	-	-
PFML [1]*	R512	70.6	80.6	87.5	92.5	89.2	93.7	96.2	97.8	79.8	91.2	96.5	98.6
NPFML	R512	**72.3**	**81.5**	**88.6**	**93.0**	**90.8**	**95.0**	**97.4**	**98.5**	80.6	91.4	**96.7**	**98.9**

Tables 1 and 2. To ensure fairness, we strictly followed the hyperparameter configurations of the official PFML implementation for reproduction. Specifically, NPFML achieves a performance of 72.3% on CUB200 and 90.8% on CARS196, but underperforms compared to HIST on SOP. However, the NMI on the CUB and SOP datasets is lower than that of the HIST method. By incorporating an additional hypergraph neural network, HIST achieves targeted advantages in feature optimization, leading to improved NMI performance. The results show that constructing adaptive weights for each dimension can enhance the discriminative power of the potential field model. In contrast, our method solely adopts a lightweight SEN(Structure Embedding Network) module, which dynamically adjusts the scale weights of each dimension based on the local data structure to enhance feature discriminability. Nevertheless, our core insight of dynamic dimension-wise scale weight adjustment based on local data structure can be extended to other traditional Euclidean distance-based metric learning methods.

Table 2. NMI metric comparison on CUB200, Cars196, and SOP datasets.

Method	Arch/Dim	CUB	Cars	SOP
		NMI	NMI	NMI
Proxy-NCA++ [22] ECCV2022	R512	73.9	73.8	-
MS+DAS [13] ECCV2022	R512	69.1	70.8	90.4
HIST [12] CVPR2022	R512	**74.1**	75.2	**92.8**
DIVA [14] ECCV2020	BN512	69.6	68.1	90.1
SoftTriplet [17] CVPR2019	BN512	69.3	70.1	-
Anti-Collapse [7] TMM2024	R512	73.2	75.6	91.6
PA+NIR [18] CVPR2022	R512	72.5	75.0	90.6
NPFML	R512	73.3	**77.5**	91.6

3.4 Ablation Studies

To further demonstrate the contribution of each design in our NPFML method, we conducted ablation studies and present the results in Table 3 to quantify their respective impacts. Experimental results demonstrate that both the SEN and the staged potential field with decay mechanism contribute to performance improvement independently. Specifically, integrating the enhanced Recall@K by concentrating the distance metric on discriminative feature dimensions. Meanwhile, incorporating the staged potential field further boosts performance by strengthening the attraction of semi-hard positive samples and mitigating excessive repulsion from negative samples. Additionally, during training, we estimate the within-class and between-class distance distributions, and set threshold distances for the attraction and repulsion fields based on these statistics.

Table 3. Ablation study results on CUB200 and Cars196 datasets

Method	①	②	CUB				Cars			
			R@1	R@2	R@4	R@8	R@1	R@2	R@4	R@8
Baseline	×	×	70.6	80.6	87.5	92.5	89.2	93.7	96.2	97.8
Ablation (Non-Euclidean Only)	✓	×	71.9	81.4	88.5	92.9	90.5	94.6	96.9	98.0
Ablation (Multi-level Only)	×	✓	70.9	80.8	87.9	92.6	89.3	94.0	96.4	98.1
Ours	✓	✓	**72.3**	**81.5**	**88.6**	**93.0**	**90.8**	**95.0**	**97.4**	**98.5**

①: Non-Euclidean ; ②: Multi-level ; ✓: Module Included; ×: Module Excluded

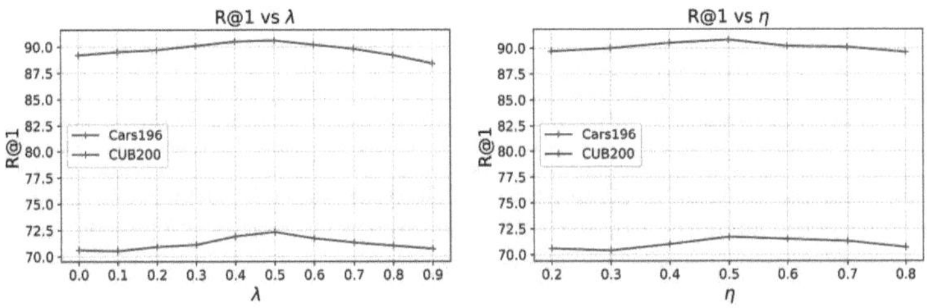

(a) R@1 VS Hyperparameter λ (b) R@1 VS Hyperparameter η

Fig. 3. Variation in Recall@1 with hyperparameter λ, η on the Cars196 and CUB200 datasets.

Effect of λ. To further validate the model's parameter sensitivity, we conducted hyperparameter experiments. Results are in Fig. 3(a). When λ is set to 0.5, the model achieves the best performance on both CUB200 and Cars196, regardless of whether segmented potential fields are applied. Essentially, our method emphasizes discriminative dimensions and suppresses redundant ones to fit the true manifold structure of the data. However, over-reliance on this mechanism causes uncontrolled dimensional adjustment, preventing the model from balancing intra-class compactness and inter-class separation and ultimately degrading performance.

Effect of η. Results are shown in Fig. 3(b). We verified the results of η in the range [0.2, 0.8] for the MSM layer. When η is set to 0.5, the model achieves the best performance.

Visual Results. As shown in Fig. 4, the T-SNE visualization results illustrate the distribution of learned feature embeddings, providing insights into how our method clusters semantically similar samples. Additionally, the top-K retrieval results in Fig. 5 demonstrate that our method achieves strong performance on both CUB200 and CARS196 datasets. Our method achieves top-ranked low-distance same-class samples (green) and lower-ranked larger-distance different-class samples (red), showcasing robust fine-grained retrieval.

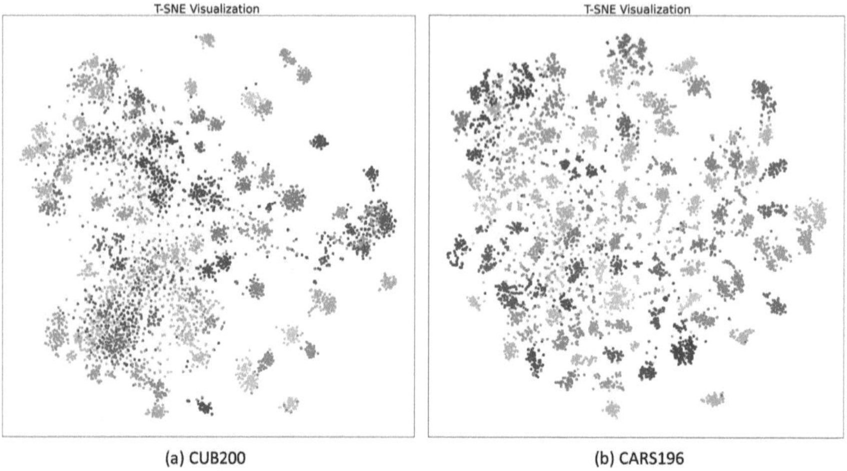

Fig. 4. The T-SNE visualization depicts the feature distributions of the CUB200 and CARS196 test sets, with results for CUB200 on the left and CARS196 on the right. Distinct classes are represented by different colors.

Fig. 5. Top-k retrieval results on CUB200 (first two rows) and CARS196 (last two rows). Query images are black-framed. Same-class images are green-highlighted, and different-class ones are red-highlighted. (Color figure online)

4 Conclusion and Limitations

In this work, we have introduced a novel deep metric learning framework, NPFML, which significantly enhances the discriminative power of the embedding space through a dual-branch module and a multi-stage hierarchical potential field. The dual-branch module assigns adaptive weights to each dimension, capturing variations in data across different dimensions, while the multi-stage hierarchical potential field provides fine-grained control over sample distances, effectively handling complex data structures.

Although NPFML performs strongly, there are still a few notable limitations. First, the staged potential field and structural embedding network introduce additional computational overhead, with our training time requiring only approximately 8% more than the baseline. Second, while our method effectively emphasizes discriminative feature dimensions, it relies on careful tuning of hyperparameters for the potential field decay distance. Finally, like PFML, NPFML depends on extensive category proxies for potential field modeling, requiring more time or hardware resources for large-scale training.

Acknowledgments. This work was supported by the Nature Science Foundation of Hubei Province (No. 2025AFB056), the Key Project of the Science Research Plan of the Hubei Provincial Department of Education (No. D20241103), the Hubei Province Key Laboratory of Intelligent Information Processing and Real-time Industrial System (Wuhan University of Science and Technology) (No. ZNXX2023QNO3), the Fund of Hubei Key Laboratory of Inland Shipping Technology and Innovation (No. NHHY2023004), the Key Laboratory of Social Computing and Cognitive Intelligence (Dalian University of Technology), Ministry of Education (No. SCCI2024YB02), the Research Project of the Hubei Provincial Department of Science and Technology (No. 2024CSA075), and the Entrepreneurship Fund for Graduate Students of Wuhan University of Science and Technology (No. JCX2023049, JCX2023160).

References

1. Bhatnagar, S., Ahuja, N.: Potential field based deep metric learning. In: Proceedings of the IEEE/CVF Conference on Computer Vision and Pattern Recognition, pp. 25549–25559 (2025)
2. Brattoli, B., Tighe, J., Zhdanov, F., Perona, P., Chalupka, K.: Rethinking zero-shot video classification: end-to-end training for realistic applications. In: Proceedings of the IEEE/CVF Conference on Computer Vision and Pattern Recognition, pp. 4613–4623 (2020)
3. Deng, J., Guo, J., Xue, N., Zafeiriou, S.: Arcface: Additive angular margin loss for deep face recognition. In: Proceedings of the IEEE/CVF Conference on Computer Vision and Pattern Recognition, pp. 4690–4699 (2019)
4. Duan, Y., Zheng, W., Lin, X., Lu, J., Zhou, J.: Deep adversarial metric learning. In: Proceedings of the IEEE/CVF Conference on Computer Vision and Pattern Recognition, pp. 2780–2789 (2018)
5. Hadsell, R., Chopra, S., LeCun, Y.: Dimensionality reduction by learning an invariant mapping. In: Proceedings of the IEEE/CVF Conference on Computer Vision and Pattern Recognition. vol. 2, pp. 1735–1742. IEEE (2006)

6. Hoffer, E., Ailon, N.: Deep metric learning using triplet network. In: Similarity-Based Pattern Recognition: Third International Workshop, pp. 84–92. Springer (2015)
7. Jiang, X., et al.: Anti-collapse loss for deep metric learning. IEEE Trans. Multimedia (2024)
8. Kim, S., Kim, D., Cho, M., Kwak, S.: Proxy anchor loss for deep metric learning. In: Proceedings of the IEEE/CVF Conference on Computer Vision and Pattern Recognition, pp. 3238–3247 (2020)
9. Kim, S., Seo, M., Laptev, I., Cho, M., Kwak, S.: Deep metric learning beyond binary supervision. In: Proceedings of the IEEE/CVF Conference on Computer Vision and Pattern Recognition, pp. 2288–2297 (2019)
10. Krause, J., Stark, M., Deng, J., Fei-Fei, L.: 3D object representations for fine-grained categorization. In: Proceedings of the IEEE/CVF International Conference on Computer Vision, pp. 554–561 (2013)
11. Li, P., Li, Y., Xie, H., Zhang, L.: Neighborhood-adaptive structure augmented metric learning. In: Proceedings of the AAAI Conference on Artificial Intelligence, vol. 36, pp. 1367–1375 (2022)
12. Lim, J., Yun, S., Park, S., Choi, J.Y.: Hypergraph-induced semantic tuplet loss for deep metric learning. In: Proceedings of the IEEE/CVF Conference on Computer Vision and Pattern Recognition, pp. 212–222 (2022)
13. Liu, L., et al.: Das: Densely-anchored sampling for deep metric learning. In: European Conference on Computer Vision, pp. 399–417. Springer (2022)
14. Milbich, T., et al.: Diva: Diverse visual feature aggregation for deep metric learning. In: European Conference on Computer Vision, pp. 590–607. Springer (2020)
15. Movshovitz-Attias, Y., Toshev, A., Leung, T.K., Ioffe, S., Singh, S.: No fuss distance metric learning using proxies. In: Proceedings of the IEEE/CVF International Conference on Computer Vision, pp. 360–368 (2017)
16. Oh Song, H., Xiang, Y., Jegelka, S., Savarese, S.: Deep metric learning via lifted structured feature embedding. In: Proceedings of the IEEE/CVF Conference on Computer Vision and Pattern Recognition, pp. 4004–4012 (2016)
17. Qian, Q., et al.: Softtriple loss: deep metric learning without triplet sampling. In: Proceedings of the IEEE/CVF Conference on Computer Vision and Pattern Recognition, pp. 6450–6458 (2019)
18. Roth, K., Vinyals, O., Akata, Z.: Non-isotropy regularization for proxy-based deep metric learning. In: Proceedings of the IEEE/CVF Conference on Computer Vision and Pattern Recognition, pp. 7420–7430 (2022)
19. Schroff, F., Kalenichenko, D., Philbin, J.: Facenet: a unified embedding for face recognition and clustering. In: Proceedings of the IEEE/CVF Conference on Computer Vision and Pattern Recognition, pp. 815–823 (2015)
20. Sohn, K.: Improved deep metric learning with multi-class n-pair loss objective. Adv. Neural Inf. Process. Syst. **29** (2016)
21. Sun, Y., et al.: Circle loss: a unified perspective of pair similarity optimization. In: Proceedings of the IEEE/CVF Conference on Computer Vision and Pattern Recognition, pp. 6398–6407 (2020)
22. Teh, E.W., DeVries, T., Taylor, G.W.: Proxynca++: revisiting and revitalizing proxy neighborhood component analysis. In: European Conference on Computer Vision, pp. 448–464. Springer (2020)
23. Wah, C., Branson, S., Welinder, P., Perona, P., Belongie, S.: The caltech-ucsd birds-200-2011 dataset (2011)

24. Wang, X., Zhang, H., Huang, W., Scott, M.R.: Cross-batch memory for embedding learning. In: Proceedings of the IEEE/CVF Conference on Computer Vision and Pattern Recognition, pp. 6388–6397 (2020)
25. Xu, X., Yuan, X., Wang, Z., Zhang, K., Hu, R.: Rank-in-rank loss for person re-identification. ACM Trans. Multimed. Comput. Commun. Appl. **18**(2s), 1–21 (2022)
26. Yang, B., et al.: HSE: Hybrid species embedding for deep metric learning. In: Proceedings of the IEEE/CVF International Conference on Computer Vision, pp. 11047–11057 (2023)
27. Yuan, X., et al.: Searching parameterized retrieval and verification loss for re-identification. IEEE J. Sel. Top. Signal Process. **17**(3), 560–574 (2023)
28. Zhai, A., Wu, H.Y.: Classification is a strong baseline for deep metric learning. arXiv preprint arXiv:1811.12649 (2018)
29. Zheng, W., Chen, Z., Lu, J., Zhou, J.: Hardness-aware deep metric learning. In: Proceedings of the IEEE/CVF Conference on Computer Vision and Pattern Recognition, pp. 72–81 (2019)
30. Zhu, Y., Yang, M., Deng, C., Liu, W.: Fewer is more: a deep graph metric learning perspective using fewer proxies. Adv. Neural. Inf. Process. Syst. **33**, 17792–17803 (2020)

Robust Ensemble of GNNs with Adaptive Graph Structure Learning

Ruichao Ren, Yiqi Wang, Jiaxin Zhang, Wen Yin, Yong Guo, and Xiaoling Li(✉)

College of Computer Science and Technology, National University of Defense Technology, ChangSha, China
{renruichao23,lixiaoling}@nudt.edu.cn

Abstract. In recent years, Graph Neural Networks (GNNs) have emerged as powerful tools for analyzing graph-structured data, demonstrating remarkable success in various applications. However, most existing GNN models inherently assume that the input graph data is clean and accurate, which often does not hold in real-world scenarios due to the presence of noise and adversarial attack. These imperfections can significantly degrade the quality of the graph, leading to erroneous predictions and reduced classification accuracy. To address these challenges, it is crucial to develop robust algorithms capable of mitigating adversarial perturbations over graph structure. In this work, we propose a novel robust graph structure learning framework called Robust Adaptive Graph Neural Network (RA-GNN), which is designed to enhance the robustness of GNNs against structural perturbations. Specifically, we introduce a novel loss function considering the dynamic structural perturbations, and a learnable combined model consisting of a multi-layer perceptron and a GNN. RA-GNN adaptively recovers the unperturbed graph structure from noisy graphs with various perturbation rate during training. Extensive experiments demonstrate that our method achieves superior robustness across varying levels of noise intensity, effectively eliminating perturbations, and significantly improving node classification performance.

Keywords: Graph Neural Networks · Node Classification · Graph Robust Learning

1 Introduction

Graph Neural Networks (GNNs) have achieved notable success in the field of graph data processing, demonstrating promising application prospects [8,11,13,17,19]. However, the robustness of GNNs poses a significant challenge in practical applications, with many empirical observations indicating that model performance is highly susceptible to structural perturbations [6]. In response, extensive research has focused on enhancing GNN robustness against structural attacks [12,16,22].

J. Lokoč et al. (Eds.): MMM 2026, LNCS 16413, pp. 164–176, 2026.
https://doi.org/10.1007/978-981-95-6957-1_12

Under the aforementioned scenario, existing methods for enhancing the robustness of GNNs [2,9,15,20] usually set certain criteria to detect potentially perturbed edges, and then purify the graph structure by reducing the weights of these edges or directly removing them. These criteria often rely on predefined and static rules, such as measuring node similarity based on the number of common neighbors and node feature similarity to identify adversarial links, or employing random propagation strategies with fixed probabilities [20]. Most existing defense strategies use the same denoising constraint across varying levels of perturbation intensity, limiting their application in scenarios with significant changes in perturbation intensity and potentially leading to unnecessary purification operations [2]. However, in reality, structure-oriented attacks and perturbations are diverse and vary in intensity [15]. Predefined static rules are typically tailored to attacks of specific magnitudes, and their defense effectiveness may diminish when confronted with attacks that are either more severe or more subtle, as they lack the adaptability to dynamically adjust to the fluctuating nature of adversarial perturbations.

To address the aforementioned challenges, this paper proposes a Robust Adaptive Graph Neural Network (RA-GNN), a novel framework designed to enhance the robustness of GNNs against adversarial structural perturbations. As illustrated in Fig 1, RA-GNN incorporates two core mechanisms to achieve adaptive and reliable performance. First, it introduces a flexible norm constraint that dynamically adjusts the strength of structural purification based on the detected perturbation intensity. This allows the model to impose stricter constraints under mild perturbations and more relaxed ones under severe attacks, thereby optimizing the purification process across varying adversarial conditions. Second, recognizing the potential negative impact of unreliable graph structures on GNN performance, RA-GNN integrates an MLP branch that operates independently of graph topology. This branch is combined with the GNN module through a learnable coefficient, enabling joint feature learning and adaptive weighting between structure-dependent and structure-independent representations. Such a design ensures that the model can rely more on node features via the MLP when the graph structure is highly perturbed, while still leveraging purified structural information when it is credible. The main contributions of this work are summarized as follows:

- We propose RA-GNN, a robustness enhancement method that adaptively purifies graph structures under varying perturbation intensities through a flexible norm constraint.
- We design a dynamic fusion mechanism that combines GNN and MLP outputs using a learnable coefficient, enhancing prediction reliability by leveraging both structural and feature information.
- Extensive experiments under diverse attack scenarios demonstrate the superiority of RA-GNN in node classification tasks compared to existing methods.

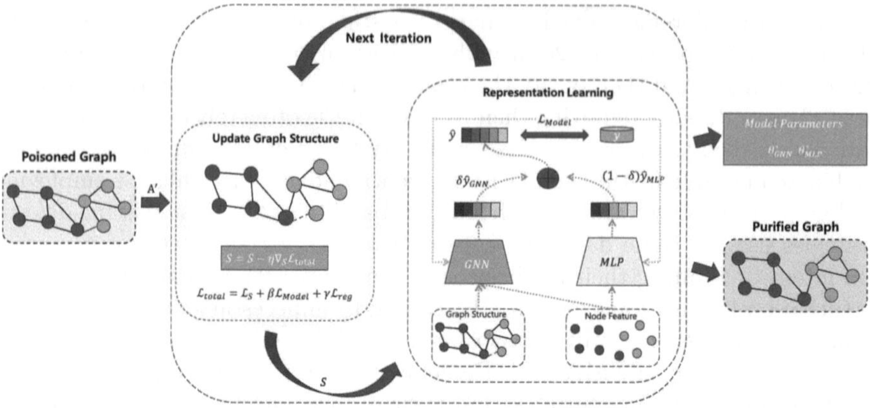

Fig. 1. Our model takes as input a poisoned graph A'. First,we jointly optimize a constraint loss and the classification loss to purify the structure; Next, a GNN and a MLP collaboratively extract node representations, and their outputs are fused via a learnable parameter δ.

2 Related Work

2.1 Adversarial Perturbations in Graph Data

Adversarial perturbations in graph-structured data, such as social networks, knowledge graphs, and molecular graphs, refer to the subtle and often imperceptible modifications made by attackers. These modifications can lead to incorrect predictions in downstream tasks such as node classification [10], link prediction [18], and graph classification [17]. Poisoning attacks, a primary form of adversarial attack, involve making minor perturbations to the training dataset to disrupt model training, thereby degrading the model's performance during inference.

2.2 Attack Methods

Nettack [21] was the first to propose a poisoning attack against Graph Neural Networks (GNNs). This method jointly optimizes minor perturbations in graph topology and node features to misclassify target nodes during training while preserving the overall statistical characteristics of the graph data. Nettack employs a greedy-discrete projection strategy to optimize the attack's effectiveness against a surrogate model. Building on Nettack's research, Metattack [22] innovatively introduced meta-learning to the graph poisoning scenario, addressing Nettack's limitation of targeting only a few nodes and its lack of scalability. Unlike Nettack, which targets specific nodes, Metattack aims at the entire graph. This method can significantly reduce the model's overall accuracy across the graph without accessing the target node labels by perturbing graph structure and node features, and its effectiveness is independent of the specific GNN architecture. To apply attacks in scenarios with stricter budget constraints, Dai et al. [3]

proposed UGBA (Unnoticeable Graph Backdoor Attack), a graph backdoor poisoning method under low-budget constraints. UGBA adaptively selects a few key nodes as poisoning anchors and then injects subgraph triggers with fake graph structures and labels into the training set, significantly reducing the scale of modifications required for the attack.

2.3 Defense Strategies

As adversarial attacks evolve, defense strategies against them are also being proposed. These defense strategies focus on different aspects to improve the robustness of GNNs. Adversarial training strategies typically involve modifications to the GNN training process. For example, AdvT [4] introduces a local regularization strategy into the adversarial training process, perturbing node feature vectors and encouraging the model to make similar predictions for perturbed and clean samples, thereby enhancing model robustness. Another approach to enhancing robustness is to purify the graph structure by reducing the weights of suspected perturbed edges or removing them entirely. GNN-Jaccard [14], based on the homogeneity assumption of graphs (i.e., nodes with similar features are more likely to be connected), calculates the Jaccard similarity of feature vectors between adjacent nodes before GNN training. It then removes edges from the adjacency matrix with similarity below a threshold, thus filtering out suspected perturbed edges. Pro-GNN [9] designs constraint terms based on the unique properties of clean graphs and reconstructs the adjacency matrix while training the GNN model. Additionally, some methods explore other defense strategies; beta-GNN [2] combines MLP with GNN models through a learnable parameter β to jointly predict perturbed graphs; RUNG [7] investigates the estimation bias brought by the l1 constraint in robustness enhancement and proposes an unbiased estimation aggregation function; Noisy-GNN [5] ensures the expected robustness of GNNs in the worst-case scenario by introducing random Gaussian noise sampled in the GNN hidden layers. Despite these defense strategies enhancing the robustness of GNNs, there are still issues in practical applications. For example, these methods are still not flexible enough to face attacks of varying perturbation intensities. Moreover, such robustness enhancement strategies can also easily cause unnecessary structural modifications, which can damage the original graph data structure.

3 Method

In this section, we first introduce some concepts and notations to define the problem and then describe the model we propose.

We apply the α-norm to the constraints, which more consistently enhances robustness and adaptively responds to perturbations of varying intensity owing to the flexibility of the α-norm Based on the homogeneous graph assumption, we construct a feature smoothing constraint term $\mathcal{L}_f$ to ensure that there is a higher probability of connections between nodes with similar feature vectors. In order to minimize the negative impact of unreliable graph structure information on model performance, RA-GNN uses an MLP-assisted GNN for feature learning, and dynamically adjusts the weights of GNN and MLP model outputs through a learnable coefficient δ.

3.1 Notation

Let $G = (V, E)$ be a graph, where $V = \{v_1, v_2, \ldots, v_N\}$ is the set of nodes in G and E is the set of edges in G. The graph structure can be represented in the form of an adjacency matrix, denoted as $A \in \mathbb{R}^{N \times N}$, where A_{ij} represents the connection between node v_i and node v_j. Additionally, $X = [x_1, x_2, \ldots, x_N]^T \in \mathbb{R}^{N \times d}$ is the node feature matrix, where X_i represents the feature vector of node v_i. The Frobenius norm of a matrix can be represented as the sum of the squares of all its elements, i.e., $\|S\|_F^2 = \sum_{i,j} S_{ij}^2$. The l_1 norm of a matrix is represented as $\|S\|_1 = \sum_{i,j} |S_{ij}|$, and the nuclear norm of a matrix is represented as the sum of its singular values, i.e., $\|S\|_* = \sum_{i=1}^{\text{rank}(S)} \sigma_i$, where σ_i is the i-th singular value of S.

For the node classification problem, only a subset of nodes V_L in graph $G = (E, V)$ have labels Y_L. The goal is to learn a classifier f_θ that can correctly map V_L to Y_L so that it can predict labels for unlabeled nodes. The optimization objective of f_θ can be expressed as:

$$\min_\theta \mathcal{L}_{\text{GNN}}(\theta, A, X, Y_L) = \sum_{v_i \in V_L} \ell(f_\theta(X, A)_i, y_i), \quad (1)$$

where θ are the parameters of the classifier f_θ, $f_\theta(X, A)_i$ is the predicted label for node v_i, and $\ell(\cdot, \cdot)$ is a loss function that calculates the difference between f_θ and the true labels.

3.2 The Proposed Framework

Jin et al. [9]'s research indicates that low-rank and sparse constraints can effectively remove adversarial edges while preserving normal edges when reconstructing the adjacency matrix. Additionally, considering the stealthiness of adversarial attacks, it is necessary to limit the difference between the reconstructed matrix S and the perturbed matrix A'. Existing methods generally use the Frobenius norm of the matrix to construct constraint terms. However, Xie et al. [15]'s research, using the Frobenius norm can lead to poor defense performance under high perturbation intensity because the square summation relationship makes it very sensitive to noise changes under high-intensity attacks, resulting in excessive constraint strength. Inspired by the α-norm proposed by Xie et al., defined as:

$$g(\alpha) = \|S_{i,\cdot} - S_{j,\cdot}\|_\alpha = \frac{(1+\alpha)\|S_{i,\cdot} - S_{j,\cdot}\|_2^2}{\|S_{i,\cdot} - S_{j,\cdot}\|_2 + \alpha}, \quad (2)$$

the α-norm behaves similarly to the Frobenius norm when α is large and similarly to a linear function when α is small. Therefore, when α is large, it constrains high-intensity perturbations too strongly and low-intensity perturbations too loosely, indicating that a smaller α should be chosen for constraint. To this end, we apply α-norm to the structural constraint term to achieve an appropriate regularization effect under different perturbation intensities.

RA-GNN uses the following constraint term when reconstructing the adjacency matrix S:

$$\arg\min_{S \in \mathcal{S}} L_0 = \|A - S\|_\alpha + \omega_1 \|S\|_1 + \omega_* \|S\|_*, \text{ s.t. } S = S^T, \tag{3}$$

where using the α-norm instead of the Frobenius norm can more stably enhance robustness and adaptively cope with varying-intensity perturbations. Additionally, we impose an l_1 norm on S to encourage sparsity and a nuclear norm l_* to encourage low-rank structure.

Existing research by Kipf & Wellin [10] has shown has shown that many real-world clean graphs are homogeneous. Therefore, we can construct a feature smoothing constraint term L_f to ensure that nodes with similar feature vectors have a higher probability of existing links, defined as:

$$\mathcal{L}_f = \frac{1}{2} \sum_{i,j=1}^{N} S_{ij} (\mathbf{x}_i - \mathbf{x}_j)^2, \tag{4}$$

which is equivalent to:

$$\mathcal{L}_f = \text{tr}(\mathbf{X}^\top \mathbf{L} \mathbf{X}), \tag{5}$$

where L represents the Laplacian matrix, defined as $L = D - A$, and D is a diagonal matrix, defined as the degree matrix of A, i.e., $D_{ii} = \sum_j A_{ij}$.

To avoid the negative impact of unreliable graph structural information on model performance, RA-GNN uses an MLP to assist the GNN in feature learning and dynamically adjusts weights of the GNN and MLP outputs through a learnable coefficient δ. The model's output can be expressed by the following formula:

$$\hat{y} = \delta \cdot \hat{y}_{GNN} + (1 - \delta) \cdot \hat{y}_{MLP} \tag{6}$$

At this point, the model's classification loss is:

$$\mathcal{L}_{Model} = l(\hat{Y}, Y_L) \tag{7}$$

Finally, combining the structural constraint with the classification loss, we obtain the overall optimization objective of RA-GNN:

$$\arg\min_{S \in \mathcal{S}} L_0 = \|A - S\|_\alpha + \alpha \|S\|_1 + \beta \|S\|_* + \text{tr}(\mathbf{X}^\top \mathbf{L} \mathbf{X}) + \mathcal{L}_{\text{Model}}, \quad \text{s.t. } S = S^T. \tag{8}$$

We have summarized the specific implementation details of RA-GNN into the pseudocode presented in algorithm 1.

Algorithm 1. RA-GNN

Require: Poisoned matrix A', Attribute matrix X, Labels Y_L
Ensure: Learned adjacency S, GNN parameters θ_{gnn}, MLP parameters θ_{mlp}

Define: $\hat{y} = \delta \cdot f_{\text{gnn}}(S, X) + (1 - \delta) \cdot f_{\text{mlp}}(X)$
$\mathcal{L}_{\text{Model}} = \mathcal{L}_{\text{Model}}(\hat{y}, y)$
$\mathcal{L}_{\text{Total}} = \mathcal{L}_{\text{Model}} + \beta \mathcal{L}_S + \gamma \mathcal{L}_{\text{reg}}$
Initialize $S = A'$
Randomly initialize $\theta_{\text{gnn}}, \theta_{\text{mlp}}$
for $i = 1$ to epoch **do**
    $S = S - \eta \nabla_S \mathcal{L}_{\text{Total}}$
    for $j = 1$ to instep **do**
        check(S)
        $g_{\text{gnn}} = \nabla_{\theta_{\text{gnn}}} \mathcal{L}_{\text{Model}}$
        $g_{\text{mlp}} = \nabla_{\theta_{\text{mlp}}} \mathcal{L}_{\text{Model}}$
        $\theta_{\text{gnn}} = \theta_{\text{gnn}} - \eta' g_{\text{gnn}}$
        $\theta_{\text{mlp}} = \theta_{\text{mlp}} - \eta' g_{\text{mlp}}$
        $\delta = \delta - \eta' \frac{\partial \mathcal{L}_{\text{Model}}}{\partial \delta}$
    end for
end for
return $S, \theta_{\text{gnn}}, \theta_{\text{mlp}}$

4 Experiment

In this section, we first assess the performance of RA-GNN on semi-supervised learning tasks by conducting comparative experiments with state-of-the-art defense approaches against Metattack. Subsequently, we performed thorough analytical investigations of the experimental results.

4.1 Experimental Settings

Datasets. We evaluate the performance of our proposed method on four well-established node classification datasets, including three citation networks (i.e., Cora, Citeseer, ACM) and one blog network (i.e., Polblogs) [9]. For all datasets, we adopt the experimental setup of Pro-GNN, where 10% of nodes are randomly sampled for the training set and 10% for the validation set. The remaining 80% are reserved for evaluating non-targeted adversarial attacks. To mitigate the effects of randomness, we conduct each experiment 5 times with distinct random seeds and report the average results with standard deviation. The statistics of the datasets are presented in Table 1. It is worth noting that **Polblogs lacks node features**, which prevents the use of the MLP module in RA-GNN. Pro-GNN [9] also lacks results on the Polblogs dataset. As an alternative, we adopt RA-GNN-α(see Sect. 4.3) as a substitute for reporting results on Polblogs.

Baselines. To verify the performance of the RA-GNN model, we compared it with state-of-the-art GNN models and defense strategies, this comparison being implemented through the DeepRobust adversarial attack and defense library.

Table 1. The statistics of four datasets.

Datasets	Nodes	Classes	Edges	Features
Cora	2485	7	5096	1433
Citeseer	2110	6	3668	3703
Polblogs	1222	2	16714	/
ACM	3025	3	13128	1870

- **GCN** [10]: The GCN learns node embeddings by aggregating information from their neighbors, which have been widely applied to node classification tasks.
- **Pro-GNN** [9]: The Pro-GNN analyzes the properties of graph data, which inherently exhibit low-rank and sparse characteristics. It leverages these properties to reconstruct the graph structure and performs joint optimization of the graph structure reconstruction and model training processes.
- **β-GNN** [2]: The β-GNN combines the GNN model with an MLP for joint feature learning and prediction. It dynamically adjusts the weights of these two components through a learnable parameter β.
- **NoisyGCN** [5]: Noisy-GCN samples noise from a random Gaussian distribution and adds this noise to the output of the first layer of the GCN hidden layer. This tightens the upper bound of the model's error in the worst-case scenario, thereby enhancing the robustness of the GCN.

Implementation Details. Following the original implementation, Meta-attack is employed using its default parameter configuration Notably, for our experiments across all three datasets, we selected the most destructive attack variant - Meta-Self [22] - for testing.

All model training and testing were conducted on the same operating system configuration and within the same Anaconda virtual environment.Training and test phase of all model were conducted on the same configuration. The physical configuration of the computer comprises a 12-core CPU 12700 operating at 3.6GHz, 48GB RAM, and an NVidia 4090D GPU. The versions of the key packages in the virtual environment of the computer were Torch 2.3.0, torch_geometrics 2.6.1 and CUDA 11.8. Each experiment was repeated five times with different initial random seeds to ensure the robustness of the results. For all perturbation rates below 15%, we set the value of the parameter α for the α-norm to 5; for perturbation rates greater than or equal to 15%, we set the value of α to 0.1. The impact of the α value on the model's performance will be discussed in Sect. 4.4. We employ a grid search [1] to precisely select the optimal hyperparameter combination. The search space was defined as follows: {learning rate(η'), [1e-4, 1e-2]}, {learning rate adj(η), [1e-4, 5e-2]},{ω_1, [2e-4, 6e-2]}, {ω_*, [0, 5]}, and {γ, [6e-2, 2e1]}.

4.2 Results and Analysis

Table 2. Performance comparison on different datasets and under different perturbation rates. RA-GNN-α is a variant of RA-GNN (see Sect. 4.3).

Dataset	Ptb(%)	GCN	NoisyGCN	β-GNN	Pro-GNN	RA-GNN	RA-GNN-α
Cora	0	83.50±0.44	83.07±0.30	84.17±0.19	82.98±0.23	**84.93±0.11**	84.19±0.38
	5	76.55±0.79	79.55±0.45	79.97±0.42	82.27±0.45	**82.93±0.42**	82.49±0.77
	10	70.39±1.28	73.98±0.81	76.04±0.83	79.03±0.59	**79.38±0.65**	79.31±0.31
	15	65.10±0.71	69.66±0.42	73.24±1.18	76.40±1.27	**78.88±0.80**	77.46±0.45
	20	59.56±2.72	59.29±0.91	69.52±2.22	73.32±1.56	**74.14±1.02**	73.75±0.36
	25	47.53±1.96	53.79±1.14	68.46±0.60	69.72±1.69	70.68±0.46	**71.13±0.36**
Citeseer	0	71.96±0.55	72.65±0.50	73.89±0.16	73.28±0.69	**75.32±0.55**	74.11±0.30
	5	70.88±0.62	72.49±1.05	73.57±0.25	73.09±0.34	**74.85±0.38**	73.23±0.77
	10	67.55±0.89	70.18±0.35	72.78±0.67	72.51±0.75	**74.42±0.23**	73.27±0.39
	15	64.52±1.11	67.17±0.35	71.09±0.24	72.03±1.11	**74.57±0.81**	73.18±1.01
	20	62.03±3.49	59.32±1.66	69.00±0.17	70.02±2.28	**71.34±1.16**	70.35±0.92
	25	56.94±2.09	59.81±0.80	68.63±0.41	68.95±2.78	**71.87±0.96**	71.06±0.53
Polblogs	0	**95.69±0.38**	95.21±0.15	94.95±0.08	-	-	95.34±0.05
	5	73.07±0.80	77.98±0.33	75.07±0.89	-	-	**93.58±0.89**
	10	70.72±1.13	74.03±1.22	75.24±0.41	-	-	**90.72±0.45**
	15	64.96±1.91	69.37±0.15	70.06±0.56	-	-	**87.28±1.03**
	20	51.27±1.23	63.35±0.77	61.37±0.46	-	-	**82.25±0.10**
	25	49.23±1.36	62.17±4.71	56.93±1.51	-	-	**74.34±0.15**
ACM	0	70.88±1.62	70.85±1.24	70.50±1.04	70.12±0.53	**71.39±1.13**	70.27±1.16
	5	66.79±0.63	67.51±0.34	66.51±0.60	66.02±0.24	**68.18±0.36**	66.13±0.34
	10	61.89±0.24	62.89±0.25	62.11±0.21	62.08±0.11	**63.93±0.11**	62.49±0.27
	15	58.23±0.63	59.27±0.25	58.86±0.26	59.83±0.18	**60.48±0.32**	59.92±0.32
	20	53.43±0.51	55.49±0.34	54.15±0.56	56.01±0.59	**59.78±4.12**	55.82±0.38
	25	52.62±0.32	54.48±0.47	53.31±0.31	54.26±0.14	**56.84±1.25**	53.47±0.16

Performance of RA-GNN. Table 2 compares the performance of our method with existing robust learning methods on four datasets under different perturbation levels (0%–25%). It can be observed that compared with the baseline methods, RA-GNN achieves significant improvements in classification performance. On the Citeseer dataset, our method achieves the highest classification accuracy across all perturbation rates. Although RA-GNN does not consistently achieve the best performance across all perturbation rates on the Cora dataset, its overall performance remains competitive with state-of-the-art methods such as Pro-GNN. Notably, RA-GNN shows particularly significant performance improvements under high perturbation rates (25%). Since the Polblogs dataset lacks meaningful node features and RA-GNN incorporates an MLP module, it is infeasible to conduct RA-GNN experiments on this dataset. As an alternative, we report the experimental results of the RA-GNN-α variant (see Sect. 4.3) in the table. Specifically, under no perturbation and low perturbation levels (less than or equal to 10%), its classification accuracy is significantly higher than that of the traditional non-robust GCN, indicating that RA-GNN can effectively utilize graph structural information.

Across all tested scenarios, our method consistently outperforms robustness-enhanced methods, demonstrating that the introduction of the MLP in RA-GNN can avoid the negative impact of unreliable graph structures. In particular, at the high perturbation rate of 25%, the accuracy improves by 9%, showing statistical significance. It is worth noting that the introduction of the α-norm constraint results in a lower standard deviation in five repeated runs, which fully validates the effectiveness of this method in enhancing output stability and yielding smoother and more reliable predictions.

4.3 Ablation Study

Table 3. Classification performance of RA-GNN-α , RA-GNN-MLP and RA-GNN on Cora dataset

Ptb Rate (%)	0	5	10	15	20	25
RA-GNN-α	84.19	82.49	79.31	77.46	73.30	71.13
RA-GNN-MLP	83.42	82.50	79.33	78.02	72.97	69.89
RA-GNN	**84.93**	**82.92**	**79.38**	**78.88**	**74.14**	**70.68**

To comprehensively assess the contributions of individual components within the RA-GNN architecture to its adversarial defense capabilities, an ablation study was conducted.

The RA-GNN model comprises two principal elements: the application of the α-norm within the structural constraint term and an MLP with adaptive learning weights. In the ablation study, one component was systematically removed while the other was maintained to evaluate the resultant changes in model performance. This approach yielded two variants: RA-GNN-α and RA-GNN-MLP. Specifically, RA-GNN-α signifies the exclusion of the MLP module from RA-GNN, retaining solely the application of the α-norm. In contrast, RA-GNN-MLP only integrates the MLP without employing the α-norm constraint. Experimental outcomes were reported on the Cora dataset, where analogous patterns were discernible across other datasets as well. As demonstrated in Table 3, RA-GNN-α demonstrated superior performance over RA-GNN-MLP under both clean and high-intensity-perturbation settings. This improvement can be explained by the following points:

- **1.** Under unperturbed conditions, the graph's structural information is deemed reliable; however, the MLP alone fails to leverage this reliable information, thereby limiting its ability to assist the GNN in making accurate predictions.
- **2.** Under conditions of high perturbation intensity ($\geq 20\%$), the graph's structural information necessitates purification. Models employing the α-norm can more effectively accomplish the process of graph structure purification, thus yielding superior performance.

The findings from the ablation study indicate that the two modules within the RA-GNN model serve distinct functions, and the removal of either consistently leads to degraded in classification performance. By integrating both modules, RA-GNN achieves enhanced defensive efficacy against adversarial attacks.

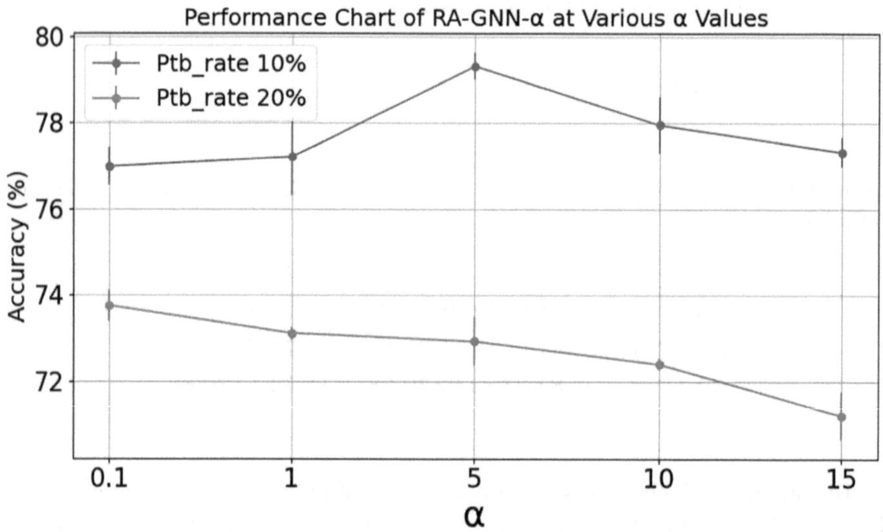

Fig. 2. Performance Chart of RA-GNN-α at Various α Values

4.4 Parameter Analysis

In this subsection, we investigate the impact of the parameter α in the α-norm on model performance. To isolate the influence of the MLP, we conducted experiments using the RA-GNN-α variant. We confined the value of α to one of the value set $\{0.1, 1, 5, 10, 15\}$ and observed the model's behavior under each possible value. We report the results for the Cora dataset under perturbation rates of 10% and 20%, as similar trends were observed in other scenarios. The performance variation of RA-GNN-alpha is depicted in Fig. 2. At lower perturbation intensities (10%), selecting a larger value of α (≥ 5) is generally more effective than choosing a smaller α value. However, an excessively large α can also limit the model's performance, as overly stringent constraints can negatively affect the purification of the graph structure. Similarly, at higher perturbation intensities (20%), smaller α value is more beneficial for the model. This is because high-intensity perturbations tend to modify the graph structure more extensively, and a smaller α value allows for a more relaxed constraint, encouraging more purification operations. This aligns with the theoretical motivation presented earlier.

5 Conclusion

Graph Neural Networks are particularly vulnerable to disruptions from graph adversarial attacks, with perturbation intensities often exhibiting variability in practical applications. In this study, we propose a robust Graph Neural Network, termed RA-GNN, which integrates a Multi-Layer Perceptron to jointly learn feature information and adaptively recovers graph structure in accordance with the magnitude of perturbations. The experimental findings indicate that RA-GNN effectively diminishes the influence of adversarial edges and surpasses many current robustness-enhancement approaches, especially under high-perturbation conditions. Future work will focus on improving RA-GNN's efficiency and reducing its computational cost.

Acknowledgements. This work was supported by Research Foundation from NUDT (Grant No.ZK2023-12).

References

1. Akiba, T., Sano, S., Yanase, T., Ohta, T., Koyama, M.: Optuna: a next-generation hyperparameter optimization framework. In: Proceedings of the 25th ACM SIGKDD International Conference on Knowledge Discovery and Data Mining (2019)
2. Aslan, H.I., Wiesner, P., Xiong, P., Kao, O.: β-gnn: a robust ensemble approach against graph structure perturbation. In: Proceedings of the 5th Workshop on Machine Learning and Systems, pp. 168–175 (2025)
3. Dai, E., Lin, M., Zhang, X., Wang, S.: Unnoticeable backdoor attacks on graph neural networks. In: Proceedings of the ACM Web Conference 2023, pp. 2263–2273 (2023)
4. Dai, Q., Shen, X., Zhang, L., Li, Q., Wang, D.: Adversarial training methods for network embedding. In: The World Wide Web Conference, pp. 329–339 (2019)
5. Ennadir, S., Abbahaddou, Y., Lutzeyer, J.F., Vazirgiannis, M., Boström, H.: A simple and yet fairly effective defense for graph neural networks. In: Proceedings of the AAAI Conference On Artificial Intelligence. vol. 38, pp. 21063–21071 (2024)
6. Entezari, N., Al-Sayouri, S.A., Darvishzadeh, A., Papalexakis, E.E.: All you need is low (rank) defending against adversarial attacks on graphs. In: Proceedings of the 13th International Conference On Web Search And Data Minin, pp. 169–177 (2020)

7. Hou, Z., Feng, R., Derr, T., Liu, X.: Robust graph neural networks via unbiased aggregation. Adv. Neural. Inf. Process. Syst. **37**, 110097–110130 (2024)
8. Hu, S., Zeng, W., Zhang, P., Tang, J.: Neural graph similarity computation with contrastive learning. Appl. Sci. **12**(15), 7668 (2022)
9. Jin, W., et al.: Graph structure learning for robust graph neural networks. In: Proceedings of the 26th ACM SIGKDD International Conference On Knowledge Discovery And Data Mining, pp. 66–74 (2020)
10. Kipf, T.: Semi-supervised classification with graph convolutional networks. arXiv preprint arXiv:1609.02907 (2016)
11. Si, J., Gan, X., Xiao, T., Yang, B., Dong, D., Pang, Z.: Stegnn: Spatial-temporal embedding graph neural networks for road network forecasting. In: 2022 IEEE 28th International Conference on Parallel and Distributed Systems (ICPADS), pp. 826–834. IEEE (2023)
12. Takahashi, T.: Indirect adversarial attacks via poisoning neighbors for graph convolutional networks. In: 2019 IEEE International Conference on Big Data (Big Data), pp. 1395–1400. ieee (2019)
13. Velickovic, P., et al.: Graph attention netw. stat **1050**(20), 10–48550 (2017)
14. Wu, H., et al.: Adversarial examples on graph data: deep insights into attack and defense. arXiv preprint arXiv:1903.01610 (2019)
15. Xie, X., Chen, W., Kang, Z.: Robust graph structure learning under heterophily. Neural Netw. **185**, 107206 (2025)
16. Xu, K., et al.: Topology attack and defense for graph neural networks: an optimization perspective. arXiv preprint arXiv:1906.04214 (2019)
17. Yin, W., et al.: Enhancing graph neural networks for graph classification via structure-guided adaptor. In: 2025 28th International Conference on Computer Supported Cooperative Work in Design (CSCWD), pp. 1992–1997. IEEE (2025)
18. Ying, R., et al.: Graph convolutional neural networks for web-scale recommender systems. In: Proceedings of the 24th ACM SIGKDD International Conference On Knowledge Discovery and Data Mining, pp. 974–983 (2018)
19. Zhang, J., et al.: Test-time training on graphs with large language models (LLMS). In: Proceedings of the 32nd ACM International Conference on Multimedia, pp. 2089–2098 (2024)
20. Zhu, D., Zhang, Z., Cui, P., Zhu, W.: Robust graph convolutional networks against adversarial attacks. In: Proceedings of the 25th ACM SIGKDD International Conference On Knowledge Discovery And Data Mining, pp. 1399–1407 (2019)
21. Zügner, D., Akbarnejad, A., Günnemann, S.: Adversarial attacks on neural networks for graph data. In: Proceedings of the 24th ACM SIGKDD International Conference On Knowledge Discovery and Data Mining, pp. 2847–2856 (2018)
22. Zügner, D., Günnemann, S.: Adversarial attacks on graph neural networks via meta learning (2024). https://arxiv.org/abs/1902.08412

Enhancing Vision Transformer with Multiple Fractional-Order Differential Operators for Image Desnowing

Yuxuan Li[1(✉)] and Yuning Ren[2]

[1] The University of Sydney, Sydney, NSW 2006, Australia
yuxuan.li.usyd.au@gmail.com
[2] China University of Political Science and Law, Beijing 100088, China

Abstract. Image desnowing aims to eliminate the complex visual degradation caused by snowflake noise and is an important branch of image restoration. In this paper, we consider the self-similar complex edges and rough translucent structures of snowflake noise, which can be characterized by fractal dimension. We use multiple fractional-order differential operators to model fractals, thereby enhancing the Vision Transformer (ViT), and propose MF-ViT. MF-ViT is a dedicated deep learning desnowing model based on the specific prior modeling of the fractal features of snowflake noise. Specifically, to enhance fractal feature representation ability, we incorporate fractional differential operators of different orders into the attention and feedforward networks of ViT, which help to handle fractal features. We empirically evaluate the proposed MF-ViT on five benchmark public desnowing datasets. The results show that MF-ViT achieves state-of-the-art performance in both simulation and real-world images with snowflake noise. This paper also provides new model improvement ideas for other machine vision pattern analysis tasks with fractal dimension features.

Keywords: Image desnowing · Image restoration · Fractional-order differential operators · Fourier transform · Fractal modeling

1 Introduction

Snowfall can cause visual information degradation such as blurring, occlusion and artifacts, significantly affecting computer vision tasks such as object detection and tracking [1]. Desnowing is one of the image restoration techniques, aiming to eliminate the visual degradation caused by snowflake noise and restore high-quality images [2]. Early image desnowing algorithms are often based on pure prior knowledge [3,4]. With the development of machine intelligence, desnowing methods based on deep learning have stronger generalization capabilities and can handle more complex and diverse snowflake noises.

The deep learning desnowing model DesnowNet [5] adopts the following mathematical modeling. Suppose observed image with snowflake noise at pixel position x, $K(x)$, is mathematically expressed as:

J. Lokoč et al. (Eds.): MMM 2026, LNCS 16413, pp. 177–189, 2026.
https://doi.org/10.1007/978-981-95-6957-1_13

$$K(x) = J(x)(1 - Z(x)) + C(x)Z(x), \tag{1}$$

where $J(x)$ is the clean image, $Z(x)$ is the snow mask, $C(x)$ denotes the chromatic aberration map for snow images. JSTASR [6] has made a further modeling of snow-covered images, denoted as $\mathcal{I}(x)$ which can be represented as

$$\mathcal{I}(x) = \mathcal{K}(x)\mathcal{T}(x) + \mathcal{A}(x)(1 - \mathcal{T}(x)), \tag{2}$$

where $\mathcal{T}(x)$ represents the transmission map, $\mathcal{A}(x)$ represents the effect of snow on the color and intensity of the light, $\mathcal{K}(x)$ represents the veiling effect-free but snowy image, which can be further decomposed into:

$$\mathcal{K}(x) = \mathcal{J}(x)(1 - \mathcal{Z}(x)\mathcal{R}(x)) + \mathcal{C}(x)\mathcal{Z}(x)\mathcal{R}(x), \tag{3}$$

where $\mathcal{J}(x)$ is the clean image, $\mathcal{R}(x)$ is a binary mask that represents the area covered by snow, $\mathcal{Z}(x)$ represents the color distortion caused by snow, $\mathcal{C}(x)$ represents the color properties, taking into account the tint variation of the snow. Deep learning-based desnowing models have also proposed other modeling methods [7,8].

We further consider the modeling and representation problem of snowflake noise. Snowflake noise has complex edges and self-similar structures at multiple observation scales, possessing fractal features. Tafti P.D. et al. demonstrate that the fractional Laplacian operator, implemented in the frequency domain as a multiplier, serves as an effective fractional-order differential operator for modeling fractals [9]. At the same time, we also consider that early machine vision deep learning models are mainly based on convolution and their abilities to understand global features are limited. The Vision Transformer (ViT) has effectively solved this problem by using multi-head attention [10]. The multi-scale self-similar characteristics of snowflake noise also have global features, so the structure based on ViT is an effective solution.

In this paper, we implement **M**ultiple **F**ractional-Order differential operators using the frequency-domain multiplier after Fourier transform, incorporate them into **ViT**, and propose **MF-ViT**. MF-ViT is a dedicated desnowing model, which uses convolutional networks to generate tokens and reconstruct images (motivated by [11]). MF-ViT uses the ViT structure for feature processing and complementarily processes fractal features by using fractional-order differential operators of different orders in the attention and feedforward networks. As shown in Fig. 1, experiments have verified that proposed MF-ViT can achieve state-of-the-art performance in the desnowing task.

The main contributions of this paper can be summarized as follows: (1) To the best of our knowledge, we are the first to consider fractal features in desnowing models. (2) We use fractional differential operators to represent fractals in ViT and propose a desnowing model MF-ViT. (3) We empirically verify that MF-ViT has reached state-of-the-art level, especially in the real-world snowfall scenario.

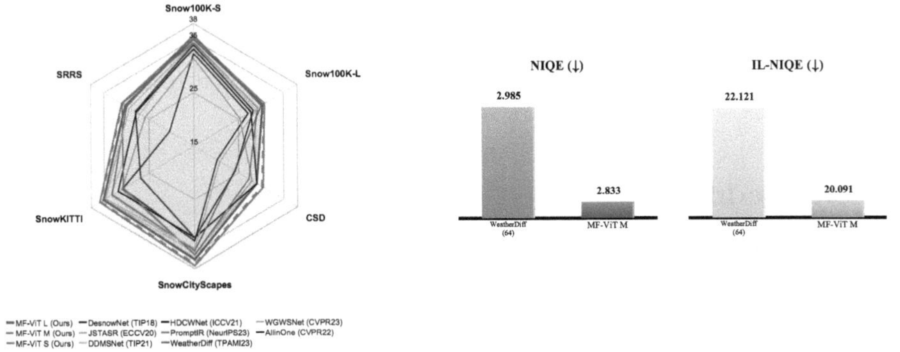

(a) (b)

Fig. 1. The comparison between MF-ViT and the existing state-of-the-art methods on five public benchmark datasets, where Snow100K is divided into two difficulty levels: S and L. (a) PSNR comparison, higher is better. (b) Snow100K real-world metrics comparison, lower is better.

2 Related Work

2.1 Image Desnowing

Deep learning technology enables image desnowing model to obtain snowflake feature knowledge, thereby improving the adaptability and generalization ability of the model. Y. Liu et al. divide snowflake noise into different attributes and design a multi-level network DesnowNet to successively remove snowflakes [5]. Chen et al. add the shadowing effect to further model the snowflake, and proposed a new desnowing algorithm named JSTASR [6] with the fusion understanding of size and transparency. W. Chen et al. use a new hierarchical decomposition of snowflake features and a loss function to design the HDCWNet [8] that can effectively deal with snow particles of different sizes. K. Zhang et al. design a deep learning desnowing network DDMSNet [7] that can adaptively judge the semantic depth using multi-scale fusion understanding.

These models have achieved excellent results by adopting different modeling methods and design concepts. However, the lack of fractal feature consideration in existing works limits the further breakthrough in model performance.

2.2 Multi-task Image Restoration

At present, some works have implemented one deep learning framework to carry out various image restoration tasks. Models can take into account the characteristics of multiple weather conditions simultaneously to complete various tasks [12,13]. Some models do not rely on specific modeling, thus enabling unified handling of all adverse weather conditions [14–17].

In recent years, with the development of the representation capabilities of deep learning models, models that can simultaneously perform all image restoration tasks have also been designed, rather than being limited to weather noise [2,18–21].

3 Methodology

3.1 Overall Pipeline

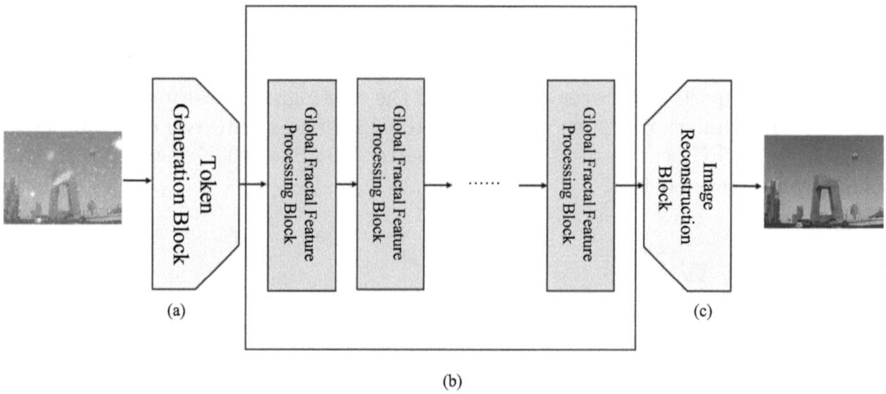

Fig. 2. The architecture of MF-ViT. (a) Tokens are encoded using a convolutional network, and the feature dimension is enhanced through an embedding layer. (b) The Global Fractal Feature Processing Block incorporates Transformer-network enhanced with fractional-order differential operators of different orders, details in Sect. 3.2. (c) The final output image is reconstructed by the reverse process of token-encoded.

The pipeline of MF-ViT is illustrated in Fig. 2, where the token generation method is mainly inspired by [11]. Global Fractal Feature Processing Blocks use different fractional-order differential operators to complementarily handle fractal features in multi-head attention and feedforward networks. Based on different hyperparameter settings, we design three variants of MF-ViT: S, M and L, details in Table 1.

Table 1. Configurations of different MF-ViT variants.

Variant	Number of Global Fractal Blocks	Embedding Dimension	Attention Heads
MF-ViT-S	8	512	4
MF-ViT-M	10	512	4
MF-ViT-L	12	1024	8

For the input snowy images, Fig. 3 describes the token generation process, where C is set to 16. We process the input snowy images using convolutions with different configurations, then make tokens through an embedding layer as following:

$$X = \text{reshape}(X,\ B, \frac{H}{8} \times \frac{W}{8}, 8C)W_E \in \mathbb{R}^{B \times N \times D} \tag{4}$$

where $N = \frac{H}{8} \times \frac{W}{8}$ is the sequence length, $W_E \in \mathbb{R}^{B \times 8C \times D}$ is a linear layer, D is the embedding dimension in Table 1.

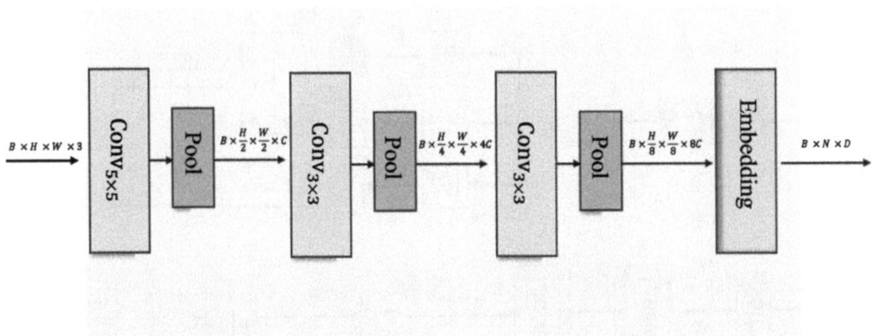

Fig. 3. Detailed Architecture of Token Generation Block.

As shown is Fig. 4, for the output tensor of global fractal feature processing blocks, image reconstruction block has a dual structure to output the clean image. The linear layer is used to reduce the feature dimension back to $8C$.

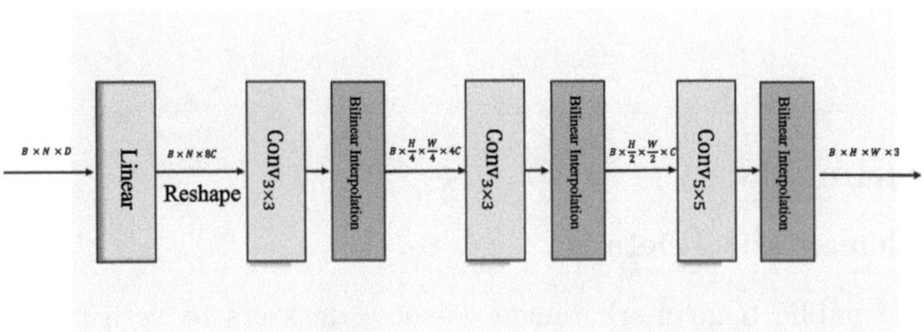

Fig. 4. Detailed Architecture of Image Reconstruction Block.

3.2 Global Fractal Feature Processing Block

The global fractal feature processing block is the main module for representing global features with fractal characteristics. This is a structure similar to

ViT, consisting of fractal multi-head attention and fractal feedforward networks, details in Fig. 5. Along the sequence axis, we perform an FFT (ω: angular frequency), multiply each frequency bin by a power-law weight of order α, and apply the inverse FFT; this is—up to a fixed 2π unit conversion—the Riesz-type fractional Laplacian on the sequence. Equivalently, we realize the symmetric fractional derivative obtained by averaging the left and right Riemann–Liouville derivatives (see [9]).

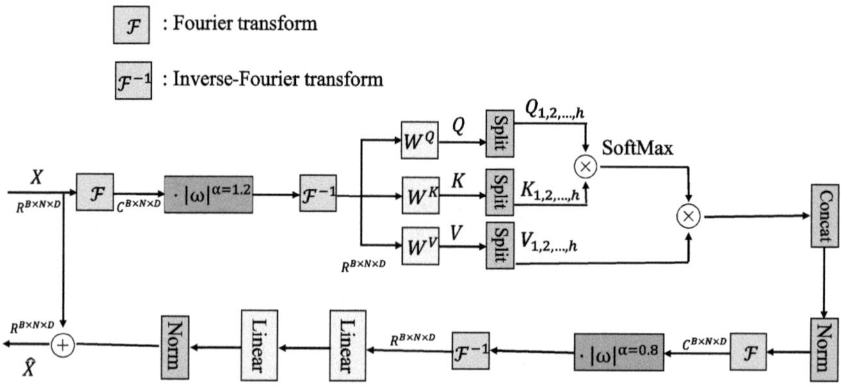

Fig. 5. Global fractal feature processing block alternately uses improved fractal multi-head attention and fractal feedforward networks to achieve multi-level integration and understanding of the fractal features of snowflake noise. In the multi-head attention, Fourier transform is applied, followed by $\alpha = 1.2$ order differentiation in the frequency domain. After mapping back to the time-domain features, conventional multi-head attention is performed. Similar operations are performed in the fractal feedforward network, with the differentiation order changed to $\alpha = 0.8$. The hidden dimension in the middle of 2 linear layers is $2D$.

4 Experiments

4.1 Implementation Details

We use five public benchmark image de-snow datasets to verify the effect of MF-ViT, including Snow100K [5], CSD [8], SnowCityScapes [7], SnowKitty [7] and SRRS [6]. In the experiments on the Snow100K dataset, we adopt two test difficulties, S and L, and simultaneously verify the practical performance in the real-world images it provided. We implement our model using the deep learning framework PyTorch [22]. The experimental configurations are summarized in Table 2.

Table 2. Experiments settings.

Settings	Details
Batch Size	16
Learning Rate	1×10^{-4}
Optimizer	Adam [23] ($\beta_1 = 0.01$, $\beta_2 = 0.99$)
Input Resolution	512×512
Output Resolution	512×512
Loss Function	PSNR Loss [24]
Total Epochs	15,000
GPU	$2 \times$ Nvidia Tesla V100 32G

4.2 Main Results

Quantitative evaluations between ground truth and desnowed images are performed via the SSIM [25] and PSNR [26] (higher is better). For real-world snowy images, we adopt NIQE [27] and IL-NIQE [28] (lower is better). Table 3 shows that on the synthetic datasets, MF-ViT has achieved state-of-the-art performance.

Table 3. Quantitative comparisons in terms of PSNR and SSIM with state-of-the-art image desnowing and restoration methods. Some good results (mainly best and second-best) are highlighted in green.

Type	Method	Venue	Snow100K-S		Snow100K-L		CSD		SnowCityScapes		SnowKITTI		SRRS	
			PSNR	SSIM	PSNR	SSIM	PSNR	SSIM	PSNR	SSIM	PSNR	SSIM	PSNR	SSIM
Snow-specific Desnowing	DesnowNet [5]	TIP'18	32.33	0.95	27.17	0.90	20.13	0.90	32.81	0.95	30.90	0.94	20.38	0.84
	JSTASR [6]	ECCV'20	31.40	0.90	25.32	0.81	27.96	0.89	35.32	0.95	31.09	0.94	25.82	0.89
	DDMSNet [7]	TIP'21	34.34	0.94	28.85	0.88	28.79	0.91	35.29	0.97	33.48	0.96	27.03	0.91
	HDCW-Net [8]	ICCV'21	33.21	0.96	27.80	0.92	29.06	0.92	32.21	0.95	31.87	0.94	27.78	0.92
Multi-task Image Restoration	All-in-One [16]	CVPR'22	33.98	0.95	28.33	0.88	24.23	0.87	32.24	0.95	26.90	0.89	27.98	0.88
	WeatherDiff [15]	TPAMI'23	35.83	0.96	30.09	0.90	24.10	0.94	34.61	0.96	33.23	0.95	29.90	0.93
	WGWSNet [14]	CVPR'23	34.14	0.95	29.42	0.90	20.73	0.94	34.23	0.96	34.70	0.95	29.31	0.92
	PromptIR [18]	NIPS'23	34.40	0.96	29.30	0.91	27.28	0.97	35.45	0.97	30.71	0.97	30.91	0.97
This Paper	**MF-ViT S**	—	35.12	0.97	30.41	0.91	29.02	0.97	36.25	0.97	35.13	0.97	30.29	0.97
	MF-ViT M	—	35.43	0.97	30.72	0.91	30.17	0.97	37.03	0.96	35.83	0.96	30.82	0.97
	MF-ViT L	—	35.38	0.97	30.85	0.91	30.47	0.97	37.40	0.96	35.99	0.97	31.01	0.97

More important is the real world desnowing effect, we train all models on Snow100K training set and test them on Snow100K real-world snowy images.

Table 4 and Fig. 6 show that, MF-ViT M can effectively remove snowflake noise and retain the original details.

Table 4. Quantitative comparisons with state-of-the-art models on Snow100K real-world data.

Method	NIQE	IL-NIQE
Snow-specific Desnowing		
DesnowNet [5]	3.384	23.8
JSTASR [6]	3.151	23.2
DDMSNet [7]	3.042	22.7
HDCW-Net [8]	3.057	22.7
Multi-task Image Restoration		
All-in-One [16]	3.073	22.1
WeatherDiff [15]	2.964	21.9
WGWSNet [14]	3.149	22.3
PromptIR [18]	3.024	22.2
This Paper		
MF-ViT S	2.912	21.4
MF-ViT M	2.833	20.1
MF-ViT L	2.952	22.1

Fig. 6. Intuitive visual effect comparisons with state-of-the-art models on Snow100K real-world data.

4.3 Ablation Experiment

Effectiveness of Fractional-Order Differential Operator. Compared to the standard ViT, MF-ViT incorporates multiple fractional-order differential operators in the frequency domain features, allowing it to complementarily learn the fractal geometric characteristics of snowflakes, making it an advanced image desnowing model. We use $\alpha = 1.2$ before attention (stronger whitening) and $\alpha = 0.8$ before the feedforward (weaker fractional differentiation) to obtain complementary learning of snowflake-like fractal features. In ablation experiments, we verify the impact of fractional-order differential operators on desnowing performance. As shown in Tables 5 and 6, the full MF-ViT with both operators consistently outperforms all variants, confirming their effectiveness in enhancing desnowing performance. Especially when dealing with real-world data, fractional differential operators can significantly improve the performance of MF-ViT.

Table 5. Ablation experimental results on the synthetic datasets.

Models	CSD	Snow100K-L
Without α	27.61	27.13
Fixed $\alpha = 0.8$	29.32	28.99
Fixed $\alpha = 1.2$	28.80	29.35
Full MF-ViT (M)	**30.17**	**30.72**

Table 6. Snow100K-Real ablation on NIQE and IL-NIQE.

Models	NIQE	IL-NIQE
Without α	3.16	22.1
Fixed $\alpha = 0.8$	2.95	21.4
Fixed $\alpha = 1.2$	2.95	21.0
Full MF-ViT (M)	**2.833**	**20.091**

Feature Visualization. To further illustrate the improvement of learning abilities of snowflake features by the proposed method, we use T-SNE to reduce the dimension and visualize the features following approach of J. Feng et al. [29]. We select the features of the Snow100K training set as the source domain and real-world data as the target domain. Figure 7 shows that the added differential operators can reduce domain shift, thereby enhancing the generalization ability in real-world applications.

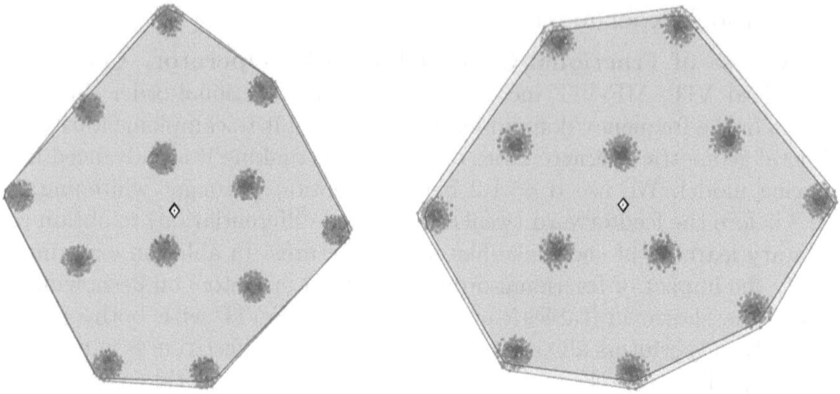

(a) (b)

Fig. 7. T-SNE visualization of features learned by MF-ViT M with (a)/without (b) added differential operators, where the blue dots represent the source domain features, and the orange dots represent the target domain features. (a) exhibits features with a higher degree of overlap, while (b) shows a high degree of feature aggregation in the source domain, with a relatively large shift from the target domain. (Color figure online)

5 Conclusion

To address the issue that existing image desnowing models do not consider snowflake fractal features, this paper introduces multiple fractional-order Riemann–Liouville differential operators in ViT, proposing MF-ViT. Experiments show that MF-ViT has achieved state-of-the-art performance and is particularly suitable for handling snowflake noise in the real world.

Discussion on Limitations. The selection of α is an empirical result rather than one obtained through neural network learning, so adaptability may be further enhanced. In the future, we hope to achieve the automatic adaptation of α and further enhance the fractal modeling and representation capabilities.

A Appendix

A.1 Ablation Experiment on Attention Distance

As a supplement to the ablation experiment, for each query token, the pixel-wise distances to all key tokens are weighted by their attention coefficients and then averaged across the image. Figure 8 shows that MF-ViT can establish long-range dependencies earlier than the version with the fractional differential operators removed.

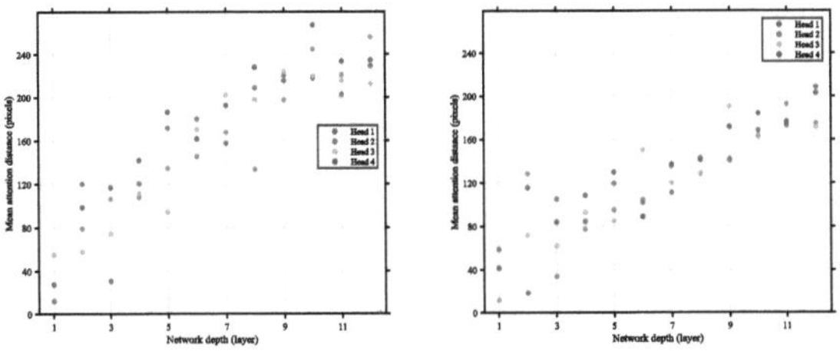

(a) (b)

Fig. 8. Attention areas by MF-ViT with (a)/without (b) added differential operators. Taking MF-ViT M as an example, the full version can achieve the improvement of attention distance earlier and more fully, indicating that it can learn global snowflake noise features more effectively, which further illustrates the effectiveness of the proposed method.

A.2 Resolution Robustness

Figure 9 shows the desnowing results of MF-ViT on the Snow100K dataset at different resolutions. The changes are not significant, indicating that MF-ViT can adapt to the requirements of multiple resolutions.

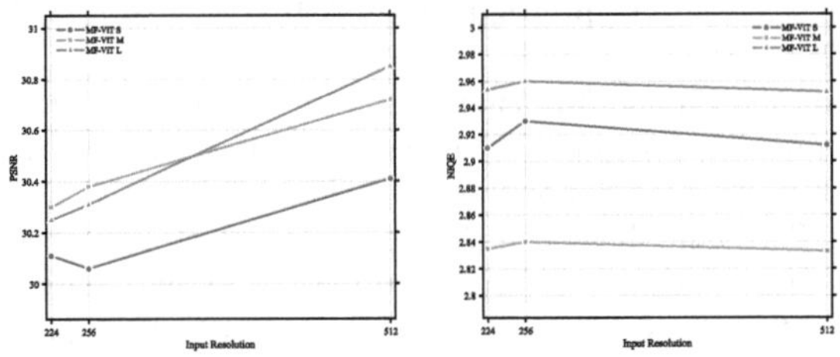

(a) (b)

Fig. 9. Desnowing evaluation results at different resolutions on Snow100K. (a) Snow100K-L. (b) Snow100K real world.

References

1. Chen, H., et al.: Snow removal in video: a new dataset and a novel method. In: 2023 IEEE/CVF International Conference on Computer Vision (ICCV), pp. 13165–13176. IEEE, Piscataway (2023). https://doi.org/10.1109/ICCV51070.2023.01215
2. Cui, Y., Knoll, A.: Enhancing local–global representation learning for image restoration. IEEE Trans. Industr. Inf. **20**(4), 6522–6530 (2024). https://doi.org/10.1109/TII.2023.3345464
3. Bossu, J., Hautière, N., Tarel, J.-P.: Rain or snow detection in image sequences through use of a histogram of orientation of streaks. Int. J. Comput. Vision **93**(3), 348–367 (2011). https://doi.org/10.1007/s11263-011-0421-7

4. Wang, Y., Liu, S., Chen, C., Zeng, B.: A hierarchical approach for rain or snow removing in a single color image. IEEE Trans. Image Process. **26**(8), 3936–3950 (2017). https://doi.org/10.1109/TIP.2017.2708502
5. Liu, Y.-F., Jaw, D.-W., Huang, S.-C., Hwang, J.-N.: DesnowNet: context-aware deep network for snow removal. IEEE Trans. Image Process. **27**(6), 3064–3073 (2018). https://doi.org/10.1109/TIP.2018.2806202
6. Chen, W.-T., Fang, H.-Y., Ding, J.-J., Tsai, C.-C., Kuo, S.-Y.: JSTASR: joint size and transparency-aware snow removal algorithm based on modified partial convolution and veiling effect removal. In: Vedaldi, A., Bischof, H., Brox, T., Frahm, J.-M. (eds.) ECCV 2020. LNCS, vol. 12366, pp. 754–770. Springer, Cham (2020). https://doi.org/10.1007/978-3-030-58589-1_45
7. Zhang, K., Li, R., Yu, Y., Luo, W., Li, C.: Deep dense multi-scale network for snow removal using semantic and depth priors. IEEE Trans. Image Process. **30**, 7419–7431 (2021). https://doi.org/10.1109/TIP.2021.3104166
8. Chen, W.-T., et al.: ALL snow removed: single image desnowing algorithm using hierarchical dual-tree complex wavelet representation and contradict channel loss. In: 2021 IEEE/CVF International Conference on Computer Vision (ICCV), pp. 4176–4185. IEEE, Piscataway (2021). https://doi.org/10.1109/ICCV48922.2021.00416
9. Tafti, P.D., Van De Ville, D., Unser, M.: Invariances, Laplacian-like wavelet bases, and the whitening of fractal processes. IEEE Trans. Image Process. **18**(4), 689–702 (2009). https://doi.org/10.1109/TIP.2008.2011451
10. Dosovitskiy, A., et al.: An image is worth 16x16 words: transformers for image recognition at scale. arXiv preprint arXiv:2010.11929 (2021). https://arxiv.org/abs/2010.11929
11. Wang, L., Cao, M., Zhong, Y., Yuan, X.: Spatial-temporal transformer for video snapshot compressive imaging. IEEE Trans. Pattern Anal. Mach. Intell. **45**(7), 9072–9089 (2023). https://doi.org/10.1109/TPAMI.2022.3225382
12. Wen, Y., Gao, T., Zhang, K., Cheng, P., Chen, T.: Restoring vision in rain-by-snow weather with simple attention-based sampling cross-hierarchy Transformer. Pattern Recogn. **156**, 110743 (2024). https://doi.org/10.1016/j.patcog.2024.110743
13. Ali, A., Sarkar, R., Chaudhuri, S.S.: Wavelet-based auto-encoder for simultaneous haze and rain removal from images. Pattern Recogn. **150**, 110370 (2024). https://doi.org/10.1016/j.patcog.2024.110370
14. Zhu, Y., et al.: Learning weather-general and weather-specific features for image restoration under multiple adverse weather conditions. In: 2023 IEEE/CVF Conference on Computer Vision and Pattern Recognition (CVPR), pp. 21747–21758. IEEE, Piscataway (2023). https://doi.org/10.1109/CVPR52729.2023.02083
15. Özdenizci, O., Legenstein, R.: Restoring vision in adverse weather conditions with patch-based denoising diffusion models. IEEE Trans. Pattern Anal. Mach. Intell. **45**(8), 10346–10357 (2023). https://doi.org/10.1109/TPAMI.2023.3238179
16. Li, R., Tan, R.T., Cheong, L.-F.: All in one bad weather removal using architectural search. In: 2020 IEEE/CVF Conference on Computer Vision and Pattern Recognition (CVPR), pp. 3172–3182. IEEE, Piscataway (2020). https://doi.org/10.1109/CVPR42600.2020.00324
17. Wang, T., et al.: GridFormer: residual dense transformer with grid structure for image restoration in adverse weather conditions. Int. J. Comput. Vision **132**(10), 4541–4563 (2024). https://doi.org/10.1007/s11263-024-02056-0

18. Potlapalli, V., Zamir, S.W., Khan, S.H., Khan, F.S.: PromptIR: prompting for all-in-one image restoration. In: Oh, A., Naumann, T., Globerson, A., Saenko, K., Hardt, M., Levine, S. (eds.) Advances in Neural Information Processing Systems (NeurIPS), vol. 36, pp. 71275–71293. Curran Associates, Inc., Red Hook (2023). https://proceedings.neurips.cc/paper_files/paper/2023/file/e187897ed7780a579a0d76fd4a35d107-Paper-Conference.pdf
19. Wang, Z., Cun, X., Bao, J., Zhou, W., Liu, J., Li, H.: Uformer: a general U-shaped transformer for image restoration. In: 2022 IEEE/CVF Conference on Computer Vision and Pattern Recognition (CVPR), pp. 17662–17672. IEEE, Piscataway (2022). https://doi.org/10.1109/CVPR52688.2022.01716
20. Cho, H., Shin, H.K., Jang, Y., Ko, S.J., Jung, S.W.: PD-CR: patch-based diffusion using constrained refinement for image restoration. IEEE Signal Process. Lett. **31**, 949–953 (2024). https://doi.org/10.1109/LSP.2024.3381908
21. Wang, L., Guo, Y., Wang, Y., Dong, X., Xu, Q., Yang, J., An, W.: Unsupervised degradation representation learning for unpaired restoration of images and point clouds. IEEE Trans. Pattern Anal. Mach. Intell. **47**(1), 1–18 (2025). https://doi.org/10.1109/TPAMI.2024.3471571
22. Paszke, A., et al.: PyTorch: an imperative style, high-performance deep learning library. arXiv preprint arXiv:1912.01703 (2019). https://arxiv.org/abs/1912.01703
23. Kingma, D.P., Ba, J.: Adam: a method for stochastic optimization. arXiv preprint arXiv:1412.6980 (2017). https://arxiv.org/abs/1412.6980
24. Chen, L., Lu, X., Zhang, J., Chu, X., Chen, C.: HINet: half instance normalization network for image restoration. In: 2021 IEEE/CVF Conference on Computer Vision and Pattern Recognition Workshops (CVPRW), pp. 182–192. IEEE, Piscataway (2021). https://doi.org/10.1109/CVPRW53098.2021.00027
25. Wang, Z., Bovik, A.C., Sheikh, H.R., Simoncelli, E.P.: Image quality assessment: from error visibility to structural similarity. IEEE Trans. Image Process. **13**(4), 600–612 (2004). https://doi.org/10.1109/TIP.2003.819861
26. Huynh-Thu, Q., Ghanbari, M.: Scope of validity of PSNR in image/video quality assessment. Electron. Lett. **44**(13), 800–801 (2008). https://doi.org/10.1049/el:20080522
27. Mittal, A., Soundararajan, R., Bovik, A.C.: Making a "completely blind" image quality analyzer. IEEE Signal Process. Lett. **20**(3), 209–212 (2013). https://doi.org/10.1109/LSP.2012.2227726
28. Zhang, L., Zhang, L., Bovik, A.C.: A feature-enriched completely blind image quality evaluator. IEEE Trans. Image Process. **24**(8), 2579–2591 (2015). https://doi.org/10.1109/TIP.2015.2426416
29. Feng, J., et al.: S4DL: shift-sensitive spatial–spectral disentangling learning for hyperspectral image unsupervised domain adaptation. IEEE Trans. Neural Netw. Learn. Syst. 1–15 (2025). https://doi.org/10.1109/TNNLS.2025.3556386

VENUS: Visual Editing with Noise Inversion Using Scene Graphs

Thanh-Nhan Vo[1,2], Trong-Thuan Nguyen[1,2(✉)], Tam V. Nguyen[3], and Minh-Triet Tran[1,2]

[1] University of Science, VNU-HCM, Ho Chi Minh City, Vietnam
tmtriet@fit.hcmus.edu.vn
[2] Vietnam National University, Ho Chi Minh City, Vietnam
{vtnhan,ntthuan}@selab.hcmus.edu.vn
[3] University of Dayton, Dayton, USA
tamnguyen@udayton.edu

Abstract. State-of-the-art text-based image editing models often struggle to balance background preservation with semantic consistency, frequently resulting either in the synthesis of entirely new images or in outputs that fail to realize the intended edits. In contrast, scene graph-based image editing addresses this limitation by providing a structured representation of semantic entities and their relations, thereby offering improved controllability. However, existing scene graph editing methods typically depend on model fine-tuning, which incurs high computational cost and limits scalability. To this end, we introduce **VENUS** (***V***isual ***E***diting with ***N***oise inversion ***U***sing ***S***cene graphs), a training-free framework for scene graph-guided image editing. Specifically, VENUS employs a split prompt conditioning strategy that disentangles the target object of the edit from its background context, while simultaneously leveraging noise inversion to preserve fidelity in unedited regions. Moreover, our proposed approach integrates scene graphs extracted from multimodal large language models with diffusion backbones, without requiring any additional training. Empirically, VENUS substantially improves both background preservation and semantic alignment on PIE-Bench, increasing PSNR from 22.45 to 24.80, SSIM from 0.79 to 0.84, and reducing LPIPS from 0.100 to 0.070 relative to the state-of-the-art scene graph editing model (SGEdit). In addition, VENUS enhances semantic consistency as measured by CLIP similarity (24.97 vs. 24.19). On EditVal, VENUS achieves the highest fidelity with a 0.87 DINO score and, crucially, reduces per-image runtime from 6–10 min to only 20–30 s. Beyond scene graph-based editing, VENUS also surpasses strong text-based editing baselines such as LEDIT++ and P2P+DirInv, thereby demonstrating consistent improvements across both paradigms.

Keywords: Image Editing · Scene Graphs · Multimodal Large Language Models · Diffusion Models

J. Lokoč et al. (Eds.): MMM 2026, LNCS 16413, pp. 190–203, 2026.
https://doi.org/10.1007/978-981-95-6957-1_14

1 Introduction

Editing visual scenes in a controllable and efficient manner remains a fundamental challenge in computer vision. Broadly, two paradigms have been explored: text-driven diffusion editing and scene graph-based editing. On the one hand, text-only methods [1,10] offer a convenient and intuitive interface, making them highly accessible and flexible across diverse editing scenarios. On the other hand, scene graph-based approaches [30] provide more precise control over object relations and spatial layout, underscoring the importance of structured scene representations for reliable visual editing. However, despite these advantages, text-driven methods often suffer from semantic imprecision, leading to object misplacement and background drift. Conversely, scene graph-based methods typically require costly training or inference-time fine-tuning, resulting in long editing latencies (6–10 min per image). Therefore, the fundamental trade-off between semantic consistency and background fidelity remains unresolved.

Contributions of this Work. We propose **VENUS** (***V***isual ***E***diting with ***N***oise inversion ***U***sing ***S***cene Graphs), a training-free framework for scene graph-guided image editing. VENUS introduces a split prompt conditioning strategy that disentangles the target object to be edited from the preserved background, thereby enabling diffusion models to perform controllable semantic editing without compromising background fidelity. Furthermore, by leveraging multimodal large language models (MLLMs) for automatic scene graph extraction and refinement, VENUS integrates seamlessly with existing diffusion backbones and significantly reduces editing latency. Notably, without requiring any additional training or inference-time fine-tuning, VENUS outperforms SGEdit [30] on the PIE-Bench dataset [10], reducing LPIPS from 0.100 to 0.070, increasing PSNR from 22.45 to 24.80, improving SSIM from 0.790 to 0.837, and raising the CLIP from 24.19 to 24.97. Beyond outperforming scene graph-based methods, VENUS also consistently surpasses strong text-based baselines, including LEDIT++ [1] and P2P+DirInv [10], demonstrating its effectiveness across editing paradigms.

2 Related Work

2.1 Scene Graph to Image Synthesis and Editing

Scene graphs represent visual scenes as nodes (objects) and edges (spatial or semantic relations), which have been successfully applied to image captioning [16], cross-modal retrieval [28], image generation [9], and image editing [30]. For visual synthesis, early work such as SG2IM [9] employed a two-stage layout refinement pipeline, whereas SATURN [23] transformed scene graphs into structured textual prompts, thereby leveraging the capabilities of modern text-to-image diffusion models without requiring additional training. For image editing, SGEdit [30] integrates scene graph parsing, concept grounding, and large language model–guided prompting into diffusion models, achieving strong accuracy and fidelity. Taken together, these works highlight the role of scene graphs as a structured and interpretable interface that is particularly well-suited for both generation and editing within contemporary diffusion-based frameworks.

2.2 LLM-Based Image Synthesis

In recent years, LLM-based methods have emerged as highly effective solutions. By leveraging the multimodal reasoning capabilities of LLMs across text, visual inputs, and structured representations such as scene graphs, many methods [17,20] have achieved remarkable performance in producing high-quality, semantically aligned images. Unlike traditional pipelines that rely solely on text encoders, LLMs can serve as high-level planners, determining scene layouts and object compositions prior to the generation stage. For example, [24,30] employ LLMs to generate structured layouts or scene graphs, which in turn guide diffusion or autoregressive models, yielding more controllable and faithful synthesis.

2.3 Diffusion-Based Image Editing

DDPM [8] and Stable Diffusion [19] have greatly enhanced the ability to synthesize diverse, photorealistic images with high detail and naturalness. As a result, diffusion-based models have been widely adopted across domains, including medicine, the arts, and media. Meanwhile, image editing is particularly notable, as it demands a fine-grained understanding of both natural language and visual semantics. Existing methods can be broadly categorized into three main types. The first category consists of training-based methods, such as DiffusionCLIP [12] and InstructPix2Pix [2], which train models from scratch on large-scale, language-guided datasets. The second category includes test-time fine-tuning methods, such as StyleDiffusion [25] and Imagic [11], which adapt pretrained models during inference to maximize representational capacity. Finally, there are training-free and fine-tuning-free methods, including LEDITS++ [1], Plug-and-Play (PnP) [22], Prompt-to-Prompt (P2P) [6], and MasaCtrl [3], which operate on pretrained models, typically by modifying latent representations.

2.4 Discussion

Limitations of Previous Methods. As discussed in Sects. 2.1, 2.2, and 2.3, existing methods face significant challenges in jointly preserving background fidelity and maintaining semantic consistency during editing. On the one hand, text-driven diffusion models often introduce background drift or semantic misalignment, producing outputs that deviate from the intended edits. On the other hand, scene graph-based methods, while offering greater controllability, typically require additional training or fine-tuning at inference time [11,30]. This requirement increases computational and reproduction cost and leads to substantial editing latency, with processing times commonly ranging from six to ten minutes per image. Specifically, these limitations severely restrict scalability and hinder practicality in real-world scenarios. Consequently, the *trade-off between semantic alignment and background preservation* remains unresolved in prior work, motivating the need for a more efficient and training-free solution.

Advantages of Our Approach. VENUS addresses the aforementioned limitations by *leveraging scene graphs as structural priors while preserving the accessibility of text-based editing.* In particular, our approach integrates scene graph guidance directly into the diffusion process through a training-free noise inversion mechanism, which ensures that edits are accurately localized while leaving unedited regions unaffected. This design *retains the fine-grained relational control afforded by scene graphs*, while simultaneously benefiting from the high visual fidelity of state-of-the-art diffusion models. Furthermore, VENUS eliminates the need for additional training or inference-time fine-tuning, thereby *reducing editing latency* from several minutes to only a few seconds. Consequently, our approach enables controllable, efficient, and high-quality scene graph–based image editing, while remaining fully compatible with existing diffusion backbones.

3 Problem Formulation

In this work, we address the problem of editing images from scene graphs, where a scene graph is defined as a set of triplets, each augmented with bounding boxes and object categories. Formally, we define the scene graph as in Eq. (1).

$$\mathcal{G} = \{(s_i, r_i, o_i)\}_{i=1}^{N}, \quad N = |\mathcal{G}| \tag{1}$$

where s_i, r_i and o_i denote the subject, relation, and object of the i-th triplet.

Given a source image $\mathcal{I}$, our approach first detects objects and relations to construct an initial scene graph $\mathcal{G}$. The user then interactively modifies $\mathcal{G}$ (by selecting objects, editing attributes, or altering relations) to obtain a target graph $\mathcal{G}^*$. We mathematically express the interactive editing task in Eq. (2).

$$(\mathcal{I}, \Delta\mathcal{G}) \xrightarrow{f_\theta} \hat{\mathcal{I}} \tag{2}$$

where $\Delta\mathcal{G} = \mathcal{G}^* - \mathcal{G}$ represents the user-specified changes, and θ denotes the parameters of the model. The editor generates the edited image $\hat{I}$ by applying the modifications in $\Delta\mathcal{G}$ to $\mathcal{I}$, while preserving visual fidelity in unchanged regions.

4 Our Proposed Approach

Figure 1 presents our proposed approach, termed VENUS, which leverages multimodal LLM to extract and refine scene graphs from an image. We then parse the graph, prune irrelevant relations, and preserve key ones, using the result as a conditioning signal for precise and controllable diffusion-based image editing.

4.1 Scene Graph Construction and Refinement via MLLM

To leverage the spatial understanding capabilities of scene graphs, we employ Qwen-VL-2.5-7B, an advanced lightweight MLLM, to extract the scene graph $\mathcal{G}$ from an input image $\mathcal{I}$. Typically, after parsing the scene graph with an MLLM, a user may manually manipulate it. Alternatively, the MLLM itself can be employed to perform automated edits. In the automatic scenario, we generate an edited scene graph $\mathcal{G}'$ conditioned on an instruction $\mathcal{T}$, as defined in Eq. (3).

$$\mathcal{G} = \text{MLLM}(\mathcal{I}), \quad \mathcal{G}' = \text{MLLM}(\mathcal{I}, \mathcal{G}, \mathcal{T}) \tag{3}$$

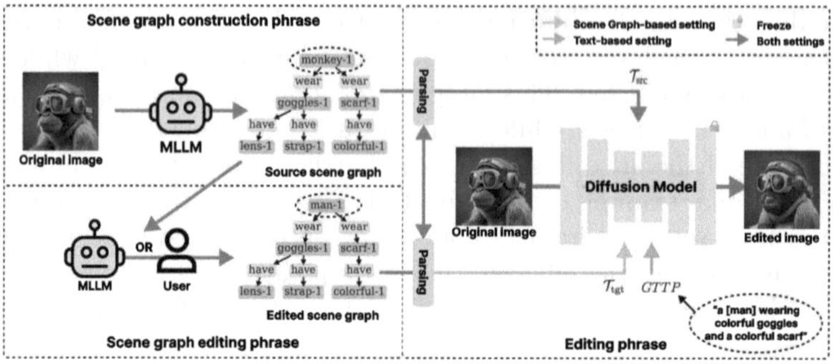

Fig. 1. An illustration of our VENUS approach. A scene graph is constructed from the input image, then edited by either the MLLM or the user. The edited graph is parsed into structured prompts, which condition a frozen diffusion model to edit the image. **(Best viewed in color and with zoom.)**

In the Eq. (3), the two-step process ensures that the original scene graph $\mathcal{G}$ and the edited scene graph $\mathcal{G}'$ remain structurally aligned. Thereby, this process preserves objects, subjects, and unaffected relations while simultaneously capturing the semantic modifications specified by the editing instruction.

4.2 Target Prompt and Source Prompt Construction

Target Prompt Construction. We first parse scene graphs into textual descriptions for text-to-image models. Unlike the previous method [23], which focuses on image generation, our goal is editing, thus we construct two separate prompts: one for edited content $\mathcal{T}_{\mathcal{G}_{\text{new}}}$ and one for preserved background $\mathcal{T}_{\mathcal{G}_{\text{bgd}}}$. We adopt a similar strategy by converting each relation triplet into a concise textual phrase of the form $t = (s, r, o)$ where s, r, and o denote the subject, relation, and object, respectively. These phrases are then concatenated to form a raw caption, which is defined as $\mathcal{T}_{\mathcal{G}} = [t_1, t_2, \ldots, t_N], \quad t_i = (s_i, r_i, o_i) \in \mathcal{G}$ For example, the triplet ("dog", "sitting on", "bench") becomes the phrase "dog sitting on bench". To better guide the image generation process, we further decompose the target caption $\mathcal{T}_{\text{tgt}}$ into two components: $\mathcal{T}_{\mathcal{G}_{\text{new}}}$ and $\mathcal{T}_{\mathcal{G}_{\text{bgd}}}$, defined in Eq. (4).

$$\mathcal{T}_{\text{tgt}} = \mathcal{T}_{\mathcal{G}_{\text{new}}} + \mathcal{T}_{\mathcal{G}_{\text{bgd}}} \tag{4}$$

where $\mathcal{T}_{\mathcal{G}_{\text{new}}}$ denotes the set of relation phrases that appear exclusively in $\mathcal{G}'$ but not in $\mathcal{G}$, and $\mathcal{T}_{\mathcal{G}_{\text{bgd}}}$ represents the set of relations shared by both $\mathcal{G}$ and $\mathcal{G}'$. The corresponding subgraphs $\mathcal{G}_{\text{new}}$ and $\mathcal{G}_{\text{bgd}}$ are formally defined in Eq. (5).

$$\mathcal{G}_{\text{new}} = \mathcal{G}' \setminus \mathcal{G}, \quad \mathcal{G}_{\text{bgd}} = \mathcal{G} \cap \mathcal{G}' \tag{5}$$

In the Eq. (5), the separation emphasizes novel or edited content during image synthesis and mitigates potential information loss caused by token truncation (e.g., exceeding the 77-token limit of CLIP's text encoder [18]).

Source Prompt Construction. Our approach supports the diffusion model in inverting image noise while preserving the vanilla scene structure, we do not utilize the full scene graph $\mathcal{G}$. Instead, we selectively retain only the subset of relation triplets that are common to both the vanilla and edited graphs, i.e., $\mathcal{G} \cap \mathcal{G}'$. These relations form the background caption, denoted as $\mathcal{T}_{\mathcal{G}_{\text{bgd}}}$. Accordingly, the source caption is defined as $\mathcal{T}_{\text{src}} = \mathcal{T}_{\mathcal{G}_{\text{bgd}}}$. This design choice ensures that the source prompt encodes only the stable semantic components, thereby facilitating faithful reconstruction of the input image during inversion.

4.3 Image Editing

Our approach can integrate with various diffusion-based editing backbones. In this work, we choose LEDIT++ [1] due to its strong performance in semantic editing and background preservation (see Table 1). In addition, our approach enables unified conditioning on both scene graphs and textual inputs.

Scene Graph-Based Editing. In our approach, we leverage the LEDIT++ [1] framework for scene graph-based image editing, in which the source $\mathcal{T}_{\text{src}}$ and target prompts $\mathcal{T}_{\text{tgt}}$, are used to guide the diffusion model throughout both the inversion and editing phases. Unlike vanilla LEDIT++ that utilize editing prompts in the form of short attribute-level instructions (e.g., "+glasses", "-hat"), we intentionally omit such simplistic expressions. Instead, we incorporate structurally rich prompts derived from scene graphs to provide more comprehensive semantic control over the editing process, as formally defined in Eq. (6).

$$\mathcal{I}_{\text{edited}} = \mathcal{M}(\mathcal{T}_{\text{src}}, \mathcal{T}_{\text{tgt}}) \tag{6}$$

where $\mathcal{M}$ denotes a text-guided image editing model, and $\mathcal{I}_{\text{edited}}$ is the output image after applying semantic edits guided by both source and target prompts.

Text-Based Editing. For text-based image editing, we also use LEDIT++ as the backbone of our framework. We observe that existing benchmarks predominantly evaluate edited images based on their CLIP similarity with the target prompt. To align with this evaluation, we utilize the ground-truth target prompt (GTTP) as input during editing. In addition, we incorporate the source prompt $\mathcal{T}_{\text{src}}$, which is constructed from the original scene graph, as auxiliary guidance during inversion and generation, as formally defined in Eq. (7).

$$\mathcal{I}_{\text{edited}} = \mathcal{M}(\mathcal{T}_{\text{src}}, \text{GTTP}) \tag{7}$$

where $\mathcal{M}$ denotes a text-guided image editing model, and $\mathcal{I}_{\text{edited}}$ is the output image after applying semantic edits guided by both source and target prompts.

Classifier-Free Guidance. Prompts are injected during both inversion and editing through classifier-free guidance (CFG) [7], ensuring that T_{src} and T_{tgt}

continuously influence the internal representations and effectively guide the image generation process. Specifically, this mechanism leads to edits that are more consistent and of higher visual quality. The inference-time CFG is defined as $\epsilon_{\text{pred}} = \epsilon_{\text{null text}} + s(\epsilon_{\text{text}} - \epsilon_{\text{null text}})$, where ϵ_{pred} denotes the final predicted noise, ϵ_{text} the prediction conditioned on the prompt, $\epsilon_{\text{null text}}$ the unconditioned prediction, and $s > 1$ the guidance scale controlling the strength of conditioning.

Table 1. Comparison of different models on background preservation and semantic consistency. **GTTP** denotes the *Ground-Truth Target Prompt* used during editing and evaluation. Best score in **bold**, second best in underline.

Method	Backbone	PSNR ↑	SSIM ↑	LPIPS ↓	CLIP Whole ↑
Text-based Editing Methods					
P2P [6]	–	17.87	0.7114	0.2088	25.01
Pix2Pix-Zero [2]	–	20.44	0.7467	0.1722	22.80
MasaCtrl [3]	–	22.17	0.7967	0.1066	23.96
PnP [22]	–	22.28	0.7905	0.1134	25.41
PnP + DirInv [10]	–	22.46	0.7968	0.1061	25.41
P2P + DirInv [10]	–	**27.22**	**0.8476**	**0.0545**	25.02
LEDIT++$_{\text{(only GTTP)}}$ [1]	SDv2.1	23.18	0.8218	0.0855	**26.85**
VENUS w/GTTP	LEDIT++	23.54	0.8288	0.084	**26.89**
VENUS w/GTTP	P2P+DirInv	**27.25**	**0.853**	**0.05**	25.99
Scene graph-based Editing Methods					
DiffSG [27]	–	9.35	0.41	0.55	12.45
SIMSG [5]	–	19.48	0.70	0.40	20.40
SGEdit [30]	SDv2.1	**22.45**	**0.79**	**0.10**	**24.19**
VENUS (Ours)	LEDIT++	**24.80**	**0.837**	**0.070**	**24.97**

5 Experiment Results

5.1 Implementation Details

Model Configuration. Our experiments build upon the LEDIT++ framework [1], employing Stable Diffusion v2.1 (SDv2.1) [19] as the generative backbone to enhance both visual fidelity and editability. Unless otherwise stated, we set the number of sampling steps to 50, the skip parameter to 25, and the random seed to 42. For semantic understanding and editing guidance, we use Qwen-VL 2.5 (7B) [21] as the multimodal language model to extract scene graphs from images and suggest structural edits. To prevent overly long prompts and token truncation in the text encoder, the number of extracted relations is capped at 15. All experiments are conducted on the NVIDIA A100 GPU with 80GB.

Dataset and Metric. We evaluate on PIE-Bench [10], a text-based benchmark designed for diverse image editing tasks. In addition, we report performance using PSNR [13], SSIM [26], and LPIPS [29] for background preservation, and CLIP similarity [18] for semantic alignment with the target prompt.

5.2 Quantitative Results

Table 1 reports a comparison between our approach and state-of-the-art baselines across four key metrics. Compared to the strongest scene graph-based model (SGEdit), our approach improves PSNR from 22.45 to 24.80 and SSIM from 0.79 to 0.84, while reducing LPIPS from 0.100 to 0.070. The CLIP similarity also increases from 24.19 to 24.97, indicating improved semantic alignment without compromising content fidelity. In the text-based setting, VENUS surpasses strong baselines, such as P2P+DirInv and LEDIT++, by a significant margin. With the LEDIT++ backbone under the GTTP configuration, we obtain a CLIP score of 26.89 (+0.04 over LEDIT++), along with measurable improvements in background preservation. Using the P2P+DirInv backbone, our approach yields a +0.97 increase in the CLIP score and further improves preservation metrics. This demonstrates that our split-prompt design forces the model to respect edited relations while avoiding catastrophic drift in background context, a balance that previous direct-prompting approaches fail to achieve. We emphasize that conditioning on the GTTP in PIE-Bench is not an unfair bias but mirrors the standard operating mode of conventional text-based image editing, where the user-provided target prompt directly guides the edit. Our text-based setting therefore faithfully simulates real-world usage, ensuring a fair, apples-to-apples comparison with existing text-based baselines (e.g., LEDIT++, P2P+DirInv, SGEdit). This result highlights that VENUS is not restricted to scene graph editing: even in the purely text-based setting, VENUS achieves the highest CLIP similarity (26.89) and improved fidelity, surpassing previous models.

Table 2. Comparison on EditVal. Accuracy is measured using OwL-ViT [15], and fidelity is measured using DINO [4] (higher is better). Reported runtime corresponds to the average editing time per image. Best score in **bold**, second best in <u>underline</u>.

Method	Accuracy	Fidelity	Time per image
	OwL-ViT ↑	DINO ↑	
Scene graph-based Editing Methods			
SIMSG	0.11	0.57	–
DiffSG	0.01	0.13	–
SGEdit	**0.53**	<u>0.83</u>	<u>6–10 m</u>
VENUS (Ours)	<u>0.32</u>	**0.87**	**20–30 s**

Additionally, we evaluate our approach on the EditVal benchmark using OwL-ViT [15] and DINO [4] as automatic metrics in Table 2. Our approach attains an OwL-ViT accuracy of 32% and a DINO fidelity score of 87%. While the fidelity surpasses all baselines, the accuracy falls short of SGEdit (53%). We attribute this gap to two primary factors: (1) EditVal is built on MS-COCO [14], which features diverse aspect ratios and resolutions, whereas our pipeline is based on Stable Diffusion v2.1 [19] without specific adaptation for such variations; and (2) EditVal emphasizes precise spatial localization of edits, a criterion not explicitly optimized by LEDIT++ backbones. We note that this limitation is common across many diffusion-based models. Moreover, *without any fine-tuning phase, VENUS significantly reduces editing time to just 20–30 s per image.*

Table 3. Comparison of editing performance with constructed and ground-truth target prompts. Scores are ImageReward (higher is better) and FID (lower is better), evaluated on PIE-Bench. Best score in **bold**, second best in underline.

Method	Backbone	ImageReward ↑		FID ↓
		Constructed	GT	
LEDIT++ (only GTTP)	SDv2.1	0.514	**0.636**	53.01
VENUS w/GTTP	LEDIT++	0.363	0.624	53.54
VENUS (Ours)	LEDIT++	**0.716**	0.071	**47.38**

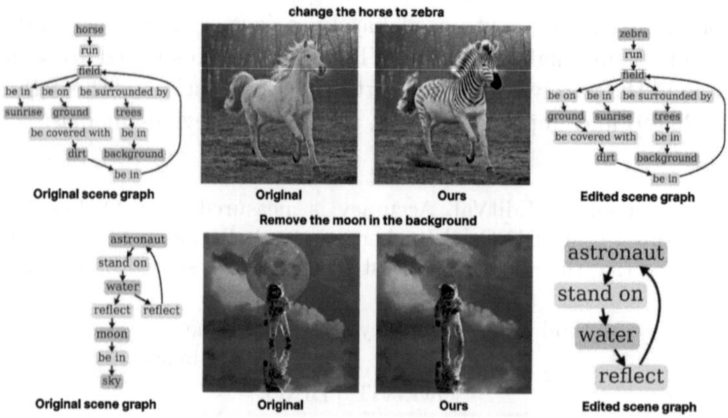

Fig. 2. Examples of scene graph guided image editing. Top row: changing the horse into a zebra by updating the corresponding node in the scene graph. Bottom row: removing the moon in the background by deleting its associated nodes and relations. **(Best viewed in color and with zoom.)**.

Table 3 compares VENUS with text-based editing baselines using the ImageReward and FID metrics. With our constructed target prompts, the scene graph-based setting attains the highest ImageReward score of 0.716, outperforming both the text-based variant (0.363) and the LEDIT++ baseline with GTTP only (0.514). When evaluated using ground-truth prompts, LEDIT++ (GTTP only) achieves the highest score (0.636), with our text-based variant following closely at 0.624. In terms of FID, our setting achieves the best visual quality (47.38), outperforming LEDIT++ (53.01) and our text-based variant (53.54).

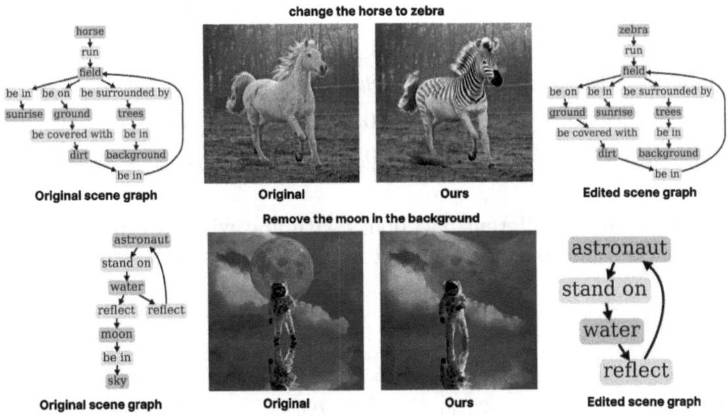

Fig. 3. Comparison of editing results across a variety of tasks. Compared to LEDIT++, P2P-DirInv, and PnP-DirInv, VENUS (ours) produces edits that are both semantically accurate and visually consistent. **(Best viewed in color and with zoom.)**.

5.3 Qualitative Results

To demonstrate the effect of incorporating scene graph inputs into the generated outputs, Fig. 2 showcases VENUS's ability to edit images based on complex scene graph modifications. The edited results exhibit a strong correspondence between the manipulated scene graphs and the visual outputs. With the proposed split prompt strategy, our approach can effectively handle structural edits, such as object removal. For example, in the second case, the moon is successfully removed from the background while preserving other scene elements.

In addition, we conduct a qualitative assessment to visually analyze the strengths and limitations of our approach. As illustrated in Fig. 3, VENUS produces edited images with superior background preservation and semantic consistency compared to state-of-the-art text-based editing models [1,10]. The proposed approach more accurately focuses on the intended editing regions while minimizing unintended changes, thereby enhancing background fidelity.

Table 4. Evaluating the contribution of the Src prompt to background preservation and semantic consistency. w/o$_{\text{Src prompt}}$ removes the source prompt during inversion; VENUS w/$_{\text{GTTP}}$ replaces our constructed target prompt with the ground-truth target prompt from PIE-Bench. Best score in **bold**, second best in <u>underline</u>.

Method	PSNR ↑	SSIM ↑	LPIPS ↓	CLIP Whole ↑
Scene Graph-based Editing Approaches				
VENUS	**24.80**	**0.837**	**0.070**	**24.97**
w/o$_{\text{Src prompt}}$	<u>23.42</u>	<u>0.820</u>	<u>0.082</u>	<u>24.96</u>
Text-based Editing Approaches				
VENUS w/$_{\text{GTTP}}$	**23.54**	**0.829**	**0.084**	**26.89**
LEDIT++ [1]	<u>23.18</u>	<u>0.822</u>	<u>0.086</u>	<u>26.85</u>

Table 5. Comparison of different backbones with and without our text-based setting approach for background preservation and semantic consistency.

Method	Backbone	PSNR ↑	SSIM ↑	LPIPS ↓	CLIP Whole ↑
Text-based Editing Methods					
PnP + DirInv [10]	–	22.46	0.7968	0.1061	<u>25.41</u>
P2P + DirInv [10]	–	**27.22**	**0.8476**	**0.0545**	25.02
LEDIT++$_{\text{(only GTTP)}}$	–	<u>23.18</u>	<u>0.8218</u>	<u>0.0855</u>	**26.85**
Our Approach					
VENUS w/$_{\text{GTTP}}$	PnP+DirInv	22.35 (-0.11)	0.8019 (+0.0051)	0.1066 (+0.0005)	<u>26.21</u> (+0.80)
VENUS w/$_{\text{GTTP}}$	P2P+DirInv	**27.25** (+0.03)	**0.853** (+0.0054)	**0.05** (-0.0045)	25.99 (+0.97)
VENUS w/$_{\text{GTTP}}$	LEDIT++	<u>23.54</u> (+0.36)	<u>0.8288</u> (+0.007)	<u>0.084</u> (-0.0015)	**26.89** (+0.04)

5.4 Ablations Study

Enhancing Text-Based Image Editing Models. Table 5 reports that the effect of scene graph–based split prompt conditioning consistently improves text-based image editing across backbones (PnP+DirInv, P2P+DirInv, LEDIT++). The largest gain is observed in P2P+DirInv (PSNR 27.25, SSIM 0.853, LPIPS 0.05, +0.97 CLIP), indicating stronger semantic alignment with high fidelity. LEDIT++ achieves modest but improvements in fidelity and context retention.

Effectiveness and Efficiency. As shown in Table 4, when the source prompt is removed, the performance drops significantly in terms of scene integrity, with a slight decrease in semantic consistency in both settings. Specifically, this indicates that the source prompt effectively improves background fidelity. Moreover, scene graph-based methods better maintain background details than text-based methods, highlighting the value of structured representations for coherent edits.

Table 6. Comparison of different MMLs and the improvements achieved by VENUS in both text-based settings and scene graph-based settings for background preservation and semantic consistency. GTTP denotes the *Ground-Truth Target Prompt* used during editing and evaluation. Best score in **bold**, second best in underline.

Method	Multimodal LLM	Para.	PSNR ↑	SSIM ↑	LPIPS ↓	CLIP Whole ↑
Text-based Editing Approaches						
VENUS w/GTTP	Qwen2.5-VL-instruct	7B	23.54	0.8288	0.084	26.89
VENUS w/GTTP	Phi-3.5-vision-instruct	4.15B	**23.58**	**0.8256**	**0.081**	26.87
VENUS w/GTTP	InternVL2.5	4B	23.24	0.8217	0.085	**26.92**
Scene graph-based Editing Approaches						
VENUS	Qwen2.5-VL-instruct	7B	24.80	**0.837**	**0.070**	**24.97**
VENUS	Phi-3.5-vision-instruct	4.15B	24.62	0.834	0.073	24.84
VENUS	InternVL2.5	4B	**25.08**	0.835	0.073	23.65

Effect of Different MLLM Backbones. We evaluate our split-prompt strategy in both text-based and scene-graph-based settings using Qwen2.5-VL-instruct (7B), Phi-3.5-vision-instruct (4.15B), and InternVL2.5 (4B). As shown in Table 6, performance is stable across backbones, with larger models (Qwen2.5-VL) tending to yield higher CLIP, while smaller models (InternVL2.5) excel in background preservation. These results indicate that our approach is robust to the choice of MLLM and allows trade-offs between semantic alignment and fidelity.

6 Conclusion

In this work, we introduced VENUS, a training-free approach for image editing guided by scene graphs. By leveraging a split-prompt strategy that disentangles edited content from preserved background, VENUS enables diffusion models to achieve semantic controllability and background fidelity, overcoming a limitation of prior methods. Beyond scene graph-based methods, VENUS also enhances prominent text-based editing backbones, consistently improving performance.

For future work, we plan to further investigate the potential of lightweight multimodal LLMs, which our experiments suggest can achieve performance comparable to that of larger counterparts. In addition, we aim to explore the use of richer scene graph annotations, including bounding boxes/masks, to enhance position editing accuracy by more precisely delineating true object boundaries.

Acknowledgment. This research is funded by Vietnam National University HoChiMinh City (VNU-HCM) under a project within the framework of the Program titled “Strengthening the capacity for education and basic scientific research integrated with strategic technologies at VNU-HCM, aiming to achieve advanced standards comparable to regional and global levels during the 2025–2030 period, with a vision toward 2045”.

References

1. Brack, M., et al.: Ledits++: limitless image editing using text-to-image models. In: Proceedings of the IEEE/CVF Conference on Computer Vision and Pattern Recognition, pp. 8861–8870 (2024)
2. Brooks, T., Holynski, A., Efros, A.A.: Instructpix2pix: learning to follow image editing instructions. In: Proceedings of the IEEE/CVF Conference on Computer Vision and Pattern Recognition, pp. 18392–18402 (2023)
3. Cao, M., Wang, X., Qi, Z., Shan, Y., Qie, X., Zheng, Y.: Masactrl: tuning-free mutual self-attention control for consistent image synthesis and editing. In: Proceedings of the IEEE/CVF International Conference on Computer Vision, pp. 22560–22570 (2023)
4. Caron, M., et al.: Emerging properties in self-supervised vision transformers. In: Proceedings of the IEEE/CVF International Conference on Computer Vision, pp. 9650–9660 (2021)
5. Dhamo, H., et al.: Semantic image manipulation using scene graphs. In: Proceedings of the IEEE/CVF Conference on Computer Vision and Pattern Recognition, pp. 5213–5222 (2020)
6. Hertz, A., Mokady, R., Tenenbaum, J.M., Aberman, K., Pritch, Y., Cohen-Or, D.: Prompt-to-prompt image editing with cross attention control. arXiv abs/2208.01626 (2022). https://api.semanticscholar.org/CorpusID:251252882
7. Ho, J.: Classifier-free diffusion guidance. arXiv abs/2207.12598 (2022). https://api.semanticscholar.org/CorpusID:249145348
8. Ho, J., Jain, A., Abbeel, P.: Denoising diffusion probabilistic models. Adv. Neural. Inf. Process. Syst. **33**, 6840–6851 (2020)
9. Johnson, J., Gupta, A., Fei-Fei, L.: Image generation from scene graphs. In: Proceedings of the IEEE Conference on Computer Vision and Pattern Recognition, pp. 1219–1228 (2018)
10. Ju, X., Zeng, A., Bian, Y., Liu, S., Xu, Q.: PNP inversion: boosting diffusion-based editing with 3 lines of code. In: International Conference on Learning Representations (ICLR) (2024)
11. Kawar, B., et al.: Imagic: text-based real image editing with diffusion models. In: Proceedings of the IEEE/CVF Conference on Computer Vision and Pattern Recognition, pp. 6007–6017 (2023)
12. Kim, G., Kwon, T., Ye, J.C.: Diffusionclip: text-guided diffusion models for robust image manipulation. In: Proceedings of the IEEE/CVF Conference on Computer Vision and Pattern Recognition, pp. 2426–2435 (2022)
13. Korhonen, J., You, J.: Peak signal-to-noise ratio revisited: is simple is beautiful? (2012). https://doi.org/10.1109/QoMEX.2012.6263880
14. Lin, T.Y., et al.: Microsoft coco: common objects in context. In: European Conference on Computer Vision, pp. 740–755. Springer (2014)
15. Minderer, M., et al.: Simple open-vocabulary object detection. In: European Conference on Computer Vision, pp. 728–755. Springer (2022)
16. Nguyen, K., Tripathi, S., Du, B., Guha, T., Nguyen, T.Q.: In defense of scene graphs for image captioning. In: Proceedings of the IEEE/CVF International Conference on Computer Vision (ICCV), pp. 1407–1416 (2021)
17. Qin, J., et al.: Diffusiongpt: LLM-driven text-to-image generation system. arXiv preprint arXiv:2401.10061 (2024)
18. Radford, A., et al.: Learning transferable visual models from natural language supervision. In: International Conference on Machine Learning, pp. 8748–8763. PmLR (2021)

19. Rombach, R., Blattmann, A., Lorenz, D., Esser, P., Ommer, B.: High-resolution image synthesis with latent diffusion models. In: Proceedings of the IEEE/CVF Conference on Computer Vision and Pattern Recognition, pp. 10684–10695 (2022)
20. Sun, P., et al.: Autoregressive model beats diffusion: llama for scalable image generation. arXiv preprint arXiv:2406.06525 (2024)
21. Team, Q.: Qwen2.5-vl (2025). https://qwenlm.github.io/blog/qwen2.5-vl/
22. Tumanyan, N., Geyer, M., Bagon, S., Dekel, T.: Plug-and-play diffusion features for text-driven image-to-image translation. In: Proceedings of the IEEE/CVF Conference on Computer Vision and Pattern Recognition, pp. 1921–1930 (2023)
23. Vo, T.N., Nguyen, T.T., Nguyen, T.V., Tran, M.T.: Saturn: autoregressive image generation guided by scene graphs. In: Proceedings of the International Conference on Multimedia Analysis and Pattern Recognition (MAPR) (2025)
24. Wang, Z., Li, A., Li, Z., Liu, X.: Genartist: multimodal LLM as an agent for unified image generation and editing. Adv. Neural. Inf. Process. Syst. **37**, 128374–128395 (2024)
25. Wang, Z., Zhao, L., Xing, W.: Stylediffusion: controllable disentangled style transfer via diffusion models. In: Proceedings of the IEEE/CVF International Conference on Computer Vision, pp. 7677–7689 (2023)
26. Wang, Z., Bovik, A., Sheikh, H., Simoncelli, E.: Image quality assessment: from error visibility to structural similarity. IEEE Trans. Image Process. **13**(4), 600–612 (2004). https://doi.org/10.1109/TIP.2003.819861
27. Yang, L., et al.: Diffusion-based scene graph to image generation with masked contrastive pre-training. arXiv preprint arXiv:2211.11138 (2022)
28. Yoon, S., et al.: Image-to-image retrieval by learning similarity between scene graphs. In: Proceedings of the AAAI Conference on Artificial Intelligence, vol. 35, pp. 10718–10726 (2021)
29. Zhang, R., Isola, P., Efros, A.A., Shechtman, E., Wang, O.: The unreasonable effectiveness of deep features as a perceptual metric. In: CVPR (2018)
30. Zhang, Z., Chen, D., Liao, J.: Sgedit: bridging LLM with text2image generative model for scene graph-based image editing. arXiv preprint arXiv:2410.11815 (2024)

Noise Scale Controllable Anomaly Synthesis Strategy for Industrial Anomaly Detection and Localization

Yuchen Deng[1], Hongyou Chen[1](✉), Lingfeng Qu[2], Yong Jiang[1], and Yong Fan[1]

[1] Sichuan Big Data and Intelligent System Engineering Technology Research Center, Southwest University of Science and Technology, Mianyang 621010, China
chy2019@foxmail.com

[2] Cyberspace Institute of Advanced Technology, Guangzhou University, Guangzhou 510006, China

Abstract. In the surface anomaly detection of industrial products, due to the scarcity of abnormal samples, unsupervised methods have become mainstream approaches, and the detection ability is enhanced by introducing anomaly synthesis strategy. Currently, the method of adding Gaussian noise to normal features to simulate abnormal features has achieved excellent results, but the control of noise scale has become a difficult problem. Therefore, we propose a noise scale controllable anomaly feature synthesis strategy (NSCAS). The strategy consists of three parts: noise scale control, abnormal mask generation and abnormal feature synthesis. The NSCAS model achieves advanced detection results on MVTec-AD (I-AUROC of 99.6%), VisA (P-AUROC of 98.4%) datasets. Its effectiveness and generalization ability have been further verified in the self-built speaker shell anomaly dataset (SS-AD).

Keywords: Deep learning · Industrial anomaly detection · Anomaly synthesis · Noise scale control · Unsupervised learning

1 Introduction

Surface anomaly detection and location aims to identify the anomaly on the surface of the object and locate the area where the anomaly is located. Due to the small number of abnormal samples in the industrial production process and the high cost of sample labeling, the use of supervised methods is limited. Therefore, unsupervised anomaly detection technology has been widely used in industrial detection scenarios [2,9,12]. The existing anomaly detection methods are roughly divided into three categories: reconstruction-based methods [1,8,23], embedding-based methods [7,14], and synthesis-based methods [18,22].

Reconstruction-based methods [1,23] aim to reconstruct the input image from the latent space, and then detect anomalies by analyzing the errors before and after reconstruction. However, these methods rely heavily on the quality of the

J. Lokoč et al. (Eds.): MMM 2026, LNCS 16413, pp. 204–218, 2026.
https://doi.org/10.1007/978-981-95-6957-1_15

reconstructed image. Embedding-based methods [7]use pre-trained convolutional neural networks to extract features, where anomalous features are clearly separated from normal clusters in the feature space. However, the features extracted using pre-trained networks can be biased. Synthesis-based methods introduce discriminative information through anomaly synthesis based on reconstruction and embedding frameworks [10,16], but such anomalies may deviate too far from normal patterns to provide meaningful learning information.

Recently, hybrid methods based on synthesis and embedding have achieved high-level results [5,6,11]. ImageNet pre-trained backbone network (such as ResNet) is used to extract features, and then feature-level anomalies are synthesized by adding Gaussian noise to normal features based on feature-level anomaly synthesis strategy [5,13,20,21]. This simple synthesis strategy avoids repeated feature extraction, efficiently generating and utilizing diverse anomalies. The similarity between synthetic and real anomaly features depends on the noise scale, which controls their distance from normal features. Excessive σ value will lead to a high Euclidean distance between abnormal features and normal features. Training on too large σ will lead to loose decision boundaries, resulting in high false negatives. On the contrary, if σ is too small, the training process will become unstable, and the discriminator cannot be well generalized to normal features [10].

To alleviate the above issues, a controllable anomaly synthesis strategy is proposed. Firstly, in order to control the size of the noise scale, the distribution difference between the synthetic anomaly and the real anomaly feature is as small as possible. We control the distance between synthesized anomalous features and normal features in the feature space using a noise scale controller. Secondly, the anomaly mask is generated by the binarization of Perlin noise, so as to limit the range of noise added on the normal feature map to simulate the real abnormal geometric features. Finally, in order to reduce the interference of irrelevant information such as background, shadow and edge, the detection accuracy of the discriminator is improved by adaptive channel weighting.

2 Related Work

Reconstruction-based methods are widely used in the field of industrial anomaly detection. Its core is to assume that the model can learn effective reconstruction patterns from normal data. Therefore, the normal data can be well reconstructed, and the reconstruction error is small. However, the unknown abnormal data cannot be effectively reconstructed by the model, and the reconstruction error is large. Then the anomaly is detected by analyzing the reconstruction error. Autoencoder (AE) is the most typical method of reconstruction methods [8,24,25]. The core idea is to map the high-dimensional input into a low-dimensional potential space through an encoder. Then the potential representation is reconstructed back to the original space through a decoder. However, if only normal samples are used for training, the details of normal data may be over-fitted during the training process, resulting in the accurate reconstruction of abnormal regions.

The embedding-based method shows excellent performance in anomaly detection and localization tasks. These methods use ImageNet pre-training network for feature extraction. By embedding normal features into the compressed space, the abnormal features are far away from the normal sample clustering center in the feature space. The Patchcore model [14] and the EfficientAD model [3] extract patch features through the pre-training model to construct a representative normal image patch feature memory library, and combine the nearest neighbor search to calculate the abnormal score. This kind of method scores the input features by calculating the Mahalanobis distance or the maximum feature distance between the patch and the normal feature patch.

The synthesis-based method treats the generation of anomalies as data augmentation of normal samples by controlling the injection or synthesis of pseudo anomalies on normal data. CutPaste [10] randomly cuts a patch on the image, and then pastes it to another random position of the image to generate a pseudo-abnormal sample. Then a neural network is trained to distinguish between normal samples and abnormal samples. [18,20,22] Use Perlin noise binarization to generate a more natural binary mask, which is used to capture various abnormal shapes, and sample abnormal texture synthesis anomalies from the abnormal source image data set unrelated to the input image.

Image-level anomaly synthesis provides detailed anomalous textures but lacks diversity, whereas feature-level anomaly synthesis is more efficient but faces challenges in controllability. Therefore, this paper proposes a controllable anomaly synthesis strategy.

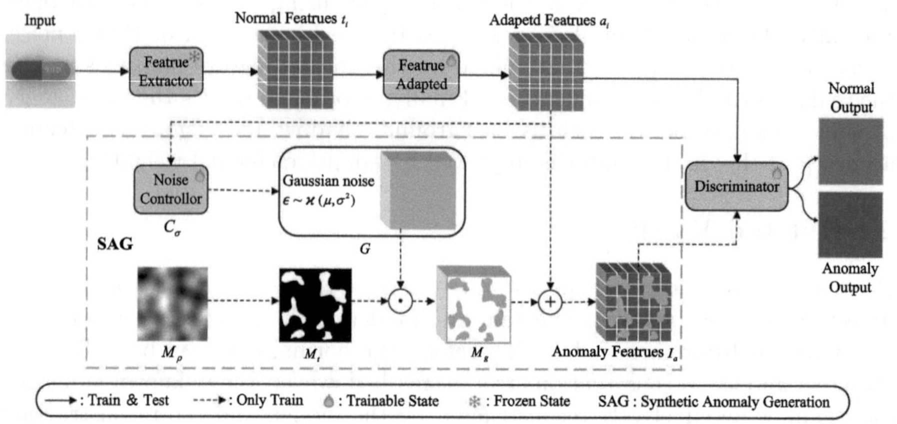

Fig. 1. NSCAS architecture description. In the training phase, the abnormal features are synthesized by SAG. The anomaly is synthesized by adding Gaussian noise to the feature space, and the scale of the added noise is controlled. The range of noise addition is constrained by the anomaly mask, and finally the anomaly feature map is obtained. And are input into the discriminator as positive samples and negative samples respectively to train the discriminator. Remove SAG when reasoning.

3 Proposed Method

The proposed framework is shown in Fig. 1. In the training phase, it consists of a feature extractor E_ϕ, a feature adapter A_θ, a noise scale controller C_σ, an abnormal feature generator G and a discriminator D_ψ. The backbone of E_ϕ is pre-trained on ImageNet and kept frozen, and A_θ is to reduce the potential domain bias caused by the pre-trained backbone network. C_σ maintains an appropriate distance between the generated abnormal features and the normal features in the feature space by controlling the standard deviation σ of Gaussian noise. G synthesizes abnormal features by adding Gaussian noise to normal features. The classification of normal and abnormal images is completed by a discriminator D_ψ, which is trainable. In the inference phase, only the feature extractor E_ϕ, the feature adapter A_θ, and the discriminator D_ψ are used.

3.1 Feature Extractor and Feature Adaptor

The pre-trained network is used to extract multi-scale and multi-channel features. In this paper, the ResNet backbone network ϕ, which is pre-trained and frozen parameters on ImageNet, is used as the feature extractor. The normal image x_i first outputs the feature tensor $\phi_j(x_i) \in \mathbb{R}^{H_j \times W_j \times C_j}$ with shape $H_j \times W_j \times C_j$ at the j-th layer through the feature extractor E_ϕ. The vector at position (h, w) is denoted by $\phi_{i,j}^{(h,w)} \in \mathbb{R}^{C_j}$. The channel vector at (h,w) is denoted by $\phi_{(i,j)}^{(h,w)} \in \mathbb{R}^{C_j}$ The domain features are aggregated by adaptive average pooling. The domain set centered on (h, w) with size p is:

$$N_p^{h,w} = \left\{(a,b)\middle| a \in \left[h - \lfloor\frac{p}{2}\rfloor, \cdots, h + \lfloor\frac{p}{2}\rfloor\right], b \in \left[w - \lfloor\frac{p}{2}\rfloor, \cdots, w + \lfloor\frac{p}{2}\rfloor\right]\right\} \quad (1)$$

Then the domain aggregation feature vector at (h,w) is expressed as $s_{i,j}^{h,w} \in \mathbb{R}^{C_j}$. To fuse low-level and high-level feature information at the same time, a multi-level feature cascade function f_{concat} is defined, and the aggregation results of each layer are adjusted to the maximum resolution (H_m, W_m) of the lowest level through the upsampling function $f_{resize}^{H_m,W_m}$. To maintain the versatility of the features, we only use two intermediate layers j and $j+1$. Thus, the feature mapping of the image x_i is expressed as:

$$t_i = E_\phi(x_i) = f_{concat}\left(\left\{f_{resize}^{H_m,W_m}(s_{i,j})\middle| j \in J\right\}\right) \quad (2)$$

where J denotes the selected hierarchy set.

We use A_θ to map the features extracted by the pre-trained and frozen E_ϕ to a low-dimensional representation space that is more suitable for the current task, thereby alleviating the feature distribution deviation caused by the pre-trained ResNet. It is realized as a fully connected layer. Generate adaptive local features, denoted as $a_i^{h,w}$.

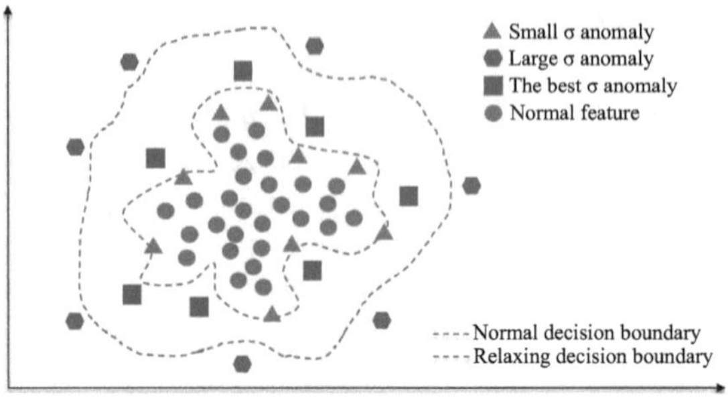

Fig. 2. In the feature space, the green circle is a normal feature, and the brown triangle is an abnormal feature obtained by adding a small σ Gaussian noise to the normal feature. The red hexagon is the anomaly characteristic corresponding to the larger σ. The blue square is the best σ corresponding anomaly feature. (Color figure online)

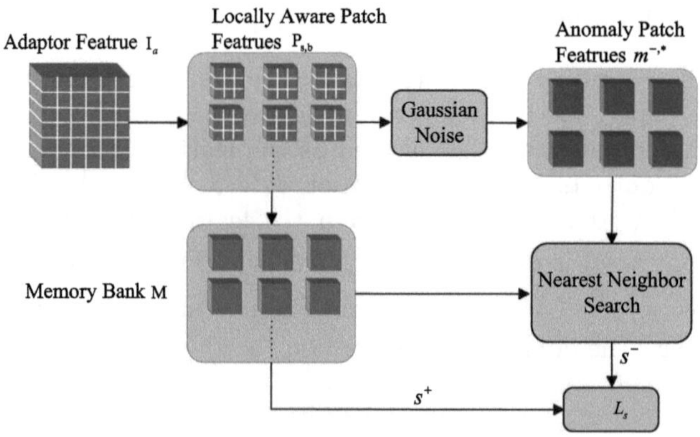

Fig. 3. Noise Controllor architecture description. The adapted normal feature map I_a is decomposed into a patch-level feature repository M with domain-aware capabilities. Randomly select k normal patch features and add Gaussian noise to synthesize abnormal patch features $g^j_{h,w}, (j \in k)$. The abnormal fraction of $g^j_{h,w}$ is calculated.

3.2 Noise Scale Controller

Normal feature points will form compact clusters in the feature space [15]. The scale σ of Gaussian noise controls the degree of deviation between the synthesized abnormal features and the normal feature clusters. As shown in Fig. 2, the small value of σ will make the synthetic abnormal features within or on the decision boundary, resulting in unstable training. The larger σ value will make the abnormal feature too far away from the decision boundary, resulting in boundary relaxation [11]. To ensure that the synthesized anomalies are positioned at an appropriate distance from the boundary, inspired by PatchCore [14], we measure the degree of anomaly by calculating the distance between the synthesized anomalous feature vectors and the normal feature vectors in the memory bank. The anomaly degree of the outermost features in the normal feature cluster is then used as the threshold. Then, the threshold is used to guide the adjustment of the σ value, so as to control the abnormal synthesis. The framework of the noise scale controller is shown in Fig. 3.

Normal Feature Memory Bank: The normal feature memory bank is composed of patch-level features. The adaptive feature map $a_i^{(h,w)} \in \mathbb{R}^{H\times W\times C}$ is obtained after the feature map of image x_i passes through the feature adapter A_θ, where H, W and C are the height, width and channel size of the feature map respectively. We use a $b\times b$ sliding window to divide the feature map into several patches of $b\times b$ size, and the step size is s, the step size parameter is optional, and the default setting is 1. For the feature map tensor a_i, its local perceptual patch feature set $P_{s,b}(a_i)$ is defined as:

$$P_{s,b}(a_i) = \left\{a_i\left(N_b^{h,w}\right) \,\middle|\, h < H,\, w < W,\, h, w \in \mathbb{N}\right\} \tag{3}$$

Finally, the normal feature memory M is defined as:

$$\mathrm{M} = \bigcup_{x_i\in\chi_n} P_{s,b}\left(a_j(x_i)\right) \tag{4}$$

where χ_n is the first n pictures in the training sample.

Feature-Level Patch Exception Generation: According to the methods of References [5,11,21], abnormal features are generated by adding Gaussian noise to normal features. Specifically, a Gaussian noise vector $\varepsilon_i^{h,w} \in \mathbb{R}^c$ is sampled, where each element is independently drawn from a Gaussian distribution $\varepsilon_i^{h,w} \sim \kappa(\mu_g, \sigma_g)$. The normal feature $a_i^{h,w}$ is then perturbed to obtain the Gaussian anomaly feature $g_i^{h,w}$:, which is expressed as:

$$g_i^{h,w} = a_i^{h,w} + \varepsilon_i^{h,w} \tag{5}$$

Distance Control: In order to control the distance between the synthesized abnormal features and the normal features, we first randomly select k feature patch sets in the normal feature memory library M, and then use the above feature-level patch anomaly generation method to obtain k abnormal patch feature sets $m^{-,*}$. Initially, the noise scale σ is large, so $m^{-,*}$ is far from normal features. We calculate the anomaly score $s^- \in \mathbb{R}$ of $m^{-,*}$ and the normal score s^+ of the corresponding normal patches. The anomaly score loss function L_s is constructed using s^- and s^+ to guide anomaly synthesis. The anomaly score s^* is defined as the minimum L_2 distance between a generated abnormal patch feature $m_i^{-,*}$ and its nearest neighbor m_j^* in M:

$$m_j^* = \arg\min_{m\in M} \|m_i^{-,*} - m\|_2 \tag{6}$$

$$s^* = \|m_i^{-,*} - m_j^*\|_2 \tag{7}$$

If the feature m_j^* closest to $m_i^{-,*}$ in M is far away from its adjacent feature points, it is necessary to reduce its abnormal score to make the abnormal score more representative. For this reason, we introduce the scaling factor ω. The final score s^- is:

$$s^- = \left(1 - \frac{\exp\left(\|m_i^{-,*} - m_j^*\|_2\right)}{\sum_{m \in \mathcal{Z}_b(m_j^*)} \exp\left(\|m_i^{-,*} - m\|_2\right)}\right) \cdot s^* \tag{8}$$

The abnormal score of s^+ for the normal feature m_j^* is:

$$s^+ = \frac{1}{b} \sum_{m \in \mathcal{Z}_b(m_j^*)} \|m_j^* - m\|_2 \tag{9}$$

$Z_b(m^*)$ denotes the b nearest neighbor patch feature from the patch feature m_j^* in M. The abnormal score s^+ represents the distribution of normal samples near the decision boundary. If the abnormal score s^+ of the normal patch feature is large, it indicates that the normal feature distribution near the patch is sparse, so the distance between the normal feature and the abnormal feature is large. On the contrary, the distance is small. Therefore, we set the threshold according to s^+ to control the distance between the synthetic anomaly and the decision boundary. The threshold is usually set to $2s^+$. Abnormal score loss function L_s:

$$L_s = \left[\max(0, s^- - 2\,s^+)\right]^2 = \begin{cases} (s^- - 2\,s^+)^2, & s^- > 2\,s^+ \\ 0, & s^- \le 2\,s^+ \end{cases} \tag{10}$$

L_s provides gradient guidance to adjust the noise scale σ, ensuring that the synthesized anomaly is close to the decision boundary but does not cross it.

3.3 Anomaly Feature Generator

Firstly, the performance cons binary anomaly mask M_t is created by binarizing the randomly generated Perlin noise image M_ρ, following a process similar to References [22] and [18]. Then, sampled Gaussian noise ε is added only within the effective range defined by M_t to obtain M_g. Finally, M_g is added to the adaptive feature a_i to generate the perturbation feature map I_a. The abnormal feature generation is defined as:

$$M_g = M_t \odot \varepsilon \tag{11}$$

$$I_a = M_g + a_i \tag{12}$$

Through this anomaly feature generation strategy, the generated anomalies are more real, more diverse and more natural.

3.4 Discriminator

The discriminator D_ψ, acting as a normality scorer, evaluates each position (h, w) on the input feature map. It maximizes scores for normal features and minimizes scores for synthetic abnormal features to distinguish between normal and abnormal samples. The input to the discriminator consists of the multi-channel feature map a_i and the synthesized abnormal feature map I_a from each training image x_i after feature extraction and adaptation. To enhance inter-channel dependency and representation, adaptive channel weighting is applied. Given input feature $X \in \mathbb{R}^{C\times H\times W}$, we first perform global average pooling on each channel to get the channel descriptor $z \in \mathbb{R}^C$:

$$z_c = \frac{1}{H \times W}\sum_{i=1}^{H}\sum_{j=1}^{W} X_{c,i,j} \tag{13}$$

Then, the channel weights $s \in [0,1]^C$ are learned via two fully connected layers:

$$s = \lambda(W_2 \cdot \mathrm{ReLU}(W_1 z)) \tag{14}$$

where $\lambda(\cdot)$ is the element-by-element Sigmoid. Finally, the weight vector s is multiplied by the original feature channel to complete the channel weighting, which is defined as:

$$\tilde{X}_c = s_c \cdot X_c \tag{15}$$

The discriminator uses a two-layer multi-layer perceptron (MLP) as the classifier, and the output is $D_\psi(g_{h,w}) \in \mathbb{R}$, indicating the normality of the position. In order to make the discriminator not only distinguish between normal samples and abnormal samples, but also prevent excessive punishment of extreme scores, we use bilateral truncation L_1 loss:

$$\ell^i_{h,w} = \max(0, \mathrm{th}^+ - D_\psi(g^i_{h,w})) + \max(0, -\mathrm{th}^- + D_\psi(g^{i-}_{h,w})) \tag{16}$$

The truncation terms $\mathrm{th}^+ = 0.5$ and $\mathrm{th}^- = -0.5$. The total objective function of the training phase is as follows:

$$L = \min_{\theta,\psi} \sum_{x^i \in \mathcal{X}_{train}} \sum_{h,w} \frac{l^i_{h,w}}{H_0 * W_0} + L_s \tag{17}$$

L_s is derived from formula 10.

3.5 Inference and Abnormal Scoring

As shown in Fig. 3, the inference process excludes both the noise controller and the anomaly feature generator. Given a test image $x_i \in X_{\text{test}}$, the adaptive feature map $a_i \in \mathbb{R}^{C\times H\times W}$ is obtained by sequentially applying the pretrained feature extractor E_ϕ and the feature adapter A_θ. Next, each C-dimensional feature vector $a_i^{(h,w)} \in \mathbb{R}^C$ at position (h, w) is fed into the discriminator D_ψ, producing a local anomaly score $s_{h,w}^i$.

The anomaly score map in the inference phase is defined as:

$$S_A(\mathbf{x}_i) := \{s_{h,w}^i \mid \forall(h,w) \in \{1,...,H\} \times \{1,...,W\}\} \tag{18}$$

The anomaly map $S(x_i)$ is then upsampled to the original input image resolution and smoothed with a Gaussian filter (standard deviation $\sigma = 4$) to obtain the final refined anomaly map. Finally, the maximum value of the anomaly score in the anomaly map is used as the overall anomaly score for the image.

Table 1. The performance of NSCAS is compared with different methods on the Mvtec AD dataset.image-level AUROC (I-AUROC) and pixel-level AUROC (P-AUROC) are displayed in each entry as I-AUROC%/P-AUROC%. The best results are highlighted in bold.

Category	Reconstruction-based		Synthesizing-based		Embedding-based			Our
	PNPT [19]	DBPI [17]	DRAEM [22]	CutPaste [10]	SimpleNet [11]	CRAS [6]	SSN [13]	NSCAS
carpet	98.0/97.5	99.0/98.4	97.0/95.5	93.9/98.3	98.5/97.3	98.9/98.6	98.5/98.1	**99.8/98.6**
grid	98.3/96.9	98.4/95.6	99.9/**99.7**	**100**/97.5	99.0/98.5	99.3/98.1	99.1/97.7	**99.9**/98.7
leather	98.5/97.7	**100**/97.1	**100**/98.6	**100/99.5**	99.1/99.0	**100**/99.1	**100**/98.9	**100**/99.1
tile	97.0/93.8	**100/99.3**	99.6/99.2	94.6/90.5	98.5/95.1	99.8/96.3	99.7/95.0	**100**/96.1
wood	98.1/94.7	99.6/95.1	99.1/**96.4**	99.1/95.5	97.1/93.5	99.6/93.5	98.6/92.8	**100**/94.5
bottle	98.5/96.8	99.4/98.0	99.2/99.1	98.2/97.6	**100**/97.1	**100/98.7**	**100**/97.7	**100**/98.1
cable	96.6/95.9	95.9/93.6	91.8/94.7	81.2/90.0	97.7/97.0	96.7/98.2	97.9/97.0	**98.9/98.7**
capsule	91.8/97.0	97.1/98.6	**98.5**/94.3	98.2/97.4	96.2/**99.2**	96.2/98.8	98.2/98.7	98.2/99.0
hazelnut	98.5/97.4	98.8/98.7	**100/99.7**	98.3/97.3	**100**/96.7	99.8/98.3	99.9/98.0	**100**/98.0
metal_nut	98.1/94.3	100/98.2	98.7/**99.5**	99.9/93.1	98.0/97.4	99.6/98.6	99.7/98.6	**100**/98.3
pill	94.2/94.2	95.2/97.4	**98.9**/97.6	94.9/95.7	97.3/97.6	96.3/**98.6**	98.1/98.3	**98.9**/98.5
screw	94.4/98.0	98.1/**99.1**	93.9/97.6	88.7/96.7	96.2/98.1	88.9/97.8	84.3/98.6	**98.3/99.1**
toothbrush	98.5/97.5	**100/99.0**	**100**/98.1	99.4/98.1	98.7/98.2	**100/99.0**	91.4/98.3	**100**/98.5
transistor	98.2/95.8	99.7/95.6	93.1/90.9	96.1/93.0	**100**/96.3	**100/97.1**	**100**/95.1	**100**/96.9
zipper	95.9/96.8	93.8/98.5	**100**/98.8	99.9/**99.3**	98.9/97.7	99.7/98.9	99.3/98.3	**100**/98.8
Avg	97.0/96.3	98.3/97.5	98.0/97.3	96.1/96.0	98.3/97.2	98.3/98.0	98.0/97.3	**99.6/98.1**

4 The Experiment and Analysis

4.1 Datasets

MVTec AD: The dataset [4] is the first comprehensive, multi-target, multi-defect anomaly detection dataset with pixel-level accurate labels. It contains 5 textures and 10 industrial items, with a total of 5,354 high-resolution images, including more than 70 defect types. VisA: The Visual Anomaly [26] dataset is one of the largest datasets for industrial anomaly detection, containing 12 object categories and a total of 10,821 color images. SS-AD: The abnormal data set of the speaker shell is a self-built data set. The data is collected by a linear array camera, and a total of 5,894 color images are collected. The training set contains 4,126 normal images, and the test set contains 806 normal images and 962 abnormal images.

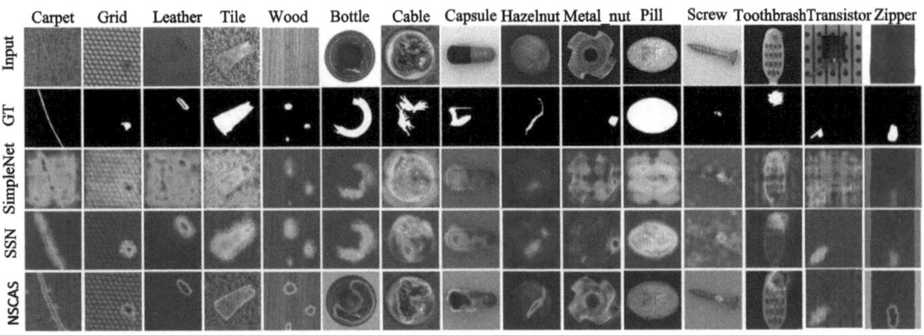

Fig. 4. The NSCAS and SimpleNet methods are qualitatively compared in different categories of MVTec AD datasets. 'GT' denotes Ground Truth.

4.2 Implementation Details

Experimental settings. We use WideResNet50 as the backbone network for the feature extractor E_ϕ, and fuse features from level 2 and level 3. The domain size p is set to 3. All input images are resized and center-cropped to 256×256 pixels. The discriminator consists of a linear layer, a batch normalization layer, a Leaky ReLU (slope 0.2), and another linear layer. We train both the feature adapter and the discriminator using the Adam optimizer. The learning rates for the feature adapter and discriminator are set to 0.0001 and 0.0002, respectively. The training is run for 300 epochs with a batch size of 16. All methods were evaluated on the same hardware, consisting of an NVIDIA TITAN V and an Intel(R) Xeon(R) CPU E5-2650 v4 @ 2.20GHz. The inference time is 12 ms, with a throughput of 213 images per second.

Evaluation indicators. Area Under the Receiver Operating Characteristic Curve (AUROC) is a commonly used evaluation metric in anomaly detection, we use it to evaluate the discriminative ability of models at image and pixel levels.

4.3 Comparative Experiments

To evaluate our proposed method, we conducted comparative experiments with various representative and state-of-the-art approaches. For the reconstruction method, we selected the PNPT [19], DBPI [17]models. For the synthesis, we choose DRAEM [22] and CutPaste [10]. For the embedding method, we choose CRAS [6], RD++ [16], SimpleNet [11] and SuperSimpleNet(SSN) [13].

MVTec AD dataset: As shown in Table 1, NSCAS achieves 100% image level AUROC(I-AUROC) in 9 of 15 categories, significantly exceeding other models in quantity. NSCAS enables the generated anomalous features to exhibit a concentrated intra-class distribution. The average of 99.9% I-AUROC and 99.3% pixel-level AUROC(P-AUROC) on MVTec AD. As shown in Fig. 4. There is a problem of low confidence in SimpleNet [11], but NSCAS can effectively locate anomalies in each category of Mvtec AD.

Table 2. The performance of NSCAS is compared with different methods on the VisA dataset.

Metrics	Reconstruction-based		Synthesizing-based	Embedding-based			Ours
	PNPT	DBPI	DRAEM	SimpleNet	CRAS	SSN	NSCAS
I-AUROC	92.2	85.4	88.7	89.2	93.5	92.3	**93.7**
P-AUROC	97.0	90.9	93.5	96.6	97.7	98.0	**98.4**

VisA dataset: As shown in Table 2, NSCAS exhibits excellent anomaly detection and localization performance on the VisA dataset, achieving an average I-AUROC of 93.7% and an average P-AUROC of 98.4%. Compared with the second-best method, I-AUROC and P-AUROC are improved by 0.2% and 1.4%, respectively. As shown in Fig. 5, The results show that NSCAS still exhibits excellent detection performance in the face of small anomalies.

SS-AD dataset: As shown in Table 3, NSCAS shows good performance on the SS-AD dataset, with an average I-AUROC of 98.3% and an average P-AUROC of 93.0%. As shown in Fig. 6, although the size of the anomaly is small, the model also shows good performance in fine-grained positioning.

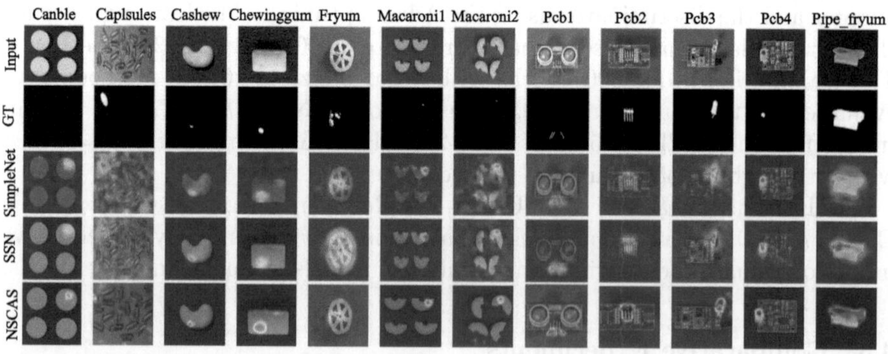

Fig. 5. The NSCAS and SimpleNet methods are qualitatively compared in different categories of VisA datasets. 'GT' denotes Ground Truth.

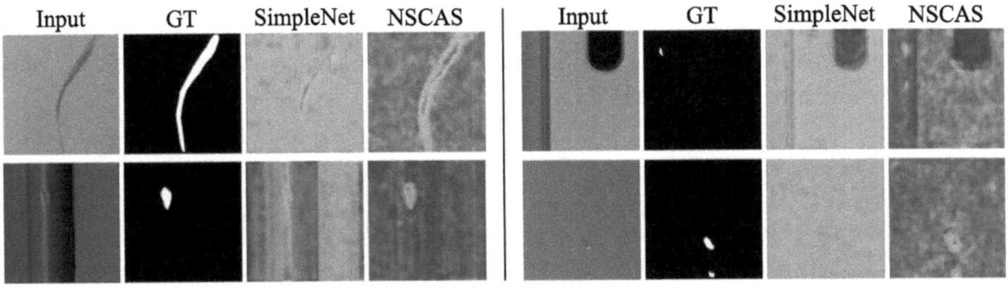

Fig. 6. The NSCAS and SimpleNet methods are qualitatively compared in different categories of SS-AD datasets. 'GT' denotes Ground Truth.

Table 3. The performance of NSCAS is compared with SimpleNet on the SS-AD dataset. The best results are highlighted in bold.

Metrics	SimpleNet [11]	NSCAS
I-AUROC	96.2	**98.3**
P-AUROC	89.6	**95.0**

4.4 Ablation Study

Ablation studies on the MVTec AD dataset (Table 4) validate the contribution of each component in NSCAS. We added three components: Noise Controller (NC), Anomaly Mask Generation (AMG), and Channel-wise Adaptive Weighting (CAW). On MVTec AD, results with added components were compared to SimpleNet (SN). The experiments demonstrate that the newly introduced components help improve detection performance. Table 4 clearly shows that by adding NC, AMG, and CAW, the evaluation metrics improve progressively.

Notably, the inclusion of NC led to substantial performance improvements, with I-AUROC and P-AUROC increasing by 0.7% and 0.5%, respectively, underscoring its critical role in the model. AMG primarily enhances pixel-level localization accuracy, enabling P-AUROC to achieve near-optimal performance when

Table 4. Performance ablation of different components in NSCAS on MVTec AD, as measured by I-AUROC/% and P-AUROC/%. 'NC', 'AMG' and 'CAW' correspond to noise control, anomaly mask generation and adaptive channel weighting, respectively.

SN	NC	AMG	CAW	I-AUROC	P-AUROC
✓				98.3	97.2
✓	✓			99.0	97.7
✓		✓	✓	98.6	97.5
✓	✓		✓	99.1	98.0
✓	✓	✓		99.3	97.9
✓	✓	✓	✓	**99.6**	**98.1**

combined with NC. Conversely, CAW strengthens the model's capability to assess image-level abnormalities, allowing I-AUROC to reach near-optimal performance when integrated with NC.

5 Conclusion

In this work, we propose a new controllable anomaly synthesis strategy based on Euclidean distance to enhance anomaly detection performance. This method uses the noise scale controller and the anomaly mask obtained by perlin noise binarization to control the synthesis of artificial anomalies. And through adaptive channel weighting to reduce the attention of discriminator to background and shadow. NSCAS has significantly improved the performance of industrial anomaly detection and localization. This method has been tested on several industrial datasets and has achieved favorable results across all of them.ability.

Acknowledgment. This work was supported in part by the National Natural Science Foundation of China (NSFC) under Grant 62402125, Major Program of Natural Science Foundation of Sichuan Province, China under Grant 2025ZNSFSC0005, Central Government Guides Local Projects of China under Grant 2025ZYDF105 and Excellent Talent Cultivation Foundation Guided Cultivation Project of Southwest University of Science and Technology under Grant 25ZX7128.

References

1. Akcay, S., Atapour-Abarghouei, A., Breckon, T.P.: Ganomaly: semi-supervised anomaly detection via adversarial training. In: Asian Conference on Computer Vision, pp. 622–637. Springer (2018)
2. Baitieva, A., Hurych, D., Besnier, V., Bernard, O.: Supervised anomaly detection for complex industrial images. In: Proceedings of the IEEE/CVF Conference on Computer Vision and Pattern Recognition, pp. 17754–17762 (2024)
3. Batzner, K., Heckler, L., König, R.: Efficientad: accurate visual anomaly detection at millisecond-level latencies. In: Proceedings of the IEEE/CVF Winter Conference on Applications of Computer Vision, pp. 128–138 (2024)
4. Bergmann, P., Fauser, M., Sattlegger, D., Steger, C.: Mvtec ad–a comprehensive real-world dataset for unsupervised anomaly detection. In: Proceedings of the IEEE/CVF Conference on Computer Vision and Pattern Recognition, pp. 9592–9600 (2019)
5. Chen, Q., Luo, H., Lv, C., Zhang, Z.: A unified anomaly synthesis strategy with gradient ascent for industrial anomaly detection and localization. In: European Conference on Computer Vision, pp. 37–54. Springer (2024)
6. Chen, Q., et al.: Center-aware residual anomaly synthesis for multi-class industrial anomaly detection. arXiv preprint arXiv:2505.17551 (2025)
7. Deng, H., Li, X.: Anomaly detection via reverse distillation from one-class embedding. In: Proceedings of the IEEE/CVF Conference on Computer Vision and Pattern Recognition, pp. 9737–9746 (2022)

8. Gong, D., et al.: Memorizing normality to detect anomaly: memory-augmented deep autoencoder for unsupervised anomaly detection. In: Proceedings of the IEEE/CVF International Conference on Computer Vision, pp. 1705–1714 (2019)
9. Hyun, J., Kim, S., Jeon, G., Kim, S.H., Bae, K., Kang, B.J.: Reconpatch: contrastive patch representation learning for industrial anomaly detection. In: Proceedings of the IEEE/CVF Winter Conference on Applications of Computer Vision, pp. 2052–2061 (2024)
10. Li, C.L., Sohn, K., Yoon, J., Pfister, T.: Cutpaste: self-supervised learning for anomaly detection and localization. In: Proceedings of the IEEE/CVF Conference on Computer Vision and Pattern Recognition, pp. 9664–9674 (2021)
11. Liu, Z., Zhou, Y., Xu, Y., Wang, Z.: Simplenet: a simple network for image anomaly detection and localization. In: Proceedings of the IEEE/CVF Conference on Computer Vision and Pattern Recognition, pp. 20402–20411 (2023)
12. Rolih, B., Ameln, D., Vaidya, A., Akcay, S.: Divide and conquer: high-resolution industrial anomaly detection via memory efficient tiled ensemble. In: Proceedings of the IEEE/CVF Conference on Computer Vision and Pattern Recognition, pp. 3866–3875 (2024)
13. Rolih, B., Fučka, M., Skočaj, D.: Supersimplenet: unifying unsupervised and supervised learning for fast and reliable surface defect detection. In: International Conference on Pattern Recognition, pp. 47–65. Springer (2025)
14. Roth, K., Pemula, L., Zepeda, J., Schölkopf, B., Brox, T., Gehler, P.: Towards total recall in industrial anomaly detection. In: Proceedings of the IEEE/CVF Conference on Computer Vision and Pattern Recognition, pp. 14318–14328 (2022)
15. Ruff, L., et al.: Deep one-class classification. In: International Conference on Machine Learning, pp. 4393–4402. PMLR (2018)
16. Tien, T.D., et al.: Revisiting reverse distillation for anomaly detection. In: Proceedings of the IEEE/CVF Conference on Computer Vision and Pattern Recognition, pp. 24511–24520 (2023)
17. Wang, S., Lv, C., Zhang, Z., Wei, X.: Dual-branch learning with prior information for surface anomaly detection. IEEE Trans. Instrum. Meas. **72**, 1–11 (2023)
18. Yang, M., Wu, P., Feng, H.: Memseg: a semi-supervised method for image surface defect detection using differences and commonalities. Eng. Appl. Artif. Intell. **119**, 105835 (2023)
19. Yao, H., Cao, Y., Luo, W., Zhang, W., Yu, W., Shen, W.: Prior normality prompt transformer for multiclass industrial image anomaly detection. IEEE Trans. Ind. Inform. (2024)
20. Yao, H., Yu, W., Wang, X.: A feature memory rearrangement network for visual inspection of textured surface defects toward edge intelligent manufacturing. IEEE Trans. Autom. Sci. Eng. **20**(4), 2616–2635 (2022)
21. You, Z., et al.: A unified model for multi-class anomaly detection. Adv. Neural. Inf. Process. Syst. **35**, 4571–4584 (2022)
22. Zavrtanik, V., Kristan, M., Skočaj, D.: Draem-a discriminatively trained reconstruction embedding for surface anomaly detection. In: Proceedings of the IEEE/CVF International Conference on Computer Vision, pp. 8330–8339 (2021)
23. Zavrtanik, V., Kristan, M., Skočaj, D.: Reconstruction by inpainting for visual anomaly detection. Pattern Recogn. **112**, 107706 (2021)
24. Zhou, Y.: Rethinking reconstruction autoencoder-based out-of-distribution detection. In: Proceedings of the IEEE/CVF Conference on Computer Vision and Pattern Recognition, pp. 7379–7387 (2022)

25. Zong, B., et al.: Deep autoencoding gaussian mixture model for unsupervised anomaly detection. In: International Conference on Learning Representations (2018)
26. Zou, Y., Jeong, J., Pemula, L., Zhang, D., Dabeer, O.: Spot-the-difference self-supervised pre-training for anomaly detection and segmentation. In: European Conference on Computer Vision, pp. 392–408. Springer (2022)

Enhancing Image Generation of Diffusion Models with Structural Image Guidance

Wei Wang and JiaYi Hu(✉)

School of Electronic and Information Engineering, Liaoning Technical University, Huludao 125105, China
939769714@qq.com

Abstract. ControlNet incorporates structural control images into diffusion models, enhancing controllability in text-to-image generation. However, it requires training on millions of paired samples over days or even months, consuming substantial resources, which poses challenges for ordinary users to develop new model types. Moreover, ControlNet exhibits limited generalization across different control types. Based on this, this paper proposes StrControl, which fine-tunes a Scribble model pre-trained on a large-scale dataset. Channel attention SECA is introduced into the middle layer to enhance the perception of structural information, while zero-convolution layers in the original ControlNet are replaced with cross-normalization to improve the robustness of shallow feature extraction, thereby stabilizing the structural enhancement effect of SECA. Building on these structural optimizations, StrControl further integrates LoRA into the linear layers of the attention modules, enabling lightweight adaptation and enhancing flexibility in handling complex control images. Experimental results demonstrate that the proposed method maintains high image quality under limited-sample conditions while effectively highlighting the subject and aligning with semantic content.

Keywords: Diffusion Models · LoRA · Image Generation

Fig. 1. The images generated by StrControl

1 Introduction

The availability of large-scale "Text-Image" datasets and conditional guidance techniques has enabled Text-To-Image (T2I) generation [1, 2]. Notably, Stable Diffusion [3] employs a latent diffusion model (LDM) to accelerate generation by operating in a compressed feature space. To enhance control, methods like T2I-Adapter and ControlNet [4, 5] introduce structural conditions (e.g., sketches or edge maps) to guide the generation process more effectively.

J. Lokoč et al. (Eds.): MMM 2026, LNCS 16413, pp. 219–231, 2026.
https://doi.org/10.1007/978-981-95-6957-1_16

In current controllable image generation methods, structural conditions are typically processed using a parallel branch or adapter [4, 5]. For example, T2I-Adapter introduces an external adapter module to encode control images and align their features with Stable Diffusion; ControlNet replicates Stable Diffusion's UNet backbone and inserts control branches to achieve controllable generation. Although ControlNet performs well with various control signals, it struggles with cross-type generalization and structural consistency. A single ControlNet model often fails to adapt to different input styles—for example, an edge-based model may lose critical structures when processing sketch inputs. Training a versatile ControlNet that supports multiple control types requires extensive aligned annotation data, leading to high computational costs.

To address these limitations, we propose an enhanced ControlNet-based image generation model that leverages the pre-trained Scribble model to guide diffusion models using edge maps and sketches as structural conditions. Our approach ensures text-aligned generation while emphasizing key features and suppressing redundant artifacts.(1)We adapt the pre-trained Scribble ControlNet using LoRA (Low-Rank Adaptation), compressing original parameters into low-rank matrices for parameter-efficient updates.(2)The intermediate layers of ControlNet process both UNet-encoded control images and text features, augmented by our proposed SECA attention [6, 7]. (3)We replacing Zero Convolutions in ControlNet, this method mitigates early-stage information conflicts caused by distribution mismatches between pre-trained and new modules. It accelerates convergence and enhances training stability by normalizing shallow features, reducing interference from noisy edges.

The proposed StrControl framework achieves more stable training and higher fidelity in structure-preserving generation compared to standard ControlNet implementations.

Fig. 1 **and** 2 demonstrate the images generated by our proposed StrControl framework.

Fig. 2. The images generated by StrControl

2 Related Work

2.1 Diffusion and ControlNet

Diffusion models first proposed by Sohl-Dickstein et al. (2015) [8], they were later refined by Ho et al. (2020) into DDPM, using fixed noise schedules and simplified loss functions [9]. Dhariwal et al. (2021) then enhanced their architecture, surpassing GANs in performance [10].To enhance the expressive power of diffusion models, researchers proposed conducting the diffusion process in a latent space, which achieves breakthroughs in conditional image generation. Following this, large-scale text-to-image models such as DALL-E 2 [11] 、Stable Diffusion and SD1V5 [12] emerged, further advancing the field.

To improve controllability, recent work integrates spatial conditioning [5, 13], addressing the limitations of text-only guidance. ControlNet introduces an auxiliary network to process conditional images, integrating it into Stable Diffusion as a plugin to incorporate conditional control images. However, training a separate ControlNet for each type of conditional image requires large-scale data and time, consuming significant resources. To address this, methods like ControlNetXt modify the network architecture to reduce model scale [6, 14, 15]. ControlNetXt adapts the pretrained large model by removing the control branch and replacing it with a lightweight convolutional module composed of multiple ResNets, significantly cutting learnable parameters [15, 16].To solve these issues, StrControl performs lightweight fine-tuning on pretrained models. Pretrained models, having learned robust knowledge from large datasets, offer strong generalization [17–19].Leveraging this, StrControl adjusts the existing knowledge structure, speeding up convergence

2.2 Low-Rank Adaptation(LoRA)

With the rapid advancement of AIGC (Artificial Intelligence Generated Content), there is a growing demand for personalized and specialized applications. LoRA (Low-Rank Adaptation) [20] is an efficient fine-tuning technique for large language models and image generation modelsInstead of directly updating the original model weights, LoRA freezes the pre-trained parameters and introduces trainable low-rank matrices, significantly reducing both computational cost and model file size.

Recent studies have applied LoRA techniques to enhance image generation capabilities. Yarden Frenkel et al. proposed B-Lora, which jointly trains LoRA weights for two Transformer blocks in Stable Diffusion XL (SDXL) to decouple style and content [21] . DiffLora employs a diffusion-based hypernetwork to predict personalized LoRA weights from reference images, enabling zero-shot customized generation [22]. LoRA has also been adapted to improve ControlNet architectures [23–25]. ZKJ Hartley et al. integrated LoRA with ControlNet for domain adaptation, achieving style transfer on botanical datasets [23]. LooseControl employs LoRA to fine-tune ControlNet for achieving generalized depth control, effectively mitigating the issue of color loss commonly observed in standard fine-tuning and thereby ensuring more stable outputs [24]. CtrLoRA integrates LoRA into the linear layers of the pretrained ControlNet model, reducing the number of trainable parameters while accelerating the training process [25].

3 Method

3.1 Preliminaries

Diffusion model consists of forward (diffusion) process and reverse (generation) process, both implemented as parameterized Markov chains. The forward process can be confined as follows:

$$x_t = \sqrt{\overline{\alpha}_t}x_0 + \sqrt{1 - \overline{\alpha}_t}\varepsilon, \quad \varepsilon \sim \mathcal{N}(0, I) \tag{1}$$

$x_0$is inputs data, ϵ is The noise mapping follows a Gaussian sampling process. $\overline{\alpha}_t := \prod_{s=0}^{t} \alpha_s$ ∘ $\overline{\alpha}_t$ =1- β_t is a differentiable function of the timestep. The loss function of diffusion models is the MSE loss function:

$$L = w \cdot \mathbb{E}_{x_0,t,c_t,\varepsilon\sim N(0,1)}\left[\left\|x - \int_{\theta}(x_t, t, c_t)\right\|^2\right] \tag{2}$$

c_t is text prompts, w denotes the weight of the loss function.ControlNet introduces controllable generation by integrating conditional controls. It calculates the loss function as:

$$\begin{aligned} L = w \cdot \mathbb{E}_{x_0,t,c_t,\varepsilon\sim N(0,1)}\left[\left\|x - \int_{\theta}(x_t, t, c_t)\right\|^2\right] = \\ w \cdot \mathbb{E}_{x_0,t,c_t,c_f,\varepsilon\sim N(0,1)}\left[\left\|x - \int_{\theta}(x_t, t, c_t, c_f)\right\|^2\right] \end{aligned} \tag{3}$$

3.2 SECA

The Efficient Channel Attention (ECA) and Squeeze-and-Excitation (SE) modules offer complementary approaches to enhancing model expressiveness through channel relationships. ECA utilizes efficient 1D convolution for local interaction, while SE employs fully connected layers for global modeling and suppression of irrelevant features. By combining them in series, we establish a synergistic attention flow from global to local. This allows ECA to perform local refinement conditioned on SE's global attention, enhancing spatial detail without added computation. In contrast to parallel structures that extract multi-branch information independently, this progressive serial structure more effectively captures the contextual information of related features.

Input feature image $X \in \mathbb{R}^{B \times C \times H \times W}$ enter the global attention stage of the SE module to obtain the SE attention weights w_{SE}:

$$w_{SE} = \sigma(W_2 \cdot \delta(W_1 \cdot GAP(X))) \tag{4}$$

GAP(.) is global average pooling; W_1 、 W_2 is fully connected layer weights; δ is ReLU function; σ is Sigmoid function. ECA adjusts local cross-channel on the SE attention weights:

$$w_{ECA-SE} = HardSigmoid(K \cdot w_{SE}) \tag{5}$$

$K \in \mathbb{R}^k$ is *1Cond*, $w_{ECA-SE} \in \mathbb{R}^{B \times C \times 1 \times 1}$。Final output is:

$$Y = X \cdot w_{ECA-SE} \tag{6}$$

SECA achieves Global—Local collaborative optimization: SE first select important channel, then ECA precisely adjust SE's output. This design mitigates the issues of over-smoothing or excessive localization. Fig. 3 is SECA structure.

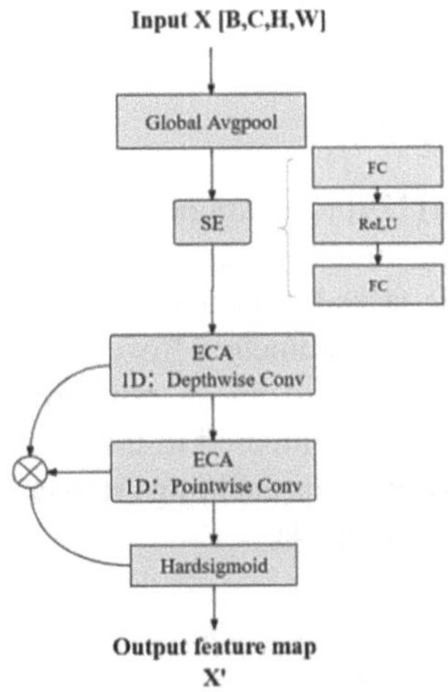

Fig. 3 SECA structure diagram

3.3 Cross Normalization

The central issue in fine-tuning pre-trained models involves how to properly introduce extra parameters and components. The newly added modules or parameters often lead to training instability and poor convergence due to distribution misalignment and incompatibility with the pre-trained model's data distribution.

Normalization methods [26–28], like batch normalization and graph normalization, standardize layer inputs by normalizing them to zero mean and unit variance, thereby enhancing training stability and convergence speed. Cross-normalization is a method that fuses backbone and conditional features, enhancing multimodal task performance by harmonizing different data modalities in feature space. Unlike ControlNet's zero-convolution approach for conditional injection, cross-normalization aligns image and condition features without compromising the pretrained Stable Diffusion model. Experiments demonstrate that replacing zero-convolutions with cross-normalization in intermediate layers yields optimal image quality with improved training stability, particularly in few-shot task. We represent the denoised features and controlled features as x_d and x_c, x_m and $x_c \in \mathbb{R}^{H \times W \times C}$ use the mean and variance calculated from the main branch x_m to normalize the control features x_c,First, calculate the mean and variance of the denoising features:

$$\mu_m = \frac{1}{n}\sum_{i=1}^{n} x_{m,i} \tag{7}$$

$$\sigma_m^2 = \frac{1}{n} \sum_{i=1}^{n} \left(x_{m,i} - \mu_m\right)^2 \tag{8}$$

Then, we normalize the control features using the mean and variance of the denoising features, γ is the model scales the normalized values with learnable parameters.:

$$\hat{X}_c = \frac{X_c - \mu_m^{\ *}}{\sqrt{\sigma_m^2}} \gamma \tag{9}$$

3.4 Low-Rank Adaptation(LoRA)

LoRA (Low-Rank Adaptation) [29] is a parameter-efficient fine-tuning technique used for large language models and image generation models. Its core principle involves optimizing models for specific tasks by introducing small trainable low-rank matrices while keeping the original model parameters frozen. LoRA introduces trainable low-rank matrix decompositions at selected linear transformations within the model $\Delta W = BA$, $B \in \mathbb{R}^{h \times j}$, $A \in \mathbb{R}^{j \times h}$,$j \ll h$. Compared to traditional full-parameter fine-tuning, LoRA needs fewer trainable parameters, enabling rapid task-switching capabilities ideal for multi-task learning scenarios (Fig. 4).

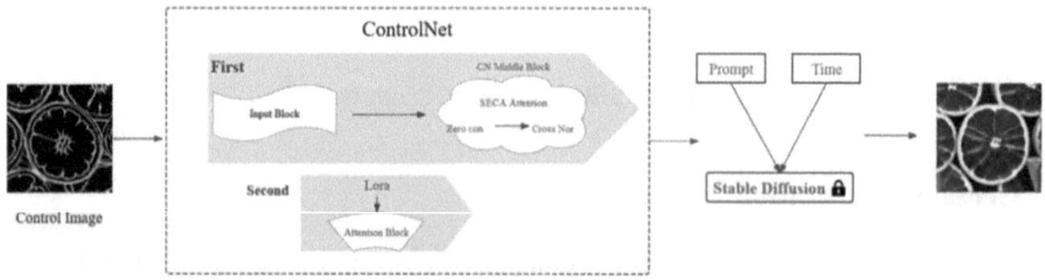

Fig. 4 Model Training Process. StrControl keeps Stable Diffusion frozen during training. The key architectural innovations include: **Block Enhancement:** we replace ControlNet's third middle block layer with SECA Attention (attention combining SE and ECA)and substitute zero-convolutions with cross-normalization in middle block outputs. **Two-Phase Training:**(1).Trains only intermediate modules to adapt to new image styles.(2).Full-model fine-tuning with LoRA integration. We replace QKV projection layers in self-attention with LoRA's 1D convolutions and substitute output linear layers with LoRA's 2D convolutions

4 Experiment

4.1 Dataset and Setting

We collect dataset proposed in SmartControl [30] with all images resized to 512×512 and select 1000 image samples per task for training. We choose the following state-of-the-art models in controllable image generation as competing methods:T2I-Adapter [31], ControlNet and UniControl [32].

4.2 Experiment Implement

We use ControlNet as the base model. The training configuration employs: Optimizer: AdamW; Learning rate:1e-5;Batch size: 16. We implement a two-phase training strategy:

First: Integrating SECA Attention into intermediate layers and replacing zero-convolutions with cross-normalization. Enabling rapid model adaptation to new sample patterns through intermediate layer training.

Second: Unfrozen the remaining modules and finetuned with LoRA, which is applied to the linear layers of the original model's attention mechanism for efficient parameter adaptation.

4.3 Result Analyse

Evaluation Metrics: We use FID、CLIP-score and SSIM.FID assesses generation quality by comparing feature distributions between synthetic and real images; CLIP-score measures text-image alignment through semantic similarity; SSIM quantifies structural similarity in luminance, contrast, and texture.

Table 1. HED Dataset

Method	FID↓	SSIM↓	CLIP-score↑
T2I-Adapter(SD1.5)	—	0.106	0.261
Uni-Control(SD1.5)	17.08	**0.082**	0.227
ControlNet(SD1.5)	15.41	0.123	0.264
StrControl(SD1.5)	**15.00**	**0.096**	**0.289**

Table 2. SKETCH Dataset

Method	FID↓	SSIM↓	CLIP-score↑
T2I-Adapter(SD1.5)	26.94	0.139	0.267
Uni-Control(SD1.5)	24.95	0.116	0.265
ControlNet(SD1.5)	26.88	0.125	0.269
StrControl(SD1.5)	**25.30**	**0.110**	**0.282**

Comparison of Controllability with State-of-the-Art Methods:
The experimental results are shown in **Table 1** and **2**, which can be summarized as the following observations:

For edge-aware tasks, StrControl achieves superior FID scores compared to existing methods, though its structural similarity (SSIM) is slightly lower than Uni-Control. This occurs because StrControl actively eliminates distracting noise patterns to enhance subject prominence—a critical capability for foreground extraction tasks. Notably, StrControl still maintains higher SSIM values than the original ControlNet.

The CLIP-score metric, which evaluates semantic alignment between generated images and text descriptions, demonstrates StrControl's superior performance across all three experimental tasks compared to baseline models. The fine-tuned StrControl exhibits enhanced capability to capture nuanced textual details. SECA improves cross-modal attention and enables finer-grained text-visual feature interaction. LoRA-based Specialized Tuning optimizes semantic sensitivity via low-rank adaptations and preserves pretrained knowledge while adapting to textual nuances.

The proposed model demonstrates superior performance compared to existing conditional image generation approaches, achieving better text-semantic alignment while more effectively highlighting key subjects and rendering finer details in the generated images. Fig. 5 presents a comparative visualization of images generated by StrControl versus baseline methods using identical text prompt. StrControl demonstrates optimal performance across color texture Fidelity、prompt adherenceand subject prominence, achieving superior overall image quality.

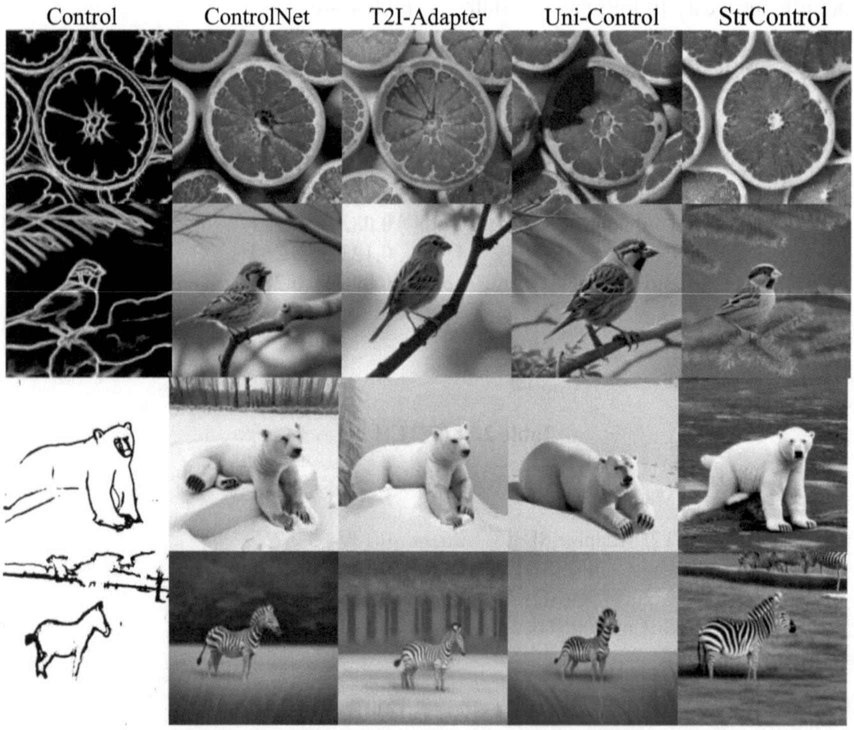

Fig. 5 The results of the comparative experiment image generation

4.4 Ablation Study

To evaluate the improvements of StrControl over the base ControlNet architecture, this study conducts ablation experiments by incrementally integrating the proposed modules into the ControlNet.

SECA Attention

We introduce the SECA into the ControlNet middle layer. This module combines the fused attention of SE and ECA, designed to enhance the model's perception of local details and channel importance, thereby improving the conversion quality from importance and structural control maps (e.g., sketches, edge maps) to images. The SE significantly enhances the model's semantic structure awareness through global channel modeling of intermediate feature maps, leading to notable CLIP-score improvements. However, due to its reliance on fully-connected networks for global modeling, SE demonstrates limited capability in capturing local features. As illustrated in Fig. 6, the standalone SE fails to adequately preserve fine details in certain cases. By further incorporating the ECA module with its 1D convolution-guided local channel interaction, the combined approach effectively compensates for SE's deficiencies in local feature modeling.

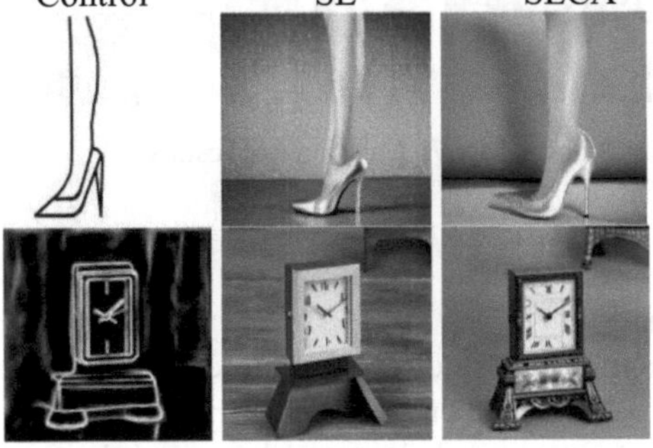

Fig. 6 Ablation test results

Table 3. Ablation test results

Method	FID↓		SSIM↓		CLIP-score↑	
	Hed	Sketch	Hed	Sketch	Hed	Sketch
ControlNet(CN)	16.41	26.88	0.123	0.125	0.264	0.256
CN + SE	16.00	26.20	0.119	0.121	0.268	0.273
CN + SECA	15.82	26.03	0.115	0118	0.273	0.276
CN + SECA+Cross	15.30	25.50	0.102	0.113	0.280	0.279
+Lora	**15.00**	**25.30**	**0.096**	**0.110**	**0.289**	**0.282**

LoRA

LoRA (Low-Rank Adaptation) serves as a lightweight adjustment mechanism, modifying the attention layers with minimal additional parameters to enhance training efficiency and stability.Building upon the SECA module and cross-normalization, LoRA further improves the model's adaptability, enabling more flexible handling of complex control maps. To evaluate its effectiveness, we conduct an ablation study by applying LoRA to the linear layers of the attention mechanism and comparing it with the baseline without LoRA. The experimental results are presented in Table 3, while Fig. 7 illustrates the comparative effects before and after LoRA adaptation.

Control No Lora With Lora

Fig. 7 Ablation test results

Cross Normalization

We introduce cross-normalization to replace zero convolution in ControlNet, aligning data distributions for a more efficient and stable training process. Zero convolution is located in two key parts: the encoded control layer and the final layer where control features are fused with backbone features. To evaluate whether cross-normalization improves training stability, we analyze the loss curve (per step) during training, particularly focusing on fluctuations within the first 10 epochs. As shown in Fig. 8, the loss curve with cross-normalization exhibits smoother convergence and reduced oscillations, confirming its stabilizing effect on training.

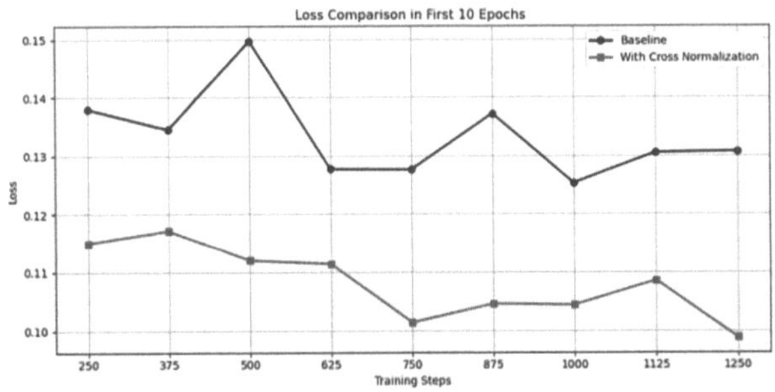

Fig. 8 Loss curve graphs of the first 10 epochs before and after introducing cross normalization

5 Conclusion

This paper proposes StrControl, a novel image generation model for structural control. Built upon ControlNet, StrControl leverages large-scale pretrained weights and enhances key feature extraction through an inserted SECA module. It replaces zero convolutions with cross-normalization in the middle layers to stabilize training. Furthermore, we employ LoRA to efficiently fine-tune the attention mechanisms with minimal parameters. Experimental results show that StrControl maintains high generation quality while robustly extracting structural information from complex sketches, achieving superior robustness and object composition capabilities.

References

1. Dhariwal, P., Nichol, A.: Diffusion models beat GANs on image synthesis. Adv. Neural Inf. Proces. Syst. **34**, 8780–8794 (2021)
2. Guo. Y., Yang. C., Rao. A., et al.: Animatediff: Animate your personalized text-to-image diffusion models without specific tuning. arXiv preprint arXiv:2307.04725, (2023).
3. Rombach, R., Blattmann, A., Lorenz, D., et al.: High-resolution image synthesis with latent diffusion models. In: Proceedings of the IEEE/CVF Conference on Computer Vision and Pattern Recognition, pp. 10684–10695 (2022)
4. Zhang, L., Rao, A., Agrawala, M., Adding Conditional Control to Text-to-Image Diffusion Models, (2023).
5. Mou, C., Wang, X., Xie, L., et al. T2i-adapter: learning adapters to dig out more controllable ability for text-to-image diffusion models. Proceedings of the AAAI Conference on Artificial Intelligence, 2024, 38(5): 4296–4304.
6. Hu, J., Shen, L., Sun, G.: Squeeze-and-excitation networks. In: Proceedings of the IEEE Conference on Computer Vision and Pattern Recognition, pp. 7132–7141 (2018)
7. Wang, Q., Wu, B., Zhu, P., et al.: ECA-net: efficient channel attention for deep convolutional neural networks. In: Proceedings of the IEEE/CVF Conference on Computer Vision and Pattern Recognition, pp. 11534–11542 (2020)
8. Sohl-dickstein, J., Weiss, E., Maheswaranathan, N., et al.: Deep unsupervised learning using nonequilibrium thermodynamics. In: International Conference on Machine Learning. PMLR, pp. 2256–2265 (2015)
9. Ho, J., Jain, A., Abbeel, P.: Denoising diffusion probabilistic models. Adv. Neural Inf. Proces. Syst. **33**, 6840–6851 (2020)

10. Dhariwal, P., Nichol, A.: Diffusion models beat gans on image synthesis. In: Proceeding of the 34th Advances in Neural Information Processing Systems, vol. 34, pp. 8780–8794. MA: MIT, Cambridge (2021)
11. Ramesh, A., Dhariwal, P., Nichol, A., et al.: Hierarchical text-conditional image generation with clip latents. arXiv preprint arXiv:2204.06125, (2022)
12. Runwayml Stable Diffusion v1.5: Stable Diffusion v1–5 Model Card [EB/OL], (2023-08-24) [2024-06-30] https://huggingface.co/runwayml/stable-diffusion-v1-5
13. Li, Y., Liu, H., Wu, Q., et al.: Gligen: open-set grounded text-to-image generation. In: Proceedings of the IEEE/CVF Conference on Computer Vision and Pattern Recognition, pp. 22511–22521 (2023)
14. Zavadski, D., Feiden, J.F., Rother, C.: Control net-XS: rethinking the control of text-to-image diffusion models as feedback-control systems. In: European Conference on Computer Vision, pp. 343–362. Springer Nature Switzerland, Cham (2024)
15. Peng, B., Wang, J., Zhang, Y., et al.: Controlnext: Powerful and efficient control for image and video generation. arXiv preprint arXiv:2408.06070, (2024)
16. Hecong, W.: Control LoRA: A Lightweight Neural Network To Control Stable Diffusion Spatial Information [EB/OL]. (2023)
17. Blattmann, A., Dockhorn, T., Kulal, S., et al.: Stable video diffusion: Scaling latent video diffusion models to large datasets. arXiv preprint arXiv:2311.15127, (2023)
18. Podell, D., English, Z., Lacey, K., et al.: SDXL: Improving latent diffusion models for high-resolution image synthesis. arXiv preprint arXiv:2307.01952, (2023)
19. Zhang, Y.C., Qian, S.J., Peng, B.H., et al.: Prompt highlighter: interactive control for multi-modal llms. In: Proceedings of the IEEE/CVF Conference on Computer Vision and Pattern Recognition, p. 1321513224 (2024)
20. Hu, E.J., Shen, Y., Wallis, P., et al. Lora: Low-rank adaptation of large language models. arXiv preprint arXiv:2106.09685, (2021)
21. Frenkel, Y., Vinker, Y., Shamir, A., et al.: Implicit style-content separation using b-lora. In: European Conference on Computer Vision, pp. 181–198. Springer Nature Switzerland, Cham (2024)
22. Wu, Y., Shi, Y., Wei, J., et al.: Difflora: Generating personalized low-rank adaptation weights with diffusion. arXiv preprint arXiv:2408.06740, (2024)
23. Hartly, Z.J.K., Lind, R.J., Pound, M.P., et al.: Domain targeted synthetic plant style transfer using stable diffusion LoRA and control net. In: Proceedings of the IEEE/CVF Conference on Computer Vision and Pattern Recognition, pp. 5375–5383 (2024)
24. Bhat, S.F., Mitra, N., Wonka, P.: Loosecontrol: lifting controlnet for generalized depth conditioning [C]//ACM SIGGRAPH. Conf. Papers. **2024**, 1–11 (2024)
25. Xu, Y., He, Z., Shan, S., et al.: Ctr LoRA: An Extensible and Efficient Framework for Controllable Image Generation. arXiv preprint arXiv:2410.09400, (2024)
26. Ba, J.K., Kiros, J.R.: G.E. Hinton Layer normalization. arXiv preprint arXiv: 1607.06450, (2016).
27. Ioffe, S., Szegedy, C.: batch normalization: accelerating deep network training by reducing internal covariate shift. In: International Conference on Machine Learning, pp. 448–456. PMLR (2015)
28. Wuand, Y.X., He, K.M.: Group normalization. In: Proceedings of the European Conference on Computer Vision (ECCV), pp. 3–19 (2018)
29. Hu, E.J., Shen, Y., Wallis, P., et al.: Lora: low-rank adaptation of large language models. ICLR. **1**(2), 3 (2022)
30. Liu, X., Wei, Y., Liu, M., et al.: Smartcontrol: enhancing controlnet for handling rough visual conditions. In: European Conference on Computer Vision, pp. 1–17. Springer Nature Switzerland, Cham (2024)

31. Mou, Chong, et al.: T2i-adapter: learning adapters to dig out more controllable ability for text-to-image diffusion models. Proceedings of the AAAI Conference on Artificial Intelligence. 38. 5 (2024)
32. Qin, C, Zhang, S, Yu, N, et al.: Unicontrol: A unified diffusion model for controllable visual generation in the wild. arXiv preprint arXiv:2305.11147 (2023)

HCFFPN: Hierarchical Cross-Scale Feature Fusion Pyramid Network for Small Target Detection in Unmanned Aerial Vehicle Images

Lin Wang[1], Tiansong Li[1], Guofen Wang[1(✉)], Shaoguo Cui[1], Hongkui Wang[2], and Li Yu[3]

[1] Computer and Information Science, Chongqing Normal University, Chongqing 401331, China
wanggf@cqnu.edu.cn

[2] Communication Engineering, Hangzhou Dianzi University, Hangzhou 310018, China

[3] Electronic Information and Communications, Huazhong University of Science and Technology, Wuhan 430074, China

Abstract. In complex backgrounds, the precise detection of small targets in unmanned aerial vehicle (UAV) images remains a crucial yet highly challenging task, especially considering that objects in natural scenes often display significant scale variations. These variations can lead to incomplete feature extraction and consequently hinder accurate recognition. As a common solution, Feature Pyramid Networks (FPNs) have been introduced to enhance multi-scale feature representation. However, existing FPNs typically employ simple feature concatenation and addition operations that are insufficient for capturing rich semantic information, thus constraining their effectiveness in handling large scale variations within complex scenes. To address this issue, we propose a novel Hierarchical Cross-scale Feature Fusion Pyramid Network (HCFFPN) that captures semantic relationships across adjacent feature levels via a Multi-level Feature Extractor (MLFE) to enhance inter-level correlation, and integrates high- and low-frequency information through a Deep Feature Aggregation Module (DFAM), resulting in more robust feature representations. To verify the effectiveness of the proposed HCFFPN, we conducted a large number of experiments and comprehensive evaluations on the VisDrone and UAVDT datasets. The experimental results show that, compared with existing FPN-based architectures, the proposed HCFFPN demonstrates significant improvements in both detection accuracy and robustness for small targets in complex UAV scenes.

Keywords: Unmanned aerial vehicle images · Small target detection · Hierarchical feature pyramid fusion · Multi-scale features

J. Lokoč et al. (Eds.): MMM 2026, LNCS 16413, pp. 232–246, 2026.
https://doi.org/10.1007/978-981-95-6957-1_17

1 Introduction

In recent years, unmanned aerial vehicles (UAVs) have been widely applied in target detection for remote sensing, traffic monitoring, search and rescue, and security applications (e.g., [30]). The key to the successful implementation of these application scenarios lies in whether the target objects can be quickly and accurately identified from the images obtained by UAVs. Nevertheless, target detection technology for UAVs faces a series of distinctive challenges. Small targets in images have limited pixel representation and insufficient feature information, which makes them difficult to detect. This is further complicated by occlusion and overlap caused by dense target distribution, image blur or distortion due to UAV movement, and the limited hardware resources available on UAV platforms. Consequently, achieving a balance between recognition accuracy and detection speed remains a critical research focus in this domain.

As a widely recognized classic model architecture in small target detection tasks, the core advantage of Feature Pyramid Network (FPN) [12] as a traditional neck network is that it can construct a multi-scale feature representation system by fusing multi-level features in the backbone network, as shown in Fig. 1. It significantly improves the model's detection ability for targets of different sizes by combining deep abstract semantic features with shallow high-resolution detail features. However, while the fusion method of the neck network based on FPN, such as CEASC [3], CGMDet [31] and SOD-UAV [11], improves multi-scale target detection by upscaling low-resolution feature maps and fusing them with adjacent high-level features at the channel level, it fails to effectively capture the intrinsic relationships across different feature levels. This oversight leads to insufficient utilization of feature information, leaving room for performance improvements, especially in small target detection.

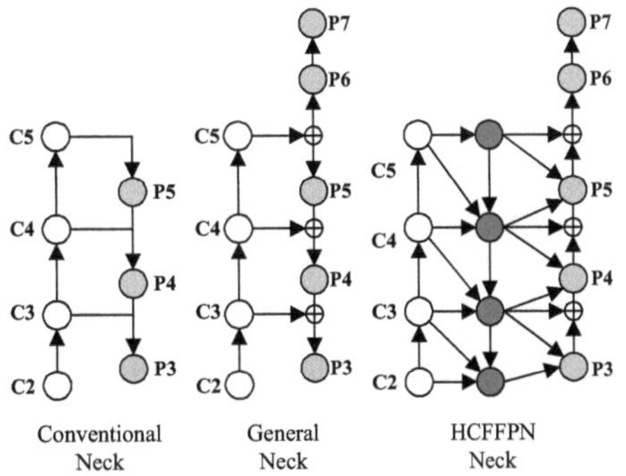

Fig. 1. The left, middle, and right panels compare conventional [12], recent general [3], and our proposed HCFFPN neck networks, respectively. Unlike prior designs, HCFFPN introduces hierarchical structural optimization for multi-scale integration and an adaptive fusion layer (where C: backbone features, P: fused neck outputs).

To address the above issues, we propose a Hierarchical Cross-Scale Feature Fusion Pyramid Network (HCFFPN). This architecture enhances semantic coherence across different feature levels by integrating spatial context information with cross-layer feature commonality, thereby significantly improving the model's ability to learn discriminative multi-scale representations. Within HCFFPN, an Multi-Level Feature Extractor (MLFE) is introduced to refine and align hierarchical features through a set of purposefully engineered feature interaction strategies. These operations enable MLFE to capture and harmonize multi-scale semantic relationships, enhancing both feature expressiveness and fusion accuracy. To further improve the quality of fused features, we propose a Deep Feature Aggregation Module (DFAM) that addresses feature redundancy and information loss in conventional fusion pipelines. DFAM achieves selective feature refinement by decomposing feature representations into high- and low-significance subspaces, guided by an adaptive gating strategy that emphasizes informative patterns while suppressing less relevant ones. Additionally, it incorporates feature reconstruction to mitigate semantic misalignment between shallow and deep features, thereby improving the fidelity and robustness of the final representation. Overall, the synergy between MLFE and DFAM enables HCFFPN to optimize the multi-scale feature fusion process effectively. As a result, the proposed framework achieves significant performance improvements, particularly in small target detection, offering a more reliable solution for object detection in complex scenarios.

There are three contributions in this paper:

1) We propose a hierarchical cross-scale feature fusion pyramid network HCFFPN for target detection in UAV images. This approach not only effectively detects small targets but also serves as a plug-and-play model, significantly enhancing the detection accuracy of the baseline model.
2) We propose an Multi-Level Feature Extractor (MLFE) module that pioneers a "split-interact-recalibrate" paradigm, enabling fine-grained spatial-channel feature calibration through group-wise decomposition and self-calibrating attention mechanisms, moving beyond the coarse branch-level selection of SKNet [9]-like methods.
3) A DFAM is built in this study to further process high-frequency and low-frequency information, maximizing the retention of small target feature information in the deep network and enhancing the representational capacity of small targets.

2 Related Work

2.1 General Object Detection

Object detection methods are generally categorized into two-stage (e.g., Faster R-CNN [17]) and single-stage (e.g., YOLO [21]) approaches. Two-stage detectors achieve higher accuracy but are computationally expensive, while single-stage detectors are efficient but less accurate. Although recent advances such as RetinaNet's Focal Loss and anchor-free methods (e.g., FCOS [20]) have mitigated some issues, detecting densely distributed small objects in drone imagery remains challenging due to insufficient feature representation.

2.2 Small Object Detection

Small object detection faces inherent challenges from limited pixel resolution. Current approaches employ data augmentation, high-resolution features QueryDet [23], attention mechanisms (e.g., CEASC [3]), and transformer architectures

MUPT-Net [25] to improve performance. However, existing methods primarily optimize detection heads or loss functions, overlooking critical inefficiencies in multi-scale feature fusion. Our work specifically addresses this bottleneck through neck structure optimization for better small-object localization.

2.3 Feature Fusion for Multi-Scale Detection

Building upon FPN's top-down architecture, subsequent improvements introduced bidirectional connections BiFPN [19], automated pathway design Nas-FPN [6], and attention mechanisms A2-FPN [7]. Recent approaches like SGGF-Net incorporate global context but increase computational overhead. Our HCFFPN optimizes cross-scale fusion efficiency while maintaining accuracy, specifically designed for drone-based small object detection.

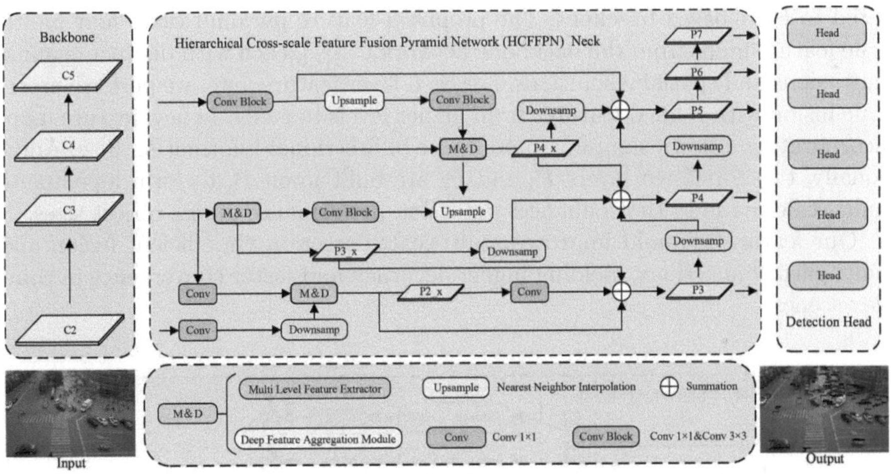

Fig. 2. The Hierarchical Cross-Scale Feature Fusion Pyramid Network HCFFPN employs a multi-level feature extractor and a deep feature aggregation module in the neck structure to replace the traditional fusion method. HCFFPN is suitable for any detector based on the FPN structure and improves feature integration through its advanced fusion strategy.

3 Proposed Method

This section elaborates our proposed HCFFPN, which is designed to systematically optimize cross-layer feature integration. The architecture comprises two parallel core components: the MLFE and the DFAM, working collaboratively to overcome conventional FPN limitations. The MLFE enhances semantic correlation between adjacent hierarchical features and strengthens local-global information interaction. In parallel, the DFAM intensifies both high-frequency and low-frequency information in multi-level feature fusion, utilizing convolutional modules to ensure efficient fusion and transformation. HCFFPN precisely represents hierarchical feature correlations while preserving crucial frequency information in the deep fusion network.

3.1 Overall Network Architecture

While FPN has served as the cornerstone for multi scale object detection, its linear feature concatenation limits inter-layer information flow. We propose HCFFPN, a hierarchical cross-layer fusion pyramid network, to overcome this bottleneck. Unlike FPN's single-path fusion, HCFFPN (Fig. 2) introduces adaptive transformation modules for backbone feature recalibration and multi-branch cross connections enabling bidirectional feature propagation. This architecture enhances multi-scale fusion efficiency while maintaining real-time inference speeds, achieving superior feature richness and detection accuracy compared to FPN-based baselines. The proposed feature pyramid takes four multi-scale feature maps from the backbone network,$C_{i\in(1,5)}$each with distinct channel dimensions and spatial resolutions. For each layer feature map, we perform cross-scale fusion with refined features from higher levels to generate new feature map, incorporating a down sampling module for bidirectional information flow. Additionally, two top-down layers P_6 and P_7 are built upon P_5 to form a complete multi-scale pyramid that enhances detection performance across object sizes.

Our feature pyramid improves multi-scale detection via efficient fusion and multi-path connections, yielding higher accuracy and faster convergence in complex scenes.

Fig. 3. Multi-level Feature Extractor(MLFE)

3.2 Multi-level Feature Extractor

To effectively integrate multi-level features from the backbone network (e.g., $C_i \in R^{c\times h\times w}$ and $C_{i+1} \in R^{c\times h\times w}$), we propose a novel Multi-level Feature Extractor(MLFE). This module not only achieves deep fusion of multi-scale features but also significantly enhances the representational power of the fused features through a fine-grained group-wise attention mechanism.

As shown in Fig. 3, the MLFE module first element-wise sums the input features F_1 and F_2 from adjacent layers to obtain a coarsely fused feature F_{global}. Instead of employing multi-branch convolutions to capture different receptive fields as in SKNet, we innovatively introduce a Global Attention Gate. This gate generates a global attention map F_{map} by applying a convolution and a Sigmoid activation to F_{global}, which is then used to scale the fused feature:

$$F = F_{map} \otimes F_{global}, \tag{1}$$

thereby suppressing less useful information and highlighting critical features.

The core innovation of our module lies in the subsequent Group Interaction and Recalibration step. We split the preliminarily weighted feature F into G groups ($G = 4$) along the channel dimension. Different from SKNet, which uses a global channel attention vector to weight entire branch outputs, we independently process each group feature $F[g]$ as follows:

Interactive Convolution: Each group feature $F[g]$ is first processed by a shared InteractConv layer, designed to capture local context and interactions within the group.

$$F[g] = InteractConv(F[g]) \tag{2}$$

Intra-Group Spatial Attention: We generate a unique spatial attention map for the interacted features. Specifically, we reshape the feature map and calculate its ratio to the mean value, followed by a Softmax function to obtain a normalized attention distribution that precisely indicates the importance of each spatial location within the group.

$$SpatialAttention[g] = M\&S(F[g]) \tag{3}$$

Feature Reweighting: The output of the group is finally obtained by element-wise multiplying the original group feature $F[g]$ with the computed spatial attention map.

$$Out[g] = F[g] \otimes SpatialAttention[g] \tag{4}$$

Afterwards, all group outputs are concatenated and fed into a Group Normalization layer. This layer standardizes the concatenated features across groups and introduces learnable affine parameters γ and β to restore their representational capacity, outputting the enhanced fused feature F_{out}.

$$Out[g] = Concat(Out[g_s]) \tag{5}$$

$$F_{out} = GNWA(Out[g]) \tag{6}$$

where $InteractConv$ represents the local feature interaction convolution, which aims to capture more complex feature representations and is capable of extracting higher-order features or creating associations between features. $Concat$ refers to concatenation along the channel, $M\&S$ indicates taking the mean of each element and then performing normalization and $Softmax$ operations, and $GNWA$ represents global normalization operation and affine transformation.

Comparison with SKNet: The essence of SKNet is to generate features using multiple convolutional kernels and adaptively select receptive fields through global channel attention. In contrast, our MLFE module takes a different approach. Its core idea is to group the fused single feature and perform fine-grained spatial attention recalibration and interaction within each group. This method avoids the parameter and computational overhead associated with multiple branches and achieves more efficient and precise feature fusion through a finer-grained spatial attention mechanism and explicit feature interaction, as shown in Table 3.

3.3 Deep Feature Aggregation Module

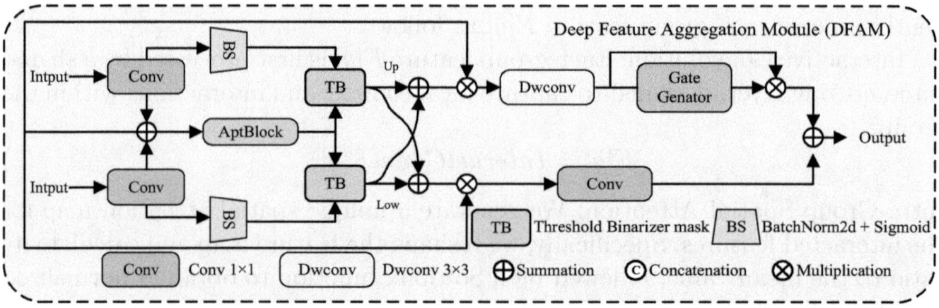

Fig. 4. Deep Feature Aggregation Module (DFAM).

To address the loss of high-frequency features and weak feature correlations in deep networks, we propose a Deep Feature Aggregation Module (DFAM), which decouples multi-scale backbone features into high/low-frequency components for separate processing. Unlike conventional fusion schemes, DFAM employs frequency-aware decomposition to preserve critical low-frequency contexts, High-frequency enhancement branches to amplify subtle patterns (e.g., small-object edges). This dual-path strategy mitigates information degradation during feature propagation, particularly for low-saliency small targets prone to spatial suppression. As visualized in Fig. 4, DFAM maintains feature integrity throughout the processing chain.

Specifically, as previously stated, upon receiving the features $C_i \in R^{c \times h \times w}$ and $C_{i+1} \in R^{c \times h \times w}$ from adjacent levels that have been aligned at different scales, they will respectively undergo convolution operations followed by element-wise addition operations. The resulting features $F_{c_{i \in (i,i+1))}}$ and F_a can be expressed as:

$$F_{c(i)} = Conv(C_i) \tag{7}$$

$$F_a = F_{c(i)} \bigoplus F_{c(i+1)} \tag{8}$$

where $\bigoplus$ represents element-wise summation, and $Conv$ denotes a 1×1 convolutional layer.

Next, we apply the adaptive average pooling module to obtain the feature channel information weight w_a of F_a. Meanwhile, the features $F_{c(i)}$ and $F_{c(i+1)}$ pass through the BS module to generate their respective different feature information weights $w_{c(i)}$ and $w_{c(i+1)}$, which can be expressed as:

$$w_a = AptBlock(F_a) \tag{9}$$

$$w_{c(i)} = BS(F_{c(i)}) \tag{10}$$

$$w_{c(i+1)} = BS(F_{c(i+1)}) \tag{11}$$

where $AptBlock$ indicates the adaptive average pooling block which consists of multiple average poolings and $Sigmoid$ functions. BS is composed of batch normalization operations and $Sigmoid$ functions.

The preliminary high-frequency and low-frequency information F_{up} and F_{low} of w_a are obtained by comparing it with the feature weights of two distinct levels respectively. Subsequently, the high-frequency and low-frequency information of the two different levels are element-wise summed and multiplied by F_a to guarantee the non-loss of global information. This can be represented as:

$$F_{up_{c(i)}}, F_{low_{c(i)}} = TB(w_a, w_{c(i)}) \tag{12}$$

$$F_{up} = F_a \bigotimes (F_{up_{c(i)}} \bigotimes F_{up_{c(i+1)}}) \tag{13}$$

$$F_{low} = F_a \bigotimes (F_{low_{c(i)}} \bigotimes F_{low_{c(i+1)}}) \tag{14}$$

where $\bigotimes$ represents the element-wise product. $F_{up_{c(i)}}$, $F_{low_{c(i)}}$, $F_{up_{c(i+1)}}$, and $F_{low_{c(i+1)}}$ respectively denote the high-frequency and low-frequency information of $w_{c(i)}$ and $w_{c(i+1)}$. TB is a binary processing operation for comparing different weight information. Subsequently, a depthwise separable convolution operation is conducted on F_{up}, and more refined high-frequency information is generated through a gating mechanism and multiplied with the features after the separable convolution. This is carried out to better retain that relatively weak high-frequency information during feature transformation. F_{low} merely requires a single convolution operation to better preserve the original low-frequency information. Ultimately, the high- and low-frequency information that has undergone secondary processing is element-wise added to obtain the final feature aggregation output as F_{out}. It can be represented as:

$$F'_{up} = Gate(Dwconv(F_{up})) \bigotimes Dwconv(F_{up}) \tag{15}$$

$$F'_{low} = Conv(F_{low}) \tag{16}$$

$$F_{out} = F'_{up} \bigoplus F'_{low} \tag{17}$$

where $Dwconv$ represents a $3 \otimes 3$ depthwise separable convolution, F'_{up} and F'_{low} respectively denote the processed high-frequency and low-frequency information, $Conv$ indicates a $1 \otimes 1$ convolution, and $Gate$ refers to a gated generator that maps the high-frequency information to a single channel to generate more refined high-frequency information.

4 Experiments

We evaluate the effectiveness of HCFFPN by comparing it to the existing approaches and conducting comprehensive ablation studies.

4.1 Datasets and Metrics

The VisDrone [5] dataset contains 10,209 high-resolution (2000×1500) aerial images with many small objects across 10 categories. Following prior work, we use 6,471 images for training and 548 for testing. The UAVDT [4] dataset includes 23,258 training and 15,069 testing images (1080×540 resolution), annotated for cars, trucks, and buses. We evaluate accuracy using mAP and AP, and efficiency using GFLOPs, Params and Fps.

Table 1. Comparisons with the existing approaches on VisDrone dataset. "-" indicates that the result is not reported or not publicly available.

Base Detector	Method	Backbone	mAP	AP50	AP75	GFLOPs	Params	Fps
YOLOv3-SPP3 [28]	Baseline	Darknet-53	26.4	-	-	284.10	63.9M	-
ClusDet [24]	Baseline	ResNeXt-101	28.4	56.2	31.6	-	-	-
Improved CascadeNet [29]	Baseline	ResNet-50	28.8	47.1	29.3	-	-	-
DMNet [8]	Baseline	ResNeXt-101	29.4	49.3	30.6	-	-	-
YOLC [14]	Baseline	ResNeXt-101	29.7	52.4	29.2	-	-	-
FSAF [5]	Baseline	ResNet18	**26.3**	50.3	**23.7**	518.25	42.22M	18.84
	CEASC [3]	ResNet18	25.0	48.9	22.0	153.92	42.76M	23.71
	HCFFPN(Ours)	ResNet18	25.8	**51.3**	22.8	**131.98**	**40.25M**	**24.64**
RetinaNet [13]	Baseline	ResNet18	21.8	39.3	21.1	529.81	42.74M	24.37
	QueryDet [23]	ResNet50	19.6	35.7	19.0	-	-	-
	QueryDet-CSQ [23]	ResNet50	19.3	35.0	18.9	-	-	-
	CEASC [3]	ResNet18	21.6	39.6	20.6	157.41	43.28M	23.59
	HCFFPN(Ours)	ResNet18	**23.0**	**41.6**	**22.3**	**136.47**	**40.76M**	**25.32**
GFL [10]	Baseline	ResNet18	28.4	50.0	27.8	524.95	42.52M	21.46
	MobileNet V2 [18]	MobileNet V2	28.5	50.2	28.1	491.47	-	-
	ShuffleNet V2 [15]	ShuffleNet V2	26.2	46.6	25.7	488.94	-	-
	CGMDet [31]	CGAM-Net	29.3	50.9	**29.4**	-	-	-
	SOD-UAV [11]	ELAN-Net	26.3	45.7	26.8	**126.2**	**32.2M**	-
	CEASC [3]	ResNet18	28.7	50.7	28.4	150.18	43.06M	21.55
	HCFFPN (Ours)	ResNet18	**29.6**	**52.0**	**29.4**	128.58	40.54M	**22.73**

4.2 Implementation Details

Our experiments use PyTorch [16] and MMDetection [1] on an NVIDIA RTX 4080 Super GPU. VisDrone: Models are trained for 15 epochs (batch size = 2, LR = 0.01) with 1333×800 input. The LR follows linear warm-up and decays at epochs 11/14. UAVDT: Training runs for 6 epochs (LR = 0.01, decays at epochs 4/5) with 1024×540 input. We adopt GFL, RetinaNet and FSAF as base detectors. Neck output channels are set to 512 for VisDrone, with other hyperparameters aligned with CEASC.

4.3 Results on VisDrone Dataset

Evaluation on Different Detectors. Our HCFFPN is plug-and-play and can be seamlessly integrated into various FPN-based detectors. To verify its effectiveness, we evaluate it with three widely used baseline detectors: GFL [10], RetinaNet [13], and FSAF [5]. As shown in Table 1, the integration of HCFFPN reduces the GFLOPs of all basic detectors by at least 10%, the Parameters are reduced by about 2M, and the Fps is also improved; in this case, the mAP is also improved, indicating its effectiveness and universality in accelerating object detection without sacrificing model accuracy.

Comparison With Existing Methods. We compare our approach with existing GFL, RetinaNet and FSAF-based methods. At the same data enhancement settings, our method achieved higher detection accuracy, and GFLOPS, parameters and fps were also significantly improved. As shown in Table 1 GFL detector, our method outperforms CEASC, CGMDet, and SOD-UAV in terms of mAP and AP metrics while reducing the size of the model. Specifically, it improved mAP, AP50 by 0.3% and 1.1%, respectively. These gains are attributed to the enhanced multi-layer feature fusion in our method, which effectively balances model complexity with detection performance.

Visualization of Detection Results. Figure 5 compares feature heat maps with CEASC, demonstrating that our HCFFPN enlarges the receptive field for small objects, particularly in dense and distant regions, thereby improving detection. This enhancement stems from our feature fusion strategy, which preserves small-object information while strengthening feature correlations.

4.4 Results on UAVDT Dataset

Comparison With Existing Methods. The UAVDT dataset, with its abundance of tiny targets and challenging conditions like low-light and complex backgrounds, is ideal for evaluating small-target detection. As shown in Table 2, our method outperforms CEASC, APNet, and (GFL + EFC) in mAP. Compared to CEASC, we achieve gains of 1.0% mAP, 1.9% AP75, and 0.6% AP50. Against APNet, we improve by 0.2% mAP and 2.1% AP50. Even compared to the 2024 state-of-the-art (GFL + EFC), our method leads by 0.1% mAP and 0.8% AP75 while maintaining comparable AP50. These results demonstrate our algorithm's robustness in standard scenarios and superior detection in challenging drone imagery.

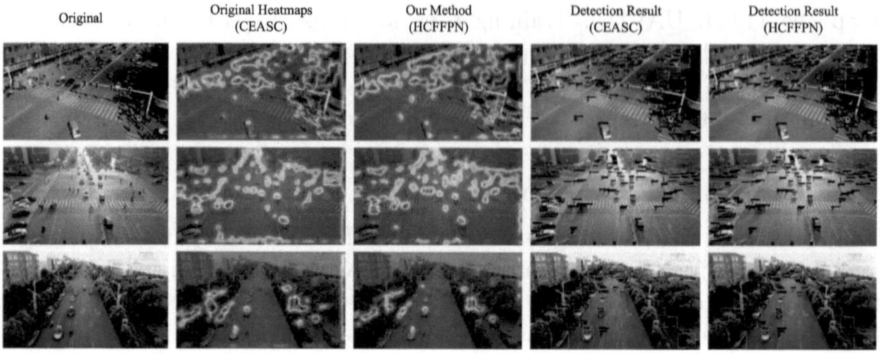

Fig. 5. Attention heatmaps and detection results on VisDrone: HCFFPN vs. CEASC. Highlighted regions denote network focus areas, validating HCFFPN's enhanced small-object localization capability.

Table 2. Comparisons with existing detectors on the UAVDT dataset. "-" indicates that the result is not reported or not publicly available

Model	Backbone	mAP	AP50	AP75
ClusDet [24]	ResNet-50	13.7	26.5	12.5
GLSAN [2]	ResNet-50	17.0	28.1	18.8
DREN [26]	ResNet-50	15.1	-	-
GFL [10]	ResNet-18	16.9	29.5	17.9
APNet [27]	ResNet-18+APDC	17.9	29.4	-
CEASC [3]	ResNet-18	17.1	30.9	17.8
GFL+EFC [22]	ResNet-18	18.0	**31.5**	18.9
HCFFPN	ResNet-18	**18.1**	**31.5**	**19.7**

Visualization Performance. As shown in Fig. 6 our HCFFPN effectively integrates limited small-target information to enhance feature representation, outperforming methods like CEASC and GFL+EFC in detecting small, background-similar targets under weak lighting.

4.5 Ablation Study

In this section, we conduct comprehensive ablation studies to evaluate the effectiveness of the key components in MLFE, SKNet and DFAM. All experiments are performed on the VisDrone dataset, using GFL with a ResNet18 backbone as the baseline. To assess the contribution of each component in our proposed HCFFPN, we replace the conventional FPN and PAFPN structures with our method and analyze MLFE, SKNet and DFAM individually and jointly. As shown in Table 3, both modules significantly improve detection performance over the baselines. When using only MLFE, all metrics outperform the FPN base-

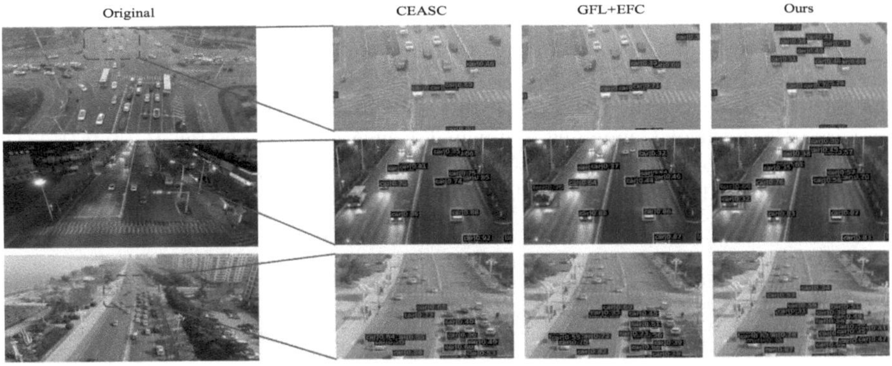

Fig. 6. Detection results on UAVDT under challenging low-light and similar-background conditions. Left: Original image with ground truth (red); Right: CEASC, GFL+EFC vs. Ours (enhanced accuracy). (Color figure online)

line. Similarly, DFAM alone surpasses the PAFPN baseline across all evaluation criteria. The best performance is achieved when both modules(MLFE, DFAM) are combined, demonstrating their complementary effects. Of course, in order to effectively compare MLFE with SKNet, when we replace MLFE with SKNet and DFAM, the overall performance will be reduced. Therefore, our MLFE is very effective compared to SKNet.

Table 3. Ablation study of MLFE, SKNet and DFAM detector on the VisDrone dataset

MLFE	SKNet	DFAM	mAP	AP50	AP75
FPN			28.4%	50.0%	27.8%
PAFPN			28.6%	50.9%	28.6%
√			28.6%	50.6%	28.1%
		√	29.3%	51.4%	28.9%
	√	√	28.5%	51.2%	28.6%
√		√	**29.6%**	**52.0%**	**29.4%**

5 Conclusion

In this study, an HCFFPN strategy using MLFE and DFAM is proposed for small target detection in unmanned aerial vehicle images. It is worth noting that the proposed HCFFPN method can be flexibly integrated into the underlying FPN. In addition, a large number of experimental results on VisDrone and UAVDT datasets show that HCFFPN shows remarkable effectiveness in small target detection and effectively reduces the computing resource requirements of the neck network.

Acknowledgement. This work was supported by the National Natural Science Foundation of China (Grant 62202134); the "Pioneer" and "Leading Goose" R&D Program of Zhejiang Province (Grants 2025C01035, 2025C01006); the Natural Science Foundation of Chongqing (Grant CSTB2024NSCQ-MSX0726); the Science and Technology Research Program of Chongqing Municipal Education Commission (Grants KJZD-K202500510, KJQN202400507); the Talents Fund of Chongqing Normal University (Grants 21XLB031, 23XLB028); and the Dazhou Key Laboratory of Government Data Security (Grant ZSAQ202403).

References

1. Chen, K., et al.: Mmdetection: open mmlab detection toolbox and benchmark. arxiv 2019. arXiv preprint arXiv:1906.07155 (1906)
2. Deng, S., et al.: A global-local self-adaptive network for drone-view object detection. IEEE Trans. Image Process. **30**, 1556–1569 (2020)
3. Du, B., Huang, Y., Chen, J., Huang, D.: Adaptive sparse convolutional networks with global context enhancement for faster object detection on drone images. In: Proceedings of the IEEE/CVF Conference on Computer Vision and Pattern Recognition, pp. 13435–13444 (2023)
4. Du, D., et al.: The unmanned aerial vehicle benchmark: object detection and tracking. In: Proceedings of the European Conference on Computer Vision (ECCV), pp. 370–386 (2018)
5. Du, D., et al.: Visdrone-det2019: the vision meets drone object detection in image challenge results. In: Proceedings of the IEEE/CVF International Conference on Computer Vision Workshops (2019)
6. Ghiasi, G., Lin, T.Y., Le, Q.V.: NAS-FPN: learning scalable feature pyramid architecture for object detection. In: Proceedings of the IEEE/CVF Conference on Computer Vision and Pattern Recognition, pp. 7036–7045 (2019)
7. Hu, M., Li, Y., Fang, L., Wang, S.: A2-FPN: attention aggregation based feature pyramid network for instance segmentation. In: Proceedings of the IEEE/CVF Conference on Computer Vision and Pattern Recognition, pp. 15343–15352 (2021)
8. Li, C., Yang, T., Zhu, S., Chen, C., Guan, S.: Density map guided object detection in aerial images. In: Proceedings of the IEEE/CVF Conference on Computer Vision and Pattern Recognition Workshops, pp. 190–191 (2020)
9. Li, X., Wang, W., Hu, X., Yang, J.: Selective kernel networks. In: Proceedings of the IEEE/CVF Conference on Computer Vision and Pattern Recognition, pp. 510–519 (2019)
10. Li, X., et al.: Generalized focal loss: learning qualified and distributed bounding boxes for dense object detection. Adv. Neural. Inf. Process. Syst. **33**, 21002–21012 (2020)
11. Li, Y., Wang, Y., Ma, Z., Wang, X., Tang, Y.: Sod-UAV: small object detection for unmanned aerial vehicle images via improved yolov7. In: ICASSP 2024-2024 IEEE International Conference on Acoustics, Speech and Signal Processing (ICASSP), pp. 7610–7614. IEEE (2024)
12. Lin, T.Y., Dollár, P., Girshick, R., He, K., Hariharan, B., Belongie, S.: Feature pyramid networks for object detection. In: Proceedings of the IEEE Conference on Computer Vision and Pattern Recognition, pp. 2117–2125 (2017)

13. Lin, T.Y., Goyal, P., Girshick, R., He, K., Dollár, P.: Focal loss for dense object detection. In: Proceedings of the IEEE International Conference on Computer Vision, pp. 2980–2988 (2017)
14. Liu, C., Gao, G., Huang, Z., Hu, Z., Liu, Q., Wang, Y.: Yolc: you only look clusters for tiny object detection in aerial images. IEEE Trans. Intell. Transp. Syst. (2024)
15. Ma, N., Zhang, X., Zheng, H.T., Sun, J.: Shufflenet v2: practical guidelines for efficient CNN architecture design. In: Proceedings of the European Conference on Computer Vision (ECCV), pp. 116–131 (2018)
16. Paszke, A., et al.: Pytorch: an imperative style, high-performance deep learning library. In: Advances in Neural Information Processing Systems, vol. 32 (2019)
17. Ren, S., He, K., Girshick, R., Sun, J.: Faster R-CNN: towards real-time object detection with region proposal networks. IEEE Trans. Pattern Anal. Mach. Intell. **39**(6), 1137–1149 (2016)
18. Sandler, M., Howard, A., Zhu, M., Zhmoginov, A., Chen, L.C.: Mobilenetv2: inverted residuals and linear bottlenecks. In: Proceedings of the IEEE Conference on Computer Vision and Pattern Recognition, pp. 4510–4520 (2018)
19. Tan, M., Pang, R., Le, Q.V.: Efficientdet: scalable and efficient object detection. In: Proceedings of the IEEE/CVF Conference on Computer Vision and Pattern Recognition, pp. 10781–10790 (2020)
20. Tian, Z., Chu, X., Wang, X., Wei, X., Shen, C.: Fully convolutional one-stage 3D object detection on lidar range images. Adv. Neural. Inf. Process. Syst. **35**, 34899–34911 (2022)
21. Wang, C.Y., Bochkovskiy, A., Liao, H.Y.M.: Yolov7: trainable bag-of-freebies sets new state-of-the-art for real-time object detectors. In: Proceedings of the IEEE/CVF Conference on Computer Vision and Pattern Recognition, pp. 7464–7475 (2023)
22. Xiao, Y., Xu, T., Yu, X., Fang, Y., Li, J.: A lightweight fusion strategy with enhanced inter-layer feature correlation for small object detection. IEEE Trans. Geosci. Remote Sens. (2024)
23. Yang, C., Huang, Z., Wang, N.: Querydet: cascaded sparse query for accelerating high-resolution small object detection. In: Proceedings of the IEEE/CVF Conference on Computer Vision and Pattern Recognition, pp. 13668–13677 (2022)
24. Yang, F., Fan, H., Chu, P., Blasch, E., Ling, H.: Clustered object detection in aerial images. In: Proceedings of the IEEE/CVF International Conference on Computer Vision, pp. 8311–8320 (2019)
25. Yin, J., et al.: MUPT-Net: multi-scale u-shape pyramid transformer network for infrared small target detection. Displays **83**, 102681 (2024)
26. Zhang, J., Huang, J., Chen, X., Zhang, D.: How to fully exploit the abilities of aerial image detectors. In: Proceedings of the IEEE/CVF International Conference on Computer Vision Workshops (2019)
27. Zhang, P., Zhang, G., Yang, K.: Apnet: accurate positioning deformable convolution for UAV image object detection. IEEE Lat. Am. Trans. **22**(4), 304–311 (2024)
28. Zhang, P., Zhong, Y., Li, X.: Slimyolov3: narrower, faster and better for real-time UAV applications. In: Proceedings of the IEEE/CVF International Conference on Computer Vision Workshops (2019)
29. Zhang, X., Izquierdo, E., Chandramouli, K.: Dense and small object detection in UAV vision based on cascade network. In: Proceedings of the IEEE/CVF International Conference on Computer Vision Workshops (2019)

30. Zhang, Y., Ye, M., Zhu, G., Liu, Y., Guo, P., Yan, J.: FFCA-YOLO for small object detection in remote sensing images. IEEE Trans. Geosci. Remote Sens. **62**, 1–15 (2024)
31. Zhou, L., et al.: A multi-scale object detector based on coordinate and global information aggregation for UAV aerial images. Remote Sens. **15**(14), 3468 (2023)

MP-CLIP: Unlocking Long-Text Understanding in CLIP via Multi-paragraph Encoding

Zhaofu Zeng[1] and Jian Xing[2](✉)

[1] Laboratory of Space-Air-Ground-Ocean Integrated Network Security, School of Cyberspace Security, Hainan University, Haikou, China

[2] Laboratory of Space-Air-Ground-Ocean Integrated Network Security, School of Cyberspace Security, Hainan University. Hainan Provincial Engineering Research Center of Cryptology and Cross-Border Data Security, Haikou, China

xingjian@hainanu.edu.cn

Abstract. Contrastive Language-Image Pre-training (CLIP) has demonstrated strong performance across various downstream tasks. However, its text encoder is limited to a maximum of 77 tokens, making it incapable of handling longer textual inputs. Although some recent multimodal models have addressed this length limitation, they still struggle to capture structural semantics. As shown in Figs. 1 and 2, Long-CLIP struggle to distinguish between structurally correct and incorrect text descriptions based on the given image, as it tend to focus on learning general representation across varying text lengths while neglecting the modeling of structured representation. In this paper, We propose MP-CLIP, a fine-tuned CLIP variant that supports longer inputs and enhances structured representation for better understanding of both short and long descriptions. MP-CLIP introduces two key components: (1) Multi-Segment Aggregation Encoder, and (2) Multi-Granularity Matching strategy. MP-CLIP achieves an average 3% improvement over Long-CLIP on long-text image retrieval tasks, and surpasses CLIP on short-text retrieval benchmarks such as MSCOCO and Flickr30k. It also maintains comparable performance to CLIP on zero-shot image classification, and achieves superior results on fine-grained tasks, outperforming Long-CLIP by 7% on VG-Relation. Moreover, MP-CLIP can be seamlessly integrated into image generation pipelines to guide high-quality image synthesis from detailed descriptions.

Keywords: Multimodality · Text-Image Retrieval · Text-to-Image Generation · Fine-Grained Representation

1 Introduction

Contrastive Language-Image Pre-training (CLIP) [19] is a widely adopted vision-language pretraining model composed of an image encoder and a text encoder.

J. Lokoč et al. (Eds.): MMM 2026, LNCS 16413, pp. 247–261, 2026.
https://doi.org/10.1007/978-981-95-6957-1_18

It aligns visual and textual modalities via contrastive learning to build a unified embedding space, achieving strong zero-shot performance across tasks such as image classification [6,15], image-text retrieval [7,24], and image generation [22, 25].

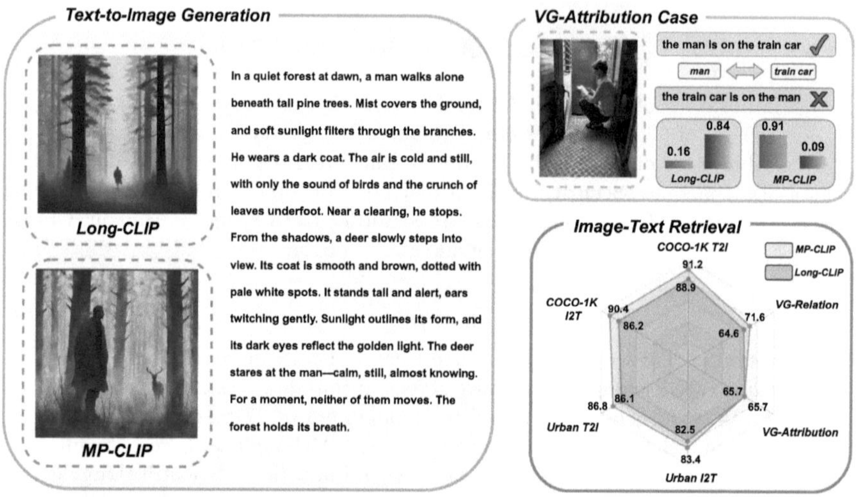

Fig. 1. Usage of Our Proposed MP-CLIP. MP-CLIP can directly replace the original CLIP in generative tasks. Compared to Long-CLIP, it generates images that better align with long descriptions (left side). MP-CLIP also demonstrates a stronger ability to detect semantic perturbations (top right). Furthermore, it significantly outperforms Long-CLIP on long-text and fine-grained tasks (bottom right).

However, CLIP's text input is constrained to a maximum of 77 tokens due to its positional encoding design, which imposes a strict upper bound on input length. This limitation hampers its ability to process and understand long descriptions. Moreover, the mismatch between the rich semantics of images and the relatively sparse information in short texts forces CLIP to compress visual representations to fit the textual input, leading the image encoder to prioritize only key elements and ignore fine-grained details. Therefore, extending CLIP's capability to handle longer texts is essential.

A straightforward approach is to interpolate the positional embeddings to support longer sequences and fine-tune CLIP on image-text pairs with long descriptions. However, such naive interpolation significantly harms performance on short-text tasks. Alternatively, replacing the original text encoder with a large language model requires expensive pretraining on massive datasets.

To address these challenges, Long-CLIP [32] retains the original positional embeddings for the first 20 positions and interpolates the remaining ones to support up to 248 tokens. This approach maintains the integrity of pretrained positions while extending input length. To preserve short-text performance, Long-CLIP aligns compressed, coarse-grained image features with short texts, helping

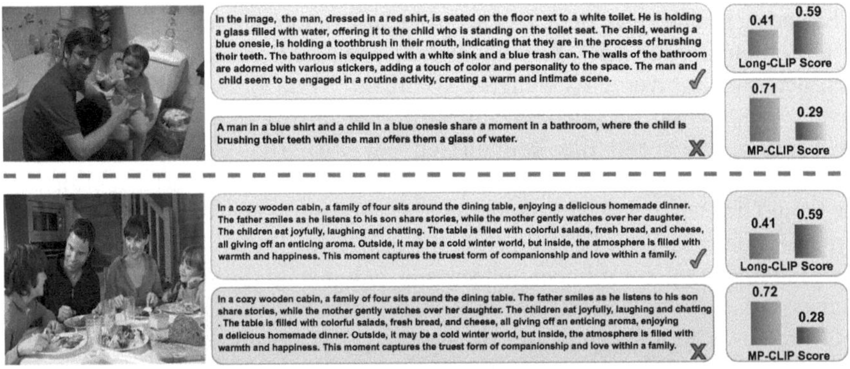

Fig. 2. The Long-CLIP scores between images and true/false captions (normalized between the two results). The results indicate that Long-CLIP still struggles to understand the structure and semantics of long texts.

the model focus on key visual components while still capturing details. However, as noted in Winoground [27], CLIP processes text primarily as a bag-of-words approach, largely ignoring syntactic structure and word order. Long-CLIP inherits this limitation in its understanding of both short and long texts. For short texts, as shown in Fig. 1, Long-CLIP fails to identify structural perturbations. Specifically, it considers the structurally incorrect sentence "the train car is on the man" to be a better match for the image. For long texts, as shown in Fig. 2, Long-CLIP suffers from the same issue, assigning higher similarity scores to shuffled long texts than to correctly ordered ones. In addition, as shown in Fig. 2, Long-CLIP assigns a higher similarity score between the image and an incorrect short text than between the image and its corresponding correct long description. These observations indicate that Long-CLIP still struggles with fine-grained understanding across both short and long texts.

To overcome these limitations, We introduce MP-CLIP, which employs two novel strategies to not only improve understanding of both long and short texts but also enhance structured representation.(1) Multi-Paragraph Aggregation Encoder (MPAE): We first split long texts into semantically coherent segments using the Segment Processor, each shorter than 77 tokens. Each segment is encoded using CLIP's original text encoder to obtain a set of segment embeddings. A lightweight multi-paragraph aggregation module then fuses these embeddings into a single long-text representation. This method extends the input length while preserving CLIP's pretrained knowledge and enhances the model's structured representation. (2) Multi-Granularity Matching (MGM): To enhance fine-grained discrimination between short and long texts, we adopt contrastive losses with different temperature parameters for each, allowing the model to better capture nuanced semantic differences. Furthermore, we introduce a semantic consistency loss to preserve alignment between long and short descriptions of the same image, preventing the model from over-separating them during training.

Experimental results demonstrate that our model significantly outperforms the CLIP and Long-CLIP. Specifically, for long-text image retrieval, our approach achieves an average 4% higher recall rate than Long-CLIP. In short-text image retrieval tasks, it substantially surpasses the original CLIP. For zero-shot classification, the recall rate remains on par with the original CLIP without degradation. On fine-grained tasks, our model exhibits superior performance over Long-CLIP, particularly achieving a 7% higher on the VG-Relation dataset. Moreover, our model can be seamlessly integrated into image generation pipelines as a plug-and-play component, effectively enhancing the quality of generated images.

Overall, our contributions are three-fold: (1) We propose a Multi-Paragraph Aggregation Encoder that enables CLIP to encode long texts while enhancing its structured representation. (2) We design a multi-granularity contrastive strategy that captures fine-grained semantic differences across varying text lengths while enforcing semantic consistency. (3) Extensive experiments show that MP-CLIP surpasses Long-CLIP in long-text and fine-grained understanding tasks, retains strong performance on short-text benchmarks, and proves effective in generative tasks. Codes are released at https://github.com/Omntrix/MP-CLIP.

2 Related Works

2.1 Vision-Language Pretraining

Vision-Language Models (VLMs) have rapidly advanced in learning robust cross-modal representations for diverse downstream tasks. Among them, Contrastive Language-Image Pre-training (CLIP) [19] aligns visual and textual modalities via contrastive learning on large-scale image-text pairs, achieving strong zero-shot generalization across detection [5,8], segmentation [12,33], video understanding [21,28], and image generation [20,26].

However, some subsequent works [11] have noted that CLIP struggles to capture fine-grained semantic information. To address this, they adopt contrastive learning strategies that align individual text tokens with corresponding image regions. Other studies [10,30] further reveal that CLIP interprets text in a bag-of-words manner, neglecting word order and syntactic structure. To overcome this, they improve the quality of negative samples to help the model distinguish hard negatives and better capture fine-grained textual semantics. However, these approaches still struggle to model fine-grained information effectively in long caption scenarios.

2.2 Long-Text Semantic Representation

Standard Transformer architectures struggle with modeling texts that exceed their maximum input length or encompass extensive semantic spans. To handle longer sequences, models like Longformer [1] and BigBird [31] introduce sparse attention mechanisms, enabling the processing of longer text inputs. To overcome CLIP's limitation on input text length, one straightforward solution is to replace its text encoder with a language model capable of processing longer sequences. However, this approach requires extensive pretraining. Instead, Long-CLIP [32] offers an efficient fine-tuning strategy using knowledge-preserved stretching and primary component matching to enhance long-text semantic representation.

2.3 Vision-Language Datasets

Traditional multimodal datasets like ImageNet [4] and Cifar10 [14], limited to fixed categories, are becoming inadequate for open-world vision-language applications. To address this, image-text datasets such as Visual Genome [13], MSCOCO [16], Flickr30k [29], Conceptual-12M [2], and LAION-5B [23] have been developed, improving models' real-world understanding and generalization. However, their captions are typically short and low in information detail, which limits their ability to capture fine-grained visual semantics. ShareGPT4V [3] addresses this by providing 1.2 million image long-text pairs, with an average of over 800 characters per caption. The texts offer rich information, including object attributes, spatial relationships, and background knowledge, making it well-suited for training models to understand fine-grained visual semantics.

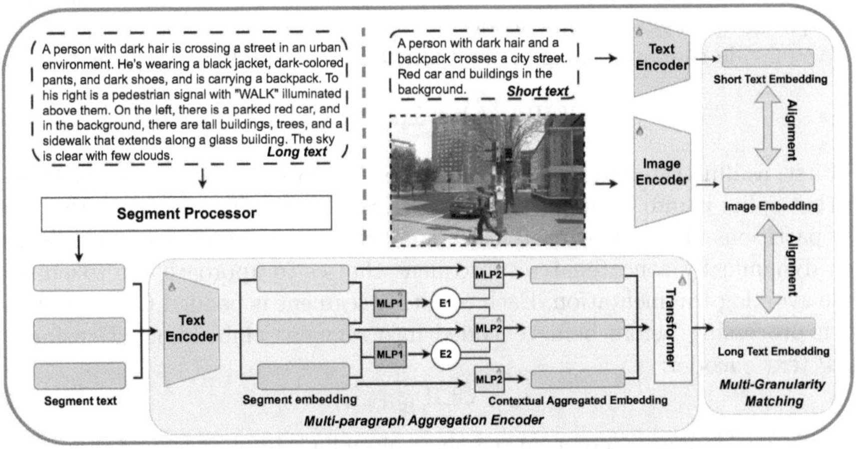

Fig. 3. Overview of MP-CLIP. We encode long text using a multi-segment aggregation encoder without modifying the structure of the original CLIP text encoder. The model is trained with a Multi-Granularity Matching. MLP1 denotes the Compatibility MLP, and MLP2 denotes the Update MLP.

3 Method

The overview of MP-CLIP is illustrated in Fig. 3, We propose MP-CLIP to enhance CLIP's long-text comprehension, improve fine-grained understanding, and maintain strong performance on short texts, achieved through a multi-paragraph aggregation encoder and a multi-granularity matching strategy. The multi-paragraph aggregation encoder is described in Sect. 3.1, and the multi-granularity matching strategy is detailed in Sect. 3.2.

3.1 Multi-paragraph Aggregation Encoder

As shown in Fig. 2, Long-CLIP often mismatches images with misleading short texts or long texts with disordered token sequences, making it difficult to align with the correct descriptions. This indicates its limitations in capturing the semantics and structural order of long texts. To address this issue, we propose a multi-paragraph aggregation encoder that preserves the original CLIP text encoder architecture. Specifically, a Segment Processor divides the long text into shorter, coherent segments compatible with the CLIP encoder. Each segment is independently encoded, and the resulting embeddings are aggregated into a unified representation. Moreover, prior work [32] has shown that CLIP's positional encoding primarily focuses on the first 20 tokens, while the remaining positions are underutilized. Our segmentation strategy generates segments close to the 77-token limit, fully activating these underused positional embeddings and enhancing long-text understanding. In the following, we detail the multi-paragraph aggregation encoder.

We encode the input image I_i and text C_i using the following equations.

$$V_i = \text{CLIP}_{\text{vis}}(I_i), \quad T_i = \text{CLIP}_{\text{text}}(C_i). \tag{1}$$

First, as illustrated in Fig. 3, the long text is processed by a Segment Processor that splits it into coherent segments $S_1, S_2, \ldots, S_N$. The Segment Processor first partitions the text at sentence boundaries to preserve semantic integrity, then dynamically concatenates subsequent clauses to approach the token limit while avoiding fragmentation. Each resulting segment is padded to 77 tokens for batch processing before being encoded into segment embeddings T_i using the CLIP text encoder:

$$T_i = \text{CLIP}_{\text{text}}(S_i). \tag{2}$$

In natural language, adjacent segments often have implicit logical connections (e.g., causality, contrast, elaboration) that cannot be captured by independent segment embeddings T_i. To extract such inter-segment relations, we employ a compatibility MLP (denoted as MLP1 in the Fig. 3) to obtain a logic-aware embedding e_i between each adjacent pair:

$$e_i = \text{MLP}_{\text{compat}}([T_i; T_{i+1}]). \tag{3}$$

To enhance each segment's awareness of its surrounding context, we apply an update MLP (denoted as MLP2 in the Fig. 3) to fuse its original representation with logical information from neighboring segments:

$$p_i = \text{MLP}_{\text{update}}([T_i; e_{i-1}; e_i]). \tag{4}$$

To aggregate the contextual aggregated embeddings $\mathbf{p}_1, \mathbf{p}_2, \ldots, \mathbf{p}_N$ into a unified representation for the entire long text, we design a memory-token-based Transformer aggregation module. Inspired by the [CLS] token used in BERT, we introduce a learnable memory token $\mathbf{m} \in \mathbb{R}^D$, where D is the dimensionality of the segment embeddings. This memory token serves as a global semantic carrier to integrate information across all segments.

The memory token is prepended to the sequence of contextual aggregated embeddings, forming the input $[\mathbf{m}, \mathbf{p}_1, \mathbf{p}_2, \ldots, \mathbf{p}_N]$, which is then fed into a multi-layer Transformer encoder. Let the output of the Transformer be $[\mathbf{h}_0, \mathbf{h}_1, \ldots, \mathbf{h}_N]$, where $\mathbf{h}_0$ corresponds to the memory token:

$$[\mathbf{h}_0, \mathbf{h}_1, \ldots, \mathbf{h}_N] = \text{Transformer}([\mathbf{m}, \mathbf{p}_1, \mathbf{p}_2, \ldots, \mathbf{p}_N]), \tag{5}$$

Finally, we take the output at the memory token position, $\mathbf{h}_0$, as the unified representation $\mathbf{m}'$ of the entire long text:

$$\mathbf{m}' = \mathbf{h}_0. \tag{6}$$

3.2 Multi-granularity Matching

To enhance fine-grained alignment between images and texts of different lengths, we propose a Multi-Granularity Matching strategy, which jointly optimizes three objectives: (1) image-to-long-text contrastive loss, (2) image-to-short-text contrastive loss, and (3) semantic consistency loss. Each component is weighted appropriately to reflect their granularity and semantic role.

Contrastive Loss with Granularity-Specific Temperature. Given an image embedding $\mathbf{v}_i$ and its corresponding short and long text embeddings $\mathbf{t}i^{\text{short}}$ and $\mathbf{t}i^{\text{long}}$, we adopt two separate contrastive objectives with different temperature scaling factors τshort and τlong, inspired by the observation that long texts provide dense fine-grained descriptions requiring stricter matching, while short texts are more abstract and tolerant.

The contrastive loss for long-text alignment is defined as:

$$\mathcal{L}\text{long} = \frac{1}{2}\left(\text{CE}\left(\frac{\mathbf{V}\mathbf{T}\text{long}^\top}{\tau_{\text{long}}}\right) + \text{CE}\left(\frac{\mathbf{T}\text{long}\mathbf{V}^\top}{\tau\text{long}}\right)\right), \tag{7}$$

Similarly, the short-text contrastive loss is:

$$\mathcal{L}\text{short} = \frac{1}{2}\left(\text{CE}\left(\frac{\mathbf{V}\mathbf{T}\text{short}^\top}{\tau_{\text{short}}}\right) + \text{CE}\left(\frac{\mathbf{T}\text{short}\mathbf{V}^\top}{\tau\text{short}}\right)\right), \tag{8}$$

here, $\text{CE}(\cdot)$ denotes the cross-entropy loss with label smoothing, and the temperature parameters $\tau_{\text{short}} > \tau_{\text{long}}$ allow for more lenient matching with short text.

Semantic Consistency Loss. Since the long text semantically encompasses its corresponding short text, their embeddings should reside in close proximity within the embedding space. To ensure that the long and short text embeddings remain semantically aligned. we introduce a semantic consistency loss:

$$\mathcal{L}\text{sim} = \frac{1}{N} \sum i = 1^N \left[\max \left(0, \delta - \left| \mathbf{t}_i^{\text{long}} - \mathbf{t}_i^{\text{short}} \right|_2 \right) \right]^2, \tag{9}$$

where δ is a margin parameter that controls the minimum similarity between the long and short text embeddings. This term acts as a regularizer to prevent the two representations from drifting too far apart during optimization.

Final Objective. The final loss is a weighted sum of the three components:

$$\mathcal{L}\text{total} = \mathcal{L}\text{long} + \alpha \cdot \mathcal{L}\text{short} + \lambda\text{sim} \cdot \mathcal{L}_{\text{sim}}, \tag{10}$$

where α and λ_{sim} are tunable hyperparameters that control the importance of short-text alignment and semantic consistency, respectively.

4 Experiments

4.1 Experiment Setting

Training Setting. All experiments are implemented using PyTorch and conducted on a single NVIDIA A6000 GPU. The paragraph-level aggregation encoder consists of a two-layer Transformer and two three-layer MLP modules, referred to as the compatibility MLP and the update MLP, respectively. We use the ShareGPT4V [3] as our training set, which contains approximately 1.2 million (long text, image) pairs. Among them, 400k pairs are randomly selected for training. During training, we initialize our model with a pre-trained CLIP checkpoint and train it for 5 epochs with a batch size of 128. We adopt the AdamW optimizer with a weight decay of 0.1 and an initial learning rate of 4e-6. The hyperparameters are set as follows: δ is set to 0.1, α is set to 1, and λ_{sim} is set to 0.1.

Evaluation Datasets. To evaluate the model's ability to understand long texts, we employ two long-text benchmarks: (1) the Urban-1k dataset introduced in Long-CLIP [32], which includes 200 urban scene images and their corresponding long descriptions; and (2) We construct a new benchmark by randomly sampling 1k images from the test split of the MSCOCO2017 dataset [16] and generating detailed long captions using LLaVA [17]. For short-text evaluation, we use MSCOCO2017 [16] and Flickr30k [29]. Specifically, we use the standard 5k validation set for MSCOCO2017, and the entire 30k image-text pairs in Flickr30k, instead of only the 1k test subset, to increase difficulty and avoid benchmark saturation. For zero-shot image classification. We evaluated the zero-shot classification capability of CLIP on Cifar10 [14] and ImageNet-O [9]. To further assess the model's sensitivity to fine-grained semantics and syntactic structure, we evaluate on two challenging benchmarks [30]: VG-Relation and VG-Attribution. Each test instance consists of an image and two captions—one semantically aligned and one adversarially altered. The model must identify the correctly aligned description.

Table 1. R@1 of long-caption text-image retrieval on MSCOCO-1K and Urban-1k.

	MSCOCO-1K		Urban-1k	
	Image-to-Text	Text-to-Image	Image-to-Text	Text-to-Image
ViT-L/14				
CLIP	76.8	78.0	47.0	47.0
Direct Fine-tuning	84.1	85.4	78.0	76.5
Long-CLIP	86.2	88.9	82.5	86.1
MP-CLIP	**90.4**	**91.2**	**83.4**	**86.8**

Table 2. R@K of short-caption text-image retrieval on MSCOCO and Flickr30k.

	MSCOCO				Flickr30k			
	Image-to-Text		Text-to-Image		Image-to-Text		Text-to-Image	
	R@1	R@5	R@1	R@5	R@1	R@5	R@1	R@5
ViT-L/14								
CLIP	56.1	79.5	35.4	60.1	38.0	65.3	28.0	49.3
Direct Fine-tuning	37.9	63.1	23.1	45.1	26.3	48.3	17.9	34.9
Long-CLIP	**62.8**	**85.1**	**46.3**	**70.8**	**53.4**	77.5	**41.2**	64.1
MP-CLIP	59.8	84.6	44.6	69.8	53.2	**77.6**	40.5	**64.3**

Table 3. Results of zero-shot image classification and fine-grained benchmarks.

	Cifar10	ImageNet-O	VG-Attribution	VG-Relation
ViT-L/14				
CLIP	**95.5**	31.9	61.1	61.5
Long-CLIP	95.3	**33.7**	**65.7**	64.6
MP-CLIP(ours)	93.4	32.88	**65.7**	**71.6**

4.2 Overall Results

We evaluate MP-CLIP against CLIP and Long-CLIP on four tasks: long-text retrieval, short-text retrieval, short-text classification, and fine-grained recognition. For fair comparison, CLIP is fine-tuned on long-text tasks using the input extension strategy proposed by Long-CLIP to support longer sequences. To the best of our knowledge, Long-CLIP is one of the latest and few existing models that enhance CLIP's long-text understanding through fine-tuning, making it a strong and representative baseline for our evaluations.

As shown in Table 1, MP-CLIP outperforms Long-CLIP by approximately 4% on MSCOCO-1K and 1% on Urban-1k in long-text retrieval. In Table 2, Our method significantly surpasses CLIP in short-text retrieval and outperforms Long-CLIP by +0.2% in R@5 T2I on Flickr30K. Additionally, in Table 3,

Table 4. Ablation study on the effectiveness of our two proposed strategies.

MPAE	MGM	VG-Att	VG-Rel	Flickr I2T R@1	MSCOCO-1K T2I R@1	Urban-1k T2I R@1
✗	✗	61.1	61.5	38.0	85.4	76.5
✓	✗	64.9	68.9	40.6	89.1	83.2
✗	✓	65.4	66.9	49.1	86.6	80.5
✓	✓	**65.7**	**71.6**	**53.2**	**91.2**	**86.8**

Table 5. Performance comparison of long-text processing strategies.

Strategy	VG-Att	VG-Rel	MSCOCO T2I R@1	MSCOCO-1K T2I R@1	Urban-1k T2I R@1
Transformer	64.5	63.2	37.6	84.4	78.6
Transformer + Token	64.9	66.9	38.9	89.1	83.2
Ours	**65.7**	**71.6**	**44.6**	**91.2**	**86.8**

MP-CLIP maintains competitive performance in short-text classification without noticeable degradation. In fine-grained tasks, as shown in Table 3, MP-CLIP improves over Long-CLIP by 7% on the VG-Relation task and achieves comparable performance on VG-Attribution. These results indicate that our method enhances structured representations.

Overall, experimental results demonstrate that MP-CLIP not only lifts the length limitation of CLIP by enabling effective alignment between long texts and images, achieving better performance than Long-CLIP on long-text tasks, but also enhances structured representation. In addition, it maintains strong performance on short-text retrieval benchmarks.

4.3 Ablation Studies

Effectiveness of Core Components. MP-CLIP integrates two essential components. The first is the Multi-paragraph Aggregation Encoder, which enhances long-text structured representation by modeling inter-paragraph relations without modifying the original CLIP architecture. The second is the Multi-Granularity Matching strategy, which guides the model to capture fine-grained differences and commonalities between short and long texts.

To evaluate the effectiveness of these components, we conduct ablation studies on several CLIP variants, each enabling or disabling specific strategies. When only the Multi-paragraph Aggregation Encoder is used with a standard contrastive loss and without the semantic consistency objective, as shown in Table 4, row 2, performance on short-text retrieval tasks drops significantly. This highlights the importance of the Multi-Granularity Matching strategy in preserving short-text retrieval capabilities. Conversely, when only the Multi-Granularity Matching strategy is used and segment-level features are simply aggregated via a vanilla transformer, as shown in Table 4, row 3, the model struggles to represent long texts effectively. These results indicate that both components are critical: removing either one leads to substantial performance degradation on both short-text and long-text tasks.

Strategies for Long-Text Representation. To build effective long-text representation and structured representation, we employ two MLP modules to capture logical relationships between segments and incorporate contextual information into each paragraph embedding. The enriched segment embeddings are then aggregated using a transformer with memory token to produce the final long-text embedding. There are various ways to aggregate paragraph embeddings for long-text encoding, including using only partial components of our Multi-paragraph Aggregation Encoder. To evaluate the effectiveness of the full combination, we implement and compare three strategies: (1) aggregating paragraph embeddings using only a vanilla transformer, (2) using a transformer with memory token, and (3) using the complete Multi-paragraph Aggregation Encoder.

As shown in Table 5, removing the MLP modules significantly degrades performance on the VG-Relation task, indicating their crucial role in modeling inter-paragraph logic and enhancing structured representation. Similarly, omitting memory token leads to substantial drops in long-text retrieval accuracy, suggesting its importance in maintaining semantic completeness. These results confirm that both the MLP-based logic modeling and memory-enhanced fusion are essential for robust long-text representation and improving structured representation.

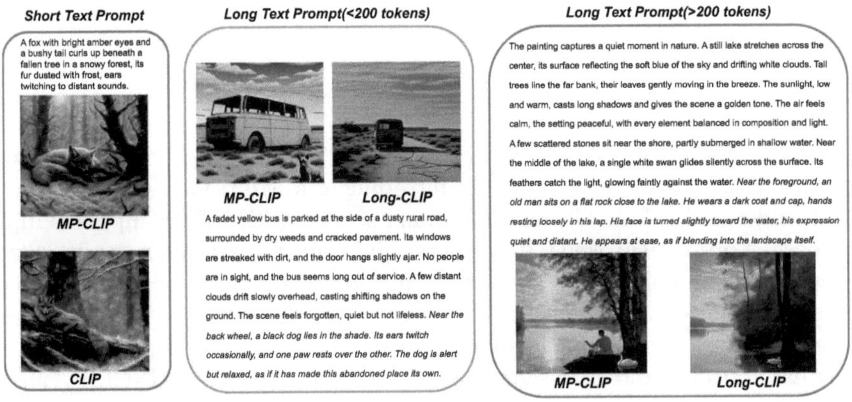

Fig. 4. Application of MP-CLIP in Image Generation. From left to right, the length of the text prompts increases. As the prompt length grows, MP-CLIP consistently captures fine-grained information in long texts, enabling the generation model to produce high-quality images.

4.4 Application of MP-CLIP in Image Generation

CLIP has been widely used in generative tasks such as image segmentation [12], 3D reconstruction [25], and image synthesis [18]. Particularly in image generation, the pretrained CLIP text encoder is often used to extract text features that guide the generation process. The quality of the generated images largely depends on the model's ability to understand the input text. However, the original CLIP model imposes a strict limit on input length, which restricts its effectiveness in handling long-text prompts. Thanks to the Multi-paragraph Aggregation Encoder and the Multi-Granularity Matching, our MP-CLIP model significantly enhances long-text representations while maintaining strong short-text understanding. MP-CLIP can be seamlessly integrated into existing generative pipelines, effectively lifting the input length constraint and enabling higher-quality image generation. For example, the original CLIP text encoder in SDXL [18] can be directly replaced with MP-CLIP without any additional training, resulting in improved visual quality.

As illustrated in Fig. 4, MP-CLIP demonstrates strong long-text modeling capabilities. For short prompts, MP-CLIP produces images highly consistent with those generated by CLIP, indicating that our model preserves CLIP's short-text understanding. For long prompts, MP-CLIP generates images that faithfully reflect the detailed content across the entire prompt. In contrast, when prompt length increases further or when key entities appear near the end of the text, Long-CLIP often fails to capture those details, resulting in incomplete or less faithful image content. This contrast highlights MP-CLIP's ability to effectively aggregate cross-segment information and maintain semantic completeness, enabling finer-grained representations of long-form textual inputs.

5 Conclusion

We present MP-CLIP, a model that enhances CLIP's long-text understanding and structured representation by introducing a Multi-Paragraph Aggregation Encoder and a Fine-Grained Matching strategy. MP-CLIP supports input sequences of up to 308 tokens. It achieves significant improvements on long-text retrieval tasks and shows strong performance on fine-grained benchmarks, outperforming Long-CLIP by 7% on VG-Relation. Moreover, MP-CLIP maintains competitive performance on short-text tasks, matching the baseline CLIP in short-text retrieval and zero-shot classification. In addition, MP-CLIP can be seamlessly integrated into image generation pipelines, enabling higher-quality image generation with longer text prompts.

Scalability. Both MP-CLIP and Long-CLIP are trained on ShareGPT-4V, where the average text length is about 250 tokens, and currently no publicly available image-text dataset exceeds 500 tokens, which inherently limits model input lengths. Critically, scaling MP-CLIP to 500 tokens only requires aggregating three additional short-text segments, which is far more efficient than Long-CLIP that would need expensive retraining for hundreds of extra tokens.

Moreover, MP-CLIP is trained on only about half of the ShareGPT-4V dataset, yet it still surpasses Long-CLIP on long-text tasks, underscoring its scalability and potential for further gains with larger datasets.

Acknowledgements. Supported by Hainan Provincial Natural Science Foundation of China:825RC742, Hainan University Scientific Research Fund Supported Project: KYQD(ZR)23182.

References

1. Beltagy, I., Peters, M.E., Cohan, A.: Longformer: the long-document transformer. arXiv preprint arXiv:2004.05150 (2020)
2. Changpinyo, S., Sharma, P., Ding, R.: Conceptual 12m: pushing web-scale image-text pre-training to recognize long-tail visual concepts. In: Proceedings of the IEEE/CVF Conference on Computer Vision and Pattern Recognition, pp. 3558–3568 (2021)
3. Chen, L., et al.: Sharegpt4v: improving large multi-modal models with better captions. In: European Conference on Computer Vision, pp. 370–387. Springer (2024)
4. Deng, J., Dong, W., Socher, R., Li, L.J., Li, K., Fei-Fei, L.: Imagenet: a large-scale hierarchical image database. In: 2009 IEEE Conference on Computer Vision and Pattern Recognition, pp. 248–255. IEEE (2009)
5. Esmaeilpour, S., Liu, B., Robertson, E., Shu, L.: Zero-shot out-of-distribution detection based on the pre-trained model clip. In: Proceedings of the AAAI Conference on Artificial Intelligence, vol. 36, pp. 6568–6576 (2022)
6. Fan, L., Krishnan, D., Isola, P., Katabi, D., Tian, Y.: Improving clip training with language rewrites. In: Advances in Neural Information Processing Systems, vol. 36 (2024)
7. Gao, P., Geng, S., Zhang, R., Ma, T., Fang, H., Qiao, Y.: Clip-adapter: better vision-language models with feature adapters. Int. J. Comput. Vision **132**(2), 581–595 (2024)
8. Gu, X., Lin, T.Y., Kuo, W., Cui, Y.: Open-vocabulary object detection via vision and language knowledge distillation. arXiv preprint arXiv:2104.13921 (2021)
9. Hendrycks, D., Zhao, K., Basart, S., Steinhardt, J., Song, D.: Natural adversarial examples. In: Proceedings of the IEEE/CVF Conference on Computer Vision and Pattern Recognition, pp. 15262–15271 (2021)
10. Huang, Y., et al.: Structure-clip: towards scene graph knowledge to enhance multi-modal structured representations. In: Proceedings of the AAAI Conference on Artificial Intelligence, vol. 38, pp. 2417–2425 (2024)
11. Kim, D., Angelova, A., Kuo, W.: Region-aware pretraining for open-vocabulary object detection with vision transformers. In: Proceedings of the IEEE/CVF Conference on Computer Vision and Pattern Recognition, pp. 11144–11154 (2023)
12. Kirillov, A., Mintun, E., Ravi, N., Mao, H., Rolland, W.Y., et al.: Segment anything. In: Proceedings of the IEEE/CVF International Conference on Computer Vision, pp. 4015–4026 (2023)
13. Krishna, R., Zhu, Y., Groth, O., Johnson, J., Hata, D.A., et al.: Visual genome: connecting language and vision using crowdsourced dense image annotations. Int. J. Comput. Vision **123**, 32–73 (2017)

14. Krizhevsky, A., Hinton, G., et al.: Learning multiple layers of features from tiny images (2009)
15. Li, J., Li, D., Xiong, C., Hoi, S.: Blip: bootstrapping language-image pre-training for unified vision-language understanding and generation. In: International Conference on Machine Learning, pp. 12888–12900. PMLR (2022)
16. Lin, T.-Y., et al.: Microsoft COCO: common objects in context. In: Fleet, D., Pajdla, T., Schiele, B., Tuytelaars, T. (eds.) ECCV 2014. LNCS, vol. 8693, pp. 740–755. Springer, Cham (2014). https://doi.org/10.1007/978-3-319-10602-1_48
17. Liu, H., Li, C., Li, Y., Lee, Y.J.: Improved baselines with visual instruction tuning. In: Proceedings of the IEEE/CVF Conference on Computer Vision and Pattern Recognition, pp. 26296–26306 (2024)
18. Podell, D., English, Z., Lacey, K., Blattmann, A., Dockhorn, T., Müller, R.: SDXL: improving latent diffusion models for high-resolution image synthesis. arXiv preprint arXiv:2307.01952 (2023)
19. Radford, A., Kim, J.W., Hallacy, C., Ramesh, A., Goh, J., et al.: Learning transferable visual models from natural language supervision. In: International Conference on Machine Learning, pp. 8748–8763. PmLR (2021)
20. Ramesh, A., Dhariwal, P., Nichol, A., Chu, C., Chen, M.: Hierarchical text-conditional image generation with clip latents. arXiv preprint arXiv:2204.06125 **1**(2), 3 (2022)
21. Rasheed, H., Khattak, M.U., Maaz, M., Khan, S., Khan, F.S.: Fine-tuned clip models are efficient video learners. In: Proceedings of the IEEE/CVF Conference on Computer Vision and Pattern Recognition, pp. 6545–6554 (2023)
22. Rombach, R., Blattmann, A., Lorenz, D., Esser, P., Ommer, B.: High-resolution image synthesis with latent diffusion models. In: Proceedings of the IEEE/CVF Conference on Computer Vision and Pattern Recognition, pp. 10684–10695 (2022)
23. Schuhmann, C., Beaumont, R., Vencu, R., Gordon, C., Wightman, M., et al.: Laion-5b: an open large-scale dataset for training next generation image-text models. Adv. Neural. Inf. Process. Syst. **35**, 25278–25294 (2022)
24. Sun, Q., Fang, Y., Wu, L., Wang, X., Cao, Y.: Eva-clip: improved training techniques for clip at scale. arXiv preprint arXiv:2303.15389 (2023)
25. Sun, Z., et al.: Alpha-clip: a clip model focusing on wherever you want. In: Proceedings of the IEEE/CVF Conference on Computer Vision and Pattern Recognition, pp. 13019–13029 (2024)
26. Tao, M., Bao, B.K., Tang, H., Xu, C.: Galip: generative adversarial clips for text-to-image synthesis. In: Proceedings of the IEEE/CVF Conference on Computer Vision and Pattern Recognition, pp. 14214–14223 (2023)
27. Thrush, T., Jiang, R., Bartolo, M., Singh, A., Williams, A., Kiela, C.: Winoground: probing vision and language models for visio-linguistic compositionality. In: Proceedings of the IEEE/CVF Conference on Computer Vision and Pattern Recognition, pp. 5238–5248 (2022)
28. Wasim, S.T., Naseer, M., Khan, S., Khan, F.S., Shah, M.: Vita-clip: video and text adaptive clip via multimodal prompting. In: Proceedings of the IEEE/CVF Conference on Computer Vision and Pattern Recognition, pp. 23034–23044 (2023)
29. Young, P., Lai, A., Hodosh, M., Hockenmaier, J.: From image descriptions to visual denotations: new similarity metrics for semantic inference over event descriptions. Trans. Assoc. Comput. Linguist. **2**, 67–78 (2014)
30. Yuksekgonul, M., Bianchi, F., Kalluri, P., Jurafsky, D., Zou, J.: When and why vision-language models behave like bags-of-words, and what to do about it? In: The Eleventh International Conference on Learning Representations (2023). https://openreview.net/forum?id=KRLUvxh8uaX

31. Zaheer, M., Guruganesh, G., Dubey, K.A., Ainslie, L., et al.: Big bird: transformers for longer sequences. Adv. Neural. Inf. Process. Syst. **33**, 17283–17297 (2020)
32. Zhang, B., Zhang, P., Dong, X., Zang, Y., Wang, J.: Long-clip: unlocking the long-text capability of clip. In: European Conference on Computer Vision, pp. 310–325. Springer (2024)
33. Zhou, Z., Lei, Y., Zhang, B., Liu, L., Liu, Y.: Zegclip: towards adapting clip for zero-shot semantic segmentation. In: Proceedings of the IEEE/CVF Conference on Computer Vision and Pattern Recognition, pp. 11175–11185 (2023)

Token-Based Multi-condition Autoregressive Diffusion for Lung CT Image Generation

Bo Wu(✉)

The University of Edinburgh, Edinburgh, UK
bowu29473@gmail.com

Abstract. Medical image generation faces challenges due to data scarcity, high annotation costs, and the need for precise structural fidelity. While diffusion models excel at generating high-quality images, they suffer from slow sampling speeds. Conversely, autoregressive models, although efficient in sequence modeling, struggle with continuous visual signals. In this work, we propose a diffusion-to-autoregressive framework for lung CT image synthesis, where the autoregressive model learns the temporal denoising dynamics of a diffusion process. Specifically, we introduce a sequential diffusion tokenizer that discretizes intermediate diffusion latents into structured token sequences, enabling the AR transformer to perform next-step predictions on diffusion trajectories rather than raw images. This formulation preserves the coarse-to-fine generation behavior of diffusion models while substantially accelerating sampling. Additionally, we incorporate a multi-condition guidance mechanism that utilizes segmentation masks and anatomical structures to enhance generation controllability. Experiments on a low-dose lung CT dataset show that our method achieves outstanding performance in FID, IS, and SSIM metrics. Furthermore, a Visual Turing Test conducted with radiologists confirms the perceptual realism of the generated images. Our approach has potential for extension to multi-task and multi-modality medical imaging applications.

Keywords: Conditional diffusion models · Autoregressive modeling · Lung CT synthesis · Medical image generation

1 Introduction

Deep learning has made significant progress in medical image processing, with widespread applications in tasks such as segmentation, classification, and detection, driving the rapid development of computer-aided diagnosis [1]. However, one of the primary challenges is the scarcity of high-quality datasets. Due to stringent legal and ethical restrictions on patient data, acquiring data is often difficult [2]. Additionally, the annotation of medical images typically requires specialists with advanced knowledge, making the process both time-consuming

J. Lokoč et al. (Eds.): MMM 2026, LNCS 16413, pp. 262–275, 2026.
https://doi.org/10.1007/978-981-95-6957-1_19

and costly. Furthermore, medical images exhibit considerable heterogeneity, as data from different devices and patient populations differ in distribution [3].

To address the challenges, researchers have explored the use of generative models for image generation, enhancement, and reconstruction in recent years. Generative Adversarial Networks (GANs) and Variational Autoencoders (VAEs) were among the earliest methods applied to medical image synthesis. As shown in Fig. 1, GANs establish an adversarial mechanism between a generator and a discriminator, enabling the generator to learn the data distribution and generate images similar to real samples. This approach has achieved notable results in various image domains, including CT, MRI, and ultrasound [4]. VAEs achieve generative modeling by maximizing the variational lower bound, enabling them to capture the data distribution in latent space and generate medical images with good continuity [5]. However, GANs often fail to capture the complete data distribution and are susceptible to issues such as training instability and mode collapse, which are particularly pronounced in medical scenarios with limited data. The images generated by VAEs tend to be blurry and struggle to accurately represent intricate and detailed anatomical features of the structure [6].

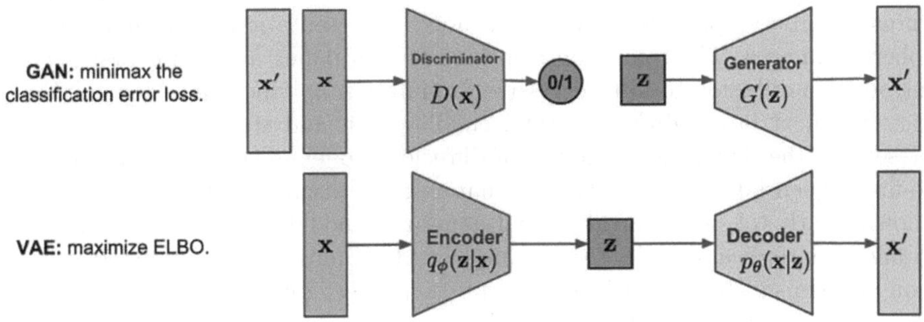

Fig. 1. An overview of the architectures and workflows of GANs and VAEs, illustrating the primary components of each model, including the generator and discriminator in GANs and the encoder and decoder in VAEs.

In recent years, denoising diffusion probabilistic models (DDPM) have become one of the leading techniques in image generation tasks [7]. This method begins with Gaussian noise and recovers images through a stepwise reverse denoising process, thereby generating high-fidelity images. Compared to GANs, diffusion models can capture the entire data distribution and generate more diverse images [8]. In addition, unlike VAEs, this method can also generate high-quality images. In medical imaging, unconditional diffusion models have been successfully applied to synthesize lung X-ray images [9]. Meanwhile, conditional diffusion models further incorporate structural guidance information, such as text, segmentation maps, and modality conversion, to enhance the medical relevance and specificity of the generated images. This method produces stable and high-quality images that are particularly suitable for medical imaging tasks requiring high texture and detail fidelity. However, the reverse diffusion process typically operates densely and requires a large number of sampling steps, leading to slow sampling speeds and computationally expensive issues.

In contrast, autoregressive (AR) models, which are based on 'next-token prediction,' have become the foundational paradigm in natural language processing (NLP). The success of large language models (LLMs) have led to widespread adoption of the autoregressive modeling approach [10]. This paradigm has also been progressively extended to visual generation tasks, modeling images through sequences of pixels, patches, or tokens to achieve image generation. However, images do not inherently possess a linear sequence structure, and directly applying autoregressive mechanisms is challenging. For example, different pixel scanning orders can lead to significant performance differences in models, and training complexity is high while generation speed is slow [11].

To address these challenges, this paper integrates autoregressive modeling with diffusion processes for lung CT image synthesis. Unlike prior approaches that treat diffusion and autoregression as separate paradigms, our method reformulates the diffusion trajectory itself as a next-step prediction problem that can be learned by an autoregressive transformer. By discretizing intermediate diffusion latents into structured token sequences, the AR model learns how the diffusion process progressively refines noisy representations into anatomically coherent images. The main contributions of this paper are as follows: (1) We introduce a diffusion-to-autoregressive formulation in which the AR model predicts the next latent diffusion state, enabling fast and stable generation while preserving the denoising dynamics of diffusion models. (2) We design a coarse-to-fine sequential diffusion tokenizer that extracts discretized latent trajectories, allowing early tokens to encode global structures and later tokens to capture fine-grained anatomical details. (3) We implement a multi-condition-guided generation mechanism that can utilize segmentation maps, anatomical structure maps, etc., to generate high-quality lung CT images. (4) The generated images exhibit visual performance across multiple Hounsfield Unit (HU) windows (e.g., lung window, bone window, soft tissue window), showing excellent fidelity and structural authenticity, with some synthetic samples capable of 'fooling' professional doctors.

2 Related Works

2.1 Diffusion Models in Medical Image Generation

Diffusion models have attracted attention in the field of medical imaging, early research primarily focused on unconditional generation. Wolleb et al. [9] used the unconditional diffusion model to achieve high-fidelity image details in brain lesion detection and reconstruction. Furthermore, diffusion models have made substantial advances across multiple modalities, including ultrasound image synthesis and fundus image generation [12]. Conditional diffusion models improve the refinement of generated images by incorporating information such as segmentation masks, anatomical structures, or texts. Song et al. [13] integrated conditional information into the diffusion model denoising process, enabling semantically guided lesion region generation. This approach helps to train segmentation networks and improves segmentation accuracy. However, the reverse diffusion process requires hundreds of steps, resulting in slow generation speeds [14]. To address this issue, researchers have proposed strategies such as few-step sampling methods and dynamic sampling strategies [15] to improve speed. However, these methods may lead to a degradation in image quality, which can adversely affect the diversity and generalization.

2.2 Visual Autoregressive and Diffusion Models

Combining the advantages of autoregressive and diffusion models has become an important research direction. Unlike the discrete and linear structure of text, images are continuous and two-dimensional information, which requires gradual modeling of the highly correlated pixel information within this two-dimensional space. This complexity often results in slow training and sampling processes [16]. Researchers have proposed image discretization, using vector quantization autoencoders (VQ-VAE) to encode images and convert them into processable discrete token sequences [17]. Furthermore, Esser et al. [18] employ a raster scan order to transform two-dimensional images into one-dimensional sequences, and leverage transformer architectures for autoregressive modeling. However, quantization introduces information loss issues. Ramesh et al. [19] use discrete tokens as conditions to guide the diffusion process during the token decoding stage. The Imagen model [20] successfully generates semantically consistent, high-resolution images through multi-stage token-conditional diffusion decoding. Additionally, DDT-LLaMa [21] attempted to generate image tokens sequentially via the diffusion process. However, its generation order was reversed, which limited the autoregressive capabilities of the tokens. The method proposed in this paper integrates the entire diffusion decoding process into the tokenizer and trains it using the standard autoregressive next-token approach.

3 Methods

To address the limitations of existing generative models regarding detail fidelity, structural consistency, and generation efficiency, we propose a novel framework that integrates discrete diffusion modeling with an autoregressive prediction. This approach formulates the diffusion process as a structured token sequence generation problem, leveraging the advantages of autoregressive modeling in capturing long-range dependencies and sequence generation, while preserving the stability and high fidelity characteristic of diffusion models in terms of generation quality. In Sect. 3.1, we first introduce the sequential diffusion tokenizer. Then, in Sect. 3.2, we present the autoregressive process. Finally, we describe the multi-condition diffusion guidance mechanism in Sect. 3.3

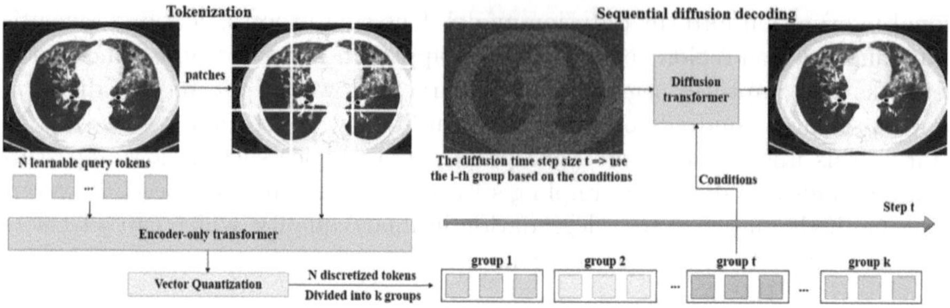

Fig. 2. Overview of the Sequential Diffusion Tokenizer. The input image is encoded into a sequence of discrete tokens via a transformer encoder and vector quantization. These tokens are grouped and temporally aligned to guide the reverse diffusion process progressively from coarse semantics to fine details.

3.1 Diffusion Images Tokenizer

The diffusion tokenizer encodes consecutive images into a discrete token sequence with temporal structure, enabling effective modeling by autoregressive models while also facilitating gradual denoising from semantic content to fine details during the diffusion decoding process. As shown in Fig. 2, the process can be divided into two stages: image tokenization and sequential diffusion decoding.

During the tokenizer stage, the image is divided into fixed-size patches, then fed into a transformer encoder $E(\cdot)$. To extract features with strong semantic representation, we introduce a set of learnable query vectors $[q_1, q_2, \ldots, q_N]$ that extract N latent image representation vectors. This process outputs a continuous sequence of vectors, which are subsequently discretized through vector quantization to form the final token sequence $z = [z_1, z_2, \ldots, z_N] = \text{quant}(E(I, [q_1, \ldots, q_N]))$. Each token corresponds to a specific region or semantic component within the image, collectively providing a structural description of the entire image. In the sequential diffusion decoding stage, the generation process is based on the token sequence. This process is trained using the Flow Matching Loss method [22] to learn image interpolation trajectories over the time interval $t \in [0, 1]$. Given the initial noise image $x_0 \sim \mathcal{N}(0, I)$ and the target image x_1, intermediate images are constructed as $x_t = t \cdot x_1 + (1-t) \cdot x_0$, and the true direction vector is defined as $v_t = x_1 - x_0$. The diffusion decoder D_{FM} takes x_t, the current time t, and the conditional token $c(t)$ as inputs and outputs the predicted direction vector $\hat{v}_t$. The training objective is to minimize the flow matching loss:

$$\ell_{\text{fm}} = \mathbb{E}_{t, x_0, x_1} \left[\| v_t - D_{\text{FM}}(x_t, t, c(t)) \|^2 \right] \tag{1}$$

To better align with the multi-step process of the diffusion model, the token sequence z is divided into K periodic subsequences $\{g_1, g_2, \ldots, g_K\}$, each containing $\frac{N}{K}$ tokens. We introduce a reparameterization function $t' = \frac{t}{t + \frac{1}{\beta}(1-t)}$ to adjust the token groups used at each time step dynamically, ultimately defining $c(t) = g_{\lceil t' \cdot K \rceil}$. This mechanism ensures that low-frequency tokens are used in the early stages to establish the global structure, while high-frequency tokens are gradually introduced in later stages to restore image details.

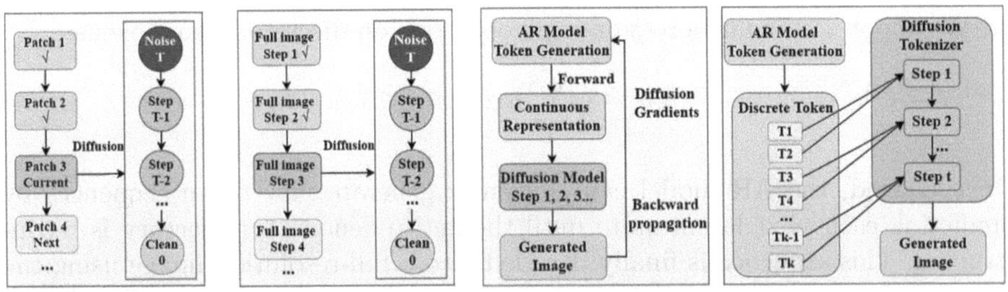

(a) Patch-level Diffusion (b) Image-level Diffusion (c) AR Training via Diffusion Gradients (d) Discrete AR with Diffusion Tokenizer

Fig. 3. Paradigms of Integrating Diffusion with Autoregressive Models. (a) applies a complete patch-level diffusion denoising process at each autoregressive generation step to handle continuous image content [23]; (b) predicts the entire image at each step, using an AR transformer conditioned on the sequence of previously denoised images [24]; (c) utilizes diffusion gradients to train AR models via continuous representations passed through diffusion steps [25]; (d) our method generates discrete tokens using vanilla AR models and reconstructs images with a diffusion-based tokenizer and decoder.

3.2 Autoregressive Process

Traditional visual sequence models require an explicit definition of the scanning order of the image pixels and adjustments to accommodate the two-dimensional nature of images. However, due to the diffusion tokenizer mentioned above, images have been discretized and linearized into a semantically consistent one-dimensional discrete token sequence $z = [z_1, z_2, \ldots, z_N]$. This sequence can be directly input into standard autoregressive language modeling frameworks for training. As shown in Fig. 3, there are currently several methods that combine AR with diffusion processes. Compared to these methods, our model generates the current group token zi at each step based solely on the previously generated token sequence. Instead of directly predicting final image tokens, our autoregressive model is designed to learn the *temporal dynamics* of the diffusion process, we treat the sequence of intermediate diffusion states as a learnable trajectory, and formulate the transition from x_t to x_{t-1} as a next-step prediction problem:

$$p(x_{t-1} \mid x_t) = \text{AR}(x_t). \tag{2}$$

To construct training data for this process, we employ a pretrained diffusion transformer (DiT) as a teacher model. During its denoising trajectory, we record all intermediate latent representations (patch token embeddings):

$$\mathcal{T} = \{x_T, x_{T-1}, \ldots, x_0\}, \tag{3}$$

where x_T is the initial noise and x_0 is the final clean image. Each pair (x_t, x_{t-1}) represents one step in the diffusion trajectory and serves as a supervised training example for the AR model. Given this sequence, the autoregressive transformer is trained to predict the next latent state x_{t-1} conditioned on the current one x_t. We adopt a standard cross-entropy objective on discrete latent tokens:

$$L_{\text{AR}} = -\sum_{t} \log p(x_{t-1} \mid x_t). \tag{4}$$

Once trained, the AR model can efficiently generate new token sequences by predicting each next latent state until the entire denoising trajectory is reconstructed. This sequence is finally decoded into a full-resolution image using the diffusion decoder introduced in Sect. 3.1. Compared to running a full diffusion sampling process, this autoregressive surrogate significantly improves inference efficiency while maintaining high structural fidelity and detail consistency.

3.3 Multi-condition Diffusion Guidance

To further enhance the structural controllability of image generation, we introduce a multi-condition diffusion guidance mechanism during the diffusion decoding stage to fully utilize anatomical structure maps, adjacent CT slices, segmentation maps, etc., to guide the generation process. Unlike the DDPM, which uses continuous image space-based conditional guidance, our mechanism primarily operates within the diffusion decoder, guiding each diffusion sampling step to progressively align the decoded image with the low-frequency distributions of multiple structural reference maps. Specifically, during the diffusion decoding process, we extract low-frequency structural information from each reference image y_s using a downsampling operation $\phi_{n_s}(\cdot)$ and align it with the low-frequency components of the current reconstructed image x_{t-1} to form a structural guidance error. After each diffusion sampling step, we add these low-frequency differences to the current image representation, thereby gradually aligning the generated image with the reference conditions at every step. The process can be formally represented as:

$$x_{t-1} = x_{t-1} + \sum_{s=1}^{M} (\phi_{n_s}(y_s) - \phi_{n_s}(x_{t-1})) \tag{5}$$

where M denotes the number of reference images, and ϕ_{ns} represents the downsampling operation applied for the s-th condition, which controls its guidance strength. Since we use a discrete token sequence-driven diffusion decoding, this can be seamlessly integrated into the image reconstruction stage corresponding to each group of token decoding, gradually aligning reference information from coarse structures to fine details. Unlike traditional diffusion models, our approach generates tokens and performs diffusion sampling simultaneously. During the generation process, each time a set of tokens is produced, a diffusion decoding step can be executed, thereby improving generation efficiency. Additionally, the autoregressive and diffusion models can be trained or fine-tuned independently, providing strong generalization capabilities and modular scalability. For example, in the future, this framework could be extended to multi-task applications, such as combining image generation with segmentation.

To ensure that the model effectively learns to leverage conditional information, we adopt a training strategy that randomly drops condition signals with a small probability (e.g., 10–15%). This intentional removal of conditions forces the model to simultaneously learn both conditional and unconditional generation, preventing it from over-relying on explicit conditioning. In practice, this is realized via the *classifier-free guidance* (CFG) mechanism [26]. During inference, both unconditional and conditional predictions are obtained, and the final denoising prediction is computed as:

$$\hat{\epsilon}_{\text{CFG}} = \hat{\epsilon}_{\text{uncond}} + \gamma \cdot (\hat{\epsilon}_{\text{cond}} - \hat{\epsilon}_{\text{uncond}}), \tag{6}$$

where $\hat{\epsilon}_{\text{uncond}}$ and $\hat{\epsilon}_{\text{cond}}$ denote the unconditional and conditional denoising outputs, respectively, and γ is the guidance scale that controls the trade-off between condition adherence and generation diversity. A larger γ enforces stronger alignment with conditions, while a smaller γ encourages more diverse outputs. In our multi-condition setting, this mechanism extends to multiple structural references, further enhancing controllability while maintaining flexibility.

4 Implementations

This section describes the implementation and configurations of each module. The training dataset comprises 4,388 low-dose CT images of the lungs. All images have undergone standardized 2,000 HU full-window processing to optimize tissue contrast visualization. These cross-sectional images capture the bilateral complete lung fields and key anatomical structures near the heart.

4.1 Experiments Setup

In the tokenizer, we divide the image into 16×16 patches, which are then processed by eight Transformer layers with a dimensionality of 768, resulting in a one-dimensional token sequence of length 256. We then apply a causal mask to the query token, and the quantization employs VQ-VAE with a codebook size of 16,384 and an embedding dimension of 8. During training, we use flow matching loss to guide the learning of the diffusion direction, introduce token retention loss to prevent information loss, and weight-combine vector quantisation error with LPIPS perceptual loss:

$$\ell_{\text{total}} = \ell_{\text{fm}} + \ell_{\text{VQ}} + 0.5 \cdot \ell_{\text{PIPS}} + 0.5 \cdot \ell_{\text{repa}} \tag{7}$$

The tokenizer was trained using the Adam optimiser, with an initial learning rate of 0.0001 and a batch size of 512. The model was trained on the University of Edinburgh's high-performance computing platform Cirrus for 130,000 steps.

The Diffusion Decoder is implemented using a diffusion transformer structure, consisting of 12 transformer layers with a patch size of 8. During training, the model receives tokens from the Tokenizer output and introduces conditional token information through Cross-Attention and Adaptive LayerNorm. The optimization objective is to predict the direction vector $v_t = x_1 - x_0$, thereby enabling path learning from noisy images to target images.

In the autoregressive part, we employ a standard decoder-only Transformer, utilizing RMSNorm normalization, SwiGLU activation functions, and 1D RoPE positional encoding to accommodate the linear structure of token sequences. Additionally, we introduce a classifier-free guidance mechanism to enhance generative diversity. During training, we use the AdamW optimizer with an initial learning rate of 0.0002, $\beta_1 = 0.9$, $\beta_2 = 0.95$, a batch size of 512, and 250 epochs. We also adopt an exponential moving average (EMA) of the denoising network weights with a decay of 0.9999.

4.2 Evaluation Metric

We adopt three metrics to evaluate the model's performance.

Fréchet Inception Distance (FID). measures the distributional discrepancy between generated images and real images in the feature space of an Inception-v3 network. It is defined as:

$$\text{FID} = \|\mu_r - \mu_g\|^2 + \text{Tr}(\Sigma_r + \Sigma_g - 2(\Sigma_r \Sigma_g)^{1/2}) \quad (8)$$

where μ_r, Σ_r are the mean and covariance of real image features, and μ_g, Σ_g are those of generated images. Lower FID indicates better distribution alignment.

Inception Score (IS). evaluates both the semantic clarity and the diversity of generated images. It is computed based on the KL divergence between the conditional label distribution $p(y|x)$ and the marginal distribution $p(y)$:

$$\text{IS} = \exp\left(\mathbb{E}_x \left[\text{KL}(p(y|x)\|p(y))\right]\right) \quad (9)$$

Higher IS suggests the model generates more diverse and recognizable images.

Set-level Structural Similarity Index Measure (SSIM). quantifies the perceptual similarity between generated and ground truth images by comparing luminance, contrast, and structure:

$$\text{SSIM}(x, \hat{x}) = \frac{(2\mu_x \mu_{\hat{x}} + C_1)(2\sigma_{x\hat{x}} + C_2)}{(\mu_x^2 + \mu_{\hat{x}}^2 + C_1)(\sigma_x^2 + \sigma_{\hat{x}}^2 + C_2)} \quad (10)$$

We report the average SSIM over the entire evaluation set (set-level SSIM), where higher values (closer to 1) indicate better preservation of structural information.

Table 1. Performance comparison of generative models at 256×256 and 512×512 resolutions; models include GAN-based (StyleGAN2), diffusion-based (DDPM unguided/guided, DiT), masked generative (MaskGIT), visual autoregressive (VAR-d24), and our model combining sequential diffusion with autoregressive decoding.

Model	256×256			512×512		
	FID ↓	IS ↑	SSIM ↑	FID ↓	IS ↑	SSIM ↑
StyleGAN2 [27]	10.25	162.39	0.51	22.14	147.20	0.45
DDPM (unguided)	21.73	97.28	0.31	46.27	54.07	0.24
DDPM (guided)	7.96	174.17	0.57	14.22	182.31	0.49
DiT [28]	4.28	205.69	0.65	9.61	198.21	0.55
MaskGIT [29]	14.37	141.73	0.39	29.47	113.92	0.32
VAR-d24 [30]	**3.08**	206.13	0.68	6.27	199.76	0.56
Ours	3.41	**221.27**	**0.71**	**5.94**	**214.63**	**0.59**

5 Results

Under the full HU window condition (−1000 to 1000 HU), we generated a total of 8,000 lung CT images and compared with several state-of-the-art generative models. All models were trained on the same dataset, as shown in Table 1.

For the FID, our method outperforms StyleGAN2 and DDPM at both resolutions, demonstrating stable and high-fidelity image generation. It also approaches the visual autoregressive model VAR-d24, indicating that introducing a sequential diffusion can alleviate the accumulation of blurring in long sequence generation. Furthermore, for the IS, our method achieves the highest score, indicating that the generated images not only exhibit greater confidence in local quality but also display richer and more consistent global category distributions. Our model also performs excellently in set-level SSIM, demonstrating stronger structural alignment capabilities. For lung images of highly regular anatomical structures, where structural consistency reflects whether the model has captured the true anatomical distribution. It is worth noting that as image resolution increases, the performance of MaskGIT and DDPM (unguided) declines significantly, whereas our method and VAR-d24 maintain good stability. Although VAR-d24 has an autoregressive advantage, it is slightly lacking in semantic control. In contrast to these models, our approach not only focuses on generating raw data but also on its annotations. It is scalable and can be easily extended to generate CT scans with annotated pathologies.

In terms of inference speed, StyleGAN2 remains the fastest due to its feed-forward design, whereas DDPM and DiT are the slowest because of their multi-step denoising. MaskGIT and VAR-d24 achieve moderate efficiency, with the latter slowed by fully autoregressive sampling. Our method provides a favorable balance: it is substantially faster than diffusion-only models while being only slightly slower than MaskGIT and VAR-d24, yet achieving clearly superior image fidelity and structural coherence.

To further evaluate the perceptual realism of our generated lung CT images, we conducted a Visual Turing Test with three experienced radiologists. A balanced dataset containing 40 samples (20 real and 20 generated) was prepared. In each trial, a single CT slice was randomly selected and presented in a random order. The radiologists were asked to classify each image as either "Real" or "Fake". The confusion matrices in Fig. 4 report the detailed responses. These results indicate that the majority of generated images were consistently perceived as authentic by the radiologists. In particular, the high false positive rates (generated → real) suggest that our method produces images that closely resemble true CT scans in terms of structural and textural fidelity. Collectively, the outcomes demonstrate that our framework has successfully passed the Visual Turing Test, confirming its capability to generate realistic medical images.

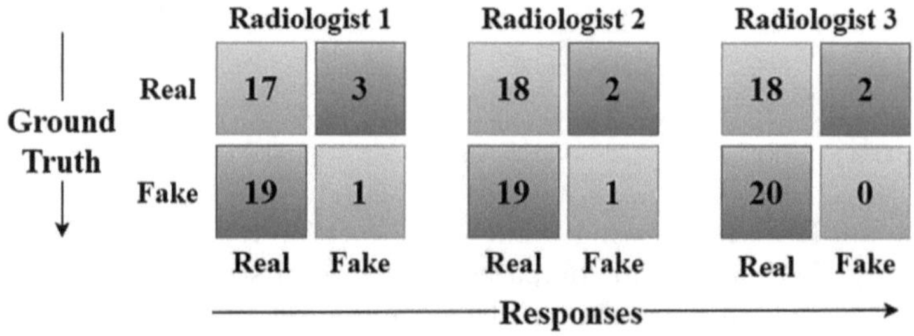

Fig. 4. Visual Turing Test results with three radiologists. Each confusion matrix shows their ability to distinguish real from generated CT scans. Notably, 95% of generated images were judged as "Real" by Radiologist 1, 95% by Radiologist 2, and 100% by Radiologist 3, indicating that our framework produces highly realistic medical images.

6 Ablation Study

We investigate the impact of the coarse-to-fine token grouping strategy by varying the number of token groups $K \in \{1, 2, 4, 8, 16, 32\}$, with results summarized in Table 2. A clear performance trend emerges as K increases. When $K = 1$, the model degenerates to a single-stage decoding process, yielding the worst generation quality across all metrics, indicating that the absence of a hierarchical structure severely limits the model's ability to recover anatomical details. Increasing K from 1 to 8 progressively improves performance, leading to substantial reductions in FID and consistent gains in IS and SSIM. This shows that introducing a moderate number of hierarchical groups enables the model to first reconstruct global structural patterns and then refine local details more effectively. Interestingly, the performance begins to degrade when K is further increased beyond 8. Both FID and SSIM deteriorate for $K = 16$ and $K = 32$, indicating that excessively fine-grained grouping oversplits the token space and introduces instability in the denoising trajectory. In such cases, early-stage groups receive insufficient semantic context, making the reconstruction process noisier and less coherent. In addition, the inference time increases monotonically with larger K, as more decoding stages must be executed sequentially. A larger K also results in smaller token subsets per stage, reducing the degree of parallelism and causing each denoising step to incur proportionally higher computational overhead. Overall, $K = 8$ provides the best trade-off between efficiency and quality, suggesting that a moderate number of coarse-to-fine levels offers the optimal balance between global consistency and fine-detail synthesis.

Table 2. Ablation study on the number of token groups K in the coarse-to-fine diffusion decoding pipeline for generating 256×256 CT images.

K	FID	IS	SSIM	Time (s/img)
1	15.61	155.14	0.35	3.35
2	12.53	178.31	0.48	3.58
4	8.32	194.28	0.57	3.64
8	**3.41**	**221.27**	**0.71**	4.05
16	5.33	201.98	0.64	4.48
32	9.41	191.49	0.53	5.87

7 Conclusion

In this work, we presented a generative framework that integrates diffusion processes with autoregressive modeling for lung CT image synthesis. By reformulating diffusion trajectories as sequential token prediction, our method leverages the long-range dependency modeling capabilities of autoregressive transformers while preserving the coarse-to-fine denoising behavior of diffusion models. The proposed sequential diffusion tokenizer effectively discretizes intermediate diffusion latents, enabling efficient next-step prediction, and the multi-condition guidance mechanism further enhances controllability by incorporating segmentation maps, anatomical structures, and other clinical priors. The experimental results demonstrate that our method outperforms StyleGAN2, DDPM, DiT, MaskGIT, and VAR-d24. Notably, the generated images successfully passed the Visual Turing Test with experienced radiologists, validating their clinical authenticity across different Hounsfield Unit windows. Despite these promising results, several limitations remain. The scale and diversity of the generated images are still relatively limited, which may restrict their application scope in rare pathological situations. Future work will involve expanding the dataset to include multi-center, multi-modality, and pathological cases to enhance generalization and robustness. Additionally, we plan to extend the framework to multi-task settings, such as simultaneous generation and segmentation.

References

1. He, K., et al.: Transformers in medical image analysis. Intell. Med. **3**(1), 59–78 (2023)
2. Kaissis, G. A., Makowski, M. R., Rückert, D., Braren, R. F.: Secure, privacy-preserving and federated machine learning in medical imaging. Nature Mach. Intell., 2(6),305–311 (2020)
3. Zech, J. R., et al.: Variable generalization performance of a deep learning model to detect pneumonia in chest radiographs: a cross-sectional study. PLoS medicine, 15(11),e1002683 (2018)
4. Khosravi, B., et al. Creating high fidelity synthetic pelvis radiographs using generative adversarial networks: unlocking the potential of deep learning models without patient privacy concerns. J. Arthroplasty, 38(10),2037–2043 (2023)
5. Gan, M., Wang, C.: Esophageal optical coherence tomography image synthesis using an adversarially learned variational autoencoder. Biomed. Opt. Express **13**(3), 1188–1201 (2022)
6. Khader, F., et al.: Denoising diffusion probabilistic models for 3d medical image generation. Sci. Rep., 13(1), 7303 (2023)
7. Ho, J., Jain, A., Abbeel, P.: Denoising diffusion probabilistic models. Adv. Neural. Inf. Process. Syst. **33**, 6840–6851 (2020)
8. Bayat, R.: A study on sample diversity in generative models: Gans vs. diffusion models (2023)
9. Wolleb, J., Bieder, F., Sandkühler, R., Cattin, P.C.: Diffusion models for medical anomaly detection. In: International Conference on Medical image computing and computer-assisted intervention, pp. 35–45. Springer (2022)
10. Touvron, H., et al.: Llama: open and efficient foundation language models. arXiv preprint arXiv:2302.13971 (2023)
11. Menick, J., Kalchbrenner, N.: Generating high fidelity images with subscale pixel networks and multidimensional upscaling. arXiv preprint arXiv:1812.01608 (2018)
12. Go, S., Ji, Y., Park, S. J., Lee, S.: Generation of structurally realistic retinal fundus images with diffusion models. In: Proceedings of the IEEE/CVF conference on computer vision and pattern recognition, pp. 2335–2344 (2024)
13. Song, Y., et al.: Score-based generative modeling through stochastic differential equations. arXiv preprint arXiv:2011.13456 (2020)
14. Song, J., Meng, C., Ermon, S.: Denoising diffusion implicit models. arXiv preprint arXiv:2010.02502 (2020)
15. Cheng, L., et al.: DPM-solver: a fast ode solver for diffusion probabilistic model sampling in around 10 steps. Adv. Neural. Inf. Process. Syst. **35**, 5775–5787 (2022)
16. Chen, M., et al.: Generative pretraining from pixels. In: International conference on machine learning, pp. 1691–1703. PMLR (2020)
17. Razavi, A., den Oord, A.V., Vinyals, O.: Generating diverse high-fidelity images with vq-vae-2. Adv. Neural Inf. Process. Syst. 32 (2019)
18. Esser, P., Rombach, R., Ommer, B.: Taming transformers for high-resolution image synthesis. In: Proceedings of the IEEE/CVF conference on computer vision and pattern recognition, pp. 12873–12883 (2021)
19. Ramesh, A., Dhariwal, P., Nichol, A., Chu, C., Chen, M.: Hierarchical text-conditional image generation with clip latents. arXiv preprint arXiv:2204.06125, 1(2),3 (2022)
20. Saharia, C., et al.: Photorealistic text-to-image diffusion models with deep language understanding. Adv. Neural Inf. Process. Syst., 35, 36479–36494 (2022)

21. Pan, K., et al.: Generative multimodal pretraining with discrete diffusion timestep tokens. In: Proceedings of the Computer Vision and Pattern Recognition Conference, pp. 26136–26146 (2025)
22. Lipman, Y., TQ Chen, R., Ben-Hamu, H., Nickel, M., Le, M.: Flow matching for generative modeling. arXiv preprint arXiv:2210.02747 (2022)
23. Deng, C., Zhu, D., Li, K., Guang, S., Fan, H.: Causal diffusion transformers for generative modeling. arXiv preprint arXiv:2412.12095 (2024)
24. Gu, J., et al.: Denoising autoregressive transformers for scalable text-to-image generation. In: The Thirteenth International Conference on Learning Representations (2025)
25. Pan, X., et al.: Transfer between modalities with metaqueries. arXiv preprint arXiv:2504.06256 (2025)
26. Ho, J., Salimans, T.: Classifier-free diffusion guidance. arXiv preprint arXiv:2207.12598 (2022)
27. Karras, T., et al.: Analyzing and improving the image quality of stylegan. In: Proceedings of the IEEE/CVF conference on computer vision and pattern recognition, pp. 8110–8119 (2020)
28. William Peebles and Saining Xie. Scalable diffusion models with transformers. In: Proceedings of the IEEE/CVF international conference on computer vision, pp. 4195–4205, 2023
29. Chang, H., Zhang, H., Jiang, L., Liu, C., Freeman, W. T.: Maskgit: masked generative image transformer. In: Proceedings of the IEEE/CVF conference on computer vision and pattern recognition, pp. 11315–11325 (2022)
30. Tian, K., Jiang, Y., Yuan, Z., Peng, B., Wang, L.: Visual autoregressive modeling:scalable image generation via next-scale prediction. Adv. Neural. Inf. Process. Syst. **37**, 84839–84865 (2024)

DAHM: A Dual-Stream Attention Fusion Model for Hate Content Detection

Qingguan Li, Jiawei Cong, and Kai Zhao(✉)

School of Computer Science and Technology, Xinjiang University, Urumqi 830046, China
zhawkk@xju.edu.cn

Abstract. The rapid proliferation of social media has underscored the critical need for effective detection of hate content disseminated online, particularly as such content often comprises both textual and visual modalities. Existing multimodal detection methods frequently suffer from inadequate complementarity between text and image modalities, leading to an imbalance wherein one modality disproportionately influences the classification outcome. To address these limitations, we propose the Dual-Stream Attention Fusion Model (DAHM), specifically designed for hate content detection. DAHM employs a dual-stream architecture, wherein textual and visual features are extracted independently before undergoing an attention-based fusion process. This mechanism enhances the interaction between the two modalities, ensuring that both contribute meaningfully to the final classification. The model further integrates the outputs from each stream through a weighted aggregation strategy, thereby mitigating the risk of over-reliance on a single modality and enhancing overall detection performance. Experimental results show our model surpasses baselines on critical metrics, achieving a higher F1-score and a lower MMAE on challenging datasets. This underscores its effectiveness in delivering fair, robust, and reliable multimodal hate content detection.

Keywords: Hate content detection · Multimodal analysis · Dual-Stream network architecture · Attention fusion

1 Introduction

Hate content usually refers to malicious attacks, insults, and discriminatory expressions against specific individuals or groups, often on sensitive topics such as race, gender, religion, and nationality [16]. Three pieces of data with hate content are shown in Fig. 1. The text "Me sending my dog out for supplies since he can't contract COVID-19" in Fig. 1(a), combined with the image of the dog sitting in the driver's seat holding the steering wheel reflects, in a humorous way, people's banter about quarantine measures during the epidemic. The image labeled "Asians" in Fig. 1(b), showing a big cat stepping on a kitten labeled "Asians, The entire fuckin pkanat" is an example of hate content, serving as a

J. Lokoč et al. (Eds.): MMM 2026, LNCS 16413, pp. 276–289, 2026.
https://doi.org/10.1007/978-981-95-6957-1_20

metaphor for hostility and oppression towards Asians, with blatant racial hatred. Figure 1(c), although the text "me when the vaccine drops on the NHS" is ostensibly related to vaccination, the image expresses fear and mistrust of vaccines through the apathetic expression and the image of multiple injections, implying resistance to public health policies. These examples demonstrate that the covert and complex nature of hate content makes it difficult to accurately recognize whether hateful sentiments are also present in a single image or text. The intertwining of different modal messages creates a more complex communication mechanism, and traditional unimodal detection methods are overstretched in this context.

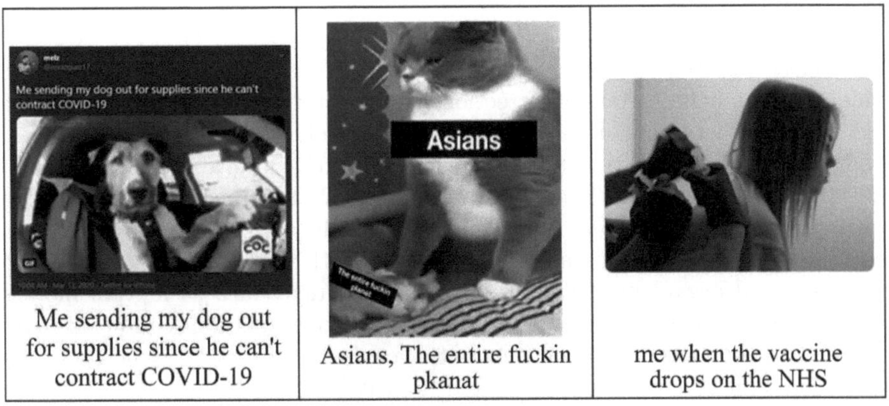

(a) (b) (c)

Fig. 1. Example with hate content data.

In recent years, the introduction of attention mechanisms has demonstrated excellent performance in various multimodal tasks. Whether in visual quizzing [7] or sentiment analysis [26], the attention mechanism effectively enhances the model's ability to capture important information by adaptively assigning attention weights to each modality. In multimodal tasks, the attentional mechanism plays a crucial role by enabling fine-grained feature extraction within each modality and establishing complex semantic links, thereby uncovering complementary information across modalities.

Inspired by this insight, this paper presents the Dual-Stream Attention Fusion Model (DAHM) for hate content detection. The model employs a dual-stream architecture to process text and image modalities separately, with a focus on enhancing inter-modal information interaction and fusion through attention fusion model. Specifically, text and image features are first extracted using RoBERTa-BiLSTM for text and ViT for images, generating feature representations for each modality. The attention fusion model then models the interactions between these modalities, enabling effective fusion of textual and visual information. To capture the shared information between the two modalities while preserving their unique features, the model employs MultiHeadAttention [37], which adaptively focuses on important aspects of both modalities. In the classification phase, DAHM classifies text and image features independently using two separate classifiers. To balance the contributions from both modalities, the outputs of these classifiers are fused using a weighted aggregation strategy, leading to the final classification prediction. This fusion approach minimizes modality imbalance and enhances the interaction of text and image features, resulting in improved performance in hate content detection tasks.

2 Related Work

2.1 Text Hate Content Detection

The widespread popularity of social media and various social platforms has made automated hate content detection an important research direction. This field spans data mining, information retrieval, and natural language processing (NLP) [21]. To address this challenge, researchers have proposed several text-specific hate content detection datasets [10,32,39], which provide valuable resources for model training and evaluation. Early research relied heavily on traditional machine learning methods [6,8,31,38,41], which typically focused on extracting and organizing low-level features from text, such as n-gram models and sentiment features [28]. Although these traditional methods can be effective for hate content detection to some extent, their feature extraction and representation capabilities are relatively limited, and it is not easy to handle complex semantic information and contextual relationships.

Subsequently deep learning based techniques [3,5,12,15,17,29,45] and strategies for fine-tuning large pre-trained models have been introduced into the field of hate content detection. These methods significantly improve the detection performance by automatically learning high-level semantic features and complex text patterns. In particular, large pre-trained language models such as BERT [11], RoBERTa [25] and GPT-3 [4] have been widely applied to this task in recent years with remarkable results [36]. Furthermore, Min et al. [30] proposed the EHSor method, which is based on self-trained multi-label learning combined with emotion detection and utilizes the correlation between emotional states and hate content to improve the effectiveness of hate content detection innovatively. These deep learning models can capture subtle semantic differences in text and enhance the accuracy and robustness of detection through multi-level feature representation.

2.2 Multimodal Hate Content Detection

Multimodal hate content detection integrates textual and visual information to enable a more comprehensive understanding and identification of complex online hate content. In 2020, Kiela et al. [19] launched the "Hateful Memes Challenge", significantly advancing research in this field. Lippe et al. [24] proposed a multimodal framework based on integrated learning, achieving strong performance in this Challenge. Building upon these early efforts, subsequent studies introduced more sophisticated architectures. Lee et al. [21] proposed a framework that improves classification accuracy and interpretability by deconstructing target entities within multimodal posts. Yang et al. [43] later introduced a cross-domain knowledge transfer approach to enhance the categorization of different types of hate speech.

As research progressed, more advanced methodologies were developed to better capture implicit and contextual cues. Zhang et al. [44] proposed a method based on optimal transport theory to improve the detection of implicitly harmful content. Chhabra and Vishwakarma [9] introduced an approach combining multi-scale visual features with knowledge distillation to enhance the recognition of subtle offensive content. Dwivedy and Roy [14] improved classification accuracy by leveraging deep feature fusion and transfer learning strategies. Yang et al. [42] using graph neural networks to generate both shared and modality-specific representations, thereby facilitating more effective inter-modal interactions.

More recently, further advancements were made. Arya et al. [1] utilized the CLIP model to extract and fuse image and text features for improved hate content classification. Wu et al. [40] proposed a method that integrates primary semantics with contextual information to enhance understanding of the underlying meaning in terse multimodal content. Ayetiran et al. [2] introduced a deep learning approach based on a unified modal attention mechanism to improve prediction performance. Meanwhile, Lin et al. [23] enhanced the interpretability of detection models by generating adversarial explanations through multimodal dialogues involving large language models.

3 Methodology

The DAHM model for multimodal hate content detection, as shown in Fig. 2.

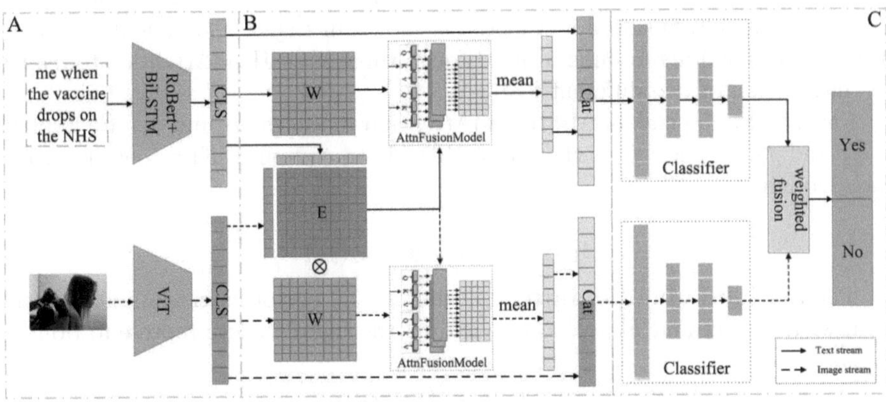

Fig. 2. Architecture of DAHM.

3.1 Problem Statement

In the hate content detection task, each piece of data consists of a piece of text and an image, representing the text modality and the image modality, respectively, and the corresponding label for that piece of data is y. The combination of these two modalities provides a comprehensive description of the hate content. Specifically, the model's output is a probability distribution containing two values, where y_0 represents the predicted probability that the data does not include hate content, and y_1 represents the predicted probability that it contains hate content. If $y_1 > y_0$, the data is expected to be harmful; otherwise, it is harmless.

3.2 Multimodal Feature Representation

This study employs RoBERTa-BiLSTM for text processing and ViT [13] for image processing to meet the feature extraction requirements of both modalities, as illustrated in Fig. 2(A). Specifically, RoBERTa is first utilized to generate contextualized word representations, which are then fed into a BiLSTM layer to further capture bidirectional sequential dependencies. The output of the BiLSTM serves as the final text feature representation, thereby compensating for potential limitations in RoBERTa's sequence modeling capability.

Finally, both text and image features are represented using CLS tokens extracted from their respective models. Let T denote the text feature and I the image feature; both are used in subsequent stages to integrate and process multimodal information.

3.3 Attentional Fusion Model

Fusing the main features in text and images effectively is crucial in the hate content detection task. To achieve this goal, a feature fusion model (AttnFusionModel) based on MultiHeadAttention mechanism is designed and implemented in this study, AttnFusionModel model is shown in Fig. 3. The model gradually fuses features of different modalities through three MultiHeadAttention to realize the interaction of two different features.

First, two independent MultiHeadAttention modules, a_1 and a_2, are applied to f_1 and f_2, respectively, where $Q = K = V$ to compute self-attention within each modality:

$$t_1 = \mathrm{a}_1(f_1), \quad t_2 = \mathrm{a}_2(f_2) \tag{1}$$

Then, a third MultiHeadAttention module a_3 performs cross-modal attention, with $Q = t_1$ and $K = V = t_2$, yielding the fused feature representation:

$$t_3 = \mathrm{a}_3(t_1, t_2) \tag{2}$$

This design ensures intra-modal contextual encoding while enabling intermodal fusion to capture complementary information, and finally obtain the fused feature t_3.

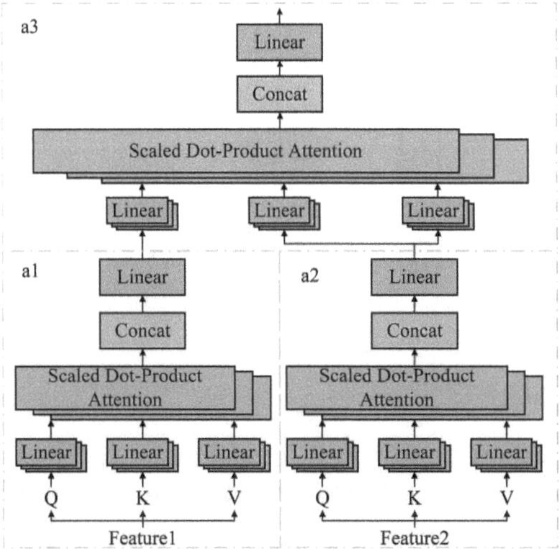

Fig. 3. Architecture of AttnFusionModel.

3.4 Feature Interaction And Fusion

As shown in Fig. 2(B), a feature interaction mechanism based on the outer product of text and image CLS tokens is designed to achieve cross-modal fine-grained relationship modeling and initial fusion.

$$E = \mathrm{T} \otimes \mathrm{I} \tag{3}$$

where E represents the interaction between text and image features, this tensor can be regarded as containing all possible combinations of features between the two modalities, capturing the fine-grained interaction information between them.

To enhance the fusion between features, this study introduces a learnable projection matrix W to project text features and image features into the shared space separately:

$$G_T = TW, \quad G_I = IW \tag{4}$$

where W is a learnable weight matrix, and G_T and G_I denote the projected text and image features.

To further optimize the fusion process, this study uses AttnFusionModel for the projected text features and image features to interact with each other separately based on the interaction matrix E to obtain a new weighted feature representation:

$$F_T = \text{AttnFusionModel}_1(G_T, E), \quad F_I = \text{AttnFusionModel}_2(G_I, E) \tag{5}$$

The role of the AttnFusionModel here is to dynamically adjust the importance of each feature by aligning it with the interaction feature E, thus increasing the model's focus on important information. For further processing, this study obtains a one-dimensional feature representation by averaging operations on these features in the second dimension:

$$\bar{F_T} = \frac{1}{\mathrm{d}}\sum_{i=1}^{\mathrm{d}} F_T, \quad \bar{F_I} = \frac{1}{\mathrm{d}}\sum_{i=1}^{\mathrm{d}} F_I \tag{6}$$

where d represents the length of the sequence. Finally, the dimensionality-reduced text features $\bar{F_T}$ and image features $\bar{F_I}$ are concatenated with the original text features T and image features I, respectively, to form the final multimodal feature representation:

$$F_T^{\text{final}} = \text{Concat}(T, \bar{F_T}), \quad F_I^{\text{final}} = \text{Concat}(I, \bar{F_I}) \tag{7}$$

Through this splicing operation, the model retains the features of the original modality while introducing multimodal interaction information during fusion.

3.5 Dual-Stream Architecture

This study uses a dual-stream architecture to process the modal information of the text stream and image stream separately. In this architecture, the features F_T^{final} and F_I^{final} obtained from the text and image streams are passed through two independent classifiers, as illustrated in Fig. 2(C). Each classifier is implemented as a multi-layer perceptron (MLP), which maps the modality-specific features to the corresponding prediction space:

$$P_T = \text{MLP}_1(F_T^{\text{final}}), \quad P_I = \text{MLP}_2(F_I^{\text{final}}) \tag{8}$$

In this study, the prediction results of the text stream P_T and the image stream P_I are integrated using a geometric mean strategy. Specifically, the final category probability distribution is defined as:

$$P = \frac{\sqrt{P_T \cdot P_I}}{\sum_k \sqrt{P_{T,k} \cdot P_{I,k}}} \tag{9}$$

$P = (y_0, y_1)$ represents the final category probability distribution. This fusion method highlights categories that receive strong support from both modalities, thereby improving the reliability of the final prediction.

4 Experiments

4.1 Datasets

This study uses three publicly available hate content detection datasets: Harm-C [33], Harm-P [34], and MultiOFF [35]. The MultiOFF dataset comprises 743 samples, which are divided into 445 for training, 149 for validation, and 149 for testing. The Harm-C dataset contains 3,544 samples, with 3,013 allocated to training, 177 to validation, and 354 to testing. Similarly, the Harm-P dataset consists of 3,552 samples, including 3,020 for training, 177 for validation, and 355 for testing.

4.2 Settings

Experiments were conducted on a server equipped with 30GB RAM and an NVIDIA A10 GPU (24 GB) running Ubuntu 22.04. The AdamW optimizer was employed to ensure stable and efficient training. The learning rate was set to 3×10^{-5}, with a batch size of 32 and 30 training epochs.

4.3 Evaluation Metrics

In this study, for the MultiOFF dataset, following the evaluation method in Suryawan shi et al. [35], F1, Precision, and Recall are selected as the primary evaluation metrics. For the Harm-C and Harm-P datasets, following the method in Pramanick et al. [34], Acc, F1, and MMAE are used as the evaluation metrics, and for Acc and F1, the larger the value, the better, while for MMAE, the smaller the value, the better.

4.4 Baselines

For the MultiOFF dataset, the following multimodal baseline models were selected: StackedLSTM+VGG16, BiLSTM+VGG16, CNNText+VGG16 [35], DisMultiHate [21], MSKAV+KDAC [9], and UIMAF [2]. For the Harm-C and Harm-P datasets, the following multimodal baseline models were selected: MMBT [18], ViLBERT [27], Visual BERT [22], CLIP [20], MOMENTA [34], TOT [44], and ISM(BLIP) [42]. In this study, a comparative test was conducted using the above baseline model with the DAHM model, and the experimental results are shown in Table 1 and Table 2.

Table 1. Comparison of MultiOFF performance (%)

Model	MultiOFF		
	Precision↑	F1↑	Recall↑
StackedLSTM+VGG16	46.3	37.3	61.1
BiLSTM+VGG16	48.0	48.6	58.4
CNNText+VGG16	46.3	37.3	61.1
DisMultiHate	64.5	64.6	65.1
MSKAV+KDAC	67.0	67.9	69.0
UIMAF	70.3	70.2	70.0
DAHM	**74.6**	**74.2**	**73.9**

Table 2. Comparison of Harm-C and Harm-P performance

Model	Harm-C			Harm-P		
	Acc↑(%)	F1↑(%)	MMAE↓	Acc↑(%)	F1↑(%)	MMAE↓
MMBT	73.48	67.12	0.3258	82.54	80.23	0.1413
ViLBERT	78.53	78.06	0.1881	87.25	86.03	0.1276
Visual BERT	81.36	80.13	0.1857	86.80	86.07	0.1318
CLIP	78.10	77.64	0.2010	84.02	83.85	0.1508
MOMENTA	83.82	82.80	0.1743	89.84	88.26	0.1314
TOT	87.01	85.93	0.1634	91.55	91.29	0.1245
ISM(BLIP)	87.08	86.23	0.1585	**92.29**	91.61	0.1085
DAHM	**87.19**	**86.99**	**0.1483**	91.93	**91.88**	**0.1040**

4.5 Results Analysis

The performance of the proposed DAHM model was systematically evaluated across three datasets. On the MultiOFF and Harm-C datasets, DAHM consistently achieved the best results, demonstrating its strong ability to balance positive-class recognition with overall classification accuracy.

Evaluation on the Harm-P dataset reveals that while DAHM's Accuracy is slightly behind the baseline, it demonstrates superior performance on more robust classification metrics. Specifically, DAHM surpasses the ISM (BLIP) model with a higher F1 score and a lower MMAE. This indicates that DAHM, despite a marginal dip in overall accuracy, is more adept at handling class imbalance and provides more reliable, well-calibrated predictions.

As shown in Fig. 4, the experimental results reveal a distinct performance profile for the DAHM model. The model demonstrates high sensitivity, consistently and accurately identifying positive samples across all datasets, which validates its capability for effective semantic feature extraction. However, a notable and recurring limitation is the model's tendency to generate false positives. This asymmetric error pattern suggests that while the model has learned the features of the positive class well, it struggles to definitively reject challenging or ambiguous negative instances. Therefore, future improvements should prioritize enhancing the model's specificity to reduce these misclassifications and improve its overall robustness.

4.6 Ablation Study

To evaluate the multimodal fusion approach proposed in this study in the hate content detection task, three ablation experiments were designed. The performance of these ablation models was compared with the full model DAHM, and the experimental results are shown in Table 3. Removing the interaction feature E, DAHM(*E), leads to performance degradation, highlighting its role

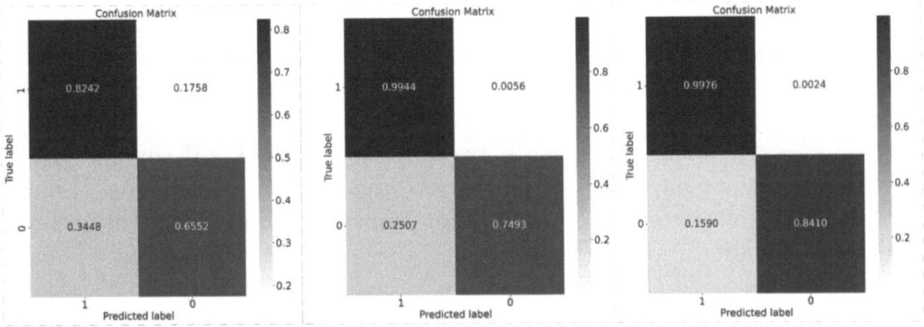

Fig. 4. Experimental results (confusion matrix) for different datasets. (a): for the MultiOFF dataset. (b): for the Harm-C dataset. (c): for the Harm-P dataset.

in capturing fine-grained text-image interactions. Eliminating the AttnFusionModel, DAHM(*MA), also causes a significant drop, confirming its importance in enhancing inter-modal alignment and feature representation. Lastly, removing the feature concatenation, DAHM(*CAT), reduces performance, indicating that combining raw and attention-processed features helps preserve critical information for classification. Overall, each component plays a vital role in the model's effectiveness.

Table 3. Ablation study results for MultiOFF, Harm-C, and Harm-P

Model	MultiOFF (%)			Harm-C			Harm-P		
	Pre↑	F1↑	Recall↑	Acc↑(%)	F1↑(%)	MMAE↓	Acc↑(%)	F1↑(%)	MMAE↓
DAHM	**74.6**	**74.2**	**73.9**	**87.19**	**86.99**	**0.1483**	**91.93**	**91.88**	**0.1040**
DAHM(*E)	72.5	72.0	71.6	85.29	83.86	0.1614	86.20	86.19	0.1376
DAHM(*MA)	72.5	72.7	72.9	83.89	83.00	0.1536	87.04	87.04	0.1289
DAHM(*CAT)	60.0	60.1	60.2	70.89	69.68	0.2871	88.73	88.72	0.1130

4.7 Case Study

As shown in Fig. 5, three representative samples from the dataset are presented, along with SHAP-based visual and textual attributions for DAHM's predictions. In the first meme, the text appears neutral, welcoming someone to a "circus" and "party", but the image depicts a clown stylized as a political figure, implicitly mocking the Democratic Party. The model correctly identifies the implicit harm by relying on visual sarcasm despite the benign language. In the second meme, the text mentions criminal charges such as "manslaughter" and "charged", yet the tone remains objective and the image is emotionally neutral. The model misclassifies this sample as harmless, indicating difficulty in detecting harmful intent expressed through factual or context-dependent language. In the third meme, a humorous caption referring to a dog as "President Woof" is paired with an innocuous, patriotic image. The model correctly classifies it as harmless, supported by the clear alignment between visual and textual content. These cases illustrate the model's strengths in capturing implicit harmful cues through multimodal interactions, as well as its limitations when harmful intent is subtle or context-dependent.

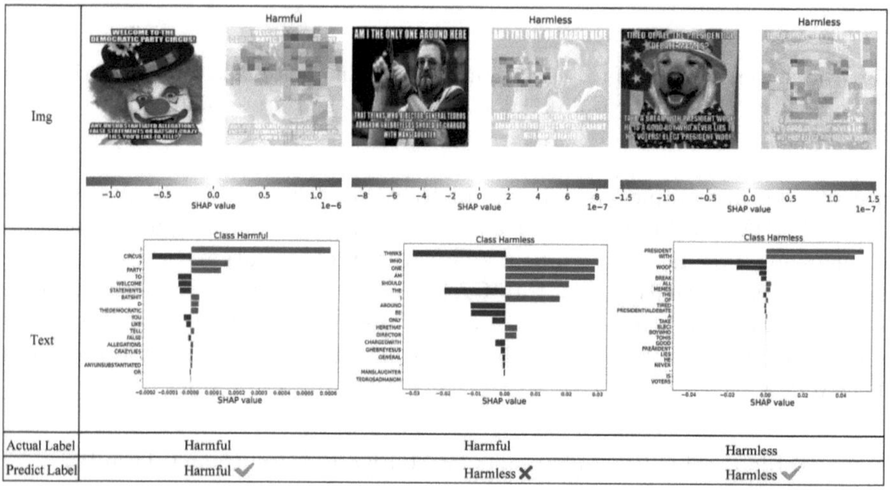

Fig. 5. Case studies of memes.

5 Conclusions

The DAHM model proposed in this paper offers an effective solution to the multimodal hate content detection problem. By leveraging both text and image modalities, the model demonstrates significant improvements in processing complex multimodal data and effectively capturing the complementary information from both sources. While the model shows promising results in experiments, there remains room for further refinement, particularly in specific performance metrics. The model's performance across various datasets indicates the potential of its design and provides valuable insights into multimodal hate content detection. Future research will focus on optimizing the model's performance, with an emphasis on improving accuracy, efficiency, and generalization.

Acknowledgments. This work is supported by the Key R&D Program of Xinjiang Uygur Autonomous Region (2022B01046).

References

1. Arya, G., et al.: Multimodal hate speech detection in memes using contrastive language-image pre-training. IEEE Access (2024)
2. Ayetiran, E.F., Özgöbek, Ö.: An inter-modal attention-based deep learning framework using unified modality for multimodal fake news, hate speech and offensive language detection. Inf. Syst. **123**, 102378 (2024)
3. Badjatiya, P., Gupta, S., Gupta, M., Varma, V.: Deep learning for hate speech detection in tweets. In: Proceedings of the 26th International Conference on World Wide Web Companion, pp. 759–760 (2017)
4. Brown, T.B.: Language models are few-shot learners. arXiv preprint arXiv:2005.14165 (2020)
5. Cao, R., Lee, R.K.W., Hoang, T.A.: Deephate: hate speech detection via multi-faceted text representations. In: Proceedings of the 12th ACM Conference on Web Science, pp. 11–20 (2020)
6. Chatzakou, D., Kourtellis, N., Blackburn, J., De Cristofaro, E., Stringhini, G., Vakali, A.: Mean birds: detecting aggression and bullying on twitter. In: Proceedings of the 2017 ACM on Web Science Conference, pp. 13–22 (2017)
7. Chen, C., Han, D., Chang, C.C.: CAAN: context-aware attention network for visual question answering. Pattern Recogn. **132**, 108980 (2022)
8. Chen, Y., Zhou, Y., Zhu, S., Xu, H.: Detecting offensive language in social media to protect adolescent online safety. In: 2012 International Conference on Privacy, Security, Risk and Trust and 2012 International Conference on Social Computing, pp. 71–80. IEEE (2012)
9. Chhabra, A., Vishwakarma, D.K.: Multimodal hate speech detection via multi-scale visual kernels and knowledge distillation architecture. Eng. Appl. Artif. Intell. **126**, 106991 (2023)
10. Davidson, T., Warmsley, D., Macy, M., Weber, I.: Automated hate speech detection and the problem of offensive language. In: Proceedings of the International AAAI Conference on Web and Social Media, vol. 11, pp. 512–515 (2017)
11. Devlin, J.: Bert: pre-training of deep bidirectional transformers for language understanding. arXiv preprint arXiv:1810.04805 (2018)
12. Djuric, N., Zhou, J., Morris, R., Grbovic, M., Radosavljevic, V., Bhamidipati, N.: Hate speech detection with comment embeddings. In: Proceedings of the 24th International Conference on World Wide Web, pp. 29–30 (2015)
13. Dosovitskiy, A., et al.: An image is worth 16×16 words: transformers for image recognition at scale. arXiv preprint arXiv:2010.11929 (2020)
14. Dwivedy, V., Roy, P.K.: Deep feature fusion for hate speech detection: a transfer learning approach. Multimed. Tools Appl. **82**(23), 36279–36301 (2023)
15. Gambäck, B., Sikdar, U.K.: Using convolutional neural networks to classify hate-speech. In: Proceedings of the First Workshop on Abusive Language Online, pp. 85–90 (2017)
16. Govers, J., Feldman, P., Dant, A., Patros, P.: Down the rabbit hole: detecting online extremism, radicalisation, and politicised hate speech. ACM Comput. Surv. **55**(14s), 1–35 (2023)
17. Gröndahl, T., Pajola, L., Juuti, M., Conti, M., Asokan, N.: All you need is "love" evading hate speech detection. In: Proceedings of the 11th ACM Workshop on Artificial Intelligence and Security, pp. 2–12 (2018)
18. Kiela, D., Bhooshan, S., Firooz, H., Perez, E., Testuggine, D.: Supervised multimodal bitransformers for classifying images and text. arXiv preprint arXiv:1909.02950 (2019)

19. Kiela, D., et al.: The hateful memes challenge: detecting hate speech in multimodal memes. In: Advances in Neural Information Processing Systems, vol. 33, pp. 2611–2624 (2020)
20. Kumar, G.K., Nandakumar, K.: Hate-clipper: multimodal hateful meme classification based on cross-modal interaction of clip features. In: 2nd Workshop on NLP for Positive Impact, NLP4PI 2022 Held in Conjunction with the 2022 Conference on Empirical Methods in Natural Language Processing, EMNLP 2022, pp. 171–183. Association for Computational Linguistics (ACL) (2022)
21. Lee, R.K.W., Cao, R., Fan, Z., Jiang, J., Chong, W.H.: Disentangling hate in online memes. In: Proceedings of the 29th ACM International Conference on Multimedia, pp. 5138–5147 (2021)
22. Li, L.H., Yatskar, M., Yin, D., Hsieh, C.J., Chang, K.W.: Visualbert: a simple and performant baseline for vision and language. arXiv preprint arXiv:1908.03557 (2019)
23. Lin, H., Luo, Z., Gao, W., Ma, J., Wang, B., Yang, R.: Towards explainable harmful meme detection through multimodal debate between large language models. In: Proceedings of the ACM on Web Conference 2024, pp. 2359–2370 (2024)
24. Lippe, P., et al.: A multimodal framework for the detection of hateful memes. arXiv preprint arXiv:2012.12871 (2020)
25. Liu, Y.: Roberta: a robustly optimized BERT pretraining approach. arXiv preprint arXiv:1907.11692 (2019)
26. Liu, Z., Yang, T., Chen, W., Chen, J., Li, Q., Zhang, J.: Sentiment analysis of social media comments based on multimodal attention fusion network. Appl. Soft Comput. **164**, 112011 (2024)
27. Lu, J., Batra, D., Parikh, D., Lee, S.: Vilbert: pretraining task-agnostic visiolinguistic representations for vision-and-language tasks. In: Advances in Neural Information Processing Systems, vol. 32 (2019)
28. Malmasi, S., Zampieri, M.: Challenges in discriminating profanity from hate speech. J. Exp. Theor. Artif. Intell. **30**(2), 187–202 (2018)
29. Mehdad, Y., Tetreault, J.: Do characters abuse more than words? In: Proceedings of the 17th Annual Meeting of the Special Interest Group on Discourse and Dialogue, pp. 299–303 (2016)
30. Min, C., et al.: Finding hate speech with auxiliary emotion detection from self-training multi-label learning perspective. Inf. Fusion **96**, 214–223 (2023)
31. Nobata, C., Tetreault, J., Thomas, A., Mehdad, Y., Chang, Y.: Abusive language detection in online user content. In: Proceedings of the 25th International Conference on World Wide Web, pp. 145–153 (2016)
32. Park, J.H., Fung, P.: One-step and two-step classification for abusive language detection on twitter. arXiv preprint arXiv:1706.01206 (2017)
33. Pramanick, S., et al.: Detecting harmful memes and their targets. In: Findings of the Association for Computational Linguistics: ACL-IJCNLP 2021, pp. 2783–2796 (2021)
34. Pramanick, S., Sharma, S., Dimitrov, D., Akhtar, M.S., Nakov, P., Chakraborty, T.: Momenta: a multimodal framework for detecting harmful memes and their targets. In: Findings of the Association for Computational Linguistics: EMNLP 2021, pp. 4439–4455 (2021)
35. Suryawanshi, S., Chakravarthi, B.R., Arcan, M., Buitelaar, P.: Multimodal meme dataset (multioff) for identifying offensive content in image and text. In: Proceedings of the Second Workshop on Trolling, Aggression and Cyberbullying, pp. 32–41 (2020)

36. Tekiroğlu, S.S., Chung, Y.L., Guerini, M.: Generating counter narratives against online hate speech: data and strategies. In: Proceedings of the 58th Annual Meeting of the Association for Computational Linguistics, pp. 1177–1190 (2020)
37. Vaswani, A.: Attention is all you need. In: Advances in Neural Information Processing Systems (2017)
38. Waseem, Z.: Are you a racist or am i seeing things? Annotator influence on hate speech detection on twitter. In: Proceedings of the First Workshop on NLP and Computational Social Science, pp. 138–142 (2016)
39. Waseem, Z., Hovy, D.: Hateful symbols or hateful people? Predictive features for hate speech detection on twitter. In: Proceedings of the NAACL Student Research Workshop, pp. 88–93 (2016)
40. Wu, F., et al.: Fuser: an enhanced multimodal fusion framework with congruent reinforced perceptron for hateful memes detection. Inf. Process. Manag. **61**(4), 103772 (2024)
41. Xiang, G., Fan, B., Wang, L., Hong, J., Rose, C.: Detecting offensive tweets via topical feature discovery over a large scale twitter corpus. In: Proceedings of the 21st ACM International Conference on Information and Knowledge Management, pp. 1980–1984 (2012)
42. Yang, C., Zhu, F., Han, J., Hu, S.: Invariant meets specific: a scalable harmful memes detection framework. In: Proceedings of the 31st ACM International Conference on Multimedia, pp. 4788–4797 (2023)
43. Yang, C., Zhu, F., Liu, G., Han, J., Hu, S.: Multimodal hate speech detection via cross-domain knowledge transfer. In: Proceedings of the 30th ACM International Conference on Multimedia, pp. 4505–4514 (2022)
44. Zhang, L., et al.: Tot: topology-aware optimal transport for multimodal hate detection. In: Proceedings of the AAAI Conference on Artificial Intelligence, vol. 37, pp. 4884–4892 (2023)
45. Zhang, Z., Robinson, D., Tepper, J.: Detecting hate speech on twitter using a convolution-GRU based deep neural network. In: Gangemi, A., Navigli, R., Vidal, M.-E., Hitzler, P., Troncy, R., Hollink, L., Tordai, A., Alam, M. (eds.) ESWC 2018. LNCS, vol. 10843, pp. 745–760. Springer, Cham (2018). https://doi.org/10.1007/978-3-319-93417-4_48

TTEdit: Cross-Modal Fusion with Diffusion Models for Detail-Aware Fashion Editing

Feng Zhang[1,3,5], Junliang Tan[1,2], Zhenming Chen[1], Hao Feng[1], Biao Guo[1], Junyan Chen[1], Yao Lu[1], and Ming Jiang[1,3,4](✉)

[1] School of Computer Science and Information Security, Guilin University of Electronic Technology, Guilin 541004, China
zhangfeng@guet.edu.cn
[2] Guangxi Academy of Artificial Intelligence, Nanning 530201, China
[3] Guangxi Metaverse Application Engineering Center, Guangxi Institute of Digital Technology, Nanning 530000, China
mjiang@guet.edu.cn
[4] School of Information Engineering, Guangxi Polytechnic of Construction, Nanning 530003, China
[5] Guangxi Guanyang Digital Technology, Nanning 530000, China

Abstract. Recent advances in text-driven image editing using visual language models demonstrate considerable potential. However, purely text-based methods face persistent challenges in fashion image editing, where a semantic gap hinders the accurate expression of complex garment details and the generation of visuals faithful to textual descriptions. While FashionTex pioneered the use of texture images as auxiliary inputs with GANs for local editing, inherent limitations of GANs constrain generative quality. To address these issues, we propose **TTEdit**, a novel diffusion model-based framework for fashion image editing. TTEdit innovatively integrates dual-modality information from text descriptions and texture images to achieve precise, localized garment edits. Our approach consists of two core components: (1) **MCNet**, a lightweight and accurate localization module built upon mainstream segmentation models to identify editing regions, and (2) the **Texture-Adapter**, which injects fabric texture features into the diffusion model via a cross-modal attention mechanism. Extensive experiments demonstrate that TTEdit significantly outperforms existing methods, achieving state-of-the-art results with an FID of 11.55, SSIM of 0.871, and CLIP-I of 0.83. These metrics surpass the previous best method, Insert Anything (FID 11.98, SSIM 0.848, CLIP-I 0.81), confirming our framework's superior generative quality, fidelity, and controllability.

Keywords: Fashion Image Editing · Multimodality · Diffusion Model

1 Introduction

Image editing is a key technology for digital content creation. It involves semantic modification, quality enhancement, and content regeneration of existing images

J. Lokoč et al. (Eds.): MMM 2026, LNCS 16413, pp. 290–303, 2026.
https://doi.org/10.1007/978-981-95-6957-1_21

[1]. The scope ranges from color correction to semantic reconstruction, supporting applications such as beautification, style transfer, and content completion. At the intersection of computer vision and graphics, deep learning methods provide users with creative freedom and precise control [2].

Fig. 1. TTEdit. Users can edit clothing regions by painting with a reference texture. Our method automatically adapts the reference image and seamlessly integrates it into the source image, enabling high-quality editing results.

Generative models have transformed image editing. Early methods based on GANs [3] achieved initial success but suffered from unstable training, mode collapse, and artifacts. Diffusion models [4], in contrast, generate high-fidelity images through a progressive denoising process, offering better stability.

Vision–language models further enabled text-driven image editing. By leveraging cross-modal pretraining such as CLIP [5], image editing can be guided by natural language. This paradigm reduces the technical barrier and allows non-experts to engage in creative tasks. Despite these advantages [6–8], text-driven methods remain limited in fashion editing. They struggle with precise region localization and often misinterpret fine-grained semantics, leading to inaccurate visual outputs.

FashionTex [9] addresses semantic limitations with reference textures, improving semantic–visual alignment. However, as a GAN-based method, it still suffers from mode collapse and low image quality. More critically, it lacks the ability for precise local editing without disturbing unrelated regions.

We propose TTEdit to overcome these challenges. The MCNet module, a lightweight dual-branch image–text encoder, collaborates with a segmentation model to locate target regions accurately. In the rendering stage, the Texture-Adapter employs multi-scale feature injection, dynamically aligning texture features with UNet layers. This ensures consistency from local details to global structure. Experiments demonstrate that TTEdit produces superior image quality, realistic textures, and smooth editing boundaries. Our main contributions are:

- We propose the TTEdit editing framework, which combines text with texture images and diffusion models, effectively solving the problems of insufficient text description representation and inability to accurately locate the editing region.
- In order to improve the editing accuracy, we designed MCNet, which can assist the segmentation model to locate the garment regions that need to be edited and provide precise guidance for subsequent image modification.
- To effectively utilize texture details as conditional guidance for diffusion-based generation, we introduce a Texture-Adapter module. Furthermore, we improve the formulation of texture loss to produce more realistic and visually consistent results.
- Experimental results show that our method outperforms the existing methods in a number of metrics, including image fidelity, local consistency, and user judgments, and achieves a more realistic and detailed fashion image editing effect.

2 Related Work

Text-Driven Image Editing. With the rapid progress of text-driven image generation, models such as DALL-E 2 [10] and Stable Diffusion [11] enable image creation directly from natural language. InstructPix2Pix [13] further extends this paradigm to instruction-based editing, allowing images to be modified according to natural language guidance. In fashion editing, Texfit [14] explores modifying clothing styles, patterns, and colors through textual input, supporting flexible personalization. To improve semantic alignment, methods such as GLIDE [15], Textual Inversion [16], and T2I-Adapter [17] strengthen the mapping between language and images. While these approaches are effective in generating diverse content, they still struggle with fine-grained control, particularly for complex textures and material details. To address this, we introduce reference texture images as additional conditions, enhancing texture representation and yielding more realistic editing results.

Reference-Based Fashion Editing. A large body of work focuses on transferring target clothing onto user images. TryOnGAN [18] applies layered interpolation to improve pose consistency, while PASTA-GAN++ [19] leverages spatial guidance and Texture-Adapter modules for more realistic try-on results. FashionTex [9] combines texture references with text prompts to enhance visual fidelity and detail control. Despite these advances, most approaches rely on GANs, which face limitations in fine-grained texture modeling and handling complex geometric deformations. As a result, synthesized results often exhibit misaligned patterns, stretched fabrics, or unnatural folds [20,21], reducing overall visual consistency. In this work, we leverage reference texture information within diffusion models, achieving more controllable, high-quality, and natural fashion image generation.

3 Method

To achieve personalized garment editing, we propose a method that combines prompt-based conditional control with reference texture images. In Fig. 2, this is the architecture diagram of our model, given a full-body image I, our approach employs MCNet to use textual prompts to assist the general segmentation models in accurately selecting the mask M of the target garment region corresponding to the description of the user. Subsequently, the selected region is rendered using the reference texture image in conjunction with the garment prompt. The implementation details are presented below.

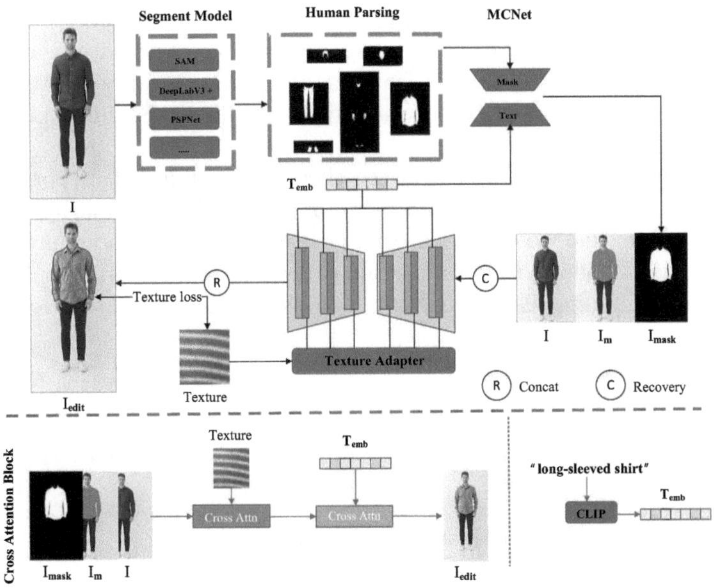

Fig. 2. Overview of our pipeline.

3.1 Region Extraction for Editing

Given that mainstream segmentation models (e.g., SAM [22], DeepLabV3+ [23], and PSPNet [24]) have demonstrated strong performance in human parsing, we use these models to parse the input image I, obtaining a detailed segmentation of the human body and the corresponding segmentation results.

To further refine the selection of the target garment mask, we design MCNet, a network that assists the segmentation model in mask selection. MCNet consists of two types of encoder: a mask image encoder E_{mask} and a text encoder T_{emb}. These encoders extract features from two different modalities—visual mask images and textual garment descriptions.

During training, we introduce a multi-modal semantic alignment loss, which aims to align the semantic space of the garment prompts with that of the corresponding mask features. This encourages the model to better understand the correspondence between natural language descriptions and visual regions. The alignment loss function is defined as follows:

$$\mathcal{L}_{\text{alignment}} = -\sum_{i=1}^{N} \log \left(\frac{\exp\left(\text{sim}\left(f_I\left(\text{aug}_I(M_i^+)\right), f_T(T_i^+)\right)/\tau\right)}{\sum_{j=1}^{N} \exp\left(\text{sim}\left(f_I\left(\text{aug}_I(M_i^+)\right), f_T(T_j)\right)/\tau\right)} \right) \tag{1}$$

where $f_I(\cdot)$ denotes the image encoder used to extract features from visual inputs, $\text{aug}_I(\cdot)$ represents the data augmentation function applied to the mask regions, M^+ is the positive sample feature extracted from the target mask region, T^+ is the positive sample feature corresponding to the associated garment text prompt, T_j refers to the j-th text feature among all candidate textual descriptions, and τ is the temperature parameter controlling the smoothness of the similarity distribution.

We encourage the similarity between the mask region and its corresponding ground-truth text to be maximized, while minimizing its similarity with other non-matching textual descriptions. In Fig. 3, the image encoder processes the input conditions, followed by multi-scale processing with downsampling operations between different scales, enabling high-quality texture extraction and generation.

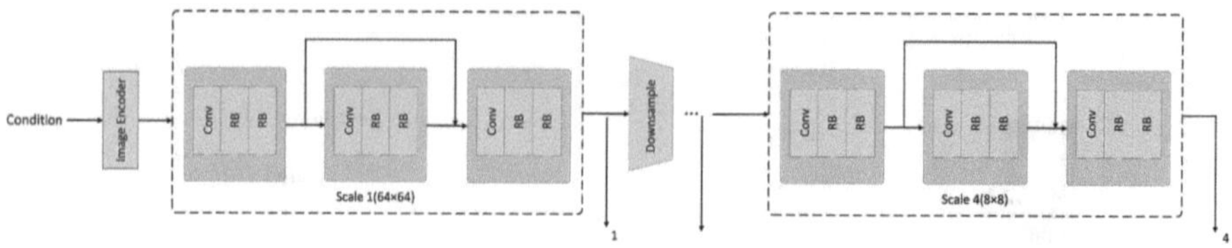

Fig. 3. Texture-Adapter Network Architecture.

During inference, we utilize the trained MCNet to evaluate the segmented mask regions produced by the parsing model. MCNet automatically selects the target mask region M_T that is semantically most relevant to the input prompt. This region represents the garment area to be edited, and is subsequently used for texture generation or image editing tasks. This enables more accurate garment manipulation and texture replacement. The selection process can be formulated as follows:

$$M^* = \arg\max_{M_i} \text{sim}\left(\phi_v(I \odot M_i), \phi_t(T)\right) \tag{2}$$

where $\phi_v(\cdot)$ denotes the image encoder used to extract features from visual inputs, $I \odot M_i$ represents the masked image obtained by applying mask M_i to the original image I, $\text{sim}(\cdot,\cdot)$ is the similarity function that measures the closeness between two feature vectors, M^* is the selected target mask region most semantically relevant to the input text prompt, $\phi_t(\cdot)$ denotes the text encoder used to extract features from the textual description T.

3.2 Content Generation

Due to the insufficient control over the generated content by text description alone, we leverage reference texture images to assist generation. The texture prompts mainly provide the target image with surface material, detail distribution, and style attributes, characterized by strong high-frequency information and strict spatial consistency requirements. To effectively extract semantic information from the texture images, we employ a Texture-Adapter to extract multi-level features. These texture semantic-rich features are injected into the cross-attention modules of the U-Net. Specifically, for each query feature Z in the U-Net module, we use the texture feature $c_{\mathcal{T}}$ as a key and value to perform the cross-attention computation:

$$Z_{\text{texture}} = Softmax\left(\frac{QK^T}{\sqrt{d}}\right)V \tag{3}$$

$$Q = ZW_q, K = c_{\mathcal{T}}W_k, V = c_{\mathcal{T}}W_v \tag{4}$$

where W_q, W_k, W_v are the learnable weights trained during the process, $c_{\mathcal{T}}$ denotes the projected texture prompt sequence.

This mechanism endows the diffusion process with texture-aware capability, enabling precise control over material and detail distribution in image generation.

To incorporate the semantic guidance of clothing-related textual prompts, we use a pre-trained CLIP text encoder to encode the input prompt and obtain a global text embedding. This embedding is then projected through a trainable projection network to generate a sequence of text embeddings c_{text}, which are subsequently injected into the cross-attention modules of the U-Net:

$$Z_{\text{text}} = \text{Softmax}\left(\frac{Q_{text}K_{text}^{\text{T}}}{\sqrt{d}}\right)V_{\text{text}} \tag{5}$$

$$Q_{text} = ZW_q^{\text{text}}, K_{text} = c_{text}W_k^{\text{text}}, V_{text} = c_{text}W_v^{\text{text}} \tag{6}$$

where $W_q^{\text{text}}, W_k^{\text{text}}, W_v^{\text{text}}$ are learnable parameters during training, and c_{text} is the projected text embedding sequence. The final fused output is defined as:

$$Z_{\text{new}} = Z + Z_{\text{texture}} + Z_{\text{text}} \tag{7}$$

During the training stage, we jointly train the Texture-Adapter and the U-Net network. The training dataset consists of paired textual prompts, reference texture images, and target images. The training objective is to minimize the diffusion reconstruction error using the standard denoising loss function:

$$\mathcal{L}_{\text{simple}} = \mathbb{E}_{x_0,\epsilon,c_\mathcal{T},c_{\text{text}},t} \left\|\epsilon - \epsilon_\theta(x_t, t, c_\mathcal{T}, c_{\text{text}})\right\|^2 \tag{8}$$

To preserve structural consistency while effectively transferring style characteristics (e.g., texture, detail, material) from the reference texture image to the target image, and simultaneously ensuring semantic consistency with the textual prompt, we employ the following style transfer loss:

$$\mathcal{L}_{\text{style}} = \sum_{l \in C} \lambda_l \cdot \|G_l(I_g) - G_l(I_s)\|_F^2 \tag{9}$$

where λ_l is the weight hyperparameter for each layer to balance contributions across layers, and $\|\cdot\|_F$ denotes the Frobenius norm, which measures the difference between two Gram matrices. $G_l(I_s)$ and $G_l(I_g)$ represent the style features of the reference and generated images at the l-th layer, respectively.

To achieve high-quality and personalized clothing texture editing, we jointly optimize multiple objectives during training. Specifically, we adopt the standard diffusion reconstruction loss $\mathcal{L}_{\text{simple}}$ to guide the overall image generation. Meanwhile, to ensure consistency in both texture appearance and semantic alignment, we define a texture loss $\mathcal{L}_{\text{texture}}$ that combines style transfer loss $\mathcal{L}_{\text{style}}$ and semantic alignment loss $\mathcal{L}_{\text{alignment}}$:

$$\mathcal{L}_{\text{texture}} = \beta_1 \cdot \mathcal{L}_{\text{style}} + \beta_2 \cdot \mathcal{L}_{\text{alignment}} \tag{10}$$

The total training objective is then defined as:

$$\mathcal{L}_{\text{total}} = \alpha \cdot \mathcal{L}_{\text{simple}} + \mathcal{L}_{\text{texture}} \tag{11}$$

where α is the weight coefficient for the loss of diffusion reconstruction, while β_1 and β_2 control the importance of style fidelity and semantic alignment in the texture region.

In our experiments, we set $\alpha = 1$, $\beta_1 = 25$, and $\beta_2 = 20$ to ensure high-quality texture transfer and precise semantic control [25].

During the inference stage, given a texture image and a text prompt as input, the model is capable of generating image results with consistent material style and detailed texture.

4 Experiments

4.1 Experimental Settings

Dataset. In this study, the Deepfashion-MultiModal dataset [26] is used for the training and evaluation of the methods. This dataset is a multi-modal fashion dataset containing 12,701 high-quality full-body human images, each labeled with (1) detailed text descriptions, (2) 24 categories of human parsing labels, and (3) human keypoint detection data. To ensure the reliability of the experiments,

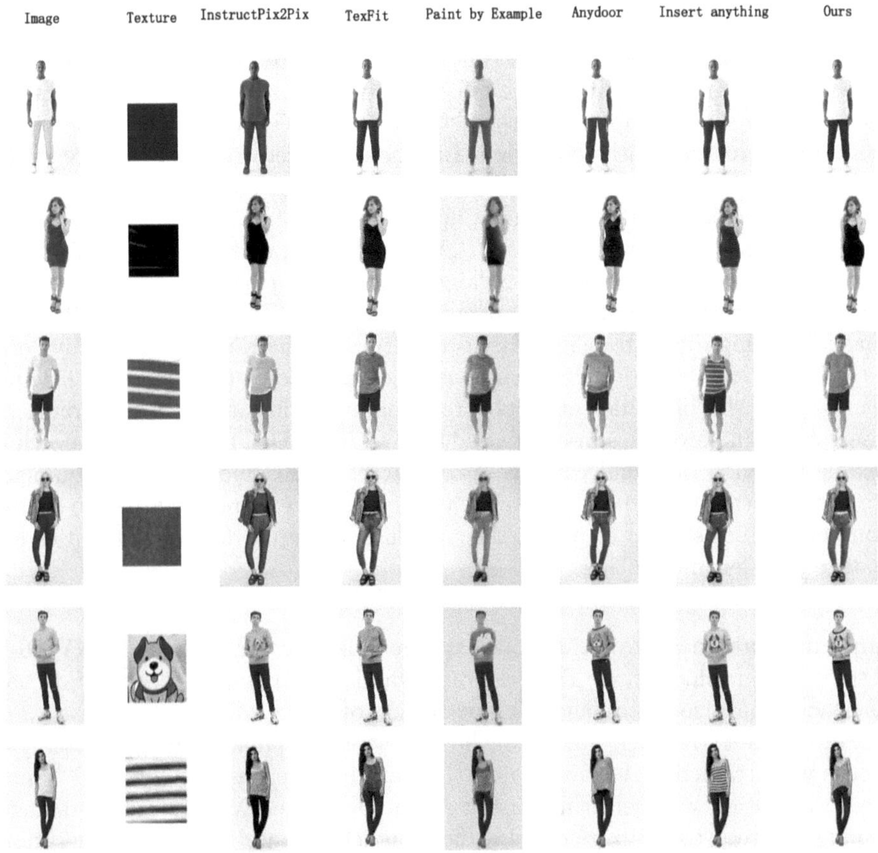

Fig. 4. Qualitative comparison with other approaches. Our method can generate results that are semantically consistent with the input reference images in high perceptual quality.

we divided the dataset into a training set (11,260 images) and a test set (1,441 images) according to standard protocols. In addition, considering the specific needs of this study, we additionally integrated local texture samples from the FashionTex dataset as supplementary reference data to enhance the model's ability to perceive garment texture features.

Baselines. To comprehensively evaluate the effectiveness of our proposed framework, we select five representative diffusion-based editing methods as baselines, covering both text-driven and reference-driven paradigms. The comparison setting is carefully designed to ensure fairness and reproducibility.

InstructPix2Pix [13] and TexFit [14] are representative text-guided editing approaches based on diffusion models. InstructPix2Pix learns to follow natural language editing instructions for general image manipulation, while TexFit specifically targets fashion image editing with text-driven control. Since both methods rely primarily on textual conditions, we first employ BLIP [12] to generate descriptive captions from the reference texture images. These captions are then used as textual inputs to InstructPix2Pix and TexFit, ensuring that the models are provided with semantic instructions derived from the texture itself. This setting allows the text-driven models to generate outputs consistent with the intended clothing patterns and styles.

In contrast, Paint by Example [27], AnyDoor [28], and Insert Anything [29] adopt reference-driven paradigms. Paint by Example performs exemplar-based editing by conditioning the diffusion process on the appearance of a reference image, enabling direct transfer of texture details to the target region. AnyDoor extends this idea to object-level customization in a zero-shot manner, allowing seamless transfer of visual concepts across contexts without additional training. Insert Anything leverages in-context editing within a DiT backbone, supporting flexible insertion of visual content into a target scene guided by an exemplar. For these methods, we directly provide the reference texture image as the condition, which allows them to utilize their intrinsic exemplar-based capabilities for texture transfer and editing.

Implementation Details. All our experiments were run on a single NVIDIA Tesla V100 graphics card. To accelerate training and improve computational efficiency, we resized all images to a resolution of 512×256.

For MCNet training, we used the following parameter configuration: 50 epochs were trained, batch size was 8, the learning rate was set to 1×10^{-4}, and the optimization was performed using Adam Optimizer. We adopt a two-stage training strategy to optimize the classification network and the image generation model separately.

In the first stage, we fine-tune MCNet, a classification network built on the CLIP architecture. To better adapt to our specific dataset and task, both the CLIP text encoder and image encoder are fine-tuned during this stage. The training parameters are set as follows: a total of 150k fine-tuning steps, a batch size of 8, and a learning rate of 5×10^{-6} to maintain the semantic alignment of the pre-trained CLIP model. Due to memory constraints, we employ mixed-precision training and adopt a gradient accumulation strategy with 4 accumulation steps.

In the second stage, we fine-tune the image generation model based on Stable Diffusion v1.5. Specifically, both the U-Net module and the Texture-Adapter are jointly trained, while all other components remain frozen to preserve the pre-trained knowledge. The training configuration remains consistent with the first stage, including 150k fine-tuning steps, a batch size of 1, and a learning rate of 1×10^{-5}. Similarly, mixed-precision training and gradient accumulation with 4 accumulation steps are applied to enhance training efficiency.

In the inference stage, we set the generation parameters to 50 iterations and the classifier-free guidance coefficient to 7.5.

4.2 Comparative Results

To evaluate the effectiveness of our method, we compare it with representative texture transfer approaches, as shown in Fig. 4. The results demonstrate that our method achieves more stable performance in texture migration, accurately preserving the fine-grained details, material properties, and style consistency of the reference textures.

For text-driven editing methods, a common pipeline first applies BLIP [12] to the texture reference image to generate a textual description containing material, pattern, and color information. This description is then fed into frameworks such as InstructPix2Pix [13] and TexFit [14], whose performance, however, deteriorates with complex prompts, producing less accurate texture reproduction. Paint by Example [27] tends to lose fine texture details, leading to overly smoothed or faded appearances. AnyDoor [28] better maintains global style but suffers from distortions and pattern inconsistencies in complex regions. Insert Anything [29] focuses on local insertions yet often introduces texture blending artifacts or geometric mismatches. In contrast, our approach consistently generates results that are both visually realistic and semantically consistent.

For quantitative evaluation, Table 1 reports results on four widely used metrics: Fréchet Inception Distance (FID) for measuring realism, Structural Similarity Index (SSIM) for structural fidelity, Learned Perceptual Image Patch Similarity (LPIPS) for perceptual consistency, and CLIP-I for semantic alignment. In addition, we conducted a user study in which 30 human annotators were asked to evaluate 100 generated images based on realism and texture consistency.

Table 1. Quantitative comparison with diffusion-based reference-guided methods.

Methods	FID ↓	SSIM ↑	LPIPS ↓	CLIP-I ↑	User Study ↑
InstructPix2Pix [13]	13.88	0.769	0.141	0.61	53
TexFit [14]	12.49	0.829	0.129	0.68	62
Paint by Example [26]	16.34	0.788	0.138	0.72	60
AnyDoor [27]	12.77	0.833	0.125	0.74	73
Insert Anything [28]	11.98	0.848	**0.119**	0.81	71
TTEdit (Ours)	**11.55**	**0.871**	0.123	**0.83**	**80**

4.3 Ablation Studies

In order to deep analyze the core mechanism of our method and verify the effectiveness of each key component, we designed two sets of systematic ablation experiments to examine (1) the effect of prompt conditional control and (2) the effect of the texture loss function.

The Effect of Prompt Conditional Control. In Fig. 5, the comparison experiment clearly demonstrates the influence of cue word condition control on the generation effect. The cue word control can effectively help the model to understand the relationship with the mask region, in order to effectively generate the dress corresponding to the texture.

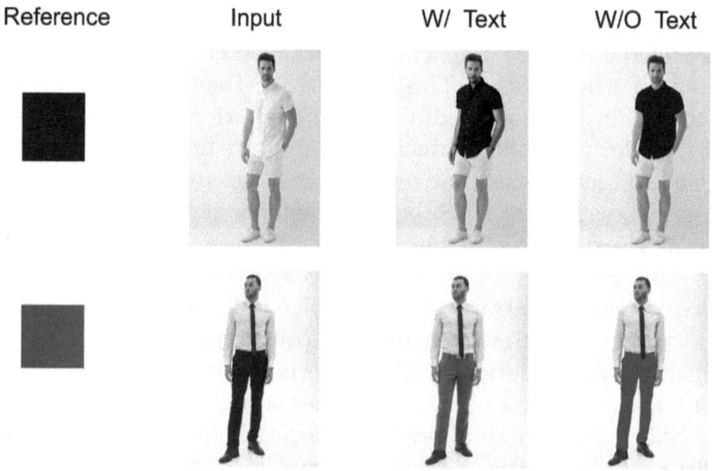

Fig. 5. Results of ablation experiments with the cue word conditional control module. Each row represents a set of dress style migration experiments, with the first column being the color cue for the reference image, the second column being the input character image, the third column being the results after the introduction of the cue word conditional control module (W/ Text), and the fourth column being the results without the introduction of the module (W/O Text).

The Effect of Texture Loss Functions. In Fig. 6, The introduction of the texture loss results in a significant performance improvement. As shown in Table 2, the adoption of the proposed loss results in notable enhancements across key metrics including FID, SSIM, LPIPS, CLIP-I and User Study. This demonstrates that the texture loss effectively strengthens the model's understanding of texture semantics by optimizing semantic alignment in the feature space, thereby enabling the generation of high-quality garments that better match the expected visual patterns.

Table 2. Ablation study on the effect of texture loss.

Methods	FID ↓	SSIM ↑	LPIPS ↓	CLIP-I ↑	User Study ↑
W/O Texture Loss	12.34	0.802	0.177	0.77	62
W/ Texture Loss	**11.55**	**0.871**	**0.123**	**0.83**	**80**

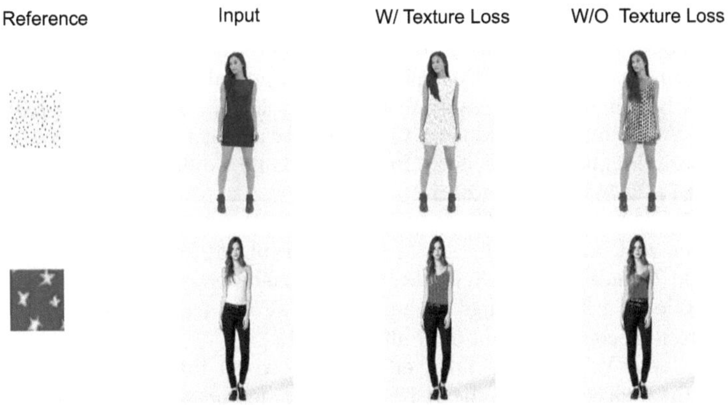

Fig. 6. Ablation study on the effect of texture loss for semantic alignment in texture synthesis.

5 Conclusion

In this paper, we presented TTEdit, a diffusion-based framework for fashion image editing that achieves high texture fidelity, accurate region localization, and strong semantic consistency. We proposed two key modules: a Mask Classification Network (MCNet) for precise editable region localization, and a Texture-Adapter for multi-scale texture injection through cross-modal attention. Comprehensive experiments demonstrate that TTEdit outperforms state-of-the-art methods in both objective metrics (FID, SSIM, LPIPS, CLIP-I) and subjective evaluations. It delivers realistic local edits, faithful texture reproduction, and robust performance across complex garment structures and materials. This work highlights the potential of diffusion models for fine-grained and controllable fashion image editing.

Acknowledgements. This work is supported by Guangxi Key R&D Program (AB25069302), the Nanning Yongjiang Talent Program (No. 2023018, No. 2019004), Guilin S&T Program (20230119-1), Graduate Innovation Project of Guangxi Academy of Artificial Intelligence and Guangxi Natural Science Foundation (No. 2025GXNS-FAA069684).

References

1. Shuai, X., Ding, H., Ma, X., et al.: A survey of multimodal-guided image editing with text-to-image diffusion models. arXiv preprint arXiv:2406.14555 (2024)
2. Zhan, F., Yu, Y., Wu, R., et al.: Multimodal image synthesis and editing: the generative AI era. IEEE Trans. Pattern Anal. Mach. Intell. **45**(12), 15098–15119 (2023)
3. Goodfellow, I., Pouget-Abadie, J., Mirza, M., et al.: Generative adversarial nets. In: Advances in Neural Information Processing Systems, vol. 27 (2014)
4. Ho, J., Jain, A., Abbeel, P.: Denoising diffusion probabilistic models. In: Advances in Neural Information Processing Systems, vol. 33, pp. 6840–6851 (2020)
5. Radford, A., Kim, J.W., Hallacy, C., et al.: Learning transferable visual models from natural language supervision. In: International Conference on Machine Learning, pp. 8748–8763. PMLR (2021)
6. Baykal, A.C., Anees, A.B., Ceylan, D., et al.: CLIP-guided StyleGAN inversion for text-driven real image editing. ACM Trans. Graph. **42**(5), 1–18 (2023)
7. Kawar, B., Zada, S., Lang, O., et al.: Imagic: text-based real image editing with diffusion models. In: Proceedings of the IEEE/CVF Conference on Computer Vision and Pattern Recognition, pp. 6007–6017 (2023)
8. Lin, Y., Chen, Y.W., Tsai, Y.H., et al.: Text-driven image editing via learnable regions. In: Proceedings of the IEEE/CVF Conference on Computer Vision and Pattern Recognition, pp. 7059–7068 (2024)
9. Lin, A., Zhao, N., Ning, S., et al.: FashionTex: controllable virtual try-on with text and texture. In: ACM SIGGRAPH 2023 Conference Proceedings, pp. 1–9 (2023)
10. Rombach, R., Blattmann, A., Lorenz, D., et al.: High-resolution image synthesis with latent diffusion models. In: Proceedings of the IEEE/CVF Conference on Computer Vision and Pattern Recognition, pp. 10684–10695 (2022)
11. Ramesh, A., Dhariwal, P., Nichol, A., et al.: Hierarchical text-conditional image generation with CLIP latents. arXiv preprint arXiv:2204.06125 (2022)
12. Li, J., Li, D., Xiong, C., Hoi, S.C.H.: BLIP: bootstrapping Language-Image Pre-training for unified vision-language understanding and generation. In: Proceedings of the 39th International Conference on Machine Learning (ICML 2022), PMLR, vol. 162, pp. 12888–12900 (2022)
13. Brooks, T., Holynski, A., Efros, A.A.: InstructPix2Pix: learning to follow image editing instructions. In: Proceedings of the IEEE/CVF Conference on Computer Vision and Pattern Recognition (CVPR), pp. 18392–18402 (2023)
14. Wang, T., Ye, M.: TexFit: text-driven fashion image editing with diffusion models. In: Proceedings of the AAAI Conference on Artificial Intelligence, vol. 38, no. 9, pp. 10198–10206 (2024)
15. Nichol, A., Dhariwal, P., Ramesh, A., et al.: Glide: towards photorealistic image generation and editing with text-guided diffusion models. arXiv preprint arXiv:2112.10741 (2021)
16. Gal, R., Alaluf, Y., Atzmon, Y., et al.: An image is worth one word: personalizing text-to-image generation using textual inversion. arXiv preprint arXiv:2208.01618 (2022)
17. Mou, C., Wang, X., Xie, L., et al.: T2I-adapter: learning adapters to dig out more controllable ability for text-to-image diffusion models. In: Proceedings of the AAAI Conference on Artificial Intelligence, vol. 38, no. 5, pp. 4296–4304 (2024)
18. Lewis, K.M., Varadharajan, S., Kemelmacher-Shlizerman, I.: TryOnGAN: body-aware try-on via layered interpolation. ACM Trans. Graph. (TOG) **40**(4), 1–10 (2021)

19. Xie, Z., Huang, Z., Zhao, F., et al.: PASTA-GAN++: a versatile framework for high-resolution unpaired virtual try-on. arXiv preprint arXiv:2207.13475 (2022)
20. Zhang, J., Li, K., Lai, Y.K., et al.: PISE: person image synthesis and editing with decoupled GAN. In: Proceedings of the IEEE/CVF Conference on Computer Vision and Pattern Recognition, pp. 7982–7990 (2021)
21. Raffiee, A.H., Sollami, M.: GarmentGAN: photo-realistic adversarial fashion transfer. In: 2020 25th International Conference on Pattern Recognition (ICPR), pp. 3923–3930. IEEE (2021)
22. Kirillov, A., et al.: Segment anything. In: Proceedings of the IEEE/CVF International Conference on Computer Vision (ICCV), pp. 4015–4026 (2023)
23. Xiao, J., Wang, Z.: DeepLabv3+: encoder-decoder with atrous separable convolution for semantic image segmentation. In: Proceedings of the IEEE Conference on Computer Vision and Pattern Recognition (CVPR), pp. 1–5 (2019)
24. Zhao, H., Shi, J., Qi, X., Wang, X., Jia, J.: Pyramid scene parsing network. In: Proceedings of the IEEE Conference on Computer Vision and Pattern Recognition (CVPR), pp. 2881–2890 (2017)
25. Gatys, L.A., Ecker, A.S., Bethge, M.: Image style transfer using convolutional neural networks. In: Proceedings of the IEEE Conference on Computer Vision and Pattern Recognition, pp. 2414–2423 (2016)
26. Jiang, Y., Yang, S., Qiu, H., Wu, W., Loy, C.C., Liu, Z.: Text2Human: text-driven controllable human image generation. ACM Trans. Graph. (Proceedings of SIGGRAPH) **41**(4), 1–11 (2022)
27. Yang, B., Gu, S., Zhang, B., et al.: Paint by example: exemplar-based image editing with diffusion models. In: Proceedings of the IEEE/CVF Conference on Computer Vision and Pattern Recognition, pp. 18381–18391 (2023)
28. Chen, X., Huang, L., Liu, Y., et al.: AnyDoor: zero-shot object-level image customization. In: Proceedings of the IEEE/CVF Conference on Computer Vision and Pattern Recognition, pp. 6593–6602 (2024)
29. Song, W., Jiang, H., Yang, Z., et al.: Insert anything: image insertion via in-context editing in DiT. arXiv preprint arXiv:2504.15009 (2025)

UCAMNet: HVI Color Space Based Unsupervised Low-Light Enhancement via Uncertainty Constraint and Attention Mechanism

Jingshuo Guan[1], Na Qi[1(✉)], Qing Zhu[1], and Liang Chen[2]

[1] Beijing University of Technology, Beijing, China
guanjingshuo@emails.bjut.edu.cn, {qina,ccgszq}@bjut.edu.cn
[2] Fujian Normal University, Fuzhou, China

Abstract. Low-light image enhancement is a computer vision task that aims to improve the visual perceptual quality of images captured in poorly illuminated scenes. At present, low-light enhancement methods based on deep learning can obtain high-quality enhanced images. However, these methods fail to fully account for the statistical characteristics of different regions, with the uncertainty of images remaining poorly underutilized. Furthermore, most existing methods perform enhancement directly in the sRGB color space, where the strong coupling between color and brightness often gives rise to issues such as texture distortion, noise amplification, and color shifts. To address these issues, we propose a novel network with **U**ncertainty **C**onstraint and **A**ttention **M**echanism (**UCAMNet**), an unsupervised framework based on the HVI color space, to improve the performance of low-light enhancement. UCAMNet first transforms the unpaired low/normal-light images from the sRGB color space to the HVI color space, which allows for a more thorough decoupling of color and brightness information, thereby decomposing the images into HV color and intensity components. Then a generative adversarial network guided by uncertainty constraint is proposed to enhance the intensity component with the guidance of variance estimation. Additionally, we further introduce a channel-spatial guided attention network to address the color distortion and noise problem, thus making the network better preserve the texture details. Extensive experimental results demonstrate that UCAMNet outperforms the state-of-the-art methods in terms of visual quality and quantitative metrics.

Keywords: Low-light enhancement · GAN · uncertainty-driven · attention module · HVI color space

1 Introduction

Due to challenging environmental conditions, images captured under low-light scenarios exhibit significant detail loss and reduced contrast [1]. As a basic computer vision task, low-light image enhancement (LLIE) not only improves subjective visual perception, but also plays an important role in high-level tasks such as semantic segmentation [2,3], face recognition [4,5] and object detection [6,7].

J. Lokoč et al. (Eds.): MMM 2026, LNCS 16413, pp. 304–318, 2026.
https://doi.org/10.1007/978-981-95-6957-1_22

Existing LLIE methods are broadly categorized into two parts: traditional methods [8–10] and deep learning-based methods [11–13]. Traditional LLIE methods include histogram equalization and Retinex methods, which mostly rely on handcrafted priors or statistical assumptions, leading to limited adaptability, color distortion, amplified noise, and inconsistent enhancement across diverse scenes. In contrast, deep learning-based methods leverage data-driven priors and deep neural networks, demonstrating superior performance in most scenarios. However, most LLIE methods require paired training datasets, which are difficult to acquire in real-world settings. To address this problem, unsupervised GAN-based methods have emerged as promising solutions [14, 15]. For instance, EnlightenGAN [14] utilizes a global-local discriminator, a self-regularized perceptual loss, and an attentional mechanism to enhance low-light images effectively. Moreover, UTrCGAN [15] decomposes images into reflectance and illumination based on Retinex-based model, incorporates uncertainty constraints into cycle-consistent GAN framework, enabling adaptive regional illumination adjustment. However, UTrCGAN ignores noise suppression in the reflectance component and lacks full bidirectional cycle consistency on low/normal-light images.

Traditional LLIE methods typically decompose images into reflectance and illumination components in the sRGB color space based on Retinex model [11, 16]. However, the strong coupling between brightness and color channels in sRGB space often causes color distortion during enhancement. Inspired by Horizontal/Vertical-Intensity (HVI) [17], a new color space minimizing noise via reduced Euclidean distance between similar colors, we propose transforming images from sRGB to HVI space for more effective enhancement. The HV color component contains rich intrinsic texture and color information but is often corrupted by noise in low-light conditions. Traditional CNN-based denoising methods [18, 19] struggle with long-range dependencies and non-local self-similarities, leading to color artifacts and detail loss. While Transformer architectures [20] have demonstrated superior capability in modeling long-range dependencies, leading to significant improvements in LLIE [21–23]. However, existing Transformer-based methods primarily focus on global image enhancement and do not specifically address noise and texture detail loss in the HV color component. The potential of Transformer for noise suppression and detail preservation in LLIE tasks remains largely unexplored. Thus, we propose a Channel-Space Guided Attention Network (CSGAN) to reduce noise in the HV color component, preserve the texture details and adaptively enhance the quality.

Our main contributions can be summarized as follows:

- We propose an unsupervised LLIE framework that transforms sRGB to HVI color space for image decomposition, then establishes the cycle-consistent GANs on low/normal-light images with uncertainty-guided constraints, enabling paired-data-free enhancement while preserving image quality.
- We propose a CSGAN by introducing a channel-spatial guided attention mechanism to jointly addresses noise suppression and detail preservation in the HV color component, overcoming limitations of existing Transformer-based approaches.

– Experimental results demonstrate that our proposed UCAMNet outperforms the state-of-the-art methods in terms of subjective and objective quality.

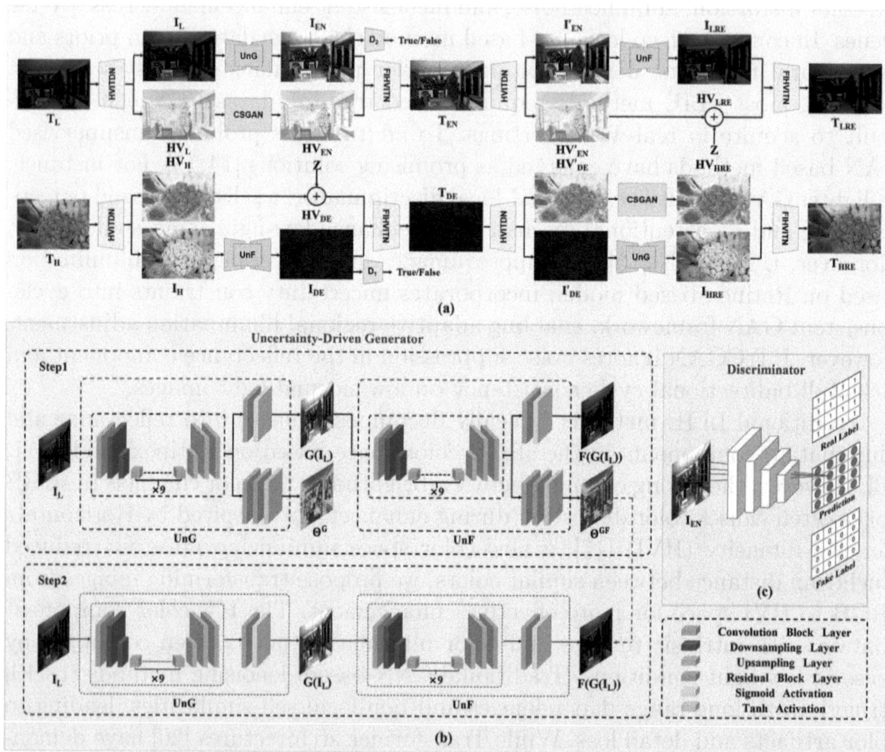

Fig. 1. Overview of the UCAMNet framework. (a) The network takes unpaired low-light images T_L and normal-light images T_H for training, employing HVITDN for HVI decomposition, CSGAN for HV denoising, uncertainty-driven generators (UnG, UnF) for illumination transfer with discriminators (D_1, D_2), and FIHVITN for reconstruction. (b) The training process of uncertainty-driven generator involves two steps: In 'step1', the mean values of the enhanced intensity component $G(I_L)$ and the recovered intensity component $F(G(I_L))$ are estimated, along with uncertainty parameters Θ^G and Θ^{GF}. In 'step2', Θ^{GF} is utilized to further guide network optimization.

2 Methodology

2.1 Overall Network Architecture

We propose a dedicated unsupervised learning method to enhance low-light images with regions of disparate exposure levels. $P_{data}(T_L) = \{T_{Lk}|k = 1, ..., M\} \subset LL$ and $P_{data}(T_H) = \{T_{Hk}|k = 1, ..., N\} \subset HL$ are unpaired images, where M and N denote the numbers of low-light and normal-light images, and LL and HL denote the sets of the low-light and normal-light images, respectively. To address the challenges of traditional enhancement methods that often produce unnatural color shifts and texture degradation, we incorporate the color space transformation into the image decomposition, apply the cycle-consistent GANs to the low/normal-light images, and propose an attention network to reduce the noise and preserve the texture details. The proposed architecture of UCAMNet for LLIE is shown in Fig. 1(a). To mitigate the inherent coupling between color and luminance while effectively suppressing noise artifacts, we first transform sRGB images into the HVI color space [17]. This transformation enables the decomposition of images into two distinct components: the intensity component containing scene brightness information and the HV color component preserving chromatic and structural details. Subsequently, the LLIE task is decoupled into two subtasks: illumination enhancement and noise removal. We utilize the GANs to learn non-linear mapping between low-light and normal-light domains in the intensity maps as well as propose a channel-spatial guided attention network (CSGAN) to suppress noise in HV color maps.

As shown in Fig. 1(a), T_L and T_H are firstly transformed into the HVI color space through the HVI transformation decomposition network (HVITDN) [17], yielding their corresponding HV color maps HV_L, HV_H, and intensity maps I_L, I_H. Given training samples $P_{data}(I_L) = \{I_{Lk}|k = 1, \ldots\ldots, M\} \subset IL$ and $P_{data}(I_H) = \{I_{Hk}|k = 1, \ldots\ldots, N\} \subset IH$, we aim to learn the mapping function between IL and IH. We formulate the illumination enhancement as learning a mapping $G : I_L \rightarrow I_H$ implemented by UnG, such that the enhanced output $I_{EN} = G(I_L)$ is statistically indistinguishable from the distribution I_H through adversarial training. To address the inherent ambiguity in unpaired learning, we couple this with an inverse mapping $F : I_H \rightarrow I_L$ implemented by UnF and introduce a cycle consistency loss to enforce $F(G(I_L)) \approx I_L$. The uncertainty-driven generators (UnG and UnF) share the same architecture. Thus, G and F should be inverses of each other, and both mappings should be bidirectional. Furthermore, to account for spatially varying illumination characteristics, we incorporate uncertainty constraint for adaptive illumination enhancement, focusing on high-variance regions. In addition, we use 70×70 PatchGANs [24] as discriminators D_1 and D_2 to classify 70×70 overlapping patches of generated intensity maps I_{EN} and real intensity maps I_H as real or fake. The CSGAN restores natural color and removes noise from HV_L and HV_{DE}, producing HV_{EN} and HV_{HRE} respectively. Since HV_H and HV_{EN} are noise-free with natural colors, random noise maps Z sampled from a predefined Gaussian distribution are added to them, where μ and σ^2 obey the uniform distributions $U(-0.3, -0.1)$ and $U(0.0001, 0.001)$ [25], respectively. Finally, HV_{EN} and I_{EN} are concatenated and transformed back to the sRGB color space via the Fusion Inverse HVI Transformation Network (FIHVITN), thereby obtaining the enhanced output T_{EN}.

2.2 HVI Transformation Decomposition Network

The HVI color space [17] is specifically designed to address limitations of traditional color spaces (e.g., sRGB, HSV) in LLIE by reducing color artifacts and noise amplification. The transformation from sRGB involves a nonlinear mapping from sRGB to HSV and an optimization-based projection to HVI color space. The input sRGB images T_L are first transformed to HSV color space, which separate brightness (Value, V) from chromatic information (Hue, H; Saturation, S). The V component is derived from the maximum intensity across sRGB channels, while S reflects color purity relative to brightness, and H encodes the dominant wavelength, determined by the sRGB channel contributing most to the Value. Then, the HSV color space is subsequently optimized to derive the HVI representation, addressing two principal limitations: spectral discontinuity in the Hue component and noise amplification in low-intensity regions. The Hue discontinuity is resolved through polarization into orthogonal horizontal (H') and vertical (V') components using trigonometric decomposition, which enforces smooth transitions across critical wavelength boundaries. Noise suppression is achieved via an intensity collapse function applied to the Value channel V, which simultaneously preserves structural details in high-luminance regions while attenuating noise in dark areas. The final HVI representation is constructed by first deriving the horizontal map $\hat{H}$ and vertical map $\hat{V}$ via element-wise multiplication of H' and V' with S and collapse function, respectively, followed by concatenation of these resulting $\hat{H}$, $\hat{V}$, and the optimized intensity channel map. Thus, T_L are decomposed into HV color component maps HV_L and intensity component maps I_L via HVITDN, enabling targeted enhancement in subsequent processing.

2.3 Uncertainty-Driven Generator

To leverage regional characteristics and prior information during the enhancement, we introduce uncertainty modeling into the generator and propose an uncertainty-driven framework with a two-step training process, aiming to improve the enhancement quality of the intensity component.

The training process of the UCAMNet's generator is divided into two steps, as shown in Fig. 1(b). Based on the cycle consistency loss, the uncertainties Θ^G and Θ^{GF} are estimated in 'step1' of uncertainty-driven generator, and the latter is utilized in the 'step2' to optimize the final mean $F(G(I_L))$. More specifically, in 'step1', the intensity maps of low-light images I_L sequentially pass through the UnG and UnF, so that the mean value of the enhanced intensity maps $G(I_L)$ and the recovered intensity maps $F(G(I_L))$ are accurately estimated, respectively. On the another branch of the network followed by $Tanh$ activation, the variances Θ^G and Θ^{GF} are estimated at the same time as $G(I_L)$ and $F(G(I_L))$, respectively. Subsequently, the cycle consistency loss function can be further constrained by using the variance as the weight. In the 'step2', the variance maps Θ^{GF} serve as the guidance signal to compute a new adaptive weighted cycle consistency loss, refining the network training. Thus, the network focuses on high-variance regions and is more inclined to reduce the prediction errors in these regions to enhance the performance. Notably, the network in 'step2' is initialized using the weights from the pre-trained network of 'step1', and the parameters of other sub-networks are frozen during the training in 'step2'.

2.4 Channel-Spatial Guided Attention Network

To address noise in the decomposed HV color component and enhance image quality, we propose a channel-spatial guided attention network (CSGAN) that integrates Transformer and CNN architectures to capture global and local information. By leveraging the channel-spatial attention, it focuses on critical image regions, using the extracted features to model long-range dependencies.

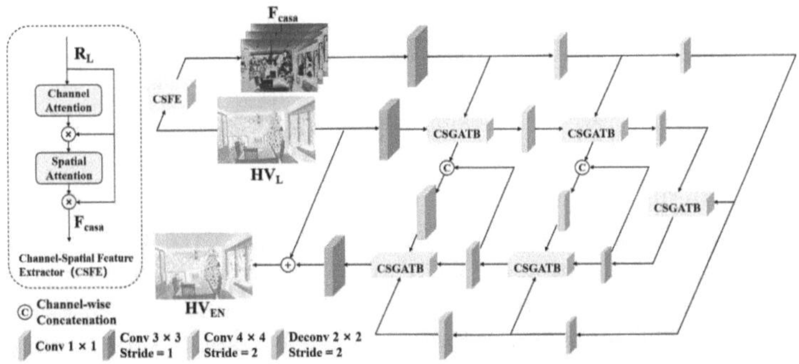

Fig. 2. Illustration of the CSGAN framework. The HV_L inputs are fed into a channel-spatial feature extractor (CSFE) to obtain F_{casa}, which are imported to U-shaped architecture with channel-spatial guided attention Transformer block (CSGATB) to obtain the denoised HV color component maps HV_{EN}.

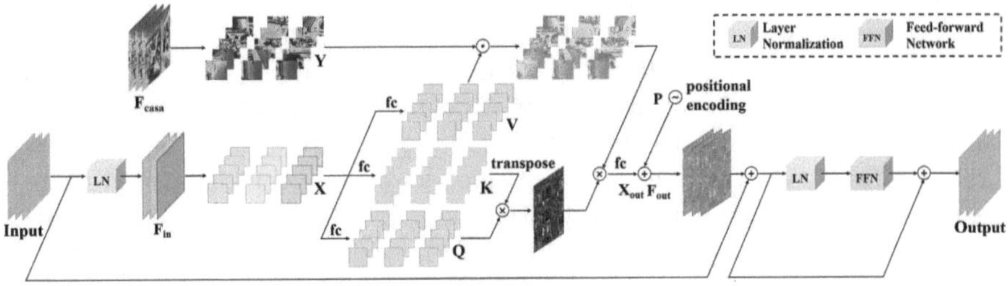

Fig. 3. Illustration of the CSGATB framework. CSGATB is composed of two layer normalization (LN), a feed-forward network (FFN) and a module that uses the features F_{casa} to guide the multi-head self-attention computation.

As shown in Fig. 2, CSGAN takes the HV color maps HV_L from HVITDN as inputs and outputs the denoised map HV_{EN}. The network employs a three-scale U-shaped architecture with skip connections to minimize the information loss. Additionally, the Channel-Spatial Feature Extractor (CSFE) extracts supplementary features F_{casa} from HV_L. These features F_{casa} are then provided as input to the corresponding Channel-Spatial Guided Attention Transformer Block (CSGATB). Within the CSGATB module (Fig. 3), input features first undergo layer normalization (LN) and are then reshaped into tokens X, with each X being divided into j heads: $X = [X_1, X_2, \ldots, X_j]$. Figure 3 shows the case of $j = 1$. For each $head_i$, where $i = 1, 2, \ldots, j$, three unbiased fully connected (fc) layers are utilized to linearly project X_i into query elements Q_i, key elements K_i, and value elements V_i, enabling multi-head self-attention to model long-range dependencies. This mechanism dynamically adjusts weights across sequence positions by computing similarities among Q_i, K_i, and V_i. Meanwhile, F_{casa} are also reshaped into tokens Y and split into j heads: $Y = [Y_1, Y_2, \ldots, Y_j]$, to guide the self-attention computation for HV color map restoration. The self-attention for each $head_i$ is computed as follows:

$$Attention(Q_i, K_i, V_i, Y_i) = (Y_i \odot V_i) softmax(\frac{{K_i}^T Q_i}{\alpha_i}), \tag{1}$$

where $\odot$ denotes the element-wise multiplication, Y_i is the weighted element and α_i is a learnable parameter for adaptive scaling matrix multiplication. Subsequently, j heads are concatenated to be processed by an fc layer, combined with positional encoding and reshaped to produce the output features F_{out}. Finally, the output of the CSGATB is obtained by LN, FFN and residual connection.

2.5 Fusion Inverse HVI Transformation Network

The enhanced intensity component maps I_{EN} and denoised HV color component maps HV_{EN} are first channel-wise concatenated to form the enhanced image in the HVI color space. To convert this image back to sRGB color space, an inverse HVI transformation is applied through the following steps. Intermediate variables are obtained by normalizing the horizontal and vertical components of HVI with the previously used intensity collapse function in HVITDN. These normalized values then enable to estimate the corresponding HSV components. Hue (H) is calculated via the arctangent of the ratio between the normalized vertical and horizontal components. Saturation (S) is derived from the magnitude of these normalized components. The Value (V) component is adjusted using a linear parameter to fine-tune overall brightness, building on the intensity information preserved during the forward transformation. Finally, the derived HSV component maps are converted back to sRGB to obtain T_{EN} using standard HSV-to-sRGB conversion [26].

2.6 Loss Function

In our UCAMNet, we utilize the loss of image recovery, brightness enhancement, and noise removal as the total loss.

a) Loss of Image Recovery: To leverage the complementary information in both HVI and sRGB color spaces, the total image recovery loss combines cycle consistency losses across both domains. It can be expressed as:

$$L_{IR} = \phi_{sRGB} L_{cycT}^{sRGB} + \phi_{HVI} L_{cycT}^{HVI}, \tag{2}$$

where ϕ_{sRGB} and ϕ_{HVI} are weights to balance the respective contributions. We denote the compound loss function $L_{CS}(a, b)$ as the sum of the Charbonnier loss $||\cdot||_{char}$ and SSIM loss $||\cdot||_{ssim}$ as follows,

$$L_{CS}(a, b) = ||a - b||_{char} + ||a - b||_{ssim}, \tag{3}$$

where a and b denote two images. The cycle consistency loss in each color space is then expressed as,

$$L_{cycT}^{sRGB} = L_{CS}(T_{LRE}, T_L) + L_{CS}(T_{HRE}, T_H), \tag{4}$$

$$L_{cycT}^{HVI} = L_{CS}(T_{LRE}^{HVI}, T_L^{HVI}) + L_{CS}(T_{HRE}^{HVI}, T_H^{HVI}), \tag{5}$$

where T_{LRE}^{HVI}, T_L^{HVI}, T_{HRE}^{HVI} and T_H^{HVI} are the corresponding HVI images obtained by applying HVI transformation to the sRGB images T_{LRE}, T_L, T_{HRE} and T_H, respectively.

b) Loss of Brightness Enhancement: The brightness enhancement loss consists of an adversarial loss L_{adv}, an uncertainty-driven cycle consistency loss L_{ucyc} and an uncertainty-driven identity loss L_{uidt}. For the generator mapping functions G, F and their respective discriminators D_1, D_2, the total brightness enhancement loss is formulated as,

$$L_{BE} = w_{adv} L_{adv} + w_{ucyc} L_{ucyc} + w_{uidt} L_{uidt}, \tag{6}$$

where w_{adv},w_{ucyc} and w_{uidt} are weight parameters to balance the respective contributions.

We apply adversarial losses to both mapping functions $G : I_L \rightarrow I_H$ and $F : I_H \rightarrow I_L$. These loss can be respectively expressed as follows,

$$L_{adv}^{G} = (D_2(I_{EN}) - 1)^2, \tag{7}$$

$$L_{adv}^{D_2} = D_2(I_{EN})^2 + (D_2(I_H) - 1)^2, \tag{8}$$

$$L_{adv}^{F} = (D_1(I_{DE}) - 1)^2, \tag{9}$$

$$L_{adv}^{D_1} = D_1(I_{DE})^2 + (D_1(I_L) - 1)^2. \tag{10}$$

Traditionally, the cycle consistency loss is computed via the L_1 norm, which treats each pixel equally. However, we enforce uncertainty constraint to assign higher weights to regions with high variance, the training process is divided into two steps. The corresponding L_{ucyc} is expressed as the following forms:

$$\begin{aligned} L_{ucyc}^{(1)} &= \left(\frac{1}{M}\sum_{k=1}^{M} exp(-u_k^{GF})||I_{LRE k} - I_{L k}||_{char} + 2u_k^{GF}\right) \\ &+ \left(\frac{1}{N}\sum_{k=1}^{N} exp(-u_k^{FG})||I_{HRE k} - I_{H k}||_{char} + 2u_k^{FG}\right), \end{aligned} \tag{11}$$

and

$$L_{ucyc}^{(2)} = \frac{1}{M}\sum_{k=1}^{M} v_k^{GF}||I_{LRE k} - I_{L k}||_{char} + \frac{1}{N}\sum_{k=1}^{N} v_k^{FG}||I_{HRE k} - I_{H_k}||_{char}, \tag{12}$$

where $u_k^{GF} = ln\Theta_k^{GF}$ and $u_k^{FG} = ln\Theta_k^{FG}$, $v_k^{GF} = u_k^{GF} - min(u_k^{GF})$ and $v_k^{FG} = u_k^{FG} - min(u_k^{FG})$ are the non-negative linear scaling functions.

In addition, in order to ensure that the generator's output remains unchanged when the input image belongs to the target domain, we define the corresponding L_{uidt} for the 'step1' and 'step2' of the uncertainty-driven generator as follows,

$$\begin{aligned} L_{uidt}^{(1)} = &(\frac{1}{M}\sum_{k=1}^{M} exp(-u_k^{FL})||F(I_{L k}) - I_{L k}||_{char} + 2u_k^{FL}) \\ &+ (\frac{1}{N}\sum_{k=1}^{N} exp(-u_k^{GH})||G(I_{H k}) - I_{H k}||_{char} + 2u_k^{GH}), \end{aligned} \tag{13}$$

and

$$L_{uidt}^{(2)} = \frac{1}{M}\sum_{k=1}^{M} v_k^{FL}||F(I_{L k}) - I_{L k}||_{char} + \frac{1}{N}\sum_{k=1}^{N} v_k^{GH}||G(I_{H k}) - I_{H k}||_{char}, \tag{14}$$

where $u_k^{FL} = ln\Theta_k^{FL}$ and $u_k^{GH} = ln\Theta_k^{GH}$, Θ_k^{FL} and Θ_k^{GH} denote uncertainty maps from feeding I_{Lk} into UnF and I_{Hk} into UnG, respectively. The terms $v_k^{FL} = u_k^{FL} - min(u_k^{FL})$ and $v_k^{GH} = u_k^{GH} - min(u_k^{GH})$ represent non-negative linear scaling functions.

c) Loss of Noise Removal: The loss function of noise removal consists of the denoising loss $L_{denoise}$ and the color constancy loss L_{color}. The total loss function for the CSGAN is formulated as,

$$L_{NR} = \psi_{denoise} L_{denoise} + \psi_{color} L_{color}, \tag{15}$$

where $\psi_{denoise}$ and ψ_{color} are weight parameters to balance the respective contributions. The denoising loss $L_{denoise}$ aims to ensure the generated HV color map HV_{HRE} resembles the ground truth HV color map, which is formulated as,

$$L_{denoise} = L_{CS}(HV_{HRE}, HV_H). \tag{16}$$

In addition, the color constancy loss L_{color} enforces color consistency between the denoised HV color map and the input HV color map, formulated as,

$$\begin{aligned} L_{color} = &||VGG(Map(HV_{EN})) - VGG(Map(HV_L))||_2 \\ &+ ||VGG(Map(HV_{HRE})) - VGG(Map(HV_{DE}))||_2, \end{aligned} \tag{17}$$

where $VGG(\cdot)$ denotes the extraction of multi-scale features from the input image using a specific layer of the pre-trained vgg19 network, $Map(\cdot)$ represents the channel mapping function. For the HV color component $HV \in \mathbb{R}^{B\times 2\times H\times W}$, $Map(HV) = [HV^H, HV^V, 0] \in \mathbb{R}^{B\times 3\times H\times W}$, where HV^H and HV^V denote the horizontal and the vertical channels of HV, respectively.

3 Experiments

3.1 Experiment Setting

We implement our framework with PyTorch on the NVIDIA RTX 3090 GPU. The model is trained with the Adam optimizer for 100 epochs each step. The learning rate is initially set to 1×10^{-4} and then linearly decays to 0. As for the regularization parameters, we set $\phi_{sRGB} = 10$, $\phi_{HVI} = 5$, $w_{adv} = 1$, $w_{ucyc} = 10$, $w_{uidt} = 5$, $\psi_{denoise} = 1$ and $\psi_{color} = 1$, respectively.

We evaluate our model on three standard LLIE benchmarks, including LOL-v1 dataset (485 image pairs for training and 15 pairs for testing) [11], LOL-v2-

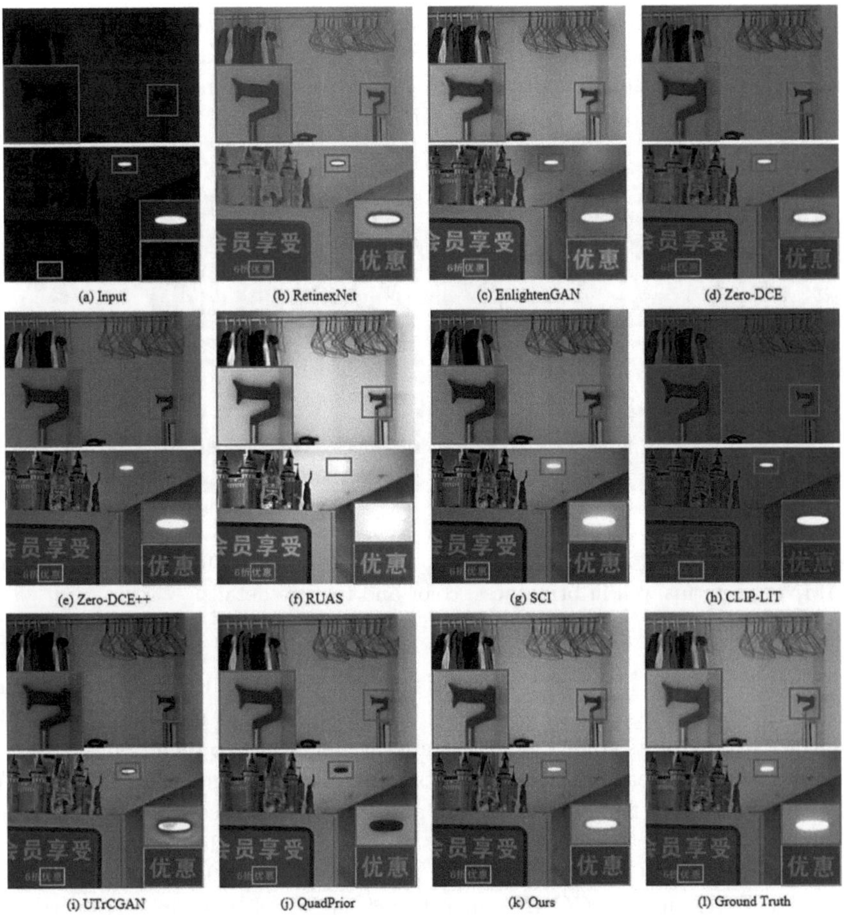

Fig. 4. Visual comparisons on the low-light images from the LOL-v1 dataset (the first row) and LOL-v2-real dataset (the second row). While other methods either generate noisy images, or introduce black spot corruptions, UCAMNet can remove the noise and reconstruct well-exposed image details.

real dataset (689 images for training and 100 pairs for testing) [27] and LOL-v2-syn dataset (900 images for training and 1000 pairs for testing) [27]. It is important to note that while trained on the paired dataset, our model does not require paired low/normal-light images as input during training.

We adopt PSNR, SSIM and No Reference Quality Evaluation (NIQE) as evaluation metrics. The higher PSNR and SSIM values, the better performance, as well as the lower NIQE values, the better performance.

3.2 Comparison with Current Methods

In order to evaluate the performance of our method in LLIE task, quantitative and qualitative results are presented in comparison with several state-of-the-art methods including RetinexNet [11], EnlightenGAN [14], Zero-DCE [12], Zero-DCE++ [28], RUAS [29], SCI [5], CLIP-LIT [30], UTrCGAN [15] and QuadPrior [31], respectively. We obtain the results of other methods by adopting official pre-trained models without retraining from scratch.

Fig. 5. Visual comparisons on the low-light images from the LOL-v2-syn dataset. Our UCAMNet performs well in brightness, color and texture details.

Figure 4 and Fig. 5 present comparisons between the UCAMNet and nine state-of-the-art methods on three paired datasets. It can be seen that the results of RetinexNet exhibits color deviations and artifacts due to its emphasis on texture enhancement. Zero-DCE++ exhibits apparent color distortion, while RUAS and CLIP-LIT suffer from overexposure and severe noise, respectively. Although EnlightenGAN preserves texture details well in LOL-v1 dataset, it introduces visible artifacts in LOL-v2-real dataset and color bias in LOL-v2-syn dataset. Both Zero-DCE and UTrCGAN generate artifacts and noise in certain regions of both LOL-v1 and LOL-v2-real datasets. SCI results in significant overexposure in LOL-v2-syn dataset. While QuadPrior achieves superior performance in texture preservation and noise removal across all datasets, it still presents illumination inconsistencies in LOL-v2-real dataset. Overall, our method performs well on all three datasets in different aspects, including texture details preservation, brightness and color restoration, as well as suppressing artifacts and noise.

The quantitative comparison results are listed in Table 1. As shown in the table, the average performance of our proposed UCAMNet method has almost reached the optimal level when compared with all competing methods.

Table 1. Quantitative results on LOL-v1, LOL-v2-real, and LOL-v2-syn datasets. The best and second best scores are **highlighted** and underlined, respectively.

Methods	LOL-v1			LOL-v2-real			LOL-v2-syn		
	PSNR↑	SSIM↑	NIQE↓	PSNR↑	SSIM↑	NIQE↓	PSNR↑	SSIM↑	NIQE↓
RetinexNet [11]	16.77	0.425	9.728	17.67	0.640	5.090	17.30	0.738	4.778
EnlightenGAN [14]	17.48	0.652	4.888	18.64	0.677	5.504	16.57	0.772	3.831
Zero-DCE [12]	14.86	0.562	8.223	18.06	0.580	8.767	17.76	**0.814**	4.357
Zero-DCE++ [28]	14.71	0.485	5.900	17.96	0.583	6.145	17.58	0.811	4.259
RUAS [29]	16.41	0.503	5.927	15.33	0.493	6.172	13.40	0.640	5.092
SCI [5]	14.78	0.525	8.374	17.30	0.540	8.739	15.43	0.744	4.652
CLIP-LIT [30]	12.39	0.493	8.791	15.15	0.533	9.222	16.19	0.772	4.690
UTrCGAN [15]	18.56	0.689	4.774	19.74	0.701	5.078	15.99	0.750	4.707
QuadPrior [31]	18.33	0.767	4.642	20.53	0.797	4.821	16.08	0.756	4.150
Ours	**21.32**	**0.790**	**3.516**	**21.30**	**0.809**	**4.185**	**18.85**	0.799	**3.423**

3.3 Ablation Study

We conduct an ablation study for our proposed network on the LOL-v1 dataset to further validate the effectiveness of uncertainty constraint (UC) and CSGAN. As listed in Table 2, we consider ablation setups including 'M0', 'M1', 'M2' and ours, which denote as 'UCAMNet w/o both UC and CSGAN', 'UCAMNet w/o CSGAN', 'UCAMNet w/o UC', and our UCAMNet, respectively. Qualitative results of all settings are presented in Fig. 6. It can be observed that the 'M0' produces results with noticeable noise and loss of detail in HV color component map, as well as low contrast and blur in the intensity component map. The brightness contrast and the denoising effect can be improved by the 'M1' and 'M2' via UC and CSGAN, respectively. Our UCAMNet can effectively suppress the noise, preserve structural details, recover the color fidelity and improve the brightness due to the integration of the uncertainty constraint and CSGAN.

Table 2. Quantitative evaluation on the enhanced results obtained from different settings. The best scores are **highlighted**.

Model	UC	CSGAN	PSNR↑	SSIM↑	NIQE↓
M0			19.25	0.624	8.187
M1	✓		19.98	0.671	7.803
M2		✓	20.16	0.730	5.296
Ours	✓	✓	**21.32**	**0.790**	**3.516**

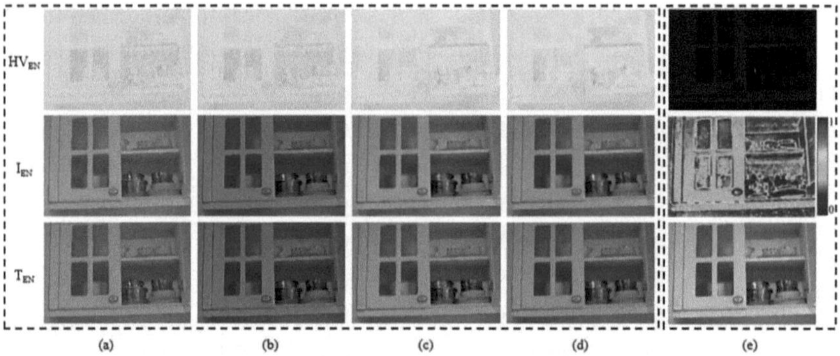

Fig. 6. Visual results of ablation study on the LOL-v1 dataset. (a)M0, (b)M1, (c)M2, and (d)ours. The rows of (a) to (d) sequentially show the results of HV color component HV_{EN}, intensity component I_{EN}, and image T_{EN}, respectively. From top to bottom, (e) shows the input, our uncertainty (variance) map, and the ground truth, respectively.

4 Conclusion

In this paper, we propose UCAMNet, an unsupervised low-light image enhancement framework designed to address two major limitations of existing methods, including inadequate region-wise uncertainty modeling and color artifacts/noise caused by coupling of color and brightness in the sRGB color space. By converting images from sRGB to HVI color space, our UCAMNet effectively decouples color and brightness information, thereby improving the illumination while maintaining the color authenticity. We incorporate an uncertainty constraint to enhance the intensity component, leveraging variance estimation to enhance adaptability and robustness. In addition, we introduce CSGAN to suppress noise and preserve the details. Extensive experiments demonstrate that the UCAMNet achieves competitive performance in both visual quality and objective metrics.

Acknowledgement. This work was supported by Beijing Natural Science Foundation (Project No. 4252028, 4232017), the National Natural Science Foundation of China under Grant 61906009, and the NSF of Fujian Province under Grant No. 2023J01083.

References

1. Jingchun, Z., Su, G.E., Sunar, M.S.: Low-light image enhancement: a comprehensive review on methods, datasets and evaluation metrics. J. King Saud Univ.-Comput. Inf. Sci. **36**(10), 102234 (2024)
2. Chen, L.-C., Zhu, Y., Papandreou, G., Schroff, F., Adam, H.: Encoder-decoder with atrous separable convolution for semantic image segmentation. In: Proceedings of the European Conference on Computer Vision (ECCV), pp. 801–818 (2018)
3. Islam, M.J., et al.: Semantic segmentation of underwater imagery: dataset and benchmark. In: 2020 IEEE/RSJ International Conference on Intelligent Robots and Systems (IROS), pp. 1769–1776. IEEE (2020)
4. Deng, J., Guo, J., Xue, N., Zafeiriou, S.: Arcface: additive angular margin loss for deep face recognition. In: Proceedings of the IEEE/CVF Conference on Computer Vision and Pattern Recognition, pp. 4690–4699 (2019)
5. Ma, L., Ma, T., Liu, R., Fan, X., Luo, Z.: Toward fast, flexible, and robust low-light image enhancement. In: Proceedings of the IEEE/CVF Conference on Computer Vision and Pattern Recognition, pp. 5637–5646 (2022)
6. Cui, Z., Qi, G.-J., Gu, L., You, S., Zhang, Z., Harada, T.: Multitask aet with orthogonal tangent regularity for dark object detection. In: Proceedings of the IEEE/CVF International Conference on Computer Vision, pp. 2553–2562 (2021)
7. Du, Z., Shi, M., Deng, J.: Boosting object detection with zero-shot day-night domain adaptation. In: Proceedings of the IEEE/CVF Conference on Computer Vision and Pattern Recognition, pp. 12666–12676 (2024)
8. Pizer, S.M., et al.: Adaptive histogram equalization and its variations. Comput. Vis. Graph. Image Process. **39**(3), 355–368 (1987)
9. Guo, X., Li, Y., Ling, H.: Lime: low-light image enhancement via illumination map estimation. IEEE Trans. Image Process. **26**(2), 982–993 (2016)
10. Li, M., Liu, J., Yang, W., Sun, X., Guo, Z.: Structure-revealing low-light image enhancement via robust retinex model. IEEE Trans. Image Process. **27**(6), 2828–2841 (2018)
11. Wei, C., Wang, W., Yang, W., Liu, J.: Deep retinex decomposition for low-light enhancement, arXiv preprint arXiv:1808.04560 (2018)
12. Guo, C., et al.: Zero-reference deep curve estimation for low-light image enhancement. In: Proceedings of the IEEE/CVF Conference on Computer Vision and Pattern Recognition, pp. 1780–1789 (2020)
13. Wu, W., Weng, J., Zhang, P., Wang, X., Yang, W., Jiang, J.: Uretinex-net: retinex-based deep unfolding network for low-light image enhancement. In: Proceedings of the IEEE/CVF Conference on Computer Vision and Pattern Recognition, pp. 5901–5910 (2022)
14. Jiang, Y., et al.: Enlightengan: deep light enhancement without paired supervision. IEEE Trans. Image Process. **30**, 2340–2349 (2021)
15. Guan, J., Qi, N., Zhu, Q., Chen, L.: Utrcgan: uncertainty-driven cycle-consistent generative adversarial network for low-light image enhancement. In: 2024 IEEE International Conference on Image Processing (ICIP), pp. 1473–1479. IEEE (2024)
16. Zhang, Y., Zhang, J., Guo, X.: Kindling the darkness: a practical low-light image enhancer. In: Proceedings of the 27th ACM International Conference on Multimedia, pp. 1632–1640 (2019)
17. Yan, Q., et al.: HVI: a new color space for low-light image enhancement. In: Proceedings of the Computer Vision and Pattern Recognition Conference, pp. 5678–5687 (2025)

18. Guo, S., Yan, Z., Zhang, K., Zuo, W., Zhang, L.: Toward convolutional blind denoising of real photographs. In: Proceedings of the IEEE/CVF Conference on Computer Vision and Pattern Recognition, pp. 1712–1722 (2019)
19. Tian, C., Xu, Y., Zuo, W.: Image denoising using deep CNN with batch renormalization. Neural Netw. **121**, 461–473 (2020)
20. Vaswani, A., et al.: Attention is all you need. In: Advances in Neural Information Processing Systems, vol. 30 (2017)
21. Xu, X., Wang, R., Fu, C.-W., Jia, J.: SNR-aware low-light image enhancement. In: Proceedings of the IEEE/CVF Conference on Computer Vision and Pattern Recognition, pp. 17714–17724 (2022)
22. Wang, T., Zhang, K., Shen, T., Luo, W., Stenger, B., Lu, T.: Ultra-high-definition low-light image enhancement: a benchmark and transformer-based method. In: Proceedings of the AAAI Conference on Artificial Intelligence, vol. 37, pp. 2654–2662 (2023)
23. Cai, Y., Bian, H., Lin, J., Wang, H., Timofte, R., Zhang, Y.: Retinexformer: one-stage retinex-based transformer for low-light image enhancement. In: Proceedings of the IEEE/CVF International Conference on Computer Vision, pp. 12504–12513 (2023)
24. Isola, P., Zhu, J.-Y., Zhou, T., Efros, A.A.: Image-to-image translation with conditional adversarial networks. In: Proceedings of the IEEE Conference on Computer Vision and Pattern Recognition, pp. 1125–1134 (2017)
25. Wu, K., Huang, J., Ma, Y., Fan, F., Ma, J.: Cycle-retinex: unpaired low-light image enhancement via retinex-inline cyclegan. IEEE Trans. Multimed. **26**, 1213–1228 (2023)
26. Foley, J.D., Van Dam, A.: Fundamentals of Interactive Computer Graphics. Addison-Wesley Longman Publishing Co., Inc. (1982)
27. Yang, W., Wang, W., Huang, H., Wang, S., Liu, J.: Sparse gradient regularized deep retinex network for robust low-light image enhancement. IEEE Trans. Image Process. **30**, 2072–2086 (2021)
28. Li, C., Guo, C., Loy, C.C.: Learning to enhance low-light image via zero-reference deep curve estimation. IEEE Trans. Pattern Anal. Mach. Intell. **44**(8), 4225–4238 (2021)
29. Liu, R., Ma, L., Zhang, J., Fan, X., Luo, Z.: Retinex-inspired unrolling with cooperative prior architecture search for low-light image enhancement. In: Proceedings of the IEEE/CVF Conference on Computer Vision and Pattern Recognition, pp. 10561–10570 (2021)
30. Liang, Z., Li, C., Zhou, S., et al.: Iterative prompt learning for unsupervised backlit image enhancement. In: Proceedings of the IEEE/CVF International Conference on Computer Vision, pp. 8094–8103 (2023)
31. Wang, W., Yang, H., Fu, J., et al.: Zero-reference low-light enhancement via physical quadruple priors. In: Proceedings of the IEEE/CVF Conference on Computer Vision and Pattern Recognition, pp. 26057–26066 (2024)

DPC-FCNet: A Dual-Channel Cross-Modality Person re-Identification Network with Enhanced Multi-Level Feature Correlation

Guang Huo and Yue Wang(✉)

Northeast Electric Power University, Jilin Jilin 132000, China
2202301036@neepu.edu.cn

Abstract. Cross-modality person re-identification aims to match the same person captured by different sensor cameras, but heterogeneous images exhibit significant modality differences. To address the problem of insufficient correlation between different network layers and inconsistent discriminative feature regions in cross-modality feature extraction, this paper proposes a Dual-Path Cross-Level Feature Correlation Network (DPC-FCNet) that enhances multi-level feature correlation. Under the synergistic effect of shallow independent modeling and deep-level fusion, the network significantly improves the discriminative power and correlation of cross-modality features. A dual-modal collaborative enhancement strategy was adopted to improve the adaptability of the model to cross-modal color shifts. Regarding the network design, independent convolution kernels in the dual-path architecture extract modality-specific features at shallow levels, retaining differentiated information. At deeper levels, a channel-space dual attention module was introduced to strengthen cross-modality semantic correlations in a hierarchical manner. Simultaneously, learnable gating parameters dynamically balance the original features and attention features, with residual connections ensuring a stable gradient propagation. The experimental results show significant improvements in the rank and mAP values on the benchmark datasets SYSU-MM01 and RegDB. This design provides a new approach for cross-modality interaction and opens new research directions in the field of multi-modal joint representation, with its dual-path associative learning and dynamic fusion strategy.

Keywords: Cross-modality · Person re-identification · Feature correlation

1 Introduction

As a cutting-edge intelligent analysis technique in computer vision, Person Re-identification (ReID) achieves precise identification of target subjects across non-overlapping camera networks through feature matching in visual data streams [1]. With the accelerated advancement of smart city infrastructures, this technology has evolved into a fundamental component of modern security ecosystems, driving innovations in three critical domains: large-scale video analytics, smart surveillance configuration, and cross-system trajectory reconstruction.

J. Lokoč et al. (Eds.): MMM 2026, LNCS 16413, pp. 319–331, 2026.
https://doi.org/10.1007/978-981-95-6957-1_23

In nighttime scenarios, visible light cameras struggle to capture effective feature information, while infrared cameras can capture pedestrian features in low-light environments. As a result, many researchers have started to explore how to match and retrieve across two modalities, leading to the emergence of cross-modal person re-identification [2]. Cross-modality re-identification faces more complex challenges, as it not only needs to address issues such as low resolution, occlusion [3], and viewpoint changes within each modality but also needs to cope with the significant differences between images captured by the two modalities.

Currently, the main method for reducing modality differences in cross-modal person re-identification is to extract modality-shared features. This is achieved by designing feature learning networks to learn the shared features between infrared and visible light images. To this end, Park et al. [4] processed cross-modality differences at the pixel level to more effectively suppress modality-related features from person representations, enhancing pixel-level associations between cross-modality local features, thus further promoting feature learning for VI-reID. Murali et al. [5] designed a two-stage feature compensation network that uses adversarial learning to generate basic features containing both modalities. Song et al. [6] proposed the Channel-Enhanced Cross-Modality Relationship Network (CECR-Net), which designed two relationship network-based modules. These two modules capture the structural relationships between different body parts, breaking the isolation between local features.

However, due to the significant modality differences between VIS and IR images, existing methods do not fully mine the shared information between the two modalities. To address the issue of ineffective integration of multi-level channel feature information and spatial feature information in existing methods, this paper proposes a Dual-Path Cross-Level Feature Correlation Network (DPC-FCNet), which enhances multi-level feature correlations. The specific contributions are as follows:

- ·By utilizing a dual-modal collaborative enhancement strategy, the discriminatory dimensions of cross-modality features are effectively enriched, providing important support for improving cross-domain retrieval accuracy.
- ·A dual-path co-evolutionary network architecture is first designed. The core breakthrough of this architecture lies in revealing the collaborative optimization mechanism of shallow independent modeling and deep association fusion: the shallow network focuses on the refined extraction of modal-specific features, while the deep network vigorously digs and strengthens the deep and highly discriminating internal association between cross-modal features through innovative association fusion mechanism. This mechanism fundamentally overcomes the key bottleneck that it is difficult to optimize the correlation and discrimination of cross-modal features in the existing methods, and realizes the revolutionary improvement of cross-modal identity recognition accuracy.
- ·A dual-dimensional feature extraction module, Dual Focus Attention (DFA), is proposed. It achieves fine-grained feature interaction through a dual-dimensional channel-space attention mechanism and adaptive gating, thus improving the model's feature extraction ability.

2 Related Works

2.1 Single-Modal Person Re-Identification.

Initial investigations in person re-identification predominantly relied on hand-engineered feature descriptors. Farenzena et al. [7] introduced a biomechanical symmetry-guided region decomposition framework, which minimized background interference through precise foreground-background isolation using anthropometric priors. Nevertheless, constrained representation capacity of manual features led to inevitable performance plateau. The paradigm shift towards deep learning architectures has revolutionized retrieval paradigms through structural innovation and loss function optimization. Pioneering this transition, Sun et al. [8] developed a Component-Aware Neural Architecture (PCB) incorporating multi-granularity feature alignment mechanisms. Specifically, their designed recurrent part refinement (RPP) module enabled adaptive spatial-semantic consistency across fragmented body regions, substantially improving discriminative feature learning robustness.

2.2 Cross-Modality Person Re-Identification.

Visible-infrared cross-modal person re-identification has emerged as a pivotal research direction in intelligent surveillance systems. Ye et al. [9] developed a Hierarchical Feature Learning Framework (DDAG) that implements region-aware attention mechanisms through feature space decomposition, enhancing discriminative capability by capturing modality-specific regional patterns. To address cross-modal discrepancies, Wu et al. [10] presented a Dual-alignment Network (MPANet) that simultaneously optimizes modality consistency while preserving discriminative cross-spectrum characteristics through contrastive embedding learning. Furthermore, Wu et al. [11] designed a Cross-dimensional Interaction Module (FAM) incorporating spatial-semantic dependency modeling and channel-wise correlation enhancement, establishing robust feature associations across heterogeneous modalities.

3 Proposed Method

3.1 Overall Solution of DPC-FCNet.

The proposed DPC-FCNet network consists of four main parts: data augmentation, dual-stream input, feature extraction, and loss functions. The overall network architecture is shown in Fig. 1.

The single-branch network design performs cross-modal feature extraction by mixing infrared and visible light images. This indiscriminate processing approach, while constructing common features between modalities, inevitably weakens the modality-specific discriminative features during the feature extraction process. The lack of these low-level features negatively impacts the performance of the network.

Therefore, this paper designs a dual-channel feature extraction network that enhances multi-level feature correlations to separately process the data streams of visible light images and infrared images. First, the images undergo data augmentation and then enter

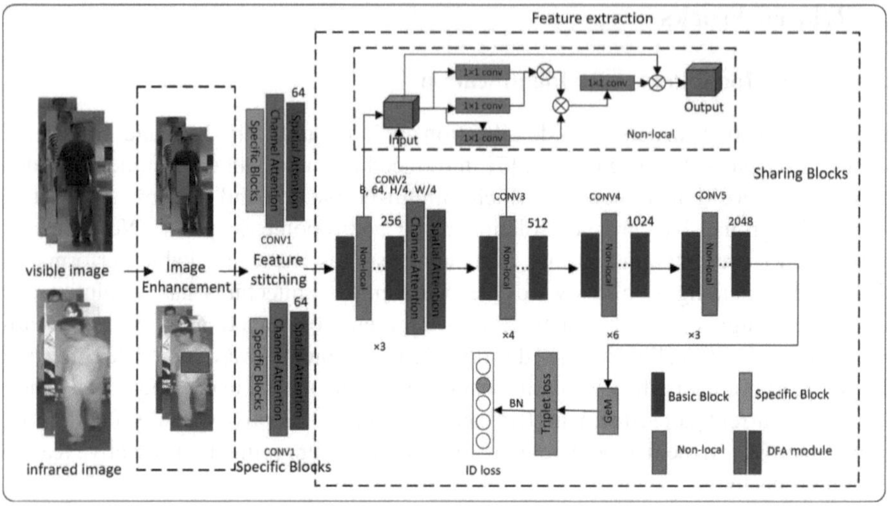

Fig. 1. Overall network architecture diagram.

the dual-stream network. In the proposed Feature Correlation Network (DPC-FCNet), a Dual-Path Independent Kernels design is adopted to separately extract the unique shallow features of infrared and visible light modalities, allowing the initial convolutional layer (Conv1) to preserve modality-specific information while laying the foundation for differentiated features in subsequent layers. The initial convolutional layer is shown in Fig. 2.

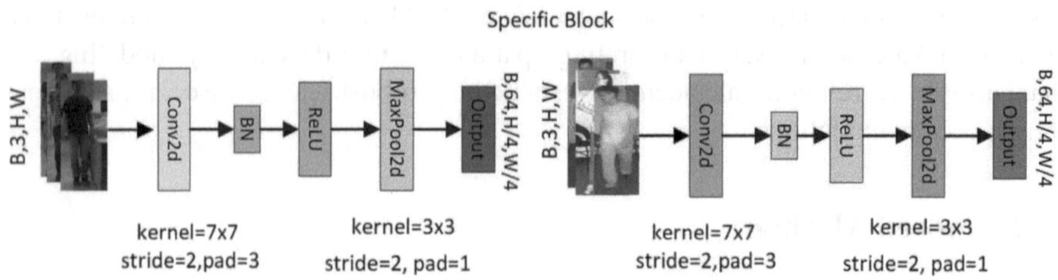

Fig. 2. Schematic of the initial convolution layer.

The second layer (Conv2) to the fifth layer (Conv5) achieves multi-level feature interaction through shared cross-channel convolution modules (Cross-Channel Shared Convolution Blocks). However, due to the significant differences between the infrared and visible light modalities, it is insufficient to solely rely on the network's learning capability to explore the relationship between the infrared and visible light modalities. To address this issue, this paper proposes the Dual Focus Attention (DFA) module, which is added after the initial convolution layer and the second layer of the shared network. This design not only facilitates feature alignment between the two modalities in the deep semantic space but also enhances the cross-level correlation constraints (achieved through the DFA module and enhanced cross-modality gating), allowing the network to adaptively capture complementary features between the different modalities. To enable the model to capture more detailed information, non-local blocks are inserted into the last two residual blocks of the second layer and the last three residual blocks of the third layer. The structure combines shallow independent modeling with deep relational fusion, significantly improving the correlation and discriminability of cross-modality features while preserving modality-specific information. After feature extraction, adaptive triplet loss and id loss are used for end-to-end training of the model. The id loss, as shown in Formula 1, is used for classification tasks to optimize identity consistency across modalities.

$$L_{id} = -\frac{1}{N}\sum_{i=1}^{N} y_i \log(p_i) \tag{1}$$

Here, y_i represents the encoded true label, and p_i is the predicted probability distribution by the model (normalized by softmax).

TripletLoss_ADP is shown in Formula 2.

$$L_{tri} = \frac{1}{N}\sum_{i=1}^{N}\left[\max\left(\| f_a - f_p \|^2 - \| f_a - f_n \|^2 + \alpha \cdot \left(\frac{\| f_a - f_p \|}{\| f_a - f_n \|}\right)^{\gamma}, 0\right)\right] \tag{2}$$

Here, *fa* is the feature of the anchor sample, *fp* is the feature of the positive sample, and *fn* is the feature of the negative sample. α and γ are hyperparameters used to dynamically adjust the weight of hard samples. The final loss is the weighted sum of the above two losses and the KL divergence loss.

$$L_{kl} = \frac{1}{N}\sum_{i=1}^{N}\sum_{c=1}^{C} P_c(x_i)\log\left(\frac{P_c(x_i)}{Q_c(x_i)}\right) \tag{3}$$

The KL divergence loss is shown in Formula 3. Here, P and Q are the probability distributions output by the two modalities (which need to be normalized through softmax). The weight is controlled by args.kl. The final loss is the weighted sum of the three losses mentioned above.

3.2 Data Augmentation.

In the task of cross-modal pedestrian recognition, data enhancement has been proved to be an important link to improve the model feature expression ability. Occlusion is inevitable in the real scene and color deviation is caused by illumination and other transformations. In order to reduce the big error in the final matching result caused by the model's excessive attention to the sample color and the contour characteristics of the occlusion, this paper adopts a dual-mode collaborative enhancement strategy: firstly, the visible light image is enhanced by dynamic region shielding in the preprocessing stage, and the complex occlusion scene is simulated by randomly erasing local texture information, and the erasing schematic diagram is shown in Fig. 3; At the same time, the single-channel infrared image is extended to three channels to adapt to the input dimension of the model. In particular, by adopting the channel enhancement joint learning strategy [12] and dynamically adjusting the color channel response characteristics, not only the adaptability of the model to cross-modal color shift is significantly improved, but also a variety of modal transfer sample pairs are constructed at the training sample level.

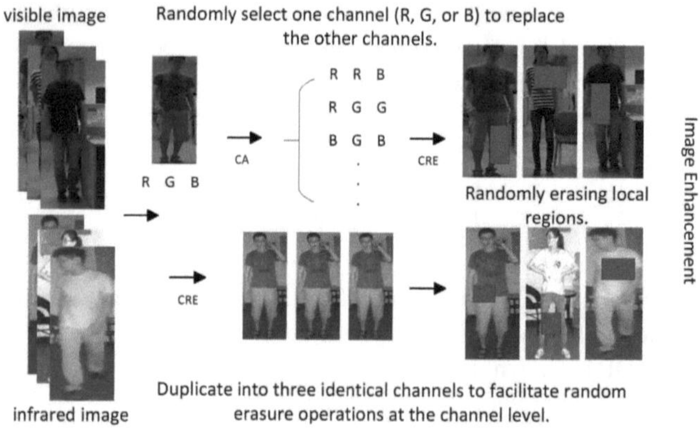

Fig. 3. Data augmentation diagram.

3.3 Dual Focus Attention Module.

In the task of cross-modal pedestrian re-recognition, due to the significant differences (texture details, edge sharpness, etc.) between infrared mode and visible mode, as well as the changes of target scale, posture and background complexity under different shooting conditions, the baseline network framework often faces the problem of insufficient modal-specific feature extraction.

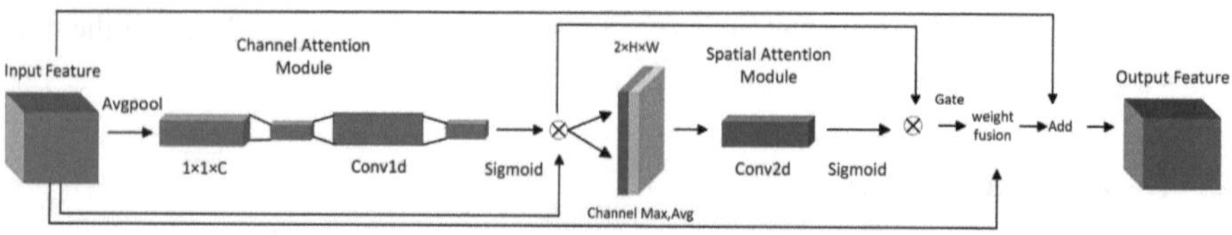

Fig. 4. Structure diagram of the DFA module.

In order to obtain more comprehensive image feature information of the target at different levels and realize the integration of context information, this paper designs and proposes a dual-focus attention module (DFA) for the first time, which can extract and enhance the fine-grained features of the sample image in two dimensions of channel and space. After adding it to the first layer of the special layer of the backbone architecture and the first layer of the shared layer, it is found that the network can better capture the fine-grained features of the sample image and improve the detection and matching accuracy of the model. The structure diagram of DFA module is shown in Fig. 4.

The advantages of the DFA module compared with other mainstream feature extraction modules: 1) A 1D convolution (kernel_size = 3) is used to replace the fully connected layer, significantly reducing the number of parameters while maintaining the adaptive channel interaction capability and balancing local and global information. The channel attention calculation formula is shown in Eq. 4.

$$\mathbf{y}_c = \sigma\left(\mathrm{Conv1D}_{k=3}\left(\mathrm{GAP}(\mathbf{X})^{\top}\right)\right)^{\top} \tag{4}$$

In the formula, GAP(·) represents global average pooling, Conv1Dk = 3 is the 1D convolution with a kernel size of 3, and σ is the Sigmoid activation function.

2. The dual-path (average pooling + max pooling) feature fusion is maintained, using a 7 × 7 convolution to capture large-range spatial context, and Xavier initialization is employed to improve stability. The spatial attention calculation is shown in Eq. 5. In the formula, AvgPoolc and MaxPoolc represent the average pooling and max pooling along the channel dimension, respectively, [·;·] denotes channel concatenation, and Conv2D7 × 7 is the 7 × 7 convolution kernel.

$$\mathbf{y}_s = \sigma\left(\mathrm{Conv2D}_{7\times7}\left(\left[\mathrm{AvgPool}_c(\mathbf{X});\mathrm{MaxPool}_c(\mathbf{X})\right]\right)\right) \tag{5}$$

3. Learnable parameters are used to dynamically balance the original features and attention features, achieving efficient and stable cross-modality feature fusion. The gating fusion formula is shown in Eq. 6. In the formula, λ is the learnable gating parameter, and Xatt represents the features processed by attention.

$$\mathbf{X}' = \lambda \cdot \mathbf{X}_{\mathrm{att}} + (1-\lambda)\cdot \mathbf{X} \tag{6}$$

λ enables the network to automatically adjust the optimal fusion ratio of attention features and original features (usually evolving from the initial value of 0.5 to the task optimal value between 0.3 and 0.8) according to the task requirements, data characteristics and hierarchical depth. Compared with fixed λ, it improves the model capacity and generalization ability and reduces the cost of hyperparameter tuning.

4 Experiments

4.1 Datasets and Evaluation Indicators.

All experiments in this study were conducted under the same environmental setup, using an RTX 3090 Ti GPU. The input image size was adjusted to 288 × 144 [12]. The model training employed the Stochastic Gradient Descent (SGD) optimizer, with an initial learning rate of 0.1 for both datasets. The learning rate decayed to 0.01 at the 20th epoch and to 0.001 at the 50th epoch, with a total of 100 epochs for training. Each training batch was constructed by randomly sampling 8 pedestrian identities, with 4 visible light images and 4 infrared images selected to ensure modality balance.

The datasets used in this study are the widely recognized SYSU-MM01 and RegDB datasets in the field of cross-modality person re-identification. The SYSU-MM01 dataset was collected by surveillance cameras deployed in various environments. Cameras 1, 2, 4, and 5 capture visible light images, while cameras 3 and 6 capture infrared images. The RegDB dataset was organized and released by Tien et al. This dataset contains images of 412 pedestrians collected in open-scene settings.

The evaluation metrics mainly include the Rank-K metric and Mean Average Precision (mAP). The Rank-K evaluation metric calculates the feature similarity between the query pedestrian image and all samples in the database, ranks the candidate results by matching scores from high to low, and then computes the probability that the target pedestrian is included in the top n returned results. Rank-K can be calculated using the following formula:

$$Rank\text{-}K = \frac{1}{N_q}\sum_{i=1}^{N_q} r_i(k) \tag{7}$$

4.2 Analysis of Ablation Experimental Results.

To visually demonstrate the effect of data augmentation, a comparative experiment was conducted, and the results are shown in Table 1. Among them, CA represents the random channel swapping augmentation, and CRE represents the random channel erasure experiment. The experimental results indicate that the data augmentation significantly improves the retrieval performance of cross-modal person re-identification.

Table 1. Ablation Experiment of Data Augmentation.

Methods	Rank-1	Rank-10	Rank-20	mAP
AGW[12]	47.5%	-	-	47.6%
AGW + CRE	60.1%	89.6%	95.0%	58.6%
AGW + CA	54.2%	-	-	51.5%
AGW + CA + CRE	**69.2%**	**96.4%**	**99.5%**	**66.7%**

Due to the significant differences between the infrared modality and the visible light modality, relying solely on the learning capability of the basic network to explore the relationship between the infrared and visible light modalities is insufficient. To demonstrate the effectiveness of the DFA module, horizontal experiments were conducted by embedding different modules. All experiments were performed under the same configuration environment, with the modules embedded in the first layer of the shared layer. The Baseline refers to the performance results reproduced under the same device environment. Table 2 provides a detailed comparison of the performance of different modules.

To visually demonstrate the role of the DFA module, Grad-Cam technology is used to generate heatmaps. The darker the color on the heatmap, the higher the model's attention to the features in that region. To further validate the effectiveness of integrating

Table 2. Performance Comparison of Baseline with Different Embedded Modules.

Methods	Rank-1	Rank-10	Rank-20	mAP
Baseline	69.2%	96.4%	99.5%	66.7%
Baseline+SENet	69.6%	95.7%	98.4%	65.8%
Baseline+CBAM	64.1%	94.2%	98.0%	60.9%
Baseline+EMA	68.6%	94.4%	97.9%	66.1%
Baseline + DFA	**72.2%**	**97.6%**	**99.7%**	**68.7%**

the DFA mechanism into the Baseline(AGW + Data Augmentation) [9] network for recognizing body parts, images were selected from the test set for evaluation, as shown in Fig. 5. Experimental results show that after incorporating the DFA module, the network significantly enhanced its focus on the pedestrian torso region. This effectively improved the comprehensiveness of feature extraction and strengthened the ability to learn discriminative features.

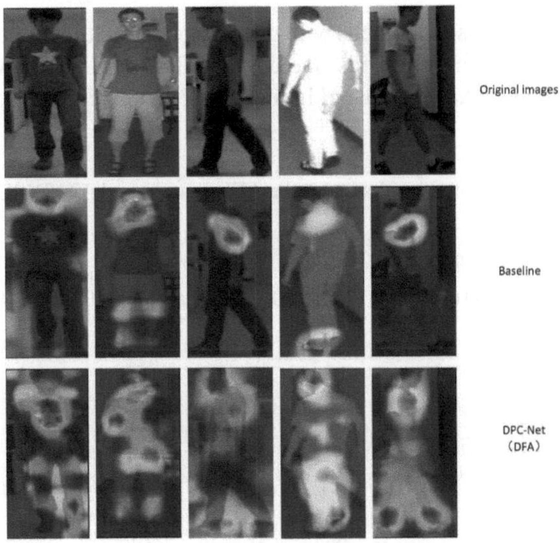

Fig. 5. Comparison image after integrating the module.

Ablation studies validate the role of multi-level feature correlation and optimize the embedding position of the DFA module. As shown in Table 3, the best retrieval performance is achieved when the module is embedded after the initial convolution layer (Conv1) and the second layer of the shared network (Conv2). Since the performance of conv4 and conv5 has declined, no comparison will be made between them. The improvements made at various stages of the network have enhanced the retrieval capability of cross-modality person re-identification, thereby improving the model's retrieval accuracy.

Table 3. Optimization of DFA Module Embedding Position (1–5 correspond to conv1-conv5).

1	2	3	4	5	Rank-1	Rank-10	Rank-20	mAP
√					70.6%	95.8%	98.5%	67.3%
	√				70.2%	95.6%	98.3%	66.7%
		√			66.9%	95.3%	98.2%	64.5%
	√			√	63.4%	94.2%	97.9%	61.2%
	√	√			67.8%	95.4%	98.4%	64.9%
√	√				**74.9%**	**98.3%**	**99.7%**	**70.8%**

4.3 Analysis of Comparison Experimental Results.

The performance comparison methods with the network structure proposed in this paper include seven deep learning methods, including AGW. The SYSU-MM01 dataset is evaluated in two modes: All-Search and Indoor-Search, with comparison data shown in Table 4 and Table 5.

Table 4. Experimental results under the ALL-search mode on SYSU-MM01.

Methods	All Search			
	Rank-1(%)	Rank-10(%)	Rank-20(%)	mAP(%)
AGW[13]	47.5	84.3	92.1	47.6
CM-NAS[14]	60.8	92.1	96.8	58.9
PMT[15]	67.5	95.3	98.6	64.9
HAT[16]	55.2	92.1	97.3	53.8
SMCL[17]	67.4	92.9	96.8	61.8
DART[18]	68.7	96.4	99.0	66.3
CAJ[9]	68.8	95.1	98.3	66.8
DPC-FCNet	**74.9**	**98.3**	**99.7**	**70.8**

As shown in Table 4, the proposed network model achieves 74.9% on the evaluation metric rank-1, with all rank values in the experiment being the average of ten tests. Compared to the best method in the table, rank-1 improved by 6.1%, while the mean Average Precision (mAP) increased by 4%, indicating the effectiveness of the proposed network in retrieval accuracy.

Table 5. Experimental results under the Indoor-search mode on SYSU-MM01.

Methods	Indoor Search			
	Rank-1(%)	Rank-10(%)	Rank-20(%)	mAP(%)
AGW[13]	54.1	91.1	95.9	62.9
CM-NAS[14]	68.0	94.8	97.9	52.4
PMT[15]	71.6	96.7	99.2	76.5
HAT[16]	62.1	95.7	99.2	69.3
SMCL[17]	68.8	96.6	98.8	75.6
DART[18]	72.5	97.8	99.5	78.2
CAJ[9]	76.2	95.1	98.3	80.3
DPC-FCNet	**82.3**	**98.5**	**99.8**	**84.7**

As shown in Table 5, in the indoor mode of the SYSU-MM01 dataset, DPC-FCNet achieves 81.3% on the rank-1 evaluation metric, the highest among the listed methods. Compared to the best method in the table, rank-1 improved by 5.7%, and mAP increased by 4.4%.

Due to the simplicity of the indoor environment and minimal lighting changes, performance in indoor search is superior to that in outdoor scene search. The experimental results in this paper further demonstrate that the DPC-FCNet significantly enhances the discriminability and relevance of cross-modality features through the synergistic effect of shallow independent modeling and deep relational fusion.

Table 6. Comparison of experimental results on the RegDB dataset.

Methods	Rank-1(%)	Rank-10(%)	Rank-20(%)	mAP(%)
AGW[13]	70.0	86.2	91.5	66.3
CM-NAS[14]	82.8	95.1	97.7	79.3
PMT[15]	84.8	-	-	76.5
HAT[16]	71.8	87.1	92.1	67.5
SMCL[17]	83.9	-	-	79.8
DART[18]	83.6	-	-	75.7
CAJ[9]	83.5	94.5	96.8	76.8
DPC-FCNet	**91.4**	**97.4**	**98.6**	**86.3**

The comparison of the RegDB dataset is shown in Table 6. From the rank values and mAP of DPC-FCNet, rank-1 improves by 6.6% and mAP improves by 6.5% compared to the best method in the table. The experimental results outperform other mainstream methods, indicating that the proposed network further enhances the effectiveness of cross-modality feature extraction. Due to the complex and diverse shooting conditions of the SYSU-MM01 dataset, the retrieval accuracy is lower than that on RegDB.

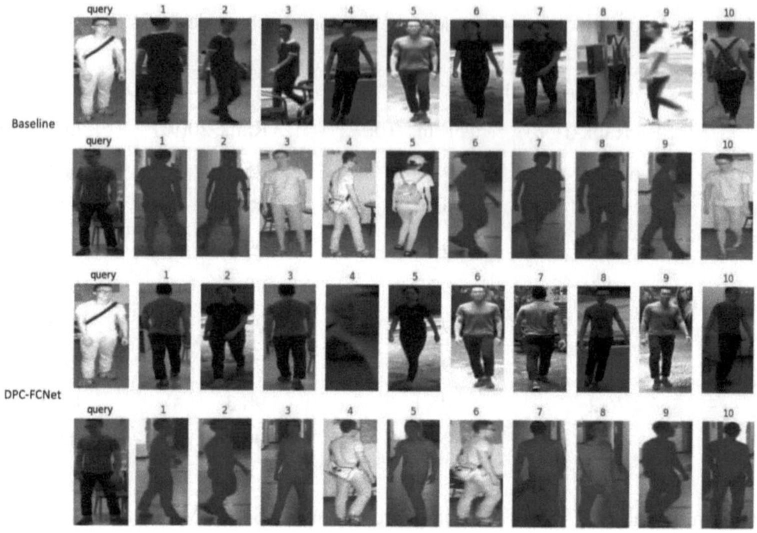

Fig. 6. Comparison image after integrating the module.

To demonstrate the effectiveness of the DPC-FCNet network, some retrieval results on SYSU-MM01 are visualized, as shown in Fig. 6. Green numbers indicate correct retrievals, while red numbers indicate incorrect ones. It can be seen that the retrieval results of this network are significantly better.

5 Conclusion

The proposed dual-channel network for enhancing multi-level feature correlation effectively addressed the key challenge of insufficient modality-shared feature extraction. The network achieves a balance between retaining modality-specific information and shared features through a collaborative mechanism of shallow independent modeling and deep associative fusion. The experimental results demonstrate that DPC-FCNet significantly outperforms the existing methods on two benchmark datasets.

The multi-level feature association and dynamic balance mechanism in this paper provide important insights into the modality interaction problem in the field of multimodal learning. It shows great potential in extended applications such as cross-domain identification and cross-modal retrieval, and promotes the development and progress of related fields.

References

1. Abed, A., Akrout, B., Amous, I.: Deep learning-based few-shot person re-identification from top-view RGB and depth images. Neural Comput. Appl. (Prepublish). **36**, 1–18 (2024)
2. Yunshang, L., Ying, C.: Infrared-visible cross-modal person re-identification via dual-attention collaborative learning. Signal Process. Image Commun. **109**, 116868 (2022)

3. Bai Shutao, Chang Hong, and Ma Bingpeng: Incorporating texture and silhouette for video-based person re-identification (2024)
4. Liu, Y., Ge, H., Luo, Y., et al.: Find hidden modality divergence: adversarial aware learning for unsupervised visible-infrared person re-identification. IEEE Trans. Neural Networks Learn. Syst. **36**, 19879 (2025)
5. Murali, N., Mishra, D.: Feature compensation network for prototype-based cross-modal person re-identification. IEEE Access. **12**, 117994 (2024)
6. Song, W., et al.: Channel enhanced cross-modality relation network for visible-infrared person re-identification. Appl. Intell. **55**, 1 (2025)
7. Fan, H., Chen, Y., Chen, Y., et al.: Wavelet-based multi-level information compensation learning for visible-infrared person re-identification. Digit. Signal Process. **168**(PA), 105471–105471 (2026)
8. Tao, M., Long, H., Kong, G., et al.: Combining information augmentation aggregation and dual-granularity feature fusion for visible-infrared person re-identification. SIViP. **19**(1), 163–163 (2024)
9. Ye, M., Shen, J., Crandall, D.J. et al.: Dynamic dual-attentive aggregation learning for visible-infrared person re-identification. In: Eur. Conf. Comput. Vis., pp. 229–247. Springer, Cham (2020)
10. Wu, Q., Dai, P., Chen, J., et al.: Discover cross-modality nuances for visible-infrared person re-identification. In: Proceedings of the IEEE/CVF Conference on Computer Vision and Pattern Recognition, pp. 4330–4339. IEEE/CVF (2021)
11. Wu, B., Feng, Y., Sun, Y., et al.: Feature Aggregation Via Attention Mechanism for Visible-Thermal Person Re-Identification, pp. 140–144. IEEE (2023)
12. Shi, L., Ma, Y., Tao, Y., et al.: Bridging modal gaps: a cross-modal feature complementation and feature projection network for visible-infrared person re-identification. Neurocomputing. **642**, 130367–130367 (2025)
13. Wang J,Gao X,Niu S, et al.: Learning discriminative features via deep metric learning for video-based person re-identification. Expert Syst. Appl., 286 128123–128123.(SS2025)
14. Fu, Chaoyou, et al.: CM-NAS: Cross-Modality Neural Architecture Search for Visible-Infrared Person re-Identification. (2021).
15. Lu, H., Zou, X., Zhang, P.: Learning progressive modality-shared transformers for effective visible-infrared person re-identification. In: Proceedings of the AAAI Conference on Artificial Intelligence (AAAI 2023), pp. 1835–1843. AAAAI Press (2023)
16. Ye, M., Shen, J., Shao, L.: Visible-infrared person re-identification via homogeneous augmented trimodal learning. IEEE Trans. Inf. Forensics Secur. **16**, 728–739 (2020)
17. Wei, Z., Yang, X., Wang, N., et al.: Syncretic modality collaborative learning for visible infrared person re-identification. In: Proc. IEEE/CVF Int. Conf. Comput. Vis, pp. 225–234. IEEE/CVF (2021)
18. Yang, M., Huang, Z., Hu, P., et al.: Learning with twin noisy labels for visible-infrared person reidentification. In: Proc. IEEE/CVF Conf. Comput. Vis. Pattern Recognit, pp. 14308–14317. IEEE/CVF (2022)

Surface Defect Detection of Photovoltaic Panels Based on Deep Learning and Electroluminescent Images

Xiaoqiang Wang, Liurui Zhao(✉), and Yanjie Wang

College of Intelligent Science and Technology, Inner Mongolia University of Technology, Hohhot, China
zhaoliurui99@163.com

Abstract. In the quality inspection of photovoltaic (PV) modules, defect detection methods that combine electroluminescence (EL) imaging with deep learning have attracted considerable attention. However, EL images often suffer from complex background interference, minor target defects, and weak feature boundaries, which hinder the accuracy and robustness of existing detection methods. To address these challenges, this paper proposes an improved YOLOv9s algorithm, PV-YOLOv9s, to enhance the performance of defect detection in complex scenarios. First, wavelet gated convolution (WGConv) is proposed and the WGRepNC-SPELAN4 module is further constructed to achieve the capture of multi-scale and multi-frequency information, which effectively enhances the background suppression and feature expression. Second, the Multi-Scale Dual Attention (MSDA) module is incorporated to promote multi-scale feature fusion and improve the model's ability to detect defects of varying sizes. Furthermore, the integration of the Coordinate Attention (CoordAtt) mechanism strengthens the model's joint perception of spatial positions and channel features, thus improving localization accuracy. Finally, the InnerEIoU loss function is employed to accelerate convergence and enhance bounding box regression precision. Experimental results demonstrate that the proposed PV-YOLOv9s achieves a mean Average Precision (mAP50:95) of 74.2% on a PV module EL image defect detection dataset, outperforming the latest YOLOv13s by 2.1%.

Keywords: Photovoltaic Panels · Electroluminescence Image · Defect Detection · YOLOv9 · Attention Mechanism

1 Introduction

Driven by the "dual carbon" goals, the global photovoltaic (PV) industry has entered a phase of rapid development. According to statistics from the International Renewable Energy Agency (IRENA), as of 2024, the global installed capacity of photovoltaics has exceeded 1.2 terawatts, with an average annual growth rate of over 20%. However, during the manufacturing, transportation,

J. Lokoč et al. (Eds.): MMM 2026, LNCS 16413, pp. 332–345, 2026.
https://doi.org/10.1007/978-981-95-6957-1_24

and prolonged outdoor operation of PV modules, surface defects such as cracks, broken grid lines, and thick fingers may arise due to material flaws, process deviations, or environmental stress. These subtle defects not only reduce the power generation efficiency of PV systems but may also trigger hotspot effects, accelerate power degradation, and even pose fire hazards.

Currently, mainstream imaging techniques for PV panel surface defect detection include infrared thermography, photoluminescence (PL), and electroluminescence (EL) imaging. Infrared thermography identifies defects via temperature anomalies but is highly sensitive to ambient conditions. PL imaging uses external light to excite fluorescence based on carrier lifetime differences; however, image contrast depends heavily on illumination uniformity, making it prone to artifacts. In contrast, EL imaging applies a forward bias to stimulate carrier recombination, producing light correlated with local current density—directly revealing internal electrical and structural defects. Puranik et al. [1] proposed the generalized quantitative electroluminescence (g-QUEL) method, integrating EL images, datasheet data, and voltage measurements for quantitative assessment. Lin et al. [2] developed a pseudo-coloring technique to enhance defect visibility in EL images. Harsh Rajesh Parikh et al. [3] extracted statistical features from EL image histograms for training machine learning models.

However, the multi-scale morphological variations of defects and complex noise in EL images pose challenges for traditional image processing algorithms in accurately extracting defect features. With recent advances in deep learning, many studies have combined EL imaging with deep learning to enhance PV panel surface defect detection. Chen et al. [4] proposed an improved Faster R-CNN method incorporating the Convolutional Block Attention Module (CBAM) and a clustering algorithm to enhance crack detection and model convergence. Pan et al. [5] embedded an Adaptive Complementary Fusion (ACF) module into YOLOv5, boosting both detection accuracy and speed. Zhu et al. [6] introduced the C2DEM-YOLO network, which features a C2Dense module for multi-level feature extraction, the EMA attention mechanism, and the InnerCIoU loss function, effectively addressing detection challenges in complex EL image backgrounds.

In summary, defect detection methods that integrate EL imaging with deep learning have significantly advanced their application in the field of PV defect inspection. However, the aforementioned approaches still face considerable challenges when detecting defects in EL images under complex background conditions. To address these issues, this paper proposes an improved YOLOv9s-based algorithm, referred to as PV-YOLOv9s. The main contributions of this work are summarized as follows:

(1) To address the challenges of complex background information and defect feature extraction in EL images under intricate scenarios, a novel Wavelet-Gated Convolution (WGConv) is proposed to capture multi-scale and multi-frequency feature information. Building upon this, a WGRepNCSPELAN4 module is designed by replacing specific standard convolutional layers in the original RepNCSPELAN4 with WGConv. This modification effectively

suppresses complex background interference in EL images and enhances the model's feature extraction capability.
(2) To address the difficulty of capturing multi-scale defects, the Multi-Scale Dual Attention (MSDA) module is employed to effectively capture and utilize multi-scale feature information, enabling accurate defect recognition.
(3) To mitigate the problem of false detections, the Coordinate Attention (CoordAtt) mechanism is introduced to fuse spatial coordinate information with channel features, thereby enhancing the model's perception of target location information and reducing false positives.
(4) To address the issue of insufficient detection accuracy, the InnerEIoU loss function is adopted to replace the CIoU loss function, thereby improving the model's convergence speed and localization precision.

2 Method

2.1 PV-YOLOv9s

YOLOv9s, as a model within the YOLO series, builds upon the core architecture of YOLOv9 [7] with further optimizations. The photovoltaic panel surface defect detection model proposed in this paper is based on the YOLOv9s algorithm, as illustrated in Fig. 1. The PV-YOLOv9s network consists of three components: the backbone network, the neck network, and the detection head. To overcome the limitations of conventional convolution operations in capturing multi-scale and multi-frequency features, a WGConv is proposed and used to replace standard convolutions at key positions within the RepNCSPELAN4 module, thereby enhancing the model's ability to represent multi-scale defect features of photovoltaic modules. Furthermore, the MSDA module is introduced after the auxiliary training branch detection head to improve the model's capability in detecting multi-scale targets. Subsequently, the CoordAtt module is incorporated into the neck network to effectively enhance the model's perception of target location information by fusing spatial coordinate information with channel features. Finally, the original loss function of YOLOv9s is improved by adopting the InnerEIoU loss function, which further boosts the detection accuracy of the model.

2.2 Wavelet Gated Convolution (WGConv)

To address the limitations of traditional convolution operations in capturing multi-scale and multi-frequency features, Finder et al. [8] proposed WTConv, which utilizes wavelet transforms for multi-scale feature extraction. Building upon this, we further propose WGConv. The core idea of WGConv is to perform multi-scale decomposition of the input image using wavelet transform to capture features at different frequency bands, and to dynamically ad-just the contribution of each band through a gating mechanism. As illustrated in Fig. 2(a), the working principle is as follows:

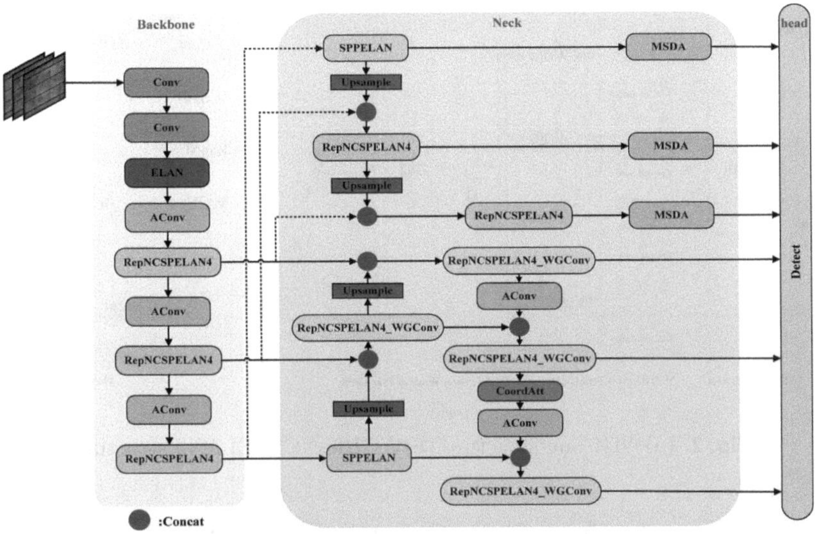

Fig. 1. Architecture of the PV-YOLOv9s network.

Wavelet Decomposition: The input image is first decomposed into low and high frequency components through wavelet transform. The low-frequency component preserves the global structural information of the image, while the high-frequency components capture image details and edge information.

Convolution Operation: Convolution is performed separately on the decomposed low-frequency and high-frequency components to extract more informative features. The specific computation of the convolution operation is as follows:

$$Y_{LL}, Y_{LH}, Y_{HL}, Y_{HH} = Conv(W, X_{LL}, X_{LH}, X_{HL}, X_{HH}) \tag{1}$$

Here, W denotes the convolution kernel weights; X_{LL}, X_{LH}, X_{HL} and X_{HH} represent the different frequency band components obtained from wavelet decomposition; and Y_{LL}, Y_{LH}, Y_{HL}, and Y_{HH} represent the outputs after convolution.

Gating Mechanism: To adaptively weight different frequency bands, we introduce a gating function G. First, the convolved features are fused to obtain a joint representation:

$$X_{fusion} = Fuse(Y_{LL}, Y_{LH}, Y_{HL}, Y_{HH}) \tag{2}$$

where Fuse(·) denotes channel-wise concatenation followed by 1×1 convolution for dimensionality alignment. The gating weights are computed as:

$$G = \sigma(Conv_{gate}(W_{gate}, X_{fusion})) \tag{3}$$

Here, σ is the sigmoid function, $Wgate$ represents the convolutional weights of the gating mechanism, and $G \in [0, 1]$ modulates the contribution of each frequency band. Feature Fusion and Inverse Wavelet Transform: The weighted convolutional features are split into two parts:

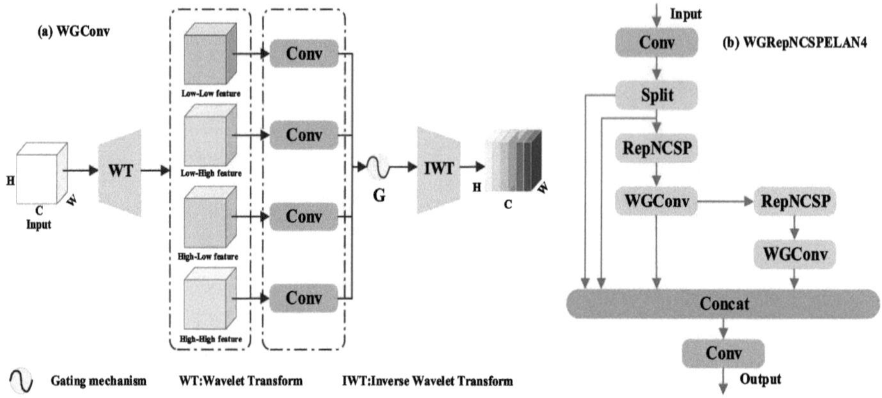

Fig. 2. (a)WGConv structure (b)WGRepNCSPELAN4 structure.

Y_{base}: features from the low-frequency branch (Y_{LL}). Y_{tag}: features from the high-frequency branches (Y_{LH}, Y_{HL}, Y_{HH}) fused together.

The final output is reconstructed via inverse wavelet transform (IWT):

$$\gamma = IWT(G \odot Y_{base} + (1 - G) \odot Y_{tag}) \tag{4}$$

Here, γ denotes the final output feature map, $\odot$ is element-wise multiplication, and the IWT reconstructs the multi-frequency information back into the spatial domain.

2.3 WGRepNCSPELAN4

In photovoltaic defect detection, surface flaws on panels often present multi-scale patterns, subtle details, and are easily affected by background noise, placing high demands on the feature extraction capabilities of existing detection models. To address these challenges, this paper proposes the WGRepNCSPELAN4 module, which integrates WGConv into the RepNCSPELAN4 architecture to combine frequency-domain enhancement, gated feature selection, and multi-branch aggregation, achieving a balance between feature extraction efficiency and computational cost.

As shown in Fig. 2(b), WGRepNCSPELAN4 replaces conventional convolutions with WGConv at key positions. During feature extraction, WGConv performs wavelet decomposition and gated convolution to obtain multi-scale features rich in detail and semantic information. These features are then fused via CSP connections and ELAN aggregation, enabling interaction across scales and forming a hierarchical, information-rich feature pyramid. The resulting feature maps preserve essential details while enhancing the model's representation of multi-scale photovoltaic defect features.

2.4 Multi-Scale Dilated Attention (MSDA)

In the field of deep learning, especially in computer vision tasks, effectively capturing and utilizing multi-scale feature information has long been a key research focus and challenge. The MSDA module [9], as an innovative attention mechanism, significantly enhances the model's capability to detect multi-scale objects.

As shown in Fig. 3(a), the MSDA module adopts a hierarchical attention head architecture, where the input feature map $X \in \mathbb{R}^{C \times H \times W}$ is divided along the channel dimension into n attention heads that are processed in parallel. Each attention head i independently performs the following operations:

$$h_i = SWDA(Q_i, K_i, V_i, r_i), 1 \leq i \leq n \tag{5}$$

where $Q_i, K_i, V_i \in \mathbb{R}^{\frac{C}{n}} \times H \times W$ are the query, key, and value vectors of the ith header, and r_i denotes the expansion rate parameter specific to that header.

This design allows each attention head to capture semantic information at different scales. For each position (i,j) in the feature map, the Sliding Window Dilated Attention (SWDA) sparsely selects keys and values within a dilation-based sliding window centered at (i,j). The output value X_{ij} is computed as follows:

$$X_{ij} = Attention(q_{ij}, K_r, V_r) = Softmax(\frac{q_i K_r^T}{\sqrt{d_k}})V_r \tag{6}$$

Here, q_{ij} denotes the query vector, K_r and V_r represent the key and value matrices sampled via dilated convolution and d_k is the dimension of the attention head. This design enables a single MSDA module to cover multi-scale features ranging from local details to global context.

After processing by SWDA, the output features h_i from each attention head are fused through concatenation, expressed by the following formula:

$$X = Linear(Concat\,[h_1, \cdots, h_n]) \tag{7}$$

The linear layer enables cross-channel information interaction through the weight matrix $W \in \mathbb{R}^{C \times H \times W}$, ultimately producing the enhanced feature $X \in \mathbb{R}^{C \times H \times W}$.

2.5 Coordinate Attention (CoordAtt)

Complex lighting conditions, varying defect scales, and background noise often hinder deep learning models from accurately localizing defects and reducing false detections. CoordAtt [10], as an efficient attention mechanism, enhances the model's spatial awareness by integrating coordinate information with channel features, offering a promising solution to these challenges.

As illustrated in Fig. 3(b), for the input feature map $X \in \mathbb{R}^{C \times H \times W}$, the CoordAtt structure first performs one-dimensional average pooling along the horizontal and vertical directions to obtain two direction-aware feature maps:

$$z_c^h(h) = \frac{1}{W} \sum_{0 \leq i < W} x_c(h, i), z_c^w(w) = \frac{1}{H} \sum_{0 \leq j < H} x_c(j, w) \tag{8}$$

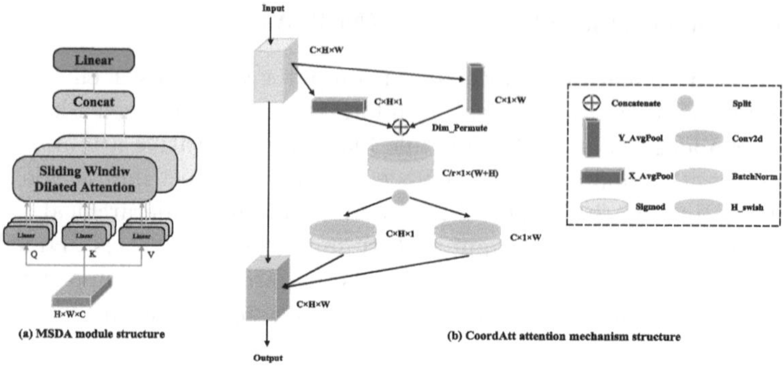

Fig. 3. (a)MSDA module structure (b)CoordAtt attention mechanism structure.

Here, $Z_c^h \in \mathbb{R}^{H\times 1\times C}$ and $Z_c^w \in \mathbb{R}^{1\times W\times C}$ encode global channel information along the vertical and horizontal directions, respectively, effectively retaining positional cues often lost in conventional global pooling.

After concatenating the feature maps from the two directions, a convolution layer F_1 is applied to reduce the dimensionality, followed by a nonlinear activation function δ, resulting in an intermediate feature $f \in \mathbb{R}^{1\times(H+W)\times\frac{C}{r}}$:

$$f = \delta(F_1([z^h, z^w])) \tag{9}$$

The feature f is split into horizontal $f \in \mathbb{R}^{1\times H\times C/r}$ and vertical $f^w \in \mathbb{R}^{1\times W\times C/r}$ components, each processed by convolution layers F_h and F_w to recover the channel dimension. A Sigmoid activation is then applied to produce direction-aware attention weights $g^h \in \mathbb{R}^{H\times 1\times C}$ and $g^w \in \mathbb{R}^{1\times W\times C}$:

$$g^h = \sigma(F_h(f^h)), g^w = \sigma(F_w(f^w)) \tag{10}$$

Here, σ denotes the Sigmoid function, which normalizes the weights into the range of (0, 1).

The generated horizontal and vertical attention weights are multiplied with the original feature map to enhance the features of the target regions:

$$y_c(i, j) = x_c(i, j) \times g_c^h(i) \times g_c^w(j) \tag{11}$$

This operation enhances the channel responses of the target regions, enabling the model to precisely focus on defect features.

2.6 InnerEIoU Loss Function

In photovoltaic defect detection, accurate bounding box regression is crucial for precise localization. InnerEIoU introduces auxiliary boxes to refine geometric constraints, enabling more precise gradient flow and significantly improving convergence speed and localization accuracy.

InnerEIoU consists of two components: the improved EIoU [11] loss and the auxiliary bounding box IoU [12] loss. Its mathematical formulation is as follows:

$$L_{InnerEIoU} = L_{EIoU} + IoU - IoU^{inner} \quad (12)$$

where L_{EIoU} denotes the extended EIoU loss, and IoU^{inner} represents the IoU value of the auxiliary bounding box.

EIoU, based on CIoU [13], replaces the abstract aspect ratio metric with an ex-plicit width and height difference loss. The formula is defined as:

$$L_{EIoU} = L_{LoU} + L_{dis} + L_{asp} = 1 - IoU + \frac{\rho^2(b, b^{gt})}{c^2} + \frac{\rho^2(w, w^{gt})}{c_W^2} + \frac{\rho^2(h, h^{gt})}{c_h^2} \quad (13)$$

Here, L_{IoU} is the overlap loss, L_{dis} measures the center point distance, and L_{asp} captures aspect ratio differences. ρ denotes the Euclidean distance, while c, c_w, and c_h refer to the diagonal length, width, and height of the smallest enclosing box covering both prediction and ground truth.

To speed up convergence for high-IoU samples, InnerEIoU introduces a scaling ratio to create scaled auxiliary boxes, known as Inner Boxes. These are defined for both the ground truth and predicted bounding boxes as follows:

$$b_l^{gt} = x_c^{gt} - \frac{w^{gt} \cdot ratio}{2}, b_r^{gt} = x_c^{gt} + \frac{w^{gt} \cdot ratio}{2} \quad (14)$$

$$b_t^{gt} = y_c^{gt} - \frac{h^{gt} \cdot ratio}{2}, b_b^{gt} = y_c^{gt} + \frac{h^{gt} \cdot ratio}{2}, b_l = x_c - \frac{w \cdot ratio}{2} \quad (15)$$

$$b_r = x_c + \frac{w \cdot ratio}{2}, b_t = y_c - \frac{h \cdot ratio}{2}, b_b = y_c + \frac{h \cdot ratio}{2} \quad (16)$$

The IoU of the auxiliary bounding boxes is calculated as follows:

$$inter = (min(b_r^{gt}, b_r) - max(b_l^{gt}, b_l)) \cdot (min(b_b^{gt}, b_b) - max(b_t^{gt}, b_t)) \quad (17)$$

$$IoU^{inner} = \frac{inter}{(w^{gt} \cdot h^{gt} + w \cdot h) \cdot ratio^2 - inter} \quad (18)$$

By adjusting the ratio, the gradient magnitude of the auxiliary box becomes greater than that of the original box, thereby accelerating model convergence.

3 Experiments

3.1 Experimental Environment

The experiments were conducted on the Ubuntu 16.04.6 operating system, equipped with an Nvidia Tesla V100S PCIe GPU with 32 GB of memory and an Intel(R) Xeon(R) Gold 5218R. The Python version used was 3.8.0, with CUDA version 11.6, and the deep learning framework was PyTorch 1.13.1. The experimental hyperparameters were configured as follows: the model was trained for 200 epochs with data augmentation enabled, a batch size of 48, and an input image size of 640 × 640. The optimizer used was SGD with an initial learning rate of 0.01, momentum of 0.937, and a weight decay of 0.0005.

3.2 Datasets

The PVEL-AD dataset [14], jointly released by Hebei University of Technology and Beihang University, is designed to provide high-quality image data support for research on PV module defect detection. The publicly accessible portion of the dataset contains 36,543 EL images of PV modules, among which 4,500 are annotated with various internal defects. In this study, we construct a PV surface defect detection dataset by combining 4,149 EL images from the PVEL-AD dataset with 550 private EL images of PV module surface defects, resulting in a total of 4,699 images categorized into six classes, as shown in Fig. 4. The dataset is split into training, validation, and test sets in a 7:1.5:1.5 ratio and annotated using the Labelimg tool. Additionally, we evaluate the generalization capability of the PV-YOLOv9s network using the NEU-DET dataset [15] and the PCB dataset [16].

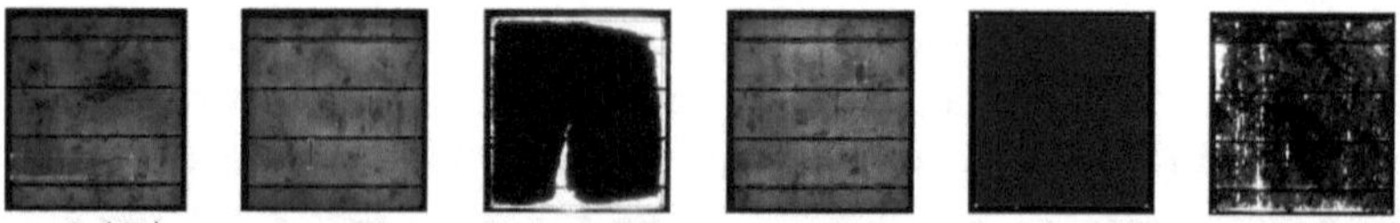

Fig. 4. Types of surface defects in photovoltaic modules.

3.3 Evaluation Metrics

The following evaluation metrics are used in this study as assessment criteria:

$$Precision = \frac{TP}{TP + FP}, Recall = \frac{TP}{TP + FN}, F_1 = 2 \times \frac{precision \times recall}{precision + recall} \tag{19}$$

$$AP = \int_0^1 precision(recall)d(recall), mAP = \frac{1}{N}\sum_0^N AP \tag{20}$$

where TP denotes true positives, FP denotes false positives, and FN denotes false negatives. Precision(P), Recall(R), and F_1 Score(F_1) are used to measure classification performance. Average Precision(AP) evaluates the detection accuracy of the model for a single class, while mean Average Precision(mAP) represents the average of AP across multiple classes and is used to assess the model's overall detection performance across all categories.

3.4 Ablation Experiments

To evaluate the effectiveness of the proposed improvements for photovoltaic defect detection, ablation studies were conducted using YOLOv9s as the baseline. The WGConv module, MSDA module, CoordAtt mechanism, and InnerEIoU loss were incrementally integrated under identical experimental settings. Detailed results are presented in Table 1.

Ablation experiments indicate that integrating WGConv alone achieves the highest Recall (90.20%) among all single-module settings. Incorporating MSDA increases mAP50 and mAP50:95 by 0.7% and 0.3%, respectively, while reducing Precision by 0.8% and increasing Recall by 1.2%, reflecting a trade-off among these metrics. CoordAtt brings minor overall gains, while InnerEIoU boosts Precision by 1.4%. Integrating all four modules achieves a balanced Precision of 90.00% and Recall of 90.10%, with mAP50, mAP50:95, and F_1 reaching their peak values, demonstrating effective synergy and complementary strengths.

Table 1. Ablation experiments of the PV-YOLOv9s algorithm.

YOLOv9s	WGConv	MSDA	CoordAtt	InnerEIoU	P%	R%	mAP50%	mAP50:95%	F_1%
√					90.10	88.60	92.40	72.80	89.34
√	√				87.70	**90.20**	93.10	73.20	88.93
√		√			89.30	89.80	93.10	73.10	89.55
√			√		89.50	88.60	92.50	72.90	89.05
√				√	91.50	87.70	92.80	73.30	89.56
√	√	√			89.60	88.60	92.60	73.80	89.10
√	√		√		90.70	87.80	92.40	73.00	89.23
√	√			√	89.50	88.60	93.10	73.30	89.05
√		√	√		90.50	88.60	92.90	73.60	89.54
√		√		√	87.70	89.20	92.70	73.50	88.44
√			√	√	90.30	88.40	93.00	72.90	89.34
√	√	√	√		90.90	88.40	93.00	73.60	89.63
√	√		√	√	**91.60**	86.10	92.80	72.00	88.76
√		√	√	√	89.60	89.80	92.80	74.10	89.70
√	√	√	√	√	90.00	90.10	**93.40**	**74.20**	**90.05**

3.5 Comparative Experiments

The comparative experimental results, shown in Table 2, demonstrate that the PV-YOLOv9s algorithm exhibits significant advantages in photovoltaic module surface defect detection. The model excels at feature extraction from electroluminescence images, particularly achieving a notable 48.80% AP95 detection accuracy in crack defect identification, which validates its effectiveness in accurately detecting cracks in EL images. Notably, the algorithm attains 98.30% detection accuracy for short circuit defects and 53.70% for broken grid defects, confirming the robustness of its multi-scale feature fusion mechanism under complex backgrounds. Compared to the latest YOLOv12s and YOLOv13s baselines,

PV-YOLOv9s outperforms across multiple defect types, especially in recognizing cracks, thick lines, and short circuits, establishing a clear technical advantage. This performance breakthrough demonstrates strong applicability for high-precision defect identification, providing an effective solution for photovoltaic panel quality inspection.

Table 2. Comparative experiments of network models.

Model	AP50:95%						mAP50%	mAP50:95%
	Ck	Fr	BC	TL	SC	MD		
Faster R-CNN	30.92	49.10	90.63	36.06	91.63	93.81	87.50	65.40
SSD	26.02	42.56	93.32	40.41	97.85	98.59	87.30	66.50
RetinaNet	33.79	42.98	95.44	37.47	97.94	**100.00**	88.40	67.90
YOLOv8s	43.90	53.20	96.00	46.80	96.10	99.40	92.20	72.60
RT-DETR	42.00	51.00	94.30	43.60	81.70	96.90	89.50	68.30
YOLOv9s	46.30	52.20	95.10	**49.00**	94.30	99.50	92.40	72.80
YOLOv10s	39.80	**53.70**	96.70	48.00	96.20	99.40	91.30	72.30
YOLOv11s	41.60	53.00	96.60	46.40	95.90	99.40	91.20	72.10
YOLOv12s	42.30	52.70	96.20	48.00	92.10	99.40	91.90	71.80
YOLOv13s	39.50	**53.70**	**96.90**	47.50	95.60	99.30	91.90	72.10
PV-YOLOv9s	**48.80**	**53.70**	96.10	48.70	**98.30**	99.40	**93.40**	**74.20**

3.6 Generalization Experiment

The evaluation results on two representative industrial defect datasets, NEU-DET and PCB, demonstrate the excellent cross-dataset generalization ability of PV-YOLOv9s, as detailed in Table 3. The algorithm achieves leading detection performance across various categories and scenarios, notably attaining the highest AP50 scores in multiple key classes, with strong overall mAP50 results. The consistent performance across datasets indicates that PV-YOLOv9s overcomes dataset-specific limitations, exhibiting stable and superior defect recognition capabilities in diverse industrial inspection contexts, thereby validating its robust model generalization.

3.7 Visual Analytics

To highlight the proposed method's improvements, representative test images were selected for heatmap comparisons. As shown in Fig. 5, the proposed PV-YOLOv9s algorithm effectively integrates global information, precisely localizes fine-grained defects like cracks, and maintains strong performance in detecting large-scale anomalies. These results collectively validate the effectiveness of PV-YOLOv9s in photovoltaic defect detection.

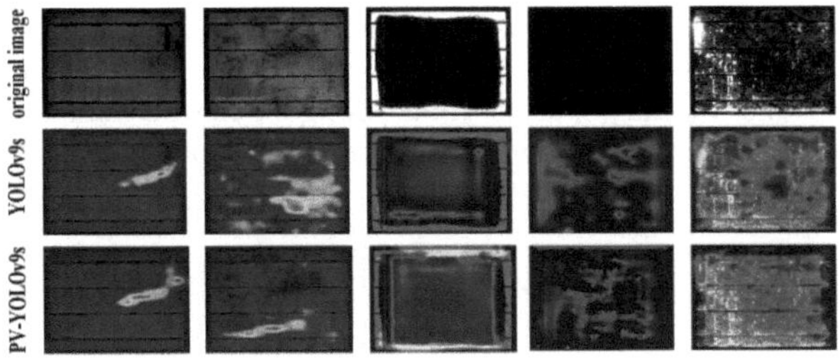

Fig. 5. Heatmap Comparison.

Table 3. Generalization experiments on the NEU-DET and PCB dataset

Dataset	Method(year)	AP50%						mAP50%
		Cr	In	Pa	PS	RS	Sc	
NEU-DET	Faster R-CNN [17] (2015)	44.20	81.90	92.00	78.40	63.10	95.20	75.80
	YOLOv5 [17] (2021)	48.50	85.10	91.60	76.90	66.40	93.10	77.00
	YOLOv7 [17] (2022)	49.60	83.60	81.90	83.40	59.80	**95.70**	75.70
	EDDNet [17] (2023)	41.70	76.30	86.30	85.10	58.10	85.60	72.40
	Improved YOLOX [18] (2024)	**55.10**	83.00	**93.60**	86.10	59.70	84.20	77.00
	Improved YOLOv5n [19] (2025)	50.40	**90.50**	84.50	86.10	54.30	89.70	75.30
	PV-YOLOv9s(Ours)	48.10	84.80	92.30	**86.40**	**70.50**	90.80	**78.80**
		MH	MB	OC	SH	SP	SC	
PCB dataset	YOLOv5 [17] (2021)	98.60	85.00	87.50	95.70	73.80	85.80	87.80
	YOLOv7 [17] (2022)	89.60	87.90	87.50	86.30	90.40	87.70	88.20
	LDD-Net [20] (2023)	90.10	**98.73**	**98.58**	97.91	88.72	92.36	95.90
	PMSA-DyTr [17] (2024)	97.80	97.30	96.80	92.70	92.50	94.90	95.30
	YOLOv8-PCB [21] (2025)	**99.50**	93.50	95.70	**98.00**	**98.20**	91.60	96.10
	PV-YOLOv9s(Ours)	**99.50**	95.40	95.90	95.40	97.10	**95.40**	**96.50**

4 Conclusion

This paper presents PV-YOLOv9s, an enhanced YOLOv9s algorithm for PV defect detection, addressing challenges such as complex EL backgrounds, weak feature extraction, and false detections. The model integrates WGConv for multi-scale feature learning, WGRepNCSPELAN4 for noise suppression and feature enhancement, MSDA for small-defect detection, CoordAtt for precise localization, and InnerEIoU for faster convergence and improved bounding box accuracy. Experiments on EL datasets show superior performance over existing methods, though low-contrast and high-noise conditions still cause weak feature representation and blurred boundaries. Future work will optimize the architecture, strengthen multi-scale learning, and refine attention mechanisms to overcome these limitations.

Acknowledgments. This research was supported by the Inner Mongolia Natural Science Foundation (Grant No. 2023MS06021), the National Natural Science Foundation of China (Grant No. 62165011), and the Inner Mongolia Science and Technology Breakthrough and Leadership Project (Grant No. 2025KJTW0025-04).

References

1. Puranik, V.E., Kumar, R., Gupta, R.: Generalized quantitative electroluminescence method for the performance evaluation of defective and unevenly degraded crystalline silicon photovoltaic module. Prog. Photovoltaics Res. Appl. **31**(3), 269–282 (2023)
2. Lin, K.M., Lin, H.H., Dandage, H.K., et al.: Pseudo colorization of electroluminescence images of multi-crystalline silicon solar cells for defect inspection. Mod. Phys. Lett. B **33**(14n15), 1940010 (2019)
3. Parikh, H.R., Buratti, Y., Spataru, S., et al.: Solar cell cracks and finger failure detection using statistical parameters of electroluminescence images and machine learning. Appl. Sci. **10**(24), 8834 (2020)
4. Chen, A., Li, X., Jing, H., et al.: Anomaly detection algorithm for photovoltaic cells based on lightweight multi-channel spatial attention mechanism. Energies **16**(4), 1619 (2023)
5. Pan, W., Sun, X., Wang, Y., et al.: Enhanced photovoltaic panel defect detection via adaptive complementary fusion in YOLO-ACF. Sci. Rep. **14**(1), 26425 (2024)
6. Zhu, J., Zhou, D., Lu, R., et al.: C2DEM-YOLO: improved YOLOv8 for defect detection of photovoltaic cell modules in electroluminescence image. Nondestructive Test. Eval. **40**(1), 309–331 (2025)
7. Wang, C.Y., Yeh, I.H., Liao, H.Y.M.: Yolov9: learning what you want to learn using programmable gradient information. In: European Conference on Computer Vision (ECCV 2024), pp. 1–21. Springer, Cham (2024)
8. Finder, S.E., Amoyal, R., Treister, E., et al.: Wavelet convolutions for large receptive fields. In: European Conference on Computer Vision (ECCV 2024), pp. 363–380. Springer, Cham (2024)
9. Jiao, J., Tang, Y.M., Lin, K.Y., et al.: Dilateformer: multi-scale dilated transformer for visual recognition. IEEE Trans. Multimed. **25**, 8906–8919 (2023)
10. Hou, Q., Zhou, D., Feng, J.: Coordinate attention for efficient mobile network design. In: Proceedings of the IEEE/CVF Conference on Computer Vision and Pattern Recognition (CVPR 2021), pp. 13713–13722 (2021)
11. Zhang, Y.F., Ren, W., Zhang, Z., et al.: Focal and efficient IOU loss for accurate bounding box regression. Neurocomputing **506**, 146–157 (2022)
12. Yu, J., Jiang, Y., Wang, Z., et al.: Unitbox: an advanced object detection network. In: Proceedings of the 24th ACM International Conference on Multimedia (MM 2016), pp. 516–520 (2016)
13. Rezatofighi, H., Tsoi, N., Gwak, J.Y., et al.: Generalized intersection over union: a metric and a loss for bounding box regression. In: Proceedings of the IEEE/CVF Conference on Computer Vision and Pattern Recognition (CVPR 2019), pp. 658–666 (2019)
14. Su, B., Zhou, Z., Chen, H.: PVEL-AD: a large-scale open-world dataset for photovoltaic cell anomaly detection. IEEE Trans. Industr. Inf. **19**(1), 404–413 (2022)
15. Song, K., Yan, Y.: A noise robust method based on completed local binary patterns for hot-rolled steel strip surface defects. Appl. Surf. Sci. **285**, 858–864 (2013)

16. Ding, R., Dai, L., Li, G., et al.: TDD-net: a tiny defect detection network for printed circuit boards. CAAI Trans. Intell. Technol. **4**(2), 110–116 (2019)
17. Su, J., Luo, Q., Yang, C., et al.: PMSA-DYTR: prior-modulated and semantic-aligned dynamic transformer for strip steel defect detection. IEEE Trans. Industr. Inf. **20**(4), 6684–6695 (2024)
18. Li, C., Xu, A., Zhang, Q., et al.: Steel surface defect detection method based on improved YOLOX. IEEE Access (2024)
19. Tian, J.H., Feng, X.F., Li, F., et al.: An improved YOLOv5n algorithm for detecting surface defects in industrial components. Sci. Rep. **15**(1), 9756 (2025)
20. Zhang, L., Chen, J., Chen, J., et al.: LDD-Net: lightweight printed circuit board defect detection network fusing multi-scale features. Eng. Appl. Artif. Intell. **129**, 107628 (2024)
21. Wang, J., Xie, X., Liu, G., et al.: A lightweight PCB defect detection algorithm based on improved YOLOv8-PCB. Symmetry **17**(2), 309 (2025)

Low-Dimension Representation Estimation in Principal Component Analysis Under Missing Data

Thanh Tu Do[3(✉)], Van Hua[1], Uyen Dang[3], Thu Nguyen[2], Steven Hicks[2], Pål Halvorsen[2], Michael A. Riegler[2], and Binh T. Nguyen[3,4,5]

[1] University of Medicine and Pharmacy at Ho Chi Minh City, Ho Chi Minh City, Vietnam
htpvan@ump.edu.vn
[2] Simula Metropolitan, Oslo, Norway
{thu,paalh,michael}@simula.no
[3] University of Science, Ho Chi Minh City, Vietnam
tu.dothanh1906@gmail.com, ngtbinh@hcmus.edu.vn
[4] Vietnam National University, Ho Chi Minh City, Vietnam
[5] AISIA Research Lab, Ho Chi Minh City, Vietnam

Abstract. Principal Component Analysis (PCA) is well known for its effectiveness in dimensionality reduction and denoising. However, the presence of missing data in real-world scenarios often presents a formidable challenge to its application. Practitioners are usually faced with the choice of either imputing the missing data before implementing PCA on the restored dataset or attempting to extract the principal components directly from the incomplete dataset without any prior imputation. In this paper, we present a novel approach known as ***PCA for missing data (PCAM)***. This innovative technique is specifically designed to derive principal components directly from missing value datasets, reducing the need for preliminary imputation. Experimental results indicate that PCAM showcases competitive performance when juxtaposed with state-of-the-art methods from both imputation-first and direct extraction strategies, thereby demonstrating its potential to revolutionize the approach to handling missing data in PCA.

1 Introduction

In data analysis and machine learning, dimensionality reduction is useful since it enables the simplification of complex datasets while preserving their intrinsic information. In particular, Principal Component Analysis (PCA) has emerged as an important tool, renowned for its capabilities in both dimensionality reduction and denoising. Its applications span a myriad of fields, from bioinformatics to finance, image processing, social sciences, etc. However, PCA's effectiveness is often tested by the ubiquitous challenge of missing data. Missing data, whether due to sensor failures, human errors, or other unforeseen circumstances, can

J. Lokoč et al. (Eds.): MMM 2026, LNCS 16413, pp. 346–360, 2026.
https://doi.org/10.1007/978-981-95-6957-1_25

introduce biases, reduce statistical power, and lead to potentially misleading results.

There have been two dominant strategies to deal with missing data for PCA. The first, and perhaps the most intuitive, involves imputing the missing values by some imputation techniques such as DIMV [28], SOFT-IMPUTE [16], etc. By filling in the gaps, we can then apply PCA in a straightforward manner to the imputed data set, which is complete. However, each imputation method can also come with its own set of assumptions and potential pitfalls that may affect the principal components computed from the imputed data.

The second strategy is to modify the PCA algorithm to work with incomplete data, thereby extracting principal components without the need for any imputation. This approach, while promising, may also come with certain drawbacks. For example, experiments in [11] demonstrated that GowPCoA could rarely be run for a high level of missing values and a low number of features, and PairCor could rarely be run for a high level of missing values and a low number of individuals. The paper also evaluates the advantages and drawbacks of some other approaches to extract principal components without imputation. Therefore, in this paper, we propose the **PCA for missing data (PCAM)** algorithm, a novel method to derive principal components directly from missing value datasets, as an effort to address these issues. The basic idea of PCAM is to estimate the covariance matrix directly from the input with missing data and then use that covariance matrix to find the projection matrix for PCA and estimate a representation of an input observation in the reduced dimension directly. As will be illustrated from the experiments, PCAM is competitive compared to state-of-the-art techniques in accuracy.

Our contributions are as follows: (i) we propose a novel method, namely *PCA for missing data* (PCAM), to derive principal components directly from datasets with missing values without any preliminary imputation; (ii) we conduct various experiments to compare PCAM with various state-of-the-art methods; (iii) our experiments provide a comparison on the effects of various imputation techniques to PCA; and (iv) we provide direction for potential future work.

The rest of this paper is organized as follows. Section 2 shows general works related to the research problem. Then, we review the fundamentals of the Gaussian conditional formula and the DPER algorithm in Sect. 3. Next, we detail the approach for estimating missing values of two strategies in Sect. 4, before conducting experiments and comparing PCAM's estimation of lower dimensional embedding against other state-of-the-art methods in Sect. 5. Finally, the paper ends with our conclusion and points to future work in Sect. 6.

2 Related Works

There are two dominant ways of dealing with PCA under missing data: impute the data before subsequently applying PCA to the imputed data, or to adapt PCA to be able to work directly with missing data.

The first strategy relies on imputation to make a dataset complete. The research community has developed many imputation techniques so far. Some traditional ones [14] include mean/ mode/ regression imputation and K-nearest neighbor imputation (KNNI), to name a few. For the mean method, the average attribute value for all observed data is used to fill in any missing attributes. On the other hand, the mode method fills in the missing attribute values by using the attribute value that appears the most frequently in all of the observed data. Next, an example of a learning technique is KNNI [21], where values calculated from the k nearest observed data are used to impute missing values. Various distance metrics can be used, and the method has been shown to be particularly effective when the data have an underlying structure. In addition, MissForest is a nonparametric missing value imputation using Random Forest [9,22]. It can be used with mixed-type data and does not require distributional assumptions. It has demonstrated high imputation accuracy with various missing rates. Furthermore, matrix factorization techniques aim to fill in missing values by factorizing the observed data into lower-dimensional matrices. SOFT-IMPUTE [16], in particular, leverages singular value decomposition (SVD) and has been shown to be effective for large-scale datasets with missing values.

In recent years, deep learning techniques [4,5] have also shown powerful imputation power. The Generative Adversarial Imputation Nets (GAIN) [30] is a method that imputes missing data by adapting the Generative Adversarial Nets (GAN) framework. Some components of a real data vector are observed by the generator (G). Then, it imputes the missing components conditioned on what is observed and outputs a completed vector. Next, a completed vector was chosen and the discriminator (D) tried to determine the components that were imputed and those that were truly observed. They gave D some additional information in the form of a hint vector to make sure D forces G to learn the distribution. Furthermore, Conditional Distribution-Based Imputation of Missing Values with Regularization (DIMV) is a method to fill in missing values that was proposed by the authors [28]. Finding the conditional distribution of a feature with missing entries is how DIMV functions, and then it uses the information from the fully observed values as a basis. The experimental result showed that it provides a confidence region and effectively handles multicollinearity in imputation of missing values. In addition, Deep Belief Networks (DBN) technique [15], whether discretization is done in the first or second step, the performance using DBN for imputation is significantly better than other imputation methods, such as mean, KNN, CART, and SVM. Furthermore, DNN-based MICE and gradient-boosting MICE plus CISCL (GB-MICE-CISCL) [24] are the imputation algorithm for high filling efficiency for a tabular data set depending on the kind and percentage of missing values. Recently, a technique that makes use of score-based diffusion models conditioned on observed data is called the Conditional Score-based Diffusion model (CSDI) [26]. This conditional diffusion model can take advantage of the correlations between the observed values because it has been trained for imputation. Moreover, probabilistic forecasting and time-series interpolation can both be done with CSDI.

For the second direction of adapting PCA to work on incomplete data, some algorithms work by skipping missing values. For example, GowPCoA [8] deals with missing entries by using pairwise deletion to remove missing values when computing dissimilarities. Meanwhile, another set of algorithms works by adapting the PCA to consider the missing values. For example, NIPALS [29] sequentially provides the PCA scores and loadings for complete data, and it is easily extended to incomplete data by using weighted regressions with null weights for the missing values. Meanwhile, iterative PCA [12] is an iterative procedure that provides the scores and loadings that minimize the least squares criterion in the observed entries.

Although the aforementioned methods have made significant strides in addressing the challenge of missing data in PCA, each comes with its own set of assumptions, computational complexities, and potential limitations.

3 Preliminaries

3.1 Gaussian Conditional Formula

First, we have the following well-known statement [10];

Lemma 1. *Suppose* $\mathbf{y} \sim N_p(\boldsymbol{\gamma}, \boldsymbol{\Sigma})$. *Let* $\mathbf{y} = \begin{bmatrix} \mathbf{y}_1 \\ \mathbf{y}_2 \end{bmatrix}$ *with sizes* $\begin{bmatrix} q \times 1 \\ (p-q) \times 1 \end{bmatrix}$, *and we partition accordingly:* $\boldsymbol{\gamma} = \begin{bmatrix} \gamma_1 \\ \gamma_2 \end{bmatrix}, \boldsymbol{\Sigma} = \begin{bmatrix} \boldsymbol{\Sigma}_{11} & \boldsymbol{\Sigma}_{12} \\ \boldsymbol{\Sigma}_{21} & \boldsymbol{\Sigma}_{22} \end{bmatrix}$. *Then, the distribution of* $\mathbf{y}_1$ *conditional on* $\mathbf{y}_2 = a$ *follows multivariate normal distribution with mean* $\bar{\gamma} = \gamma_1 + \boldsymbol{\Sigma}_{12}\boldsymbol{\Sigma}_{22}^{-1}(\mathbf{a} - \gamma_2)$, *and covariance matrix* $\overline{\boldsymbol{\Sigma}} = \boldsymbol{\Sigma}_{11} - \boldsymbol{\Sigma}_{12}\boldsymbol{\Sigma}_{22}^{-1}\boldsymbol{\Sigma}_{21}$.

3.2 Direct Parameter Estimation for Randomly Missing Data (DPER Algorithm)

DPER [19] is an approach that directly estimates parameters from incomplete data, allowing fast and highly accurate estimation. It estimates the mean and covariance matrices by using pairs of features based on the assumption that each arbitrary pair of features follows a bivariate normal distribution. Since the scope of this paper is for PCA, we are interested in estimating the covariance matrix only. Therefore, we use DPER* [28], a slight modification of DPER, instead of the original DPER algorithm for single-class data.

In the following, we outline the DPER* algorithm. For each pair of features, we can arrange the data into the following pattern

$$\boldsymbol{x} = \begin{pmatrix} x_{11} & \dots & x_{1\,m} & x_{1\,m+1} & \dots & x_{1n} & * & \dots & * \\ x_{21} & \dots & x_{2\,m} & * & \dots & * & x_{2n+1} & \dots & x_{2l} \end{pmatrix}.$$

So, each column represents an observation, and $x_{ij} \in \mathbb{R}$ is an entry at the ith feature and the jth observation.

Assume that we have a set of i.i.d. observations from a bivariate normal distribution $\boldsymbol{\mu} = (\mu_1, \mu_2)^T$ and covariance matrix $\boldsymbol{\Sigma} = \begin{pmatrix} \sigma_{11} & \sigma_{12} \\ \sigma_{21} & \sigma_{22} \end{pmatrix}$. Then, let L be the likelihood of the data and $s_{11} = \sum_{j=1}^{m} x_{1j}^2$, $s_{12} = \sum_{j=1}^{m} x_{2j}x_{1j}$, $s_{22} = \sum_{j=1}^{m} x_{2j}^2$.

Then, the resulting estimators obtained by maximizing L w.r.t $\mu_1, \sigma_{11}, \mu_2, \sigma_{22}$, and σ_{12} are:

$$\hat{\sigma}_{11} = \frac{\sum_{j=1}^{n} x_{1j}}{n}, \; \hat{\sigma}_{22} = \frac{\sum_{j=1}^{m} x_{2j}^2 + \sum_{j=n+1}^{l} x_{2j}^2}{m+l-n}$$

and $\hat{\sigma}_{12}$, where $\hat{\sigma}_{12}$ is the maximizer of:

$$\eta = C - \frac{1}{2} m \log\left(\sigma_{22} - \frac{\sigma_{12}^2}{\sigma_{11}}\right) - \frac{1}{2}\left(s_{22} - 2\frac{\sigma_{12}}{\sigma_{11}} s_{12} + \frac{\sigma_{12}^2}{\sigma_{11}^2} s_{11}\right)\left(\sigma_{22} - \frac{\sigma_{12}^2}{\sigma_{11}}\right)^{-1}.$$

Algorithm 1. DPER*

Input: ***centered*** input data $\mathcal{X}$ of p features.
Ouput: $\hat{\mathbf{\Sigma}} = (\sigma_{ij})_{i,j=1}^{p}$.

1. Estimate the diagonal elements of $\mathbf{\Sigma}$: $\hat{\sigma}_{ii}$ is the uncorrected sample variance of all the available entries in the ith feature.
2. For each $\hat{\sigma}_{ij}$ that lies below the diagonal of $\mathbf{\Sigma}$, compute $\hat{\sigma}_{ij}$ based on Equation (??). If two solutions maximize the likelihood function, choose the one closest to the estimate based on case deletion.
3. $\hat{\mathbf{\Sigma}} \leftarrow \frac{1}{2}(\hat{\mathbf{\Sigma}} + \hat{\mathbf{\Sigma}}^T)$
4. diag $(\hat{\mathbf{\Sigma}}) \leftarrow$ diag $(\hat{\mathbf{\Sigma}})/2$

4 Methodology

In this section, we detail the proposed *PCA for missing data* (PCAM) algorithm, which can derive principal components directly from datasets with missing values without any preliminary imputation.

4.1 Algorithm Structure Overview

The PCAM algorithm, described in algorithm 2, consists of two methods: FIT and TRANSFORM. The FIT method requires the estimation of the covariance matrix $\mathbf{\Sigma}$, the precision matrix $\mathcal{P}$, and the eigenvectors $\mathbf{V}$. Therefore, this section is organized as follows: Subsects. 4.2, 4.3, and 4.4 describe the estimation of the covariance matrix, the precision matrix, and the projection matrix, respectively; Subsects. 4.5 and 4.6 describe the transformation from input space into a lower dimensional-space using a linear transformation. The estimation of the three matrices is enveloped within the FIT method, while the transformation described in Sects. 4.5 and 4.6 is implemented by the TRANSFORM method of the algorithm 2. Finally, Subsect. 4.7 discusses the overall implementation of the algorithm.

4.2 Estimating the Covariance Matrix.

Assume that the data are centered so that each column mean is zero, the algorithm first estimates $\hat{\mathbf{\Sigma}}_D$ of $\mathbf{\Sigma}$ from the normalized input $\mathcal{X}$ using DPER* [28]. However, when the number of variables is not small compared to the number of samples, the estimate of the non-shrinkage covariance matrix is not necessarily well conditioned for inversion [13], which is needed later. Therefore, we used a linear shrinkage estimate [13], which shrinks the estimated covariance $\hat{\mathbf{\Sigma}}_D$ toward the identity matrix to minimize the expected quadratic loss,

$$\min_{c} \mathbb{E}[||\hat{\mathbf{\Sigma}} - \mathbf{\Sigma}||] \quad \text{such that} \quad \hat{\mathbf{\Sigma}} = c\boldsymbol{\mu}\mathbf{I} + (1-c)\hat{\mathbf{\Sigma}}_D. \tag{1}$$

Here, $\mathbf{I}$ is the identity matrix; c is the linear shrinkage intensity; $\boldsymbol{\mu} = \frac{\text{Tr}(\mathbf{\Sigma};\mathbf{I})}{p}$, where p is the number of features; $\hat{\mathbf{\Sigma}}_D$ is DPER estimation of the covariance matrix; and $\mathbf{\Sigma}$ is the true unobserved covariance. In this study, we choose the shrinkage intensity by grid searching for several intensity values. Then we choose the intensity that results in the best classification accuracy, as described in algorithm 3. The result of this step is the estimation $\hat{\mathbf{\Sigma}}$ of the covariance matrix.

4.3 Estimating the Precision Matrix.

We can simply estimate the precision matrix using the inverse matrix of the estimated covariance matrix $\hat{\mathbf{\Sigma}}$. However, it is not always the case that the best estimation in the inversion of the covariance matrix is the best estimation of the inversion of the covariance matrix, especially for high-dimensional data [7]. Hence, we used Graphical Lasso [7] to enforce sparsity in the estimation of the precision matrix. We estimate the precision matrix $\mathcal{P}$ by minimizing

$$\log\det\mathcal{P} - \text{Tr}(\mathbf{S}\mathcal{P}) - \alpha||\mathcal{P}||_1, \tag{2}$$

where $||.||_1$ is the L_1 norm - the sum of the absolute value of elements of the precision matrix, and α is the L_1 penalty. The L_1 penalty for the Graphical Lasso method by grid searching over several values. During the experiments, we found that when the shrunk covariance is not employed, the Graphical Lasso method sometimes fails to compute because the estimated covariance is numerically ill-conditioned. So, we employed Shrunk Covariance to estimate $\mathbf{\Sigma}$ based on the resulting estimate from DPER* before applying Graphical Lasso on that estimate for the estimation of the precision matrix.

4.4 Finding the Projection Matrix and Getting the Lower-Dimensional Representation for Fully Observed Samples

Now that $\mathbf{\Sigma}$ is estimated, suppose that eigen decomposition on $\hat{\mathbf{\Sigma}}$ gives the sorted eigenvalue-eigenvector pairs $(\lambda_1, \mathbf{v}_1), ..., (\lambda_p, \mathbf{v}_p)$ so that $\lambda_1 \geq \lambda_2 \geq ... \geq \lambda_p$. In addition, assume that the first r eigenvalues are used for dimension reduction. Then, let

$$\mathbf{V} = (\mathbf{v}_1, ..., \mathbf{v}_r) \tag{3}$$

To get the lower-dimensional representation for fully observed samples, we can simply perform a linear transformation using the eigenvectors to get the lower-dimensional representation for a fully observed sample $\mathbf{x}$.

$$\mathbf{r} = \mathbf{V}^T \mathbf{x}. \tag{4}$$

4.5 Matrix Partitioning for the Derivation of the Lower-Dimensional Representation of a Sample with Missing Entries

Now, consider samples with missing entries. Suppose that we have a sample $\mathbf{x}$ that can be partitioned into

$$\mathbf{x} = \begin{pmatrix} \mathbf{x}_o \\ \mathbf{x}_m \end{pmatrix}, \tag{5}$$

where $\mathbf{x}_o$ corresponds to the observed entries in $\mathbf{x}$ and $\mathbf{x}_m$ corresponds to the missing entries in $\mathbf{x}$.

Next, partition $\mathbf{V}$ into $\mathbf{V} = \begin{pmatrix} \mathbf{V}_o \\ \mathbf{V}_m \end{pmatrix}$, where $\mathbf{V}_o$ corresponds to observed features in $\mathbf{x}$ (i.e., the features in $\mathbf{x}_o$) and $\mathbf{V}_m$ corresponds to missing features in $\mathbf{x}$ (i.e., the features in in $\mathbf{x}_m$). In addition, we also partition the estimated covariance matrix $\widehat{\mathbf{\Sigma}}$ and precision matrix $\hat{\mathcal{P}}$ into

$$\widehat{\mathbf{\Sigma}} = \begin{pmatrix} \widehat{\mathbf{\Sigma}}_o & \widehat{\mathbf{\Sigma}}_{om} \\ \widehat{\mathbf{\Sigma}}_{mo} & \widehat{\mathbf{\Sigma}}_m \end{pmatrix} \tag{6}$$

and

$$\hat{\mathcal{P}} = \begin{pmatrix} \hat{\mathcal{P}}_o & \hat{\mathcal{P}}_{om} \\ \hat{\mathcal{P}}_{mo} & \hat{\mathcal{P}}_m \end{pmatrix} \tag{7}$$

where $\widehat{\mathbf{\Sigma}}_o, \hat{\mathcal{P}}_o$ and $\widehat{\mathbf{\Sigma}}_m, \hat{\mathcal{P}}_m$ corresponding to the observed and missing part of $\mathbf{x}$, respectively.

4.6 Deriving an Estimate for the Representation of an Input Observation in the Lower-Dimension Space

Given the three estimated parameters $\hat{\mathbf{\Sigma}}, \hat{\mathcal{P}}$, and $\mathbf{V}$ as estimated in Sect. 4.5, we are now able to estimate the projection of each sample $\mathbf{x}$ onto a lower dimensional space, as described in function TRANSFORM in algorithm 4. The function transforms each sample $\mathbf{x}$ by first partitioning the sample and estimated parameter into a continuously observed partition $\mathbf{x}_o$ and a missing partition $\mathbf{x}_m$. Then, if we assume each sample $\mathbf{x} \sim \mathcal{N}(\mathbf{0}, \hat{\mathbf{\Sigma}})$, the missing partition $\mathbf{x}_m$ can be estimated by the conditional Gaussian expectation formula 1 given the observed partition $\mathbf{x}_o$.

$$\hat{\mathbf{x}}_m = \mathbb{E}[\mathbf{x}_m] + \mathbf{\Sigma}_{om}\mathbf{\Sigma}_o^{-1}(\mathbf{x}_o - \mathbb{E}[\mathbf{x}_o]) \tag{8}$$

Recall that $\mathbf{x}$ is centered, so that $\mathbb{E}[\mathbf{x}_o]$ and $\mathbb{E}[\mathbf{x}_m]$ are $\mathbf{0}$, hence,

$$\hat{\mathbf{x}}_m = \mathbf{\Sigma}_{om}\mathbf{\Sigma}_o^{-1}\mathbf{x}_o. \tag{9}$$

Finally, the projection can be estimated by transforming sample $\mathbf{x}$ by a linear transformation $\mathbf{V}^\top$

$$\begin{aligned}\mathbf{V}^\top\mathbf{x} &= (\mathbf{V}_o^\top, \mathbf{V}_m^\top)\begin{pmatrix}\mathbf{x}_o\\ \mathbf{x}_m\end{pmatrix}\\ &= \mathbf{V}_o^\top\mathbf{x}_o + \mathbf{V}_m^\top\mathbf{x}_m\\ &= \mathbf{V}_o^\top\mathbf{x}_o + \mathbf{V}_m^\top\mathbf{\Sigma}_{om}\mathbf{\Sigma}_o^{-1}\mathbf{x}_o\\ &= (\mathbf{V}_o^\top + \mathbf{V}_m^\top\mathbf{\Sigma}_{om}\mathbf{\Sigma}_o^{-1})\mathbf{x}_o.\end{aligned} \tag{10}$$

Note that we replaced $\mathbf{x}_m$ by $\hat{\mathbf{x}}_m$ in Eq. 9, hence the third equality. Now, replace $\mathbf{\Sigma}_{om}$ by $\hat{\mathbf{\Sigma}}_{om}$ and $\mathbf{\Sigma}_o^{-1}$ by $\hat{\mathbf{\Sigma}}_o^{-1}$ gives the following estimate for the projection $\mathbf{V}^\top\mathbf{x}$ of $\mathbf{x}$

$$\mathbf{r} = (\mathbf{V}_o^\top + \mathbf{V}_m^\top\hat{\mathbf{\Sigma}}_{om}\hat{\mathcal{P}})\mathbf{x}_o. \tag{11}$$

4.7 PCAM Implementation

The PCAM algorithm consists of two methods, namely FIT and TRANSFORM, as presented in the algorithm 2.

Algorithm 2. PCAM algorithm

Input: Centered input $\mathcal{X}$, labels y, classifier $\mathcal{F}$
Output: The estimation of $\mathcal{X}$'s embedding in the lower-dimensional space $\mathcal{R}$.

```
1: function PCAM(𝒳, ℱ)
2:     (Σ̂, 𝒫̂, V) ← FIT(𝒳, ℱ)
3:     ℛ ← TRANSFORM(𝒳, Σ̂, 𝒫̂, V)
4:     return ℛ
5: end function
```

The FIT method, as detailed in algorithm 3, addresses the problem of estimating the covariance matrix $\hat{\mathbf{\Sigma}}$ and $\hat{\mathcal{P}}$ and $\mathbf{V}$, and the practical problem which is choosing shrinkage intensity c^* and L_1 penalty α^*. The FIT method starts with obtaining the DPER* estimation of the covariance matrix $\hat{\mathbf{\Sigma}}_D$. Next step, to grid search for the pair of parameters (c, α), we split the dataset into train set $\mathbf{X}_t$ and validation set $\mathbf{X}_v$; and used a variable S to keep track of the best accuracy of the classification task on the validation set. For each of the pair of parameters (c, α), we shrank the DPER's estimation of covariance $\hat{\mathbf{\Sigma}}_D$ toward the identity matrix by linear shrinkage intensity c, resulting in covariance $\hat{\mathbf{\Sigma}}_c$.

Subsequently, we performed an Eigen decomposition on $\hat{\mathbf{\Sigma}}_c$, resulting in the projection matrix $\mathbf{V}$. In the next step, we perform Graphical Lasso on $\hat{\mathbf{\Sigma}}_c$ with a L_1 penalty of α, resulting in the estimation $\hat{\mathcal{P}}_\alpha$ of the precision matrix. Given the estimated covariance matrix $\hat{\mathbf{\Sigma}}_c$, projection matrix $\mathbf{V}$, and estimated precision matrix $\hat{\mathcal{P}}_\alpha$; we employed the TRANSFORM method to estimate the lower-dimensional representation $\mathbf{Z}_t$ and $\mathbf{Z}_v$ of the train and validation set, as described in algorithm 4. We can then train the classifier $\mathcal{F}$ in the training set consisting of $\{\mathbf{Z}_t, \mathbf{y}_t\}$ and evaluate the classification accuracy on the validation set $\{\mathbf{Z}_v, \mathbf{y}_v\}$. The accuracy on validation set is used to select the best pair of parameter (c^*, α^*). Finally, the FIT method returns a triplet $(\hat{\mathbf{\Sigma}}^*, \hat{\mathcal{P}}^*, \mathbf{V}^*)$. In which: $\hat{\mathbf{\Sigma}}^* = \text{shrunk\_covariance}(\hat{\mathbf{\Sigma}}_D, c^*)$, $\hat{\mathcal{P}}^* = \text{graphical\_lasso}(\hat{\mathbf{\Sigma}}^*, \alpha^*)$, and $\mathbf{V}^* = \text{eigen\_decomposition}(\hat{\mathbf{\Sigma}}^*)$.

The FIT method searched for (c, α) over a set $\{0.0, 0.1, \cdots 0.9\} \times \{10^{-i} | i \in \{0, 1, 2, 3\}\}$, where $\{0.1, \cdots 0.9\}$ is the set of possible shrinkage intensity c and $\{10^{-i} | i \in \{0, 1, 2, 3\}\}$ is the set of possible L_1 penalty α.

Algorithm 3. PCAM's FIT method

Input: Centered train and validation sets $\{X_t, y_t\}, \{X_v, y_v\}$, classifier $\mathcal{F}$
Output: Estimated covariance matrix $\hat{\mathbf{\Sigma}}$, estimated precision matrix $\hat{\mathcal{P}}$, projection matrix $\mathbf{V}$

function FIT($\{X_t, y_t\}, \{X_v, y_v\}, \mathcal{F}$)
  $\hat{\mathbf{\Sigma}}_D = \text{DPER}^*(\mathcal{X})$
  Score S = 0 ▷ Keep track of best accuracy
  for $(c, \alpha) \in \{0.1, \cdots, 0.9\} \times \{10^{-3}, 10^{-2}, 10^{-1}, 1\}$ **do**
    $\hat{\mathbf{\Sigma}}_c = \text{shrunk\_covariance}(\hat{\mathbf{\Sigma}}_D, c)$ ▷ Shrunk covariance estimation
    $(\lambda_1, \mathbf{v}_1), ..., (\lambda_p, \mathbf{v}_p) \leftarrow$ sorted eigenvalue-eigenvector pairs of $\hat{\mathbf{\Sigma}}_c$, so that $\lambda_1 \geq \lambda_2 \geq ...\lambda_p$
    $\mathbf{V} = (\mathbf{v}_1, ..., \mathbf{v}_r)$
    $\hat{\mathcal{P}}_\alpha = \text{graphical\_lasso}(\hat{\mathbf{\Sigma}}_c, \alpha)$ ▷ Estimation of inverse covariance matrix using Graphical Lasso
    $Z_t = \text{TRANSFORM}(X_t, \hat{\mathbf{\Sigma}}_c, \hat{\mathcal{P}}_\alpha, V)$
    $Z_v = \text{TRANSFORM}(X_v, \hat{\mathbf{\Sigma}}_c, \hat{\mathcal{P}}_\alpha, V)$
    $\mathcal{F} \leftarrow \text{Train}(\mathcal{F}, \{Z_t, y_t\})$
    Accuracy $s = \text{Evaluate}(\mathcal{F}, \{(Z_v, y_v\})$
    if $s > S$ **then**
      $S \leftarrow s$
      $(c^*, \alpha^*) \leftarrow (c, \alpha)$
    end if
  end for ▷ Grid search for parameter c^* of Shrunk Covariance and parameter α^* Graphical Lasso
  $\hat{\mathbf{\Sigma}}^* \leftarrow \text{shrunk\_covariance}(\hat{\mathbf{\Sigma}}, c^*)$
  $\hat{\mathcal{P}}^* \leftarrow \text{graphical\_lasso}(\hat{\mathbf{\Sigma}}, \alpha^*)$
  $\mathbf{V}^* \leftarrow \text{eigen\_decomposition}(\hat{\mathbf{\Sigma}}^*)$
  return $\hat{\mathbf{\Sigma}}^*, \hat{\mathcal{P}}^*, \mathbf{V}^*$
end function

The TRANSFORM method as described in algorithm 4, requires four arguments, namely a normalized input $\mathcal{X}$, an estimated covariance matrix $\hat{\mathbf{\Sigma}}$, a projection matrix $\mathbf{V}$, and an estimated precision matrix $\hat{\mathcal{P}}$. For each sample x in $\mathcal{X}$, we can rearrange the entry and then partition the sample, the projection matrix, the estimated covariance matrix, and the precision matrix into contiguous partitions of observed and missing entry as described in algorithm 4. The projection of sample $\mathbf{x}$ is given by Eq. 11.

Algorithm 4. PCAM's TRANSFORM method

Input: Normalized input $\mathcal{X}$, estimated covariance matrix $\hat{\mathbf{\Sigma}}$, estimated precision matrix $\hat{\mathcal{P}}$, projection matrix $\mathbf{V}$
Output: the representation of $\mathcal{X}$ in the lower-dimensional space $\mathcal{R}$.

function TRANSFORM($\mathcal{X}, \hat{\mathbf{\Sigma}}, \hat{\mathcal{P}}, \mathbf{V}$)
  $\mathcal{R} \leftarrow \{\}$
  for sample $\mathbf{x}$ in $\mathcal{X}$ **do**
    Partition
    - $\mathbf{x} = \begin{pmatrix} \mathbf{x}_o \\ \mathbf{x}_m \end{pmatrix}$
    - $\mathbf{V} = \begin{pmatrix} \mathbf{V}_o \\ \mathbf{V}_m \end{pmatrix}$
    - $\hat{\mathbf{\Sigma}} = \begin{pmatrix} \hat{\mathbf{\Sigma}}_o & \hat{\mathbf{\Sigma}}_{om} \\ \hat{\mathbf{\Sigma}}_{mo} & \hat{\mathbf{\Sigma}}_m \end{pmatrix}$
    - $\hat{\mathcal{P}} = \begin{pmatrix} \hat{\mathcal{P}}_o & \hat{\mathcal{P}}_{om} \\ \hat{\mathcal{P}}_{mo} & \hat{\mathcal{P}}_m \end{pmatrix}$
    $\mathcal{R} \leftarrow \mathcal{R} \cup (\mathbf{V}_o + \mathbf{V}_m^\top \hat{\mathbf{\Sigma}}_{om} \hat{\mathcal{P}}_o)\mathbf{x}_o$
  end for
  return $\mathcal{R}$
end function

5 Experiments

5.1 Experiment Settings

In this study, we compare PCAM's estimation of lower dimensional embedding against other alternative methods by comparing the accuracy of downstream classification tasks. The alternative methods can be grouped into two categories. The first category utilizes the strategy of first imputing the missing data and then using PCA to reduce the dimension of the imputed data such that the explained variance ratio is greater than or equal to 95%. Five methods included in this category are GAIN [30], KNN-Imputation, MICE [27], Miss Forest [25], SOFT-IMPUTE [16]. The second category directly reduces the dimensionality of the missing data; the example of this category is the TSR algorithm [6]. The explained variance ratio 95% threshold is applied for both TSR and our proposed method. Here, we refer to KNN-PCA as the approach where KNN is used for imputation prior to PCA dimension reduction. It is also similar for GAIN-PCA, MICE-PCA, Miss Forest-PCA, and SOFT IMPUTE-PCA.

Here, we use classification accuracy rather than MSE as our primary metric because PCAM's internal use of a sparse inverse for estimating principal components leads to fundamentally different properties than traditional sparse PCA methods. While the sparse PCA baselines (MICE+SparsePCA, SOFT+SparsePCA, KNNI+SparsePCA) directly optimize for sparse loadings with good reconstruction, PCAM employs sparse inverse computation as an intermediate step to estimate non-sparse PCA under missing data conditions.

We conducted experiments on four datasets, namely, Ecoli [17], Frogs [2], Heart [1], and Ionosphere [3]. For each dataset, we emulate the missing entries completely at random (MCAR) with the probability from 10% to 70% with a step size of 20%. A summary of the number of observations for each dataset and the dimension for each observation is given in table 1.

Table 1. Dataset descriptions.

	Ecoli	Frogs	Heart	Ionosphere
# Instances	336	7195	267	351
# Features	7	22	44	32

For the implementation of GAIN, we used the official code provided at the author's repository[1]. For MICE and KNN Imputer, we employed scikit-learn [20] implementation. We used *fancyimpute* [23] package for Soft Impute, and finally, we used *missingpy* package for the implementation of Miss Forest. We used three classifiers for downstream classification tasks: K-Nearest Neighbors (KNN), Support Vector Machine (SVM), and Multi-layer Perceptron Classifier (NN) using scikit-learn [20]. All experiments were carried out on a machine running Ubuntu 22.04LTS with AMD 3.9Ghz Ryzen 5600G chipset and 16G RAM.

Table 2. A comparison of KNN classification accuracy between PCAM and other methods across multiple missing rates.

Dataset	Missing rate Method	10.0%	30.0%	50.0%	70.0%
Ecoli	PCAM	94.12 ± 5.96	88.99 ± 5.01	**86.16 ± 2.48**	78.52 ± 5.06
	GAIN-PCA	92.39 ± 2.84	81.58 ± 2.26	77.91 ± 4.03	72.19 ± 9.63
	KNN-PCA	95.73 ± 1.67	91.02 ± 2.73	84.77 ± 2.77	80.56 ± 3.32
	MICE-PCA	95.18 ± 1.25	**92.90 ± 2.02**	85.95 ± 2.32	80.68 ± 3.67
	Miss Forest-PCA	**95.77 ± 1.36**	91.56 ± 4.42	85.10 ± 3.10	**81.10 ± 4.14**
	Soft Impute-PCA	95.11 ± 1.96	91.39 ± 2.23	85.51 ± 3.26	77.80 ± 6.41
	TSR	95.46 ± 1.23	92.57 ± 0.62	85.52 ± 4.42	78.04 ± 5.32
Frogs	PCAM	98.11 ± 0.47	95.36 ± 0.69	93.31 ± 0.52	88.86 ± 1.10
	GAIN-PCA	98.00 ± 0.43	94.22 ± 0.41	87.58 ± 0.98	70.64 ± 2.62
	KNN-PCA	**99.17 ± 0.34**	**98.62 ± 0.24**	93.28 ± 0.37	81.97 ± 0.83
	MICE-PCA	99.12 ± 0.23	98.18 ± 0.11	95.82 ± 0.39	88.17 ± 0.79
	Miss Forest-PCA	99.00 ± 0.16	98.52 ± 0.13	95.80 ± 0.59	89.22 ± 0.82
	Soft Impute-PCA	98.73 ± 0.48	97.60 ± 0.19	94.09 ± 0.56	88.39 ± 0.95
	TSR	99.12 ± 0.20	98.58 ± 0.10	**96.28 ± 0.40**	**90.31 ± 0.76**
Heart	PCAM	75.91 ± 4.32	**77.82 ± 6.11**	**78.26 ± 4.94**	74.29 ± 3.16
	GAIN-PCA	**78.72 ± 5.49**	75.14 ± 2.03	71.54 ± 6.29	**79.18 ± 4.30**
	KNN-PCA	75.25 ± 4.91	74.42 ± 4.78	75.92 ± 2.32	72.37 ± 7.42
	MICE-PCA	76.14 ± 3.35	76.77 ± 3.22	6 9.14 ± 5.30	77.28 ± 3.47
	Miss Forest-PCA	75.35 ± 5.20	73.80 ± 2.99	76.64 ± 4.36	75.03 ± 3.57
	Soft Impute-PCA	74.54 ± 5.75	76.40 ± 4.46	73.50 ± 5.28	75.64 ± 5.95
	TSR	74.74 ± 4.58	72.56 ± 7.33	70.79 ± 5.52	69.61 ± 2.25
Ionosphere	PCAM	**86.59 ± 1.92**	85.18 ± 4.93	**84.36 ± 3.67**	77.42 ± 3.90
	GAIN-PCA	84.90 ± 2.32	84.53 ± 2.64	80.56 ± 3.95	60.63 ± 3.31
	KNN-PCA	84.24 ± 2.43	84.67 ± 2.80	82.99 ± 3.70	**79.71 ± 2.75**
	MICE-PCA	84.39 ± 2.80	**85.26 ± 3.36**	83.31 ± 3.11	77.66 ± 5.29
	Miss Forest-PCA	84.06 ± 2.15	83.14 ± 1.79	80.64 ± 3.74	75.35 ± 4.03
	Soft Impute-PCA	85.46 ± 1.67	83.47 ± 1.40	82.02 ± 3.06	74.65 ± 1.49
	TSR	82.35 ± 1.21	78.86 ± 3.43	79.01 ± 3.83	69.71 ± 3.67

[1] https://github.com/jsyoon0823/GAIN.

5.2 Results and Analysis

The detailed experimental results of KNN classification and SVM classification accuracy between PCAM and other methods across multiple missing rates on four datasets (such as Ecoli, Frogs, Hearts, and Ionosphere) are shown in the tables 2 and 3.

From table 2, we can see the promising performance of PCAM compared to the other techniques when the downstream classifier is KNN. For example, for the Heart dataset, under missing rate 50%, the accuracy of PCAM is 78.26, while for GAIN-PCA, KNN-PCA, MICE-PCA, Miss Forest - PCA, Soft Impute - PCA, TSR, it is 71.54 %, 75.92 %, 69.14 %, 76.64 %, 73.50 %, 70.79 %, respectively. As another example, on the Ionosphere dataset, at 50 % missing rate, the accuracy of PCAM is 84.36, while the accuracy for the remaining methods (GAIN-PCA, KNN-PCA, MICE-PCA, Miss Forest - PCA, Soft Impute - PCA, TSR) is 80.56 %, 82.99 %, 83.31 %, 80.64 %, 82.02 %, 79.01 %, respectively.

Next, when comparing the SVM classification accuracy between PCAM and other methods, the table 3 also shows that the PCAM method gives promising results compared to other methods. For example, at a missing rate of 10% for the Frogs dataset, the SVM classification accuracy of PCAM is 97.99 %, a 4.01 percent improvement compared to GAIN-PCA, 3.38 percent improvement com-

Table 3. A comparison of SVM classification accuracy between PCAM and other methods across multiple missing rates.

Dataset	Missing rate Method	10.0%	30.0%	50.0%	70.0%
Ecoli	PCAM	**95.93 ± 2.89**	87.74 ± 6.23	**89.25 ± 3.65**	**82.00 ± 6.00**
	GAIN-PCA	77.46 ± 5.44	77.46 ± 5.44	77.46 ± 5.44	77.46 ± 5.44
	KNN-PCA	93.55 ± 2.24	86.83 ± 3.52	84.54 ± 3.94	77.46 ± 5.44
	MICE-PCA	93.86 ± 2.55	87.92 ± 3.41	86.29 ± 3.65	77.12 ± 4.89
	Miss Forest-PCA	94.15 ± 1.69	**89.77 ± 2.92**	88.39 ± 1.89	
	Soft Impute-PCA	93.03 ± 2.31	87.70 ± 4.04	81.16 ± 4.46	
	TSR	93.58 ± 1.91	89.47 ± 3.22	87.28 ± 2.14	77.12 ± 4.89
Frogs	PCAM	**97.99 ± 0.53**	**95.62 ± 0.56**	**93.57 ± 0.83**	**90.11 ± 0.38**
	GAIN-PCA	93.98 ± 0.14	89.40 ± 0.30	81.60 ± 0.94	71.94 ± 3.57
	KNN-PCA	94.61 ± 0.53	94.69 ± 0.30	90.23 ± 0.65	80.68 ± 0.46
	MICE-PCA	94.51 ± 0.45	93.33 ± 0.38	90.49 ± 0.67	83.91 ± 0.88
	Miss Forest-PCA	94.53 ± 0.53	94.33 ± 0.28	92.82 ± 0.73	88.36 ± 0.83
	Soft Impute-PCA	94.26 ± 0.23	92.82 ± 0.25	90.04 ± 0.86	83.31 ± 0.70
	TSR	94.49 ± 0.42	93.54 ± 0.28	91.22 ± 0.92	86.90 ± 0.93
Heart	PCAM	**79.23 ± 4.42**	**80.10 ± 7.68**	**80.80 ± 6.45**	76.32 ± 2.99
	GAIN-PCA	78.32 ± 5.66	76.84 ± 4.99	79.55 ± 6.32	79.04 ± 4.36
	KNN-PCA	78.07 ± 5.39	76.00 ± 5.30	78.75 ± 5.95	79.04 ± 4.36
	MICE-PCA	78.37 ± 5.64	77.99 ± 5.64	79.60 ± 6.47	79.04 ± 4.36
	Miss Forest-PCA	77.56 ± 5.42	77.19 ± 5.55	79.60 ± 6.47	79.04 ± 4.36
	Soft Impute-PCA	78.70 ± 5.89	77.28 ± 5.54	78.83 ± 5.88	79.04 ± 4.36
	TSR	78.32 ± 5.66	76.60 ± 5.42	77.99 ± 5.84	79.04 ± 4.36
Ionosphere	PCAM	91.15 ± 1.28	89.92 ± 3.32	**89.69 ± 4.13**	**82.72 ± 5.65**
	GAIN-PCA	89.75 ± 2.57	86.61 ± 4.54	81.18 ± 1.55	64.98 ± 8.13
	KNN-PCA	90.93 ± 1.65	88.06 ± 3.07	86.38 ± 2.25	78.58 ± 3.60
	MICE-PCA	**92.28 ± 3.58**	**91.20 ± 3.82**	87.13 ± 4.60	76.17 ± 5.12
	Miss Forest-PCA	90.02 ± 2.50	87.38 ± 2.63	85.53 ± 3.80	77.17 ± 2.05
	Soft Impute-PCA	90.93 ± 3.71	89.13 ± 4.09	86.58 ± 3.66	77.85 ± 4.04
	TSR	89.67 ± 3.16	87.38 ± 3.09	80.58 ± 4.60	72.99 ± 4.78

pared to KNN-PCA, 3.5 percent improvement compared to TSR. As another example, for the Inososphere dataset, at a missing rate of 50%, the SVM classification accuracy of PCAM is 89.69 %, an 8.51 % improvement compared to GAIN-PCA, 2.56 % improvement compared to MICE-PCA and 9.11 % improvement compared to TSR.

In general, the results show that PCAM has a competitive performance compared to the other techniques under comparison. Another interesting fact is that most of the time, PCAM, GAIN-PCA, KNN-PCA, MICE-PCA, Miss Forest-PCA, and Soft Impute-PCA outperform TSR, a method that provides the principal components directly from missing data directly without prior imputation.

6 Conclusion

In this paper, we introduced PCAM, a method to estimate the lower space presentation directly from datasets with missing values. This helps reduce the need for prior imputation in PCA. We also conducted various experiments to show that PCAM has a competitive performance compared to the other state-of-the-art methods under comparison. In addition, the experiments conducted in this study also show that compared to TSR, a method to directly obtain the principal components from missing data, the PCA based on imputed data can many times give better classification accuracies when KNN or SVM is the classifier for the downstream classification.

In the future, we plan to speed up this PCAM by parallel computing to improve the speed of the estimations. Moreover, it is well known that PCA is a useful tool for noise reduction, exploring the relationships between the features and discovering the underlying structures of the data. So, in the future, we will explore the performance of PCAM in these directions. We also plan to extend this work to monotone missing data, where instead of using DPER* for estimating the covariance matrix, we will use EPEM [18]. In addition, we will explore the use of kernels with PCAM for nonlinear data.

References

1. et al, C.: SPECTF Heart. UCI Machine Learning Repository (2001). https://doi.org/10.24432/C5N015
2. et al, C.: Anuran Calls (MFCCs). UCI Machine Learning Repository (2017). https://doi.org/10.24432/C5CC9H
3. et al, S.: Ionosphere. UCI Machine Learning Repository (1989). https://doi.org/10.24432/C5W01B
4. Almeida, R.J., Adriaans, G., Shapovalova, Y.: Graphical causal models and imputing missing data: a preliminary study. In: International Conference on Information Processing and Management of Uncertainty in Knowledge-Based Systems, pp. 485–496. Springer (2020)
5. Choudhury, S.J., Pal, N.R.: Imputation of missing data with neural networks for classification. Knowl.-Based Syst. **182**, 104838 (2019)

6. Folch-Fortuny, A., Arteaga, F., Ferrer, A.: PCA model building with missing data: new proposals and a comparative study. Chemom. Intell. Lab. Syst. **146**, 77–88 (2015)
7. Friedman, J., Hastie, T., Tibshirani, R.: Sparse inverse covariance estimation with the graphical lasso. Biostatistics **9**(3), 432–441 (12 2007). https://doi.org/10.1093/biostatistics/kxm045
8. Gower, J.: Statistical methods of comparing different multivariate analyses of the same data. Math. Archaeol. Historical Sci. **138**, 149 (1971)
9. Jin, H., Jung, S., Won, S.: missforest with feature selection using binary particle swarm optimization improves the imputation accuracy of continuous data. Genes & Genomics **44**(6), 651–658 (2022)
10. Johnson, R.A., Wichern, D.W., et al.: Applied Multivariate Statistical Analysis, vol. 5. Prentice hall Upper Saddle River, NJ (2002)
11. Josse, J., Husson, F., et al.: Multiple imputation in principal component analysis. Adv. Data Anal. Classif. **5**(3), 231–246 (2011)
12. Kiers, H.A.: Weighted least squares fitting using ordinary least squares algorithms. Psychometrika **62**, 251–266 (1997)
13. Ledoit, O., Wolf, M.: A well-conditioned estimator for large-dimensional covariance matrices. J. Multivariate Anal. **88**(2), 365–411 (2004). https://doi.org/10.1016/S0047-259X(03)00096-4, https://www.sciencedirect.com/science/article/pii/S0047259X03000964
14. Lin, W.C., Tsai, C.F.: Missing value imputation: a review and analysis of the literature (2006–2017). Artif. Intell. Rev. **53**, 1487–1509 (2020)
15. Lin, W.C., Tsai, C.F., Zhong, J.R.: Deep learning for missing value imputation of continuous data and the effect of data discretization. Knowl.-Based Syst. **239**, 108079 (2022)
16. Mazumder, R., Hastie, T., Tibshirani, R.: Spectral regularization algorithms for learning large incomplete matrices. J. Mach. Learn. Res. **11**(Aug), 2287–2322 (2010)
17. Nakai, K.: Ecoli. UCI Machine Learning Repository (1996). https://doi.org/10.24432/C5388M
18. Nguyen, T., Nguyen, D.H., Nguyen, H., Nguyen, B.T., Wade, B.A.: EPEM: efficient parameter estimation for multiple class monotone missing data. Inf. Sci. **567**, 1–22 (2021)
19. Nguyen, T., Nguyen-Duy, K.M., Nguyen, D.H.M., Nguyen, B.T., Wade, B.A.: DPER: direct parameter estimation for randomly missing data. Knowl.-Based Syst. **240**, 108082 (2022)
20. Pedregosa, F., et al.: Scikit-learn: machine learning in Python. J. Mach. Learn. Res. **12**, 2825–2830 (2011)
21. Pujianto, U., Wibawa, A.P., Akbar, M.I., et al.: K-nearest neighbor (k-nn) based missing data imputation. In: 2019 5th International Conference on Science in Information Technology (ICSITech), pp. 83–88. IEEE (2019)
22. Ramosaj, B., Pauly, M.: Predicting missing values: a comparative study on non-parametric approaches for imputation. Comput. Stat. **34**, 1741–1764 (2019)
23. Rubinsteyn, A., Feldman, S.: fancyimpute: An imputation library for python. https://github.com/iskandr/fancyimpute (2016)
24. Samad, M.D., Abrar, S., Diawara, N.: Missing value estimation using clustering and deep learning within multiple imputation framework. Knowl.-Based Syst. **249**, 108968 (2022)
25. Stekhoven, D.J., Bühlmann, P.: Missforest-non-parametric missing value imputation for mixed-type data. Bioinformatics **28**(1), 112–118 (2012)

26. Tashiro, Y., Song, J., Song, Y., Ermon, S.: CSDI: conditional score-based diffusion models for probabilistic time series imputation. Adv. Neural. Inf. Process. Syst. **34**, 24804–24816 (2021)
27. Van Buuren, S., Groothuis-Oudshoorn, K.: mice: multivariate imputation by chained equations in r. J. Stat. Softw. **45**, 1–67 (2011)
28. Vu, M.A., Nguyen, T., Do, T.T., Phan, N., Halvorsen, P., Riegler, M.A., Nguyen, B.T.: Conditional expectation for missing data imputation. arXiv preprint arXiv:2302.00911 (2023)
29. Wold, H., Lyttkens, E.: Nonlinear iterative partial least squares (nipals) estimation procedures. Bull. Int. Stat. Inst. **43**(1), 29–47 (1969)
30. Yoon, J., Jordon, J., Schaar, M.: Gain: Missing data imputation using generative adversarial nets. In: International Conference on Machine Learning, pp. 5689–5698. PMLR (2018)

On the Brittleness of CLIP Text Encoders

Allie Tran(✉) and Luca Rossetto

Dublin City University, Dublin, Ireland
{allie.tran,luca.rossetto}@dcu.ie

Abstract. Multimodal co-embedding models, especially CLIP, have advanced the state of the art in zero-shot classification and multimedia information retrieval in recent years by aligning images and text in a shared representation space. However, such modals trained on a contrastive alignment can lack stability towards small input perturbations. Especially when dealing with manually expressed queries, minor variations in the query can cause large differences in the ranking of the best-matching results. In this paper, we present a systematic analysis of the effect of multiple classes of non-semantic query perturbations in an multimedia information retrieval scenario. We evaluate a diverse set of lexical, syntactic, and semantic perturbations across multiple CLIP variants using the TRECVID Ad-Hoc Video Search queries and the V3C1 video collection. Across models, we find that syntactic and semantic perturbations drive the largest instabilities, while brittleness is concentrated in trivial surface edits such as punctuation and case. Our results highlight robustness as a critical dimension for evaluating vision-language models beyond benchmark accuracy.

Keywords: Multimodal Embedding · Multimedia Information Retrieval · Encoding Brittleness

1 Introduction

The rise of large-scale vision-language models (VLMs) has marked a significant advancement in AI, with models like CLIP, BLIP, and Flamingo demonstrating remarkable capabilities on zero-shot tasks [2,16,21,25,33,37]. These models are trained on vast datasets of images and text, learning to align visual and textual representations through contrastive learning. CLIP [21], in particular, combines a vision encoder and a text encoder, allowing it to perform zero-shot image classification by ranking images based on their semantic similarity to a given text query.

Despite their impressive performance on broad benchmarks, a critical, yet underexplored, vulnerability of these models is their surprising sensitivity to minor linguistic variations. While VLMs are expected to be robust to subtle changes in a text query, small perturbations–such as punctuation, spelling, or word order have been observed to cause large fluctuations in their output [30], severely impacting their reliability in zero-shot tasks. This phenomenon poses a

J. Lokoč et al. (Eds.): MMM 2026, LNCS 16413, pp. 361–374, 2026.
https://doi.org/10.1007/978-981-95-6957-1_26

substantial challenge for their deployment in real-world applications where user inputs are naturally noisy and varied.

To study this phenomenon rigorously, we distinguish between two related concepts. We define **instability** as the divergence between the image ranking retrieved by an original query and that retrieved by its perturbed version. Instability captures whether the system produces consistent outputs under small input changes. We then define **brittleness** as instability *normalized by embedding distance* between the original and perturbed queries. This measure accounts for the fact that some perturbations cause larger shifts in text embedding space than others: a model is brittle if even a small movement in embedding space results in disproportionately large ranking changes.

This paper provides a systematic analysis of how CLIP's zero-shot retrieval performance is affected by such perturbations. We study not only semantic changes (e.g., negation, spatial reasoning) but also meaning-preserving variations (e.g., typos, casing, paraphrases). Specifically, we address the following research questions:

1. To what extent do small perturbations to the text input affect CLIP's image rankings (*instability*)?
2. Which types of linguistic perturbations cause the most significant instabilities?
3. How can we quantify brittleness in a way that distinguishes genuine semantic variation from overreaction to small, meaning-preserving edits?

Our contributions are as follows: First, we present the first large-scale, controlled evaluation of CLIP's text encoders under structured lexical, syntactic, and semantic perturbations, spanning 190 TRECVID queries and over one million keyframes. Second, we introduce instability as a retrieval-oriented measure of robustness, quantified through overlap@k and RBO, and show that instability is systematic but varies across perturbation classes. Third, we propose a new brittleness index that normalises instability by embedding distance, revealing that CLIP overreacts most strongly to trivial edits (e.g., punctuation, case) rather than to semantic changes. Fourth, by comparing LAION-trained baselines with fine-tuned (FARE2) and alternative (EVA02-L14) text encoders, we show that architectural and training choices improve robustness but do not eliminate brittleness. Finally, we discuss the implications of these findings for real-world applications of CLIP, from retrieval to moderation and safety-critical tasks, and sketch directions for mitigation, including brittleness-aware training objectives and query-normalisation layers. The code used for data processing and analysis is available at https://github.com/allie-tran/clip-brittleness.

2 Related Work

2.1 Prior Work on CLIP's Limitations

Findings from single-modality research already warn us against over-interpreting deep models as compositional reasoners. In language, shuffling token order degrades BERT much less than one would expect, implying substantial bag-of-words behavior [9]. In vision, BagNet–a CNN constrained to local features with minimal spatial integration–performs surprisingly well on ImageNet [5]. This suggests that high accuracy can be achieved by aggregating local cues rather than building holistic scene representations, which in turn leads to confusions about object layout and to collapsing multiple objects into a single description [19,36].

Cross-modal evaluations make the weakness explicit. Winoground [28] is purpose-built so that *word order matters*: each example has two images and two captions that use the same words in different orders, with the correct image-caption pairing flipping solely because of order and role binding. CLIP-style VLMs perform only slightly above chance on this setting, indicating poor compositional reasoning even when vocabulary is held constant. Subsequent studies report sharp drops whenever genuine concept binding is required [15] or when spatial relations must be resolved, as in the What's Up benchmark [10]. The visual front-ends used in many multimodal LLMs–often CLIP-based–also exhibit systematic failures on basic visual patterns [29]. Beyond composition and relations, core linguistic operators like negation remain unreliable [3,26], and even counting is brittle [20]. Taken together, the evidence points to limited structural understanding across modalities rather than isolated edge cases.

2.2 Why This Brittleness Arises

Several strands of work trace these failures back to properties of the training signal and the learned embedding geometry. First, contrastive training does not require order-sensitive text supervision: a bag-of-words signal can suffice to train strong zero-shot visual models [27]. If the loss can be minimized without rewarding composition, models have little incentive to learn it.

Second, analyses of CLIP's joint embedding space indicate a persistent *modality gap*: text and image embeddings occupy largely disjoint regions with weak alignment [17]. Empirical procedures that reduce this gap yield consistent performance improvements [17]. Subsequent evidence attributes the gap primarily to the contrastive learning objective rather than to dataset mismatch or insufficient training [8]. Moreover, the embedding distribution concentrates on a thin, off-centre ellipsoidal shell, reflecting strong geometric biases [14]. These properties may heighten sensitivity to small, meaning-preserving textual perturbations, as minor shifts in the text embedding can traverse decision boundaries even when semantics are unchanged.

Finally, there are emerging theoretical limits. [11] shows that a single shared latent space cannot simultaneously represent, in a fully faithful way, more than one among basic descriptions, attribute binding, spatial relations, and negation. If so, the failures observed under minor text changes are not just data gaps; they are symptoms of an architectural trade-off built into CLIP's design.

2.3 CLIP's Text Encoders

CLIP's text encoders instantiate a Transformer architecture [31]: inputs are tokenised into subwords, embedded, and contextualized via self-attention. Relative to task-tuned encoders such as BERT and RoBERTa, CLIP's text encoders underperform on language understanding benchmarks like GLUE [7,18,32], while exhibiting strong cross-modal associations between lexical items and visual attributes [6]. This contrast indicates an encoder optimised for visual alignment rather than fine-grained linguistic structure. To mitigate sensitivity to textual variation, prior work augments training with paraphrases to stabilize responses to rephrasings [12] and proposes defenses against character-level adversarial noise [22]. The latter yields gains for various downstream tasks, but the injected symbols and digits are atypical of real user queries. We instead examine linguistically valid, everyday variations, such as orthographic errors, case changes, keyword-style queries, and paraphrases, and assess their impact on *ranking stability* rather than only top-1 retrieval. For deployed retrieval systems, stable rankings under meaning-preserving edits are a primary requirement. Accordingly, we ask: given the geometric and supervisory constraints of CLIP-style training, how volatile are text-to-image rankings under small, realistic changes to the text?

3 Methodology

We systematically evaluate the brittleness of CLIP's text encoders by applying a series of perturbations to the text input and measuring their impact on the model's output. First, we select a set of CLIP models for our experiments. From a set of queries, we encode the text input using the selected CLIP models and rank keyframes from a video dataset based on their similarity to the text input. The top-ranked keyframes are used as the reference to measure changes introduced by perturbations. We then apply a series of perturbations to the original queries and analyse the changes in the image ranking with respect to the original ranking.

3.1 Experimental Setup

We selected a set of CLIP models for our experiments, focusing on those that have been widely used and benchmarked in the literature [13] as well as a fine-tuned variant that has been shown to improve robustness against perturbations [22]. The models are listed in Table 1, ranging from the standard CLIP ViT-B/32 to larger models like ViT-H/14. All models are trained on the LAION-2B dataset, which provides two billion image-text pairs, with one model trained on a 400M subset for efficiency [24]. In addition, we include two recently proposed models: FARE2-H14, which is fine-tuned on character-level perturbations to enhance robustness [1], and EVA02-L14, which uses a large language model as its text encoder [35].

We use the queries from the TRECVID [4] Ad-Hoc Video Search (AVS) Task from the years 2016 to 2022 for our experiment, resulting in a total of 190 queries,

Table 1. CLIP models used in the experiments with their descriptions. Short names are used for brevity in the text.

Short Name	Description
LAION5B	`laion5b_s13b_b90k`: Multilingual CLIP ViT-B/32 + XLM-RoBERTa, trained on LAION-5B (13B samples, batch 90k)
LAION-B32	`ViT-B-32_laion2b_s34b_b79k`: CLIP ViT-B/32, English subset of LAION-5B (2B samples), ~66% zero-shot ImageNet
LAION-L14	`ViT-L-14_laion2b_s32b_b82k`: CLIP ViT-L/14, LAION-2B, balanced accuracy vs. speed
LAION-L14-400M	`ViT-L-14_laion400m_e32`: ViT-L/14 trained on 400M subset, likely lighter/faster variant
LAION-H14	`ViT-H-14_laion2b_s32b_b79k`: Large-capacity CLIP ViT-H/14, trained on LAION-2B, strong benchmark performance
FARE2-H14	`ViT-H-14_rho50_k1_constrained_FARE2`: ViT-H/14 LAION-2B fine-tuned on character-level perturbations
EVA02-L14	`Meta-Llama-3-8B-Instruct`: LLM2CLIP-EVA02-L-14-336, a large language model being aligned with CLIP's vision encoder

spanning a broad range of visual topics. Each query contains information about a place, object, action, or any combination thereof. Since these queries were explicitly designed for the evaluation of retrieval methods, they are inherently suited for our purpose.

We use the first shard of the Vimeo Creative Commons Collection (V3C1) [23] as a content collection. V3C1 contains 100 h of web-sourced video with a broad and diverse range of content. It comes pre-segmented into roughly one million video shots, each of which represented by a representative keyframe. These keyframes are used in our experiments to compare the effects of different query perturbations on the list of retrieved results.

3.2 Text Perturbations

We systematically apply various perturbations to the text input to CLIP, categorizing them into three classes:

- **Class 1:** Simple lexical perturbations, such as case changes, typos, deletions, additions, substitutions, and punctuation changes.
- **Class 2:** Syntactic perturbations, such as noun phrase extraction, keyword extraction, and reordering. This class represent how users might naturally simplify or modify queries to focus on key concepts.
- **Class 3:** Semantic perturbations, including paraphrasing and synonym replacement. These perturbations are designed to preserve the overall meaning of the query while altering its surface form.

These perturbations are designed to be small enough that they do not fundamentally change the meaning of the query but are sufficient to test the robustness of the text encoders. For example, changing the case of a word or replacing a word with its synonym should not alter the semantic meaning of the query, but it may still affect the ranking of images. Furthermore, to observe the 'Bag-of-Words' effect, we also shuffle the order of words in extractive perturbations (e.g. keyword-only, noun-only). We do not shuffle for full-sentence pertubations since the grammatical structure would be lost.

3.3 Evaluation Metrics

We evaluate robustness using two complementary dimensions: (i) the stability of retrieval rankings under perturbations and (ii) the relation between this stability and the magnitude of the perturbation in embedding space.

Text Distance. First, we measure how much a perturbation changes the text representation itself. For each query, we compute the cosine distance between the original and perturbed text vectors. Larger values indicate stronger movement in embedding space, while smaller values suggest that the perturbation leaves the representation nearly unchanged.

Ranking Stability/Instability. Next, we quantify the divergence in retrieval results. We compare the original and perturbed rankings by computing the Rank-Biased Overlap (RBO) [34], which emphasises agreement at the top of the ranking while still accounting for deeper positions. Formally, RBO is defined as:

$$\text{RBO}(p) = (1-p)\sum_{k=1}^{|A|} \frac{p^{k-1}}{k} \left(\frac{|A_k \cap B_k|}{|A_k \cup B_k|} \right), \tag{1}$$

where A_k and B_k denote the top-k items in the original and perturbed rankings, respectively. We define **instability** as:

$$\text{Instability} = 1 - \text{RBO}(p = 0.99).$$

This measure captures how much the rankings diverge, with higher values indicating greater instability.

Brittleness. Finally, we normalize instability by the corresponding intra- (perturbations within a query) and inter-query text distances to account for the fact that large embedding shifts naturally lead to ranking differences. We define the **brittleness index** as:

$$\mathcal{B} = \log \left(\frac{\text{Instability} \times \text{Inter-query Distance}}{\text{Intra-query Distance}} \right).$$

where *Inter-query Distance* is the average cosine distance between all pairs of original queries in the dataset, which normalises for the overall scale of the embedding space; and *Intra-query Distance* is the cosine distance between the original and perturbed query, which captures the magnitude of the specific perturbation. This measure highlights overreactions: cases where a small embedding change causes disproportionately large shifts in the ranking. A model with low instability but high brittleness is especially vulnerable to meaning-preserving edits such as typos or case changes.

Taken together, these metrics provide a graded view of robustness. Text distance reflects representational sensitivity, instability captures retrieval-level divergence, and brittleness disentangles disproportionate responses from genuinely large semantic changes.

4 Results

We present results in three stages: (i) direct overlap analysis at different cut-offs k, (ii) instability distributions and regression confirming differences across models and perturbation classes, and (iii) brittleness, which normalises instability by embedding distance.

4.1 Overlap Analysis

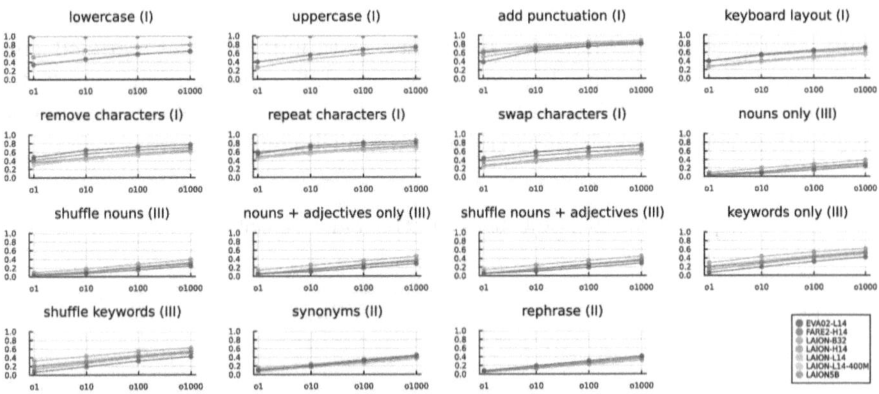

Fig. 1. Overlap@k between original and perturbed rankings for each perturbation type across models.

Figure 1 plots overlap@k between the original and perturbed rankings for each perturbation type. Several patterns emerge.

First, as k increases, overlap rises across all conditions, since agreement is easier to achieve at deeper ranks. However, the critical region for retrieval systems is the head of the ranking (e.g., $k = 10$), where differences remain substantial.

Second, most models internally normalise case, leading to identical rankings for lowercase and uppercase variants. More broadly, Class 1 (I) lexical perturbations (case, punctuation, keyboard errors, character swaps) show relatively high overlap. Yet their overlap should arguably be closer to 1.0: for example, adding a single period at the end of a query should not alter retrieval at all. The observed degradation suggests that CLIP's text encoders remain unnecessarily sensitive to surface noise. Among the tested systems, EVA02-L14 and FARE2-H14 show the highest overlap on Class 1 edits, consistent with their training for robustness against such perturbations.

Third, Class 2 (II) syntactic perturbations (keyword-only, noun/adjective-only, and word-order shuffles) considerably lower overlap. Interestingly, shuffling words alone does not produce as dramatic an effect as extracting only nouns or keywords, suggesting that CLIP exhibits bag-of-words behaviour: word order is often ignored, but loss of modifiers or function words destabilises rankings. EVA02-L14 performs consistently poorly in this class, highlighting an ongoing weakness in handling syntactic reduction.

Finally, Class 3 (III) semantic perturbations (synonym substitutions, paraphrasing) show very low overlap, indicating that CLIP's text encoders are highly sensitive to phrasing and word choice, even when overall meaning remains unchanged. This effect is least pronounced for EVA02-L14, whose LLM-based text encoder is somewhat more tolerant of paraphrastic variation than LAION-based ViTs.

4.2 Instability Across Models and Classes

While overlap curves highlight top-k stability, instability provides a graded summary across the full ranking. Figure 2 shows instability distributions aggregated across perturbations: EVA02-L14 and FARE2-H14 are consistently more stable, along with LAION5B, which was trained on multilingual texts.

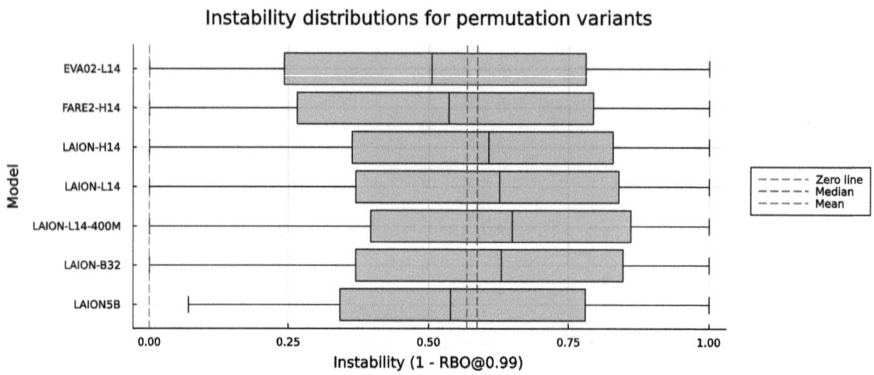

Fig. 2. Instability distributions across models. Lower values indicate greater robustness. The vertical lines show the average mean and median instability for all models.

Mixed-effects regression confirms these trends quantitatively: (i) baseline instability is substantial ($\hat{\mu} = 0.568$, $p \ll 0.001$), (ii) model-level coefficients highlight EVA02-L14, FARE2-H14, and LAION5b as significantly more stable, and (iii) perturbation-class effects are strongest for Class 2 and Class 3 variants.

4.3 Brittleness

One explanation is that instability is simply proportional to how far perturbed queries move in embedding space. Indeed, we find a strong positive correlation between instability and text distance (Fig. 3). But the relationship differs across models, and the slope of the regression line varies substantially, indicating that some models are more sensitive to small perturbations than others.

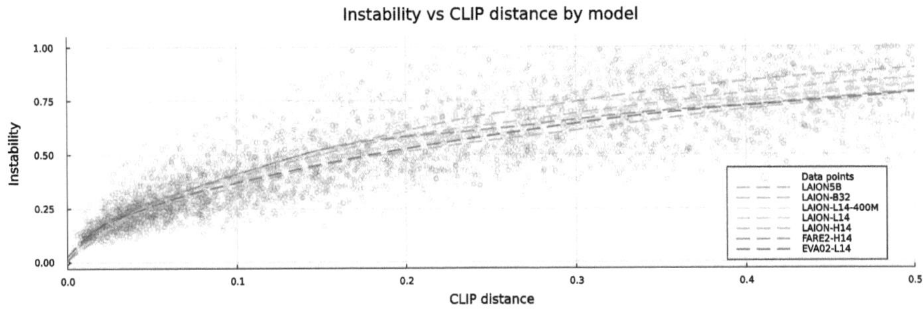

Fig. 3. Instability vs. text distance (normalised by inter-query distance) for a sample of perturbed queries, with LOESS fits per model. Slopes differ across models, indicating varying sensitivity to embedding shifts.

The brittleness heatmap (Fig. 4) highlights a different pattern than raw instability: Class 1 lexical perturbations induce the highest brittleness, since trivial edits (case, punctuation) cause ranking changes disproportionate to their minimal embedding distance. By contrast, larger embedding shifts from Class 2 syntactic or Class 3 semantic perturbations explain much of their instability, yielding lower brittleness. Depsite having lower instability, LAION5B is remarkably more brittle EVA02-L14 and FARE2-H14 consistently exhibit the lowest brittleness, suggesting that training and architecture refinements mitigate, but do not eliminate, overreaction to surface-level noise.

5 Discussion

Our analysis reveals three key aspects of CLIP's behaviour under text perturbations.

Instability is Pervasive but Structured. Across all models, even minor linguistic variations lead to non-trivial ranking changes. The overlap and instability measures show that surface edits (e.g., punctuation, casing) are not entirely ignored, while structural changes (e.g., dropping modifiers, paraphrasing) cause systematic divergence. This confirms earlier warnings that VLMs operate closer to bag-of-words matchers than to genuine compositional reasoners [9,28].

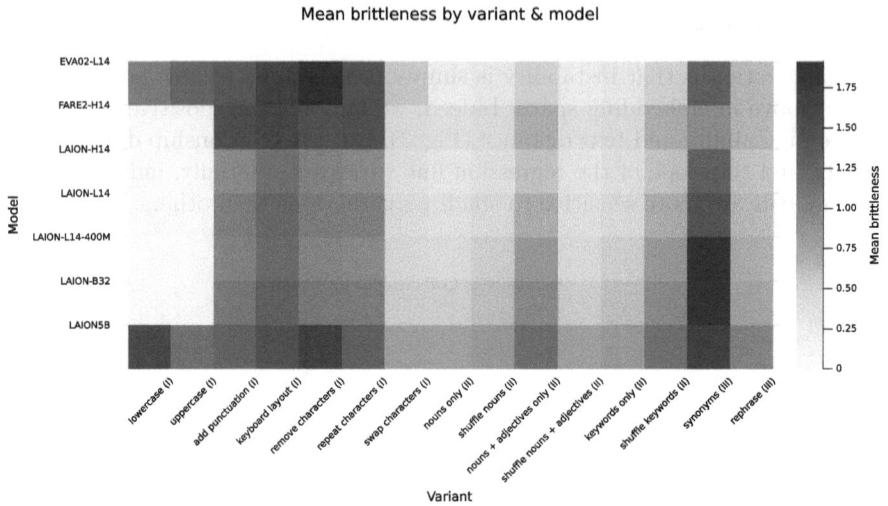

Fig. 4. Brittleness index across models and perturbation classes. Darker colours indicate higher brittleness. Lexical perturbations (Class 1) stand out as disproportionately brittle.

Brittleness is Concentrated in Trivial Edits. Normalising instability by embedding distance highlights that the greatest overreactions occur for formally small changes. Adding a period or altering character case should not affect retrieval, yet these edits yield disproportionate shifts in the ranking. This brittleness points to an undesirable property of the embedding geometry: decision boundaries are so thin that even minimal perturbations can cross them.

Model Design Matters. Comparisons across models show that architecture and training regime shape robustness. EVA02-L14 and FARE2-H14, both trained with additional constraints or alternative text encoders, exhibit consistently lower instability and brittleness. However, neither eliminates sensitivity to paraphrastic or syntactic variation. This suggests that robustness is not simply a matter of scale but also of how alignment objectives are defined.

Implications. For downstream applications, instability undermines reliability. In retrieval, unstable top-k lists reduce user trust; in moderation or safety-critical settings, brittleness could lead to inconsistent or unsafe filtering. More broadly, our results imply that evaluation of VLMs should go beyond accuracy on benchmark queries to include measures of robustness under realistic input variation.

Limitations. Our study isolates text-side perturbations. While this is appropriate for analysing CLIP's text encoders, multimodal robustness also depends on visual perturbations and their interaction with text. We evaluated a set of widely used CLIP variants, but further work should include multimodal LLMs that build on CLIP's encoders. Finally, our brittleness index is one possible normalisation; alternative formulations could capture sensitivity to perturbations differently. Furthermore, we limit the current study to queries in English to include a broader range of CLIP models for analysis. A subset of the analysed models have been trained on multilingual content. Although we have no reason to believe that the presented results are exclusive to English, no analysis of the behaviour of text encoder when faced with non-English queries has been conducted so far.

Outlook. The brittleness index suggests potential training interventions: incorporating brittleness penalties into loss functions, or learning projection layers that map noisy queries into more stable embedding regions. Both would encourage models to preserve ranking stability under small, meaning-preserving edits. Ultimately, a robust VLM must treat linguistic variation not as adversarial noise but as an expected feature of human input.

6 Conclusion

We have presented a systematic study of the robustness of CLIP's text encoders under lexical, syntactic, and semantic perturbations. Our analysis introduced two complementary notions: *instability*, the divergence in retrieval rankings caused by input variation, and *brittleness*, which normalises instability by embedding distance to highlight disproportionate overreactions. Using these measures across a large set of queries and models, we showed that instability is pervasive, with syntactic reductions and paraphrastic changes driving the largest divergences, while brittleness is concentrated in trivial edits such as punctuation or casing.

Comparisons across model families revealed that architectural and training refinements, such as FARE2-H14 and EVA02-L14, significantly reduce instability and brittleness but do not eliminate them. This suggests that robustness cannot be taken for granted, even in state-of-the-art visionlanguage models. For downstream applications, from video retrieval to content moderation, instability undermines reliability and brittleness poses safety risks, making robustness evaluation as important as accuracy benchmarks.

Looking ahead, our brittleness index offers not only an analytic tool but also a potential training signal: it could be incorporated into loss functions to penalise overreaction to meaning-preserving edits. Likewise, projection layers or query-normalisation mechanisms may mitigate brittleness by smoothing the embedding space around trivial variations. More broadly, we hope this work encourages the community to treat robustness to everyday linguistic variation as a core requirement for deploying visionlanguage models in real-world settings.

References

1. Abad Rocamora, E., Schlarmann, C., Deep Singh, N., Wu, Y., Hein, M., Cevher, V.: Robustness in both domains: clip needs a robust text encoder. arXiv e-prints pp. arXiv–2506 (2025)
2. Alayrac, J.B., et al.: Flamingo: a visual language model for few-shot learning. Adv. Neural. Inf. Process. Syst. **35**, 23716–23736 (2022)
3. Alhamoud, K., et al.: Vision-language models do not understand negation. In: Proceedings of the Computer Vision and Pattern Recognition Conference, pp. 29612–29622 (2025)
4. Awad, G., Fiscus, J., Godil, A., Diduch, L., Graham, Y., Quénot, G.: Trecvid 2024 - evaluating video search, captioning, and activity recognition. In: Proceedings of TRECVID 2024. NIST, USA (2024)
5. Brendel, W., Bethge, M.: Approximating CNNs with bag-of-local-features models works surprisingly well on imagenet. CoRR arxiv:1904.00760 (2019)
6. Chen, Z., Chen, G.H., Diao, S., Wan, X., Wang, B.: On the difference of bert-style and clip-style text encoders. arXiv preprint arXiv:2306.03678 (2023)
7. Devlin, J., Chang, M.W., Lee, K., Toutanova, K.: Bert: pre-training of deep bidirectional transformers for language understanding. In: Proceedings of the 2019 Conference of the North American Chapter of the Association for Computational Linguistics: Human Language Technologies, vol. 1 (long and short papers), pp. 4171–4186 (2019)
8. Fahim, A., Murphy, A., Fyshe, A.: It's not a modality gap: characterizing and addressing the contrastive gap. arXiv preprint arXiv:2405.18570 (2024)
9. Hessel, J., Schofield, A.: How effective is BERT without word ordering? Implications for language understanding and data privacy. In: Zong, C., Xia, F., Li, W., Navigli, R. (eds.) Proceedings of the 59th Annual Meeting of the Association for Computational Linguistics and the 11th International Joint Conference on Natural Language Processing, vol. 2: Short Papers, pp. 204–211. Association for Computational Linguistics, Online (2021)
10. Kamath, A., Hessel, J., Chang, K.W.: What's "up" with vision-language models? Investigating their struggle with spatial reasoning (2023)
11. Kang, R., Song, Y., Gkioxari, G., Perona, P.: Is clip ideal? No. can we fix it? Yes! (2025)
12. Kim, H., et al.: Fine-tuning clip text encoders with two-step paraphrasing. arXiv preprint arXiv:2402.15120 (2024)
13. Leopold, M., Schoeffmann, K.: diveXplore at ivr4b 2024. In: 2024 International Conference on Content-Based Multimedia Indexing (CBMI), pp. 1–7. IEEE (2024)
14. Levi, M.Y., Gilboa, G.: The double-ellipsoid geometry of clip. arXiv preprint arXiv:2411.14517 (2024)
15. Lewis, M., et al.: Does clip bind concepts? Probing compositionality in large image models (2024)
16. Li, J., Li, D., Xiong, C., Hoi, S.: Blip: bootstrapping language-image pre-training for unified vision-language understanding and generation. In: International Conference on Machine Learning, pp. 12888–12900. PMLR (2022)
17. Liang, V.W., Zhang, Y., Kwon, Y., Yeung, S., Zou, J.Y.: Mind the gap: understanding the modality gap in multi-modal contrastive representation learning. Adv. Neural. Inf. Process. Syst. **35**, 17612–17625 (2022)

18. Liu, Y., et al.: Roberta: a robustly optimized bert pretraining approach. arXiv preprint arXiv:1907.11692 (2019)
19. Newman, K., Wang, S., Zang, Y., Heffren, D., Sun, C.: Do pre-trained vision-language models encode object states? arXiv preprint arXiv:2409.10488 (2024)
20. Paiss, R., et al.: Teaching clip to count to ten. In: Proceedings of the IEEE/CVF International Conference on Computer Vision, pp. 3170–3180 (2023)
21. Radford, A., et al.: Learning transferable visual models from natural language supervision. In: International Conference on Machine Learning, pp. 8748–8763. PmLR (2021)
22. Rocamora, E.A., Schlarmann, C., Singh, N.D., Wu, Y., Hein, M., Cevher, V.: Robustness in both domains: clip needs a robust text encoder. arXiv preprint arXiv:2506.03355 (2025)
23. Rossetto, L., Schuldt, H., Awad, G., Butt, A.A.: V3C – a research video collection. In: Kompatsiaris, I., Huet, B., Mezaris, V., Gurrin, C., Cheng, W.-H., Vrochidis, S. (eds.) MMM 2019. LNCS, vol. 11295, pp. 349–360. Springer, Cham (2019). https://doi.org/10.1007/978-3-030-05710-7_29
24. Schuhmann, C., et al.: Laion-5b: an open large-scale dataset for training next generation image-text models. Adv. Neural. Inf. Process. Syst. **35**, 25278–25294 (2022)
25. Singh, A., et al.: FLAVA: a foundational language and vision alignment model. In: IEEE/CVF Conference on Computer Vision and Pattern Recognition, CVPR 2022, New Orleans, LA, USA, 18–24 June 2022, pp. 15617–15629. IEEE (2022)
26. Singh, J., Shrivastava, I., Vatsa, M., Singh, R., Bharati, A.: Learn "no" to say "yes" better: improving vision-language models via negations. arXiv preprint arXiv:2403.20312 (2024)
27. Tejankar, A., et al,: A fistful of words: learning transferable visual models from bag-of-words supervision. CoRR arxiv:2112.13884 (2021)
28. Thrush, T., et al.: Winoground: probing vision and language models for visio-linguistic compositionality. In: IEEE/CVF Conference on Computer Vision and Pattern Recognition, CVPR 2022, New Orleans, LA, USA, 18–24 June 2022, pp. 5228–5238. IEEE (2022)
29. Tong, S., Liu, Z., Zhai, Y., Ma, Y., LeCun, Y., Xie, S.: Eyes wide shut? Exploring the visual shortcomings of multimodal LLMs. In: Proceedings of the IEEE/CVF Conference on Computer Vision and Pattern Recognition (CVPR), pp. 9568–9578 (2024)
30. Vadicamo, L., et al.: Evaluating performance and trends in interactive video retrieval: insights from the 12th VBS competition. IEEE Access **12**, 79342–79366 (2024)
31. Vaswani, A., et al.: Attention is all you need. Adv. Neural Inf. Process. Syst. **30** (2017)
32. Wang, A., Singh, A., Michael, J., Hill, F., Levy, O., Bowman, S.R.: Glue: a multi-task benchmark and analysis platform for natural language understanding. arXiv preprint arXiv:1804.07461 (2018)
33. Wang, J., et al.: Omnivl: one foundation model for image-language and video-language tasks. Adv. Neural. Inf. Process. Syst. **35**, 5696–5710 (2022)
34. Webber, W., Moffat, A., Zobel, J.: A similarity measure for indefinite rankings. ACM Trans. Inform. Syst. (TOIS) **28**(4), 1–38 (2010)

35. Wu, A., et al.: Llm2clip: extending the capability boundaries of clip through large language models (2025)
36. Yuksekgonul, M., Bianchi, F., Kalluri, P., Jurafsky, D., Zou, J.: When and why vision-language models behave like bags-of-words, and what to do about it? (2023)
37. Zhai, X., et al.: Lit: Zero-shot transfer with locked-image text tuning. In: IEEE/CVF Conference on Computer Vision and Pattern Recognition, CVPR 2022, New Orleans, LA, USA, 18–24 June 2022, pp. 18102–18112. IEEE (2022)

DPNet: A Dual-Perception Fusion Network for Automated Coronary Artery Segmentation

Zhiting Chen[1], Jieyun Bai[1,2], Shunning Li[3], Xiaoshen Zhang[3], and Hua Lu[3(✉)]

[1] College of Information Science and Technology, Jinan University, Guangzhou, China
[2] Auckland Bioengineering Institute, Auckland, New Zealand
[3] The Department of Cardiovascular Surgery, The First Affiliated Hospital of Jinan University, Guangzhou, China
specialhual@icloud.com

Abstract. Coronary Artery Disease (CAD) is one of the leading causes of death worldwide, and its accurate diagnosis relies on precise segmentation of Coronary Arteries (CA). Although Computed Tomography Coronary Angiography (CTCA) is the clinical gold standard, manual segmentation suffers from inefficiency and subjectivity. To address the challenges of low contrast and complex vascular structures in CA segmentation, we propose DPNet, a dual-perception fusion network with parallel branches: (1) The Edge Perception Branch (EPB) extracts fine-grained vessel details through 3D convolutions, effectively handling low-contrast boundaries, and (2) The Structural Perception Branch (SPB) captures long-range dependencies via self-attention mechanisms, enhancing segmentation robustness for intricate vascular topologies. We propose two novel components: (i) The Bilateral Feature Communication Module (BFCM) that enhances cross-scale feature interaction via channel attention, and (ii) The Structural Reconciliation Module (SRM) that bridges semantic gaps between EPB and SPB to correct the shape of CA. Extensive experiments on the ImageCAS dataset demonstrate that DPNet achieves a Dice Similarity Coefficient (DSC) of 0.831, outperforming the second-best method (TransUNet) by 2%, with a 41% reduction in Hausdorff Distance (HD). Validation on ASOCA further confirms the model's generalizability. Ablation studies verify the significance of BFCM and SRM ($p<0.05$).

Keywords: Coronary Artery Segmentation · Computed Tomography Coronary Angiography · Convolutional Neural Networks · Vision Transformers

H. Lu—Supported by the Key Areas Research and Development Program of Guangdong Province (No. 2025B1111030001), entitled "Research on Clinical Novel Techniques for Transcatheter Mitral Valve Replacement and Related Device Development and Industrialization." And the First Affiliated Hospital of Jinan University (No. 2023A03J0565), which supports the project "Screening and Prediction of Protein Biomarkers for Stanford Type A Aortic Dissection Based on Differential Proteomics.".

J. Lokoč et al. (Eds.): MMM 2026, LNCS 16413, pp. 375–388, 2026.
https://doi.org/10.1007/978-981-95-6957-1_27

1 Introduction

Coronary Artery Disease (CAD) accounts for one-third of cardiovascular-related mortality globally, primarily caused by atherosclerotic plaque accumulation leading to myocardial ischemia or necrosis [25]. Computed Tomography Coronary Angiography (CTCA) serves as the gold-standard non-invasive modality for Coronary Artery (CA) assessment [23]. Current clinical workflows rely on manual stenosis grading in CTCA images, a time-consuming process (30–60 minutes/case) with notable inter-observer variability [7]. These limitations have driven the development of automated CA Segmentation (CAS) algorithms, which are increasingly recognized as essential tools for standardized, efficient diagnosis. However, owing to the intrinsic characteristics of CA, it continues to face two challenges (Fig. 1A): **(1) Low contrast:** On a local scale, CA constitute a minor fraction of the thoracic cavity(occupying $< 0.3\%$ of thoracic volume). They exhibit a low contrast with respect to the surrounding environment, and the edges of the CA are indistinct, thereby presenting a substantial hindrance to the segmentation of CA from the intricate background; and **(2) Complex structure:** On a global scale, the anatomical configuration of CA is exceedingly intricate. The tubular structures are characterized by being curved and branching, with the branches differing among individuals in terms of quantity, length, and size [19]. The distal segments of the CA tree are overly slender, which makes it highly arduous to capture complex details precisely during the segmentation process. Recent research in automatic CAS has developed specialized network architectures to address the unique structural characteristics of CA. While these methods demonstrate clinically acceptable performance, they frequently exhibit discontinuous segmentation due to insufficient consideration of overall morphological characteristics. The CAS algorithm that can simultaneously address two challenges remains an indispensable and unmet requirement in medical image analysis.

To address the aforementioned segmentation challenges, this paper proposes a Dual-Perception parallel fusion Network (DPNet) that systematically integrates edge perception and structure perception features for comprehensive characterization of CA morphology. The architecture comprises two parallel encoder-decoder pathways. An Edge-Perception Branch (EPB) employing a three-layer 3D Convolutional Neural Network (CNN) to enhance boundary delineation by resolving low contrast between CA and surrounding tissues. And a Structure-Perception Branch (SPB) utilizing Vision Transformer (ViT) to capture complex topological patterns through global contextual modeling. Two key technical innovations ensure effective multi-modal feature integration. On the one hand, the Bilateral Feature Communication Module (BFCM) establishes cross-layer information exchange between dual branches, enabling complementary fusion of local edge features and global structural representations. This mechanism effectively mitigates the inherent receptive field limitations of single-modality feature extraction. On the other hand, the Structural Reconciliation Module (SRM) performs adaptive feature recalibration through channel-wise attention mechanisms. It dynamically reconciles semantic disparities between CNN-derived edge features and ViT-based structural features while correcting vascular shape during feature fusion. The principal contributions are summarized as follows:

- A dual-perception hybrid parallel architecture that collaboratively combines edge-perception and structure-perception realizes the synchronous optimization of local boundary accuracy and global structure coherence in CAS.
- An inter-layer feature interaction framework through BFCM, enabling hierarchical fusion of multi-scale features while enhancing semantic consistency across network depths.
- A discrepancy-aware fusion mechanism via SRM, which resolves cross-branch feature conflicts through attention-guided channel recalibration and spatial refinement.

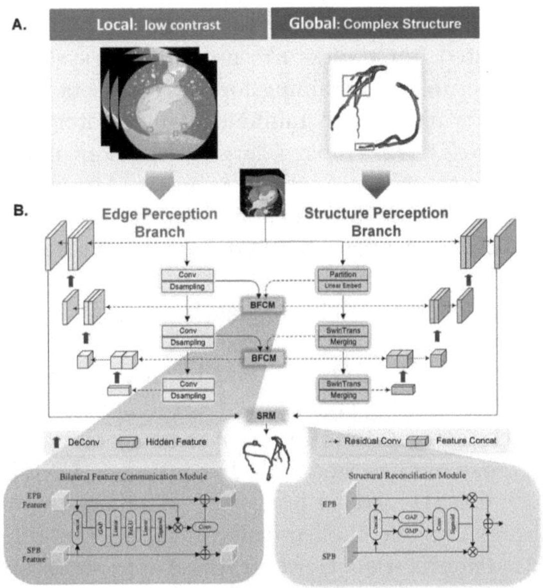

Fig. 1. Overview of DPNet. (A), Major challenges in the automatic CAS. (B), Overall architecture of the proposed dual-perception with local and global coupling (DPNet).

2 Related Work

Recent advances in deep learning have significantly enhanced medical image analysis, demonstrating superior efficiency and reliability compared to conventional methodologies. This section systematically reviews deep learning frameworks for medical image and CAS tasks.

CNNs have been widely applied to medical image segmentation, with 3D-UNet [3] extending the original UNet to volumetric data and improving capture of 3D CA structures through tailored encoderdecoder paths. Variants like UNet++ [28] introduce nested dense skip connections to enhance multi-scale feature fusion without manual tuning. To address class imbalance from small CA targets, Yu et al. [26] applied densely connected volumetric CNNs to cardiac MR images, and Huang et al. [11] segmented CA centerlines with 3D-UNet, demonstrating robustness on sparse annotations. Pan et al.'s [21] 3D Dense-U-Net integrates DenseNet blocks with 3D-UNet to improve gradient flow and feature reuse, though fine vessel details remain challenging. Attention mechanisms were later added: CS^2Net [20] employs dual spatialchannel attention to better model curved vessels, and CAS-Net [4] fuses attention-oriented feature fusion with multi scale aggregation to enrich semantic and spatial cues. Cascaded strategies–such as Kid-Net's [24] two-stage volumetric CNN and Zeng et al.'s [27] coarse-to-fine framework further refine vessel boundaries in a stepwise manner. Automated frameworks like nnUNet [12] systematically optimize preprocessing, architecture, and training for new datasets, achieving State-Of-The-Art (SOTA) performance with minimal manual intervention. Inspired by ConvNeXt [17], MedNeXt [22] employs large-kernel convolutions and inverted bottlenecks to emulate transformer inductive biases. Although UXNet [15] also utilized ConvNeXt, it did not fully exploit its structural potential. However, these convolution-Based benchmark methods remain constrained by their fixed-size receptive fields.

To capture global context, ViT [5], has been proposed. Therefore, some researchers began to turn their attention to ViT in CAS. For example, He et al. [10] further improved Zeng's [27] method by incorporating ViT, which can benefit the overall structure segmentation of CA. They used a lightweight ViT called P2T, for patch segmentation. In recent years, some researchers have proposed coupled networks of CNN and ViT. For instance, UNETR [9] and Swin-UNETR [16], which are 3D medical image segmentation architecture based on a Transformer encoder capable of effectively capturing long-range dependencies and global contextual information. However, these approaches are predominantly structured with ViT as encoder. But ViT-Based models are less effective than CNN in handling small local structures [1,17], potentially leading to inaccurate segmentation in small vascular regions. Therefore, some researchers have cascaded CNN and ViT together in encoder. For example, TransUNet [2] effectively combining the local feature extraction capabilities of CNN with the global context modeling capabilities of ViT. Experimental results demonstrated that this model outperformed the others in various medical tasks, thereby highlighting the significant potential of integrating CNN-Based and ViT-Based encoders within U-shaped medical image segmentation architectures.

The progression from CNNs to ViTs and hybrid models reveals a persistent challenge in vascular representation learning: effectively balancing local boundary ambiguity and global topological complexity in CAs. While CNNs excel at local feature extraction and ViTs model long-range dependencies, neither singular approach sufficiently addresses the coexistence of these dual characteristics. This fundamental gap motivates our dual-perception network, integrating an EPB for resolving low-contrast boundaries and a SPB for capturing anatomical topology, thereby synergizing localized precision with global coherence.

3 Method

We propose a DPNet, an end-to-end CAS framework, that integrates both local and global coupling mechanisms within a novel dual-perception parallel architecture (Fig. 1B). First, the input volume $X_{in} \in \mathbb{R}^{H \times W \times D \times C}$ (where H, W, D and C are the dimensions of the input) is fed into two branches: EPB and SPB. The former is responsible for effective extraction of vessel details (Sect. 2.1), while the latter supplements global information (Sect. 2.2). This dual-perception architecture combines CNN's local edge perception with ViT's global structural perception. A channel attention-Based BFCM filters irrelevant features while preserving diversity (Sect. 2.3), while a spatial attention-driven SRM refines segmentation accuracy through error correction (Sect. 2.4).

3.1 Edge Perception Branch

By leveraging the local segmentation capabilities of CNN operations, the intricate challenge posed by the low contrast between the CA and its surrounding tissues can be systematically addressed and effectively mitigated.

As shown in Fig. 1B, the EPB comprises three stages. During encoding phase, each stage involves a sequential application of first two $3 \times 3 \times 3$ convolutional operations, and then a ReLU activation function before a $2 \times 2 \times 2$ max pooling operation with a stride of 2 for downsampling. This sequence effectively halves the image resolution at each downsampling stage while doubling the number of feature channels. During decoding, a deconvolutional layer is utilized to upsample the resolution of the feature map by a factor of two. The output of it then fused with the corresponding output from the previous stage through skip connections, and the concatenated features are subsequently fed into another residual block for further processing.

3.2 Structure Perception Branch

Due to CNN's inherent constraint by fixed-size receptive fields, capturing long-range dependencies becomes challenging. For CA, characterized by their intricate structural complexity, this constraint hinders the effective extraction of global information. In contrast, by leveraging a self-attention mechanism, ViT is capable of establishing long-range dependencies within the input data and capturing comprehensive global information from the acquired images.

As illustrated in Fig. 1B, the SPB consists of three stages of Swin Transformer [16]. During encoding, the initial stage generates 3D tokens via a linear embedding layer. To preserve the hierarchical structure of the encoder, the resolution of the feature representation is progressively halved at the conclusion of each of the two subsequent layers via a patch merging layer. Specifically, the patch merging layer aggregates patches with a resolution of $2 \times 2 \times 2$ and concatenates them. Thereafter, a linear transformation reduces the dimensionality of the feature representations by half. In the last two stages, each transformer block is composed of two successive Swin Transformer blocks, each of which consists of a shifted window based Multi-head Self Attention (MSA) [16].

3.3 Bilateral Feature Communication Module

To enrich the features derived from both the EPB and SPB in the initial two stages, we are devised a BFCM, which integrates the coupling features stemming from the layers of these two branches through a channel attention mechanism, and subsequently injects the refined features back into their respective original branches. The primary aim of this design is to bolster the model's sensitivity to pivotal features, while retaining the diversity and richness of the feature set, and simultaneously diminishing the impact of less crucial features.

The proposed BFCM is shown in Fig. 1B. Initially, we concatenate the feature volumes f^i_{EPB} and f^i_{SPB} extracted respectively from EPB and SPB to produce the concatenated feature f^i_M, where i indicates the stage index within a dual-perception architecture that comprises three stages. Subsequently, we are employed average pooling to obtain a compact representation of the channel information. This representation is then fed to a Multi-Layer Perceptron (MLP) followed by a sigmoid activation function to perform a non-linear transformation, yielding a channel attention map f^i_{EX}. Then, we are performed a multiplication between the channel attention volume f^i_{EX}, and the fused features f^i_M, effectively assigning varying weights to different channels, and using a $3 \times 3 \times 3$ convolution layer to halve the number of channels. Finally, to ensure the retention of feature richness within the shallow layers of the network, we are introduced the original features f^i_{EPB} and f^i_{SPB} into the processing pipeline. This above process can be summarized as follows:

$$F^i_e = f^i_{EPB} + W(f^i_{EX} \odot f^i_M), F^i_s = f^i_{SPB} + W(f^i_{EX} \odot f^i_M) \quad (1)$$

where $W(\cdot)$ represents the convolutional operations using $3 \times 3 \times 3$ convolutional kernel. F^i_e and F^i_s represent the output of the EPB and SPB by BFCM, respectively.

3.4 Structural Reconciliation Module

To bridge the semantic gap existing between EPB and SPB, we are conceived and implemented a SRM based on spatial attention mechanism. After learning the position and morphology of the CA in the image, spatial fusion can precisely align the local vascular wall texture features extracted by EPB with the overall structural relationship features obtained by SPB in the spatial dimension. This alignment overcomes the semantic gap problem of the two branches, enabling the fused features to take into account both the local details and the global context of the small CA targets simultaneously, thereby correcting the boundaries and positions of the CA.

The SRM module is shown in Fig. 1B. Firstly, we concatenate the output features derived from EPB F_{EPB} with those from SPB F_{SPB}. Then, global maximum pooling and global average pooling are carried out to capture the significant features and average feature distribution on each channel. Moreover, the two pooled feature graphs are joined in the channel dimension to form a multi-channel feature volume F_{SP} which contains the importance and distribution information. Subsequently, a $3 \times 3 \times 3$ convolution layer is used to learn the spatial attention weights, followed by a Sigmoid activation function. Finally, an attention map is element-wise multiplied and added to EPB and SPB feature volumes to surpress emphasis and irrelevant information in key spatial regions. The expression formula is as follows:

$$F = F_{EPB} \odot \sigma(W(F_{SP})) + F_{SPB} \odot (1 - \sigma(W(F_{SP})) \quad (2)$$

where $W(\cdot)$ represents the convolutional operations using $3 \times 3 \times 3$ convolutional kernel. $\sigma(\cdot)$ denotes the Sigmoid activation function and F is the output of SRM.

4 Experiment

4.1 Experimental Setup

We evaluated our method on two public datasets: ImageCAS [27], encompasses a comprehensive collection of 1000 patient cases, for primary performance assessment. And ASOCA [6], which comprises 20 normal cases and 20 disease cases, to further examine the generalization capabilities of our approach. All experiments were conducted on an Ubuntu 18.04 server with dual NVIDIA RTX 3090 GPUs using PyTorch. Training utilized the Adam optimizer with an initial learning rate of 0.0001, 200 epochs, batch size 1, employing a compound Dice-CELoss. Input volumes (512×512×N voxels, N: 168334) were randomly cropped, padded, and resized to 128×128×128 voxels, augmented via rotation and noise injection. Testing employed sliding windows (128×128×128 voxels) with half-window overlap. Datasets were split into 70% training, 10% validation, and 20% testing. Segmentation accuracy was quantified using Dice Similarity Coefficient (DSC), Jaccard Index (JC), and Hausdorff Distance (HD) to measure overlap, intersection-over-union ratio, and maximum boundary deviation, respectively.

4.2 Comparisons with Other Methods on ImageCAS

We compare our method with 11 methods. These methods are divided into two main groups. One is a CNN-Based group, which contains traditional CNN-Based networks (i.e. ResNet [14], UNet [3] and UNet++ [28]), and two methods which designed explicitly for CAS (i.e., DenseVoxNet [26] and CS^{2}Net [20]), and nnUNet [13] and Transformer-inspired CNN-Based networks (i.e., UXNet [15]

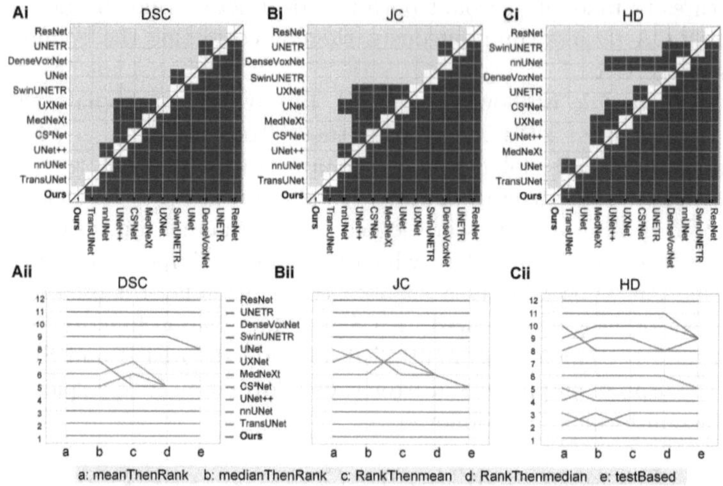

Fig. 2. Quantitative results on the ImageCAS dataset. **(A)** DSC, **(B)** JC and **(C)** HD values of 11 methods and Ours (DFNet). (**Ai**, **Bi** and **Ci**) Significance maps for visualizing ranking stability based on statistical significance (Yellow: metrics from the algorithm on the x-axis were significantly superior to the algorithm on the y-axis. Blue: indicates no significant difference); (**Aii**, **Bii** and **Cii**) Line plots for visualizing ranking robustness across different ranking methods. For each ranking method encoded on the x-axis, the height of the line represents the corresponding rank. Horizontal lines indicate identical ranks for all methods.

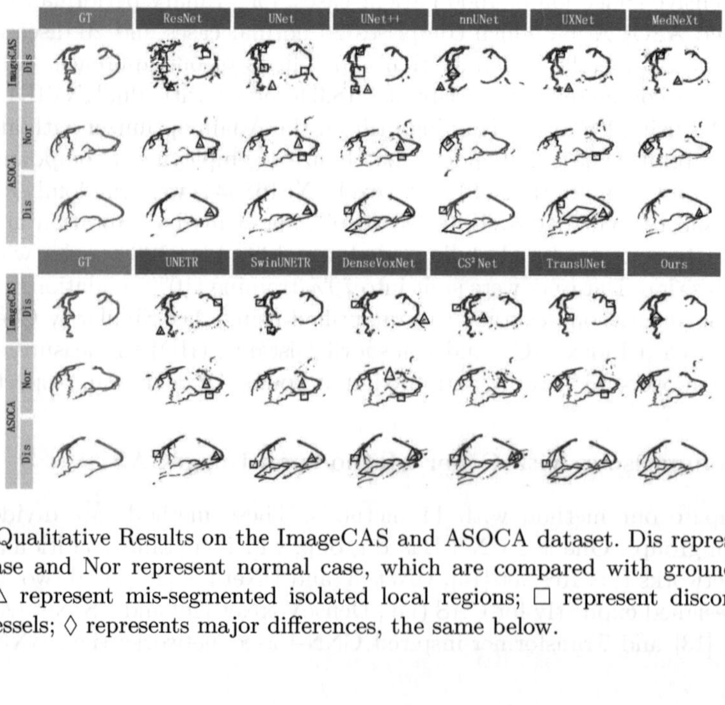

Fig. 3. Qualitative Results on the ImageCAS and ASOCA dataset. Dis represent diseased case and Nor represent normal case, which are compared with ground truths (GT). △ represent mis-segmented isolated local regions; □ represent discontinuous blood vessels; ◇ represents major differences, the same below.

Table 1. Quantitative comparison of our method and others on ImageCAS and ASOCA (mean ± standard deviation). The best results are in bold, and the second-best results are underlined.

Type		Models	ImageCAS			ASOCA			Para.
			DSC↑	JC↑	HD↓	DSC↑	JC↑	HD↓	
CNN-Based	Traditional	ResNet	0.582 (0.081)	0.415 (0.080)	60.4 (21.1)	0.736 (0.042)	0.583 (0.053)	14.5 (20.9)	46.7M
		UNet	0.759 (0.039)	0.612 (0.053)	24.3 (17.0)	0.824 (0.037)	0.690 (0.068)	19.5 (20.8)	22.6M
		UNet++	0.782 (0.039)	0.644 (0.052)	17.5 (13.7)	0.838 (0.049)	0.703 (0.059)	21.4 (24.8)	23.3M
	CAS-Based	DenseVoxNet	0.733 (0.042)	0.580 (0.052)	36.8 (22.2)	0.761 (0.037)	0.615 (0.049)	36.6 (24.5)	4.3M
		CS^{2}Net	0.772 (0.039)	0.630 (0.051)	24.4 (19.8)	0.821 (0.045)	0.699 (0.063)	25.6 (29.9)	5.8M
	Ensemble	nnUNet	0.785 (0.048)	0.685 (0.043)	35.6 (37.1)	0.838 (0.049)	0.725 (0.071)	23.3 (34.4)	88.2M
	ViT-inspired	UXNet	0.773 (0.032)	0.631 (0.042)	22.1 (18.6)	0.840 (0.032)	0.671 (0.047)	16.7 (21.6)	52.9M
		MedNeXt	0.774 (0.035)	0.633 (0.046)	18.3 (18.3)	0.854 (0.028)	0.750 (0.044)	9.0 (20.4)	31.3M
ViT-Based	ViT-Encoder	UNETR	0.721 (0.037)	0.649 (0.064)	26.1 (17.9)	0.803 (0.048)	0.673 (0.067)	19.7 (21.9)	26.1M
		SwinUNETR	0.746 (0.054)	0.562 (0.049)	38.1 (23.3)	0.846 (0.033)	0.734 (0.049)	12.5 (23.1)	61.9M
	Coupling-Encoder	TransUNet	0.813 (0.029)	0.593 (0.073)	10.5 (10.5)	0.827 (0.037)	0.707 (0.054)	14.4 (24.2)	108.8M
		Ours	**0.831(0.031)**	**0.712(0.045)**	**7.4 (6.2)**	**0.858** (0.029)	**0.752 (0.044)**	**7.8(20.2)**	**2.1M**

and MedNeXt [22]). Another one is ViT-Based group, which consists of pure ViT's encoder networks (i.e., UNETR [9] and SwinUNETR [8]) and CNN-ViT coupling encoder networks (TransUNet [2] and ours model, DPNet).

As shown in Table 1, DPNet consistently achieves the best segmentation performance on the ImageCAS dataset across all evaluation metrics. The proposed model attained the highest DSC of 0.831, and outperforming the second-best method (TransUNet) by 2%. For the JC, DPNet achieved a value of 0.712, it exhibited a 3% enhancement compared to the runner-up in nnUNet. For boundary accuracy, the proposed approach yielded the smallest HD of 7.4, indicating the most precise contour delineation; the next lowest HD was produced by TransUNet, and reducted by 41%. What's more, DPNet obtained the lowest parameter of 2.1M. These results indicate that the proposed method produces the most accurate and detailed CAS among all compared methods. Notably, for the purpose of attaining a fair and comprehensive assessment, we conducted an exhaustive evaluation of the results by implementing the Biomedical Image Analysis ChallengeS (BIAS) method (including significance maps for statistical pairwise comparisons and composite ranking strategies for aggregate evaluation.) [18]. As shown in Fig. 2, significance maps further demonstrate DPNet's marked superiority over other representative methods, with statistical significance ($p<0.05$) and consistently ranks highest in five different ranking methods overall across all metrics, showing the method's superior performance. These methods include: 1) computing the mean and subsequently ranking the aggregated scores (MeanThenRank); 2) calculating the median and then ranking the aggregated scores (MedianThenRank); 3) determining the ranking and then computing the mean of the aggregated ranks (RankThenMean); 4) calculating the ranking and then computing the median of the aggregated ranks (RankThenMedian); and 5) using the Wilcoxon signed-rank test to establish rankings based on the number of significant results (TestBased).

As shown in Fig. 3, the qualitative experimental outcomes of our proposed method and other competing approaches on typical cases. In the absence of any post-processing, a comparison has been made between the complex structures obtained in global segmentation and the details in local segmentation against the Ground Truths (GT) for Dis, from ImageCAS. The Traditional CNN-Based networks, ResNet tends to introduce significant noise (labeled △), while UNet and UNet++ effectively suppress noise but result in vascular discontinuity (labeled □). The CAS-Based networks have a good solution to vascular disconnection, but generates some noise. And Transformer-inspired CNN-Based networks such as UXNet and MedNeXt demonstrate that larger convolution kernels enhance global continuity of blood vessels but also introduce a certain degree of noise. These observations indicate that the global segmentation capability of these CNN-Based methods is limited. ViT-encoder-Based networks, namely UNETR and SwinUNETR, perform better than CNN-Based methods in maintaining vascular continuity but simultaneously introduce more noise. Therefore, both the CNN-Based method and the ViT-encoder-Based method have certain defects. Finally, coupling-encoder-Based networks, such as TransUNet and our proposed DPNet, not only reduce noise effectively (labeled △) but also ensure continuous vessel annotation (labeled □).

4.3 Generalization to ASOCA

As shown in Table 1, DPNet achieved SOTA performance on the ASOCA dataset, surpassing 11 comparison methods across DSC, JC, and HD metrics. Our method attained a mean DSC of 0.858, JC of 0.752, and HD of 7.8, outperforming the runner-up MedNeXt by margins of 0.12% (DSC), 0.22% (JC), and 15% (HD). Notably, DPNet exhibited statistically significant improvements ($p<0.05$ for DSC) over all competitors. Compared to UXNet and SwinUNETR, our approach demonstrated superior DSC (2%, 1.3%) and JC (3.5%, 2.4%) gains, along with substantially reduced HD (60–85%) against ResNet and TransUNet. Furthermore, DPNet consistently outperformed CAS-Based methods across all metrics.

We analyzed the results on the normal/diseased subgroup of ASOCA. DPNet maintained excellent performance in both subgroups. DSC/JC was higher than MedNeXt, and HD was lower than MedNeXt (DSC/JC increased by 0.9–1.4%; HD is reduced by 6.7–15%). Statistical analysis confirmed that compared with normal cases, HD in the disease subgroup was significantly improved ($p<0.05$). The qualitative experimental results of DPNet and its counterparts on the normal and disease cases are illustrated in Fig. 2. For the normal case, comparing with partly CNN-Based models, ViT-encoder-Based models improved discontinuous CA (labeled □), but produced a lot of fragments (labeled △) that do not belong to the CA. Although the CAS-Based models, ensured the continuity of CA, they generated a lot of noise at the same time. On the contrary, the proposed model, nnUNet and TransUNet gained CA details (labeled ◊), ensured the continuity of CA, and reduced noise. For the disease case, it can be seen that the disease case is generally better at processing discrete points than the normal case. UNet++, nnUNet, UXNet, SwinUNETR, CAS-Based and coupling-encoder-Based models preserved more details (labeled ◊). In the case of discontinuous CA, traditional CNN-Based models have the least ideal effect, while MedNeXt and ours have the best effect. In segmenting discontinuous CA regions, CNN architectures demonstrate limited precision, while MedNeXt and ours outperform existing approaches.

4.4 Ablation Studies

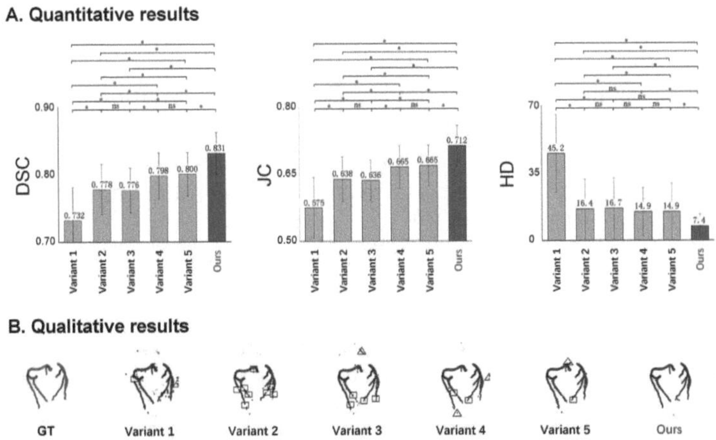

Fig. 4. Ablation studies. **A.** Quantitative and **B.** Qualitative for different variants of the proposed model The variants include: variant 1: EPB; variant 2: SPB; variant 3: EPB&SPB; variant 4: EPB&SPB&BFCM; variant 5: EPB&SPB&SRM; and Ours. An asterisk (*) indicates a statistical difference($p<0.05$), while 'ns' indicates no statistical difference($p\leq0.05$).

To evaluate the effectiveness of various modules in DPNet, we conducted a step-by-step ablation experiment on ImageCAS. In summary, we explored the effectiveness of the coupling architecture in the following variants: **Variant 1**: The EPB by a three-layered CNN network; **Variant 2**: The SPB by a three-layered Swin Transformer network; **Variant 3**: A based coupling dual-branch network architecture, comprising the EPB and SPB, is characterized by the absence of feature fusion operations across corresponding layers. Instead, the integration of these branches is achieved through an additive fusion operation at the output level; **Variant 4**: Utilize BFCM to facilitate connectivity and integration of features across the layers of the based coupling dual-branch network; **Variant 5**: Employ SRM as an alternative to the additive operation for feature interconnection and integration within the based coupling dual-branch framework.

The ablation studies demonstrate the rational design and effectiveness of our proposed DPNet architecture. As shown in Fig. 4A, Variant 3 without effective fusion modules achieves superior performance over the Variant 1 ($p<0.05$) but shows no statistically significant difference compared to the Variant 2 ($p \geq 0.05$), suggesting that mere parallel processing of local and global features through separate EPB and SPB yields suboptimal integration. This limitation is addressed through our proposed fusion strategies: Variant 4 and Variant 5 both significantly outperform Variant 3 ($p<0.05$), with BFCM enhancing cross-branch feature interaction and SRM optimizing boundary-aware fusion. Specifically, BFCM improves DSC by 2.7% and reduces HD by 10.8%, while SRM achieves 3% DSC enhancement and 10.8% HD reduction. These results validate that coordinated multi-level fusion mechanisms are essential for effectively combining CNN's local edge perception with ViT's global structure perception in medical image segmentation tasks.

From the visualization results (Fig. 4B), baseline variations 1–3 are manifested as foreign fragments and discontinuous CA regions. The noise is gradually reduced by combining Variant 4 and Variant 5 while maintaining the continuity of CA. DPNet achieves the best performance with the minimum artifact and the retention of CA structural integrity.

5 Conclusion

In summary, automatic CAS faces two critical challenges: low contrast and complex vascular topology. Our proposed DPNet addresses these through a dual-perception architecture that synergizes CNN's local edge-perception and ViT's global structure-perception. The EPB employs localized CNN to extract fine-grained textures, effectively suppressing noise artifacts in low-contrast regions, while the SPB utilizes self-attention to model long-range dependencies, resolving structural discontinuities in tortuous CA. Unlike existing methods–where CNN-Based approaches often fragment complex vasculature and ViT-Based models propagate peripheral noise–our hybrid design achieves balanced optimization via BFCM and SRM. These mechanisms enable dynamic integration of local edge precision and global structure continuity, as evidenced by qualitative improvements in noise suppression and vessel connectivity. Quantitative evaluations under the BIAS benchmark demonstrate SOTA performance, surpassing 11 methods. And validated on ASOCA further confirms superior generalizability in heterogeneous scenarios.

References

1. Bai, Y., Mei, J., Yuille, A.L., Xie, C.: Are transformers more robust than cnns? Adv. Neural. Inf. Process. Syst. **34**, 26831–26843 (2021)
2. Chen, J., Mei, J., Li, X., Lu, Y., Yu, Q., et al.: Transunet: rethinking the u-net architecture design for medical image segmentation through the lens of transformers. Med. Image Anal. **97**, 103280 (2024)
3. Çiçek, Ö., Abdulkadir, A., Lienkamp, S.S., Brox, T., Ronneberger, O.: 3D U-net: learning dense volumetric segmentation from sparse annotation. In: Ourselin, S., Joskowicz, L., Sabuncu, M.R., Unal, G., Wells, W. (eds.) MICCAI 2016. LNCS, vol. 9901, pp. 424–432. Springer, Cham (2016). https://doi.org/10.1007/978-3-319-46723-8_49

4. Dong, C., Xu, S., Dai, D., Zhang, Y., et al.: A novel multi-attention, multi-scale 3d deep network for coronary artery segmentation. Med. Image Anal. **85**, 102745 (2023)
5. Dosovitskiy, A., Beyer, L., Kolesnikov, A., Weissenborn, D., Zhai, X., et al.: An image is worth 16×16 words: transformers for image recognition at scale. arXiv preprint arXiv:2010.11929 (2020)
6. Gharleghi, R., Adikari, D., Ellenberger, K., Ooi, S.Y., Ellis, C., et al.: Automated segmentation of normal and diseased coronary arteries-the asoca challenge. Comput. Med. Imaging Graph. **97**, 102049 (2022)
7. Gharleghi, R., Chen, N., Sowmya, A., Beier, S.: Towards automated coronary artery segmentation: a systematic review. Comput. Methods Programs Biomed. **225**, 107015 (2022)
8. Hatamizadeh, A., Nath, V., Tang, Y., Yang, D., Roth, H.R., Xu, D.: Swin unetr: swin transformers for semantic segmentation of brain tumors in MRI images. In: International MICCAI Brainlesion Workshop, pp. 272–284. Springer, Heidelberg (2021). https://doi.org/10.1007/978-3-031-08999-2_22
9. Hatamizadeh, A., Tang, Y., Nath, V., Yang, D., Myronenko, A., Landman, B., et al.: Unetr: transformers for 3d medical image segmentation. In: Proceedings of the IEEE/CVF Winter Conference on Applications of Computer Vision, pp. 574–584 (2022)
10. He, S., Ji, Y., Zhang, Y., Zeng, A., Pan, D., et al.: CFNET: a coarse-to-fine framework for coronary artery segmentation. In: Proceedings of PRCV, pp. 431–442. Springer, Heidelberg (2023). https://doi.org/10.1007/978-981-99-8469-5_34
11. Huang, W., Huang, L., Lin, Z., Huang, S., Chi, Y., et al.: Coronary artery segmentation by deep learning neural networks on computed tomographic coronary angiographic images. In: In Proceedings of IEEE EMBC, pp. 608–611. IEEE (2018)
12. Isensee, F., Jaeger, P.F., Kohl, S.A., Petersen, J., Maier-Hein, K.H.: nnu-net: a self-configuring method for deep learning-based biomedical image segmentation. Nat. Methods **18**(2), 203–211 (2021)
13. Isensee, F., Wald, T., Ulrich, C., Baumgartner, M., Roy, S., et al.: nnu-net revisited: a call for rigorous validation in 3d medical image segmentation. In: International Conference on Medical Image Computing and Computer-Assisted Intervention, pp. 488–498. Springer, Heidelberg (2024). https://doi.org/10.1007/978-3-031-72114-4_47
14. Koonce, B.: Resnet 50. In: Convolutional Neural Networks with Swift for Tensorflow: Image Recognition and Dataset Categorization, pp. 63–72. Springer, Heidelberg (2021). https://doi.org/10.1007/978-1-4842-6168-2_6
15. Lee, H.H., Bao, S., Huo, Y., Landman, B.A.: 3d ux-net: a large kernel volumetric convnet modernizing hierarchical transformer for medical image segmentation. arXiv preprint arXiv:2209.15076 (2022)
16. Liu, Z., Lin, Y., Cao, Y., Hu, H., Wei, Y., et al.: Swin transformer: hierarchical vision transformer using shifted windows. In: Proceedings of the IEEE/CVF International Conference on Computer Vision, pp. 10012–10022 (2021)
17. Liu, Z., Mao, H., Wu, C.Y., Feichtenhofer, C., Darrell, T., Xie, S.: A convnet for the 2020s. In: Proceedings of the IEEE/CVF Conference on Computer Vision and Pattern Recognition, pp. 11976–11986 (2022)
18. Maier-Hein, L., Reinke, A., Kozubek, M., Martel, A.L., Arbel, T., et al.: Bias: transparent reporting of biomedical image analysis challenges. Med. Image Anal. **66**, 101796 (2020)

19. Medrano-Gracia, P., Ormiston, J., Webster, M., Beier, S., Ellis, C., et al.: A study of coronary bifurcation shape in a normal population. J. Cardiovasc. Transl. Res. **10**, 82–90 (2017)
20. Mou, L., Zhao, Y., Fu, H., Liu, Y., Cheng, J., et al.: Cs2-net: deep learning segmentation of curvilinear structures in medical imaging. Med. Image Anal. **67**, 101874 (2021)
21. Pan, L.S., Li, C.W., Su, S.F., Tay, S.Y., Tran, Q.V., Chan, W.P.: Coronary artery segmentation under class imbalance using a u-net based architecture on computed tomography angiography images. Sci. Rep. **11**(1), 14493 (2021)
22. Roy, S., Koehler, G., Ulrich, C., Baumgartner, M., Petersen, J., et al.: Mednext: transformer-driven scaling of convnets for medical image segmentation. In: Proceedings of MICCAI, pp. 405–415. Springer, Heidelberg (2023). https://doi.org/10.1007/978-3-031-43901-8_39
23. Serruys, P.W., Hara, H., Garg, S., Kawashima, H., Nørgaard, B.L., et al.: Coronary computed tomographic angiography for complete assessment of coronary artery disease: Jacc sota review. J. Am. Coll. Cardiol. **78**(7), 713–736 (2021)
24. Taha, A., Lo, P., Li, J., Zhao, T.: Kid-net: convolution networks for kidney vessels segmentation from CT-volumes. In: Frangi, A.F., Schnabel, J.A., Davatzikos, C., Alberola-López, C., Fichtinger, G. (eds.) MICCAI 2018. LNCS, vol. 11073, pp. 463–471. Springer, Cham (2018). https://doi.org/10.1007/978-3-030-00937-3_53
25. Virani, S.S., Alonso, A., Aparicio, H.J., Benjamin, E.J., Bittencourt, M.S., et al.: Heart disease and stroke statistics-2021 update: a report from the American heart association. Circulation **143**(8), e254–e743 (2021)
26. Yu, L., et al.: Automatic 3D cardiovascular MR segmentation with densely-connected volumetric ConvNets. In: Descoteaux, M., Maier-Hein, L., Franz, A., Jannin, P., Collins, D.L., Duchesne, S. (eds.) MICCAI 2017. LNCS, vol. 10434, pp. 287–295. Springer, Cham (2017). https://doi.org/10.1007/978-3-319-66185-8_33
27. Zeng, A., Wu, C., Lin, G., Xie, W., Hong, J., et al.: Imagecas: a large-scale dataset and benchmark for coronary artery segmentation based on computed tomography angiography images. Comput. Med. Imaging Graph. **109**, 102287 (2023)
28. Zhou, Z., Rahman Siddiquee, M.M., Tajbakhsh, N., Liang, J.: UNet++: a nested U-net architecture for medical image segmentation. In: Stoyanov, D., et al. (eds.) DLMIA/ML-CDS -2018. LNCS, vol. 11045, pp. 3–11. Springer, Cham (2018). https://doi.org/10.1007/978-3-030-00889-5_1

SDB: Safety Constraint Mechanism for Dual-Branch End-to-End Autonomous Driving

Hongzhi Yan and Jianmei Su(✉)

Southwest University of Science and Technology, Mianyang, Sichuan, China
sujm_swust@163.com

Abstract. Autonomous driving requires predicting a safe and compliant trajectory to facilitate reliable driving decisions. Previous end-to-end autonomous driving approaches usually research in two directions, one is based on the planned trajectory, and the other is making control decisions directly. In this paper, the proposed autonomous driving model is a two-branch combining architecture. One is waypoint prediction branch, which uses Multilayer Perceptron(MLP) prediction to generate a sequence of future waypoints to obtain a safe and reliable trajectory by using three penalty constraint. The other branch is the Proportion Integration Differentiation(PID) controller, which dynamically adjusts the decision-making control command relying on the vehicle state and trajectory point and optimal waypoints prediction branch. The experimental evaluation on the CARLA simulator shows that the proposed approach achieves the best performance on closed-loop benchmarks over state of the art approaches.

Keywords: End-to-End · Dual-Branch Structure · Dynamic BEV Feature Fusion · Penalty Constraint

1 Introduction

With the rapid development of deep learning in recent years, autonomous driving has experienced a shift from modular [20,26], handcrafted pipelines to integrated end-to-end paradigms [11–13,25]. The end-to-end framework emphasizes the concept of seamless integration of perception, decision making and control, which has gradually become a research hotspot of autonomous driving. However, there are still challenges in feasibility and security for the end-to-end planning control method. There are two main reasons: (1) the control output is susceptible to local disturbances due to the lack of forward-looking prediction and constraints on the future trajectory. (2) The lack of effective decoupling between control decision and trajectory planning makes it difficult to achieve real-time collaboration in dynamic scenarios.

For the above problems, many researchers have proposed solutions. Intermediate representations are used as auxiliary supervision signals to enhance the

J. Lokoč et al. (Eds.): MMM 2026, LNCS 16413, pp. 389–402, 2026.
https://doi.org/10.1007/978-981-95-6957-1_28

stability and generalization of the model. Bansal et al. [1] used trajectory prediction as an intermediate representation to model future trajectory generation and guide control strategies. However, early methods only generated a single trajectory, which could not cover multiple driving behaviors. Varadarajan et al. [22] addressed the issue of insufficient multi-modal coverage by introducing probabilistic anchors to generate multi-modal trajectory distributions and improve the rationality of trajectories in complex intersection scenarios. Nevertheless, traditional trajectory prediction modules often ignore compliance and safety constraints on the predicted trajectories, leading to unreasonable driving behaviors such as overlapping with other agents, exceeding the drivable area, and driving in the wrong direction.

To solve this problem, this paper proposes a Safety Constraint Mechanism for Dual-Branch End-to-End Autonomous Driving (SDB). To better integrate high-level trajectory planning with low-level motion control, a dual-branch architecture is proposed. One branch performs waypoint prediction and is used to guide the decision-making process of the other branch, enabling the model to utilize the predicted waypoint information as prior knowledge for safer and more compliant driving behavior.

In the encoder part, a gated dynamic Bird's Eye View(BEV) feature fusion method is adopted, which can adaptively fuse different modal features to improve the robustness of the model. The decoder part, multiple penalty constraints are introduced in the trajectory prediction process: Collision Penalty Constraint, Boundary Penalty Constraint and Directional Penalty Constraint. As shown in Fig. 2. Mainly to ensure the safety and compliance of the self vehicle in the process of driving. To better weight the three penalty constraint loss functions, we use the weight adaptive allocation method to dynamically adjust the weights of the three loss functions. It avoids the difficulty of manually adjusting the weights. This method has been verified on the simulation platform CARLA [9] and achieved excellent performance.

2 Related Work

End-to-End Autonomous Driving. The traditional modular paradigm has the severe disadvantages of information error accumulation and high maintenance cost. Therefore, in recent years, the end-to-end [10,11,13,25] paradigm of autonomous driving is a hot research issue. End-to-End paradigm raw sensor input can be divided into single-modal and multi-modal methods. However, the single-modal method has limited generalization ability and lacks the comprehensiveness of scene understanding. Compared with the single-modal method, the multi-modal method complements the disadvantages between different modes and has good robustness in complex or extreme scenes.

In multi-modal fusion, Vora et al. [23] append image semantic information to the point cloud. Wang et al. [24] enhance point cloud representations using image features. Chitta et al. [7] perform multi-scale feature fusion of RGB camera and LiDAR input information through a Transformer architecture. Liu et al. [17] fuse multiple modalities using a unified BEV space. Jiang et al. [13] proposed a vectorization-based, end-to-end autonomous driving model that represents the driving scene as a fully vectorized format, predicts future trajectories, and enhances the robustness of scene understanding. Zhang et al. [28] adopt an end-to-end paradigm, directly outputting the low-level control signals, including steering angle, throttle, and brake. These works promote the development of the end-to-end paradigm of autonomous driving, improve the task efficiency and scene adaptation ability of autonomous driving, and reveal future technical challenges and optimization directions.

BEV Representation in Autonomous Driving. Bird's Eye View (BEV) represents its core advantages in terms of unified 3D spatial representation, robustness to occlusion, and natural fit with downstream tasks (planning and control). Lang et al. [15] proposed a point cloud coding method based on pillars, which laid the foundation for real-time LiDAR detection. With the development of multi-modal perception, effective projection of image information from the camera perspective to BEV space has become a new research hotspot. Some methods choose to build BEV representations of images directly based on geometric projections. Philion et al. [18] use the estimated depth map to project image features into BEV space and perform spatial fusion. Hu et al. [10] proposed to use multiple frames of time-series images to extract BEV features to realize time-series modeling and future prediction.

There are also transformer-based methods [16] that interact between image features and BEV queries through a spatial cross-attention mechanism to obtain high-quality BEV representations without accurate depth estimation. Chen et al. [5] convert front-view features into BEV space using Deformable Attention, effectively addressing the geometric distortion caused by perspective projection and improving the accuracy of lane line localization in 3D space. The BEV representation also demonstrates great potential in planning and decision-making. Hu et al. [11] proposed the first BEV-based end-to-end framework that jointly optimizes perception, prediction, and planning tasks, enabling the system to complete the entire autonomous driving task directly based on the BEV representation. This work provides a valuable reference and inspiration for the development of end-to-end autonomous driving technologies.

3 Method

Overview. The overall framework of SDB is shown in Fig. 1. Image encoder (ResNet-34) and LiDAR encoder (ResNet-18) are used to extract image feature F_{img} and LiDAR feature F_{LiDAR} for RGB images and LiDAR images, respectively. In the process of extracting F_{img}, the LSS [18] algorithm is used to predict the depth distribution of image pixels and generate 3D feature point cloud along the line of sight direction, and then the point cloud is projected into the BEV grid for feature feature aggregation to generate RGB image feature $F_{\text{i\_BEV}}$ in the form of BEV. Following the practice of SECOND [27], the LiDAR encoder applies a sparse 3D convolution to the voxelized point cloud, and finally also outputs the feature $F_{\text{l\_BEV}}$ of the laser LiDAR modality in the form of BEV. In the feature fusion part, we dynamically fuse two BEV form features (paragraph 3.1).

The fused BEV feature $F_{\text{BEV}} \in \mathbb{R}^{B\times 256\times 21\times 21}$ will be spliced with the feature vector $F_{\text{c}} \in \mathbb{R}^{B\times 128}$ of the self-vehicle state information encoded by MLP to generate the fused feature $F_{\text{out}} \in \mathbb{R}^{B\times 384}$ for decision-making, which is used in the decoder for direct control branch and waypoint prediction branch. Among them, the direct control branch directly predicts the throttle value, brake value, steering value and other decision quantities. The time series information in the future position information generated by the waypoint prediction branch can make the direct control branch consider the future motion intention of the surrounding agents in the decision-making process, so as to plan in advance and avoid potential dangers. In addition, traffic constraint information such as traffic lights and stop signs will be used as additional penalty items to constraint the future path generated by the self-vehicle. When the predicted future path conflicts with traffic rules, the model will be penalized, so as to ensure that the decisions generated by the model comply with traffic rules and improve the safety of decision-making.

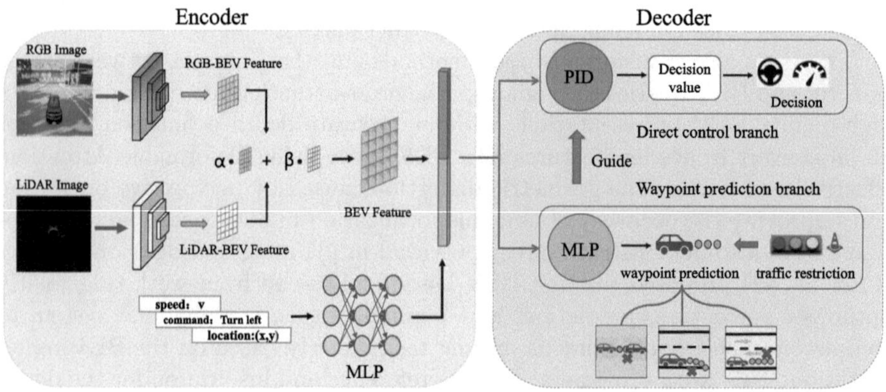

Fig. 1. Overall architecture of SDB. The overall SDB framework consists of two parts: encoder and decoder. The encoder processes images from various sensors and converts them into a Bird's Eye View (BEV), which serves as input for the decoder. The decoder consists of a direct control branch and a waypoint prediction branch, with the future trajectory information generated by the waypoint prediction branch guiding the final decision of the direct control branch.

3.1 Dynamic BEV Feature Fusion

In this study, dynamic BEV feature fusion is a multi-modal feature fusion method based on dynamic gating mechanism, which dynamically weight fuses LiDAR features in the form of BEV and RGB camera features. This fusion mechanism fully complements the advantages of both. It can also adaptively assign modal weights to the importance of input features according to the scene environmental factors. Compared with fixed weight or simple concatenation, it can effectively deal with the difference of modal feature distribution and local noise interference. For the input RGB image I_{img} and LiDAR image I_{LiDAR}, ResNet-34 and ResNet-18 were used to extract features F_{img} and F_{LiDAR}, respectively. The specific steps are as follows. First, the two features are transformed by 1×1 convolution:

$$\widetilde{F}_{\text{img}} = C_{1\times 1}(F_{\text{img}}), \quad \widetilde{F}_{\text{LiDAR}} = C_{1\times 1}(F_{\text{LiDAR}}) \tag{1}$$

where $C_{1\times 1}$ is 1×1 convolution, in this way, the expression ability of features can be improved while keeping the spatial size unchanged, for better fusion in the future. Image features and LiDAR features are concatenated to generate gating weights:

$$F_{\text{concat}} = \text{Concat}(\widetilde{F}_{\text{img}}, \widetilde{F}_{\text{LiDAR}}) \in \mathbb{R}^{B\times 2C\times H\times W} \tag{2}$$

$$G = \sigma(C_{1\times 1}(\text{ReLU}(C_{3\times 3}(F_{\text{concat}})))) \in \mathbb{R}^{B\times 2\times H\times W} \tag{3}$$

where σ is the Sigmoid function, $C_{3\times 3}$ is the 3×3 convolution, and G is the modal weight after fusion. The concatenated feature F_{concat} contains the joint information of the two modalities in space. Complete context information can be provided to the gated network. The subsequent gating network will output the fusion weights of the two channels, which can select the importance of image or LiDAR features on demand, so as to achieve high-quality modal fusion. After that, the fusion weight G is weight separated:

$$\alpha = G[:, 0, :] \qquad \beta = G[:, 1, :] \tag{4}$$

where α is the image feature weight, β is the LiDAR feature weight. Using Softmax to ensure weight normalization, that is:

$$\alpha + \beta = 1 \tag{5}$$

Finally, the RGB features and LiDAR features in BEV format are dynamically weighted fused, and the final residual connection output:

$$F_{\text{fused}} = \alpha \cdot F_{\text{i\_BEV}} + \beta \cdot F_{\text{l\_BEV}} \tag{6}$$

$$F_{\text{BEV}} = F_{\text{fused}} + F_{\text{img}} + F_{\text{LiDAR}} \tag{7}$$

Each pixel uses the adaptive weight to perform the weighted sum of the two modal features, so as to achieve dynamic fusion. The final double residual link mainly retained the feature information in the image modality and LiDAR modality to prevent the loss of detail information after fusion. Secondly, it can also slow down the vanishing gradient and increase the stability of training.

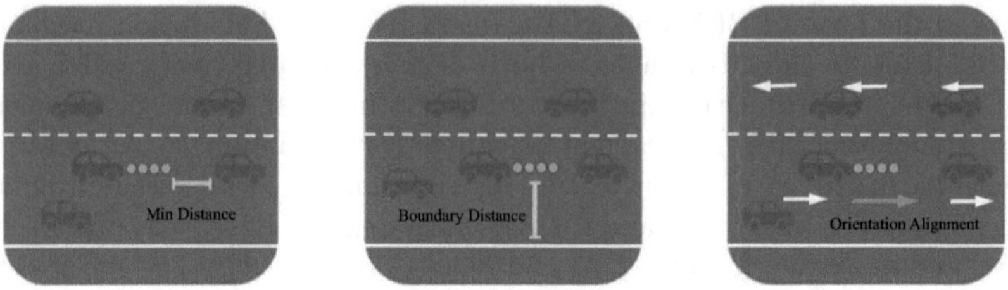

(a) Collision Penalty Constraint (b) Boundary Penalty Constraint (c) Directional Penalty Constraint

Fig. 2. Illustration of Penalty Constraint. Collision penalty constraints ensure the safety of autonomous vehicles and other intelligent entities. Boundary penalty constraints prevent future waypoints from approaching or exceeding boundaries. Directional penalty constraints ensure that the driving direction of the autonomous vehicle aligns with the lane direction.

3.2 Collision Penalty Constraint

As shown in Fig. 2 (a). The Collision Penalty Constraint method in this study designs an exponential penalty term based on the minimum distance between the sequence of waypoints $W_{t+1}, W_{t+2}, \ldots, W_{t+T} \in \mathbb{R}^{B \times T \times 2}$ generated by the model prediction (2 represents the plane coordinates (x,y) of each point) and the agent. Specifically, first SDB calculates the Euclidean distance using the predicted sequence of waypoints and the 2D coordinate positions of other dynamic and static objects:

$$D_{b,t,n} = \|W_t - O_n\|_2 \qquad t \in [1,T],\ n \in [1,N] \tag{8}$$

where b is the sample batch index. t is the waypoint time series index. n is the number of agents in the scenario. Then for the waypoint, we take the distance to the closest agent:

$$d_{b,t} = \min_n D_{b,t,n} \tag{9}$$

Finally, the penalty constraint loss is calculated:

$$\mathcal{L}_{\text{collision}} = \frac{1}{B \cdot T} \sum_{b=1}^{B} \sum_{t=1}^{T} \exp\left(-\frac{d_{bt}}{\delta}\right) \tag{10}$$

where $\exp\left(-\frac{d_{bt}}{\delta}\right)$ is the penalty term, that is, the loss is larger for closer distances and smaller for the opposite. δ is the penalty parameter (default 1.0). Finally, a scalar average penalty constraint loss is derived. The mainly penalizes the behavior of predicting waypoints close to other agents to improve the safety of autonomous driving planning and avoid collisions.

3.3 Boundary Penalty Constraint

As shown in Fig. 2 (b). The Boundary Penalty Constraint method in SDB mainly penalizes the behavior of the predicted waypoint approaching the boundary to avoid the unreasonable risk of falling on the road boundary or crossing the boundary.Specifically, Since the predicted waypoints are the real world coordinates in the actual scene, we need to look up the position value on the BEV feature map, so the real world waypoints are first mapped to the pixel coordinate system on the BEV feature map, and then the horizontal/vertical coordinates of each waypoint position are indexed:

$$(x_{b,t}, y_{b,t}) = \text{clamp}\left(\lfloor s \cdot W_{b,t} \rfloor, 0, H-1\right) \tag{11}$$

where $\lfloor \cdot \rfloor$ denotes the discrete grid coordinates of integers to BEV, and the floating-point coordinates are transformed into integer indexes; s is the scaling factor(default 0.5); H is the size of the BEV feature map; clamp(j, k, l) is a limiting function, limiting j to the $[k, l]$ interval mainly prevents the index from crossing the boundary. Find the value of the BEV boundary mask:

$$M_{b,t} = F_{\text{BEV}}(b, c, x_{b,t}, y_{b,t}) \tag{12}$$

where c is the boundary mask channel. For the waypoints are provided with how far or near the point is from the drivable area in the current BEV feature map. The margin penalty is calculated using the soft-margin strategy:

$$\mathcal{L}_{b,t} = \max(0, \text{margin} - M_{b,t}) \tag{13}$$

where margin is a security threshold (default 0.2). When the boundary mask value is less than the margin, the penalty will be generated, and the closer to the boundary, the greater the penalty. Finally, the loss of all batch and time steps is averaged to obtain the overall boundary constraint loss:

$$\mathcal{L}_{\text{bound}} = \frac{1}{B \cdot T} \sum_{b=1}^{B} \sum_{t=1}^{T} \mathcal{L}_{b,t} \tag{14}$$

This penalty mechanism aggregates the boundary penalties of all waypoints, penalizes the predicted trajectory to be close to the boundary, and enhances the robustness of the model.

3.4 Directional Penalty Constraint

As shown in Fig. 2 (c). The purpose of the Directional Penalty Constraint in this study is to make the predicted trajectory direction consistent with the geometric direction of the road, so as to ensure that the autonomous vehicle's driving trajectory conforms to the road topology (such as lane lines and lane directions). Specifically, we first obtain the direction vector of the predicted path points in the BEV feature map (the difference between adjacent path points in the physical space):

$$\boldsymbol{v}_{b,t} = W_{b,t+1} - W_{b,t}, \quad t = 0, 1, \ldots, T-2 \tag{15}$$

A total of T-1 direction vectors were obtained, after which lane direction vectors were extracted from BEV feature maps and vector normalization was performed:

$$\boldsymbol{d}_{b,t} = (d^x_{b,t}, d^y_{b,t}) \tag{16}$$

$$v_{b,t} = \frac{\boldsymbol{v}_{b,t}}{\|\boldsymbol{v}_{b,t}\|} \qquad d_{b,t} = \frac{\boldsymbol{d}_{b,t}}{\|\boldsymbol{d}_{b,t}\|} \tag{17}$$

Both the direction of the predicted point and the direction of the lane line are converted into unit vectors. The cosine similarity is then used to calculate the alignment of the two directions:

$$\cos(\theta_{b,t}) = v_{b,t} \cdot d_{b,t} \tag{18}$$

where $\cos(\theta_{b,t}) \in [-1, 1]$. A higher value indicates a more consistent direction, otherwise the opposite is true. Finally, the directional loss is averaged over all trajectory segments:

$$\mathcal{L}_{\text{dir}} = \frac{1}{B(T-1)} \sum_{b=1}^{B} \sum_{t=1}^{T} (1 - \cos(\theta_{b,t})) \tag{19}$$

When the direction is consistent, $\cos(\theta_{b,t}) = 1$, the loss is 0; when $\cos(\theta_{b,t}) = -1$ in the opposite direction, the maximum loss is 2. This penalty constraint method improves the direction compliance of the waypoints generated by the model and avoids violations such as retrograde.

3.5 End-to-End Loss

For the three penalty constraint losses in the waypoint prediction branch. Inspired by [14], we adaptively assign three kinds of loss weights. Specifically, the loss weight of each task is automatically adjusted by modeling the uncertainty of the task itself. We first initialize learnable weights, then forward compute the loss for each constraint, and automatically sum up the individual loss terms:

$$\mathcal{L}_{\text{plan}} = \sum_{i} \frac{1}{2\sigma_i^2} \mathcal{L}_i + \frac{1}{2} \log \sigma_i^2 \quad \{i \in \{\mathcal{L}_{\text{collision}}, \mathcal{L}_{\text{bound}}, \mathcal{L}_{\text{dir}}\}\} \tag{20}$$

where σ_i is a learnable parameter that's automatically optimized through back-propagation. $\log \sigma_i$ is the regularization term. $\frac{1}{2\sigma_i^2}$ is the task weight. $\mathcal{L}_{\text{plan}}$ is the waypoint prediction branch loss. For the direct control branch, the loss mainly comes from the error between the control signal output by the model (speed, throttle, steering, etc.) and the real control signal, where the control signal error is calculated by L1 loss. For the total loss, there is also a multi-task auxiliary loss, which is mainly to enhance the quality of BEV representation, traffic rule understanding, feature consistency prediction and other auxiliary goals. Although it does not directly affect the control output, it is of great help to the stability and generalization of model training. That is, the total end-to-end loss is as follows:

$$\mathcal{L} = \omega_1 \mathcal{L}_{\text{plan}} + \omega_2 \mathcal{L}_{\text{cls}} + \omega_3 \mathcal{L}_{\text{aux}} \tag{21}$$

where ω_1, ω_2, ω_3 are weight coefficients. $\mathcal{L}_{\text{cls}}$ is to control the branch loss directly. $\mathcal{L}_{\text{aux}}$ is the multi-task auxiliary loss.

4 Experiment

In this section, our method is tested in the validation of CARLA (0.9.10) driving simulator [9], using three metrics to evaluate the performance of our model: The Route Completion (RC) is used to measure the ability of the vehicle to successfully complete the specified route. Infraction Score (IS) reflects the degree of compliance of vehicles with traffic rules. Driving Score (DS), as a comprehensive performance index, is the main measurement standard, reflecting the comprehensive ability of the model to complete the route and obey the traffic rules in the driving task. And we deploy agent vehicles in 8 towns with different weather conditions of CARLA, and collect 600 000 frames of data at 10 frames per second as the dataset for model training to provide diverse and representative reference samples for model learning.

Table 1. Performance on Longest6 benchmark

RANK	Method	Driving Score↑	Route Completion↑	Infraction Score↑
1	InterFuser [21]	38.177	73.179	0.526
2	Latent TransFuser* [7]	37.31	**95.18**	0.38
3	LAV* [3]	32.74	70.36	0.51
4	WOR* [2]	20.53	48.47	0.56
5	SDB(Ours)	**46.859**	71.485	**0.692**

4.1 Comparative Experiment

In this study, a number of comparative experiments are conducted to compare and analyze the performance of various existing automated driving decision-making methods with excellent performance in two benchmark tests, Longest6 and Town05Long. Using the symbols '$*$' or '†' denotes experimental results from TransFuser [19] and InterFuser [21], respectively. The data labeled as '—' in the violation score column, indicates that the relevant literature does not provide specific results under the benchmark. The test index of '↑' indicates that the

Table 2. Performance on Town05 Long benchmark

RANK	Method	Driving Score↑	Route Completion↑	Infraction Score↑
1	InterFuser [21]	50.664	**84.934**	0.614
2	WOR* [2]	44.80	82.41	–
3	Roach* [28]	43.64	80.37	–
4	NEAT* [6]	37.72	62.13	–
5	TransFuser* [19]	33.15	56.36	–
6	LBC* [4]	7.05	32.09	–
7	CILRS* [8]	3.68	7.19	–
8	ST-P3 [10]	11.45	83.15	–
9	VAD-Base [13]	30.31	75.20	–
10	SDB(Ours)	**56.745**	81.953	**0.708**

higher the value, the better the performance. The bold result indicates that this method has achieved the best performance among the current test methods.

As shown in Table 1 and 2, our model has the highest driving score and infraction score in the autonomous driving task. In the infraction score(IS) indicator, our model shows excellent driving safety and compliance. Additionally, Comprehensive indicator driving score(DS) has also achieved optimal performance among many methods.

We also verify its robustness under different weather conditions, and compare and analyze the performance in a variety of scenarios, including night, heavy rain and sunny weather. As shown in Table 3. Comprehensive validation has demonstrated that SDB exhibits strong robustness under various complex weather conditions, while also achieving a certain degree of performance improvement.

In the comparative experiments, although our model did not achieve the best performance in the route completion (RC) metric, this precisely reflects our emphasis on safety and compliance with traffic rules. In the future, while maintaining high safety standards, we will further strive to improve route completion to achieve more comprehensive driving performance.

As shown in Fig 3. We present the visualizations of SDB under different weather conditions.

4.2 Ablation Studies

We set up a series of ablation implementations that verify the effectiveness of different component designs of our model. Where 'col', 'bound', and 'dir' represent Collision Penalty Constraint, Boundary Penalty Constraint, and Directional Penalty Constraint, respectively. 'DF' stands for dynamic BEV feature fusion.

As shown in Table 4, our designed three penalty mechanisms and dynamic BEV feature fusion have positive improvement for autonomous driving task performance.

Table 3. Performance of SDB under different weather conditions

Weather	Algorithm	Driving Score ↑	Route Completion ↑	Infraction Score ↑
CloudyNight	InterFuser	39.070	**74.982**	0.512
	SDB(Ours)	**46.108**	71.251	**0.623**
HardrainNoon	InterFuser	34.775	**73.483**	0.481
	SDB(Ours)	**43.512**	71.204	**0.642**
MidrainMorning	InterFuser	40.105	**73.133**	0.580
	SDB(Ours)	**47.073**	72.114	**0.705**

Table 4. Result of Ablation Studies

Algorithm	Driving Score↑	Route Completion↑	Infraction Score↑
w/o col&bound&dir&DF	37.849	61.263	0.603
w/o col&bound&dir	38.637	61.896	0.611
w/o bound&dir&DF	40.974	66.548	0.641
w/o col&dir&DF	41.346	68.523	0.653
w/o col&bound&DF	40.052	66.768	0.638
SDB	**46.859**	**71.485**	**0.692**

Fig. 3. Visualization of different weather conditions

5 Conclusion

In this work, we propose a Safety Constraint Mechanism for Dual-Branch End-to-End Autonomous Driving. In the feature fusion stage, we employ a dynamic fusion method that combines image features in BEV format with LiDAR features. The advantage of this approach is its ability to dynamically and adaptively accommodate multi-modal data fusion, thereby improving robustness. For the waypoint prediction branch, we propose three penalty constraints: Collision Penalty Constraint; Boundary Penalty Constraint; Directional Penalty Constraint. These three penalty constraints improve the safety and compliance of the predicted trajectory path of the model. Finally, the overall framework adopted a dual-branch structure, and the predicted future trajectory of the waypoint predicted branch output was used to guide the decision-making process of the direct control branch. In this way, the error problem of path prediction is avoided and the lack of time series information in direct control branch species is complemented. Thus, more accurate decision making and safer driving behavior can be achieved.

Acknowledgments. This paper is supported by the Major Project of Sichuan Provincial Natural Science Foundation (2025ZNSFSC0005), and the Central Government Guides Local Projects of China (2025ZYDF105).

References

1. Bansal, M., Krizhevsky, A., Ogale, A.: Chauffeurnet: learning to drive by imitating the best and synthesizing the worst. arXiv preprint arXiv:1812.03079 (2018)
2. Chen, D., Koltun, V., Krähenbühl, P.: Learning to drive from a world on rails. In: Proceedings of the IEEE/CVF International Conference on Computer Vision, pp. 15590–15599 (2021)
3. Chen, D., Krähenbühl, P.: Learning from all vehicles. In: Proceedings of the IEEE/CVF Conference on Computer Vision and Pattern Recognition, pp. 17222–17231 (2022)
4. Chen, D., Zhou, B., Koltun, V., Krähenbühl, P.: Learning by cheating. In: Conference on Robot Learning, pp. 66–75. PMLR (2020)
5. Chen, L., et al.: Persformer: 3d lane detection via perspective transformer and the openlane benchmark. In: European Conference on Computer Vision, pp. 550–567. Springer, Heidelberg (2022)
6. Chitta, K., Prakash, A., Geiger, A.: Neat: neural attention fields for end-to-end autonomous driving. In: Proceedings of the IEEE/CVF International Conference on Computer Vision, pp. 15793–15803 (2021)
7. Chitta, K., Prakash, A., Jaeger, B., Yu, Z., Renz, K., Geiger, A.: Transfuser: imitation with transformer-based sensor fusion for autonomous driving. IEEE Trans. Pattern Anal. Mach. Intell. **45**(11), 12878–12895 (2022)
8. Codevilla, F., Santana, E., López, A.M., Gaidon, A.: Exploring the limitations of behavior cloning for autonomous driving. In: Proceedings of the IEEE/CVF International Conference on Computer Vision, pp. 9329–9338 (2019)
9. Dosovitskiy, A., Ros, G., Codevilla, F., Lopez, A., Koltun, V.: Carla: an open urban driving simulator. In: Conference on Robot Learning, pp. 1–16. PMLR (2017)
10. Hu, S., Chen, L., Wu, P., Li, H., Yan, J., Tao, D.: St-p3: end-to-end vision-based autonomous driving via spatial-temporal feature learning. In: European Conference on Computer Vision, pp. 533–549. Springer, Heidelberg (2022). https://doi.org/10.1007/978-3-031-19839-7_31

11. Hu, Y., et al.: Planning-oriented autonomous driving. In: Proceedings of the IEEE/CVF Conference on Computer Vision and Pattern Recognition, pp. 17853–17862 (2023)
12. Jia, X., et al.: Think twice before driving: towards scalable decoders for end-to-end autonomous driving. In: Proceedings of the IEEE/CVF Conference on Computer Vision and Pattern Recognition, pp. 21983–21994 (2023)
13. Jiang, B., et al.: VAD: vectorized scene representation for efficient autonomous driving. In: Proceedings of the IEEE/CVF International Conference on Computer Vision, pp. 8340–8350 (2023)
14. Kendall, A., Gal, Y., Cipolla, R.: Multi-task learning using uncertainty to weigh losses for scene geometry and semantics. In: Proceedings of the IEEE Conference on Computer Vision and Pattern Recognition, pp. 7482–7491 (2018)
15. Lang, A.H., Vora, S., Caesar, H., Zhou, L., Yang, J., Beijbom, O.: Pointpillars: fast encoders for object detection from point clouds. In: Proceedings of the IEEE/CVF Conference on Computer Vision and Pattern Recognition, pp. 12697–12705 (2019)
16. Li, Z., et al.: Bevformer: learning bird's-eye-view representation from lidar-camera via spatiotemporal transformers. IEEE Trans. Pattern Anal. Mach. Intell. (2024)
17. Liu, Z., et al.: Bevfusion: multi-task multi-sensor fusion with unified bird's-eye view representation. In: 2023 IEEE International Conference on Robotics and Automation (ICRA), pp. 2774–2781. IEEE (2023)
18. Philion, J., Fidler, S.: Lift, splat, shoot: encoding images from arbitrary camera rigs by implicitly unprojecting to 3D. In: Vedaldi, A., Bischof, H., Brox, T., Frahm, J.-M. (eds.) ECCV 2020. LNCS, vol. 12359, pp. 194–210. Springer, Cham (2020). https://doi.org/10.1007/978-3-030-58568-6_12
19. Prakash, A., Chitta, K., Geiger, A.: Multi-modal fusion transformer for end-to-end autonomous driving. In: Proceedings of the IEEE/CVF Conference on Computer Vision and Pattern Recognition, pp. 7077–7087 (2021)
20. Sadat, A., Casas, S., Ren, M., Wu, X., Dhawan, P., Urtasun, R.: Perceive, predict, and plan: safe motion planning through interpretable semantic representations. In: Vedaldi, A., Bischof, H., Brox, T., Frahm, J.-M. (eds.) ECCV 2020. LNCS, vol. 12368, pp. 414–430. Springer, Cham (2020). https://doi.org/10.1007/978-3-030-58592-1_25
21. Shao, H., Wang, L., Chen, R., Li, H., Liu, Y.: Safety-enhanced autonomous driving using interpretable sensor fusion transformer. In: Conference on Robot Learning, pp. 726–737. PMLR (2023)
22. Varadarajan, B., et al.: Multipath++: efficient information fusion and trajectory aggregation for behavior prediction. In: 2022 International Conference on Robotics and Automation (ICRA), pp. 7814–7821. IEEE (2022)
23. Vora, S., Lang, A.H., Helou, B., Beijbom, O.: Pointpainting: sequential fusion for 3d object detection. In: Proceedings of the IEEE/CVF Conference on Computer Vision and Pattern Recognition, pp. 4604–4612 (2020)
24. Wang, C., Ma, C., Zhu, M., Yang, X.: Pointaugmenting: cross-modal augmentation for 3d object detection. In: Proceedings of the IEEE/CVF Conference on Computer Vision and Pattern Recognition, pp. 11794–11803 (2021)
25. Wu, P., Jia, X., Chen, L., Yan, J., Li, H., Qiao, Y.: Trajectory-guided control prediction for end-to-end autonomous driving: a simple yet strong baseline. Adv. Neural. Inf. Process. Syst. **35**, 6119–6132 (2022)
26. Xu, W., Wang, Q., Dolan, J.M.: Autonomous vehicle motion planning via recurrent spline optimization. In: 2021 IEEE International Conference on Robotics and Automation (ICRA), pp. 7730–7736. IEEE (2021)

27. Yan, Y., Mao, Y., Li, B.: Second: sparsely embedded convolutional detection. Sensors **18**(10), 3337 (2018)
28. Zhang, Z., Liniger, A., Dai, D., Yu, F., Van Gool, L.: End-to-end urban driving by imitating a reinforcement learning coach. In: Proceedings of the IEEE/CVF International Conference on Computer Vision, pp. 15222–15232 (2021)

CSQDA: A Parameter-Efficient and Memory-Efficient Tuning Method for Medical Image Classification

Yiqian Li[1] and Andy J. Ma[1,2,3](✉)

[1] School of Computer Science and Engineering, Sun Yat-sen University, No. 132 Outer Ring East Road, Guangzhou 510006, Guangdong, China
liyq263@mail2.sysu.edu.cn

[2] Guangdong Province Key Laboratory of Information Security Technology, No. 132 Outer Ring East Road, Guangzhou 510006, Guangdong, China

[3] Key Laboratory of Machine Intelligence and Advanced Computing, No. 132 Outer Ring East Road, Guangzhou 510006, Guangdong, China
majh8@mail.sysu.edu.cn

Abstract. Parameter-efficient transfer learning (PETL) has emerged as a powerful approach for adapting pre-trained models to downstream tasks while minimizing trainable parameters and computational costs. However, existing PETL methods often alter the internal structures of pre-trained models, leading to high memory usage due to intermediate feature storage during backpropagation. In this paper, we introduce a novel PETL framework that generates task-specific learning queries for pre-trained knowledge utilization, enabling efficient adaptation without extensive resource demands. The TSQG module synthesizes these queries independently of the backbone, selectively extracting informative features from intermediate representations with minimal memory overhead. To further boost performance, the extracted features are refined through a Multi-Scale Feature Enhancement module. Unlike conventional PETL approaches that rely on extensive feature propagation, our architecture achieves substantial memory savings without compromising accuracy. We validate our approach in medical image classification tasks, and extensive experiments across multiple datasets demonstrate that our method achieves state-of-the-art results with memory efficiency, making it well-suited for real-world scenarios. Our findings highlight the practicality of lightweight PETL solutions, offering a scalable and resource-efficient alternative for deploying pre-trained models in medical imaging domain.

Keywords: Parameter-efficient transfer learning · Memory-efficient transfer learning · Medical image classification

1 Introduction

With the growing success of large-scale pre-trained vision models, fine-tuning has become a standard strategy for adapting general visual representations to downstream tasks. However, full fine-tuning of these large models is computationally

J. Lokoč et al. (Eds.): MMM 2026, LNCS 16413, pp. 403–416, 2026.
https://doi.org/10.1007/978-981-95-6957-1_29

expensive and memory-intensive, especially prohibitive in resource-constrained scenarios.

Parameter-Efficient Transfer Learning (PETL) provides a beneficial substitute by updating only a small portion of parameters or inserting lightweight modules while keeping most of the pre-trained weights frozen [11]. Despite its advantages, conventional PETL approaches often integrate additional modules directly into the backbone, forcing gradients through all intermediate layers and inflating memory usage [26]. This results in the modification of every intermediate feature, which requires considerable storage during backpropagation. Furthermore, the domain gap between natural and medical images [5,16], poses an additional challenge for effective knowledge transfer and direct application of PETL methods in medical imaging.

To overcome these challenges, we propose Cross-Scale Query Driven Adaptation (CSQDA), a novel PETL framework for medical imaging that decouples task-specific learning from intermediate feature processing, significantly reducing memory overhead while maintaining fidelity and diagnostic accuracy. Instead of modifying the internal layers of the pre-trained model or propagating gradients through the entire backbone, CSQDA introduces a side-network architecture that engages the frozen backbone through query-driven interaction. CSQDA introduces two synergistic components:

First, the Task-Specific Query Generator (TSQG) generates dynamic input-conditioned queries that act as spatial and semantic anchors. These queries are designed to extract the most task-relevant features from the frozen backbone via attention mechanisms, without requiring any gradient updates to the backbone itself. Instead of initializing with a pruned sub-network derived from the pre-trained backbone as in LST [20], TSQG is trained on downstream medical datasets without relying on any pre-trained weights, ensuring that the learned queries align with clinically relevant semantics.

Second, the Multi-Scale Feature Enhancement Module (MSFEM) complements this process by hierarchically fusing features extracted from layers of the frozen backbone with the query-refined representations from TSQG. MSFEM captures spatially-aware and scale-consistent features by leveraging both low-level details and high-level semantics, allowing the model to aggregate rich contextual information without requiring full backpropagation through the backbone.

Through bidirectional feature-query interaction pathways, CSQDA facilitates efficient feature extraction and integration without backpropagating through the backbone. The two modules exchange and refine information iteratively, enabling effective adaptation, leading to a significant 65% reduction in memory consumption during training.

Our main contributions are as follows:

1. We present a new PETL paradigm for medical imaging that decouples domain-adap training from intermediate representation processing. This reduces memory costs by building a lightweight side network and avoiding gradient flow through the backbone.

2. We propose a novel module which enables query-driven feature fusion via attention mechanisms, where TSQG dynamically identifies task relevant regions guided by attention and MSFEM aggregates multi-scale context without backpropagating through the frozen LPM.
3. Extensive experiments on multiple medical imaging datasets of different modalities demonstrate that CSQDA achieves competitive performance with state-of-the-art PETL methods in both efficiency and accuracy.

2 Related Work

2.1 Parameter-Efficient Transfer Learning (PETL)

Parameter-efficient transfer learning (PETL) has emerged as a versatile strategy, demonstrating effectiveness in NLP, CV, and VL tasks by reducing adaptation costs. In NLP, partial-based tuning PETL techniques, such as BitFit [24] and Prompt Tuning [17] demonstrated the feasibility of adapting large language models by updating only bias terms or adding learnable prompts. Addition-based tuning techniques, such as LoRA [12] and Adapter [11], introduced trainable lightweight modules (e.g., low-rank matrices or bottleneck adapters), significantly reducing the number of trainable parameters without sacrificing performance.

Inspired by these advances, PETL has been extended to the vision domain, especially with the rise of Vision Transformers (ViTs) [4]. Visual Prompt Tuning (VPT) [14] introduced trainable prompts in ViTs for downstream adaptation. AdaptFormer [2] further improved efficiency by integrating lightweight adapters into feedforward network of transformer blocks. SSF [18] adjusted scale and shift factors for intermediate features of each layer, offering another direction for efficient visual adaptation. Factor-Tuning (FacT) [15] tensorizes weights of each ViT into a single 3D tensor, and decomposes their increments into lightweight factors so that only the factors need to be updated and stored during finetuning. Adapter-based approaches have also shown consistent gains in performance-efficiency trade-offs [9].

2.2 Memory-Efficient Transfer Learning and Medical Applications

Despite reducing trainable parameters, many PETL methods still incur significant memory usage during training, primarily due to activation storage for backpropagation. This poses a challenge for memory-constrained environments, especially in medical imaging scenarios.

To mitigate this, several memory-efficient PETL methods have emerged. Side-Tuning [25] introduces a lightweight side network trained in parallel with a frozen backbone, which requires fewer activations to be stored. Ladder Side Tuning [20], similarly, trains a small and separate network that takes intermediate representation as input via shortcut connections (called ladders) from backbone networks and makes predictions. VQT [22] augments the query matrix with learnable prompts to extract task-relevant representations from the backbone while keeping the intermediate features unchanged.

In medical imaging, recent efforts have increasingly explored prompt-based PETL approaches to cope with domain constraints and efficiency requirements. Dynamic Visual Prompt Tuning (DVPT) [8] learns sample- and domain-adaptive prompts by injecting parallel structures into intermediate ViT layers, but this modification significantly inflates feature-map storage and increases memory footprint. Embedded Prompt Tuning (EPT) [27] expands channels with embedded prompt vectors to improve few-shot classification accuracy, but its enlarged dimensionality also leads to inefficient memory use. To achieve memory and parameter efficiency, FPT [13] targets high-resolution medical images by freezing the main backbone and employing a low-resolution side network alongside prompt-driven fusion modules. However, when applied at standard resolutions, its selective token fusion strategy can overlook critical spatial details and yield subpar performance. Collectively, these observations highlight the need for prompt tuning approaches that balance memory efficiency with robust feature utilization.

Unlike prior prompt-based PETL approaches, our CSQDA adopts a cross-scale query-driven design to align medical semantics with frozen backbone features, achieving memory efficiency. CSQDA integrates domain-specific queries of medical datasets with a cross-scale fusion of local details and global semantic cues, resulting in scale-consistent representations aligned with clinically relevant features while remaining memory-efficient.

3 Method

3.1 Preliminaries

Transfer Learning and Intermediate Feature Utilization. To adapt pre-trained models for downstream tasks, two primary adaptation paradigms are widely adopted: linear probing, which freezes the backbone and only trains a new classification head, and fine-tuning, which updates the entire model. Recent methods such as HEAD2TOE [6], have highlighted the utility of intermediate representations in boosting downstream performance. HEAD2TOE enhances linear probing by aggregating intermediate representations at multiple checkpoints including normalization, attention, and feedforward stages. These features are pooled by token position and concatenated to form a rich representation for classification. To mitigate redundancy, a sparsity-inducing regularizer (group lasso) is employed to prune redundant features. A potential shortcoming of HEAD2TOE lies in its fixed pooling operation, which may discard task-relevant signals.

Visual Query Tuning. VQT [22] substitutes the fixed pooling with learnable query tokens that summarize information directly within the attention mechanism. Concretely, at each layer l, K learnable "query" tokens $\mathbf{P}_{l-1} \in \mathbb{R}^{K \times D}$ are introduced and undergo only the query projection $\hat{\mathbf{Q}}_l = W_q^{(l)} \mathbf{P}_{l-1}$, and are concatenated with the standard query matrix. The modified attention computation becomes:

$$[X_l \parallel \tilde{F}_l] = V_l \cdot \text{Softmax}\left(\frac{K_l^\top}{\sqrt{d}} [Q_l \parallel \hat{Q}_l] \right), \quad (1)$$

where the second term yields K new attention-summarized features $\tilde{Z}_l \in \mathbb{R}^{K \times D}$. Finally, VQT aggregates new features across all layers together with the CLS token to form the input, $\left[x_L^{(\text{CLS})} \parallel \{\tilde{F}_l\}_{l=1}^L \right]$, of the downstream classification.

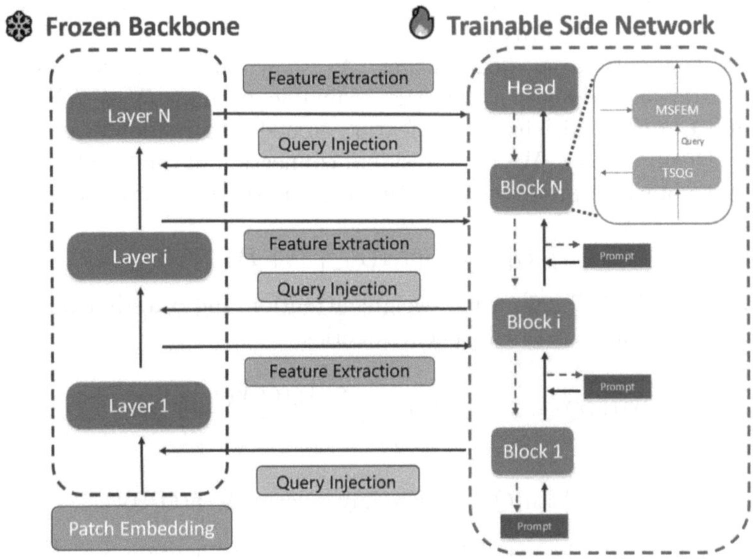

Fig. 1. The overview of the proposed framework. Our paradigm based on the side tuning architecture, where the trainable side network generates and injects query into the frozen pre-trained backbone to extract intermediate features for further enhancement. Bidirectional feature-query transfer pathways between the networks ensure that relevant feature information is iteratively aggregated through attention mechanism.

3.2 Overall Framework

Figure 1 explains our proposed framework, which adapts frozen pre-trained models to downstream medical image classification tasks through a lightweight side network that interacts with the backbone via task-specific queries and multi-scale feature enhancement. The frozen ViT backbone processes input medical images to extract intermediate features at each layer. These features are routed to a multi-scale adapter, which enhances them using parallel convolutional pathways to capture local-to-global patterns critical for medical diagnostics. The output of the adapter is fused with dynamic queries generated by the Task-Specific Query Generator (TSQG), a lightweight module with task-relevant guidance. Cross-attention then refines these features by selectively aggregating multi-scale context, which propagates to subsequent layers. The side network operates independently of the frozen backbone, requiring only 4% of its parameters, and eliminates complete backpropagation through the original backbone, which ensures both parameter efficiency and memory efficiency.

3.3 Task-Specific Query Generator

TSQG generates layer-specific queries Q_l that guide the frozen backbone to focus on diagnostically critical regions, addressing the challenge of dispersed fine-grained signals in medical images. As shown in Fig. 2, for layer l, the query generation leverages a compact trainable prompt $P_l \in R^{K*D}$ ($K \ll$ **Sequence Length**) initialized with feature embeddings randomly sampled from the downstream dataset. We use a residual connection of the fused feature from the previous block to preserve contextual continuity and a prompt adapter:

$$Q'_l = \text{Adapter}(F_{l-1}^{fused} + P_l) \tag{2}$$

Inspired by SynQT [26], a lightweight attention and a feedforward network are utilized for generating task-specific queries:

$$\text{Attn}(Q) = Q + \text{Softmax}\left(\frac{QW_q(QW_k)^\top}{\sqrt{d}}\right) QW_v \tag{3}$$

$$Q_l = \text{Attn}(Q'_l) + \text{FFN}(\text{LayerNorm}(\text{Attn}(Q'_l))) \tag{4}$$

where W_q, W_k, W_v are projection layers with hidden dimension $d \ll D$ to ensure parameter efficiency.

The output Q_l captures task-specific knowledge and then will be injected to the frozen backbone for fine-grained feature. The prompt P_l is initialized with random sampled features from the downstream medical datasets to bridge the natural-to-medical domain gap, while the residual term F_{l-1} ensures progressive refinement across layers. By decoupling task-specific learning from the frozen backbone, TSQG avoids modifying intermediate features, reducing memory overhead compared to adapter-based methods.

3.4 Multi-scale Feature Enhancement Module

The query generated by TSQG is injected to the query space of the frozen backbone to get feature F_l like VQT , and the intermediate feature X_l is extracted.

$$F'_l = \text{Attn}_l^b(Q_l) + Q_l, F_l = F'_l + \text{FFN}_l^b(\text{LayerNorm}_l^b(F'_l)) \tag{5}$$

where $\text{Attn}_l^b, \text{FFN}_l^b, \text{LayerNorm}_l^b$ are frozen blocks of layer l of the frozen backbone.

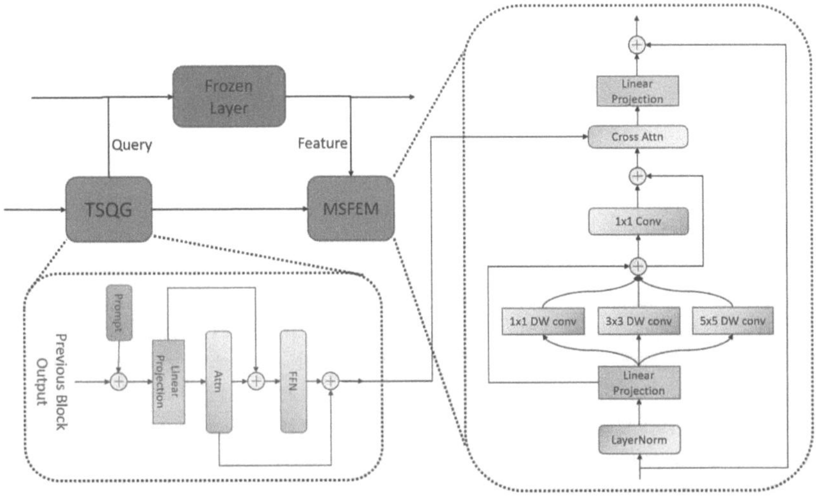

Fig. 2. Illustration of components TSQG and MSFEM of a side network block. TSQG utilizes learnable prompt and a lightweight attention for generating task-specific queries dynamically. MSFEM enhances the features extracted from the backbone through parallel convolutional pathways, capturing patterns at multiple scales.

MSFEM enhances features X_l through a multiscale enhancement adapter and cross-attention fusion, addressing the multi-scale nature of medical anomalies. The adapter processes Q_l and X_l through down projections to obtain Q_l^d and X_l^d, and further enhances X_l^d via parallel depth-wise convolutions with varying kernel sizes.

The overall calculation process can be formulated as follows:

$$X_l^{enhanced} = X_l^d + \sigma\left(f_{\text{pw}}\left(\frac{1}{3}\sum_{i=1}^{3} f_{\text{dw}^i}\left(\left(\text{LayerNorm}(X_l^d)\right)\right)\right)\right) \tag{6}$$

where f_{dw} and f_{pw} denote depth-wise and point-wise convolution. The depthwise convolution captures patterns at kernel sizes $k \in \{1, 3, 5\}$, and a pointwise convolution aggregates multiscale feature across channels. The enhanced features then interact with queries Q_l^d through cross-attention and an up projection is used to recover original dimension:

$$F_l^{\text{fused}} = F_l + \alpha \cdot \mathbf{W}_{up}\left(\text{LayerNorm}(\text{CrossAttn}(\mathbf{Q}_l^d, \mathbf{X}_l^{enhanced}))\right) \tag{7}$$

This fusion selectively amplifies task-relevant regions while suppressing background noise, which is crucial for medical images where diagnostic signals are sparse. The fused features F_l^{fused} are propagated to layer $l+1$, and help update the prompt P_{l+1} for iterative refinement.

CSQDA avoids full gradient propagation through the frozen backbone, reducing training memory by 65% compared to full-finetuning. The parallel convolutions of the multicognitive adapter mimic the multiscale analysis of radiologists. Cross-attention focus on query-guided regions, ensuring efficient adaptation without sacrificing diagnostic accuracy.

4 Experiments

4.1 Datasets and Experimental Setup

Datasets. To evaluate the effectiveness of our proposed framework, we conduct experiments on multiple medical imaging datasets from various modalities. We evaluated our model using ISIC 2016 [7], HAM10000 [21], and Messidor-2 [3]. The ISIC 2016 dataset consists of 1,279 dermoscopic images categorized as malignant or benign, used for skin cancer detection. The HAM10000 dataset includes 10,015 dermoscopic images across seven lesion types, serving as a benchmark for dermatological classification. The Messidor-2 dataset contains 1,748 retinal images for diabetic retinopathy classification, with five severity levels. For the HAM10000 dataset, we use a split ratio of 70%, 10%, 20% stratified by class label, for training, validation, and testing.

Implementation Details. In this work, we adopt the Vision Transformer (ViT) architecture, specifically ViT-B/16 [4], pre-trained on the ImageNet-21K dataset as our foundational model. We also evaluate the performance of various pre-trained backbones, including the self-supervised checkpoints, such as DINO [1] and MAE [10].

For the training procedure, we adopt the AdamW optimizer [19]. Following common practice in parameter-efficient transfer learning, the learning rate is adjusted using a cosine learning rate scheduler, with a constant batch size of 64. We use a standard cross-entropy loss for classification tasks while keeping the backbone frozen during fine-tuning. All methods are fine-tuned for 50 epochs, including 10 warm-up epochs, and hyperparameter optimization is performed via grid search on validation set to ensure fairness in comparison.

To assess the robustness of the finetuned model and mitigate overfitting, we employ several data augmentation strategies: random vertical, horizontal flipping with probability 0.5, random rotation at degree [-15, 15].

For the compared PETL methods, following the implementation of V-PETL benchmark [23], we choose four PETL methods: Adapter, BitFit, LoRA and VPT. All methods are evaluated using an input resolution of 224×224. The Top-1 accuracy and AUC score on the test sets are reported.

4.2 Experimental Results

Comparisons with State-of-the-Art Methods. We compare our framework against established baselines, including: traditional Fine-Tuning, PETL methods, and memory-efficient PETL methods. Full fine-tuning adapts all parameters of the pre-trained model, but it demands substantial GPU memory. Linear

Table 1. Results on different downstream medical datasets with pre-trained ViT-B/16 on ImageNet21K. "Mem." specifies peak GPU memory of the training process. Top-1 accuracy and AUC scores are reported.

Method	Mem. (GB)	Messidor2		ISIC2016		HAM10000		Mean	
		ACC	AUC	ACC	AUC	ACC	AUC	ACC	AUC
Traditional Fine-Tuning									
Full Fine-tuning	9.82	63.32	81.49	83.11	81.63	**86.38**	<u>96.82</u>	77.60	86.65
Linear Probing	0.74	62.75	80.41	80.73	76.57	79.10	93.29	74.19	83.42
PETL methods									
BitFit [24]	5.91	64.75	84.60	83.37	79.94	<u>85.48</u>	**97.08**	77.87	87.21
Adapter [11]	5.95	**68.19**	<u>84.81</u>	**84.69**	**82.99**	84.33	96.40	<u>79.07</u>	**88.07**
LoRA [12]	6.20	62.75	**85.85**	83.37	79.66	83.54	96.62	76.55	87.38
VPT [14]	6.32	64.46	79.10	82.58	78.75	79.20	93.76	75.41	83.87
Memory-Efficient PETL methods									
SynQT [26]	3.29	65.33	83.71	<u>83.98</u>	80.92	84.01	96.66	77.78	87.10
FPT [13]	3.49	61.03	82.50	83.11	81.39	79.00	95.82	74.38	83.66
Ours	3.51	<u>67.62</u>	83.81	**84.69**	<u>82.53</u>	<u>85.48</u>	96.13	**79.26**	<u>87.49</u>

probing, on the contrary, trains only the task-specific classification head, which is computationally inexpensive, but usually has lower adaptation capabilities.

The experimental results in Table 1 demonstrate that full fine-tuning reaches the hightest cost of 9.82GB of GPU memory during training. Other PETL methods, despite optimizing parameter efficiency, still require about 6GB of GPU memory due to intermediate feature storage during backpropagation. Although linear probing requires only 0.74GB memory, its adaptation capability is weak compared to other methods. For memory efficient PETL methods, although SynQT is not specially designed for medical image scenarios, it still outperforms many PETL methods along with memory efficiency. FPT yields subpar performance when applied at standard resolutions. We attribute this phenomenon to its fixed selective percentage of token fusion strategy, which may overlook critical spatial details for low resolution input.

In contrast, our proposed framework achieves competitive performance, securing the second-best AUC of 87.49% and the best ACC, while significantly reducing memory consumption to 3.5GB, which is only 35% of Full Fine-tuning. Furthermore, our model employs only 4% of the learnable parameters required for full fine-tuning, establishing it as a highly memory-efficient and practical solution for deploying large-scale pre-trained models in resource-constrained environments. Our results underscore the importance of efficient PETL designs, demonstrating that memory-optimized transfer learning solutions can achieve strong performance without excessive computational overhead.

Table 2. Results on different downstream medical datasets with pre-trained ViT-B/16 with different self-supervised model checkpoints. Top-1 accuracy and AUC scores are reported.

Method	Messidor2		ISIC2016		HAM10000		Mean	
	ACC	AUC	ACC	AUC	ACC	AUC	ACC	AUC
DINO backbone								
SynQT [26]	66.47	82.44	82.58	**85.38**	84.83	94.46	77.96	87.42
FPT [13]	65.90	**84.08**	83.11	81.39	77.20	94.53	75.40	86.66
Ours	**68.48**	83.04	**84.96**	84.50	**86.23**	**95.63**	**79.89**	**87.72**
MAE backbone								
SynQT [26]	64.46	81.14	82.84	**82.59**	83.99	94.88	77.09	86.20
FPT [13]	60.17	79.41	**84.16**	79.71	77.40	94.89	73.91	84.67
Ours	**64.75**	**81.26**	82.58	82.41	**84.23**	**95.11**	**77.18**	**86.26**

Performance on Different Pretrained Backbones. To further evaluate the robustness and generalizability of our approach, we conduct a comprehensive comparison against parameter- and memory-efficient PETL methods across model checkpoints of different SSL frameworks, including MAE [10], DINO [1]. As summarized in Table 2, our method consistently surpasses these baselines across multiple ViT models, demonstrating not only superior transfer performance but also enhanced adaptability to different pretraining paradigms. This highlights the compatibility of our design with task-specific query and multiscale feature enhancement techniques, reinforcing its effectiveness in low-resource fine-tuning scenarios.

4.3 Visualization Results

We present Grad-CAM visualizations to demonstrate how our proposed task-specific query generator enhances the model's attention on medically relevant details, compared to the traditional full fine-tuning (FT) approach. We use ViT-B/16 as the backbone and compare the attention focus of the FT and CSQDA models on images from the ISIC 2016, HAM10000, and Messidor-2 datasets.

The Grad-CAM heatmaps (Fig. 3) of the last layer show that the FT model tends to focus more broadly on less relevant areas of the image, while the CSQDA model, guided by task-specific queries, focuses more precisely on key regions crucial for medical diagnosis, such as lesions in skin cancer detection or retinal abnormalities (e.g., exudates) in diabetic retinopathy. These visualizations demonstrate that the task-specific queries allow the model to better prioritize medically significant features, highlighting the effectiveness of CSQDA in improving the interpretability and reliability of model predictions for medical image classification tasks.

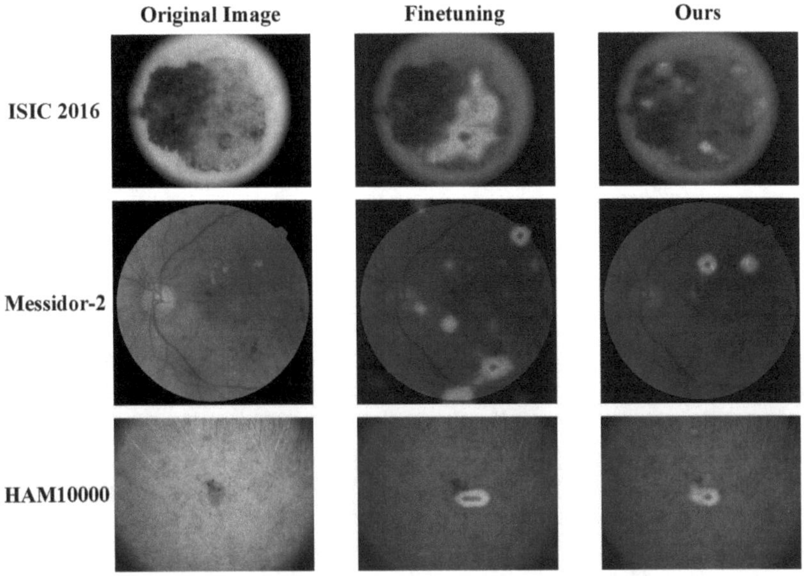

Fig. 3. Visualization results using GradCAM.

4.4 Ablation Study

Impact of Module Components. To evaluate the effectiveness of our proposed memory-efficient PETL framework, we conduct an ablation study in Table 3 by progressively incorporating components and analyzing their effects. The components we investigate include prompt adapter (projection) and prompt attention in the TSQG module and the MSFEM module.

For the TSQG module, the prompt adapter maps the learnable prompts to the appropriate space for integration with the fused residual feature, ensuring that the prompts have a meaningful impact on downstream tasks. The prompt attention mechanism enables the model to focus on studying generating queries for specific regions in the feature space relevant to the task. The MSFEM module refines the features extracted by the backbone model and enhances the task-specific information, which is beneficial to iteratively refining information. Removing this component reduces the capacity of the model to adapt the features effectively, causing performance degradation.

By progressively adding these components, we observe a consistent improvement in classification accuracy, suggesting that each module plays a crucial role in improving the memory efficiency and task adaptability of our framework.

Impact of Hyperparameters. To further evaluate the impact of different hyperparameters on model performance and memory efficiency, we conducted an analysis on two key hyperparameters: prompt length and hidden dimension, as shown in Table 4.

Table 3. Ablation Study on ISIC dataset.

Backbone	Prompt Adapter	Prompt Attn.	MSFEM	ACC (%)	AUC (%)
✓				80.47	81.86
✓	✓			82.84	82.21
✓	✓		✓	83.64	82.13
✓	✓	✓		83.90	82.10
✓	✓	✓	✓	84.69	82.53

For prompt length, we find that shorter prompts limit the model's ability to capture task complexity, while longer prompts offer richer guidance but can introduce redundancy, especially for tasks with sparse diagnostic signals. To balance efficiency and adaptability, we use a dynamic, dataset-specific prompt length with a maximum limit of 16 to avoid excessive memory overhead.

Regarding the hidden dimension of the bottleneck projection, smaller dimensions reduce the parameter count but fail to capture fine-grained features, while larger dimensions allow more complex transformations but may lead to overfitting, particularly with limited data. Our results show that a hidden dimension of 48 provides the best trade-off, balancing performance and efficiency.

Table 4. Impact of prompt length and hidden dimension on model performance.

Hyperparameter	Value	Params (M)	Mean ACC	Mean AUC
Prompt Length	1	3.60	78.20	86.93
	4	3.63	78.85	87.26
	16	3.74	78.53	87.43
Hidden Dim	32	2.81	78.44	86.90
	48	3.74	79.26	87.49
	64	4.68	78.64	87.28

5 Conclusion

In this work, we introduce CSQDA, a memory-efficient and transferable PETL framework tailored for the unique challenges of medical image analysis. By establishing an efficient bidirectional information flow between queries and features, CSQDA achieves robust adaptation while maintaining a lightweight memory footprint. Experimental results across diverse modalities and benchmarks demonstrate consistent performance gains over competitive PETL baselines, with up to 65% reduction in memory usage. These findings underscore the practical value of CSQDA in real-world clinical deployment, where scalability, accuracy, and efficiency are critical.

Acknowledgments. This work was supported partially by the National Natural Science Foundation of China (62276281) and Guangdong Basic and Applied Basic Research Foundation, China (2024A1515011882).

References

1. Caron, M., et al.: Emerging properties in self-supervised vision transformers. In: Proceedings of the IEEE/CVF International Conference on Computer Vision, pp. 9650–9660 (2021)
2. Chen, S., et al.: Adaptformer: adapting vision transformers for scalable visual recognition. Adv. Neural. Inf. Process. Syst. **35**, 16664–16678 (2022)
3. Decencière, E., et al.: Feedback on a publicly distributed image database: the messidor database. In: Image Analysis & Stereology, pp. 231–234 (2014)
4. Dosovitskiy, A., et al.: An image is worth 16 ×16 words: transformers for image recognition at scale. arXiv preprint arXiv:2010.11929 (2020)
5. Dutt, R., Ericsson, L., Sanchez, P., Tsaftaris, S.A., Hospedales, T.: Parameter-efficient fine-tuning for medical image analysis: the missed opportunity. arXiv preprint arXiv:2305.08252 (2023)
6. Evci, U., Dumoulin, V., Larochelle, H., Mozer, M.C.: Head2Toe: utilizing intermediate representations for better transfer learning. In: Proceedings of the 39th International Conference on Machine Learning. Proceedings of Machine Learning Research, vol. 162, pp. 6009–6033. PMLR (2022)
7. Gutman, D., et al.: Skin lesion analysis toward melanoma detection: a challenge at the international symposium on biomedical imaging (isbi), hosted by the international skin imaging collaboration (isic) (2016). arXiv preprint arXiv:1605.01397 (2016)
8. He, A., Wu, Y., Wang, Z., Li, T., Fu, H.: DVPT: dynamic visual prompt tuning of large pre-trained models for medical image analysis. Neural Netw. **185**, 107168 (2025)
9. He, J., Zhou, C., Ma, X., Berg-Kirkpatrick, T., Neubig, G.: Towards a unified view of parameter-efficient transfer learning. arXiv preprint arXiv:2110.04366 (2021)
10. He, K., Chen, X., Xie, S., Li, Y., Dollár, P., Girshick, R.: Masked autoencoders are scalable vision learners. In: Proceedings of the IEEE/CVF Conference on Computer Vision and Pattern Recognition, pp. 16000–16009 (2022)
11. Houlsby, N., et al.: Parameter-efficient transfer learning for NLP. In: International Conference on Machine Learning, pp. 2790–2799. PMLR (2019)
12. Hu, E.J., et al.: Lora: low-rank adaptation of large language models. In: International Conference on Learning Representations (2022)
13. Huang, Y., Cheng, P., Tam, R., Tang, X.: Fine-grained prompt tuning: a parameter and memory efficient transfer learning method for high-resolution medical image classification. In: Medical Image Computing and Computer Assisted Intervention (MICCAI), pp. 120–130. Springer, Heidelberg (2024). https://doi.org/10.1007/978-3-031-72390-2_12
14. Jia, M., et al.: Visual prompt tuning. In: European Conference on Computer Vision, pp. 709–727. Springer, Heidelberg (2022). https://doi.org/10.1007/978-3-031-19827-4_41
15. Jie, S., Deng, Z.H.: Fact: factor-tuning for lightweight adaptation on vision transformer. In: Proceedings of the AAAI Conference on Artificial Intelligence, vol. 37, pp. 1060–1068 (2023)

16. Kim, H.E., Cosa-Linan, A., Santhanam, N., Jannesari, M., Maros, M.E., Ganslandt, T.: Transfer learning for medical image classification: a literature review. BMC Med. Imaging **22**(1), 69 (2022)
17. Lester, B., Al-Rfou, R., Constant, N.: The power of scale for parameter-efficient prompt tuning. In: Proceedings of the 2021 Conference on Empirical Methods in Natural Language Processing, pp. 3045–3059 (2021)
18. Lian, D., Zhou, D., Feng, J., Wang, X.: Scaling & shifting your features: a new baseline for efficient model tuning. In: NeurIPS, vol. 35, pp. 109–123 (2022)
19. Loshchilov, I., Hutter, F.: Decoupled weight decay regularization. In: International Conference on Learning Representations (2017)
20. Sung, Y.L., Cho, J., Bansal, M.: LST: ladder side-tuning for parameter and memory efficient transfer learning. Adv. Neural. Inf. Process. Syst. **35**, 12991–13005 (2022)
21. Tschandl, P., Rosendahl, C., Kittler, H.: The ham10000 dataset, a large collection of multi-source dermatoscopic images of common pigmented skin lesions. Sci. Data **5**(1), 1–9 (2018)
22. Tu, C.H., Mai, Z., Chao, W.L.: Visual query tuning: towards effective usage of intermediate representations for parameter and memory efficient transfer learning. In: Proceedings of the IEEE/CVF Conference on Computer Vision and Pattern Recognition (CVPR) (2023)
23. Xin, Y., et al.: V-petl bench: a unified visual parameter-efficient transfer learning benchmark. In: Advances in Neural Information Processing Systems, vol. 37, pp. 80522–80535 (2024)
24. Zaken, E.B., Goldberg, Y., Ravfogel, S.: Bitfit: simple parameter-efficient fine-tuning for transformer-based masked language-models. In: Proceedings of the 60th Annual Meeting of the Association for Computational Linguistics, vol. 2: Short Papers, pp. 1–9 (2022)
25. Zhang, J.O., Sax, A., Zamir, A., Guibas, L., Malik, J.: Side-tuning: a baseline for network adaptation via additive side networks. In: Vedaldi, A., Bischof, H., Brox, T., Frahm, J.-M. (eds.) ECCV 2020. LNCS, vol. 12348, pp. 698–714. Springer, Cham (2020). https://doi.org/10.1007/978-3-030-58580-8_41
26. Zhang, T., et al.: Parameter-efficient and memory-efficient tuning for vision transformer: a disentangled approach. In: European Conference on Computer Vision, pp. 346–363. Springer, Heidelberg (2024). https://doi.org/10.1007/978-3-031-72995-9_20
27. Zu, W., Xie, S., Zhao, Q., Li, G., Ma, L.: Embedded prompt tuning: towards enhanced calibration of pretrained models for medical images. Med. Image Anal. **97**, 103258 (2024)

LCKPose: Laplacian Candidate Keypoints Modeling for 6D Object Pose Estimation

Zhiyang Mai, Yukun Qian(✉), Haitao Wang, Hejun Wu(✉), and Liangliang Zhou

School of Computer Science and Engineering, Sun Yat-sen University, Guangzhou, China
qianyk5@mail2.sysu.edu.cn, wuhejun@mail.sysu.edu.cn

Abstract. In pose estimation methods based on keypoint voting schemes, the accuracy of estimated pose depends on the quality of predicted candidate keypoints. However, due to sensor noise, occlusions and cluttered backgrounds, candidate keypoint prediction often involves aleatoric uncertainty. Existing methods typically do not explicitly account for the impact of uncertainty, which limits their performance in complex scenes. In this work, we propose **LCKPose**, a framework that quantifies the prediction uncertainty via **L**aplacian **C**andidate **K**eypoint modeling for **Pose** estimation. LCKPose employs a dual-stream hierarchical fusion network to capture fine-grained features and represents the coordinates of each candidate keypoint as a Laplace distribution. Distribution modeling enables the network to consider the uncertainty and focus on learning from reliable candidates. Moreover, on the basis of Laplace distribution modeling, LCKPose introduces a distance-based density search algorithm to locate the final predicted keypoints more effectively. Experimental results show the effectiveness of our method in reducing the adverse impact of uncertainty, achieving state-of-the-art results on the Linemod-Occlusion (89.0%) and YCB-Video (95.3%) datasets.

Keywords: 6D object pose estimation · Aleatoric uncertainty · Laplace distribution modeling

1 Introduction

6 Degrees of Freedom (6DoF) object pose estimation aims to estimate the rotation and translation of a given rigid object relative to the camera. It is a fundamental task with challenges in 3D computer vision and plays an important part in numerous real-world applications such as autonomous driving [25,34], augmented reality [29,31], and robotic grasping [5,16]. Many existing solutions [6,7,26,35] are based on keypoint voting methods, which infer candidate keypoints from visible object points and use them to establish keypoint correspondences for pose estimation. While keypoint voting methods have been widely

J. Lokoč et al. (Eds.): MMM 2026, LNCS 16413, pp. 417–431, 2026.
https://doi.org/10.1007/978-981-95-6957-1_30

adopted and shown to be effective, their performance remains limited by a variety of factors in complex scenes.

Specifically, these factors (including occlusions, cluttered backgrounds, and light variation) introduce noise to pose estimation processes and lead to error-prone predictions. Existing keypoint voting methods [6,7,26,35] typically do not involve uncertainty representation in the regression of candidate keypoints and simply constrain each of them to be as close to the ground truth as possible during training. However, the prediction of corresponding candidate keypoints from object points in the noisy regions can involve high aleatoric uncertainty [14]. During training, the model may focus too much on these points and tend to fit the noise components within them, which hinders the learning process and makes it sensitive to noise. Besides, enforcing each candidate keypoint to be accurate is unnecessary because a small number of outliers are tolerable and can be eliminated during voting. Thus, a viable approach is enabling the network to focus on learning robust features from reliable points (Fig. 1).

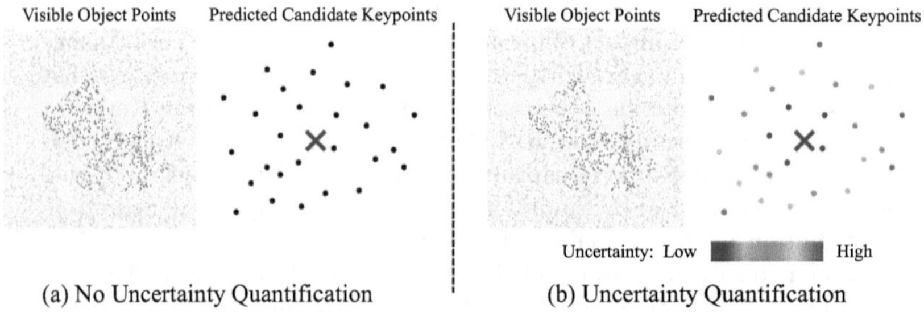

Fig. 1. Every visible object point is associated with a predicted candidate keypoint through an offset. Predicted keypoints with high uncertainty usually exhibit considerable displacements from ground truth (represented as crosses). (a) Previous methods treat each candidate keypoint equally and fail to consider uncertainty. (b) We utilize Laplace distribution modeling to quantify uncertainty of each candidate and guide network to focus on learning from reliable results.

In this work, we consider and quantify the uncertainty of predicted candidate keypoints with Laplace distribution modeling to improve model performance in complex scenes. Firstly, we construct a dual-stream hierarchical fusion network (DHF-Net) to extract high-quality point-wise features for downstream prediction tasks. DHF-Net incorporates skip connections to aggregate hierarchical features from different layers, which allows for better preservation of details. Then, to introduce uncertainty representation, we model the coordinate of each candidate as a Laplace distribution. All distributions share a common mean but may have different scale parameters. Through Laplace distribution modeling, the model is able to quantify the uncertainty of each candidate via the scale parameter and focus on learning from reliable points. Accurate candidates will also tend to cluster around the mean location and form the highest density region in the predicted candidate keypoint set. Thus, the final predicted keypoints can be obtained from candidates in this region and used for pose parameter calculation. We propose a distance-based density search (DDS) algorithm to find the highest density region. The DDS is non-iterative, fast and applicable to end-to-end training.

In summary, the main contributions of this work are:

- We construct a dual-stream hierarchical fusion network (DHF-Net) to aggregate hierarchical features, which is crucial to the preservation of details and point-wise feature extraction.
- We model the coordinate of each candidate keypoint as a Laplace distribution, which enables the model to consider aleatoric uncertainty and focus on learning from reliable points, improving prediction robustness.
- We propose a non-iterative distance-based density search (DDS) algorithm for final keypoint localization. The algorithm is fast and applicable to end-to-end training.

Extensive experiments on the Linemod-Occlusion (LM-O) [2] and YCB-V [36] datasets show that our method reduces the impact of uncertainty and improves the prediction accuracy in complex scenes.

2 Related Work

2.1 Keypoint-Based Methods for Pose Estimation

Keypoint-based methods establish correspondences of keypoints between the input data and the object CAD model as intermediate results, which are then used for subsequent pose estimation. [12,26] predict 2D keypoints in RGB images to establish 2D-3D keypoint correspondences and recover object poses using the Perspective-n-Point (PnP) algorithm. To further improve the accuracy of pose estimation, [6,7,38] additionally utilize depth information and predict 3D keypoints to establish 3D-3D keypoint correspondences. Besides, [11,19] integrate end-to-end learning into keypoint-based methods, which extend their application to some self-supervised methods.

2.2 Voting for Keypoints Prediction

Directly detecting all keypoints is difficult, especially in occluded scenes, since some keypoints of occluded objects are invisible. To address this issue, many methods [6,7,12,13,26,38] adopt voting schemes and infer keypoints from visible parts of objects. Specifically, these methods predict geometric quantities (e.g., offsets) that relate the positions of object points to 3D keypoints. Every visible object point is then associated with a predicted candidate keypoint. All these candidate keypoints are used for voting to locate the final keypoints. PVNet [26] predicts vectors that represent the directions from each object pixel in RGB images to the keypoints. Then, a RANSAC-based voting scheme is applied to locate 2D keypoints. PVN3D [7] extends the prediction to 3D space and locates 3D keypoints with MeanShift [4]. We observe that as long as most candidate keypoints are accurate, a small number of inaccurate candidates are tolerable, since these outliers can be eliminated during voting. Thus, we argue that not every predicted candidate keypoint is required to be highly accurate and that guiding the network to focus on learning from reliable points can be beneficial.

2.3 Aleatoric Uncertainty in Prediction

Aleatoric uncertainty [14] is related to the randomness and noise inherent in the observations. It is inevitable and brings challenges to learning because the task-relevant features can be severely obscured by irrelevant noise. A feasible way to reduce the impact of aleatoric uncertainty is to consider and quantify it. Uncertainty quantification can be achieved through probabilistic distribution modeling. For example, [14] captures noise with probabilistic variance to explicitly represent aleatoric uncertainty and improves performance in pixel-wise segmentation and depth regression. [24] models the uncertainty propagation relationship of the geometry projection with the Laplacian distribution for monocular 3D object detection.

3 Method

3.1 Overview

Figure 2 shows an overview of LCKPose. For each target object in an aligned RGB-D image, our goal is to estimate the transformation between the predefined object coordinate system and the camera coordinate system. This transformation consists of a rotation $\mathbf{R} \in SO(3)$ and a translation $\mathbf{t} \in \mathbb{R}^3$. Following [21,30, 33], we first detect and crop a Region of Interest (ROI) using an off-the-shelf object detector. LCKPose takes the ROI as input and extracts a point-wise RGB-D feature with DHF-Net. For each point in the ROI, LCKPose predicts a

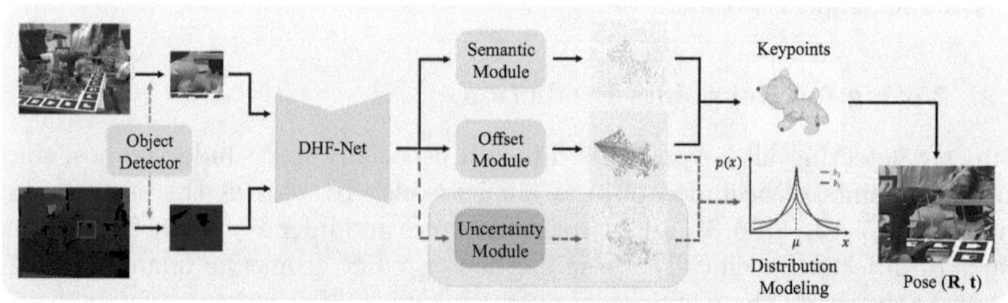

Fig. 2. Overview of LCKPose. We model the coordinate of each candidate keypoint as a Laplace distribution with a shared mean μ and a respective scale parameter b. Coordinates of points with high uncertainty are modeled using distributions with larger scale parameters.

semantic label and an offset from the RGB-D feature as in [6,7]. During training, the RGB-D feature is additionally fed into the uncertainty module to obtain a point-wise auxiliary map for Laplace distribution modeling and loss calculation. During inference, the 3D candidate keypoint set is generated using semantic labels and offsets. Then, LCKPose locates the final predicted keypoints from the set with our proposed DDS algorithm. These final predicted keypoints are used to establish keypoint correspondences and recover the object pose.

3.2 Dual-Stream Hierarchical Fusion Network

For the point-wise prediction task, a key point is to retain high-resolution spatial details. Inspired by frameworks such as U-Net [27], we introduce skip connections for both the RGB and depth streams to preserve details. However, unlike U-Net, we do not achieve cross-scale fusion by simply concatenating upsampled features with lower-level features, which may yield limited effectiveness due to potential feature misalignment [22] and heterogeneity [3] in our case. Instead, we adopt two effective cross-scale fusion modules for the RGB stream and the depth stream, respectively. The architecture of DHF-Net is shown in Fig. 3.

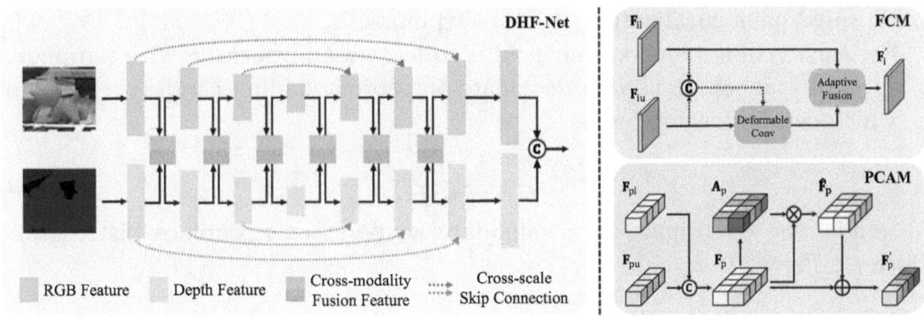

Fig. 3. Architecture of DHF-Net. To preserve details, DHF-Net incorporates skip connections within both the RGB and depth streams. Feature fusion operations for the skip connections in the RGB and depth streams are implemented by two cross-scale fusion modules, respectively. Besides, DHF-Net also adopts cross-modality fusion modules for fusion of RGB features and depth features as in [6,38].

For the RGB stream, we adopt the feature calibration module (FCM) proposed by [22]. FCM aligns the upsampled feature $\mathbf{F}_{\mathrm{iu}}$ with the lower-level feature $\mathbf{F}_{\mathrm{il}}$ through deformable convolutions and adaptively fuses them to generate the fusion feature $\mathbf{F}'_{\mathrm{i}}$. For the depth stream, we mainly focus on the feature heterogeneity between the upsampled feature $\mathbf{F}_{\mathrm{pl}}$ and the lower-level feature $\mathbf{F}_{\mathrm{pu}}$. Inspired by the feature reference module (FRM) proposed by [22] and the channel-weighted skip connection module (CWSCM) proposed by [3], we design a point-wise and channel-wise attention fusion module (PCAM). Given the concatenated feature $\mathbf{F}_{\mathrm{p}} \in \mathbb{R}^{N\times 1\times C}$, PCAM generates point-wise and channel-wise attention weights $\mathbf{A}_{\mathrm{p}} \in \mathbb{R}^{N\times 1\times C}$ with a 1×1 conv layer:

$$\mathbf{A}_{\mathrm{p}} = \mathrm{Conv}_{1\times 1}(\mathbf{F}_{\mathrm{p}}) = \mathrm{Conv}_{1\times 1}(\mathrm{Concat}(\mathbf{F}_{\mathrm{pl}}, \mathbf{F}_{\mathrm{pu}})) \tag{1}$$

The integration of channel-wise and point-wise attention enables PCAM to capture feature heterogeneity along the channel and maintain feature independence of each point. The attention-scaled feature $\hat{\mathbf{F}}_{\mathrm{p}} \in \mathbb{R}^{N\times 1\times C}$ is given by:

$$\hat{\mathbf{F}}_{\mathrm{p}} = \mathbf{F}_{\mathrm{p}} \odot \mathbf{A}_{\mathrm{p}} \tag{2}$$

The final fusion feature $\mathbf{F}'_{\mathrm{p}} \in \mathbb{R}^{N\times 1\times C'}$ is given by:

$$\mathbf{F}'_{\mathrm{p}} = \mathrm{Conv}_{1\times 1}(\mathbf{F}_{\mathrm{p}} \oplus \hat{\mathbf{F}}_{\mathrm{p}}) \tag{3}$$

3.3 Laplace Distribution Modeling for Uncertainty Quantification

We leverage Laplace distribution for modeling to consider and quantify aleatoric uncertainty. The Laplace distribution possesses many advantageous properties. Compared to the Gaussian distribution, the Laplace distribution has heavier tails, which makes it more robust to outliers. Besides, the Laplace distribution has a sharper peak at the mean. Modeling with the Laplace distribution makes accurate candidate keypoints more spatially concentrated, which brings benefits to the subsequent localization of final keypoints.

For each visible point $\mathbf{x}_i$, our goal is to learn a transform $\mathcal{F}$ with parameter $\mathbf{w}$ to predict an offset used to calculate the corresponding candidate keypoint $\mathbf{y}_i$, which can be formulated as:

$$\mathbf{y}_i = \mathcal{F}_{\mathbf{w}}(\mathbf{x}_i) \tag{4}$$

We model the coordinate of a candidate keypoint as a Laplace distribution, which can be written as:

$$p(\mathbf{y}_{gt}|\mathbf{y}_i) = Laplace(\mathbf{y}_{gt}, b(\mathbf{x}_i)) = \frac{1}{2b(\mathbf{x}_i)}\exp(-\frac{||\mathbf{y}_i - \mathbf{y}_{gt}||_1}{b(\mathbf{x}_i)}) \tag{5}$$

where $||\cdot||_1$ denotes the L1 distance function, the mean μ is set to ground truth value $\mathbf{y}_{gt}$, $b(\mathbf{x}_i)$ represents the scale parameter related to input $\mathbf{x}_i$. For each predicted result $\mathbf{y}_i$, our optimization objective is to maximize the likelihood probability $p(\mathbf{y}_{gt}|\mathbf{y}_i)$. The minimization objective can be derived from the negative log likelihood:

$$\begin{aligned} -\mathrm{log}p(\mathbf{y}_{gt}|\mathbf{y}_i) &= -\mathrm{log}(\frac{1}{2b(\mathbf{x}_i)}\exp(-\frac{||\mathbf{y}_i - \mathbf{y}_{gt}||_1}{b(\mathbf{x}_i)})) \\ &= \mathrm{log}2 + \mathrm{log}b(\mathbf{x}_i) + \frac{||\mathbf{y}_i - \mathbf{y}_{gt}||_1}{b(\mathbf{x}_i)} \end{aligned} \tag{6}$$

Let $c(\mathbf{x}_i) = \frac{1}{b(\mathbf{x}_i)}$, we have:

$$\begin{aligned} -\mathrm{log}p(\mathbf{y}_{gt}|\mathbf{y}_i) &= \mathrm{log}2 + \mathrm{log}\frac{1}{c(\mathbf{x}_i)} + c(\mathbf{x}_i)||\mathbf{y}_i - \mathbf{y}_{gt}||_1 \\ &= c(\mathbf{x}_i)||\mathbf{y}_i - \mathbf{y}_{gt}||_1 - \mathrm{log}c(\mathbf{x}_i) + \mathrm{log}2 \end{aligned} \tag{7}$$

We can see that the negative log likelihood is mainly related to two items:

$$-\log p(\mathbf{y}_{gt}|\mathbf{y}_i) \propto c(\mathbf{x}_i)||\mathbf{y}_i - \mathbf{y}_{gt}||_1 - \log c(\mathbf{x}_i) \quad (8)$$

Finally, we can define the loss function as follows to achieve Laplace distribution modeling:

$$\mathcal{L}_{\text{offset}}(\mathbf{x}_i, \mathbf{w}) = c(\mathbf{x}_i)||\mathcal{F}_{\mathbf{w}}(\mathbf{x}_i) - \mathbf{y}_{gt}||_1 - \log c(\mathbf{x}_i) \quad (9)$$

As shown in Fig. 2, we introduce the uncertainty module during training to learn the point-wise $c(\mathbf{x}_i)$ map for the calculation of $\mathcal{L}_{\text{offset}}$. To avoid a negative loss which may impair the model training in our case, we constrain the value of $c(\mathbf{x}_i)$ between 0 and 1 to ensure the logarithm term remains positive. By the way, the auxiliary uncertainty module is not necessary during inference.

The form of loss function can be explained in an intuitive way. $c(\mathbf{x}_i)$ is related to scale parameter $b(\mathbf{x}_i)$, and thus related to the variance, which reflects the uncertainty of input $\mathbf{x}_i$. A smaller $c(\mathbf{x}_i)$ generally indicates that the predicted candidate keypoint has relatively higher uncertainty and vice versa. When the error of predicted keypoint coordinate $\mathcal{F}_{\mathbf{w}}(\mathbf{x}_i)$ is large, the L1-distance term dominates and $c(\mathbf{x}_i)$ tends to be small to reduce this term, allowing the model to maintain a moderate tolerance to this predicted keypoint, which may be an outlier. When the error of $\mathcal{F}_{\mathbf{w}}(\mathbf{x}_i)$ is small, $c(\mathbf{x}_i)$ is encouraged to be large to reduce the logarithm term, which in turn increases the weight of L1-distance term and encourages the predicted result $\mathcal{F}_{\mathbf{w}}(\mathbf{x}_i)$ to be more accurate. The learnable $c(\mathbf{x}_i)$ map can be seen as a weight map of different candidate keypoints, which guides the model to focus on learning from those reliable results.

3.4 Distance-Based Density Search Algorithm

In our case, the coordinate of each point in the candidate keypoint set can be regarded as following a new unimodal distribution, which is formed by the summation of multiple Laplace distributions that share a common mean but have different scale parameters. There is only one highest density region in the candidate keypoint set. Our goal is to find this region where the ground truth keypoint is most likely located. MeanShift [4] is a universal algorithm to find the highest density region. However, it is time-consuming due to the iterative steps, which limits its applications in real-time tasks. Considering properties of the new unimodal distribution, we design a non-iterative distance-based density search algorithm.

We denote the position of the i-th candidate keypoint as k_i. For the i-th candidate keypoint, the corresponding total distance D_i to all other candidate keypoints is:

$$D_i = \sum_j ||k_i - k_j||_1, \ j \neq i \quad (10)$$

We give a brief theoretical proof in the following to demonstrate that D_i can be used to determine whether this point is located in the highest density region.

In the one-dimensional case, suppose there is a point set formed by the mixture of two sub-point sets with the same number of points. Points of these two subsets follow two Laplace distributions with the same mean μ but different scale parameters b_1 and b_2, respectively. For a point randomly sampled from the mixture set, the distribution of its position k can be written as:

$$P_{\text{dis}}(k) = \frac{1}{4b_1}\exp(-\frac{|k-\mu|}{b_1}) + \frac{1}{4b_2}\exp(-\frac{|k-\mu|}{b_2}) \tag{11}$$

The number of points $N(k)$ at position k is related to the probability density $P_{\text{dis}}(k)$ here. So the total distance $D_i(k)$ of the i-th point at position k_i to all points at position k is:

$$D_i(k) = N(k)|k_i - k| \propto P_{\text{dis}}(k)|k_i - k| \tag{12}$$

By summing the distances from the i-th point to all other points at different positions, we have:

$$D_i = \int D_i(k)dk \propto \int P_{\text{dis}}(k)|k_i - k|dk \tag{13}$$

It can be simplified as:

$$D_i \propto |k_i - \mu| + \frac{b_1}{2}\exp(-\frac{|k_i-\mu|}{b_1}) + \frac{b_2}{2}\exp(-\frac{|k_i-\mu|}{b_2}) \tag{14}$$

We can see that D_i increases with $|k_i - \mu|$, which means that if the i-th point is far from the mean μ region (the highest density region), the corresponding D_i will also be large. This can be extended to our three-dimensional case, where the predicted candidate keypoint set can be seen as a mixture of several Laplace distribution subsets with different scale parameters.

For every candidate keypoint in the set, we calculate its corresponding D. Then, we will filter those points with high D values. We define the total distance threshold as:

$$D_{\text{thres}} = \theta_{\text{thres}} \times (D_{\text{max}} - D_{\text{min}}) \tag{15}$$

where $\theta_{\text{thres}} \in (0, 1)$ is a hyper-parameter. A point is regarded as an outlier and will be filtered if the corresponding $D > D_{\text{thres}}$. The whole filtering process is non-iterative and fast. The final predicted keypoint is the centroid of retained points. The calculation process from the keypoint set to the final keypoint is differentiable, so our algorithm can be integrated into end-to-end frameworks like E2EK [19], which is crucial in some scenarios. For instance, a fully differentiable network is needed in some self-supervised methods to transmit the signal between input and the 6D pose [19,33].

4 Experiments

4.1 Datasets

Previous works have reported the evaluation accuracy over 95% on the Linemod (LM) dataset [8]. Thus, following many recent works [21,30,37], we mainly focus on using the LM-O dataset [2] and the YCB-V dataset [36]. LM-O consists of 1214 images from a sequence of LM. Texture-less objects, occluded scenes, and lighting variations make this dataset challenging and suitable for evaluating model performance under complex conditions. YCB-V is a large dataset containing many real images from videos of 21 objects with strong occlusion and clutter. In addition to the real training images, we additionally utilize publicly available synthetic physically-based rendered (PBR) images provided by the BOP challenge [9] for training on these two datasets.

4.2 Evaluation Metrics

For the LM-O dataset, we employ the common evaluation metric ADD(-S) [36] with a threshold of 10% of the object diameter. For an asymmetric object, the metric calculates the average distance of model points projected to the camera domain using the predicted pose to the same model points projected using the ground truth pose. For a symmetric object, the metric matches the closest model points instead of the exact same model points. For the YCB-V dataset, we report the area under curve (AUC) of ADD(-S) and ADD-S with a maximum threshold of 10 cm as described in [36].

4.3 Implementation Details

We apply the same dynamic zoom-in strategy as CDPN [17] to obtain noisy ROI images and generate point clouds from cropped depth ROIs as in [21]. We feed the RGB ROIs and the point clouds to the network. For the RGB stream, we use the ConvNext [23] architecture. For the depth stream (the point cloud stream), we adopt the local feature aggregation module and random sampling operation from RandLA-Net [10]. We adopt the fusion method from FFB6D [6] to achieve cross-modality fusion. The semantic, offset, and uncertainty modules are each implemented with separate MLPs. We recover the poses with the least-squares fitting algorithm [1]. During training, we use Focal Loss [20] to supervise the learning of semantic labels as in [6,7,19]. Our training loss is defined as:

$$\mathcal{L} = \alpha \mathcal{L}_{\text{semantic}} + \mathcal{L}_{\text{offset}} \tag{16}$$

where α is a weight factor between the semantic label loss and the offset loss and is set to 2 in accordance with [6,19]. We conduct our experiments on an Intel 3.70GHz CPU and an NVIDIA GeForce RTX 3090 GPU. We use the Adam [15] optimizer and train our network for 50k iterations with a batch size of 16. The learning rate is dynamically adjusted using the CyclicLR [28] policy between 2e-5 and 8e-4 and the step size is set to 5k. During inference, we utilize the detected bounding box from FCOS [32] provided by CDPNv2 [17].

4.4 Comparison with State-of-the-Art Methods

Results on LM-O Dataset. We report the ADD(-S) metric for the LM-O dataset in Table 1. Our method yields a recall of 89.0%, achieving the best mean performance compared to existing keypoint-based methods. We also show qualitative results on the LM-O dataset. As shown in Fig. 4, even facing severe occlusions, clutter in the scene, or varying environments, our LCKPose can still accurately predict keypoints and recover object poses, which demonstrates its robustness to complex scenes.

Table 2 shows an extra comparison to some of the dense correspondence-based methods. Our keypoint-based approach can also achieve state-of-the-art performance while maintaining advantages in some aspects. Taking a comparison with HiPose [21] for example, HiPose needs an additional preparation stage of binary surface encoding and is not fully differentiable. Our method is faster and can easily integrate into end-to-end frameworks, so it is still competitive.

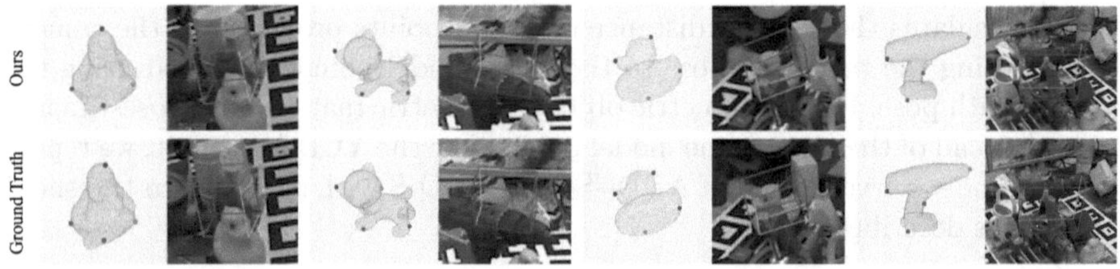

Fig. 4. Qualitative results on LM-O. We present several prediction results of the final predicted keypoints and poses, and compare them with the ground truth. Our LCKPose can achieve robust prediction in complex scenes.

Table 1. Comparison with state-of-the-art keypoint-based methods on LM-O. We report the ADD(-S) in %. (*) denotes symmetric objects. Overall best results are in bold, and the second-best results are underlined.

Method	RGB Input		RGB-D Input				
	CheckerPose [18]	6D-Diff [37]	FFB6D [6]	RCVPose [35]	E2EK [19]	DFTr [38]	Ours
ape	58.3	60.6	47.2	60.3	61.0	<u>64.1</u>	**78.0**
can	95.7	<u>97.9</u>	85.2	92.5	95.4	96.1	**98.2**
cat	62.3	<u>63.2</u>	45.7	50.2	50.8	52.2	**82.5**
driller	93.7	<u>96.6</u>	81.4	78.2	94.5	95.8	**97.3**
duck	69.9	67.2	53.9	52.1	59.6	<u>72.3</u>	**84.8**
eggbox*	70.0	73.5	70.2	<u>81.2</u>	55.7	75.3	**82.1**
glue*	86.4	<u>92.0</u>	60.1	72.1	78.3	79.3	**94.4**
holepuncher	83.8	85.5	85.9	75.2	<u>91.4</u>	86.8	**94.7**
mean	77.5	<u>79.6</u>	66.2	70.2	73.3	77.7	**89.0**

Table 2. Comparison with state-of-the-art dense correspondence-based methods on LM-O. We report the ADD(-S) in %. Overall best results are in bold, and the second-best results are underlined.

Method	GDR-Net [33]	ZebraPose [30]	Hipose [21]	Ours
ADD(-S)	62.2	76.9	**89.6**	<u>89.0</u>

Results on YCB-V Dataset. In Table 3, we report the AUC of ADD-S and AUC of ADD(-S) metrics of 21 objects and compare our method with state-of-the-art keypoint-based methods on the YCB-V dataset. Our method again achieves state-of-the-art performance.

Table 3. Comparison with state-of-the-art keypoint-based methods on YCB-V. We report the AUC of ADD-S and AUC of ADD(-S) in %. Overall best results are in bold, and the second-best results are underlined.

Method	CheckerPose [18]	6D-Diff [37]	FFB6D [6]	RCVPose [35]	E2EK [19]	DFTr [38]	Ours
AUC of ADD-S	91.3	91.5	96.6	96.6	<u>96.8</u>	96.7	**97.1**
AUC of ADD(-S)	86.4	87.0	92.7	<u>95.2</u>	94.4	94.4	**95.3**

Table 4. Ablation study on LM-O. We report the ADD(-S) in %.

Row	Method	ADD(-S)
A0	Ours	89.02
B0	A0: → w/o Skip connections	88.21
B1	A0: → w/o Skip connections (RGB)	88.66
B2	A0: → w/o Skip connections (Depth)	88.31
C0	A0: → w/o Distribution Modeling	87.96
D0	A0: DDS → MeanShift	88.88

4.5 Ablation Studies

In the following, we present ablation studies on LM-O datasets to verify the effectiveness of each design. We report the ADD(-S) metric and summarize the results in Table 4, Table 5 and Table 6.

Effectiveness of Skip Connections. We ablate skip connections in DHF-Net. As shown in Table 4, comparing A0 with B0, when adding skip connections for both the RGB stream and depth stream, the results are further improved. We also ablate the two skip connections in B1 and B2, respectively, and observe a performance drop compared to adding both of them, which indicates that the aggregation of hierarchical features can be beneficial for both parallel streams.

Ablation on Distribution Modeling. We ablate the distribution modeling as in C0 of Table 4. Comparing A0 with C0, when adopting distribution modeling for uncertainty quantification, we observed a 1.06% recall improvement. Besides, pose estimation for certain objects, such as the ape, cat and duck, remains particularly challenging, typically achieving an average accuracy of only 60% to 80%. By quantifying uncertainty, we achieve improvements of 3.11%, 2.27% and 1.34% for the ape, cat and duck, respectively. This indicates that uncertainty is indeed a critical factor limiting the accuracy of pose estimation in challenging scenes, and also demonstrates the effectiveness of distribution modeling for uncertainty quantification.

Ablation on Distance-Based Density Search Algorithm. As mentioned in Sect. 3.4, we set a threshold D_{thres} controlled by parameter θ_{thres} for filtering outliers. We conduct evaluations on keypoint set error and ADD(-S) using several different θ_{thres} values. The keypoint set error calculates the average distance from all reserved candidate keypoints to the ground truth. As presented in Table 5, we observe that the keypoint set error increases with θ_{thres}, which again confirms that points with high D values are usually far from the highest density region and have large error. Filtering those candidates with high D values can improve the accuracy of pose estimation. We finally set θ_{thres} as 0.2 for our DDS.

Table 5. Comparisons of different threshold values used in the DDS on LM-O.

θ_{thres}	0.1	0.2	0.3	0.4	0.5	0.6	0.7	0.8	0.9	1.0
Point Set Error/cm	4.26	4.29	4.31	4.32	4.33	4.35	4.36	4.38	4.40	4.46
ADD(-S)	88.93	**89.02**	88.91	88.90	88.87	88.87	88.81	88.64	88.50	87.00

In D0 of Table 4, we compare our DDS with MeanShift. We conduct additional evaluations on keypoint error and runtime to make a more thorough comparison of the two methods, as shown in Table 6. The keypoint error calculates the average distance from final predicted keypoints to the ground truth. Experiments show that there is no significant difference between our DDS and MeanShift in the accuracy of keypoint localization and pose estimation. However, our method is faster, while the computation time of MeanShift depends on the number of iterations required and can be very long in some cases.

Table 6. Comparisons of MeanShift and DDS on LM-O.

Method	MeanShift [4]	DDS
Keypoints Error/cm	4.16	4.17
Runtime/ms	6.0–10.0	4.0

5 Conclusions

In this work, we propose LCKPose. We construct a DHF-Net to extract high-quality features for downstream prediction tasks. Then, we model the coordinate of each candidate keypoint as a Laplace distribution to consider and quantify aleatoric uncertainty, which makes prediction more robust. Finally, we design a non-iterative DDS algorithm for final keypoint localization. Extensive experiments on the LM-O and YCB-V datasets show that our method reduces the impact of uncertainty and improves the overall prediction accuracy in complex scenes. In the future, we will conduct experiments on more benchmarks. We will also deploy the model on edge devices (e.g., robot arms) and measure latency and memory consumption to evaluate its real-world performance.

Acknowledgments. This work was supported by the National Key Research and Development Program of China under Grant 2023YFB4301900, and the National Natural Science Foundation of China (NSFC) (Grant No. 62272497 to Hejun Wu).

References

1. Arun, K.S., Huang, T.S., Blostein, S.D.: Least-squares fitting of two 3-d point sets. IEEE Trans. Pattern Anal. Mach. Intell. **PAMI-9**(5), 698–700 (1987)
2. Brachmann, E., Krull, A., Michel, F., Gumhold, S., Shotton, J., Rother, C.: Learning 6D object pose estimation using 3D object coordinates. In: Fleet, D., Pajdla, T., Schiele, B., Tuytelaars, T. (eds.) ECCV 2014. LNCS, vol. 8690, pp. 536–551. Springer, Cham (2014). https://doi.org/10.1007/978-3-319-10605-2_35
3. Chen, L., Xie, Y., Li, Y., Xu, Q., Dong, J.: Cwscnet: channel-weighted skip connection network for underwater object detection. IEEE Trans. Image Process. **33**, 5206–5218 (2024)
4. Comaniciu, D., Meer, P.: Mean shift: a robust approach toward feature space analysis. IEEE Trans. Pattern Anal. Mach. Intell. **24**(5), 603–619 (2002)
5. Deng, X., Xiang, Y., Mousavian, A., Eppner, C., Bretl, T., Fox, D.: Self-supervised 6d object pose estimation for robot manipulation. In: 2020 IEEE International Conference on Robotics and Automation (ICRA), pp. 3665–3671 (2020)
6. He, Y., Huang, H., Fan, H., Chen, Q., Sun, J.: Ffb6d: a full flow bidirectional fusion network for 6d pose estimation. In: CVPR, pp. 3002–3012 (2021)
7. He, Y., Sun, W., Huang, H., et al.: Pvn3d: a deep point-wise 3d keypoints voting network for 6dof pose estimation. In: CVPR, pp. 11629–11638 (2020)
8. Hinterstoisser, S., Lepetit, V., Ilic, S., Holzer, S., Bradski, G., Konolige, K., et al.: Model based training, detection and pose estimation of texture-less 3d objects in heavily cluttered scenes. In: Computer Vision – ACCV 2012, pp. 548–562 (2013)

9. Hodan, T., Sundermeyer, M., Labbé, Y., Nguyen, V.N., Wang, G., Brachmann, E., et al.: Bop challenge 2023 on detection, segmentation and pose estimation of seen and unseen rigid objects. In: 2024 IEEE/CVF Conference on Computer Vision and Pattern Recognition Workshops (CVPRW), pp. 5610–5619 (2024)
10. Hu, Q., Yang, B., Xie, L., et al.: Randla-net: efficient semantic segmentation of large-scale point clouds. In: CVPR, pp. 11105–11114 (2020)
11. Hu, Y., Fua, P., Wang, W., Salzmann, M.: Single-stage 6d object pose estimation. In: CVPR, pp. 2927–2936 (2020)
12. Hu, Y., Hugonot, J., Fua, P., Salzmann, M.: Segmentation-driven 6d object pose estimation. In: CVPR, pp. 3380–3389 (2019)
13. Huang, W.L., Hung, C.Y., Lin, I.C.: Confidence-based 6d object pose estimation. IEEE Trans. Multimedia **24**, 3025–3035 (2022)
14. Kendall, A., Gal, Y.: What uncertainties do we need in bayesian deep learning for computer vision? Adv. Neural Inf. Process. Syst. **30** (2017)
15. Kinga, D., Adam, J.B., et al.: A method for stochastic optimization. In: International Conference on Learning Representations (ICLR), California, vol. 5 (2015)
16. Li, Y., Kong, T., Chu, R., Li, Y., Wang, P., Li, L.: Simultaneous semantic and collision learning for 6-dof grasp pose estimation. In: 2021 IEEE/RSJ International Conference on Intelligent Robots and Systems (IROS), pp. 3571–3578 (2021)
17. Li, Z., Wang, G., Ji, X.: CDPN: coordinates-based disentangled pose network for real-time rgb-based 6-dof object pose estimation. In: ICCV, pp. 7677–7686 (2019)
18. Lian, R., Ling, H.: Checkerpose: progressive dense keypoint localization for object pose estimation with graph neural network. In: ICCV, pp. 13976–13987 (2023)
19. Lin, S., Wang, Z., Ling, Y., Tao, Y., Yang, C.: E2ek: end-to-end regression network based on keypoint for 6d pose estimation. IEEE Rob. Autom. Lett. **7**(3), 6526–6533 (2022)
20. Lin, T.Y., Goyal, P., Girshick, R., He, K., Dollár, P.: Focal loss for dense object detection. In: ICCV, pp. 2999–3007 (2017)
21. Lin, Y., Su, Y., Nathan, P., Inuganti, S., Di, Y., Sundermeyer, M., et al.: Hipose: hierarchical binary surface encoding and correspondence pruning for rgb-d 6dof object pose estimation. In: CVPR, pp. 10148–10158 (2024)
22. Liu, D., Liang, J., et al.: Tripartite feature enhanced pyramid network for dense prediction. IEEE Trans. Image Process. **32**, 2678–2692 (2023)
23. Liu, Z., Mao, H., Wu, C.Y., Feichtenhofer, C., Darrell, T., Xie, S.: A convnet for the 2020s. In: CVPR, pp. 11966–11976 (2022)
24. Lu, Y., Ma, X., Yang, L., Zhang, T., Liu, Y., Chu, Q., et al.: Gupnet++: geometry uncertainty propagation network for monocular 3d object detection. IEEE Trans. Pattern Anal. Mach. Intell. **47**(2), 900–915 (2025)
25. Manhardt, F., Kehl, W., Gaidon, A.: Roi-10d: monocular lifting of 2d detection to 6d pose and metric shape. In: CVPR, pp. 2064–2073 (2019)
26. Peng, S., Liu, Y., Huang, Q., Zhou, X., Bao, H.: Pvnet: pixel-wise voting network for 6dof pose estimation. In: CVPR, pp. 4556–4565 (2019)
27. Ronneberger, O., Fischer, P., Brox, T.: U-net: convolutional networks for biomedical image segmentation. In: Navab, N., Hornegger, J., Wells, W.M., Frangi, A.F. (eds.) MICCAI 2015. LNCS, vol. 9351, pp. 234–241. Springer, Cham (2015). https://doi.org/10.1007/978-3-319-24574-4_28
28. Smith, L.N.: Cyclical learning rates for training neural networks. In: 2017 IEEE Winter Conference on Applications of Computer Vision (WACV), pp. 464–472 (2017)

29. Su, Y., Rambach, J., Minaskan, N., et al.: Deep multi-state object pose estimation for augmented reality assembly. In: 2019 IEEE International Symposium on Mixed and Augmented Reality Adjunct (ISMAR-Adjunct), pp. 222–227 (2019)
30. Su, Y., Saleh, M., Fetzer, T., et al.: Zebrapose: coarse to fine surface encoding for 6dof object pose estimation. In: CVPR, pp. 6728–6738 (2022)
31. Tang, F., Wu, Y., Hou, X., Ling, H.: 3d mapping and 6d pose computation for real time augmented reality on cylindrical objects. IEEE Trans. Circuits Syst. Video Technol. **30**(9), 2887–2899 (2020)
32. Tian, Z., Shen, C., Chen, H., He, T.: FCOS: fully convolutional one-stage object detection. In: ICCV, pp. 9626–9635 (2019)
33. Wang, G., Manhardt, F., et al.: GDR-net: geometry-guided direct regression network for monocular 6d object pose estimation. In: CVPR, pp. 16606–16616 (2021)
34. Wu, D., Zhuang, Z., Xiang, C., et al.: 6d-vnet: end-to-end 6dof vehicle pose estimation from monocular RGB images. In: 2019 IEEE/CVF Conference on Computer Vision and Pattern Recognition Workshops (CVPRW), pp. 1238–1247 (2019)
35. Wu, Y., Zand, M., Etemad, A., Greenspan, M.: Vote from the center: 6 dof pose estimation in rgb-d images by radial keypoint voting. In: Computer Vision – ECCV 2022, pp. 335–352 (2022)
36. Xiang, Y., Schmidt, T., Narayanan, V., Fox, D.: Posecnn: a convolutional neural network for 6d object pose estimation in cluttered scenes. arXiv preprint arXiv:1711.00199 (2017)
37. Xu, L., Qu, H., Cai, Y., Liu, J.: 6d-diff: a keypoint diffusion framework for 6d object pose estimation. In: CVPR, pp. 9676–9686 (2024)
38. Zhou, J., Chen, K., Xu, L., Dou, Q., Qin, J.: Deep fusion transformer network with weighted vector-wise keypoints voting for robust 6d object pose estimation. In: ICCV, pp. 13921–13931 (2023)

SCP: Sinkhorn-reconciled Collaborative Prompt Learning for Vision-Language Models

Wenlong Niu and Zebao Zhang(✉)

Harbin Engineering University, Harbin 150001, China
{nwlxcdh255,zhangzebao}@hrbeu.edu.cn

Abstract. Prompt learning and Adapter applied to vision-language models are increasingly becoming a focus of research. We propose Sinkhorn-reconciled Collaborative Prompt Learning (SCP), a novel approach for fine-tuning vision-language models. Our approach aims to improve the generalization ability of the foundational vision-language models when fine-tuning downstream visual tasks in a few-shot setting. We introduce attention variants Correlative Self-Attention (CSA) and Merge Attention (MA) to design CorrAttentionMapper (CAM) and MergeBoostAdapter (MBAdapter) modules for the prompt and adapter, respectively, to enhance the prompt features expression capability and synergy between the two modalities. In addition, we reconcile the general and prompted features through Sinkhorn divergence to constrain the consistency of the model output distribution and prevent overfitting in the downstream task. Our method is fully evaluated on three benchmark tasks, including base-to-novel generalization, cross-dataset evaluation and domain generalization. Experimental results are compared with state-of-the-art prompt learning methods to demonstrate the effectiveness of the method.

Keywords: Prompt learning · Multi-modal · Adapter · Base-to-Novel generalization · Cross-dataset evaluation · Domain generalization

1 Introduction

Recent pre-trained vision-language (VLP) models, such as CLIP [21] and ALIGN [13], have demonstrated strong generalization capabilities in downstream visual tasks through training on large-scale image-text pair datasets. However, due to the large scale of these models and the lack of training data, it becomes difficult to fine-tune them for downstream tasks while maintaining a certain level of generalization ability. To address this challenge, recent studies have focused on prompt learning and adapter methods. For example, the existing multi-modal prompt learning method MaPLe [14] builds bridges between vision and language branches by setting up a prompt projection layer to facilitate collaborative interaction between the two modalities. Adapter methods [8] typically

J. Lokoč et al. (Eds.): MMM 2026, LNCS 16413, pp. 432–446, 2026.
https://doi.org/10.1007/978-981-95-6957-1_31

insert lightweight trainable networks near the prediction head of the model for fine-tuning. While these methods have shown some improvement in few-shot fine-tuning, they still fall short in maintaining the generalization ability of pre-trained models on downstream tasks. In fact, existing research has found that CoOp [31] improves the performance of a few-shots while often affecting the performance of zero-shot. This is mainly due to overfitting caused by newly introduced parameters during fine-tuning, which deviates the model from its original behavior. Therefore, further research is needed to address the above issues.

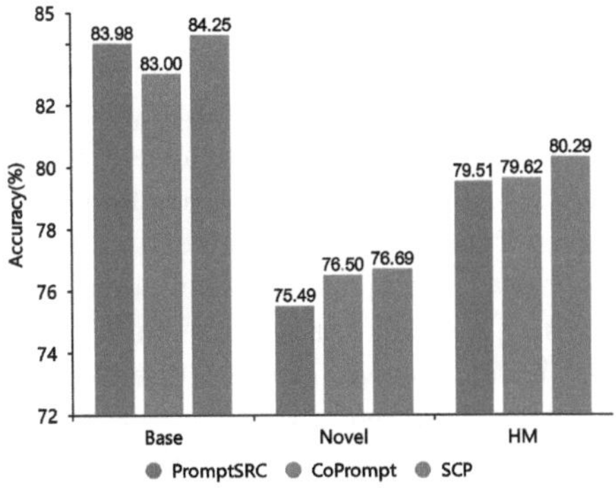

Fig. 1. Comprehensive comparison of PromptSRC, CoPrompt and SCP methods for averaging base class, novel class and harmonic mean over 11 datasets.

In this paper, we propose a novel approach to fine-tuning vision-language models, Sinkhorn-reconciled Collaborative Prompt Learning (SCP), which reduces overfitting and improves the generalization performance of models applied to downstream tasks by improving inter-modal co-consistency while keeping the distribution of the trainable model outputs consistent with that of the pre-trained models. Specifically, firstly, for prompt learning, we use the attention variant Correlative Self-Attention (CSA) [26] to design the CorrAttentionMapper (CAM) module to connect the vision-language prompts, improve the simple transformation in MaPLe using only the projection layer and further enhance the inter-modal synergy. Secondly, for the adapter, Merge Attention(MA) [29] is used to design the MergeBoostAdapter (MBAdapter) module to connect the information of the same modality, which enhances the expression of the prompted features while maintaining the general features among the same modality. Finally, the general and prompted features are reconciled by Sinkhorn divergence [24] prevent the trainable model from deviating from the semantic space of the pre-trained model. At the same time, zero-shot cross-entropy loss is introduced to participate in model training to make the model better adapted to new tasks. Overall, our approach improves the adaptability to downstream tasks while maintaining the generalization ability of pre-trained models. As shown in Fig. 1, in the base-to-novel generalization task, SCP exhibits superior performance over previous state-of-the-art methods, especially when compared to PromptSRC [15] and CoPrompt [23], which achieve an average gain of 0.78% and 0.67% in HM, respectively.

In summary, the contributions of this work are set out below:

- We propose SCP approach that uses the Sinkhorn divergence to reconcile the general and prompted features distributions to improve the model's adaptability to new tasks.

- For prompt learning, a trainable attention variant Correlative Self-Attention (CSA) is applied to design the CorrAttentionMapper (CAM) module as a bridge for vision-language prompts to further enhance inter-modal synergy.
- For the adapter, Merge Attention (MA) was used to design the MergeBoost-Adapter (MBAdapter) module to connect the same modal information to enhance the expression of prompted features.

2 Related Work

Prompt Learning in Vision-Language Models. Recently, Fundamental Vision-Language Models such as CLIP [21] and ALIGN [13] have exhibited remarkable performance improvements in various downstream visual tasks by learning from a vast number of image-text pairs. However, these models show sub-optimal performance when dealing with complex downstream tasks. To address this issue, existing research has applied prompt tuning to Pre-trained Vision-Language Models to enhance their adaptability to downstream tasks. Prompt learning usually provides auxiliary information to the model by designing appropriate prompt templates(e.g., "a photo of a <class>") based on specific tasks. Existing studies like CoOp [31] optimize prompts by replacing handcrafted ones with learnable continuous vectors. CoCoOp [30] builds on CoOp by combining image inputs to enhance prompts and boost model performance on unseen classes. MaPLe [14] uses projection to connect vision and language prompts, achieving dual-branch collaboration. PromptSRC [15] employs a self-regulating method to learn more task-agnostic knowledge, thus improving the generalization of the model. All of the above methods have demonstrated the improvement of prompt learning on model generalization performance. Our approach improves the two-branch joint prompt through the CorrAttentionMapper (CAM) module, which further improves the synergy between model branches.

Adapter. Adapter-based approaches [8], another focus in fine-tuning vision-language models, involve introducing additional trainable parameters into pre-trained models to facilitate customized learning for specific downstream tasks. Typically, adapter is placed near the prediction head of a model, where they act as learnable parameters to guide pre-trained models in learning richer features. For example, CoPrompt [23] incorporates a lightweight network adapter into the model framework to improve the performance of downstream tasks, and adapters have also been applied in image classification to increase precision. In our work, the MergeBoostAdapter (MBAdapter) is added to the model framework, connecting the general and prompted features of the same modality to strengthen downstream task capabilities.

Optimal Transport. The Optimal Transport [18] was initially introduced to address the problem of moving multiple objects at the minimum cost. Recently, optimal transport theory has regained prominence in computer vision and statistics. Due to its unique advantages in distribution matching, optimal transport has been widely applied in various fields, including multi-modal machine translation [9], graph matching for cross-domain problems [3]. In addition, Sinkhorn distances [5] simplifies the computation by introducing an entropy regularization term to optimize the optimal transport. Our method leverages the Sinkhorn divergence [24] based on Optimal Transport to reconcile the distributions of general and prompted features, thereby constraining the consistency of the output distribution.

3 Method

In this section, we present the details of the proposed SCP method. First, we review the basic requirements of the Vision-Language Models for applying prompt and the Optimal Transport fundamentals, followed by the details of our proposed SCP method.

3.1 Preliminaries

A Revisit of CLIP and MaPLe. Our approach uses CLIP [21] as a pre-trained Vision-Language Models. CLIP consists of a transformer-based text encoder $\mathcal{E}_T$ and a Vit-based image encoder $\mathcal{E}_I$. CLIP takes a hand-designed template ("a photo of a <class>") as input to the text encoder to get the text embedding W_k^c with class name, and the image X as input to the image encoder to get the image embedding $\mathcal{E}_I(X)$. Zero-sample prediction by searching for the highest similarity between the image and text embedding. That is, $p(y|x) = \frac{\exp(sim(\mathcal{E}_I(X),W_y)/\tau)}{\sum_{k=1}^{C}\exp(sim(\mathcal{E}_I(X),W_k)/\tau)}$.

Although, CLIP has a strong zero-sample prediction capability, it uses manually designed templates for text, which cannot be flexibly adapted to complex downstream tasks, therefore, there is still room for optimization. Subsequently, the text-only prompt CoOp [31] uses learnable continuous parameters based on the task text embedding instead of the original artificial template, which improves the flexibility of adapting to complex tasks. The multi-modal prompt learning method, MaPLe [14], sets up a coupling function to connect the vision-language prompts to further improve model synergy.

Specifically, for the vision branch, the image input X is segmented into blocks for projection to obtain the image embedding $\{P_n|_{n=1}^{d}\}$, which is used together with the learnable parameters to obtain the input of the image encoder $img = \{v_1, v_2, ..., v_n, p_1, p_2, ..., p_d\}$. For the language branch, The formulaic representation of the language prompts is $t_k = \{e_1, e_2, ..., e_l, c_k\}$, where c_k is the class name embedding, $e = \{e_n|_{n=1}^{l}\}$ is the learnable parameter with the same dimension as the original embedding, and l is the length of the context. The output obtained from the prompt as input to the text encoder is $\mathcal{E}_T(t_k)$, and the final prediction probability is as follows:

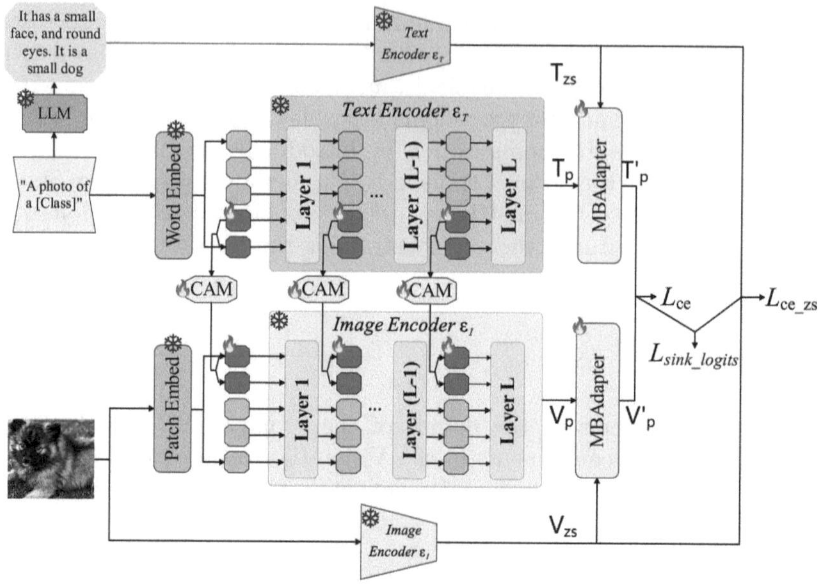

Fig. 2. Our method Sinkhorn-reconciled Collaborative Prompt Learning(SCP) reconciles the general and prompted features by Sinkhorn divergence. Simultaneously, we combine the pre-trained cross-entropy loss $\mathcal{L}_{ce\_zs}$ with the prompt cross-entropy loss $\mathcal{L}_{ce}$ to prevent the prompt model output from deviating too far from the pre-trained model output. In terms of prompt learning, we introduce the CorrAttentionMapper (CAM) module based on the Correlative Self-Attention (CSA) mechanism to further enhance model synergy. In terms of adapter, the MergeBoostAdapter (MBAdapter) module, designed based on the Merge Attention (MA) mechanism, optimizes prompt features.

$$p(y|x) = \frac{\exp\left(sim\left(\mathcal{E}_I(img), \mathcal{E}_T(t_k)\right)/\tau\right)}{\sum_{i=1}^{C}\exp\left(sim\left(\mathcal{E}_I(img), \mathcal{E}_T(t_k)\right)/\tau\right)}. \tag{1}$$

Optimal Transport. Optimal Transport(OT) is a method commonly used to compare distributions. Here, we focus on the relevant discrete case. Supposing we have two sets of points(features), the discrete distributions are equated as:

$$U = \sum_{m=1}^{M} u_m \delta_{f_m} \quad \text{and} \quad V = \sum_{n=1}^{N} v_n \delta_{g_n}, \tag{2}$$

where u and v are discrete probability vectors that sum to 1, δ_f and δ_g are the Dirac delta functions placed at the support points f and g, respectively, which lie within the embedding space. The total distance is denoted as:

$$< T, C >= \sum_{m=1}^{M}\sum_{n=1}^{N} T_{m,n} C_{m,n}. \tag{3}$$

C is called the cost matrix, where each element represents the cost between a_m and b_n, e.g. $\boldsymbol{C}_{m,n} = 1 - \sin(\boldsymbol{a}_m, \boldsymbol{b}_n)$. T is known as the transport plan and is optimized by minimizing the total distance. The optimization problem for optimal transport can be formulated as:

$$\begin{aligned} d_{\mathrm{OT}}(\boldsymbol{u}, \boldsymbol{v}|\boldsymbol{C}) &= \underset{T}{\text{minimize}} < T, C > \\ \text{subject to} \quad T\mathbf{1}_N &= \boldsymbol{u},\ T^\top \mathbf{1}_M = \boldsymbol{v},\ \boldsymbol{T} \in \mathbb{R}_+^{M\times N}. \end{aligned} \tag{4}$$

As it is time consuming to optimize the above objective directly. Therefore, we use the Sinkhorn distances [5] to accelerate the optimization by introducing entropy constraints. The entropy constrained Lagrange multiplier optimization problem is denoted as:

$$\begin{aligned} d_{\mathrm{OT},\lambda}(\boldsymbol{u}, \boldsymbol{v}|\boldsymbol{C}) &= \underset{T}{\text{minimize}} < T, \boldsymbol{C} > -\lambda h(\boldsymbol{T}) \\ \text{subject to} \quad T\mathbf{1}_N &= \boldsymbol{u}, T^\top \mathbf{1}_M = \boldsymbol{v}, \boldsymbol{T} \in \mathbb{R}_+^{M\times N}, \end{aligned} \tag{5}$$

where $h(\cdot)$ is the entropy $\lambda \geq 0$ is the hyperparameter. By iterating, we can quickly find the optimal solution :

$$T^* = \text{diag}(u^{(t)}) \exp(-C/\lambda) \text{diag}(v^{(t)}). \tag{6}$$

Here, $u^{(t)} = \boldsymbol{u} / \left(\exp(-\boldsymbol{C}/\lambda) v^{(t-1)}\right)$ and $v^{(t)} = \boldsymbol{v} / \left(\exp(-\boldsymbol{C}/\lambda)^\top \boldsymbol{u}^{(t)}\right)$, t denotes the iteration, and the initial value $v^{(0)} = 1$.

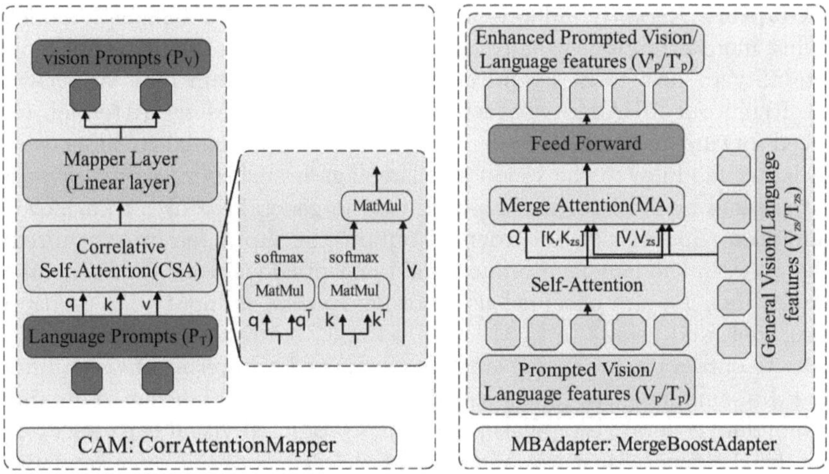

Fig. 3. (**Left**): CorrAttentionMapper(CAM) is composed of Correlative Self-Attention(CSA) and liner layer. (**Right**): MergeBoostAdapter(MBAdapter) consists of Self-Attention, Merge Attention(MA) and Feed Forward.

3.2 SCP: Sinkhorn-reconciled Collaborative Prompt Learning

We note that previous approaches either relied on prompt learning or employed adapters to adapt CLIP for downstream image recognition tasks. Inspired by related work, this paper proposes a novel fine-tuning method called SCP, which combines adapter with prompt learning and introduces the Sinkhorn divergence to reconciles the general and prompted features. The general overview diagram of SCP is shown in Fig. 2. The vision and language prompting, MBAdapter and sinkhorn divergence of our proposed SCP will be further analyzed in the following.

Vision and Language Prompting. As shown in Fig. 3: Left, our approach builds on MaPLe [14] use of linear layers for vision-language synergy, and inspired by Correlative Self-Attention (CSA), we design the CorrAttentionMapper (CAM) module to connect vision-language prompts branches to further enhance synergy. Specifically, the language prompts P_T passes through the CAM module to get the corresponding vision prompts P_V to achieve synergy between vision and language. It is specifically formulated as $P_V = CAM(P_T)$. Among them, CSA, with the help of Q,K pairwise correlation, enables each local token to prioritize its own location and its semantically similar region, enhancing the spatial consistency of features. The CSA computation method is expressed as:

$$\mathrm{CSA}_{\mathrm{Output}} = \left(\mathrm{Softmax}\left(\frac{q \cdot q^T}{\sqrt{d}}\right) + \mathrm{Softmax}\left(\frac{k \cdot k^T}{\sqrt{d}}\right)\right) \cdot v. \tag{7}$$

MBAdapter. Recently, adapter fine-tuning [8] of vision-language models is receiving more attention, usually, adapter is trainable to increase the model's adaptability to new tasks by introducing trainable parameters. As shown in Fig. 3: Right, our MBAdapter consists of self-attention, Merge Attention (MA) and feed-forward network. In order to optimize the prompted features, we add MBAdapter modules to the vision and language branches with prompt to connect the same modal information, and use the general features to achieve the optimization. Specifically, in order to obtain the input features required by MBAdapter, in the language branch, following previous work [23], general language features T_{zs} are obtained using ordinary text as input to a pre-trained text encoder, and using the LLM-enhanced text as input to the prompted text encoder to obtain the prompted language features T_p. In terms of vision branch, we use the same image as image encoder inputs in both branches to obtain the corresponding general vision features V_{zs} and prompted vision features V_p. Subsequently, the general features and prompted features of the same modality as above can be input as MBAdapter to get the enhanced prompted features which are rich in information. The MBAdapter computation can be expressed as:

$$T'_p = MBAdapter(T_p, T_{zs}), V'_p = MBAdapter(V_p, V_{zs}). \tag{8}$$

where, T'_p is the prompted language features enhanced by MBAdapter, V'_p is the prompted vision features enhanced by MBAdapter. MA calculates attention by splicing different K and V of the same modality, thus realizing to increase the diversity of prompted features information while reducing features conflicts and helping the model to better adapt to new tasks. The MA can be expressed as:

$$MA_{Output} = \text{Softmax}(\frac{Q[K, K_{\text{zs}}]^T}{\sqrt{d}})[V, V_{\text{zs}}]. \tag{9}$$

Sinkhorn Divergence. In existing studies, PLOT [2] solves the multi-prompt and vision features alignment problem by establishing OT between vision features and prompts. Our approach uses Sinkhorn divergence based on OT to establish logits level constraints between the pre-trained logits distribution and the prompted logits distribution in order to reconciling the general and prompted features, constrain the output distribution. Specifically, the pre-trained logits distribution $F = sim(V_{zs}, T_{zs})$ and the prompted logits distribution $P = sim(V_p', T_p')$ are constrained by using OT, which is the minimization of the OT distance to push F to P in order to learn the transport plan T:

$$W(F, P) = d_{\text{OT}}(\boldsymbol{u}, \boldsymbol{v}|\mathbf{1} - \boldsymbol{F}^\top \boldsymbol{P}), \tag{10}$$

Here, we use the cosine distance to represent the cost matrix $C = \mathbf{1} - (\boldsymbol{F}^\top \boldsymbol{P})$. Then we can obtain the solution of transport plan T^* as Eq. 6 and the final OT distance $W(F, P)$. Subsequently, similar to the above operation, we sequentially use OT for each of the pre-trained logits distribution F and the prompted logits distribution P to the corresponding OT distances $W(F, F)$ and $W(P, P)$. Finally, the Sinkhorn divergence can be expressed as:

$$\mathcal{L}_{sink\_logits}(F, P) = 2W(F, P) - W(F, F) - W(P, P). \tag{11}$$

Final Loss. The Sinkhorn divergence, the pre-trained cross-entropy loss $\mathcal{L}_{ce\_zs}$ and the prompted cross-entropy loss $\mathcal{L}_{ce}$ together form the final loss. Adding a balancing factor α to the Sinkhorn divergence, we obtain the final loss of the SCP is:

$$\mathcal{L} = \mathcal{L}_{ce} + \alpha \mathcal{L}_{sink\_logits} + \mathcal{L}_{ce\_zs} \tag{12}$$

4 Experiments

4.1 Experimental Evaluation

We extensively evaluated our approach in three benchmark settings.

Base-to-Novel Generalization. We follow the generic setup of Base-to-Novel Generalization, fine-tuning at few-shot and evaluating at zero-shot scenarios. The dataset is divided into Base and Novel classes, and the models are trained on Base classes and evaluated on Base and Novel classes.

Cross-Dataset Evaluation. To evaluate the generalization ability of SCP, we perform Cross-dataset Evaluation, specifically, we train the model on ImageNet and evaluate it on other datasets.

Domain Generalization. We evaluate the generalization performance of our method under out-of-domain distributions. We train the model on ImageNet and evaluate it under four other datasets with different distributions.

Datasets. In order to fully evaluate our model, we used 11 general image recognition datasets, which cover a wide range of recognition tasks. Including, ImageNet [6], Caltech101 [7], OxfordPets [20], StanfordCars [16], Flowers102 [19], Food101 [1], FGVCAircraft [17], SUN397 [28], UCF101 [25], DTD [4], EuroSAT [10]. In addition, four variant datasets are used to evaluate extraterritorial distribution experiments, include ImageNetV2 [22], ImageNet-Sketch [27], ImageNet-A [12] and ImageNet-R [11].

Table 1. Comparison with existing methods on base-to-novel generalization setting. HM: harmonic mean. '*' denote the performance obtained by our re-implementation. The underlined lines indicate sub-optimal values.

Avg	Base	Novel	HM
CLIP	69.34	74.22	71.70
CoOp	82.69	63.22	71.66
CoCoOp	80.47	71.69	75.83
MaPLe	82.28	75.14	78.55
PromptSRC*	83.98	75.49	79.51
CoPrompt*	83.00	76.50	79.62
SCP	**84.25**	**76.69**	**80.29**

ImageNet	Base	Novel	HM
CLIP	72.43	68.14	70.22
CoOp	76.47	67.88	71.92
CoCoOp	75.98	70.43	73.10
MaPLe	76.66	70.54	73.47
PromptSRC*	**77.62**	70.16	73.70
CoPrompt*	76.65	**71.26**	**73.86**
SCP	77.35	69.87	73.42

Caltech101	Base	Novel	HM
CLIP	96.84	94.00	95.40
CoOp	98.00	89.81	95.84
CoCoOp	97.96	93.81	95.84
MaPLe	97.74	94.36	96.02
PromptSRC*	98.23	94.25	96.15
CoPrompt*	**98.54**	95.20	96.84
SCP	98.45	**95.31**	**96.85**

OxfordPets	Base	Novel	HM
CLIP	91.17	97.26	94.12
CoOp	93.67	95.29	94.47
CoCoOp	95.20	97.69	96.43
MaPLe	**95.43**	97.76	96.58
PromptSRC*	95.30	97.34	96.31
CoPrompt*	95.27	97.02	96.14
SCP	95.34	**98.10**	**96.70**

StanfordCars	Base	Novel	HM
CLIP	63.37	**74.89**	68.65
CoOp	78.12	60.40	68.13
CoCoOp	70.49	73.59	72.01
MaPLe	72.94	74.00	73.47
PromptSRC*	**78.44**	74.74	**76.54**
CoPrompt*	73.25	72.20	72.72
SCP	78.17	72.73	75.36

Flowers102	Base	Novel	HM
CLIP	72.08	**77.80**	74.83
CoOp	97.60	59.67	74.06
CoCoOp	94.87	71.75	81.71
MaPLe	95.92	72.46	82.56
PromptSRC*	97.40	76.15	85.47
CoPrompt*	97.15	75.04	84.68
SCP	**98.10**	76.19	**85.77**

Food101	Base	Novel	HM
CLIP	90.10	91.22	90.66
CoOp	88.33	82.26	85.19
CoCoOp	90.70	91.29	90.99
MaPLe	90.71	**92.05**	**91.38**
PromptSRC*	90.65	91.50	91.07
CoPrompt*	90.08	91.13	90.60
SCP	**90.73**	91.83	91.28

FGVCAircraft	Base	Novel	HM
CLIP	27.19	36.29	31.09
CoOp	40.44	22.30	28.75
CoCoOp	33.41	23.71	27.74
MaPLe	37.44	35.61	36.50
PromptSRC*	**42.34**	35.53	38.64
CoPrompt*	38.12	**38.63**	38.37
SCP	41.12	37.25	**39.09**

SUN397	Base	Novel	HM
CLIP	69.36	75.35	72.23
CoOp	80.60	65.89	72.51
CoCoOp	79.74	76.86	78.27
MaPLe	80.82	78.70	79.75
PromptSRC*	**82.76**	79.04	80.86
CoPrompt*	82.30	**79.75**	**81.00**
SCP	82.32	78.31	80.27

DTD	Base	Novel	HM
CLIP	53.24	59.90	56.37
CoOp	79.44	41.18	54.24
CoCoOp	77.01	56.00	64.85
MaPLe	80.36	59.18	68.16
PromptSRC*	82.29	61.23	70.21
CoPrompt*	83.33	63.81	72.28
SCP	**83.45**	**64.13**	**72.53**

EuroSAT	Base	Novel	HM
CLIP	56.48	64.05	60.03
CoOp	92.19	54.74	68.69
CoCoOp	87.49	60.04	71.21
MaPLe	94.07	73.23	82.35
PromptSRC*	92.19	71.43	80.49
CoPrompt*	91.76	77.55	84.06
SCP	**95.56**	**79.67**	**86.89**

UCF101	Base	Novel	HM
CLIP	70.53	77.50	73.85
CoOp	84.69	56.05	67.46
CoCoOp	82.33	73.45	77.64
MaPLe	83.00	78.66	80.77
PromptSRC*	**86.69**	79.00	82.67
CoPrompt*	86.56	79.88	83.08
SCP	86.19	**80.24**	**83.11**

Implementation Details. We evaluate our proposed method, employing CLIP as a pre-trained vision-language backbone, and report average results over 3 runs. For base-to-novel generalization experiments, we fine-tune the model using a small number of settings with 16 samples selected from known classes. We train the model with 8 epochs, using the SGD optimizer with 4 batch sizes and a learning rate of 0.0035. For cross-dataset evaluation and domain generalization, we followed the MaPLe setting except for the learning rate which used 0.0015.

Baselines. Recently, CLIP [21] and CoOp-based methods have been used for comparison. For example, CoOp [31], CoCoOp [30], MaPLe [14], PromptSRC [15], CoPrompt [23].

4.2 Base-To-Novel Generalization

The results for base-to-novel generalization are shown in Table 1, SCP outperforms the state-of-the-art methods on average base and novel classes, HM on 11 datasets. Our method further increases the synergy between vision-language and enables improved modal generalization performance. Compared with PromptSRC, SCP improves 1.20% on the novel classes, and compared with CoPrompt, SCP improves 1.25% on base classes. Our method maintains the stable performance of base classes while the performance of novel classes is also improved, especially on HM, which improves 0.78% and 0.67% on PromptSRC and CoPrompt, respectively, which is a strong proof of the improvement of the overall generalization performance of the model. Notably, our SCP method outperforms existing methods in HM for 7 of the 11 datasets, and the remaining datasets have comparable performance with existing methods.

4.3 Cross-Dataset Evaluation

The results of the cross-dataset evaluation experiments are shown in Table 2, '*' denote the performance obtained by our re-implementation. SCP performs better than CoPrompt on 5 of the 10 datasets, and it outperforms PromptSRC on all datasets. In particular, compared to PromptSRC and CoPrompt, the average performance of SCP is improved by 0.84% and 0.09%, respectively.

Table 2. Comparison of SCP with existing prompt learning methods on cross-dataset evaluation. The underlined lines indicate sub-optimal values.

	Source	Target										
	ImNet	Caltech	Pets	Cars	Flowers	Food	Aircraft	SUN397	DTD	EuroSAT	UCF	*Avg*
CoOp	**71.51**	93.70	89.14	64.51	68.71	85.30	18.47	64.15	41.92	46.39	66.55	63.88
CoCoOp	71.02	**94.43**	90.14	65.32	71.88	86.06	22.94	67.36	45.73	45.37	68.21	65.73
MaPLe	70.72	93.53	**90.49**	65.57	**72.23**	86.20	**24.74**	67.01	46.49	48.06	68.69	66.30
PromptSRC*	71.44	93.67	90.02	65.20	71.05	86.36	24.21	67.01	46.16	44.46	69.07	65.72
CoPrompt*	70.60	93.75	90.42	65.84	72.19	86.24	23.94	**67.62**	**46.75**	48.13	**69.80**	66.47
SCP	70.64	93.92	90.18	**65.90**	71.62	**86.42**	24.35	67.45	46.62	**49.91**	69.21	**66.56**

4.4 Domain Generalization

The results of the domain generalization experiments are shown in Table 3, '*' denote the performance obtained by our re-implementation. Overall, the performance of SCP is competitive with all previous methods. Average performance is 0.12% lower than PromptSRC and 0.20% higher than CoPrompt.

Table 3. Comparison of performance on domaion generalization. The underlined lines indicate sub-optimal values.

	Source	Target				
	ImNet	ImNetV2	ImNetS	ImNetA	ImNetR	*Avg*
CLIP	66.73	60.83	46.15	47.77	73.96	57.18
CoOp	**71.51**	64.20	47.99	49.71	75.21	59.28
CoCoOp	71.02	64.07	48.75	50.63	76.18	59.91
MaPLe	70.72	64.07	49.15	<u>50.90</u>	76.98	60.27
PromptSRC*	<u>71.44</u>	**64.40**	**49.65**	**50.93**	**77.54**	**60.63**
CoPrompt*	70.60	64.14	49.30	50.49	77.30	60.31
SCP	70.64	<u>64.27</u>	<u>49.59</u>	50.76	<u>77.43</u>	<u>60.51</u>

Table 4. (**Left**): Ablative Analysis. (**Right**): Performance of SCP under different α values.

CSA	MBAdp.	Sinkh.	L_{zs}.	Base	Novel	HM
✓	✓	✓	✓	**84.25**	**76.69**	**80.29**
✓	✓	✓	✗	84.14	76.03	79.88
✓	✓	✗	✓	83.81	75.00	79.16
✓	✗	✓	✓	83.69	75.30	79.27
✗	✓	✓	✓	84.02	75.55	79.56
✗	✗	✓	✓	83.54	74.95	79.01
✗	✗	✗	✗	83.06	74.59	78.60

α	ImNet	Caltech	Pets	Cars	Flowers	Food	Aircraft	SUN397	DTD	EuroSAT	UCF
1.0	73.02	96.17	96.49	73.25	82.91	90.87	38.18	79.15	69.82	80.32	82.11
2.0	72.99	96.24	96.52	73.39	83.82	90.99	38.34	79.24	69.85	82.97	82.17
4.0	73.06	96.21	96.53	73.35	84.47	90.88	38.67	79.20	70.47	**86.89**	82.26
6.0	73.23	**96.85**	**96.70**	74.68	84.80	91.09	38.13	79.43	**72.53**	86.24	82.24
8.0	73.38	96.41	96.63	74.67	84.94	91.14	38.64	79.73	71.00	82.72	82.94
12.0	**73.42**	96.23	96.55	**75.36**	**85.77**	**91.28**	**39.09**	**80.27**	71.35	80.69	**83.11**
14.0	73.34	96.32	96.59	74.78	84.83	91.13	38.05	79.89	70.20	79.07	82.72

4.5 Ablative Analysis

Ablation of Key Components. In this section, we study the impact of each component by removing different components in the method. The experimental results are shown in Table 4: Left, where 'CSA', ' MBAdp.', ' Sinkh.' and ' L_{zs}.', denote Correlative Self-Attention (CSA), MergeBoostAdapter, Sinkhorn divergence and pre-trained cross-entropy loss, respectively. The best performance of SCP is obtained when all components are considered, with a HM of 80.29%. First, we remove the pre-trained cross-entropy loss, bringing the precision down to 79.88% and the performance decreases by 0.41%. This suggests that pre-trained cross-entropy loss has a positive impact on SCP performance improvement. Second, we remove Sinkhorn divergence, the accuracy drops to 79.16% and the performance decreases by 1.13%. This suggests that the moderating effect of Sinkhorn divergence is crucial for SCP. Then, we remove the MBAdapter, which lowers the accuracy to 79.27% and the performance drops 1.02%. This highlights the importance of MBAdapter in SCP. Next, we remove the CSA, which reduces the accuracy to 79.56% and the performance drops 0.73%. This demonstrates the effectiveness of the CSA module in SCP. Finally, we remove both the CSA and MBAdapter modules. The accuracy drops to 79.01% and the performance decreases by 1.28%. This suggests that their interaction can further improve the model generalization performance. Below we will analyze these important components and key parameters in detail to illustrate their importance to SCP.

Table 5. Analysis of different components of SCP.

(a) CSA replace.

CAM-attn	HM
self-attn	79.67
CSA	**80.29**

(b) MA replace.

MBAdapter-attn	HM
cross-attn	79.42
MA	**80.29**

(c) Divergence replace.

criterion	HM
KL	79.32
Sinkhorn	**80.29**

(d) Adapter choices.

MBAdapter	HM
only Text	79.40
only Image	79.55
Both	**80.29**

(e) Adapter attn analysis

Self-attn	MA	HM
✓	✗	79.72
✗	✓	79.62
✓	✓	**80.29**

(f) Adapter Layers

MBAdapter	HM
×1	**80.29**
×2	78.92
×3	76.57

Study of CSA and Sinkhorn Divergence. In this section, we examine reasonable alternatives to CSA and Sinkhorn divergence to further demonstrate the importance of the two components for SCP. For CSA, we consider the original self-attention for replacement. The results are shown in Table 5a, and we observe that the model performs best when CSA is used. For Sinkhorn divergence, we consider using KL for substitution and the results are shown in Table 5c, we observe that the model performs best when Sinkhorn divergence is used.

Study of MBAdapter. Here, we conduct a detailed study of MBAdapter to show their impact on model performance in different situations. First, we study the performance impact when adding MBAdapter to different branches, as shown in Table 5d, adding adapter to the text branch has a greater impact on the model than adding them to the image branch, and the model performs best when adapter are added to both image and text. Next, we ablated self-attention and Merge Attention (MA) in the MBAdapter to further study the impact of attention in the adapters, and the experimental results are shown in Table 5e, which results in a degradation of the model performance when we employ only self-attention or MA in the adapter layer, and the best performance of the model is obtained when both are employed. Then, we explored the effect of the number of adapter layers, and the results are shown in Table 5f, the model performance shows a decreasing trend as the number of added adapter layers increases. We attribute this to the fact that as the number of adapter layers increases. The introduction of too many attention leads to performance degradation caused by model overfitting. Finally, we considered replacing MA in the adapter with cross-attention to further explore the validity of MA. The results are shown in Table 5b, where the model performance is best when MA is used.

Sensitivity Analysis of Key Parameters. Similar to previous work [23], we analyze the sensitivity of some key parameters in this paper. Firstly, we study the balance factor α. As shown in Table 4: Right, overall, as α increases, the accuracy values show an increasing trend. The best accuracies of our 11 datasets are obtained at α of 4,6,12, among which 7 datasets achieve the best accuracy at α of 12, the accuracies on the rest of the datasets are close to the best accuracy (except EuroSAT). Factors such as different data distributions and their similarity to the pre-trained CLIP dataset can cause the optimal values from different datasets to be distributed in different α. When considering α greater than 12, the performance is not improved. Subsequently, we studied the effect of different prompt layers and epochs on the performance of SCP, as shown in Table 6a, similar to MaPLe, the prompt layers performs best when set to 9. The results of Table 6b show that the epochs perform best when set to 8.

Table 6. Sensitivity analysis.

(a) HM vs Layers.

Layers	HM
3	78.60
6	79.11
9	**80.29**
12	79.40

(b) HM vs Epochs.

Epochs	HM
3	78.03
5	79.26
8	**80.29**
10	79.69

5 Conclusion

We propose SCP, a novel approach to fine-tuning vision-language models. The method enables the model to enhance its performance on new tasks while maintaining the zero-shot generalization capability. SCP improves the overfitting problem during model fine-tuning by carefully designing the different components. We conduct extensive evaluations in three benchmark settings and demonstrate the effectiveness of SCP in Base-to-Novel Generalization, Cross-dataset Evaluation and Domain Generalization by comparing it with state-of-the-art methods. In addition, we perform detailed ablation experiments in which every important component is analyzed to demonstrate its effectiveness. We believe that our Sinkhorn-reconciled Collaborative Prompt Learning approach plays an active role in improving vision-language models.

Acknowledgments. This work is supported by the National Key R&D Program of China (2022YFC3301800) and the National Natural Science Foundation of China (Grant No. 62072135)

References

1. Bossard, L., Guillaumin, M., Van Gool, L.: Food-101-mining discriminative components with random forests. In: Computer Vision-ECCV 2014, pp. 446–461 (2014)
2. Chen, G., Yao, W., Song, X., Li, X., Rao, Y., Zhang, K.: Plot: prompt learning with optimal transport for vision-language models. In: 11th International Conference on Learning Representations, ICLR 2023 (2023)
3. Chen, L., Gan, Z., Cheng, Y., Li, L., Carin, L., Liu, J.: Graph optimal transport for cross-domain alignment. In: Proceedings of the 37th International Conference on Machine Learning. Proceedings of Machine Learning Research, vol. 119, pp. 1542–1553. PMLR (2020)
4. Cimpoi, M., Maji, S., Kokkinos, I., Mohamed, S., Vedaldi, A.: Describing textures in the wild. In: In: Proceedings of the IEEE Conference on Computer Vision and Pattern Recognition (CVPR) (2014)
5. Cuturi, M.: Sinkhorn distances: Lightspeed computation of optimal transport. In: Advances in Neural Information Processing Systems, vol. 26. Curran Associates, Inc. (2013)
6. Deng, J., Dong, W., Socher, R., Li, L., Li, K., Fei-Fei, L.: Imagenet: a large-scale hierarchical image database. In: 2009 IEEE Conference on Computer Vision and Pattern Recognition, pp. 248–255 (2009)
7. Fei-Fei, L., Fergus, R., Perona, P.: Learning generative visual models from few training examples: an incremental bayesian approach tested on 101 object categories. In: 2004 Conference on Computer Vision and Pattern Recognition Workshop, pp. 178–178 (2004)
8. Gao, P., Geng, S., Zhang, R., Ma, T., Fang, R., Zhang, Y., Li, H., Qiao, Y.: Clip-adapter: better vision-language models with feature adapters. Int. J. Comput. Vision **132**(2), 581–595 (2024)
9. Guo, W., Fang, Q., Yu, D., Feng, Y.: Bridging the gap between synthetic and authentic images for multimodal machine translation. In: Proceedings of the 2023 Conference on Empirical Methods in Natural Language Processing (2023)
10. Helber, P., Bischke, B., Dengel, A., Borth, D.: Eurosat: a novel dataset and deep learning benchmark for land use and land cover classification. IEEE J. Sel. Topics Appl. Earth Observat. Remote Sens. **12**(7), 2217–2226 (2019)
11. Hendrycks, D., et al.: The many faces of robustness: a critical analysis of out-of-distribution generalization. In: Proceedings of the IEEE/CVF International Conference on Computer Vision (ICCV), pp. 8340–8349 (2021)
12. Hendrycks, D., Zhao, K., Basart, S., Steinhardt, J., Song, D.: Natural adversarial examples. In: Proceedings of the IEEE/CVF Conference on Computer Vision and Pattern Recognition (CVPR), pp. 15262–15271 (2021)
13. Jia, C., et al.: Scaling up visual and vision-language representation learning with noisy text supervision. In: Proceedings of the 38th International Conference on Machine Learning, vol. 139, pp. 4904–4916. PMLR (2021)
14. Khattak, M., Rasheed, H., Maaz, M., Khan, S., Khan, F.: Maple: multi-modal prompt learning. In: Proceedings of the IEEE/CVF Conference on Computer Vision and Pattern Recognition (CVPR), pp. 19113–19122 (2023)
15. Khattak, M., Wasim, S., Naseer, M., Khan, S., Yang, M., Khan, F.: Self-regulating prompts: foundational model adaptation without forgetting. In: Proceedings of the IEEE/CVF International Conference on Computer Vision (ICCV), pp. 15190–15200 (2023)
16. Krause, J., Stark, M., Deng, J., Fei-Fei, L.: 3d object representations for fine-grained categorization. In: Proceedings of the IEEE International Conference on Computer Vision (ICCV) Workshops (2013)

17. Maji, S., Rahtu, E., Kannala, J., Blaschko, M., Vedaldi, A.: Fine-grained visual classification of aircraft (2013)
18. Monge, G.: M'emoire sur la th'eorie des d'eblais et des remblais. Mem. Math. Phys. Acad. Royale Sci. 666–704 (1781)
19. Nilsback, M., Zisserman, A.: Automated flower classification over a large number of classes. In: 2008 Sixth Indian Conference on Computer Vision, Graphics & Image Processing, pp. 722–729 (2008)
20. Parkhi, O.M., Vedaldi, A., Zisserman, A., Jawahar, C.V.: Cats and dogs. In: 2012 IEEE Conference on Computer Vision and Pattern Recognition, pp. 3498–3505 (2012)
21. Radford, A., et al.: Learning transferable visual models from natural language supervision. In: Proceedings of the 38th International Conference on Machine Learning, vol. 139, pp. 8748–8763. PMLR (2021)
22. Recht, B., Roelofs, R., Schmidt, L., Shankar, V.: Do imagenet classifiers generalize to imagenet? In: Proceedings of the 36th International Conference on Machine Learning, vol. 97, pp. 5389–5400. PMLR (2019)
23. Roy, S., Etemad, A.: Consistency-guided prompt learning for vision-language models. In: 12th International Conference on Learning Representations, ICLR (2024)
24. Scetbon, M., Cuturi, M.: Linear time sinkhorn divergences using positive features. In: Advances in Neural Information Processing Systems, vol. 33, pp. 13480. Curran Associates, Inc. (2020)
25. Soomro, K.: Ucf101: a dataset of 101 human actions classes from videos in the wild. arXiv preprint arXiv:1212.0402 (2012)
26. Wang, F., Mei, J., Yuille, A.: SCLIP: rethinking self-attention for dense vision-language inference. In: Computer Vision-ECCV 2024, pp. 315–332. Springer, Cham (2025). https://doi.org/10.1007/978-3-031-72664-4_18
27. Wang, H., Ge, S., Lipton, Z., Xing, E.P.: Learning robust global representations by penalizing local predictive power. In: Advances in Neural Information Processing Systems (2019)
28. Xiao, J., Hays, J., Ehinger, K., Oliva, A., Torralba, A.: Sun database: large-scale scene recognition from abbey to zoo. In: 2010 IEEE Computer Society Conference on Computer Vision and Pattern Recognition, pp. 3485–3492 (2010)
29. Zheng, B., Zhou, D.W., Ye, H.J., Zhan, D.C.: Task-agnostic guided feature expansion for class-incremental learning. In: Proceedings of the IEEE/CVF Conference on Computer Vision and Pattern Recognition (CVPR) (2025)
30. Zhou, K., Yang, J., Loy, C.C., Liu, Z.: Conditional prompt learning for vision-language models. In: Proceedings of the IEEE/CVF Conference on Computer Vision and Pattern Recognition (CVPR), pp. 16816–16825 (2022)
31. Zhou, K., Yang, J., Loy, C.C., Liu, Z.: Learning to prompt for vision-language models. Int. J. Comput. Vision **130**(9), 2337–2348 (2022)

DAGMP: A Multimodal Learning Approach Jointly Driven by Feature Fusion and Gradient Modulation

Xinying Zhou[1,2], Leixiao Li[1,2,3](✉), and Hao Lin[1,2]

[1] Inner Mongolia University of Technology, Inner Mongolia Huhhot 010080, China
llxhappy@126.com

[2] Inner Mongolia Key Laboratory of Beijiang Cyberspace Security, Inner Mongolia Autonomous Region Hohhot 010080, China

[3] Inner Mongolia Autonomous Region Software Service Engineering Technology Research Center Based on BigData, Hohhot 010080, China

Abstract. Multimodal learning holds immense potential in video understanding, semantic matching, and human-computer interaction, yet existing methods still suffer from modal information asymmetry and training imbalance issues. This paper presents DAGMP (Dual-Aware Guided Multimodal Plug-in), a plug-and-play solution for addressing modal information asymmetry and training imbalance in multimodal learning. DAGMP enhances feature alignment and noise suppression with the EEDF (Enhanced Energy-based Dynamic Fusion) strategy, and optimizes gradient consistency using the DEA (Gradient Energy with Angle) mechanism. It can be integrated into any multimodal framework without structural changes. Experiments on CREMA-D, MEAD, and AVE datasets show significant performance improvements, demonstrating its effectiveness and versatility.

Keywords: Multimodal Fusion · Cross-Modal Alignment · Plug-and-Play Multimodal Learning

1 Introduction

Despite progress in multimodal learning across numerous domains [1, 2], efficiently integrating heterogeneous modalities remains challenging [3]. Modal information asymmetry, misalignment, noise, and low-quality data [4] often undermine fusion. Static fusion (e.g., concatenation, summation) struggles with heterogeneity [5], while dynamic fusion [6, 7] may cause training instability.

Even with improved fusion strategies, multimodal models can encounter optimization bottlenecks. On MEAD and CREMA-D, they sometimes underperform unimodal models (Fig. 1[1]) due to differences in convergence, gradient magnitude, and

[1] The experiment adopts the network structure of Restnet18, maintaining the same network framework as that in the article.Present the differences in the training process through the intermediate rounds.

J. Lokoč et al. (Eds.): MMM 2026, LNCS 16413, pp. 447–460, 2026.
https://doi.org/10.1007/978-981-95-6957-1_32

directionality [8–10], causing overreliance on dominant modalities and limiting fusion expressiveness [11, 12].

To address this, we propose DAGMP, a plug-and-play fusion–modulation component. At the feature layer, EEDF achieves semantic alignment, energy-based weighting, and gated enhancement, highlighting key features and suppressing noise. At the optimization layer, DEA jointly adjusts gradient magnitude and direction, mitigating overfitting of dominant modalities and underfitting of weak ones. DAGMP can be embedded into any multimodal model without structural modification, enabling stable performance improvements.The main contributions of this paper are as follows:

(1) EEDF: Energy-driven dynamic fusion for key feature emphasis and noise suppression.
(2) DEA: Gradient adjustment for magnitude and directional consistency.
(3) DAGMP: Plug-and-play integration for universal multimodal frameworks.

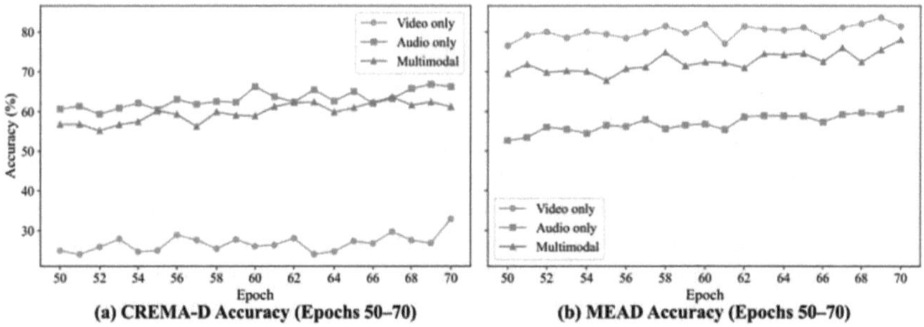

Fig. 1. CREMA-D and MEAD accuracy (epochs 50–70) for video, audio, and multimodal.

2 Related Work

2.1 Multimodal Feature Fusion Mechanism.

Multimodal learning boosts model expressiveness by integrating diverse perceptual sources, but noise, redundancy, and modality uncertainty complicate fusion. Static fusion [13] lacks adaptive weighting, while feature reconstruction and regularization [14] improve reliability but are inflexible. Attention-based dynamic fusion [15, 16] aids semantic coordination but increases parameters and struggles with noisy modalities. Energy- or uncertainty-based methods [17, 18] enhance robustness, yet estimation and fusion are often decoupled, limiting end-to-end optimization.

2.2 Imbalance Issues in Multimodal Optimization Processes

Beyond fusion, imbalance limits performance. OGM [19] balances gradient norms, GradNorm [20] normalizes multi-task gradients, and CGGM [11] aligns gradients via fixed classifiers but cannot handle dynamic imbalance. Existing methods are architecture-specific and non-modular. We propose DAGMP, a plug-and-play plugin for seamless integration into any multimodal framework.

3 Method

3.1 Overall Overview

The proposed DAGMP plugin (see Fig. 2) consists of two components: feature fusion and gradient tuning. During the forward pass, feature enhancement is performed via EEDF, while during the backpropagation phase, gradients are dynamically adjusted through DEA. Notably, DEA operates only during training, eliminating additional computational overhead during inference and thus preserving deployment costs. This integrated "fusion-tuning" design creates a closed-loop enhancement mechanism, enabling the model to benefit simultaneously at both the feature and optimization layers.

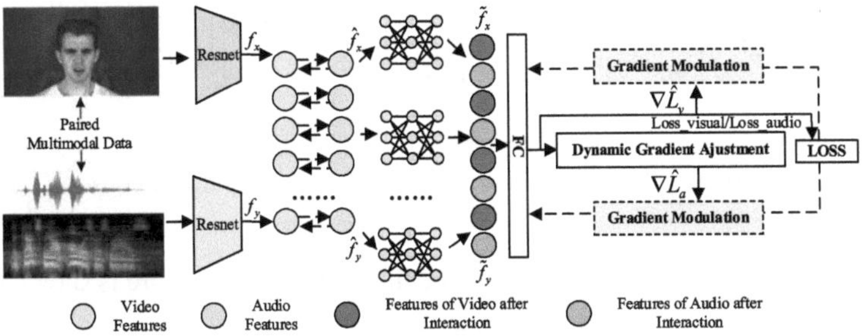

Fig. 2. DAGMP structure: visual and audio features are fused via EEDF, then forwarded, while DEA dynamically rebalances gradient magnitude and direction during backpropagation.

3.2 Enhanced Energy-Based Dynamic Fusion.

To address the issues of modal quality imbalance and unstable fusion, inspired by uncertainty perception [18], we propose the EEDF strategy (Fig. 3) based on semantic alignment, energy modeling, gated fusion, and feature enhancement to achieve dynamic modal balance and robust representation.

Semantic Alignment. First, the input multimodal features $x \in \mathbb{R}^{B \times D}$ and $y \in \mathbb{R}^{B \times D}$, each with a dimension of 512, are mapped to a unified space through two layers of linear projection and normalization operations, as shown in Eq. 1.

$$f_x = LN(\Phi(W_x x + b_x)), f_y = LN\big(\Phi\big(W_y y + b_y\big)\big) \quad (1)$$

Among these, $LN(\cdot)$ denotes the layer normalization layer, Φ represents the GELU activation function, $W_x x$ and $W_y y$ denote the linear transformation parameters, and f_x and f_y denote the mapped modal features. To enhance cross-modal semantic associations, bidirectional cross-attention is introduced, as shown in Eq. 2.

$$\hat{f}_x = CA\big(f_x, f_y\big), \hat{f}_y = CA\big(f_y, f_x\big) \quad (2)$$

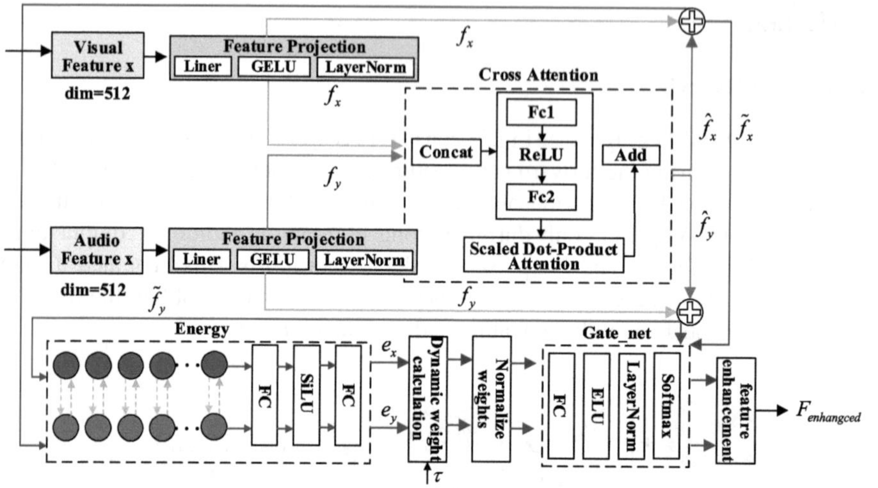

Fig. 3. EEDF flowchart: visual and audio features pass through linear projection, cross-modal attention, energy scoring, and gated fusion to produce weighted representations.

Where $CA(\cdot)$ denotes cross-attention. The final interaction feature is shown in Eq. 3:

$$\tilde{f}_x = f_x + \hat{f}_x, \tilde{f}_y = f_y + \hat{f}_y \tag{3}$$

This mechanism ensures the alignment of different modal features within a shared semantic space, preventing information fragmentation across modalities.

Energy-Score Modeling. When significant differences exist in modal quality, the key issue is how to adaptively allocate modal weights. We designed an energy function $f_e(\cdot)$ to model modal reliability, as shown in Eq. 4.

$$E_x = f_e([\tilde{f}_x, \tilde{f}_y]), E_y = f_e([\tilde{f}_y, \tilde{f}_x]) \tag{4}$$

Here, $[\cdot, \cdot]$ denotes the concatenation operation, where $f_e(\cdot)$ is a two-layer feedforward network. It applies two nonlinear transformations and outputs the energy score corresponding to each mode. The specific form is shown in Eq. 5.

$$f_e(z) = {w_2}^T \sigma(W_1 z + b_1) + b_2, z \in \mathrm{R}^{2d} \tag{5}$$

The terms $z \in \mathbb{R}^{2d}$, b_1 and b_2 represent bias terms. This formulation captures complex relationships between modalities, reflecting the relative credibility and importance of each modality within the other's context. To apply energy scores for modality fusion, a temperature parameter is introduced to control energy sensitivity. This parameter adjusts the sensitivity of energy scores, thereby influencing the model's response to different feature combinations, as shown in Eq. 6. The temperature is a hyperparameter that controls the model's sensitivity to feature combinations.

$$w_x = \frac{e^{\sigma(E_x)/\tau}}{e^{\sigma(E_x)/\tau} + e^{\sigma(E_y)/\tau}}, w_y = 1 - w_x \tag{6}$$

$\sigma(\cdot)$ denotes the *Sigmoid* function, where w_x and w_y represent weights used for weighted feature fusion, gradually decreasing during training via cosine annealing [21]. This ensures balanced modal weights in the early training stages, progressively focusing on stronger modalities for discrimination in later stages, thereby enhancing robustness.

Gate-Controlled Adaptive Fusion. To further enhance the ability to select effective modalities, we designed a gated network whose input is concatenated from raw features and weighted features. Modal fusion weights are obtained through softmax normalization, as shown in Eq. 7.

$$a = soft\max\left(g\left(\left[\tilde{f}_x, \tilde{f}_y, w_x\tilde{f}_x, w_y\tilde{f}_y\right]\right)\right), a = [a_1, a_2] \tag{7}$$

Here, a_1, a_2 represents the fusion ratio between the two modalities and satisfies the constraint $a_1 + a_2 = 1$, as shown in the abstract formula in Eq. 8.

$$a = g_\theta\left(z_{gate}\right), a \in \mathrm{R}^2 \tag{8}$$

Finally, the fused feature representation based on the weight vector a is obtained as shown in Eq. 9:

$$z = a_1\tilde{f}_x + a_2\tilde{f}_y \tag{9}$$

This gating mechanism not only utilizes interaction features but also incorporates weight-adjusted features, enabling the fusion results to achieve dynamic adaptation across different modalities.

Feature Enhancement. Features obtained through direct fusion suffer from over-smoothing or insufficient discriminative power. To address this, we further introduce a residual enhancement mechanism, first applying a nonlinear transformation to the fused features via an expansion-compression feedforward network. The formula is shown in Eq. 10.

$$\tilde{z} = \phi\left(\left[z, \tilde{f}_x + \tilde{f}_y\right]\right) \tag{10}$$

Among these, $\phi(\cdot)$ represents an expansion-compression feedforward network, which performs residual connection between the transformation result z and the input features, as shown in Eq. 11.

$$z^* = LN(\tilde{z} + z) \tag{11}$$

The residual component of $\tilde{z}+z$ embodies the superposition of input features and non-linear transformed features, ensuring no information loss while enhancing the model's representational capacity. Layer normalization further stabilizes the training process. The resulting z^* serves as the fused representation output.

3.3 Gradient Energy with Angle

To enhance fusion effectiveness, a dynamic mechanism called DEA (see Fig. 4) was designed, which integrates confidence and gradient consistency adjustment. Its core

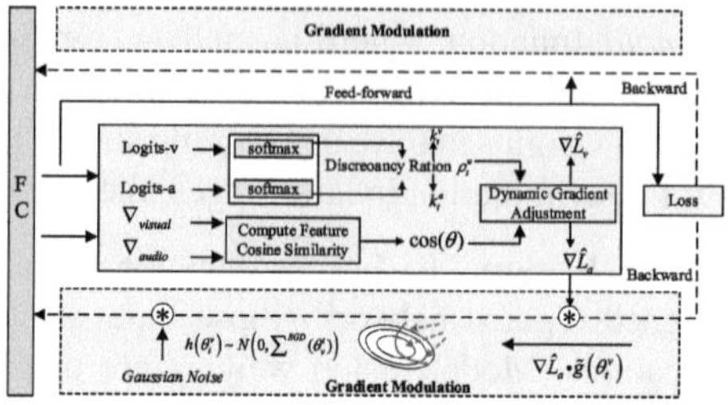

Fig. 4. DEA Mechanism: Dynamic Gradient Adjustment for Multimodal Fusion.

principle involves dynamically adjusting the gradient magnitude between audio and visual modalities during the backpropagation phase based on modal confidence differences and gradient direction consistency, thereby achieving balance and coordination between modalities.

Modal Confidence Ratio Adjustment. Inspired by OGM [19], we compute the relative importance of each modality based on its prediction confidence for the current sample during training, and dynamically scale the gradients accordingly. Specifically, let the model outputs audio modality prediction $\hat{y}^a$, visual modality prediction $\hat{y}^v$, and fused prediction $\hat{y}$, with the corresponding ground truth label y. The confidence score for each modality is defined as shown in Eq. 12.

$$s^a = \sum_{i=1}^{B} Soft\max\left(\hat{y}_i^a\right)\left[y_i\right], s^v = \sum_{i=1}^{B} Soft\max\left(\hat{y}_i^v\right)\left[y_i\right] \tag{12}$$

Here, B denotes the number of samples in the mini-batch. The Softmax function is used $[y_i]$ to convert the vector into a probability distribution. To select the probability value corresponding to the true label y_i from the Softmax output, the confidence ratio is defined as $\rho_v = \frac{s^v}{s^a}$ and $\rho_a = \frac{1}{\rho^v}$. Furthermore, a regularization factor based on hyperbolic function compression is introduced to dynamically adjust the gradient. k^a is defined as shown in Eq. 13, and k^v is defined as shown in Eq. 14.

$$k^a = \begin{cases} 1 - \tanh(a \cdot \rho_a), & \rho_a \le 1 \\ 1, & \rho_a > 1 \end{cases} \tag{13}$$

$$k^v = \begin{cases} 1 - \tanh(a \cdot \rho_v), & \rho_v > 1 \\ 1, & \rho_v \le 1 \end{cases} \tag{14}$$

Here, a serves as a temperature control hyperparameter that regulates the sensitivity of gradient adjustment. This design ensures that underperforming modes receive greater gradient update space during training.

Modal Gradient Direction Consistency Adjustment. Beyond confidence adjustment, DEA also considers the relationship between gradient directions across modalities. The consistency of optimization directions is measured by calculating the cosine similarity $cos(\theta)$ between the gradients of the two modalities in the current batch. Let the gradients of the audio and visual modalities be denoted as ∇L_a and ∇L_v, respectively. The cosine similarity is defined as shown in Eq. 15.

$$\gamma = \cos(\theta) = \frac{< \nabla L_a, \nabla L_v >}{\|\nabla L_a\| \cdot \|\nabla L_v\|} \tag{15}$$

Dual-Factor Fusion and Final Gradient. The two aforementioned adjustment mechanisms are combined to propose the following unified gradient update strategy, as shown in Eq. 16:

$$\nabla \hat{L}_u \leftarrow \gamma \cdot \left(k^u \cdot \nabla L_u + \delta_{\mathrm{u}}\right), u \in \{a, v\} \tag{16}$$

Here, $\nabla \hat{L}_u$ denotes the gradient of the loss function for model u,where k^u is a coefficient used to adjust the perturbation strength.

In implementation, we additionally inject a small Gaussian noise aa into the gradient to further enhance training robustness. This serves as an optional Gaussian perturbation term$\delta_u \sim \mathrm{N}\left(0, \sigma_u^2\right)$.

4 Experiment

4.1 Data Sets and Experimental Equipment

CREMA-D (Crowd-sourced Emotional Multimodal Actors Dataset) [22]: 91 actors, 6 emotions, 12 sentences, 7,442 audio-visual samples; MEAD (Multimodal Emotional Actor Dataset) [23]: 60 actors, 7 emotions × 3 intensities, frontal audio-visual data; AVE (Audio-Visual Event Dataset) [24]: 4,143 clips, 28 event categories with precise audio-visual annotations.

4.2 Experimental Setup

The visual encoder uses ResNet-18 for classification, with experiments on visual and audio modalities. Video frames are resized to 224 × 224, randomly cropped and horizontally flipped (50%), then normalized and standardized before 2D CNN feature extraction. Audio is sampled at 220.5 kHz, converted to 128-dim Mel spectrograms (channels expanded to three), truncated or zero-padded to 3 s. Training runs for 100 epochs using SGD with learning rate 0.005, momentum 0.9, and weight decay 1e-4 [25].

4.3 Melting Experiment

Melting of the EEDF Components. To validate the EnhancedEDF core module, we conducted ablation experiments on CREMA-D, sequentially removing Cross-Attention, energy modeling, dynamic gating, and fusion enhancement. Comparing with the full model highlights each module's contribution to cross-modal alignment, dynamic weighting, gating, and feature fusion. EEDF shows significant advantages across all ablations (mAP drop 5.4%–12.5%, $p < 0.01$; Table 1), confirming the necessity of Cross-Attention, Energy-based weights, GateNet, and FusionProj.

Table 1. Comparison of Components Within the EEDF Module

Method	CrossAttetion (A)	Energy-based Weighting(B)	GateNet(C)	FsionProj (D)	Acc	mAP
No-A	✗	✓	✓	✓	59.1%	63.2%
No-B	✓	✗(0.5/0.5)	✓	✓	58.1%	62.2%
No-C	✓	✓	✗ (aveage)	✓	64.9%	64.8%
No-D	✓	✓	✓	✗	60.9%	62.0%
Full	✓	✓	✓	✓	69.8%	74.5%

Dissolution of each Part of the DEA. To validate the role of each component in DEA, we performed ablation on the score and angle adjustment coefficients (Table 2[2]). Results on CREMA-D show that removing either ("No-score" or "No-angle") leads to clear performance drops, whereas the full DEA achieves ~2.2% and ~ 2.6% improvements, respectively. This confirms that score adjustment balances modality confidence, while angle adjustment enhances training consistency, and together they provide more stable and efficient fusion.

Table 2. Comparison of Components Within the DEA Module

Method	Score adjustment	Angle adjustment	Acc	mAP
No-score	✗	✓	66.5%	70.5%
No-Angle	✓	✗	65.6%	68.9%
Full(ours)	✓	✓	68.8%	73.3%

Dissolution of DAGMP Components. To validate the components, we conducted ablation experiments (Table 3). EEDF improves accuracy by 0.8%–4.5% via energy modeling, while DEA yields 1.6%–5.9% gains by suppressing gradient imbalance. Their combination (DAGMP) achieves 7.7% Acc and 6.0% mAP improvement on MEAD, with consistent gains on CREMA-D and AVE. Visualization in Fig. 5 shows: (a) DEA maintains positive intermodal gradient similarity and stabilizes training; (b) EEDF ensures consistency between confidence levels and allocation coefficients. These results demonstrate the effectiveness of the integrated design.

4.4 Comparative Experiment

Comparison of Integration Strategies. We compared EEDF with both traditional (baseline, summation, Film [26], Gated [27]) and emerging fusion methods (Sfusion

[2] The fusion method employs a simple concatenation approach.

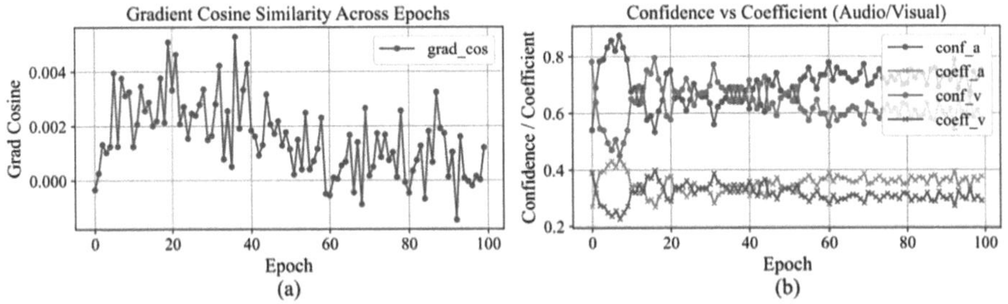

Fig. 5. DAGMP visualization. (a) Stable gradient similarity over epochs. (b) EEDF aligns modality confidence with weights.

Table 3. Melting Experiment

Dataset	MEAD		CREMA-D		AVE	
Method	Acc	mAP	Acc	mAP	Acc	mAP
Only audio	59.1%	62.8%	64.6%	66.5%	49.7%	51.5%
Only visual	83.7%	91.9%	26.7%	27.1%	34.0%	31.9%
Only EEDF	84.3%	90.1%	69.8%	74.5%	61.1%	63.3%
Only DEA	84.2%	90.7%	68.8%	73.3%	61.4%	67.9%
DAGMP	89.5%	90.1%	71.1%	75.1%	62.5%	64.9%

[28], DynMM [29], PDF [30], FuseMix [31], TelME [32]) under the same backbone (Table 4). On CREMA-D, EEDF surpasses the baseline by 9.5% in Acc and 12.7% in mAP, while on MEAD it achieves 2.6% and 4.9% gains, respectively. Overall, EEDF yields stable improvements (2.6%–12.7%) across datasets, verifying its robustness and effectiveness in multimodal fusion.

Integrate into the DEA Mechanism. Building upon Sect. 4.4.1 by incorporating the DEA mechanism, the results are presented in Table 5.[3] On the CREMA-D dataset, EED achieves a 7.3% improvement over the baseline and a 7.8% increase in mAP, maintaining its optimal performance among all compared methods. On the MEAD dataset, it achieves a 5.3% improvement over the baseline with mAP remaining largely unchanged, outperforming most other fusion methods. Overall, the DEA mechanism effectively enhances the expressive power of multimodal features, enabling EEDF to achieve stable and significant performance improvements on both datasets. This validates its effectiveness and robustness in multimodal fusion tasks.

Comparison with Other Gradient Adjustment Methods. For fair comparison, gradient adjustment methods from Ogm-ge [19], AGM [33], PMR [34], and OPM [35] were evaluated alongside EEDF (Fig. 6). While these methods partially mitigate intermodal imbalance, improvements are limited and unstable due to single-dimensional

[3] * represents integration into the DEA mechanism.

Table 4. Comparison of Fusion Methods

Dataset	CREMA-D		MEAD	
Method	Acc	mAP	Acc	mAP
Baseline(Concatenation)	60.3%	61.8%	81.7%	85.2%
Sum	61.4%	64.2%	79.9%	83.7%
Multi-Head Attention	59.7%	65.1%	76.5%	78.1%
Gated	62.0%	72.8%	74.4%	83.6%
FilM	61.6%	62.1%	79.2%	82.1%
Sfusion	67.3%	69.3%	80.3%	82.5%
DynMM	67.9%	76.7%	80.2%	87.6%
PDF	68.3%	75.3%	83.1%	89.5%
FuseMix	61.3%	65.4%	81.3%	88.7%
TelME	64.5%	66.7%	76.5%	80.1%
EEDF(ours)	69.8%	74.5%	84.3%	90.1%

Table 5. Integrate into the DEA mechanism

Dataset	CREMA-D		MEAD	
Method	Acc	mAP	Acc	mAP
Baseline(Concatenation)*	63.8%	67.3%	84.2%	90.6%
Sum*	62.5%	64.8%	85.2%	88.5%
Multi-Head Attention*	63.4%	69.7%	77.3%	79.1%
Gated*	68.0%	76.2%	76.9%	84.1%
FilM*	68.6%	76.9%	82.9%	90.1%
Sfusion*	69.2%	70.4%	85.3%	89.1%
DynMM*	68.9%	73.5%	86.6%	90.2%
PDF*	70.3%	74.6%	85.9%	89.2%
FuseMix*	65.3%	66.7%	84.4%	87.3%
TelME*	68.3%	72.3%	84.5%	88.3%
EEDF*(DAGMP)	71.1%	75.1%	89.5%	90.1%

adjustments. In contrast, DAGMP jointly modulates gradient magnitude and direction, suppressing dominant modes and enhancing weak ones, yielding superior and stable results across datasets and metrics.

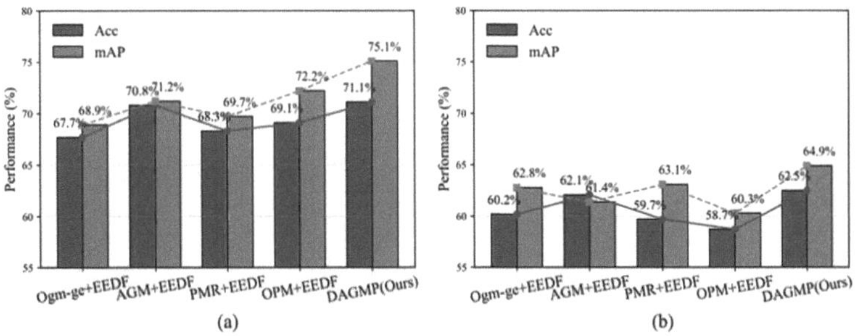

Fig. 6. Changes in (a) CREMA-D and (b) AVE performance during training with DAGMP. Results are reported on the corresponding datasets.

DAGMP provides modest but stable accuracy gains across datasets and tasks, improving performance without changing model architecture, mitigating noise or missing modalities, and yielding clearer decision boundaries, demonstrating versatility and practical applicability.

Plug-and-Play Evidence. To validate the universality of DAGMP, we conducted minimal experiments on Res-Net-18 and ViT-small (see Table 6). DAGMP consistently delivered stable improvements across different backbones and tasks: an average relative gain of 8.76% on ResNet-18, and an average relative gain of 6.29% on ViT-small, with the most significant gains observed on AVE (+13.02% and + 8.92%). Additionally, robustness experiments (see Fig. 7) demonstrate that DAGMP exhibits greater stability under missing modalities or noisy interference. Specifically, when 20% of the input modalities are missing (Miss20), DAGMP still maintains strong performance. Even with 50% of the modalities missing (Miss50), the model remains resilient, showing only a slight drop in accuracy. Furthermore, under noisy conditions with a 10 dB signal-to-noise ratio (Noisy10dB), DAGMP continues to perform effectively despite the low-quality signal and high interference. These results highlight DAGMP's robustness across various challenging scenarios, confirming its suitability for practical engineering applications.

Table 6. Backbone Network Application Comparison

Dataset	CREMA-D		AVE	
Model	Acc	mAP	Acc	mAP
ResNet-18	66.3%	71.8%	55.3%	58.9%
ResNet-18†	71.1%	75.1%	62.5%	64.9%
ViT-small	65.4%	69.7%	57.2%	59.8%
ViT-small†	69.3%	72.2%	62.3%	63.8%

† indicates the use of the DAGMP component.

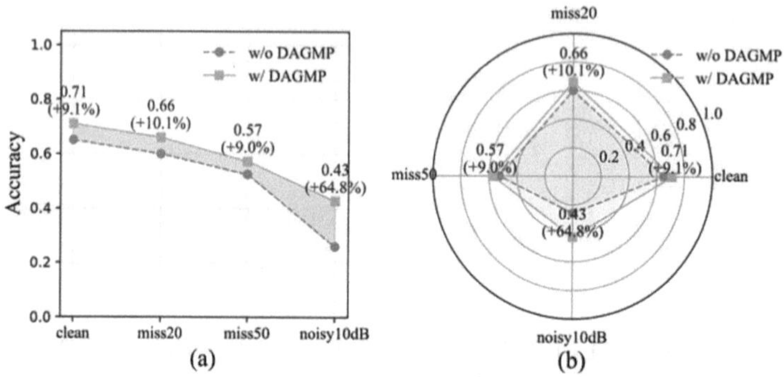

Fig. 7. Impact of DAGMP on accuracy across noise levels. Dashed lines show accuracy without DAGMP, solid lines with DAGMP.

5 Conclusion

This paper introduces DAGMP—a plug-and-play fusion-optimization integrated plugin. By incorporating the EEDF fusion strategy at the feature layer and the DEA gradient adjustment mechanism at the optimization layer, DAGMP achieves robust gains without altering the model architecture. Experiments validate the method's effectiveness and consistency across multiple datasets and robustness scenarios. Future work will explore DAGMP's potential applications in cross-task transfer learning, lightweight deployment, and multimodal large-scale models.

Acknowledgments. This work was supported by the National Natural Science Foundation of China (No.62362055), National Key Research and Development Program of China (No. 2023YFB2703900), Program for Young Talents of Science and Technology in Universities of Inner Mongolia Autonomous Region (No.NJYT22084, No.NJYT24035), Inner Mongolia Natural Science Foundation (No.2023MS06008), Scientific research projects of universities directly under the Inner Mongolia Autonomous Region (No.JY20220061, No.JY20230119, No.JY20230019), Science and Technology Plan Project of Inner Mongolia (No. 2024SKYPT0012), Ordos Key R&D Program Projects (No.YF20232328).

References

1. Baltrušaitis, T., Ahuja, C., Morency, L.P.: Multimodal machine learning: a survey and taxonomy. IEEE Trans. Pattern Anal. Mach. Intell. **41**(2), 423–443 (2019)
2. Akbari, H., Yuan, L., Qian, R., et al.: Vatt: transformers for multimodal self-supervised learning from raw video, audio and text. Published. In: The 35th Conference on Neural Information Processing Systems, vol. 34, pp. 24206–24221 (2021)
3. Tsai, Y.-H.H., Bai, S., Liang, P.P., et al.: Multimodal transformer for unaligned multimodal language sequences. In: Proceedings of the Proceedings of the Conference Association for Computational Linguistics Meeting, vol. 6558, (2019)
4. Zhang Q, Wei Y, Han Z, et al: Multimodal fusion on low-quality data: A comprehensive survey. ArXiv preprint arXiv:2404, 18947 (2024)

5. Poria, S., Cambria, E., Bajpai, R., et al.: A review of affective computing: from unimodal analysis to multimodal fusion. Inf. Fusion. **37**, 98–125 (2017)
6. Zhou P, Yang W, Chen W, et al: Modality attention for end-to-end audio-visual speech recognition. In: Proceedings of the ICASSP 2019–2019 IEEE International Conference on Acoustics, Speech and Signal Processing (ICASSP), pp. 6565–6569 (2019)
7. Chen, P., Zhao, X., Zeng, L., et al.: A review of research on SLAM technology based on the fusion of LiDAR and vision. Sensors. **25**(5), 1447 (2025)
8. Wang, W., Tran, D., Feiszli, M.: What makes training multi-modal classification networks hard? In: Proceedings of the 2020 IEEE/CVF Conference on Computer Vision and Pattern Recognition (CVPR), pp. 12695–12705 (2020)
9. Yang, Y., Wan, F., Jiang, Q.-Y., et al.: Facilitating multimodal classification via dynamically learning modality gap. Adv. Neural Inf. Proces. Syst. **37**, 62108–62122 (2024)
10. Wei, Y., Hu, D., Du, H., et al.: On-the-Fly modulation for balanced multimodal learning. IEEE Trans. Pattern Anal. Mach. Intell. **47**(1), 469–485 (2025)
11. Guo, Z., Jin, T., Chen, J., et al.: Classifier-guided gradient modulation for enhanced multimodal learning. Adv. Neural Inf. Proces. Syst. **37**, 133328–133344 (2024)
12. Kontras K, Chatzichristos C, Blaschko M, et al: Improving multimodal learning with multi-loss gradient modulation. ArXiv preprint arXiv: 2405.07930 (2024)
13. Ramachandram, D., Taylor, G.W.: Deep multimodal learning: a survey on recent advances and trends. IEEE Signal Process. Mag. **34**(6), 96–108 (2017)
14. Ma, M., Ren, J., Zhao, L., et al.: Smil: multimodal learning with severely missing modality. Proc. Proc. AAAI Conf. Artif. Intell. **35**(3), 2302–2310 (2021)
15. Alam, M.M., Dini, M.A., Kim, D.-S., et al.: TMNet: Transformer-Fused Multimodal Framework for Emotion Recognition Via EEG and Speech. ICT Express (2025)
16. Khan, M., Gueaieb, W., El Saddik, A., et al.: MSER: multimodal speech emotion recognition using cross-attention with deep fusion. Expert Syst. Appl. **245**, 122946 (2024)
17. Peng, T., Li, Z., Wang, P., et al.: A novel energy based model mechanism for multi-modal aspect-based sentiment analysis. Proc. Proc. AAAI Conf. Artif. Intell. **38**(17), 18869–18878 (2024)
18. Zhang, Q., Wu, H., Zhang, C., et al.: Provable dynamic fusion for low-quality multimodal data. In: Proceedings of the 40th International Conference on Machine Learning, pp. 1753–41769 (2023)
19. Peng, X., Wei, Y., Deng, A., et al.: Balanced multimodal learning via on-the-fly gradient modulation. In: Proceedings of the Proceedings of the IEEE/CVF Conference on Computer Vision and Pattern Recognition, pp. 8238–8247 (2022)
20. Chen, Z., Badrinarayanan, V., Lee, C.-Y., et al.: Gradnorm: gradient normalization for adaptive loss balancing in deep multitask networks. In: Proceedings of the International Conference on Machine Learning, pp. 794–803 (2018)
21. Loshchilov I, Hutter F.J.A.P.A.: Sgdr: Stochastic gradient descent with warm restarts. ArXiv preprint arXiv: 1608.03983 (2016)
22. Cao, H., Cooper, D.G., Keutmann, M.K., et al.: Crema-d: crowd-sourced emotional multimodal actors dataset. IEEE Trans. Affect. Comput. **5**(4), 377–390 (2014)
23. Wang, K., Wu, Q., Song, L., et al.: Mead: a large-scale audio-visual dataset for emotional talking-face generation. In: Proceedings of the European Conference on Computer Vision, pp. 700–717 (2020)
24. Tian, Y., Shi, J., Li, B., et al.: Audio-visual event localization in unconstrained videos. In: Proceedings of the Proceedings of the European Conference on Computer Vision (ECCV), pp. 247–263 (2018)
25. Sinha, N.K., Griscik, M.P.: A stochastic approximation method. IEEE Trans. Syst. Man Cybern. **SMC-1(4)**, 338–344 (1971)

26. Perez E, Strub F, De Vries H, et al: Film: visual reasoning with a general conditioning layer. Proceedings of the Proceedings of the AAAI Conference on Artificial Intelligence, 32 1, (2018).
27. Kiela D, Grave E, Joulin A, et al: Efficient large-scale multi-modal classification. Proceedings of the Thirty-Second AAAI Conference on Artificial Intelligence and Thirtieth Innovative Applications of Artificial Intelligence Conference and Eighth AAAI Symposium on Educational Advances in Artificial Intelligence. New Orleans, Louisiana, USA; AAAI Press. 32 1, (2018).
28. Liu, Z., Wei, J., Li, R., et al.: SFusion: self-attention based n-to-one multimodal fusion block. In: Proceedings of the International Conference on Medical Image Computing and Computer-Assisted Intervention, pp. 159–169 (2023)
29. Xue, Z., Marculescu, R.: Dynamic Multimodal Fusion. In: Proceedings of the 2023 IEEE/CVF Conference on Computer Vision and Pattern Recognition Workshops (CVPRW), pp. 17–24 (2023)
30. Cao B, Xia Y, Ding Y, et al: Predictive dynamic fusion. ArXiv preprint arXiv:2406.04802 (2024)
31. Vouitsis, N., Liu, Z., Gorti, S.K., et al.: Data-efficient multimodal fusion on a single gpu. In: Proceedings of the Proceedings of the IEEE/CVF Conference on Computer Vision and Pattern Recognition, pp. 27239–27251 (2024)
32. Yun T, Lim H, Lee J, et al: Telme: Teacher-leading multimodal fusion network for emotion recognition in conversation: ArXiv preprint arXiv:2401.12987 (2024)
33. Li, H., Li, X., Hu, P., et al.: Boosting multi-modal model performance with adaptive gradient modulation. In: Proceedings of the Proceedings of the IEEE/CVF International Conference on Computer Vision, pp. 22214–22224 (2023)
34. Fan, Y., Xu, W., Wang, H., et al.: Pmr: prototypical modal rebalance for multimodal learning. In: Proceedings of the Proceedings of the IEEE/CVF Conference on Computer Vision and Pattern Recognition, pp. 20029–22003 (2023)
35. Wei, Y., Hu, D., Du, H., et al.: On-the-fly modulation for balanced multimodal learning. IEEE Trans. Pattern Anal. Mach. Intell., 469–485 (2024)

HFS: Hierarchical Fine-Tuning for Span Detection and Aspect-Based Sentiment Analysis in the Vietnamese Language

Son T. Huynh[1,2,3], Tran Minh Huan[1,2,3], Tran Nguyen Minh Quang[1,2,3], Pham Phi Nhung[1,2,3], and Binh T. Nguyen[1,2,3](✉)

[1] Faculty of Mathematics and Computer Science, University of Science, Ho Chi Minh City, Vietnam
[2] Faculty of Mathematics and Computer Science, Vietnam National University, Ho Chi Minh City, Vietnam
[3] AISIA Research Lab, Ho Chi Minh City, Vietnam
ngtbinh@hcmus.edu.vn

Abstract. This paper proposes a novel method to enhance the performance of two key aspects of natural language processing: aspect-based sentiment analysis (ABSA) and span detection, specifically for the Vietnamese language. The proposed approach leverages the strengths of pretrained language models with a two-stage task decomposition strategy. In the first stage, the model identifies opinion spans and their associated aspects within the text, functioning as a sequence tagging task. In the second stage, it classifies the sentiment polarity related to each identified aspect, treating it as a hierarchical classification task. This approach effectively captures both linguistic dependencies and sentiment relationships within Vietnamese text by structuring the learning process in this manner. Experimental evaluations demonstrate the superiority of this approach, achieving a micro F1 score of 76.27%, which significantly outperforms prior baselines. This highlights the model's capability to handle hierarchical classification and sequence tagging tasks, particularly in low-resource language environments.

Keywords: Span detection · Aspect-based sentiment analysis · Hierarchical learning

1 Introduction

In the era of rapidly developing digital technology, along with the explosion of e-commerce, it is becoming increasingly difficult for companies or enterprises working in customer service to understand their customers. After purchasing and using the items, they can provide reviews of these products; if these reviews are used correctly and effectively, they can make both service providers and service users more understanding. If they want to get to know each other through the given customer review, sentiment analysis will be an effective way to do so,

J. Lokoč et al. (Eds.): MMM 2026, LNCS 16413, pp. 461–475, 2026.
https://doi.org/10.1007/978-981-95-6957-1_33

especially in an undertaken situation. Aspect-based sentiment analysis (ABSA) [3,4] is a subtask of sentiment analysis whose objective is to extract aspects in an unstructured text and their opinions to classify the aspect's sentiment polarity, usually into three types: positive, negative, and neutral. This subtask is widely used in understanding the writer's perspective in a given text. For example, the given text is "The design is beautiful, feels solid in hand, but the speaker is too small, the service staff is very enthusiastic", the text includes three aspects "design", "speaker", and "service staff" as well as four opinion words "beautiful", "solid", "small" and "enthusiastic" so that the polarities are POSITIVE, NEGATIVE and POSITIVE will be assigned to three aspects "design", "speaker" and "service staff", respectively.

As mentioned in [19], span detection is another task of ABSA, aiming to detect user opinions based on aspects and their corresponding sentiments. Reusing the example above, instead of giving the outputs as an ABSA task, span detection gives two shorter sentences consisting of the author's opinions: "design is beautiful, feels solid in hand", "speaker is too small," and "service staff is very enthusiastic." These opinions are then classified into the corresponding sentiment polarity. Le et al. [6] present a novel hierarchical multi-task learning model designed to simultaneously address span detection and aspect-based sentiment analysis for Vietnamese, two challenging natural language processing tasks with limited resources. This work serves as a strong foundation for leveraging multi-task hierarchical learning methods.

In this study, we investigate a new approach to the problem, with a focus on solving it using a hierarchical framework inspired by the ideas of Le et al. [6], and based on the UIT-ViSD4SA dataset [19]. In aspect-based sentiment analysis (ABSA), joint training across subtasks (e.g., aspect extraction, opinion extraction, and sentiment classification) may cause negative transfer due to conflicting learning objectives. A hierarchical approach mitigates this issue by modeling subtasks sequentially, enabling focused learning and improved overall performance. Hence, we split each task and train them independently. In each step, we will take the output from the previous task to make it the input in the next step in order to predict the new output. Our main contributions and enhancements are summarized as follows:

(i) Firstly, we start by thoroughly preprocessing the data set to ensure consistency in data quality. This step involves addressing several issues that may impact the evaluation.
(ii) Secondly, we introduce a side-by-side training process, allowing each module to be trained independently.
(iii) Finally, we propose our approach using the hierarchical method to achieve better performance at each level of the model.

The paper is organized as follows: Sect. 2 reviews the relevant literature (Related works). Section 3 presents the methodology used in this study. Section 4 presents the results and discusses the findings, followed by the conclusions in Sect. 5.

2 Related Works

2.1 Apect-Based Sentiment Analysis (ABSA)

Aspect-based sentiment Analysis (ABSA) is a crucial branch of sentiment analysis, widely used in research and real-world applications. Previous studies [1,3–5] have highlighted that it is one of the most critical and challenging tasks in natural language processing (NLP), attracting significant attention from both the research community and businesses. Previous studies emphasize its complexity due to the need to accurately detect specific aspects of products or services and then analyze the sentiment associated with each aspect. It is vital for businesses aiming to extract actionable insights from customers' feedback, enabling them to make data-driven decisions.

The SemEval datasets (2014–2016) [14–16] significantly advanced ABSA research by providing diverse, multilingual reviews across various domains, including restaurants, laptops, hotels, and electronics. SemEval 2014 introduced aspect categories (e.g., Food, Service), SemEval 2015 refined aspect definitions and removed the Conflict label, and SemEval 2016 expanded to new domains and languages (Dutch, French, Spanish). These datasets established a strong benchmark for subsequent ABSA studies.

In the context of the Vietnamese language, previous studies have focused on the problem of aspect extraction and sentiment classification, with less attention given to the span detection of user opinions. Unlike languages such as English and Chinese, which benefit from extensive resources, high-quality Vietnamese datasets remain scarce. For example, Mai et al. (2018) [12] developed a Vietnamese ABSA corpus comprising only 2,098 sentences, which concentrated on opinion target extraction (OTE) and sentiment classification. Similarly, the Vietnamese Language and Speech Processing (VLSP) community organized a shared task in 2018, but the resulting dataset mainly addressed entity-attribute aspect categories, overlooking the OTE task. In addition, the benchmark dataset for mobile e-commerce was developed and serves as a foundation for the UIT-ViSD4SA dataset. This dataset includes 35,396 spans across 11,122 smartphone feedback comments, annotated with ten aspect categories and corresponding sentiment polarities.

According to [6], the task of aspect-based sentiment analysis can be divided into three main sub-tasks. First, span extraction focuses on identifying the contiguous sequence of words within a sentence that corresponds to the opinion target, typically referring to a specific entity or feature being discussed. Second, aspect categorization involves classifying these extracted spans into a set of predefined aspect categories. Finally, polarity classification determines the sentiment orientation expressed toward each aspect, assigning it a label of positive, neutral, or negative.

2.2 Hierarchical Model in ABSA

Detecting text spans related to sentiment is a critical step in improving ABSA accuracy. Previous methods have often utilized deep learning models, such as the BiLSTM-CRF model [19], recurrent neural networks (RNNs) [10], or Transformer-based models like BERT [2], to enhance the accuracy of opinion span extraction.

Hierarchical models have demonstrated their effectiveness in the field of representation learning. Significant progress has been made in this area: [7] employed a hierarchical sequence-to-sequence model, enabling a neural auto-encoder to generate long text; [8] introduced the Feature Pyramid Network, which revolutionized object detection algorithms by leveraging the power of hierarchical representation learning. In sentiment analysis, [18] applied a hierarchical bidirectional LSTM to address the aspect-based problem. Furthermore, various studies have shown that large pre-trained language models, such as BERT and XLM, inherently learn hierarchical representations of language. For example, [9] and [17] provide evidence that BERT learns sentence representations hierarchically. In the era of deep learning, hierarchical modeling has become one of the most effective methods for representing linguistic input. These prior works have inspired our current research, which aims to maximize the potential of hierarchical models for joint task learning, specifically for span detection and aspect-based sentiment analysis.

3 Methodology

3.1 Problem Formulation

For handiness in our method, let $X = [x_1, x_2, .., x_n]$ denote a set of n review sentences. Each sentence x_i contains the tokens which will be labeled in BIO format, where $t_m = [t_{1_{x_i}}, t_{2_{x_i}}, ..., t_{m_{x_i}}]$ represents m tokens for each x_i review text and $t_{m_{x_i}} \in \mathbf{R}^d$ is a token embedding. Assume $Y = [y_1, y_2, ..., y_n]$ are the aspect term labels for the sentences review X where $y_i \in \{B, I, O\}$; The spans which are signified by $s_j = [s_{1_{x_i}}, s_{2_{x_i}}, ..., s_{j_{x_i}}]$ for j spans in every given review text. Suppose $Z = [z_{s_1}, z_{s_2}, ..., z_{s_j}]$ are the polarity sentiment labels for the spans in review x_i where $z_{s_i} \in \{POSITIVE, NEGATIVE, NEUTRAL\}$. These formulas will be used in our paper below.

3.2 Our Proposed Models

As illustrated in Fig. 1, our model includes the following components:

Aspect Architecture. The Aspect Architecture is designed as a comprehensive token classification model, consisting of several key components. First, following the approach in [6], each token t_m in a sentence is passed through an embedding layer and then encoded using the PhoBERT model [13] to produce a contextualized representation c_m. This embedding captures the semantic and syntactic context of each token within the sentence. Next, the Aspect Model processes these contextualized tokens using a pre-trained module to identify relevant aspects, with the results summarized in Table 4. To further enhance the

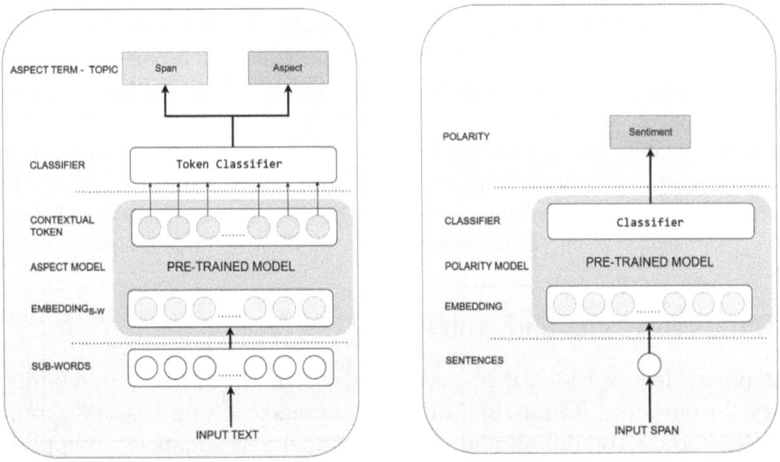

(a) Aspect Architecture (b) Polarity Architecture

Fig. 1. Illustrates the detailed structure of the model, consisting of two main components: **(a)** The Aspect Architecture processes input text by leveraging a pre-trained model to detect aspect terms and topics. **(b)** The Polarity Architecture takes the detected spans as input and classifies their sentiment using another pre-trained model and a dedicated classifier, effectively capturing sentiment polarity.

model's understanding, contextual tokens are incorporated, which significantly improve its ability to analyze sentiment by considering surrounding contextual cues. This allows for more nuanced and accurate sentiment predictions across different aspects. Finally, all contextualized tokens are fed into the PhoBERT-based classifier to perform span detection. In line with [19], a span is defined as the shortest text segment that captures the reviewer's opinion on a specific aspect category. The final output of the architecture consists of all such opinion spans identified in the input text.

Polarity Architecture. In this case, sentiment classification is a form of text classification, a fundamental task in natural language processing (NLP). Text classification involves assigning predefined labels or categories to text documents. This process enables the automated sorting and organization of textual data, facilitating the extraction of valuable information and insights from large volumes of text. Text classification is widely used in various applications, including sentiment analysis, spam detection, topic labeling, and document categorization. One of the most popular forms of text classification is sentiment analysis, where labels such as positive, negative, or neutral are assigned to a text span—in our study, this refers to a segment of text.

The overall model architecture (Fig. 3) follows a structured pipeline to extract aspects and determine their sentiment polarity from input comments. Initially, comments are processed by the Aspect module, fine-tuned using pre-trained models such as BERT Base Multilingual Cased [2], RoBERTa [11], and PhoBERT (Base and Large) [13], to identify relevant aspects like battery, price, or performance. The extracted aspect spans are then passed to the Polarity module, which uses XLNet [20] and PhoBERT Base [13] to classify sentiment as positive, negative, or neutral. Finally, the results are aggregated in the Aspect+Polarity Combination stage, producing structured outputs that link each aspect to its sentiment. This modular approach enables fine-grained sentiment analysis, offering deeper insights than traditional comment-level classification.

3.3 Data Processing and Annotation

During processing, we identified several issues in the dataset, including misspellings, inconsistent formatting, and the presence of unnecessary symbols or special characters. To address and enhance model performance, we applied specific rules to standardize the dataset (Fig. 2).

First, we performed data cleaning by removing unnecessary symbols and extra spaces. We aim for the model to be learned from natural languages that can be used in daily life. Therefore, we did not correct common spelling errors. Furthermore, we define the location where the character at the end of a span or the next character of the end word is a symbol. After that, the space could be added to these positions. This results in the data being more transparent and more reasonable to learn from. Finally, we transform the text into numerical representations using word embeddings to enhance feature extraction for aspect-based sentiment analysis.

Original Comment	Pre-processing text
Máy đẹp, chụp hình tốt. Pin sài 2 ngày hoặc có khi hơn. Sài lâu không nóng máy. Cảm ứng vân tay và mở khoa khuôn mặt nhanh. Mình dùng nhiều các phần mền ck, trả tiền qua ứng dụng như momo...mật khẩu chuyển qua cảm ứng vân tay và nhận diện khuôn mặt rất tiện lợi. Thêm nữa là chế độ bao mat khoa va an ứng dụng rất hay. Nhân viên tư vấn nhiệt tình, tận tâm.\n	Máy đẹp , chụp hình tốt . Pin sài 2 ngày hoặc có khi hơn . Sài lâu không nóng máy . Cảm ứng vân tay và mở khoa khuôn mặt nhanh . Mình dùng nhiều các phần mền ck , trả tiền qua ứng dụng như momo ... mật khẩu chuyển qua cảm ứng vân tay và nhận diện khuôn mặt rất tiện lợi . Thêm nữa là chế độ bảo mật khoá và ẩn ứng dụng rất hay . Nhân viên tư vấn nhiệt tình, tận tâm .
Beautiful device, good camera. Battery lasts 2 days or sometimes more. Long-term use does not heat up the device. Fingerprint sensor and face unlock are fast. I use a lot of transfer software(misspelling in vietnamese), pay via apps like momo ... password transfer to fingerprint sensor and face recognition is very convenient. In addition the security mode to lock and hide applications is very good. The consulting staff is enthusiastic and dedicated.\n	Beautiful device , good camera . Battery lasts 2 days or sometimes more . Long-term use does not heat up the device . Fingerprint sensor and face unlock are fast . I use a lot of transfer software(misspelling in vietnamese) , pay via apps like momo ... password transfer to fingerprint sensor and face recognition is very convenient . In addition the security mode to lock and hide applications is very good . The consulting staff is enthusiastic and dedicated .

Fig. 2. Examples illustrate the pre-processing step for aspect-based sentiment analysis in our proposal. The table compares the original comments (left column) with their pre-processed versions (right column). The pre-processing stage involves correcting misspellings **(red words)**, removing unnecessary characters **(red highlight)**, and segmenting to ensure consistency in sentiment analysis **(yellow highlight)**. Highlighted characters indicated the positions of our pre-processing. (Color figure online)

Training Process. To support our training process, we proceed to label the dataset in two tasks: span detection and text classification, each task running independently. In the span detecting task, we label the data in BIO format, and then we tokenize each sentence into tokens according to SPAN#ASPECT. In text classification, we label each span with these polarities: POSITIVE, NEGATIVE, and NEUTRAL. Table 1 describes our label in the training case:

Table 1. Training structure example

Category	Content
Given text	"Good product, smooth and worth the money, but I like the white color better. Enthusiastic staff"
Tokens	["Good", "product", ",", "smooth", "and", "worth", "the", "money", ",", "but", "I", "like", "the", "white", "color", "better", ".", "Enthusiastic", "staff"]
Labels using BIO tags	[B-GENERAL, I-GENERAL, I-GENERAL, B-PERFORMANCE, O, B-PRICE, I-PRICE, I-PRICE, I-PRICE, O, O, O, O, O, O, O, B-SER&ACC, I-SER&ACC]
Spans	["Good product,", "smooth and worth the money,", "but I like the white color better.", "Enthusiastic staff"]
Text classification	[POSITIVE, POSITIVE, NEGATIVE, POSITIVE]

These tasks will predict and give different results: (1) For span detection, our algorithms predict the specific aspect of each token in a span. Given a sentence S, for each token $t_i \in T$, we need to estimate the probability $P(a_i|t_i \in T)$ for the most likely correct aspect, where $a_i \in A$ is the aspect; (2) For text classification, this algorithm aims to determine the polarity in a span.

Note that our goal is to reach the best result in each task for better prediction. Furthermore, after training our models, we want to link these models, which inherit the best efficiency of each solution, to improve the prediction compared to [6]. To achieve this goal, we propose a hierarchical model that jointly learns span detection, aspect term extraction, and text classification linearly. Each progression gives an output; this output is then put into the input of the next model until the processing finishes.

Testing Process. In the testing phase, we evaluate the model's performance using a dataset labeled in format, following the ASPECT#POLARITY structure. This labeling scheme ensures a structured representation of aspects and their corresponding sentiment polarities, facilitating an effective evaluation of the model's ability to recognize and classify sentiment accurately. Figure 3 illustrates the overall architecture of our model, which processes input comments to perform aspect-based sentiment analysis (ABSA).

The Aspect Architecture identifies potential aspect terms within the input text, which extracts spans and opinions. The confirmed spans are subsequently

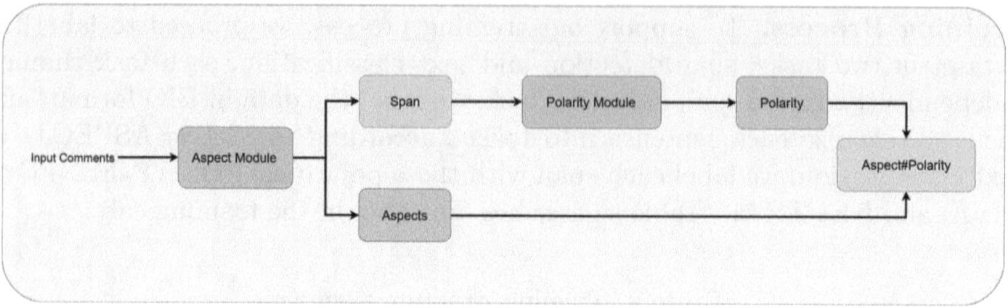

Fig. 3. The overall model proposed architecture for aspect-based sentiment analysis (ABSA), illustrating the sequential flow of processing modules. The model starts with input comments. The final classification step, Polarity, assigns sentiment labels before aggregating the results in the Aspect+Polarity output.

sent to the Polarity Module, where their sentiment polarities (e.g., positive, negative, or neutral) are determined. After sentiments are extracted, the output consolidates the aspect stored from the results of the Aspect Module, and the system outputs the final Aspect#Polarity pairs, representing each detected aspect along with its sentiment classification. Our testing example is shown as Table 2 below:

Table 2. Testing structure example

Category	Content
Given text	"Good product, smooth and worth the money, but I like the white color better. Enthusiastic staff"
Tokens	["Good", "product", ",", "smooth", "and", "worth", "the", "money", ",", "but", "I", "like", "the", "white", "color", "better", ".", "Enthusiastic", "staff"]
ASPECT#POLARITY tags	[GENERAL#POSITIVE, GENERAL#POSITIVE, GENERAL#POSITIVE, PERFORMANCE#POSITIVE, O, PRICE#POSITIVE, PRICE#POSITIVE, PRICE#POSITIVE, PRICE#POSITIVE, O, O, O, O, O, O, O, SER&ACC#POSITIVE, SER&ACC#POSITIVE]

4 Experiments

This section introduces the dataset, evaluation metrics, and results obtained from our proposed model on the UIT-ViSD4SA dataset. We first provide a detailed description of the dataset, followed by the pre-processing steps designed to enhance model performance. Our objective is to evaluate the effectiveness of the proposed approach in aspect-based sentiment analysis for Vietnamese-language e-commerce reviews.

In this study, we assess our proposal using the UIT-ViSD4SA dataset [19], a Vietnamese-language dataset containing 35,396 spans from 11,122 website comments. The dataset was gathered from an e-commerce platform in Vietnam, specifically for smartphone-related content, and was manually annotated based on 10 different aspects: SCREEN, CAMERA, FEATURES, BATTERY, PERFORMANCE, STORAGE, DESIGN, PRICE, GENERAL, and SER&ACC. The dataset is divided into three subsets: development (1,113 samples), training (7,784 samples), and testing (2,225 samples). There are 21,732 positive spans and 11,206 negative spans, while 2,214 spans are categorized as neutral. A detailed breakdown of the subsets (Train, Test, Development) can be found in Table 3.

Table 3. The overview of UIT-ViSD4SA dataset. This table provides a comprehensive summary of the UIT-ViSD4SA dataset, which is used for Vietnamese Span Detection and Aspect-Based Sentiment Analysis (ABSA). It details the distribution of data across three subsets: Train, Development, and Test, showing the number of comments, detected spans, and sentiment labels (Positive, Negative, Neutral).

Subset	Comment	Span	Positive	Negative	Neutral
Train	7,784	24,771	15,203	7,999	1,569
Development	1,113	3,582	2,296	1,062	224
Test	2,225	7,043	4,333	2,289	421
Total	11,122	35,396	21,832	11,350	2,214

4.1 Experimental Settings and Evaluation Metrics

All experiments were conducted on an RTX 4090 GPU with 24GB of VRAM. The implementation utilized the PyTorch deep learning framework, which is widely recognized for training deep learning models. The Adam optimizer was used with the following settings: a learning rate of 3e–5, β_1 set to 0.9, β_2 set to 0.999, and ϵ set to 1e–8.

We tested several transformer models for aspect detection models, including BERT Base Multilingual Cased, RoBERTa, PhoBERT Base, and PhoBERT Large. These models were configured with a linear decay learning rate scheduler, with a warmup set to 10% of the total training steps, a batch size of 32, 40 epochs for fine-tuning, gradient accumulation every four steps, a weight decay of 0.01, and a maximum input token length of 256.

We experimented with pretrained models like XLNet and PhoBERT Base for polarity classification, using hyperparameters similar to those of the aspect detection models. However, since this task's input length is shorter than the aspect detection task, we reduced the maximum sequence length from 256 to 128 tokens.

In this article, we use precision (P), recall (R), and F1 score (F1) as primary evaluation metrics. A predicted target is considered correct only if the predicted boundaries, aspect, and polarity exactly match the ground-truth target. We calculate these evaluation metrics on both the micro and macro averages to gain a wide view. Following prior studies, we adopt the micro F1-score as our standard evaluation setting.

4.2 Results and Discussions

Table 4 summarizes the overall experiments in this study, providing a comprehensive evaluation of different models applied to Aspect-Based Sentiment Analysis (ABSA) tasks. The performance of each model is assessed using multiple evaluation metrics, including micro-precision (P_{micro}), micro-recall (R_{micro}), micro-F1-score ($F1_{micro}$), macro-precision (P_{macro}), macro-recall (R_{macro}) and macro-F1-score ($F1_{macro}$).

The evaluation of our models was conducted across three subtasks: aspect extraction, polarity classification, and the combined task of aspect and polarity detection. For aspect extraction, baseline models such as BiLSTM and Hierarchical RoBERTa achieved moderate performance, with micro-F1 scores ranging from 65% to 67%. In contrast, our proposed hierarchical models, trained following the methodology introduced in this study, showed significant improvement. Notably, PhoBERT-Large outperformed all other models with a micro-F1 score of 82.63%, demonstrating its strong effectiveness in identifying aspects. For polarity classification, the XLNet-based classifier achieved the highest micro-F1 score of 82.84%, significantly surpassing the baseline models, which only achieved between 55% and 66%. PhoBERT-Base-Classification also yielded competitive results with a micro-F1 of 80.62%, marking it as one of the top-performing models in this category. Finally, in the combined task of aspect and polarity classification, baseline models like BiLSTM showed limited performance with micro-F1 scores around 61–62%. However, the combination of PhoBERT-Large and XLNet achieved the best result, reaching a micro-F1 score of 76.27%, clearly outperforming all other approaches in this more complex task.

Tables 5, 6, and 7 present the micro-F1 results for aspect classification, sentiment polarity classification, and their combination. The model excelled in Battery, Camera, and Performance (F1 $>$ 80%), but underperformed in Price and SER&ACC, with neutral sentiment proving most difficult (F1 48.76%).

For the combined aspect–polarity task, Battery and Storage achieved strong positive sentiment detection, while Price and SER&ACC struggled with negative sentiment (F1 $<$ 50%). Future work will address dataset imbalance, improve neutral sentiment recognition with contextual embeddings and sentiment-aware attention, adopt multi-task learning, and integrate external knowledge to enhance contextual understanding.

Table 4. The overall experimental results. This table presents the overall experimental results for different models on ABSA tasks, focusing on Aspect Extraction, Polarity Classification, and Combined Aspect + Polarity tasks. The performance of various models is evaluated using micro-precision, micro-recall, micro-F1, macro-precision, macro-recall, and macro-F1. The proposed models are highlighted in the table, demonstrating their superior performance compared to baseline models.

Model	P_{micro}	R_{micro}	$F1_{micro}$	P_{macro}	R_{macro}	$F1_{macro}$
	Aspect					
BiLSTM_CRF_Base	65.63%	65.15%	65.39%	62.88%	61.62%	62.17%
BiLSTM_CRF_Large	64.96%	66.85%	65.89%	62.00%	63.56%	62.76%
HierRoBERTa_SL	64.81%	67.26%	66.01%	61.69%	65.09%	63.31%
HierRoBERTa_ML	65.28%	67.50%	66.37%	63.24%	64.74%	63.91%
Our Proposal						
Bert_Base_Multilingual_Cased	79.79%	47.78%	59.92%	77.10%	47.33%	58.57%
Roberta	78.20%	77.23%	77.71%	76.36%	74.14%	75.18%
PhoBert_Base	**84.36%**	80.39%	82.32%	**82.42%**	**78.13%**	**80.18%**
PhoBert_Large	84.07%	**81.24%**	**82.63%**	81.03%	76.79%	78.89%
	Polarity					
BiLSTM_CRF_Base	54.88%	55.91%	55.39%	46.87%	46.39%	46.57%
BiLSTM_CRF_Large	56.89%	59.78%	58.30%	49.00%	50.60%	49.77%
HierRoBERTa_SL	64.64%	66.59%	65.60%	56.01%	57.47%	56.73%
HierRoBERTa_ML	65.26%	66.55%	65.90%	57.94%	57.34%	57.57%
Our Proposal						
Xlnet_Classification	**85.19%**	**80.62%**	**82.84%**	**76.00%**	70.22%	**72.92%**
PhoBert_Base_Classification	84.80%	80.25%	82.46%	75.36%	**70.36%**	72.73%
	Aspect + Polarity					
BiLSTM_CRF_Base	60.71%	61.62%	61.16%	46.18%	43.42%	44.37%
BiLSTM_CRF_Large	61.78%	62.99%	62.38%	46.84%	45.46%	45.70%
HierRoBERTa_SL	61.28%	64.01%	62.62%	50.89%	53.89%	51.66%
HierRoBERTa_ML	62.13%	64.16%	63.13%	53.91%	51.95%	52.06%
Our Proposal						
Bert_Base_Multilingual + Xlnet	68.24%	41.03%	51.25%	62.64%	32.26%	40.81%
Bert_Base_Multilingual + PhoBert	67.68%	40.70%	50.83%	55.42%	31.57%	39.42%
Roberta + Xlnet	70.79%	69.92%	70.35%	59.10%	58.34%	57.91%
Roberta + PhoBert	70.16%	69.29%	69.72%	58.42%	59.15%	57.71%
PhoBert_Base + Xlnet	**77.80%**	74.14%	75.93%	**68.63%**	62.27%	63.55%
PhoBert_Base + PhoBert	77.38%	73.74%	75.51%	67.96%	**63.47%**	**64.19%**
PhoBert_Large + Xlnet	77.60%	**74.99%**	**76.27%**	66.40%	62.35%	63.78%
PhoBert_Large + PhoBert	77.19%	74.59%	75.87%	65.53%	63.10%	63.46%

Table 5. Result per class for only aspect label.

Aspect	Precision	Recall	F1-score
BATTERY	90.45%	87.32%	88.86%
CAMERA	89.09%	87.77%	88.42%
GENERAL	80.89%	78.91%	79.89%
SER&ACC	81.23%	75.79%	78.42%
PERFORMANCE	85.46%	81.61%	83.49%
DESIGN	83.53%	85.50%	84.50%
FEATURES	83.15%	75.58%	79.18%
PRICE	62.67%	55.54%	58.89%
SCREEN	84.74%	77.26%	80.83%
STORAGE	82.99%	76.05%	79.37%

Table 6. Result per class for only sentiment polarity label.

Sentiment	Precision	Recall	F1-score
NEGATIVE	86.20%	79.09%	82.49%
NEUTRAL	53.77%	44.60%	48.76%
POSITIVE	88.02%	86.98%	87.50%

Table 7. F1score per class for aspect#polarity label.

Aspect	Negative	Neutral	Positive
BATTERY	80.62%	31.94%	81.92%
CAMERA	76.28%	51.89%	86.49%
GENERAL	63.68%	54.73%	80.77%
SER&ACC	43.57%	46.46%	85.57%
PERFORMANCE	75.96%	48.55%	82.39%
DESIGN	70.46%	41.54%	88.18%
FEATURES	75.31%	48.95%	76.26%
PRICE	49.46%	34.50%	54.74%
SCREEN	74.81%	40.00%	83.26%
STORAGE	74.38%	19.05%	84.65%

5 Conclusion

This research introduces a novel approach for joint span detection and aspect-based sentiment analysis. Using the UIT-ViSD4SA dataset, we evaluated the performance of various models for aspect-based sentiment analysis. Our results indicate that PhoBERT-based models consistently achieve the highest performance, particularly in detecting positive and negative sentiments, while BERT-based models struggle with lower recall and overall F1 scores. Despite the strong performance in some aspects, neutral sentiment detection remains a significant challenge, with considerably lower F1 scores than other sentiment classes. Our findings emphasize the importance of effective preprocessing and model fine-tuning in improving classification accuracy. Addressing these challenges could significantly enhance aspect-based sentiment analysis for Vietnamese text. Future research should explore data augmentation techniques and transformer-based architectures to further improve performance. Additionally, integrating external knowledge sources, leveraging pre-trained language models, and adopting semi-supervised or unsupervised learning approaches could help mitigate the limitations posed by data imbalance. Overall, this study demonstrates effective results, highlighting improved performance in ABSA for Vietnamese text.

Acknowledgement. Son T. Huynh was funded by the PhD Scholarship Programme of Vingroup Innovation Foundation (VINIF), code VINIF.2024.TS.006.

References

1. Chen, P., Sun, Z., Bing, L., Yang, W.: Recurrent attention network on memory for aspect sentiment analysis. In: Palmer, M., Hwa, R., Riedel, S. (eds.) Proceedings of the 2017 Conference on Empirical Methods in Natural Language Processing, pp. 452–461. Association for Computational Linguistics, Copenhagen (2017). https://doi.org/10.18653/v1/D17-1047. https://aclanthology.org/D17-1047/
2. Devlin, J., Chang, M.W., Lee, K., Toutanova, K.: BERT: Pre-training of deep bidirectional transformers for language understanding. In: Burstein, J., Doran, C., Solorio, T. (eds.) Proceedings of the 2019 Conference of the North American Chapter of the Association for Computational Linguistics: Human Language Technologies, vol. 1 (Long and Short Papers), pp. 4171–4186. Association for Computational Linguistics, Minneapolis (2019). https://doi.org/10.18653/v1/N19-1423. https://aclanthology.org/N19-1423/
3. Hu, M., Liu, B.: Mining and summarizing customer reviews. In: Proceedings of the Tenth ACM SIGKDD International Conference on Knowledge Discovery and Data Mining, KDD '04, pp. 168–177. Association for Computing Machinery, New York (2004). https://doi.org/10.1145/1014052.1014073
4. Jo, Y., Oh, A.H.: Aspect and sentiment unification model for online review analysis. In: Proceedings of the Fourth ACM International Conference on Web Search and Data Mining, WSDM '11, pp. 815–824. Association for Computing Machinery, New York (2011). https://doi.org/10.1145/1935826.1935932
5. Kiritchenko, S., Zhu, X., Cherry, C., Mohammad, S.: NRC-Canada-2014: detecting aspects and sentiment in customer reviews. In: Nakov, P., Zesch, T. (eds.) Proceedings of the 8th International Workshop on Semantic Evaluation (SemEval 2014), pp. 437–442. Association for Computational Linguistics, Dublin (2014). https://doi.org/10.3115/v1/S14-2076
6. Le, A.P., Pham, T.V., Le, T.V., Huynh, D.V., Ngo, D.: Robust hierarchical model for joint span detection and aspect-based sentiment analysis in Vietnamese. In: 2022 9th NAFOSTED Conference on Information and Computer Science (NICS), pp. 35–40 (2022). https://doi.org/10.1109/NICS56915.2022.10013463

7. Li, J., Luong, T., Jurafsky, D.: A hierarchical neural autoencoder for paragraphs and documents. In: Zong, C., Strube, M. (eds.) Proceedings of the 53rd Annual Meeting of the Association for Computational Linguistics and the 7th International Joint Conference on Natural Language Processing, vol. 1: Long Papers, pp. 1106–1115. Association for Computational Linguistics, Beijing (2015). https://doi.org/10.3115/v1/P15-1107. https://aclanthology.org/P15-1107/
8. Lin, T., Dollár, P., Girshick, R.B., He, K., Hariharan, B., Belongie, S.J.: Feature pyramid networks for object detection. CoRR arxiv:1612.03144 (2016)
9. Lin, Y., Tan, Y.C., Frank, R.: Open sesame: getting inside BERT's linguistic knowledge. In: Linzen, T., Chrupała, G., Belinkov, Y., Hupkes, D. (eds.) Proceedings of the 2019 ACL Workshop BlackboxNLP: Analyzing and Interpreting Neural Networks for NLP, pp. 241–253. Association for Computational Linguistics, Florence (2019). https://doi.org/10.18653/v1/W19-4825. https://aclanthology.org/W19-4825/
10. Liu, N., Shen, B.: Aspect-based sentiment analysis with gated alternate neural network. Know.-Based Syst. **188**(C) (2020). https://doi.org/10.1016/j.knosys.2019.105010
11. Liu, Y., et al.: Roberta: a robustly optimized bert pretraining approach (2019). https://arxiv.org/abs/1907.11692
12. Mai, L., Le, H.B.: Aspect-based sentiment analysis of Vietnamese texts with deep learning. In: Asian Conference on Intelligent Information and Database Systems (2018). https://api.semanticscholar.org/CorpusID:3750939
13. Nguyen, D.Q., Tuan Nguyen, A.: PhoBERT: pre-trained language models for Vietnamese. In: Cohn, T., He, Y., Liu, Y. (eds.) Findings of the Association for Computational Linguistics: EMNLP 2020, pp. 1037–1042. Association for Computational Linguistics, Online (2020). https://doi.org/10.18653/v1/2020.findings-emnlp.92
14. Pontiki, M., et al.: SemEval-2016 task 5: aspect based sentiment analysis. In: Bethard, S., Carpuat, M., Cer, D., Jurgens, D., Nakov, P., Zesch, T. (eds.) Proceedings of the 10th International Workshop on Semantic Evaluation (SemEval-2016), pp. 19–30. Association for Computational Linguistics, San Diego (2016). https://doi.org/10.18653/v1/S16-1002
15. Pontiki, M., Galanis, D., Papageorgiou, H., Manandhar, S., Androutsopoulos, I.: SemEval-2015 task 12: aspect based sentiment analysis. In: Nakov, P., Zesch, T., Cer, D., Jurgens, D. (eds.) Proceedings of the 9th International Workshop on Semantic Evaluation (SemEval 2015), pp. 486–495. Association for Computational Linguistics, Denver (2015). https://doi.org/10.18653/v1/S15-2082. https://aclanthology.org/S15-2082/
16. Pontiki, M., Galanis, D., Pavlopoulos, J., Papageorgiou, H., Androutsopoulos, I., Manandhar, S.: SemEval-2014 task 4: aspect based sentiment analysis. In: Nakov, P., Zesch, T. (eds.) Proceedings of the 8th International Workshop on Semantic Evaluation (SemEval 2014), pp. 27–35. Association for Computational Linguistics, Dublin (2014). https://doi.org/10.3115/v1/S14-2004. https://aclanthology.org/S14-2004/
17. Rogers, A., Kovaleva, O., Rumshisky, A.: A primer in BERTology: what we know about how BERT works. Trans. Assoc. Comput. Linguist. **8**, 842–866 (2020). https://doi.org/10.1162/tacl_a_00349. https://aclanthology.org/2020.tacl-1.54/
18. Ruder, S., Ghaffari, P., Breslin, J.G.: A hierarchical model of reviews for aspect-based sentiment analysis. In: Su, J., Duh, K., Carreras, X. (eds.) Proceedings of the 2016 Conference on Empirical Methods in Natural Language Processing, pp. 999–1005. Association for Computational Linguistics, Austin (2016). https://doi.org/10.18653/v1/D16-1103. https://aclanthology.org/D16-1103/

19. Thanh, K.N.T., Khai, S.H., Huynh, P.P., Luc, L.P., Nguyen, D.V., Van, K.N.: Span detection for aspect-based sentiment analysis in Vietnamese. In: Hu, K., Kim, J.B., Zong, C., Chersoni, E. (eds.) Proceedings of the 35th Pacific Asia Conference on Language, Information and Computation, pp. 318–328. Association for Computational Lingustics, Shanghai (2021)
20. Yang, Z., Dai, Z., Yang, Y., Carbonell, J., Salakhutdinov, R., Le, Q.V.: Xlnet: generalized autoregressive pretraining for language understanding (2020). https://arxiv.org/abs/1906.08237

Spatial-Spectral Prior Guided Mamba Network for Hyperspectral Image Super-Resolution

Guohua Miao, Zhihua Xie(✉), Haolin Chang, and Chenyu Tu

Key Lab of Advanced Electronic Materials, Jiangxi Science and Technology Normal University, Nanchang 330031, China
xie_zhihua68@aliyun.com

Abstract. Deep learning-based methods, particularly Mamba-based models, hold vast potential for hyperspectral image super-resolution (HSI-SR) reconstruction. However, the intrinsic spatial-spectral details in hyperspectral images are hard to be well represented solely utilizing the hierarchical learning manner in most Mamba structures, which results in the degradation of restoration quality on inherent fine-grain textures. To address this challenge, this paper proposes a Spatial-Spectral Prior Guided Mamba Network (SSP-Mamba) for HSI-SR, which leverages spatial-spectral priors in Wavelet domain to refine detail recovery with cross-layer embedment. Specifically, this model extracts multi-scale spatial dependences with a Wavelet transform and yields spectral responses with a spectral attention mechanism, which adaptively exploits specific bands to acquires an instructive spatial-spectral joint prior. The prior is then injected into hierarchical deep features from the backbone network—built upon a Shuffle-ReShuffle Mamba (SRM) module—through a spatial selection mechanism, thereby enabling the impressive feature representation capability. Extensive experiments on three representative datasets show that SSP-Mamba model achieves superior HSI-SR performance to other deep learning approaches in terms of the quantitative and qualitive metrics.

Keywords: Hyperspectral image super-resolution · Wavelet transform · Spatial selection mechanism · Mamba · Spatial-spectral priors

1 Introduction

Hyperspectral image (HSI), involving continuous and abundant spectral imagery with a wide coverage range, can reflect subtle response differences and material characteristics of targets across spectral bands [1], which holds significant application value in fields such as object classification [2], environmental monitoring [3], and medical imaging [4]. However, due to hardware limitations, the high spectral resolution often compromises the spatial resolution, which in turn undermines wide applications of HSI in downstream tasks. To alleviate the practical bottleneck caused by the relatively low spatial resolution, HSI super-resolution (HSI-SR) reconstruction technology has received considerable attentions in latest few decades.

Recently, the endeavors based on deep neural networks have emerged as a promising research hotspot for HSI-SR. Generally, mainstream works employ convolutional neural

J. Lokoč et al. (Eds.): MMM 2026, LNCS 16413, pp. 476–489, 2026.
https://doi.org/10.1007/978-981-95-6957-1_34

networks (CNN) or Transformer-based architectures as the backbone, such as SSPSR [5], MSFMNet [6], MSDformer [7], and CTUNet [8]. Although these solutions have shown encouraging performance, the intrinsic high dimensionality and redundancy of hyperspectral data still pose serious obstacles for the precise SR reconstruction. It is well known that CNN cannot effectively model long-range spatial and spectral dependencies. In contrast, Transformer offers strong global modeling capabilities through self-attention mechanisms. However, the quadratic computational complexity of Transformer restricts its scalability to high-dimension HSI data.

Fortunately, Mamba structure, built upon state space models, have attracted considerable attentions due to their linear computational complexity and powerful global modeling capabilities [9]. Mamba was originally designed for natural language processing, but has recently been adapted for 2D image modeling by Zhu et al. (VIM) [10] and Liu et al. (VMamba) [11], exhibiting promising merits. To alleviate the computational burden of processing HSI data, researchers have attempted to explore the application of Mamba into HSI-SR tasks. Zhi et al. [12] probed a lightweight network combining Mamba and Transformer for enhanced reconstruction performance with relatively fewer parameters. Zhu et al. [13] developed a super-resolution model combining CNN and Mamba, which effectively merges local and global information for good detail reconstruction. Xu et al. [14] designed spatial Mamba modules based on four-directional scanning and spectral Mamba modules based on bidirectional scanning, overcoming the challenge of detailed modeling in high-dimensional data. Xiao et al. [15] injected frequency-domain information via Fourier transform into Mamba framework for better feature representation. Overall, these methods typically transform HSI into a 1D sequence for modeling via bidirectional or four-directional scanning. Due to the causal nature of Mamba, these works struggle to capture comprehensive spatial dependencies [16].

Obviously, existing HSI-SR methods based on Mamba or Transformer usually exploits a cascaded feature extraction and reconstruction framework [7, 8, 13, 17], where spatial and spectral details inevitably degrades progressively as the network depth increases. As shallow structural information is gradually weakened, some high-frequency textures and edge details will be hard to maintain. Especially for Mamba network which require flattening images into 1D sequences and perform causal modeling, the original structural information is vulnerable to degradation. Based on this finding, Mamba's modeling capabilities on HSI-SR remains insufficient to ensure the completeness of fine details and the structural consistency of spatial-spectral dependences.

This paper proposes a novel HSI-SR model based on the Mamba network, named the Spatial-Spectral Prior guided Mamba Super-Resolution network (SSP-Mamba). The motivation of this approach is to interactively inject spatial-spectral priors in a cross-layer manner. Specifically, the model mainly comprises a backbone framework and a prior branch. The backbone employs a cascaded structure to progressively model deep features, utilizing Mamba modules and spectral attention mechanism to capture spatial-spectral dependencies. To alleviate the causal modeling limitations inherent in Mamba, we introduce a Shuffle operation to capture diverse spatial global dependencies. For prior information acquisition, we leverage the powerful frequency-domain separation capability of Discrete Wavelet transform (DWT) [18, 19]. By incorporating spatial and spectral attentions, the module adaptively adjusts information across different frequency bands, thereby capturing more discriminative frequency components and refining details that are easily overlooked in the hierarchical learning pipeline. Ultimately, the spatial selection mechanism is employed to fuse the prior branch features with the backbone's deep features. In summary, the main contributions of this paper are as follows:

- Built upon Mamba framework as the backbone, we incorporate spatial and spectral prior information to interactively guide the deep modeling process, facilitating the diverse details restoration.
- We design a prior extraction module that combines DWT and attention mechanism to explore spatial and spectral priors, supplementing textures that can be discarded in purely progressive learning paradigm.
- We construct a Shuffle-ReShuffle Mamba module (SRM) that integrates a Shuffle-based feature enhancement mechanism to aggregate the modeling of global spatial dependencies.

2 Methodology

2.1 Overall Framework

Most HSI-SR models employ a cascaded backbone architecture to progressively model spatial-spectral relations from local to global levels. However, such a layer-by-layer feature transmission paradigm inevitably leads to the gradual degradation of crucial but subtle details. This issue is further exacerbated in Mamba-based methods, where the transferring from1D sequence into 2D HSI may suffer from the loss of spatial-spectral structural features. To this end, our model aims to enhance the network's ability to model fine-grained details by injecting spatial-spectral priors with the help of DWT.

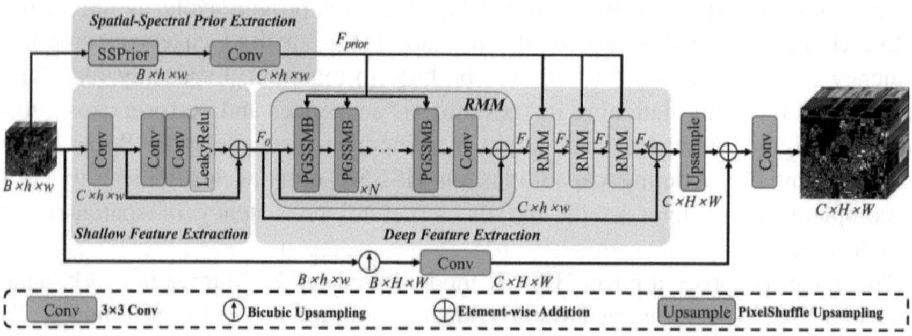

Fig. 1. Overall framework of the proposed network

The overall framework of SSP-Mamba model is illustrated in Fig. 1, which primarily includes four components: the shallow feature extraction, the spatial-spectral prior extraction, deep feature extraction, and image reconstruction.

For the convenience of description, we define the input as $X \in \mathbb{R}^{B \times h \times w}$ and the output as $Y \in \mathbb{R}^{B \times H \times W}$, where B is the number of spectral bands, $h \times w$denotes the spatial resolution of the low-resolution image, and $H \times W$ corresponds to the high-resolution output, where$H = s \times h$and$W = s \times w$, s represents the SR scale factor. The SR reconstruction process of SSP-Mamba can be formulated as:

$$Y = H_{\mathrm{SSP-Mamba}}(X) \tag{1}$$

where $H_{\mathrm{SSP\text{-}Mamba}}(\cdot)$ denotes the mapping function of the proposed scheme.

In the shallow feature extraction stage, we first apply a convolutional operation to the input low-resolution hyperspectral image X, projecting it into a feature space with C channels to enhance feature representation. Then, a residual convolutional module is used to extract shallow feature representations.

In the spatial-spectral prior extraction stage, we directly obtain feature maps from the input hyperspectral image X to retain as much of the original information as possible. The extracted prior features are then projected into C dimensional space before fed into subsequent modules.

In the deep feature extraction stage, four cascaded Residual Mamba Modules (RMM) are used to progressively mine features, with global residual connections to preserve structural information. Each RMM has two inputs, F_i, $i \in \{0, 1, 2, 3\}$ and F_{prior}.

$$F_{i+1} = H_{\mathrm{RMM}_{i+1}}\left(\ldots\left(H_{\mathrm{RMM}_1}\left(F_0, F_{\mathrm{prior}}\right)\ldots, F_{\mathrm{prior}}\right)\right) + F_0 \tag{2}$$

where $F_{i+1} \in \mathbb{R}^{C \times h \times w}$ denotes the $(i+1)$ - th output of the RMM module. Each RMM is composed of N Prior-Guided Spectral Shuffle-Mamba Block (PGSSMB)s followed by a convolutional layer, then incorporates module-level residual connections to prevent the loss of crucial information.

In the image reconstruction phase, the extracted deep features are up-sampled, and interact with the original input image via a global residual connection. Finally, a convolutional operation is applied to acquire the high-resolution output image Y.

2.2 Prior-Guided Spectral Shuffle-Mamba Block (PGSSMB)

To comprehensively fuse spatial-spectral features while effectively leveraging prior knowledge, we design the PGSSMB module, as shown in Fig. 2. Aiming to boost Mamba's capacity for capturing spatial-spectral dependencies, we incorporate targeted enhancement strategy into the Visual State Space (VSS) blocks. Although the four-directional scanning strategy partially alleviates the limitations of its receptive field and enhances contextual modeling, its ability to capture global spatial dependencies and feature interactions remains insufficient. Inspired by the endeavor in [20], we inject a spatial shuffle branch into the original Mamba in the cross-layer manner. This branch rearranges spatial features to facilitate inter-feature learning, thereby promoting the model's representation on spatial structures.

To further aggregate spectral modeling, a spectral attention mechanism (SpeA) is integrated in parallel within the core of PGSSMB. This mechanism highlights informative spectral bands, strengthening the network's capacity to model inter-band dependencies. The resulting deep features are then fed to the PFB module, where they are fused with spatial-spectral priors to cooperate in detail restoration. This detail operation is elaborated in the following section. Furthermore, to increase the network's adaptability, we introduce a residual modulation coefficient that dynamically adjusts the intensity of residual fusion.

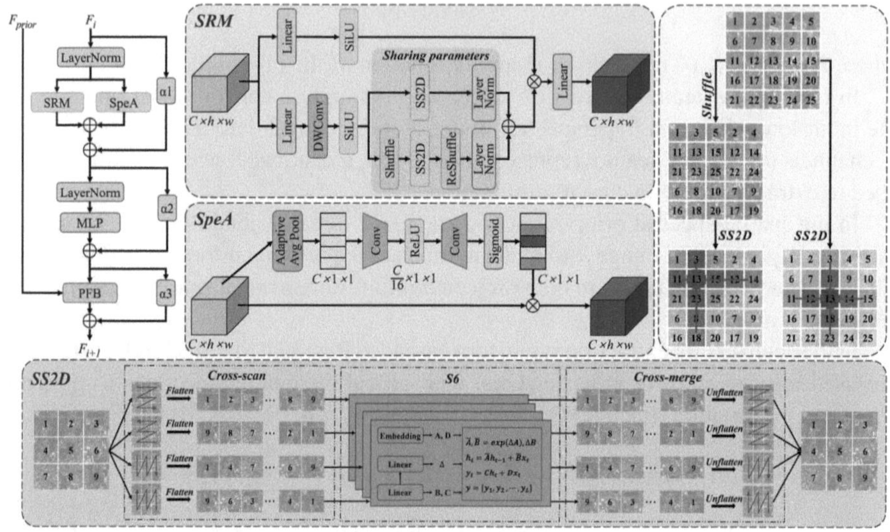

Fig. 2. The diagram of PGSSMB.

In SpeA module, the input feature F' first undergoes an adaptive average pooling along the spatial dimensions to yield a global response from each channel. This vector is then passed through a bottleneck structure composed of two fully connected layers for nonlinear transformation. Finally, a Sigmoid function generates spectral attention weights, which are multiplied channel-wise with the original feature to concentrate on informative spectral bands.

The SRM module consists of a gating-signal generation branch and a feature-extraction branch. Compared with the vanilla VSS block, the SRM module build an additional parallel feature enhancement branch based on a global cross-disruption strategy within the feature extraction pathway. This design enables the model to not only capture local relationships between adjacent pixels but also to account for long-range dependencies across distant pixels, as illustrated on the right side of Fig. 2. Under the fixed four-directional scanning scheme, correlations between the first and last elements can only be established after traversing all pixels. In contrast, the perturbation strategy allows the model to more efficiently capture potential dependencies among pixels at diverse spatial positions. Specifically, in the gating-signal generation branch, the features are first projected to a space of αC dimensions and activated to generate gating weights, where, α denotes the block expansion factor. In the feature-extraction branch, the feature is similarly projected to αC dimensions, then processed through a convolution filter, the Sigmoid-Weighted Linear Unit (SiLU) activation function, and two weight-sharing submodules to obtain global features. Among these submodules, one applies standard SS2D, while the other follows a Shuffle–SS2D–ReShuffle sequence. The outputs of the two submodules are normalized and fused via element-wise addition to yield the output of the feature-extraction branch.

2.3 Spatial-Spectral Prior (SSPrior) Block

The detail structure of the SSPrior block is depicted in Fig. 3. To make full use of the spatial-spectral nature of DWT [21], the core motivation of this block is that suitable processing strategies are undergone on specific frequency components in prior information generation. For the low-frequency components, we incorporate a spatial attention mechanism for the representation on expressive structural regions. In contrast, for the high-frequency components, convolutional operations are employed to concentrate on diverse textures and edge details, thereby accomplishing more targeted feature modeling. Additionally, learnable spectral attention weights highlight spatial features across channels, facilitating joint spatial-spectral representation. To further boost the stability and fidelity of prior representation, a global residual connection is added into the entire module to mitigate potential information loss during the modeling stage.

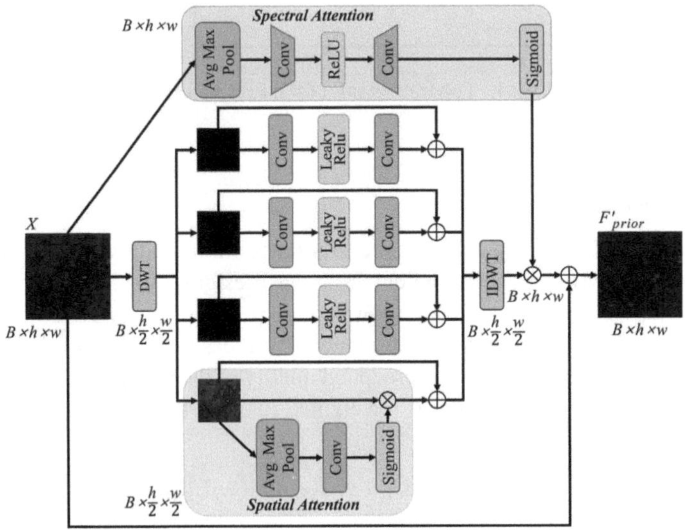

Fig. 3. The architecture of Spatial-Spectral Prior Block

2.4 Prior Fusion Block (PFB)

To adaptively select and integrate diverse information from both prior features and deep features, we design a Prior Fusion Block (PFB) inspired by the spatial selection mechanism [22], as illustrated in Fig. 4. Since prior and deep features may contain considerable redundant information, direct concatenation or addition can lead to noise accumulation. To mitigate this, the PFB first compresses the input features to extract high-level representations. To enable effective complementarity between the two feature sources, a spatial attention mechanism is introduced to adaptively modulate the high-level features. Specifically, the compressed features are concatenated and subjected to both global average pooling and max pooling to capture salient information from different statistical perspectives. The resulting descriptors are then processed by lightweight convolution layers and nonlinear mappings to generate two spatial attention maps, which are applied to the prior and deep features respectively to enhance or suppress their responses. Subsequently they are fused via element-wise addition. Finally, the fused features are aligned through a convolution layer and combined with the original deep features, preserving the semantic richness of the deep representations while enhancing spatial details, thus offering more robust and informative features for the subsequent reconstruction process.

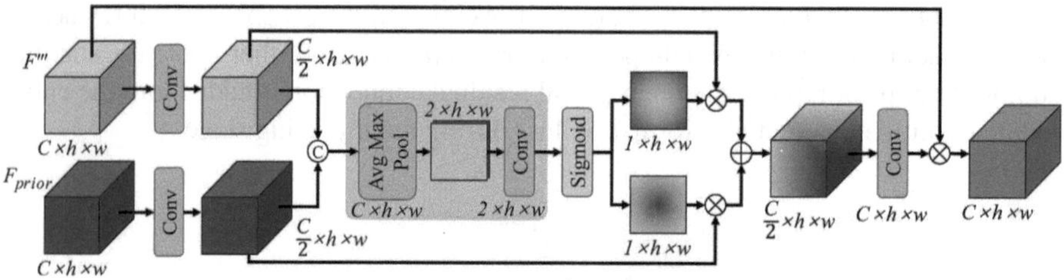

Fig. 4. The framework of the PFB module.

2.5 Loss Functions

Based on early studies [7, 8], a composite loss function, including the l_1 loss, spectral angle mapper (SAM) loss, and gradient loss, was designed to thoroughly assess the performance of SR reconstruction. Following this practice, this work also adopts the similar composite loss function to simultaneously measure the spectral consistency, spatial structural clarity, and overall perceptual quality of the restored HSI. Accordingly, the total loss function of our model is a weighted fusion from the three aspects, which is defined as follows:

$$\mathrm{L_{total}}(\theta) = \mathrm{L_1} + \lambda_1 \mathrm{L_{SAM}} + \lambda_2 \mathrm{L_{gra}} \tag{3}$$

$$\mathrm{L_1}(\theta) = \frac{1}{n}\sum_{u=1}^{n} \|Y_u - X_u\|_1 \tag{4}$$

$$\mathrm{L_{SAM}}(\theta) = \frac{1}{n}\sum_{u=1}^{n} \arccos\left(\frac{Y_u \cdot X_u}{\|Y_u\|_2 \|X_u\|_2}\right) \tag{5}$$

$$\mathrm{L_{gra}}(\theta) = \frac{1}{n}\sum_{u=1}^{n} \|M(Y_u) - M(X_u)\|_1 \tag{6}$$

where n denotes the number of images in a training batch, X_u and Y_u represent the u - th input and output hyperspectral images, respectively. $M(\cdot)$ denotes the horizontal, vertical, and spectral gradients of the input. λ_1 and λ_2 are the balance hyperparameters.

3 Experiments and Results

3.1 Datasets and Experimental Details

In this paper, three publicly hyperspectral datasets are exploited for evaluation experiments [24–26]. Following the dataset preprocessing strategy in CST [23], we crop all samples and split them into the training, validation, and test sets. For the Chikusei [24] dataset, we cropped four non-overlapping image patches with the size of $512 \times 512 \times 128$ from the top region of the image as the test set. In the remaining region, overlapping cropping was applied to increase the number of training samples. From the cropped patches, 10% were randomly selected for validation. When$s = 2$, patches with the size of $64 \times 64 \times 128$ were extracted with 32 pixels overlap, and $\times s$ bicubic interpolation down-sampling was applied to generate LR and corresponding HR image pairs. When$s = 4$, patches of size $128 \times 128 \times 128$ were extracted with 64 pixels overlap and similarly down-sampling. The Houston [25] and Pavia Center [26] datasets were processed in the similar manner. In SSP-Mamba network, the number of feature channels C in the backbone is set to 128. The backbone incorporates four RMM modules, and each module contains 3 PGSSMB blocks ($N = 3$). In the SRM module, the block expansion coefficient *expend* is set to 1 ($\alpha = 1$). Considering the loss function, we set $\lambda_1 = 0.5$ and $\lambda_2 = 0.1$. During the training stage, we employ the Adam optimizer with the batch size of 16 and train the model for 300 epochs. The initial learning rate is set to $1e^{-4}$ and is decayed to one-tenth of its original value after 150 epochs.

3.2 Comparison Results on Different Datasets

We conducted comparison experiments with several representative HSI-SR methods, including Bicubic, GDRRN [27], SSPSR [5], MSDformer [7], CST [23], MambaIRv2 [12], and FMSR [16]. For performance assessment, this work adopts popular evaluation metrics: peak signal-to-noise ratio (PSNR), structural similarity index (SSIM), and Spectral Angle Mapper (SAM). The comparison results are listed in Table 1. It can be observed that, except for the $\times 2$ upscaling case on the Houston dataset, SSP-Mamba achieves the best performance in all other experiments. Therefore, it is easy to conclude that our model consistently outperforms other competing methods in three metrics. An intuitive explanation is that our method presents significant advantages in maintaining spatial-spectral consistency, reconstructing image structure, and preserving spectral fidelity.

Additionally, to better evaluate the reconstruction quality of SSP-Mamba, we make visual comparisons with other methods under the $\times 4$ scale factor. The visualization results are depicted in Fig. 5. It is evident that the images restored by SSP-Mamba exhibit clearer edges and richer details, whereas the other methods suffer from various blurriness. Specifically, early methods such as Bicubic, GDRRN, and SSPSR produce the relatively obvious blurring effects. MSDformer, CST, MambaIRv2 and FMSR still show noticeable deficiencies in detail reconstruction, particularly in texture representation on mountains and rivers. The comparison analysis further validates the robustness and potential applicability of SSP-Mamba in realistic scenarios.

Table 1. Quantitative Comparison of Different Methods on the PaviaC, Chikusei, and Houston Datasets at ×2 and ×4 Scale Factors. (Best: Bold, Second Best: Underline)

Dataset	Scale	Metric	Bicubic	GDRRN	SSPSR	MSDformer	CST	MambaIRv2	FMSR	Ours
Chikusei	×2	PSNR	43.2125	46.6729	47.5679	47.7182	48.1442	47.6627	46.4708	**48.4770**
		SSIM	0.9721	0.9876	0.9897	0.9899	0.9909	0.9899	0.9866	**0.9915**
		SAM	1.7880	1.3509	1.2000	1.1394	1.0940	1.2984	1.4952	**1.0495**
	×4	PSNR	37.6377	39.3547	39.8927	40.0532	40.1682	39.8562	38.7227	**40.3065**
		SSIM	0.8954	0.9305	0.9384	0.9404	0.9424	0.9396	0.9171	**0.9444**
		SAM	3.4040	2.8737	2.5281	2.4019	2.3813	2.6361	3.0433	**2.2947**
Houston	×2	PSNR	50.6722	52.8606	53.1056	53.2426	**54.0285**	53.1169	51.4677	53.9736
		SSIM	0.9925	0.9957	0.9959	0.9960	**0.9967**	0.9961	0.9944	0.9967
		SAM	1.1256	0.9549	0.9344	0.9100	0.8373	1.0098	1.2353	**0.8305**
	×4	PSNR	44.1937	45.1803	46.8931	47.2795	47.7286	47.1889	46.0915	**47.7580**
		SSIM	0.9658	0.9754	0.9823	0.9832	0.9849	0.9834	0.9787	**0.9849**
		SAM	2.2090	2.3091	1.8239	1.5656	1.5178	1.7567	1.9011	**1.5048**
PaviaC	×2	PSNR	32.1036	34.8543	35.7196	35.9113	36.1186	35.7640	35.3820	**36.3016**
		SSIM	0.9122	0.9506	0.9603	0.9615	0.9629	0.9584	0.9561	**0.9646**
		SAM	4.6025	4.4787	3.6513	3.6090	3.5856	3.9388	3.8514	**3.4763**
	×4	PSNR	27.5535	28.6974	28.9923	29.0120	29.1281	28.9789	28.4594	**29.2630**
		SSIM	0.7233	0.7969	0.8124	0.8139	0.8163	0.8118	0.7868	**0.8247**
		SAM	6.5161	6.4767	5.6688	5.7576	5.7527	6.2143	6.9738	**5.5012**

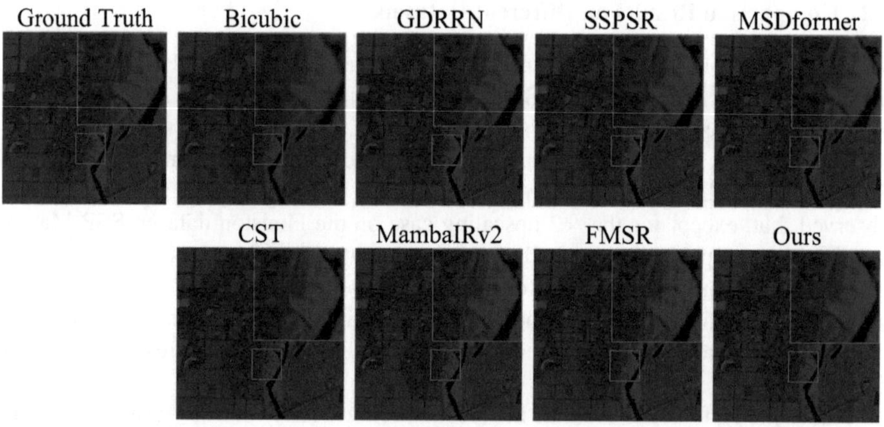

Fig. 5. Visual quality comparison on the Chikusei dataset at ×4 Scale Factor. Images are displayed with bands 43, 87, and 21 as R, G, and B channels.

3.3 Ablation Study

In this section, we conduct a series of ablation experiments based on the ×4 super-resolution task on the representative Chikusei dataset to verify the contribution of the key modules and parameters in proposed network.

The Structural Effectiveness. Considering the overall structure of SSP-Mamba, we deliberatively analysis the impact of the SRM module, prior module, and PFB module on SR performance. The quantitative results are reported in Table 2.

Table 2. Ablation Study on different modules (The bold fonts indicate the best results)

Architectures	PSNR↑	SSIM↑	SAM↓
(a) SRM w/o Shuflle	40.2443	0.9440	2.2981
(b) PGSSMB w/o PFB	40.2714	0.9437	2.2965
(c) Replacing SSPrior with Conv	40.2880	0.9442	**2.2884**
(d) Replacing PFB with Conv	40.0045	0.9404	2.3891
(e) Replacing PFB with Residual Attention	40.2954	0.9444	2.2918
(f) SSP-Mamba	**40.3065**	**0.9444**	2.2947

To ensure fairness, the comparative experiments were performed by removing or replacing the corresponding modules while keeping the rest of the architecture unchanged. Specifically, the model (a) removes the Shuffle branch in the SRM module; the structure (b) removes the PFB module, i.e., no prior information; the structure (c) replaces the spatial-spectral prior module with a standard convolution module; the structure (d) replaces the PFB module with a general convolution addition operation; the structure (e) replaces the PFB module with a residual spatial attention module; the structure (f) is the complete model with all modules. From the comparison results in Table 2, it can be seen removing or replacing these modules leads to variable degrees of performance degradation, indicating that the SRM module, prior module, and PFB module consistently make critical contributions in reconstruction process. Their collaborative design effectively improves the SR restoration quality of the overall model, confirming the rationality and effectiveness of our proposed network structure.

The Number of PGSSMB. The number of PGSSMB modules also is one of the key factors on the performance. Excessive modules significantly increase computational resource consumption and may lead to overfitting, while few modules hinder the feature extraction capability, resulting in decreased network performance. The related experimental results are reported in Table 3. It can be observed that as the number of modules increases, the evaluation metrics of the reconstructed images gradually improve. However, when the number reaches a certain point, the performance begins to decline. Considering both performance and efficiency, we set the number of PGSSMB modules to 3 to achieve an optimal balance between reconstruction quality and computational cost.

The Loss Function. To further investigate the specific contributions of each loss component during the model training stage, we conducted ablation experiments focused on the loss function. By progressively adjusting different combinations of loss functions, we inspect their impacts on the SR outcome.

Table 3. Ablation on the Number of PGSSMB (The bold fonts indicate the best results)

Parameter value	PSNR↑	SSIM↑	SAM↓
N = 1	40.1561	0.9426	2.3219
N = 2	40.2378	0.9435	2.3109
N = 3	**40.3065**	**0.9444**	**2.2947**
N = 4	40.1241	0.9429	2.3160

Table 4. Ablation Study on Loss Functions (The bold fonts indicate the best results)

$\mathscr{L}_1$	$\mathscr{L}_{SAM}$	$\mathscr{L}_{gra}$	PSNR↑	SSIM↑	SAM↓
√	×	×	40.1426	0.9426	2.3930
√	√	×	40.3036	0.9442	**2.2876**
√	×	√	40.0258	0.9414	2.4136
√	√	√	**40.3065**	**0.9444**	2.2947

Table 4 lists the ablation results on different loss functions. The experimental results indicates that integration of the SAM loss to the basic loss function significantly improves model performance, while the introduction of alone gradient loss causes a slight decline. Notably, all three loss components are ensembled, the model achieves the best performance, confirming the complementary strengths of three loss components in HSI-SR tasks.

The Residual Modulation Coefficient. To access the necessity of three residual modulation coefficients, we conduct ablation experiments by enabling them individually and in pairwise combinations. The results are listed in Table 5.

Table 5. Ablation Study on Residual Modulation Coefficient (The bold fonts indicate the best results)

$\alpha 1$	$\alpha 2$	$\alpha 3$	PSNR↑	SSIM↑	SAM↓
×	×	×	39.9179	0.9391	2.4132
√	×	×	40.0363	0.9402	2.3822
×	√	×	40.1804	0.9434	2.3336
×	×	√	39.9481	0.9399	2.4350
√	√	×	40.2577	0.9435	2.3035
√	×	√	40.1126	0.9423	2.3726
×	√	√	40.2867	0.9445	2.3035
√	√	√	**40.3065**	**0.9444**	**2.2947**

It can be observed that the network performance degrades significantly without any residual modulation coefficients. When only a single coefficient is enabled, the performance is better than that of the model without any coefficients. When adopting pairwise combinations, the network performance remains inferior to the full configuration. To sum up, the network achieves the best performance when all three coefficients are activated simultaneously.

3.4 Model Complexity Analysis

To evaluate the complexity and computational efficiency, the parameters computational cost (FLOPs) and GPU memory usage under a single batch size are visualized alongside the corresponding PSNR. Figure 6 shows the complexity and computational efficiency comparisons of different models for the ×4 super-resolution task on Chikusei dataset. From Fig. 6, GDRRN, as an earlier method, exhibits the lowest computational overhead and resource consumption but falls short in reconstruction image quality. Schemes like SSPSR, MSDformer, CST and FMSR improve performance by incorporating specific complex frameworks, at the cost of heavy resource demands. Although MambaIRv2 is lightweight in terms of computational cost as RGB image super-resolution methods, its performance remains somewhat limited in hyperspectral scenarios. SSP-Mamba achieves superior reconstruction quality while maintaining relatively low computational and memory requirements. Briefly, our model obtains a good balance between SR performance and computational efficiency.

(a) Comparison of PSNR Parameters and FLOPs (b) Comparison of PSNR Parameters and Memory Usage

Fig. 6. Comparisons of Model Complexity

4 Conclusion

This paper proposes SSP-Mamba for hyperspectral image super-resolution. We devise the SRM module by incorporating a shuffle branch into the Mamba network to improve spatial modeling, and exploit a spectral attention mechanism to fully extract spectral relations. To mitigate detail loss during feature propagation, the model leverages DWT and spectral attention to form informative priors. Furthermore, the prior fusion block is embedded to adaptively fuse prior information with deep features in cross-layer manner. Extensive experimental results demonstrate that SSP-Mamba outperforms mainstream HSI-SR methods in reconstruction accuracy, structural detail preservation, and spectral consistency, achieving the better overall SR performance.

References

1. Wang, X., Hu, Q., Cheng, Y., Ma, J.: Hyperspectral image super-resolution meets deep learning: a survey and perspective. IEEE/CAA J. Automatica Sin. **10**(8), 1668–1691 (2023)
2. Li, Y., Luo, Y., Zhang, L., Wang, Z., Du, B.: MambaHSI: spatial–spectral mamba for hyperspectral image classification. IEEE Trans. Geosci. Remote Sens. **62**, 1–16 (2024)
3. Leung, J.-H., et al.: Water pollution classification and detection by hyperspectral imaging. Opt. Express. **32**, 23956–23965 (2024)
4. Chang, Y.-C., et al.: Detection of white matter Hyperintensities in magnetic resonance imaging by hyperspectral subpixel detection. IEEE Access. **12**, 111992–112007 (2024)
5. Jiang, J., Sun, H., Liu, X., Ma, J.: Learning spatial-spectral prior for super-resolution of hyperspectral imagery. IEEE Trans. Comput. Imaging. **6**, 1082–1096 (2020)
6. Zhang, J., Shao, M., Wan, Z., Li, Y.: Multi-scale feature mapping network for hyperspectral image super-resolution. Remote Sens. **13**(20), 4180 (2021)
7. Chen, S., Zhang, L., Zhang, L.: MSDformer: multiscale deformable transformer for hyperspectral image super-resolution. IEEE Trans. Geosci. Remote Sens. **61**, 1–14 (2023)
8. Tang, Y., Yue, L., Liu, X., Li, Y., Xiao, Y., Yuan, Q.: A CNN-transformer embedded unfolding network for hyperspectral image super-resolution. IEEE Trans. Geosci. Remote Sens. **62**, 1–16 (2024)
9. Albert, G., Dao, T.: Mamba: linear-time sequence modeling with selective state spaces. 2312.00752 (2023)
10. Zhu, L., Liao, B., Zhang, Q., Wang, X., Liu, W., Wang, X.: Vision mamba: efficient visual representation learning with bidirectional state space model. In: Proceedings of the 41st International Conference on Machine Learning PMLR, vol. 235, pp. 62429–62442 (2024)
11. Liu, Y., et al.: Vmamba: Visual State Space Model. Adv. Neural Inf. Proces. Syst. **37**, 103031–103063 (2024)
12. Zhi, R., Fan, X., Shi, J.: MambaFormerSR: a lightweight model for remote-sensing image super-resolution. IEEE Geosci. Remote Sens. Lett. **21**, 1–5 (2024)
13. Zhu, Q., Zhang, G., Zou, X., Wang, X., Huang, J., Li, X.: ConvMambaSR: leveraging state-space models and CNNs in a dual-branch architecture for remote sensing imagery super-resolution. Remote Sens. **16**(17), 3254 (2024)
14. Xu, Y., et al.: MambaHSISR: mamba hyperspectral image super-resolution. IEEE Trans. Geosci. Remote Sens. **63**, 1–16 (2025)
15. Xiao, Y., Yuan, Q., Jiang, K., Chen, Y., Zhang, Q., Lin, C.-W.: Frequency-assisted mamba for remote sensing image super-resolution. IEEE Trans. Multimed. **27**, 1783–1796 (2025)
16. Guo, H., et al.: MambaIRv2: attentive state space restoration. In: Proceedings of the IEEE/CVF Conference on Computer Vision and Pattern Recognition, pp. 28124–28133 (2025)
17. Zhang, M., et al.: Spatial-spectral aggregation transformer with diffusion prior for hyperspectral image super-resolution. IEEE Trans. Circuits Syst. Video Technol. **35**(4), 3557–3572 (2025)
18. Zhao, C., Cai, W., Dong, C., Hu, C.: Wavelet-based Fourier information interaction with frequency diffusion adjustment for underwater image restoration. In: Proceedings of the IEEE/CVF Conference on Computer Vision and Pattern Recognition, pp. 8281–8291 (2024)

19. Wang, N., Mei, S., Wang, Y., Zhang, Y., Zhan, D.: WHANet: wavelet-based hybrid asymmetric network for spectral super-resolution from RGB inputs. IEEE Trans. Multimed. **27**, 414–428 (2025)
20. Deng, S.-Q., Deng, L.-J., Wu, X., Ran, R., Hong, D., Vivone, G.: PSRT: pyramid shuffle-and-Reshuffle transformer for multispectral and hyperspectral image fusion. IEEE Trans. Geosci. Remote Sens. **61**, 1–15 (2023)
21. Zou, W., Gao, H., Yang, W., Liu, T.: Wave-mamba: wavelet state space model for ultra-high-definition low-light image enhancement. In: Proceedings of the 32nd ACM International Conference on Multimedia, pp. 1534–1543 (2024)
22. Li, Y., et al.: LSKNet: a foundation lightweight backbone for remote sensing. Int. J. Comput. Vis. **133**, 1410–1431 (2025)
23. Chen, S., Zhang, L., Zhang, L.: Cross-scope spatial-spectral information aggregation for hyperspectral image super-resolution. IEEE Trans. Image Process. **33**, 5878–5891 (2024)
24. Yokoya, N., Iwasaki, A.: Airborne Hyperspectral Data over Chikusei. Space Applications Laboratory, University of Tokyo, Japan., Technical Report SAL-2016-05-27 (2016)
25. Prasad, S., Saux, B.L., Yokoya, N., Hansch, R.: 2018 IEEE GRSS data fusion challenge – fusion of multispectral LiDAR and hyperspectral data. IEEE Dataport. (2020)
26. Huang, X., Zhang, L.: A comparative study of spatial approaches for urban mapping using hyperspectral ROSIS images over Pavia City, northern Italy. Int. J. Remote Sens. **30**(12), 3205–3221 (2009)
27. Li, Y., Zhang, L., Ding, C., Wei, W., Zhang, Y.: Single hyperspectral image super-resolution with grouped deep recursive residual network. In: 2018 IEEE Fourth International Conference on Multimedia Big Data, pp. 1–4. IEEE, Xi'an (2018)

MotionSlim: A Lightweight T2M Generation Framework Based on LLM

Congrui Yu[1,2(✉)], Bo Fan[3], and Na Lyu[1,2]

[1] School of Information Science and Engineering, University of Jinan, Jinan, Shandong 250022, China
[2] Shandong Key Laboratory of Ubiquitous Intelligent Computing, Jinan, Shandong 250022, China
[3] Jinan Yinhua Information Technology Co., Ltd, Jinan, Shandong 250014, China

Abstract. In recent years, a growing number of works on human motion generation based on LLMs (Large Language Models) have emerged, achieving substantial improvements compared to traditional methods. However, due to the reliance on LLMs, most existing approaches involve massive parameter scales, which lead to high training costs, large storage overhead, and slow inference speed. To address these challenges, we propose a lightweight human motion generation framework based on LLMs, incorporating SFT (Supervised Fine-tuning) and reinforcement learning with RLOO (REINFORCE Leave-One-Out). To the best of our knowledge, this is the first work that introduces the SFT+RLOO paradigm into the LLM-based motion generation field. Our lightweight framework reduces the parameter scale to 1.5B. Experimental results demonstrate that our method effectively reduces computational cost during training and inference, significantly lowers storage requirements, and achieves performance that is comparable to or even surpasses SOTA (state-of-the-art) methods across major metrics.

Keywords: Human motion generation · Large language model · Reinforcement Learning · Lightweight Model

1 Introduction

With the rapid development of deep learning, T2M(Text-to-motion Generation) has become an important research direction in computer vision. This task aims to automatically generate realistic 3D human motion sequences from natural language descriptions, with applications spanning virtual reality, game animation, and film production.

In recent years, LLMs such as LLaMA [1] and DeepSeek [2] have made groundbreaking progress. Their strong capabilities in language understanding and reasoning have also driven advances in T2M. In particular, works such as MotionGPT [3] have leveraged the translation ability of LLMs to make significant progress in high-fidelity motion generation aligned with complex textual instructions, especially significant progress in long-sequence motion generation.

J. Lokoč et al. (Eds.): MMM 2026, LNCS 16413, pp. 490–503, 2026.
https://doi.org/10.1007/978-981-95-6957-1_35

The T2M methods based on LLM have powerful performance, while they have a large number of parameters. The existing LLM-based T2M approaches often rely on very large models (e.g., Gemma2-9B, LLaMA-13B, or even larger), which substantially increase the computational, storage, and time costs during training and inference. Simply replacing a large LLM with a smaller one may greatly reduce resource requirements but typically results in a catastrophic drop in generation quality (Sect. 4.1).

Meanwhile, in the area of LLMs, reinforcement learning, for example RLOO [4], has been shown to effectively optimize the learning of models, which enables smaller LLMs to achieve performance comparable to their larger counterparts.

Motivated by these, we propose MotionSlim, a lightweight text-to-motion generation framework that leverages a small-scale LLM fine-tuned by SFT and optimized by RLOO. By combining the lightweight advantages of small-scale LLMs, supervised fine-tuning and targeted optimization through RLOO, our experiments show that our method achieves performance close to or better than current SOTA approaches, while drastically reducing parameter size.

Our main contributions can be summarized as follows:

1) We propose MotionSlim, a novel lightweight text-to-motion generation framework that effectively leverages small-scale LLMs to achieve performance competitive with larger models, while significantly reducing computational overhead and maintaining high motion quality.

2) We design a reinforcement learning module based on the RLOO algorithm to further optimize supervised fine-tuned LLMs, effectively compensating for the performance degradation typically induced by model compression. To the best of our knowledge, this is the first successful application of RLOO in the field of text-to-motion generation.

3) We introduce a novel motion sequence length reward function that encourages the model to generate motions with temporal lengths more closely aligned with ground-truth samples. This innovation enhances the realism and temporal consistency of the generated motions.

4) Extensive experimental evaluations demonstrate that our approach achieves results comparable to or even surpassing those of large-parameter models across multiple benchmarks, all while operating with substantially fewer parameters. Furthermore, when using similarly lightweight LLMs, our approach significantly outperforms existing methods across multiple metrics, demonstrating superior effectiveness under constrained model size.

2 Related Works

2.1 Large Language Models and Reinforcement Learning

Recently, LLMs such as BERT [5], T5 [6], GPT [7], LLaMA [1], and DeepSeek [2] have developed rapidly. Traditional large-scale LLMs typically contain over 10 billion parameters, resulting in extremely high training and deployment costs. However, in recent years, a series of lightweight LLMs, such as Qwen2.5-1.5B and Hunyuan-1.8B, have been released. While their performance is somewhat reduced, they significantly lower deployment costs, requiring only a single consumer-grade GPU for fine-tuning or inference, eliminating the need for expensive multi-GPU servers and making LLMs more accessible for practical applications.

Fine-tuning is a transfer learning technique that adapts pre-trained LLMs to specific tasks by further training them on smaller, task-oriented datasets. Common fine-tuning approaches include LoRA(Low-Rank Adaptation) [8] and SFT [9], among others. LoRA avoids updating the entire model by injecting trainable low-rank matrices into linear layers, thus providing a parameter-efficient and storage-friendly alternative for large LLMs. In contrast, SFT adopts an end-to-end full-parameter update strategy: building upon pre-trained weights, it directly uses high-quality *instruction–response* pairs for multiple rounds of supervised training, guiding the model's overall parameters to converge toward human-preferred outputs via cross-entropy loss. Ouyang et al. [9] demonstrated that SFT can enable smaller LLMs to achieve performance competitive with significantly larger models.

In existing work, RL(Reinforcement Learning) is usually used to further optimize LLMs. Common RL methods include PPO [10], DPO [11], DAPO [12], GRPO [13] and RLOO [4]. PPO [10] first combined policy gradient clipping with KL(Kullback-Leibler) constraint, establishing a stable training paradigm for large-scale online policy optimization of language models; DPO [11] bypasses explicit reward models by formulating preference probabilities as differentiable losses, opening a concise path for offline one-step alignment; DAPO [12] builds on DPO [11] by introducing adaptive weighting to dynamically weaken the influence of noisy preferences on the policy; GRPO [13] replaces absolute rewards with relative rankings within groups, greatly reducing computational resource consumption. RLOO [4] is a simpler and more novel RL method. Other mainstream RL methods (e.g., those based on PPO [10] or GRPO [13]) typically treat each generated token as an action and each partial sequence as a state. In contrast, RLOO [4] is based on REINFORCE and treats the entire complete sequence as a single action, which has the advantages of simplicity and directness, avoids complex value-function estimation, saves computing resources, and better suits some end-to-end downstream tasks. Building on REINFORCE, Ahmadian et al. [4] further introduced a leave-one-out mechanism: for multiple candidate responses generated each time, the baseline for each response is set to the average score of all other responses, resulting in more accurate gradient estimates.

2.2 Text-Based Human Motion Generation

Currently, generation from textual descriptions remains the most widely studied task in the field of human motion generation. Various traditional approaches have attempted to address this challenge through different methodologies. Methods such as Text2Action [14], Ghosh et al. [15], and Qi et al. [16] utilized

GANs(Generative Adversarial Networks) for T2M generation, though their effectiveness was constrained by limitations such as insufficient motion diversity. More recently, diffusion-based models including MDM [17], PriorMDM [18], EMDM [19] and MotionDiffuse [20] have been introduced for T2M generation. These methods treat human motion as a continuous signal and employ iterative denoising processes to produce high-fidelity and diverse motion sequences. Despite their promising performance, diffusion-based approaches often depend on complex architectures and offer limited flexibility in controlling motion length. Alternatively, Guo et al. [21], Hong et al. [22], TEMOS [23], TM2T [24], and Athanasiou et al. [25] adopted VQ-VAE (Vector Quantized Variational Autoencoders) to discretize continuous motion representations into tokens via a learned codebook, thereby facilitating efficient sequence modeling and improving alignment with textual semantics.

In recent years, a growing number of LLM-based T2M methods have emerged, including MotionGPT [3], MotionAgent [26], and MotionGPT-2 [27]. Building upon the traditional VQ-VAE framework, these approaches treat each codebook index as a "motion word," thereby enabling LLMs to perform translation between natural language and a structured "motion language." MotionR1 [28] further introduced CoT (Chain-of-Thought) reasoning into T2M generation and designed a reinforcement learning optimization module based on GRPO [13]. However, these methods commonly rely on LLMs with 8B, 13B, or even larger parameter counts, resulting in significantly increased costs in training, inference, and storage. The substantial parameter scale not only heightens sensitivity to computational resources during training and fine-tuning but also raises practical barriers to deployment. Consequently, maintaining—or even enhancing—generation quality while compressing model size has become a critical and urgent research challenge.

3 Methods

Our proposed **MotionSlim** framework comprises three core modules (as illustrated in Fig. 1): (1) a VQ-VAE-based motion encoder–decoder, (2) a lightweight LLM trained with SFT, and (3) a reinforcement learning module based on the RLOO algorithm.

Specifically, the motion encoder quantizes input human motion sequences into discrete motion tokens using a learned codebook. The index of each token in the codebook is treated as a *motion ID* represented in text form. These motion IDs collectively form a "motion vocabulary," which is utilized in subsequent fine-tuning and reinforcement learning stages (see Sect. 3.1). We then employ SFT to train the lightweight LLM to perform translation between natural language text and sequences of motion IDs (see Sect. 3.2). This adapted model is referred to as the *Lightweight Motion LLM*. Furthermore, we apply reinforcement learning training to refine the model's output distribution, reducing the probability of incorrect or undesirable outputs while reinforcing factually and contextually appropriate responses (see Sect. 3.3).

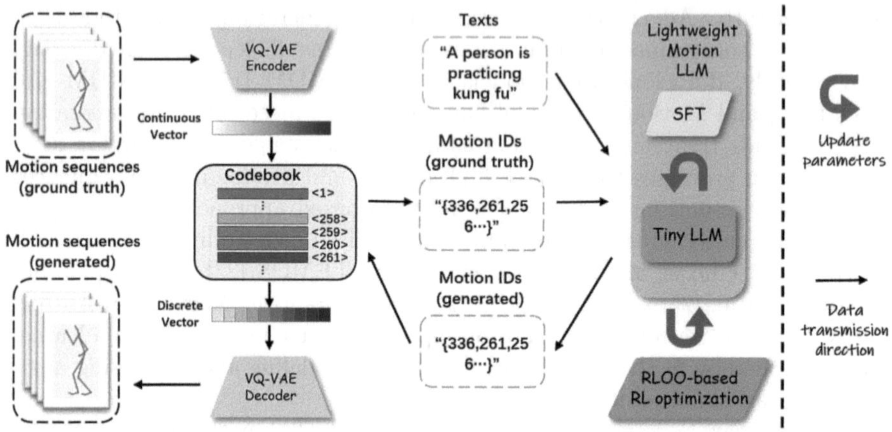

Fig. 1. The pipeline of our MotionSlim method.

3.1 Motion Encoder and Decoder

To achieve mutual conversion between human motion sequences and motion tokens, we adopt the VQ-VAE architecture, which has been widely used in prior human motion generation works such as MoMask and MotionGPT [3]. We pre-train a motion encoder and a motion decoder within this framework. The motion encoder E_m converts an input motion sequence $m_{1:T}$ into a sequence of discrete motion tokens, while the motion decoder reconstructs the motion sequence $\hat{m}_{1:T}$ from these tokens. This module enables effective bidirectional conversion between motion sequences and motion token ID sequences. The resulting motion ID sequences are subsequently used in SFT (Sect. 3.2) and RL training (Sect. 3.3).

Formally, given a motion sequence $m_{1:T} \in \mathbb{R}^{T \times D}$ as input, where T is the number of frames and D is the feature dimension per frame, the encoder E_m maps $m_{1:T}$ to a sequence of latent vectors $z_{1:T/S} \in \mathbb{R}^{T/S \times D_z}$, where S is the downsampling rate and D_z is the latent dimension. These latent vectors are then quantized using a learned codebook $\mathcal{C} = \{c_n\}_{n=1}^{N}$, where N is the codebook size and each $c_n \in \mathbb{R}^d$. During quantization, each latent vector z_i is replaced by its nearest neighbor in the codebook:

$$\hat{z}_i = Q(z_i), \quad Q(z_i) = \arg\min_{c_n \in C} \|z_i - c_n\|_2 \tag{1}$$

The motion decoder D_m reconstructs the motion sequence from the quantized latents:

$$\hat{m}_{1:T} = D_m(\hat{z}_{1:T/S}). \tag{2}$$

Following Zhang et al. [29], we train the motion encoder–decoder with a composite VQ-VAE loss:

$$\mathcal{L}_{\text{vqvae}} = \mathcal{L}_{\text{rec}} + \alpha\mathcal{L}_{\text{com}} + \beta\mathcal{L}_{\text{emb}}, \tag{3}$$

where $\mathcal{L}_{\text{rec}}$ is the reconstruction loss, $\mathcal{L}_{\text{com}}$ is the commitment loss, $\mathcal{L}_{\text{emb}}$ is the embedding loss, and α and β are hyperparameters. The loss components are defined as:

$$\mathcal{L}_{\text{rec}} = \|m - \hat{m}\|_1, \quad (4)$$

$$\mathcal{L}_{\text{com}} = \|\text{sg}(z) - \hat{z}\|_2, \quad (5)$$

$$\mathcal{L}_{\text{emb}} = \|z - \text{sg}(\hat{z})\|_2, \quad (6)$$

where m is the ground-truth motion, $\hat{m}$ is the reconstructed motion, z is the output of the encoder E_m, $\hat{z}$ is the quantized latent vector, and $\text{sg}(\cdot)$ denotes the stop-gradient operator.

As in Zhang et al. [29], we also employ EMA(Exponential Moving Average) and codebook reset techniques during training to enhance the quality of the generated motions. In our implementation, we follow some mainstream methods and set the channel dimension of the VQ-VAE latent space to 256.

3.2 Lightweight Motion LLM

Through the pretrained motion encoder–decoder in Sect. 3.1, we obtain the codebook $\mathcal{C}$ composed of motion tokens.

This codebook can be viewed as a "motion language dictionary" that contains all the motion tokens, where each token is associated with a unique motion ID (i.e., its index within the codebook). These motion IDs effectively form a "motion language vocabulary." By converting motion sequences into sequences of motion IDs, the text-to-motion generation task is reformulated as a translation task between natural language and this structured "motion language." In other words, we do not treat human motions as a specialized modality requiring task-specific architectural designs, but rather as a general-purpose "motion language" analogous to natural human languages.

Previous studies (e.g., MotionGPT [3], MotionAgent [26]) typically rely on very large LLMs. To achieve a lightweight goal, we adopt Qwen2.5-1.5B-Instruct as the backbone to handle the translation task between motion IDs and natural language. However, the drastic reduction in LLM scale inevitably leads to degraded performance. Therefore, we introduce SFT and an additional RL module (see Sect. 3.3) to compensate for this loss.

Other LLM-based T2M methods often use LoRA [8] for fine-tuning. However, when replacing large-scale LLMs with small-scale ones such as 1.5B, LoRA's computational savings from updating only a small subset of parameters become negligible compared to the overall parameter reduction, while the limited parameter updates of LoRA lead to more severe performance bottlenecks.

Ouyang et al. [9] demonstrated that SFT can enable small-scale LLMs to achieve performance comparable to much larger ones. Yet, to date, no one has applied this insight to human motion generation. Inspired by this, we employ SFT for fine-tuning our lightweight LLM. Unlike LoRA, SFT introduces explicit supervision (input–output pairs), allowing the model to achieve better performance on the specific task of motion generation.

Concretely, given the natural language descriptions of each motion sequence, denoted as $T_{cap} = \{t_{cap1}, t_{cap2}, \ldots, t_{capn}\}$, and the corresponding motion IDs, denoted as $T_{mot} = \{t_{mot1}, t_{mot2}, \ldots, t_{motn}\}$, we construct paired data in the form $T = (T_{cap}, T_{mot})$, which is used as a supervised dataset to perform SFT on the lightweight LLM.

3.3 RLOO-Based Reinforcement Learning Optimization

Due to the particularity of the T2M task, where each fine-grained motion in human actions is not always strictly aligned with fine-grained words in text, we argue that mainstream PPO-based RL methods, which treat each token as an action, may even harm T2M training. This is because in T2M, natural language prompts are usually provided for the entire motion sequence. Thus, methods like RLOO, which optimize the entire generated sequence as a whole, are more consistent with the actual scenario. Moreover, RLOO significantly reduces training time, which is in line with our lightweight objective.

Therefore, to further compensate for the performance degradation caused by reduced LLM scale, we design a reinforcement learning module based on RLOO to further optimize the model. Its framework is shown in (Fig. 2).

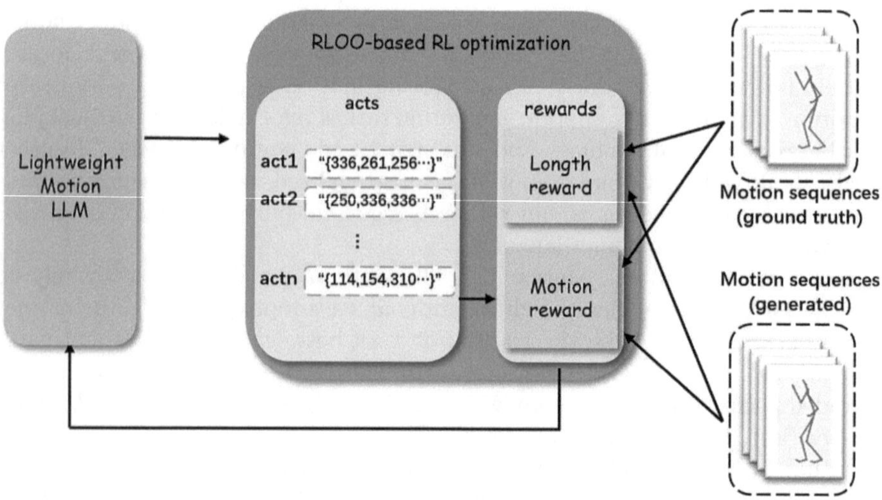

Fig. 2. The framework of our RLOO-based Reinforcement Learning Optimization.

We use the Lightweight LLM from (Sect. 3.2) as the generator. Given a natural language text prompt T_{prompt}, the generator produces k motion ID sequences $\{y(1), y(2), \ldots, y(k)\}$ as actions. For each action $y(i)$, we compute the difference between it and the average value of other actions other than it to update the gradient estimate of each sample on average, the formula can be expressed as:

$$\frac{1}{k}\sum_{i=1}^{k}\left[R(y(i),x)-\frac{1}{k-1}\sum_{j\neq i}R(y(j),x)\right]\nabla\log\pi(y(i)|x),\quad i,j\in[1,k] \tag{7}$$

where x is the input prompt, k is the number of generated actions, $y(i)$ is the i-th action, and $R(y(i),x)$ denotes its reward.

To guide the model toward better performance in terms of format, motion consistency and generated length, we design a multi-reward mechanism consisting of format reward, motion reward and length reward.

Format Reward. Since our method adopts a small-scale LLM, its instruction-following ability is weaker than that of large LLMs, thus stricter format constraints are needed. In order to ensure that the results generated by the model meet the format requirements, we design a format reward r_{for}, which checks: (1) whether the output motion IDs contain only digits; (2) whether motion IDs are within the valid codebook index range; (3) whether the output sequence not too short (less than 5); (4) whether the output is in valid JSON format. If all conditions are satisfied, $r_{for}=1$, otherwise $r_{for}=0$.

Motion Reward. To encourage reconstructed motion $\hat{m}$ to be closer to ground-truth motion m and achieve better spatio-temporal consistency, we set a motion reward r_{mot}, computed as the cosine similarity between the encoded features of $\hat{m}$ and m using the encoder E_m from (Sect. 3.1):

$$r_{mot}=\frac{E_m(\hat{m})\cdot E_m(m)}{\|E_m(\hat{m})\|_2\cdot\|E_m(m)\|_2} \tag{8}$$

Length Reward. We observe that existing generation models often produce motion ID sequences significantly longer or shorter than the ground truth. We consider that this mismatch leads to worse learning outcomes. Thus, we innovatively design a length reward r_{len} to encourage the generated length $l_{\hat{m}}$ to be closer to the ground-truth length l_m:

$$r_{len}=1-\frac{|l_{\hat{m}}-l_m|}{l_{\hat{m}}+l_m} \tag{9}$$

When $l_{\hat{m}}$ is close to l_m, r_{len} is close to 1; the larger the difference between the two, the smaller the value of r_{len} will be; when they diverge significantly, r_{len} is close to 0.

finally, our combined reward is:

$$r=\begin{cases} r_{mot}+\lambda r_{len}, & \text{if } r_{for}=1\\ 0, & \text{if } r_{for}=0\end{cases} \tag{10}$$

where λ is a weighting factor. In our experiments, λ is fixed to 0.5. We conducted preliminary trials with $\lambda\in\{0,0.2,0.5,0.7,1.0\}$ and found that $\lambda=0.5$ achieves a stable balance between the motion reward r_{mot} and the length reward r_{len}.

4 Experiments

We train our model on a server equipped with one NVIDIA A800 Tensor Core GPU and conduct all evaluations on a server with one NVIDIA GeForce RTX 3090 GPU. We evaluated our method on the most widely used dataset in human motion generation, HumanML3D [21], which contains 14,616 motions and 44,970 text descriptions. Its raw data comes from AMASS [30] and HumanAct12 [31].

4.1 Evaluation Indicators

Our evaluation metrics follow Guo et al. [21], including R-Precision@1/2/3, FID, Diversity and MM-Dist. These metrics respectively evaluate retrieval accuracy, distributional realism, sample diversity, text–motion alignment, and one-to-many generation ability. In addition, we also compare inference time, storage space, parameter count, and FLOPs to verify the significant improvement of our lightweight model in inference speed and reduction of storage costs. Our evaluation criteria are as follows:

R-Precision@k. R-Precision computes the Euclidean distance between motion and text embeddings, sorts them, and calculates the average accuracy of the top-k. Higher values indicate better text–motion matching. We evaluate $k = 1, 2, 3$.

FID(Fréchet Inception Distance). FID measures the distance between the distribution of generated motion features and ground-truth motion features, reflecting motion quality.

Diversity. It evaluates the diversity of motions generated from the same text.

MM-Dist. It measures the mean Euclidean distance between generated motions and their corresponding texts, reflecting alignment quality.

FLOPs(Floating Point Operations). FLOPs refer to the total number of floating-point operations required for a model to complete one inference. The larger the value, the greater the computational complexity, slower inference, and higher energy consumption. Conversely, the smaller the value, the lighter and more efficient it is. It can measure the computational complexity of the model. Specifically, we compared GFLOPs instead of FLOPs in Table 2, and their conversion relationship is: $1\,\mathrm{GFLOP} = 10^9\,\mathrm{FLOPs}$.

Average inference time. It represents average time per inference. We obtained this data by having each model generate same times and calculating its average time, which intuitively reflects the inference speed of the model.

Parameter count. The parameter count is the total number of learnable weights in the model, which directly determines the model size and reflects the storage and computational resources required for training and inference.

Storage space. It represents the actual disk size occupied by the model, which is a direct reflection of the storage cost of the model.

4.2 Comparisons with Other Works

We evaluate the performance of MotionSlim on HumanML3D and compare it with SOTA methods. We first compare our method with other methods under the scale of LLM claimed as suitable in their papers. We compare MotionGPT, MotionChain, MotionLLM, MotionGPT-2 and MotionR1. As shown in Table 1, our method achieves SOTA performance on FID, and also reaches close-to-SOTA performance on R-Precision, MM-Dist, and Diversity. We believe this is because our motion reward has a greater gain amplitude on FID than other indicators.

In addition, we compare FLOPs, average inference time, parameter count and storage space with several currently open-sourced LLM-based human motion generation methods. The results show that our method significantly achieves the best performance on these four indicators.

Table 1. Experimental results of our method compared to other methods with each other's LLM on the HumanML3D dataset (performance metrics)

Model	FID↓	Div↑	Top1↑	Top2↑	Top3↑	MM-Dist↓
MotionGPT [3]	0.542	9.311	0.411	0.594	0.696	3.584
MotionChain [32]	0.248	9.470	0.504	0.617	0.790	3.033
MotionAgent [26]	0.230	9.908	**0.515**	0.691	0.801	2.967
MotionGPT-2 [27]	0.191	9.860	0.496	0.691	0.782	3.080
MotionR1 [28]	0.201	**10.026**	0.515	**0.719**	**0.818**	**2.854**
Our(MotionSlim)	**0.128**	9.716	0.476	0.664	0.761	3.119

Since traditional T2M works do not compare FLOPs, average inference time and storage space, and many existing works do not provide open-source code, we reproduce available open-source methods MotionGPT [3] and MotionAgent [26] locally as baselines, and compute their FLOPs, average inference time, parameter count, and storage space for comparison. As shown in Table 2, our method significantly outperforms the others on these lightweight indicators.

Table 2. Experimental results of our method compared to other methods with each other's LLM on the HumanML3D dataset (lightweight metrics)

Model	GFLOPs↓	Time↓	Param↓	Space↓
MotionGPT [3]	1245.166	30.91s	12.991B	38.8GB
MotionAgent [26]	914.869	11.91s	9.241B	17.2GB
Our(MotionSlim)	**166.705**	**2.91s**	**1.544B**	**2.9GB**

To further verify the compensation effect of our SFT fine-tuning and RLOO reinforcement learning modules on the performance degradation caused by LLM parameter reduction, we replace the LLMs used in several currently open-sourced LLM-based human motion generation methods with the same 1.5B model as ours, and test their performance on various indicators. As shown in Table 3, our method achieves significantly better results than other methods under the 1.5B LLM, proving that our method effectively mitigates the performance degradation caused by parameter reduction.

In addition, taking MotionGPT [3] as an example, we also tested the degree of performance variation between our work and other works on LLMs of different scales. As shown in Fig. 3, our work shows that the performance degradation caused by the decrease in LLM scale is significantly smaller than that of MotionGPT on the two main indicators of FID and R-precision top 1.

Table 3. Experimental results of our method compared to other methods with 1.5B LLM on the HumanML3D dataset

Model	FID↓	Div↑	Top1↑	Top2↑	Top3↑	MM-Dist↓
MotionGPT [3]	9.771	7.406	0.148	0.261	0.350	6.057
MotionAgent [26]	95.405	0.303	0.036	0.071	0.105	9.721
Our(MotionSlim)	**0.128**	**9.716**	**0.476**	**0.664**	**0.761**	**3.119**

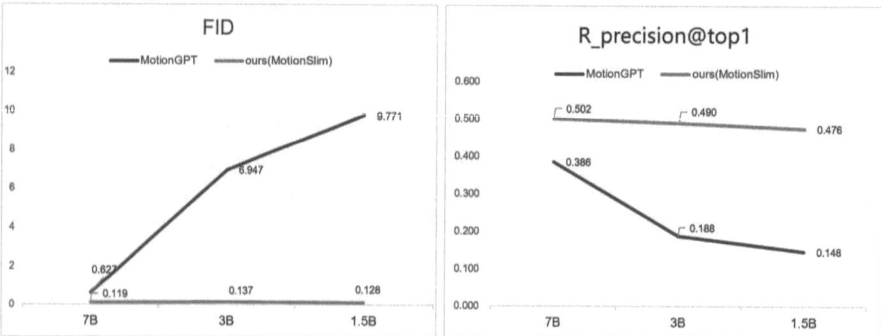

Fig. 3. Performance curve varying with LLM scale.

4.3 Ablation Studies

To verify the optimization effect of our SFT and RLOO reinforcement learning modules, we conduct ablation experiments. We compare four settings: (1) LLM without fine-tuning, (2) LLM + LoRA (widely used in other works), (3) LLM + SFT, (4) LLM + SFT + RLOO. As shown in Table 4, both our SFT fine-tuning and RLOO reinforcement learning modules improve performance.

Table 4. Experimental results of ablation experiments on our three main modules on the HumanML3D dataset

Model	FID↓	Div↑	Top1↑	Top2↑	Top3↑	MM-Dist↓
Qwen2.5-1.5B	36.523	3.271	0.032	0.068	0.128	8.452
Qwen2.5-1.5B+LoRA	9.771	7.406	0.148	0.261	0.350	6.057
Qwen2.5-1.5B+SFT	0.253	9.475	0.412	0.600	0.706	3.562
Qwen2.5-1.5B+SFT+RLOO	**0.128**	**9.716**	**0.476**	**0.664**	**0.761**	**3.119**

In addition, to verify the contribution of our two main reward functions, motion reward r_{mot} and length reward r_{len}, we compare three settings: (1) only r_{mot} enabled, (2) only r_{len} enabled, (3) both r_{mot} and r_{len} enabled. As shown in Table 5, enabling both rewards together achieves better performance than enabling either one alone.

Table 5. Experimental results of ablation experiments on motion rewards and length rewards on the HumanML3D dataset

Model	FID↓	Div↑	Top1↑	Top2↑	Top3↑	MM-Dist↓
Only r_{mot}	0.223	9.696	0.462	0.653	0.753	3.283
Only r_{len}	0.594	**9.748**	0.452	0.636	0.734	3.379
$r_{mot} + r_{len}$	**0.128**	9.716	**0.476**	**0.664**	**0.761**	**3.119**

5 Conclusion and Future

In this paper, we propose a novel lightweight text-to-motion generation framework, MotionSlim. This framework is based on a small-scale LLM and combines SFT with RLOO reinforcement learning optimization. Experimental results show that, our method can approach or even exceed SOTA in terms of generation performance while significantly reducing the number of parameters, and exhibits significant advantages in FLOPs, inference speed, and storage space. Although MotionSlim slightly underperforms large-scale models on some metrics, these trade-offs are acceptable given the drastic reduction in parameter count. When using the same small-scale LLM for generation, our method significantly outperforms current SOTA works.

Ablation studies further verify the effectiveness of our SFT and RLOO optimization modules, particularly our proposed combination of motion reward and length reward. This indicates that even under the scale of lightweight models, comparable or even superior performance to large models can be achieved through appropriate optimization strategies and reward mechanisms.

Due to lower resource consumption and faster inference speed, our method can be deployed on resource constrained devices, has broad prospects in fields such as VR and video games.

MotionSlim demonstrates the possibility of balancing lightweight design and high performance, provides new ideas for the practical application of text-to-motion generation tasks.

References

1. Touvron, H., et al.: LLaMA: Open and efficient foundation language models (2023). arXiv preprint arXiv:2302.13971
2. Bi, X., et al.: DeepSeek LLM: scaling open-source language models with longtermism (2024). arXiv preprint arXiv:2401.02954
3. Zhang, Y., et al.: MotionGPT: finetuned LLMs are general-purpose motion generators. In: Proceedings of the AAAI Conference on Artificial Intelligence, vol. 38, no. 7, pp. 7368–7376 (2024)
4. Ahmadian, A., et al.: Back to basics: revisiting reinforce style optimization for learning from human feedback in LLMs (2024). arXiv preprint arXiv:2402.14740
5. Devlin, J., Chang, M.W., Lee, K., Toutanova, K.: Bert: pre-training of deep bidirectional transformers for language understanding. In: Proceedings of the 2019 Conference of the North American Chapter of the Association for Computational Linguistics: Human Language Technologies, vol. 1 (long and short papers), pp. 4171–4186 (2019)
6. Raffel, C., et al.: Exploring the limits of transfer learning with a unified text-to-text transformer. J. Mach. Learn. Res. **21**(140), 1–67 (2020)
7. Radford, A., Narasimhan, K., Salimans, T., Sutskever, I.: Improving language understanding by generative pre-training (2018)
8. Hu, E.J., et al.: LoRA: low-rank adaptation of large language models. ICLR **1**(2), 3 (2022)
9. Ouyang, L., et al.: Training language models to follow instructions with human feedback. Adv. Neural. Inf. Process. Syst. **35**, 27730–27744 (2022)
10. Schulman, J., Wolski, F., Dhariwal, P., Radford, A., Klimov, O.: Proximal policy optimization algorithms (2017). arXiv preprint arXiv:1707.06347
11. Rafailov, R., Sharma, A., Mitchell, E., Manning, C.D., Ermon, S., Finn, C.: Direct preference optimization: your language model is secretly a reward model. Adv. Neural. Inf. Process. Syst. **36**, 53728–53741 (2023)
12. Yu, Q., et al.: DAPO: an open-source LLM reinforcement learning system at scale (2025). arXiv preprint arXiv:2503.14476
13. Shao, Z., et al.: DeepSeekMath: pushing the limits of mathematical reasoning in open language models (2024). arXiv preprint arXiv:2402.03300
14. Ahn, H., Ha, T., Choi, Y., Yoo, H., Oh, S.: Text2Action: generative adversarial synthesis from language to action. In: 2018 IEEE International Conference on Robotics and Automation (ICRA), pp. 5915–5920. IEEE (2018)
15. Ghosh, A., Cheema, N., Oguz, C., Theobalt, C., Slusallek, P.: Synthesis of compositional animations from textual descriptions. In: Proceedings of the IEEE/CVF International Conference on Computer Vision, pp. 1396–1406 (2021)
16. Qi, X., et al.: Weakly-supervised emotion transition learning for diverse 3D co-speech gesture generation. In: Proceedings of the IEEE/CVF Conference on Computer Vision and Pattern Recognition, pp. 10424–10434 (2024)

17. Tevet, G., Raab, S., Gordon, B., Shafir, Y., Cohen-Or, D., Bermano, A. H.: Human motion diffusion model (2022). arXiv preprint arXiv:2209.14916
18. Shafir, Y., Tevet, G., Kapon, R., Bermano, A.H.: Human motion diffusion as a generative prior (2023). arXiv preprint arXiv:2303.01418
19. Zhou, W., et al.: EMDM: Efficient motion diffusion model for fast and high-quality motion generation. In: Leonardis, A., Ricci, E., Roth, S., Russakovsky, O., Sattler, T., Varol, G. (eds) European Conference on Computer Vision, pp. 18–38. Springer, Cham (2024). https://doi.org/10.1007/978-3-031-72627-9_2
20. Zhang, M., et al.: Motiondiffuse: Text-driven human motion generation with diffusion model. IEEE Trans. Pattern Anal. Mach. Intell. **46**(6), 4115–4128 (2024)
21. Guo, C., et al.: Generating diverse and natural 3D human motions from text. In: Proceedings of the IEEE/CVF Conference on Computer Vision and Pattern Recognition, pp. 5152–5161 (2022)
22. Hong, F., Zhang, M., Pan, L., Cai, Z., Yang, L., Liu, Z.: AvatarCLIP: zero-shot text-driven generation and animation of 3D avatars (2022). arXiv preprint arXiv:2205.08535
23. Petrovich, M., Black, M. J., Varol, G.: TEMOS: Generating diverse human motions from textual descriptions. In: Avidan, S., Brostow, G., Cissé, M., Farinella, G.M., Hassner, T. (eds) European Conference on Computer Vision, pp. 480–497 (2022). Springer, Cham. https://doi.org/10.1007/978-3-031-20047-2_28
24. Guo, C., Zuo, X., Wang, S., Cheng, L.: TM2T: stochastic and tokenized modeling for the reciprocal generation of 3D human motions and texts. In: Avidan, S., Brostow, G., Cissé, M., Farinella, G.M., Hassner, T. (eds) European Conference on Computer Vision, vol. 13695, pp. 580–597 (2022). Springer, Cham. https://doi.org/10.1007/978-3-031-19833-5_34
25. Athanasiou, N., Petrovich, M., Black, M. J., Varol, G.: TEACH: temporal action composition for 3D humans. In: 2022 International Conference on 3D Vision (3DV), pp. 414–1423. IEEE (2022)
26. Wu, Q., Zhao, Y., Wang, Y., Liu, X., Tai, Y. W., Tang, C. K.: Motion-agent: a conversational framework for human motion generation with LLMs (2024). arXiv preprint arXiv:2405.17013
27. Wang, Y., et al.: MotionGPT-2: a general-purpose motion-language model for motion generation and understanding (2024). arXiv preprint arXiv:2410.21747
28. Ouyang, R., Li, H., Zhang, Z., Wang, X., Zhu, Z., Huang, G., Wang, X.: Motion-R1: chain-of-thought reasoning and reinforcement learning for human motion generation (2025). arXiv preprint arXiv:2506.10353
29. Zhang, J., et al.: Generating human motion from textual descriptions with discrete representations. In: Proceedings of the IEEE/CVF Conference on Computer Vision and Pattern Recognition, pp. 14730–14740 (2023)
30. Mahmood, N., Ghorbani, N., Troje, N. F., Pons-Moll, G., Black, M. J.: AMASS: archive of motion capture as surface shapes. In: Proceedings of the IEEE/CVF International Conference on Computer Vision, pp. 5442–5451 (2019)
31. Guo, C., et al.: Action2Motion: conditioned generation of 3d human motions. In: Proceedings of the 28th ACM International Conference on Multimedia, pp. 2021–2029 (2020)
32. Jiang, B., et al.: MotionChain: Conversational motion controllers via multimodal prompts. In: Leonardis, A., Ricci, E., Roth, S., Russakovsky, O., Sattler, T., Varol, G. (eds) European Conference on Computer Vision, pp. 54–74 (2024). Springer, Cham. https://doi.org/10.1007/978-3-031-73347-5_4

Boosting the Transferability of Adversarial Examples via Frequency Domain Masking and Adaptive Step Size

Jiale Yang[1], Kai Zhao[1](✉), Linlin Zhang[2], and Qingguan Li[1]

[1] School of Computer Science and Technology, Xinjiang University, Urumqi, China
zhawkk@xju.edu.cn

[2] Computer and Data Center, Xinjiang University, Urumqi, China

Abstract. Recent studies have shown that deep neural networks (DNNs) are susceptible to adversarial examples, exposing their serious vulnerabilities. This vulnerability allows adversarial examples to attack multiple models with different architectures, which is called transfer-based adversarial attacks. However, existing methods still have insufficient transferability in cross-model scenarios. To address this issue, this paper proposes enhanced frequency diversity attack (EFDA), which introduces perturbations into the image frequency domain, designs frequency domain masks combining low-frequency and mid-frequency masks to enhance specific frequency component gradients, and adopts an adaptive step size adjustment strategy to dynamically calculate the step size according to the iteration progress, thereby improving the transferability of adversarial examples. Experiments on the SVHN and CIFAR-10 datasets show that EFDA significantly reduces the classification accuracy of the target model in untargeted attacks and has better cross-model transferability than existing advanced methods.

Keywords: Neural Networks · Adversarial Attack · Frequency Domain · Transferability

1 Introduction

Recently, deep neural networks (DNNs) have made significant progress in many fields, such as image recognition and speech signal processing, thanks to their exceptional performance. This has had a profound impact on how people live and work. However, DNNs' inherent structural vulnerability has gradually been exposed in adversarial settings—attackers need only inject imperceptible changes into the original input to cause the model to misjudge critical information [1]. This hidden risk could have catastrophic consequences in safety-critical scenarios such as autonomous driving [2] and speech recognition [3].

Subsequent research has demonstrated that adversarial examples can transfer between models with different architectures and parameters [4]. Attackers can exploit similarities in feature spaces across models to bypass architectural

J. Lokoč et al. (Eds.): MMM 2026, LNCS 16413, pp. 504–518, 2026.
https://doi.org/10.1007/978-981-95-6957-1_36

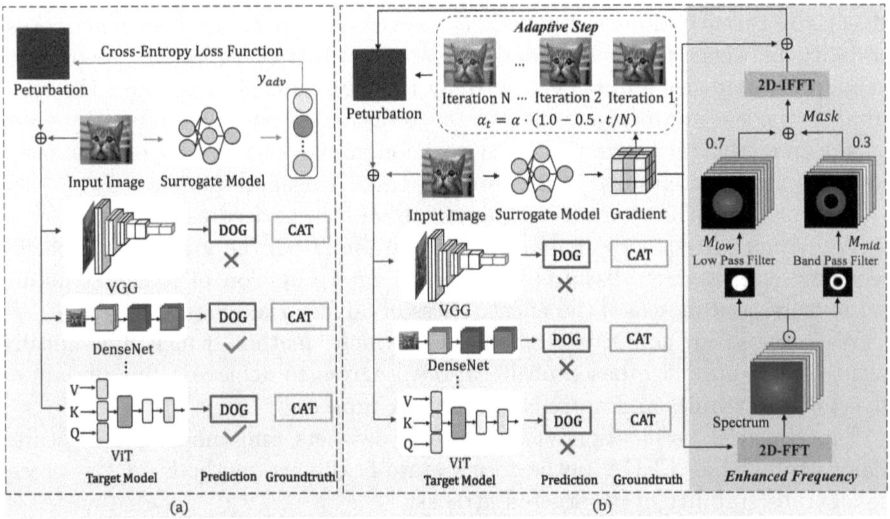

Fig. 1. (a) Existing transfer attacks target specific models (such as VGG), and shared cross-entropy loss easily leads to overfitting, making it difficult to transfer and deceive other models (such as DenseNet and ViT). (b) We use frequency domain masks and adaptive step sizes to generate adversarial examples that can effectively disrupt a variety of classification models (such as DenseNet and ViT) and have stronger transferability.

constraints and carry out cross-model attacks. This type of attack involves generating adversarial examples using a surrogate model to target an unknown architecture. While various methods have been suggested to improve the transferability of adversarial examples, most research continues to focus on spatial domain transformations, such as random translation [5,6] and scaling [7]. While studies have suggested that frequency domain transformations can increase model diversity [8], current attack methods operate only in a single domain (spatial or frequency). This severely limits the diversity of transformed images and thereby constrains the transferability of adversarial examples.

The frequency domain representation of images contains rich structural and semantic information. Low frequencies correspond to overall contours and backgrounds, while high frequencies correspond to details and edges [9]. Although deep neural networks rely on frequency domain features for image recognition, existing pixel-space attacks neglect these features, resulting in an insufficient level of cross-model transferability of adversarial examples [10]. The frequency-domain mask designed in this paper outperforms traditional single-band perturbation methods by superimposing low/mid-frequency masks with weight-optimised perturbations.

Through a simple experiment, Huang et al. [11] observed that the transferability of adversarial examples generated using traditional iterative adversarial attack methods decreases with the number of iterations. This indicates that adversarial perturbations in the vicinity of original samples have stronger transferability between different models. Inspired by this observation, we propose an adaptive step-size adjustment strategy that dynamically calculates the most suitable step size according to the current iteration progress. This strategy gradually decreases the step size with each iteration, making the update of adversarial examples more fine-grained and improving their transferability.

The main contributions of our paper can be summarised as follows:

We design a frequency domain mask that enhances the gradient of specific frequency components, disrupts the model's interpretation of image structure and texture, and increases the effectiveness of adversarial examples.

We propose an adaptive step size adjustment method which dynamically calculates the step size based on iteration progress to achieve a fine update of adversarial examples and optimise the attack process.

Experimental verification with multiple datasets and model architectures demonstrates that EFDA outperforms state-of-the-art methods in the cross-model transferability of untargeted attacks.

2 Related Work

In this section, we briefly review the literature related to adversarial attacks.

2.1 Adversarial Attacks

Adversarial attacks fall into two main categories: white-box and black-box. White-box attacks assume that the attacker has full knowledge of the target model, including its architecture, parameters and weights, and utilise gradient information to calculate adversarial perturbations. Representative methods include FGSM [4], DeepFool [12] and PGD [13]. Black-box attacks assume that the attacker has limited or no knowledge of the target model. These attacks primarily include three strategies: query-based attacks, which estimate gradients by sending queries to the target model [14]; decision-based attacks, which iteratively approximate the decision boundary using only model outputs (such as classification labels), without requiring gradients or internal information [15]; and transfer-based attacks, which use surrogate models to generate adversarial examples without querying the target model at all [16,17]. Transfer-based attacks are the most efficient and difficult to detect, making the transferability of adversarial examples a key area of research.

2.2 Frequency-Domain

Research indicates that deep neural networks (DNNs) can improve the accuracy of image processing by making use of high-frequency components that are difficult for the human visual system to perceive [9,18], and that high-frequency components are particularly important [8]. Frequency domain analysis reveals the characteristics of DNNs and offers new perspectives for improving the transferability of adversarial examples. Long et al. [19] improved the transferability of adversarial examples by applying a frequency-domain spectral transformation to the input data. Zhu et al. [20] proposed transformations in both the spatial and frequency domains to increase the diversity of adversarial examples and improve their transferability across models. Qian et al. [18] focused on accumulating high-frequency information across images during adversarial gradient computation. This yielded more stable gradient directions and significantly improved adversarial sample transferability.

2.3 Step Size Adjustment

In research aimed at improving the transferability of models against adversarial attacks, momentum-based methods update step sizes by accumulating historical gradient directions [21], essentially implementing implicit step adaptation. Shi et al. [22] balanced overfitting and effectiveness in Ada-FGSM by adapting step sizes based on gradient magnitude. Along the refinement adjustment direction, Liu et al. [23] introduced second-order momentum to dynamically adjust regional step-size optimisation perturbations. However, Tang et al. [24] incurred high computational costs when using Hessian matrix curvature estimation to control step sizes. In contrast, approaches such as DIM input transformation [25] and SGM gradient masking [26] rely on specific augmentations or model assumptions. However, step size adaptation offers greater generalisation as an internal optimisation tuning mechanism.

3 Methodology

The pipeline of our method is shown in Fig. 1 and our approach is described in detail in this section. The algorithm generates high-quality adversarial examples and improves the adversarial transferability by combining frequency domain enhancement and adaptive step size adjustment strategies.

3.1 Preliminary

The process of generating adversarial examples can essentially be modeled as a constrained optimization problem. For the parametric classification model f_θ, the objective of generating adversarial examples x_{adv} is to maximize the loss function while constraining the magnitude of the perturbation (1), i.e.:

$$\max \mathcal{J}(x_{\text{adv}}, y), \quad \text{s.t. } \|x_{\text{adv}} - x\|_p \leq \epsilon \tag{1}$$

where $\mathcal{J}(x_{\text{adv}}, y)$ is the first-order differentiable loss function (e.g., cross-entropy loss), and $\|x_{\text{adv}} - x\|_p \leq \epsilon$ indicates that the L_p-distance between the original sample x and the adversarial sample x_{adv} is restricted to a sufficiently small perturbation magnitude ϵ to ensure imperceptibility. The norm parameter p can be set to 0, 1, 2, or ∞.

In this paper, we are given an surrogate classifier f_θ parameterized by θ, and an image x, label y_{true}; a maximum perturbation radius ϵ, which restricts the maximum degree of change in the adversarial attack samples with respect to the original samples; the number of attacking steps N and the step size α, which together determine the number of iterations of sample update and the magnitude of each update; the number of orthogonal directions K, which affects the degree of multi-directional diversity; and the frequency influence coefficient λ_{freq}, which is used to balance the weights of the original gradient and the gradient after enhancement in the frequency domain.

3.2 Frequency Mask Generation

In this paper, we employ Fast Fourier Transform (FFT) and Inverse FFT (IFFT) for data conversion between frequency and spatial domains. Two critical frequency domain transformation functions are defined: FFT transformation $F(v)$ and inverse FFT transformation $F^{-1}(v)$.

Previous studies have confirmed that deep convolutional neural networks (DCNNs) exhibit an abnormal dependence on high-frequency information. Song et al. [27] found that adversarial perturbations are concentrated in high-frequency regions, with standard deviation increasing with frequency. Wang et al. [9] demonstrated that high-confidence classification can be achieved using only high-frequency components invisible to the human eye. This âĂIJoverfittingâĂİ-like dependence makes the model susceptible to being misled by high-frequency perturbations. However, image semantic understanding fundamentally relies on the synergy between low-frequency and mid-frequency components [9]. While traditional JPEG compression discards high-frequency noise, it also compromises the integrity of critical mid-frequency textures [28].

Figure 2 shows the original images and adversarial examples generated by HopSkipjump, DRA, L2T, IDAA and EFDA, as well as their spectrum diagrams. Visually, the adversarial examples generated by EFDA and other methods are highly natural, and the differences are difficult to detect. However, the spectral plots show that compared to other methods, EFDA's spectrum is richer in mid-frequency (amplitude (dB) approximately 0.2–0.4) and low-frequency (amplitude (dB) approximately 0.6–1.0) components. This unique frequency distribution indicates that EFDA can more accurately adjust specific frequencies, making adversarial examples more effective in interfering with model feature extraction and recognition.

Based on this, We designed a frequency domain mask M consisting of a low-frequency mask M_{low} and a mid-frequency mask M_{mid}.Based on the height h and width w of the image, the distance from each pixel point (i, j) to the center of the image is calculated as (2):

$$d(i,j) = \sqrt{(i - h/2)^2 + (j - w/2)^2} \tag{2}$$

Based on this distance, low-frequency mask M_{low} and medium-frequency mask M_{mid} are constructed separately, as shown (3)(4):

$$M_{\text{low}} = \exp\left(-0.5 \cdot \left(\frac{d(i,j)}{0.1h}\right)^2\right) \tag{3}$$

$$M_{\text{mid}} = \exp\left(-0.5\left(\frac{d(i,j) - 0.2h}{0.1h}\right)^2\right) \tag{4}$$

The low-frequency mask and mid-frequency mask are combined using a weighted average to form the final frequency mask $M = 0.7 \cdot M_{\text{low}} + 0.3 \cdot M_{\text{mid}}$.

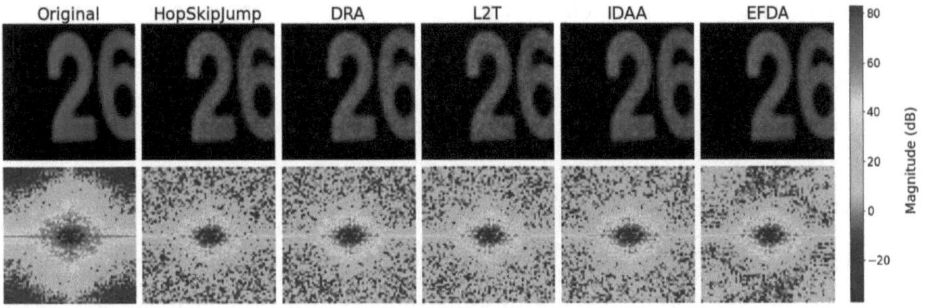

Fig. 2. Comparison of spectrum maps of different attack methods.

3.3 Orthogonal Direction Generation

To enhance the diversity and effectiveness of the attack, we introduce a multi-directional exploration strategy that uses the Gram-Schmidt orthogonalization method [29] to generate K orthogonal directions $(d_1, d_2, \cdots, d_K)$. For each new direction vector, we calculate its projection onto all previously processed directions and subtract these projections to obtain a new vector orthogonal to all previously processed directions. Each vector is normalized to unit length. The resulting K orthogonal directions provide a diverse search space for gradient updates.

First, randomly initialize K direction vectors d_i that follow a standard normal distribution $N(0, I)$. Then, starting from the second direction vector, for each $i(2 \leq i \leq K)$, orthogonally process the $j(1 \leq j \leq i-1)$ direction vectors generated previously in sequence.

3.4 Attack Process

In the untargeted attack scenario, the algorithm starts from the original input image x and initializes the adversarial examples $x_{\text{adv}} = x$. In each iteration $t(1 \leq t \leq N)$, the gradient g is computed (5):

$$g = \nabla_{x_{\text{adv}}} \mathcal{L}(z, y_{\text{true}}) \tag{5}$$

where, $z = f_\theta(x_{\text{adv}})$, which is the model's output, and $\mathcal{L}(z, y_{\text{true}})$ is the cross-entropy loss, the objective is to maximize the model's prediction error for the original label. To enhance the attack effect, the gradient (g) is transformed to the frequency domain to obtain $g_{\text{freq}} = \mathcal{F}(g)$, and the frequency domain gradient is calculated (6):

$$g_{\text{freq}}^{\text{masked}} = g_{\text{freq}} \odot M \tag{6}$$

where, $\odot$ denotes hadamard product. M is the previously computed frequency mask, and the frequency domain gradient is converted back to the spatial domain to obtain the enhanced gradient (7):

$$g_{\text{enhanced}} = \mathcal{F}^{-1}(g_{\text{freq}}^{\text{masked}}) \tag{7}$$

The composite gradient is obtained by linearly combining the original gradient and the enhanced gradient (8):

$$g_{\text{combined}} = (1 - \lambda_{\text{freq}}) \cdot g + \lambda_{\text{freq}} \cdot g_{\text{enhanced}} \tag{8}$$

In the multi-directional diversity step, for the generated K orthogonal directions $d_i (i \in \{1, 2, \ldots, K\})$, the dot product of each direction with the composite gradient g_{combined} is computed (9):

$$I_i = (d_i, g_{\text{combined}}) \tag{9}$$

The dot product I_i indicates the extent to which moving in that direction contributes to loss reduction. We retain positive influence values, i.e., $w_i = \max(I_i, 0)$, and normalize these weights (10):

$$w_i = \frac{w_i}{\sum_{j=1}^{K} w_j + \gamma} \tag{10}$$

where γ is a very small value used to avoid cases where the denominator is zero, and ultimately the final update direction is calculated based on these weighted directions (11):

$$d_{\text{final}} = \sum_{i=1}^{K} w_i \cdot d_i \tag{11}$$

Traditional iterative adversarial attacks usually use a fixed step size, which can easily lead to suboptimal solutions or convergence problems. Although the step size gradually decreases with each iteration to balance rapid convergence in the early stages with fine-tuning in the later stages, it is still not optimal. To address this issue, we propose an adaptive step size method that dynamically adjusts the step size based on the progress of the iteration.The step size adopts an adaptive strategy (12):

$$\alpha_t = \alpha \cdot (1.0 - 0.5 \cdot t/N) \tag{12}$$

As the number of iterations increases, the step size decreases gradually, making the update of adversarial examples more delicate. Update the adversarial examples according to the integrated direction and adaptive step size (13):

$$x_{\text{adv}} = x_{\text{adv}} + \alpha_t \cdot \text{sign}(d_{\text{final}}) \tag{13}$$

Meanwhile, the clipping operation ensures that the adversarial examples are within the maximum perturbation radius (14), i.e.:

$$\delta = \text{clip}(x_{\text{adv}} - x, -\epsilon, \epsilon) \tag{14}$$

Restrict the pixel values of the adversarial examples to the interval $[0, 1]$, i.e., $x_{\text{adv}} = \text{clip}(x_{\text{adv}}, 0, 1)$. After N iterations, the final generated adversarial examples x_{adv} are returned.

4 Experiment

In this section, the validity of the EFDA method is verified through extensive experiments.

4.1 Experimental Setup

Datasets. We conducted experiments on the SVHN [30] and CIFAR-10 [31] datasets. SVHN is a real-world street view digit recognition dataset containing 630,420 digit images; CIFAR-10 contains 60,000 32ÃŮ32 color images covering 10 categories.

Models. We chose 2 models as surrogate models: a DNN with skipped connections (ResNet-18 [32]) and a DNN with dense connections (DenseNet-121 [33]). We selected six deep learning models with different architectures as target models, including CNN-based VGG16 [34], ResNet-18(RN18), ResNet-34(RN34) [32], Inception [35], DenseNet-121(DN121), and Transformer-based Vision Transformer [36] (ViT).

Baseline. 3) Baseline: In order to validate the effectiveness of our method, we will be comparing it with four representative adversarial attack methods: HopSkipJump Attack [15], DRA [16], L2T [17] and IDAA [23]. These methods encompass a range of attack strategies, from random optimisation to transfer-based approaches.

Parameter and Attack Settings. In all experiments, we followed a uniform attack configuration to ensure fair comparisons. The pre-trained model is trained with batch size 128, Adam optimizer (learning rate 0.0001) for 30 epochs.EFDA is used with $\epsilon = 16/255$, 20 steps of iteration (initial $\alpha = 2/255$), frequency domain weighting 0.8, and adaptive attenuation factor 0.5. The experiments are done on NVIDIA RTX 4060 Super GPU.

4.2 The Evaluation of Untargeted Attack

We first conduct experiments to compare our EFDA with other baseline methods and show the results in Table 1. For transfer ResNet-18 $\rightarrow$ VGG16, EFDA reduces accuracy to 10.8%, outperforming DRA by 2.3%. For transfer DenseNet-121 $\rightarrow$ ResNet-18, EFDA reduces the accuracy to 13.4%, which is 24.1% better than the performance of Hopskipjump. However, it should be noted that all attack methods perform worse on ViTs than on CNNs. EFDA achieves accuracy rates of 14.1% and 13.3% on ViT, which are higher than the rates of 10.8% and 11.9% achieved by VGG16. This may be because the self-attention mechanism in ViT mitigates the effectiveness of attacks. Figure 3 shows the difference between the adversarial examples generated by DRA, IDAA and EFDA.

Table 1. Transferability against normal models: The classifier accuracy of black-box attacks on SVHN datasets. Best results are in bold.

Model	Attack	VGG16	RN18	RN34	Inception	DN121	ViT
RN18	Original	74.9	72.0	71.1	70.4	67.8	71.0
	HSJA [15]	48.8	36.3	35.9	40.1	26.8	38.7
	DRA [16]	13.1	14.7	13.9	15.3	11.1	15.1
	L2T [17]	12.1	12.8	13.4	13.0	12.0	14.7
	IDAA [23]	13.7	14.6	14.0	13.8	12.1	15.3
	EFDA	**10.8**	**12.1**	**13.1**	**12.2**	**10.4**	**14.1**
DN121	Original	75.1	71.8	71.2	69.9	67.8	71.1
	HSJA [15]	48.6	37.3	37.6	39.8	28.9	37.3
	DRA [16]	12.2	16.4	13.8	14.0	12.0	14.5
	L2T [17]	12.6	13.9	14.1	13.5	11.4	14.5
	IDAA [23]	12.9	14.2	13.2	**12.1**	12.2	14.4
	EFDA	**11.9**	**13.4**	**12.7**	12.6	**11.0**	**13.3**

4.3 Evaluation on the Other Datasets

We further validated the effectiveness of the algorithm on the CIFAR-10 dataset (Table 2). Using ResNet-18 (RN18) and DenseNet-121 (DN121) as the base models, we evaluated the performance of non-target black-box attacks (with an

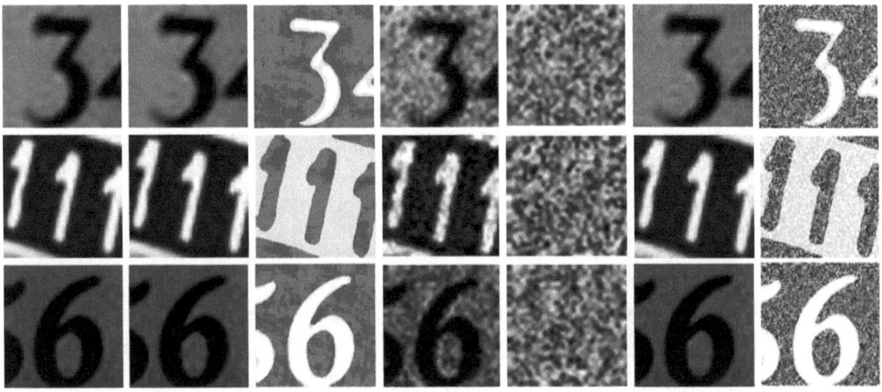

Fig. 3. Comparison between the untargeted adversarial examples generated by DRA, IDAA and our EFDA.

adaptive step size and N = 20) on six different target models with various architectures. The results show that the EFDA method performs well on most target models, especially when transferring from RN18 to ResNet-34, Inception and ViT, and from DN121 to ResNet-18, Inception and ViT. However, the L2T method outperforms EFDA on the VGG16 target model. This may be because VGG16's continuous convolutional structure is more sensitive to the specific perturbations of L2T. The adaptive step size strategy proposed in this paper has not yet been optimised sufficiently for this architecture to balance attack strength and transferability.

4.4 Ablation Study

In this section, we perform ablation analysis on the frequency domain enhancement module and adaptive step size adjustment module. In addition, separate experiments were conducted to analyze the hyperparameters λ_{freq} and frequency domain mask coefficients.

Module Analysis. The experiment verified the effect of frequency domain perturbation enhancement (Module 1) and adaptive step size adjustment (Module 2) on the transferability of adversarial examples across models(Table 3). Testing of samples generated using ResNet-18 and DenseNet-121 as surrogate models on six target models showed that when a single module was removed, the transfer accuracy from ResNet-18 to VGG16 decreased by 5.7% and 8.3%, respectively, while the transfer accuracy from DenseNet-121 to Inception increased by 1.0% to 1.7%. However, the performance improvement for ViT is limited compared to CNN, and we believe that the self-attention mechanism in ViT weakens the effectiveness of frequency-domain attacks based on local features.

Table 2. Accuracy of untargeted black box attacks on the CIFAR - 10 dataset. The best results are shown in bold.

Model	Attack	VGG16	RN18	RN34	Inception	DN121	ViT
RN18	Original	74.9	72.0	71.1	70.4	67.8	71.0
	HSJA [15]	12.1	12.8	13.4	13.0	12.0	14.7
	DRA [16]	13.7	14.6	14.0	13.8	12.1	15.3
	L2T [17]	**8.5**	10.4	10.7	10.9	10.4	12.1
	IDAA [23]	9.4	11.2	10.4	12.5	11.4	11.0
	EFDA	9.1	**8.9**	**10.0**	**10.5**	**10.2**	**9.8**
DN121	Original	75.1	71.8	71.2	69.9	67.8	71.1
	HSJA [15]	12.6	13.9	14.1	13.5	11.4	14.5
	DRA [16]	12.9	14.2	13.2	12.1	12.2	14.4
	L2T [17]	9.6	10.9	10.2	10.9	9.8	11.4
	IDAA [23]	**8.9**	10.7	9.7	11.4	10.2	11.8
	EFDA	**8.9**	**9.9**	**9.4**	**10.5**	**9.4**	**10.0**

Table 3. Accuracy of untargeted black-box attacks on SVHN datasets. Module1 refers to the frequency-domain perturbation enhancement strategy and Module2 refers to the adaptive step size adjustment strategy.

Model	Module1	Module2	VGG16	RN18	RN34	Inception	DN121	ViT
RN18	√	×	16.5	13.8	13.3	**12.9**	13.4	15.4
	×	√	19.1	14.7	13.8	14.3	12.6	15.2
	√	√	**10.8**	**13.1**	**13.1**	14.2	**10.3**	**14.1**
DN121	√	×	17.7	15.4	15.8	14.6	11.4	14.4
	×	√	19.8	15.1	15.6	14.7	12.9	14.6
	√	√	**11.9**	**13.4**	**13.4**	**13.7**	**11.0**	**13.7**

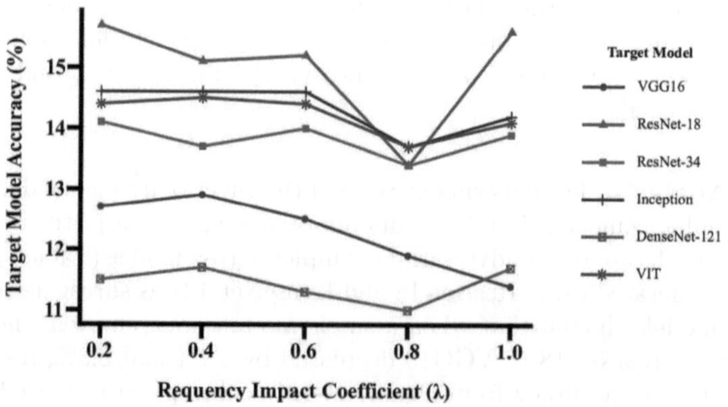

Fig. 4. Accuracy on different target models by generating adversarial examples on resnet18 for different frequency impact coefficients $\lambda (0 < \lambda \leq 1)$.

Frequency Influence Coefficient. We conducted an independent experimental analysis of the hyperparameter λ (frequency influence coefficient) in EFDA. Figure 4 shows that as λ increases, the accuracy of all target models decreases monotonically. When λ increases from 0.2 to 0.8: the accuracy of VGG16 decreases from 12.7% to 11.9%, and DenseNet-121 decreases from 11.5% to 11.0%, indicating that traditional convolutional neural networks (CNNs) are more sensitive to frequency-domain perturbations. ViT maintains an accuracy rate of 14.1% at $\lambda = 1.0$, significantly outperforming other models. When $\lambda > 0.8$, the attack effectiveness slightly weakens, so we choose $\lambda = 0.8$ to balance attack effectiveness and computational efficiency.

Analysis of Frequency Domain Mask Coefficients. As shown in Table 4, for models such as VGG16, DenseNet-121, and ViT, the optimal low-frequency to mid-frequency ratio values are 0.8:0.2, 0.4:0.6, and 0.6:0.4, respectively. The 0.7:0.3 combination achieves the best value of 12.0% in ResNet-34 and approaches optimal performance in key models such as VGG16, while maintaining suboptimal performance in other models. Based on the above considerations, we ultimately selected 0.7:0.3 as the design ratio for the frequency domain mask.

Table 4. The accuracy of the SVHN dataset created on RN18 for non-target black-box attacks, with the best results and optimal results indicated in bold.

Low	Mid	VGG16	RN18	RN34	Inception	DN121	ViT
0.1	0.9	11.0	**12.1**	12.3	12.7	10.5	**13.2**
0.2	0.8	10.8	**11.8**	**12.1**	12.5	10.7	13.5
0.3	0.7	10.7	12.7	12.5	12.5	10.7	13.3
0.4	0.6	10.7	12.4	12.9	**12.0**	**10.2**	13.6
0.5	0.5	11.0	12.8	12.6	12.5	10.5	12.6
0.6	0.4	10.6	12.5	12.2	12.3	**10.2**	12.3
0.7	**0.3**	**10.5**	**12.1**	**12.0**	12.4	**10.4**	13.5
0.8	0.2	**10.2**	12.6	**12.1**	12.5	10.8	**13.1**
0.9	0.1	10.8	12.2	12.5	**12.2**	10.7	13.4
1.0	1.0	11.2	**12.1**	12.5	12.8	10.8	14.0

5 Conclusion

This paper proposes EFDA, a black-box adversarial attack method based on transfer learning. This method uses a frequency domain mask that combines low and medium frequencies to enhance specific frequency gradients, interfering with the model's understanding of image structure and texture, and uses adaptive step size iteration to optimize adversarial examples. Experiments show that

EFDA significantly outperforms baseline methods in non-targeted attacks, but its effectiveness in attacking models with special mechanisms such as ViT needs to be improved. In the future, we will explore new architecture attack strategies to enhance transferability.

Acknowledgments. This work was supported by the National Natural Science Foundation of China under grant 62366052, 62563036 and Sichuan Province Joint Fund Project of China under grant 25QYCX0103.

References

1. Szegedy, C., Zaremba, W., Sutskever, I., Bruna, J., Erhan, D., Goodfellow, I., Fergus, R.: Intriguing properties of neural networks. CoRR (2013)
2. Wang, Y., Liu, M., Ren, Y., Zhang, X., Feng, G.: Traffic sign attack via pinpoint region probability estimation network. Pattern Recogn. **146**, 110035 (2024)
3. Chen, M., Lu, L., Yu, J., Ba, Z., Ren, K.: Advreverb: rethinking the stealthiness of audio adversarial examples to human perception. IEEE Transactions on Information Forensics and Security (2023)
4. Goodfellow, I. J., Shlens, J., Szegedy, C.: Explaining and harnessing adversarial examples. CoRR (2014)
5. Dong, Y., Pang, T., Su, H., Zhu, J.: Evading defenses to transferable adversarial examples by translation-invariant attacks. In: 2019 IEEE/CVF Conference on Computer Vision and Pattern Recognition (CVPR), pp. 4312–4321 (2019)
6. Zhang, S., Hong, H., Xie, M., et al.: Enhancing the transferability of adversarial examples with random diversity ensemble and variance reduction augmentation. IEEE Trans, Big Data (2025)
7. Wei, Z., Chen, J., Wu, Z., Zhou, Y.: Enhancing the self-universality for transferable targeted attacks. In: 2023 IEEE/CVF Conference on Computer Vision and Pattern Recognition (CVPR), pp. 12281–12290 (2023)
8. Abello, A.A., Hirata, R., Wang, Z.: Dissecting the high-frequency bias in convolutional neural networks. In: 2021 IEEE/CVF Conference on Computer Vision and Pattern Recognition (CVPR), pp. 863–871 (2021)
9. Wang, H., Wu, X., Huang, Z., Xing, E.P.: High-frequency component helps explain the generalization of convolutional neural networks. In: 2020 IEEE/CVF Conference on Computer Vision and Pattern Recognition (CVPR), pp. 8684–8694 (2020)
10. Jia, S., Ma, C., Yao, T., Chen, K., Ding, S., Yang, X.: Exploring frequency adversarial attacks for face forgery detection. In: 2022 IEEE/CVF Conference on Computer Vision and Pattern Recognition (CVPR), pp. 4103–4112 (2022)
11. Huang, Q., Lu, L., Chen, J., Zhang, Y.: Using dynamically changing step sizes to increase the success rate of adversarial attacks. In: 2024 5th IEEE Conference on Computer Vision, Image and Deep Learning (CVIDL), pp. 1071–1076 (2024)
12. Moosavi-Dezfooli, S.M., Fawzi, A., Frossard, P.: Deepfool: a simple and accurate method to fool deep neural networks. In: 2016 IEEE Conference on Computer Vision and Pattern Recognition (CVPR), pp. 2574–2582 (2016)
13. Madry, A., Makelov, A., Schmidt, L., Tsipras, D., Vladu, A.: Towards deep learning models resistant to adversarial attacks. CoRR (2017)
14. Yang, J., Jiang, Y., Huang, X., Ni, B., Zhao, C.: Learning black-box attackers with transferable priors and query feedback. In: 2020 Conference on Neural Information Processing Systems, vol. 33 (2020)

15. Chen, J., Jordan, M.I., Wainwright, M.J.: Hopskipjumpattack: A query-efficient decision-based attack. In: 2020 IEEE Symposium on Security and Privacy (sp), pp. 1277-1294 (2020)
16. Zhu, Y., et al.: Toward understanding and boosting adversarial transferability from a distribution perspective. In: 2022 IEEE Transactions on Image Processing, pp. 6487-6501(2022)
17. Zhu, R., Zhang, Z., Liang S, Liu, Z., Xu, C.: Learning to transform dynamically for better adversarial transferability. In: 2024 IEEE/CVF Conference on Computer Vision and Pattern Recognition (CVPR), PP. 24273-24283 (2024)
18. Qian, Y., et al.: Enhancing transferability of adversarial examples through mixed-frequency inputs. IEEE Transactions on Information Forensics and Security (2024)
19. Long, Y., et al.: Frequency domain model augmentation for adversarial attack. In: 2022 European Conference on Computer Vision (ECCV), pp. 549-566 (2022)
20. Zhu, P., Fan, Z., Guo, S., Tang, K., Li, X.: Improving adversarial transferability through hybrid augmentation. Comput. Secur. **139**, 103674 (2024)
21. Dong, Y., et al.: Boosting adversarial attacks with momentum. In: 2018 IEEE Conference on Computer Vision and Pattern Recognition (CVPR), pp. 9185–9193 (2018)
22. Shi, Y., Han, Y., Zhang, Q., Kuang, X.: Adaptive iterative attack towards explainable adversarial robustness. Pattern Recogn. **1**(05), 107309 (2020)
23. Liu, J, Zhang, C, Lyu, X. Boosting the transferability of adversarial examples via local mixup and adaptive step size. In: IEEE International Conference on Acoustics, Speech and Signal Processing (ICASSP). IEEE, 2025: 1-5 (2025)
24. Tang, K.: NormalAttack: curvature-aware shape deformation along normals for imperceptible point cloud attack. Secur. Commun. Netw. **2022**, 1186633 (2022)
25. Xie, C., et al.: Improving transferability of adversarial examples with input diversity. In: 2019 IEEE/CVF Conference on Computer Vision and Pattern Recognition (CVPR), pp. 2730–2739 (2019)
26. Wu, D., Wang, Y., Xia, S. T., Bailey, J., Ma, X.: Skip connections matter: On the transferability of adversarial examples generated with resnets. CoRR (2020)
27. Song, S., Chen, Y., Cheung, N. M., Kuo, C. C. J.: Defense against adversarial attacks with saak transform. CoRR (2018)
28. Wu, W., Su, Y., Lyu, M. R., King, I.: Improving the transferability of adversarial samples with adversarial transformations. In: 2021 IEEE/CVF Conference on Computer Vision and Pattern Recognition (CVPR), pp. 9024-9033 (2021)
29. Leon, S.J., Björck, Å., Gander, W.: Gram-Schmidt orthogonalization: 100 years and more. Num. Linear Algebra Appl. **20**(3), 492–532 (2013)
30. Netzer, Y., Wang, T., Coates, A., Bissacco, A., Wu, B., Ng, A.Y.: Reading digits in natural images with unsupervised feature learning. In: NIPS Workshop on Deep Learning and Unsupervised Feature Learning, 2011. 1 (2011)
31. Krizhevsky A and Hinton G.: Learning multiple layers of features from tiny images (2009)
32. He, K., Zhang, X., Ren, S., Sun, J.: Deep residual learning for image recognition. In: 2016 IEEE Conference on Computer Vision and Pattern Recognition (CVPR), pp. 770-778 (2016)
33. Huang, G., Liu, Z., Van Der Maaten, L., Weinberger, K. Q.: Densely connected convolutional networks. In: 2017 IEEE Conference on Computer Vision and Pattern Recognition (CVPR), pp. 4700-4708 (2017)
34. Simonyan, K., Zisserman, A.: Very deep convolutional networks for large-scale image recognition. CoRR (2014)

35. Szegedy, C., et al.: Going deeper with convolutions. In: 2015 IEEE Conference on Computer Vision and Pattern Recognition (CVPR), pp. 1-9 (2015)
36. Dosovitskiy, A., et al.: An image is worth 16x16 words: Transformers for image recognition at scale. CoRR (2020)

V-HOI: Velocity-Aware Human-Object Interaction Generation

Honghui Chen[1], Fan Zhou[1](✉), Ruomei Wang[1], and Baoquan Zhao[2]

[1] School of Computer Science and Engineering, Sun Yat-sen University, Guangzhou, China
chenhh53@mail2.sysu.edu.cn, isszf@mail.sysu.edu.cn
[2] School of Artificial Intelligence, Sun Yat-sen University, Zhuhai, China

Abstract. Generating realistic human-object interactions (HOI) remains challenging due to the complex temporal coordination required between human actions and object responses. Current diffusion-based methods excel at spatial modeling but fail to capture the critical temporal dynamics that govern physically plausible interactions, leading to artifacts such as premature object movement and poor action-response synchronization. To tackle this issue, we introduce V-HOI, a novel velocity-aware framework that explicitly models temporal dynamics through enhanced motion representations and adaptive constraints. A fundamental principle underlying realistic interactions is the coordinated motion dynamics where contact regions and objects exhibit coupled velocity behaviors that vary systematically across interaction phases. Our framework employs a ControlNet-inspired three-branch architecture featuring a dedicated motion branch and dual interaction branches, trained via a two-stage strategy that preserves motion generation capabilities while learning interaction dynamics. Extensive experiments on the FullBody-Manipulation dataset demonstrate substantial improvements over existing methods across both motion quality and interaction metrics, with particularly notable advances in contact modeling accuracy. Ablation studies validate the necessity of each component and confirm that our velocity-aware approach successfully generates temporally coordinated and physically plausible interaction sequences.

Keywords: Human-Object Interaction · Motion Generation · Diffusion Models · Temporal Dynamics · Velocity Modeling

1 Introduction

Generating realistic human-object interactions (HOI) represents one of the most challenging frontiers in computer vision and graphics, requiring models to synthesize motion sequences where humans and objects interact with precise temporal coordination and physical plausibility [7,22]. This capability is crucial for advancing embodied AI systems [16], creating immersive digital experiences [10], and streamlining animation workflows [20], demanding sophisticated understanding of both semantic intentions and physical constraints [6,18].

J. Lokoč et al. (Eds.): MMM 2026, LNCS 16413, pp. 519–532, 2026.
https://doi.org/10.1007/978-981-95-6957-1_37

Recent diffusion-based approaches have achieved remarkable progress in HOI generation [5,11,22], leveraging the powerful generative capabilities of diffusion models to produce diverse interaction sequences. These methods typically employ geometric constraints such as contact maps and distance fields to guide the generation process. However, they exhibit a fundamental limitation: *insufficient modeling of temporal dynamics.* Current approaches predominantly focus on static spatial relationships while overlooking the critical temporal coordination between human actions and object responses that characterizes realistic interactions.

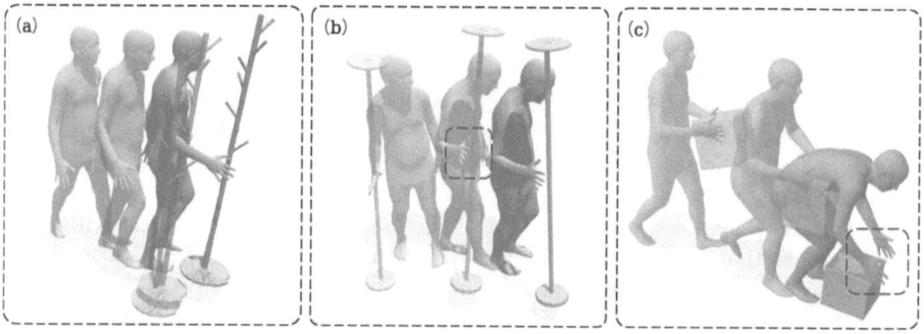

Fig. 1. Temporal misalignment artifacts in existing HOI generation methods. Common failure modes include: (a) premature object displacement occurring before physical contact, (b) unsynchronized motion patterns during single-hand manipulation tasks, and (c) poor temporal coordination in complex dual-hand interactions. These artifacts stem from insufficient temporal dynamics modeling in current approaches.

This temporal modeling deficiency manifests as systematic artifacts illustrated in Fig. 1: premature object movement before contact establishment, inconsistent manipulation patterns during interaction, and poor coordination in multi-contact scenarios. Through systematic analysis of real interaction data, we observe that successful human-object interactions exhibit distinct temporal patterns. Specifically, contact regions and objects demonstrate high velocity synchronization during active contact phases while maintaining independent motion trajectories during separation periods.

To address the temporal coordination challenge in HOI generation, we propose V-HOI, a velocity-aware framework that explicitly models temporal dynamics through three integrated technical contributions. First, we introduce a velocity-enhanced motion representation that augments standard SMPL pose parameters with explicit velocity features computed from frame-to-frame differences. This representation captures temporal dynamics directly for both human motion (joint velocities, global translation/rotation velocities) and object motion (linear/angular velocities), providing richer information for temporal coordination modeling compared to position-only approaches. Second, we design distance-adaptive velocity constraints through two complementary loss functions: an auxiliary velocity consistency loss that enforces coherence between predicted and computed velocities, and a distance-adaptive velocity coordination loss that automatically modulates interaction strength based on hand-object proximity. This approach eliminates the need for explicit contact detection while ensuring physical plausibility. Third, we develop a ControlNet-inspired three-branch architecture that strategically addresses the conflicting optimization requirements of human motion generation and interaction modeling. The framework features a pure motion branch for capturing complex human movement patterns, complemented by dual interaction branches (human and object) with bidirectional cross-attention for coordinated dynamics. A two-stage training strategy first pre-trains the motion branch, then freezes it while training the interaction network, preventing catastrophic forgetting while effectively learning interaction-specific behaviors. Comprehensive evaluation on the FullBodyManipulation dataset demonstrates substantial improvements over state-of-the-art methods: 6% improvement in contact F1-score (0.71 vs 0.67), best-in-class foot sliding (0.34) and FID scores (0.67), establishing new performance benchmarks across both interaction quality and motion fidelity metrics.

In summary, This work makes the following key contributions:

- We propose a velocity-enhanced motion representation that explicitly captures temporal dynamics for improved HOI generation.
- We design distance-based adaptive velocity constraints that ensure physical plausibility without requiring explicit contact detection.
- We develop a ControlNet-based three-branch architecture that effectively combines diffusion models with interaction-specific modules for collaborative human-object motion generation.

2 Related Work

2.1 Diffusion Models for Motion Generation

Diffusion models have emerged as the dominant paradigm for human motion synthesis, establishing new standards for both quality and controllability. Foundational works including MDM [19], MotionDiffuse [26], and MLD [3] demonstrated the effectiveness of diffusion processes for text-conditional motion generation, with subsequent advances focusing on enhanced control mechanisms and improved motion quality. Particularly relevant to our work are recent approaches that recognize the importance of temporal dynamics in motion modeling. HumanTOMATO [14] pioneered the direct incorporation of joint velocities into motion representations, demonstrating significant improvements in reconstruction quality and motion naturalness. OmniControl [21] further validated velocity modeling for enhanced controllability in complex motion sequences. These velocity-enhanced approaches consistently show superior performance in generating physically plausible motions, establishing velocity modeling as essential for realistic motion synthesis. However, despite their success in single-agent scenarios, the application of velocity-aware modeling to multi-agent interactions, particularly human-object coordination, remains largely unexplored. Our work addresses this gap by extending velocity modeling principles to the HOI generation domain.

2.2 Human-Object Interaction Generation

HOI generation has evolved from early autoregressive approaches to sophisticated diffusion-based frameworks. Pioneering work by IMoS [7] established intent-driven synthesis using conditional variational auto-regressors, demonstrating feasibility of generating diverse interaction sequences from textual instructions, though with limitations in modeling complex distributional dynamics.

The adoption of diffusion models marked a significant paradigm shift. InterDiff [22] introduced physics-informed diffusion for 3D HOI anticipation, incorporating physics constraints within diffusion processes. OMOMO [12] proposed a two-stage generation paradigm using dual diffusion models—first synthesizing hand trajectories conditioned on object motion, then generating full-body sequences. CHOIS [11] advanced controllability through keyframe constraints via sparse object waypoints.

Recognizing that generative capability alone is insufficient, researchers have increasingly incorporated explicit physical modeling. CG-HOI [5] explicitly modeled contact relationships with contact gradient guidance, while Text2HOI [2] decomposed generation into contact and motion subtasks. SemGeoMo [4] integrated multi-level geometric representations, including affordance maps and point cloud features. ROG [23] introduced Interactive Distance Fields for comprehensive geometry modeling, and ChainHOI [24] employed kinematic chain modeling for hierarchical motion structures.

Despite these advances, current HOI generation methods exhibit a fundamental limitation: they primarily focus on spatial relationships while neglecting the temporal dynamics crucial for realistic interactions. Most existing approaches lack explicit velocity modeling, resulting in temporal inconsistencies where human actions and object responses are poorly synchronized. Our work addresses this critical gap by introducing velocity-aware modeling to the HOI generation domain.

3 Methodology

Our V-HOI framework addresses the temporal coordination challenge in HOI generation through explicit velocity modeling and adaptive constraint mechanisms. The core insight driving our approach is that realistic human-object interactions exhibit predictable velocity patterns: interacting body parts and objects maintain high velocity consistency during contact phases while following independent motion trajectories during separation periods. As illustrated in Fig. 2, given text descriptions and object geometry, our framework generates realistic interaction sequences through three integrated components: velocity-enhanced

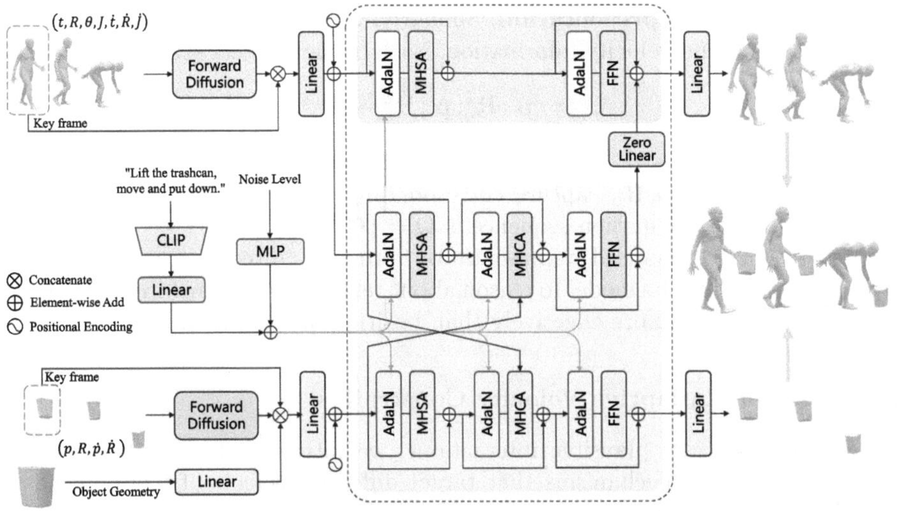

Fig. 2. Overview of the proposed V-HOI framework. Given text descriptions and object geometry, our pipeline generates realistic human-object interactions through three key components: (1) velocity-enhanced motion representation capturing explicit temporal dynamics, (2) ControlNet-based three-branch architecture with dedicated motion and interaction branches, and (3) distance-adaptive velocity constraints ensuring physical plausibility without explicit contact detection.

motion representation that captures explicit temporal dynamics, a ControlNet-inspired three-branch architecture that decouples human motion and interaction modeling, and distance-adaptive velocity constraints that ensure physical plausibility throughout the interaction sequence.

3.1 Velocity-Enhanced Motion Representation

Motivated by recent advances demonstrating the critical importance of temporal dynamics in motion modeling [8,14,21,27], we propose a velocity-enhanced motion representation that explicitly incorporates temporal information into both human and object motion parameterizations.

Human Motion Representation. We utilize the SMPL parametric model [13] as our foundation, where each frame's pose state includes global body translation $\mathbf{t} \in \mathbb{R}^3$, global body orientation $\mathbf{R}_h \in \mathbb{R}^6$ (6D rotation representation), joint rotations $\boldsymbol{\theta} \in \mathbb{R}^{21\times 6}$, and 3D joint positions $\mathbf{J} \in \mathbb{R}^{24\times 3}$ for 24 key body joints. To capture essential temporal dynamics, we augment this representation with explicit velocity features: global translation velocity $\dot{\mathbf{t}} \in \mathbb{R}^3$, global rotation velocity $\dot{\mathbf{R}}_h \in \mathbb{R}^6$, and joint position velocities $\dot{\mathbf{J}} \in \mathbb{R}^{24\times 3}$, computed as frame-to-frame differences using finite differences. For a sequence of N frames, the complete human motion representation becomes:

$$\mathbf{H} = \{\mathbf{H}^i\}_{i=1}^N, \quad \mathbf{H}^i = [\mathbf{t}^i; \mathbf{R}_h^i; \boldsymbol{\theta}^i; \mathbf{J}^i; \dot{\mathbf{t}}^i; \dot{\mathbf{R}}_h^i; \dot{\mathbf{J}}^i] \in \mathbb{R}^{288}. \tag{1}$$

Object Motion Representation. Similarly, we represent object motion with both positional and velocity information. Each frame's object state is defined as:

$$\mathbf{O}^i = [\mathbf{p}^i; \mathbf{R}_o^i; \dot{\mathbf{p}}^i; \dot{\mathbf{R}}_o^i] \in \mathbb{R}^{18}, \tag{2}$$

where $\mathbf{p} \in \mathbb{R}^3$ denotes global center position, $\mathbf{R}_o \in \mathbb{R}^6$ represents object rotation in 6D format, and $\dot{\mathbf{p}}, \dot{\mathbf{R}}_o$ capture corresponding linear and angular velocities. The complete object motion sequence is $\mathbf{O} = \{\mathbf{O}^i\}_{i=1}^N$.

This velocity-enhanced representation provides direct access to motion dynamics, enabling the model to reason about temporal relationships and enforce physical constraints more effectively than position-only representations.

3.2 Distance-Adaptive Velocity Constraints

To ensure physically plausible interactions, we introduce two complementary velocity constraint mechanisms that target different aspects of motion realism and interaction quality.

Auxiliary Velocity Consistency Loss. We enforce consistency between explicitly predicted velocity features and velocities computed from predicted positions, ensuring internal coherence in the model's temporal understanding:

$$\begin{aligned}\mathcal{L}_{aux} = &\|\dot{\hat{\mathbf{t}}} - \hat{\dot{\mathbf{t}}}\|_1 + \|\dot{\hat{\mathbf{J}}} - \hat{\dot{\mathbf{J}}}\|_1 + \|\dot{\hat{\mathbf{R}}}_h - \hat{\dot{\mathbf{R}}}_h\|_1 \\ &+ \|\dot{\hat{\mathbf{p}}} - \hat{\dot{\mathbf{p}}}\|_1 + \|\dot{\hat{\mathbf{R}}}_o - \hat{\dot{\mathbf{R}}}_o\|_1,\end{aligned} \tag{3}$$

where $\hat{\dot{(\cdot)}}$ represents model-predicted velocity features and $\dot{\hat{(\cdot)}}$ denotes velocities computed from frame-to-frame position differences.

Distance-Adaptive Velocity Coordination Loss. Our core idea lies in modeling interaction constraints through adaptive velocity coordination. Rather than requiring explicit contact detection, we leverage the key insight that hand-object proximity correlates with interaction strength. We define:

$$\begin{aligned}\mathcal{L}_{vel} = \sum_{t=1}^{T} \big[&w_{lh}^t \|\Delta\hat{\mathbf{v}}_{lh}^t - \Delta\mathbf{v}_{lh}^t\|_1 + w_{rh}^t \|\Delta\hat{\mathbf{v}}_{rh}^t - \Delta\mathbf{v}_{rh}^t\|_1 \\ &+ w_{obj}^t \|\hat{\dot{\mathbf{p}}}^t - \dot{\mathbf{p}}^t\|_2 \big],\end{aligned} \tag{4}$$

where $\Delta\hat{\mathbf{v}}_{lh}^t = \hat{\dot{\mathbf{p}}}^t - \hat{\dot{\mathbf{J}}}_{lh}^t$ represents the predicted relative velocity between object and left hand, with $\Delta\mathbf{v}_{lh}^t$ as the ground truth counterpart.

The adaptive weighting mechanism automatically modulates constraint strength based on spatial proximity:

$$w_{obj}^t = 1 - \max(w_{lh}^t, w_{rh}^t), \tag{5}$$

$$w_{lh}^t = \exp\left(-(\|\mathbf{J}_{lh}^t - \mathbf{p}^t\|_2 - d_{lh}^{\min})\right), \tag{6}$$

$$w_{rh}^t = \exp\left(-(\|\mathbf{J}_{rh}^t - \mathbf{p}^t\|_2 - d_{rh}^{\min})\right), \tag{7}$$

where $d_{lh}^{\min}$ and $d_{rh}^{\min}$ are minimum hand-object distances throughout the sequence. This design elegantly balances object motion continuity and interaction constraints: when hands are distant, objects maintain autonomous motion ($w_{obj} \to 1$); during close proximity, relative velocity constraints dominate, ensuring coordinated motion without explicit contact supervision.

3.3 ControlNet-Based Three-Branch Architecture

HOI generation involves two distinct but coupled challenges: generating natural human motion and coordinating object dynamics. Directly modeling both aspects in a single network may lead to suboptimal learning due to the different nature of human and object representations. Given that human motion exhibits significantly higher complexity and variability compared to object trajectories, we design a three-branch architecture with a dedicated pure motion branch to capture rich human movement patterns, complemented by interaction branches for human-object coordination. We address this challenge through a ControlNet-inspired three-branch architecture that strategically separates concerns while enabling coordinated learning.

Pure Motion Branch. The foundation component employs a standard diffusion transformer with L blocks, each containing adaptive layer normalization (AdaLN), self-attention, and feed-forward layers. This branch processes noised velocity-enhanced human representations $\mathbf{H}_t$ with keyframe constraints via the CHOIS masking strategy [11]. Text conditions and diffusion timesteps are integrated through AdaLN following established practices [15,17].

Dual-Branch Interaction Network. The interaction component utilizes two symmetric transformer architectures for coordinated human-object modeling. The human interaction branch processes identical noised inputs $\mathbf{H}_t$ as the motion branch, while the object branch handles noised object representations $\mathbf{O}_t$ concatenated with BPS-encoded geometry features [15]. Bidirectional cross-attention between branches facilitates information exchange for coordinated motion generation.

ControlNet Integration. Following ControlNet principles [25], we inject human interaction branch outputs into corresponding motion branch transformer blocks via zero-initialized residual connections. This block-wise fusion enables fine-grained interaction modeling while preserving pre-trained motion generation capabilities.

Two-Stage Training Strategy. Our training approach prevents catastrophic forgetting while enabling effective interaction learning:

- *Stage 1:* Pre-train the motion branch using standard human motion data with loss $\mathcal{L} + \mathcal{L}_{aux}$.
- *Stage 2:* Freeze the motion branch and train the interaction network using HOI data with loss $\mathcal{L} + \mathcal{L}_{aux} + \mathcal{L}_{vel}$.

The human interaction branch is initialized by copying pre-trained motion branch parameters, while the object branch and cross-attention modules use random initialization. This strategy ensures stable learning while adapting to interaction-specific requirements.

Training Objective. Following DDPM [9], we employ L1 reconstruction loss for direct x_0 prediction:

$$\mathcal{L} = \mathbb{E}_{\mathbf{x}_0,\epsilon,t}\|\hat{\mathbf{x}}_\theta(\mathbf{x}_t, t, \mathbf{c}) - \mathbf{x}_0\|_1, \tag{8}$$

where $\mathbf{x}_t = \sqrt{\bar{\alpha}_t}\mathbf{x}_0 + \sqrt{1-\bar{\alpha}_t}\epsilon$ represents the noised input at timestep t, and $\mathbf{c}$ denotes conditioning information (text and object geometry).

4 Experiments

We conduct a comprehensive evaluation to demonstrate the effectiveness of our velocity-aware approach across multiple dimensions of HOI generation quality.

4.1 Experimental Configuration

Dataset. We evaluate on the FullBodyManipulation dataset [12], which provides 10 h of high-fidelity human-object motion data spanning 15 object categories. Focusing on rigid object interactions, we exclude sequences with articulated objects (vacuum cleaners, mops). Following standard protocols [12], we use 15 subjects for training and 2 subjects for testing. We select this dataset over alternatives like BEHAVE [1] due to its superior motion quality, larger scale, and inclusion of locomotion patterns essential for realistic interaction modeling.

Implementation Details. Our transformer architecture employs 10 layers with 4 attention heads, feed-forward dimension of 1024, latent dimension of 512, and dropout rate of 0.1. Object geometry is encoded via 1024×3 dimensional BPS features with linear projection to latent dimension. All loss weights are set to 1.0 during training. Motion sequences undergo canonical preprocessing: rotation alignment (human left axis to +x direction), translation normalization (root joint to origin), z-score feature normalization, and padding to 120 frames maximum length.

Evaluation Metrics. We assess performance across two dimensions:

- *Motion Fidelity:* Foot sliding score (FS) quantifies unnatural foot movement, foot height (H_{feet}) measures average foot elevation, and Fréchet Inception Distance (FID) assesses distributional similarity with ground truth.
- *Interaction Quality:* Hand-object contact evaluation through binary classification using 5cm distance threshold, reporting precision (C_{prec}), recall (C_{rec}), and F1-score (C_{F1}) following established protocols [12].

Table 1. Quantitative comparison of different HOI generation methods on the FullBodyManipulation dataset. ↓ indicates lower is better, ↑ indicates higher is better. The best and second-best results are bolded and underlined, respectively.

Method	Human Motion			Interaction		
	H_{feet} ↓	FS↓	FID↓	C_{prec} ↑	C_{rec} ↑	C_{F1} ↑
Interdiff	**0.90**	0.42	208.0	0.63	0.28	0.33
MDM	6.72	0.48	6.16	0.72	0.47	0.53
OMOMO	7.08	0.40	4.19	0.73	<u>0.66</u>	0.66
CHOIS	4.20	<u>0.35</u>	<u>0.69</u>	**0.80**	0.64	<u>0.67</u>
V-HOI (Ours)	<u>2.42</u>	**0.34**	**0.67**	<u>0.76</u>	**0.67**	**0.71**

4.2 Comparison with State-of-the-Art Methods

Table 1 presents comprehensive comparison results against state-of-the-art HOI generation methods. Our V-HOI framework demonstrates superior performance across most metrics, establishing new benchmarks for HOI generation quality. For human motion fidelity, our method achieves the best foot sliding score (0.34) and FID score (0.67), indicating superior motion realism and distributional alignment with ground truth data. The second-best foot height performance (2.42) demonstrates effective foot-ground contact modeling. These results validate that our velocity-aware approach enhances motion quality while maintaining the diversity essential for realistic motion generation. Most significantly, our method achieves the highest contact F1-score (0.71), representing a 6% improvement over the previous best method (CHOIS: 0.67). This improvement stems from enhanced recall (0.67 vs 0.64), indicating more comprehensive contact detection capability. The balanced precision (0.76) demonstrates that our method achieves improved contact modeling without sacrificing accuracy. The superior interaction performance directly validates our core hypothesis: explicit velocity modeling enables better temporal coordination in HOI generation. The distance-adaptive velocity constraints successfully capture the dynamic nature of hand-object interactions, providing stronger supervision for contact modeling without requiring explicit contact annotations.

4.3 Ablation Studies

We conduct systematic ablation studies to validate each component's contribution and understand the mechanisms underlying our performance improvements.

Table 2 presents ablation results for our key components: auxiliary loss $\mathcal{L}_{aux}$, velocity consistency loss $\mathcal{L}_{vel}$, and motion branch (MB). The results demonstrate that each component contributes meaningfully to overall performance. Removing the velocity consistency loss results in noticeable decline in interaction quality, with the F1-score dropping from 0.71 to 0.67.

Table 2. Ablation study on key components. $\mathcal{L}_{aux}$: auxiliary loss, $\mathcal{L}_{vel}$: velocity consistency loss, MB: motion branch. All experiments use velocity-enhanced representation.

Components			Human Motion			Interaction		
$\mathcal{L}_{aux}$	$\mathcal{L}_{vel}$	MB	H_{feet} ↓	FS↓	FID↓	C_{prec} ↑	C_{rec} ↑	C_{F1} ↑
✓		✓	3.32	0.38	0.82	0.70	0.55	0.62
	✓	✓	2.57	0.37	2.00	0.76	0.59	0.67
✓	✓		3.91	4.64	6.28	0.63	0.32	0.43
✓	✓	✓	**2.42**	**0.34**	**0.67**	**0.76**	**0.67**	**0.71**

Table 3. Impact of transformer layer depth on performance. All experiments use the complete model with velocity-enhanced representation.

Layers	Human Motion			Interaction		
	H_{feet} ↓	FS↓	FID↓	C_{prec} ↑	C_{rec} ↑	C_{F1} ↑
4	3.66	0.42	0.89	0.70	0.53	0.60
6	2.68	0.39	0.78	0.74	0.59	0.66
8	2.46	0.36	0.74	0.72	0.62	0.66
10	**2.42**	**0.34**	**0.67**	**0.76**	**0.67**	**0.71**

The motion branch ablation provides particularly revealing insights. Without this component, performance deteriorates dramatically across all metrics—foot sliding increases from 0.34 to 4.64, and FID worsens from 0.67 to 6.28. This substantial decline validates our central design rationale: while velocity-enhanced representations effectively capture motion dynamics, they introduce considerable training complexity that impedes model convergence.

Our motion branch architecture directly tackles this challenge by providing a dedicated pathway for human motion learning. This specialized design enables stable training despite the increased representational complexity of velocity features. The ablation results confirm that without this architectural support, the model cannot effectively leverage velocity-enhanced representations, leading to poor motion synthesis and interaction modeling. These findings demonstrate that the motion branch represents a necessary architectural innovation rather than a mere performance enhancement.

Network Depth Analysis. We investigate the impact of network depth on performance. Table 3 shows results for different numbers of transformer layers. Performance consistently improves with increased depth, with 10 layers achieving optimal results. The most significant improvements occur between 4–8 layers, where H_{feet} drops by 36.8% and FID decreases by 59.3%. Though deeper networks naturally require more computational resources, the performance benefits make this a reasonable choice.

Qualitative Analysis. Figure 3 provides qualitative validation of our key design choices. We compare our complete method against ground truth and an ablated version without velocity consistency loss ($\mathcal{L}_{vel}$). Our complete method successfully generates coherent and realistic motions that closely match the ground truth, maintaining natural pose transitions throughout the interaction sequences.

Fig. 3. Qualitative ablation results comparing ground truth, our complete method, and our method without velocity consistency loss ($\mathcal{L}_{vel}$). Our complete method generates natural motions with proper hand-object coordination, while removing the velocity consistency loss leads to hand-object misalignment as highlighted in the red boxes.

The chair and floor lamp manipulation tasks demonstrate our method's ability to accurately capture complex single-handed interactions, with consistent spatial relationships between the hand and objects throughout the sequences. The red-boxed regions clearly illustrate the critical importance of the velocity consistency loss for maintaining proper hand-object synchronization. Without this component, hand-object coordination breaks down, resulting in implausible interactions where the hand fails to maintain proper contact with objects. These qualitative results provide visual confirmation of our quantitative ablation findings and demonstrate the necessity of each proposed component.

5 Conclusion

In this paper, we have presented V-HOI, a velocity-enhanced framework for human-object interaction generation. Our approach introduces explicit velocity features into motion representations and employs a ControlNet-based three-branch transformer architecture consisting of a pure motion branch and dual interaction branches with bidirectional cross-attention. Key technical contributions include a velocity-enhanced representation that incorporates motion dynamics, a contact-aware loss function for interaction modeling, and a three-branch architecture with a two-stage training strategy that prevents catastrophic forgetting while learning interaction dynamics. Experimental results on the FullBodyManipulation dataset demonstrate that our method achieves superior interaction performance and competitive motion quality metrics, representing significant improvements over existing approaches. This work advances HOI generation by explicitly modeling motion dynamics through velocity features and providing a unified framework that balances motion realism with interaction accuracy, establishing a foundation for more realistic human-object interaction modeling.

Acknowledgments. This research is supported by the National Natural Science Foundation of China (No.62172452), and in part by the Guangdong Basic and Applied Basic Research Foundation (No.2023A1515011639).

Disclosure of Interests. The authors have no competing interests to declare that are relevant to the content of this article.

References

1. Bhatnagar, B.L., Xie, X., Petrov, I.A., Sminchisescu, C., Theobalt, C., Pons-Moll, G.: Behave: dataset and method for tracking human object interactions. In: Proceedings of the IEEE/CVF Conference on Computer Vision and Pattern Recognition, pp. 15935–15946 (2022)
2. Cha, J., Kim, J., Yoon, J.S., Baek, S.: Text2hoi: Text-guided 3D motion generation for hand-object interaction. In: Proceedings of the IEEE/CVF Conference on Computer Vision and Pattern Recognition, pp. 1577–1585 (2024)
3. Chen, X., et al.: Executing your commands via motion diffusion in latent space. In: Proceedings of the IEEE/CVF Conference on Computer Vision and Pattern Recognition, pp. 18000–18010 (2023)
4. Cong, P., Wang, Z., Ma, Y., Yue, X.: Semgeomo: dynamic contextual human motion generation with semantic and geometric guidance. In: Proceedings of the Computer Vision and Pattern Recognition Conference, pp. 17561–17570 (2025)
5. Diller, C., Dai, A.: CG-HOI: contact-guided 3D human-object interaction generation. In: Proceedings of the IEEE/CVF Conference on Computer Vision and Pattern Recognition, pp. 19888–19901 (2024)
6. Fan, S., Huang, W., Cai, X., Du, B.: 3D human interaction generation: a survey. arXiv preprint arXiv:2503.13120 (2025)
7. Ghosh, A., Dabral, R., Golyanik, V., Theobalt, C., Slusallek, P.: IMoS: Intent-driven full-body motion synthesis for human-object interactions. In: Computer Graphics Forum. vol. 42, pp. 1–12. Wiley Online Library (2023)

8. Guo, C., et al.: Generating diverse and natural 3D human motions from text. In: Proceedings of the IEEE/CVF Conference on Computer Vision and Pattern Recognition, pp. 5152–5161 (2022)
9. Ho, J., Jain, A., Abbeel, P.: Denoising diffusion probabilistic models. Adv. Neural. Inf. Process. Syst. **33**, 6840–6851 (2020)
10. Kuo, C.F., Wang, M.J.J.: Motion generation and virtual simulation in a digital environment. Int. J. Prod. Res. **50**(22), 6519–6529 (2012)
11. Li, J., Clegg, A., Mottaghi, R., Wu, J., Puig, X., Liu, C.K.: Controllable human-object interaction synthesis. In: European Conference on Computer Vision, pp. 54–72. Springer (2024)
12. Li, J., Wu, J., Liu, C.K.: Object motion guided human motion synthesis. ACM Trans. Graph. (TOG) **42**(6), 1–11 (2023)
13. Loper, M., Mahmood, N., Romero, J., Pons-Moll, G., Black, M.J.: SMPL: a skinned multi-person linear model. ACM Trans. Graph. (TOG) **34**(6), 1–16 (2015)
14. Lu, S., et al.: Humantomato: text-aligned whole-body motion generation. In: International Conference on Machine Learning, pp. 32939–32977. PMLR (2024)
15. Lv, X., et al.: Himo: a new benchmark for full-body human interacting with multiple objects. In: European Conference on Computer Vision, pp. 300–318. Springer (2024)
16. Pan, L., et al.: Tokenhsi: unified synthesis of physical human-scene interactions through task tokenization. In: Proceedings of the Computer Vision and Pattern Recognition Conference, pp. 5379–5391 (2025)
17. Peebles, W., Xie, S.: Scalable diffusion models with transformers. In: Proceedings of the IEEE/CVF International Conference on Computer Vision, pp. 4195–4205 (2023)
18. Sui, K., Ghosh, A., Hwang, I., Zhou, B., Wang, J., Guo, C.: A survey on human interaction motion generation. arXiv preprint arXiv:2503.12763 (2025)
19. Tevet, G., Raab, S., Gordon, B., Shafir, Y., Cohen-or, D., Bermano, A.H.: Human motion diffusion model. In: The Eleventh International Conference on Learning Representations (2023)
20. de la Vega-Hazas, L., Calatayud, F., Iglesias, A.: Applying evolutionary computation operators for automatic human motion generation in computer animation and video games. In: International Symposium on Intelligent and Distributed Computing, pp. 247–258. Springer (2018)
21. Xie, Y., Jampani, V., Zhong, L., Sun, D., Jiang, H.: Omnicontrol: control any joint at any time for human motion generation. In: The Twelfth International Conference on Learning Representations, ICLR 2024, Vienna, Austria, May 7-11, 2024. OpenReview.net (2024)
22. Xu, S., Li, Z., Wang, Y.X., Gui, L.Y.: Interdiff: generating 3D human-object interactions with physics-informed diffusion. In: Proceedings of the IEEE/CVF International Conference on Computer Vision, pp. 14928–14940 (2023)
23. Xue, M., Liu, Y., Guo, L., Huang, S., Ding, C.: Guiding human-object interactions with rich geometry and relations. In: Proceedings of the Computer Vision and Pattern Recognition Conference, pp. 22714–22723 (2025)
24. Zeng, L.A., et al.: Chainhoi: joint-based kinematic chain modeling for human-object interaction generation. In: Proceedings of the Computer Vision and Pattern Recognition Conference, pp. 12358–12369 (2025)
25. Zhang, L., Rao, A., Agrawala, M.: Adding conditional control to text-to-image diffusion models. In: Proceedings of the IEEE/CVF International Conference on Computer Vision, pp. 3836–3847 (2023)

26. Zhang, M., et al.: Motiondiffuse: text-driven human motion generation with diffusion model. IEEE Trans. Pattern Anal. Mach. Intell. **46**(6), 4115–4128 (2024)
27. Zhao, K., Li, G., Tang, S.: Dartcontrol: a diffusion-based autoregressive motion model for real-time text-driven motion control. In: The Thirteenth International Conference on Learning Representations (2025)

WildfireX-SLAM: A Large-Scale Low-Altitude RGB-D Dataset for Wildfire SLAM and Beyond

Zhicong Sun[1], Jacqueline Lo[1(✉)], and Jinxing Hu[2(✉)]

[1] Hong Kong Polytechnic University, Hong Kong S.A.R., China
zhicong.sun@connect.polyu.hk, jacqueline.lo@polyu.edu.hk
[2] Shenzhen Institutes of Advanced Technology, Chinese Academy of Sciences, Shenzhen, China
jinxing.hu@siat.ac.cn

Abstract. 3D Gaussian splatting (3DGS) and its subsequent variants have led to remarkable progress in simultaneous localization and mapping (SLAM). While most recent 3DGS-based SLAM works focus on small-scale indoor scenes, developing 3DGS-based SLAM methods for large-scale forest scenes holds great potential for many real-world applications, especially for wildfire emergency response and forest management. However, this line of research is impeded by the absence of a comprehensive and high-quality dataset, and collecting such a dataset over real-world scenes is costly and technically infeasible. To this end, we have built a large-scale, comprehensive, and high-quality synthetic dataset for SLAM in wildfire and forest environments. Leveraging the Unreal Engine 5 Electric Dreams Environment Sample Project, we developed a pipeline to easily collect aerial and ground views, including ground-truth camera poses and a range of additional data modalities from unmanned aerial vehicle. Our pipeline also provides flexible controls on environmental factors such as light, weather, and types and conditions of wildfire, supporting the need for various tasks covering forest mapping, wildfire emergency response, and beyond. The resulting pilot dataset, WildfireX-SLAM, contains 5.5k low-altitude RGB-D aerial images from a large-scale forest map with a total size of 16 km^2. On top of WildfireX-SLAM, a thorough benchmark is also conducted, which not only reveals the unique challenges of 3DGS-based SLAM in the forest but also highlights potential improvements for future works. The dataset and code will be publicly available. Project page: https://zhicongsun.github.io/wildfirexslam.

Keywords: Wildfire · Simultaneous Localization and Mapping · Unmanned Aerial Vehicle

1 Introduction

Simultaneous Localization and Mapping (SLAM) has continually benefited from advancements in multimedia technologies. Recently, high-fidelity real-time rendering techniques, prominently represented by 3D Gaussian Splatting (3DGS),

J. Lokoč et al. (Eds.): MMM 2026, LNCS 16413, pp. 533–546, 2026.
https://doi.org/10.1007/978-981-95-6957-1_38

have significantly propelled the development of SLAM. Cutting-edge 3DGS-based SLAM, such as SplaTAM [16], MonoGS [23], and GS-SLAM [32], have achieved remarkable success in terms of mapping quality and real-time performance, especially in controlled static indoor environments. Furthermore, methods like DG-SLAM [32] address dynamic tracking challenges by filtering out moving objects, showing state-of-the-art results in dynamic indoor settings. However, despite these developments, the application of 3DGS-based SLAM in large-scale outdoor environments remains underexplored. This gap is particularly prominent in scenarios such as wildfire emergency response and forest management, where robust SLAM systems can play a transformative role.

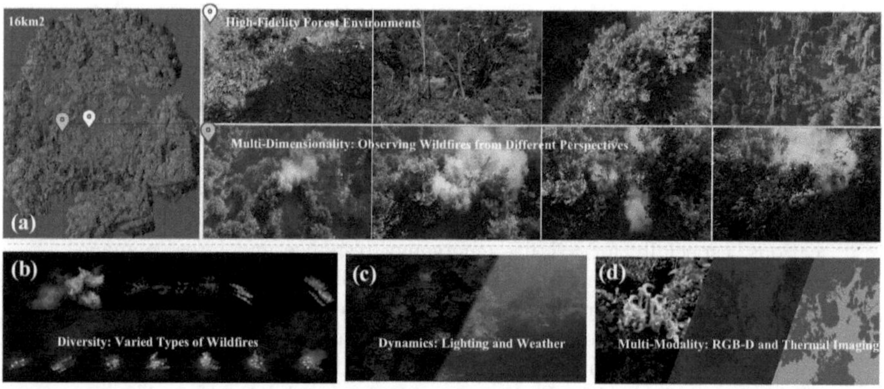

Fig. 1. Overview of *WildfireX-SLAM.* (a) We have constructed a 16 km^2 forest scene featuring wildfires, offering a high-fidelity forest environment and multi-dimensional wildfire observations. (b) Our framework enables the configuration of various types of wildfires. (c) Through diverse weather and lighting conditions, our framework provides high-fidelity and realistic scenes. (d) We deliver multi-modal data, especially RGB-D from UAVs, to facilitate tasks such as SLAM and navigation.

Existing SLAM datasets, such as KITTI [10], Oxford RobotCar [22], and EuRoC [2], have laid the groundwork for advancing SLAM techniques, but they focus on structured, man-made environments. Forest environments present a unique set of challenges due to monotonic textures, sparse structural features, and the dynamic nature of wildfires. Forest-specific datasets, such as SLOAM [5], FinnForest [1], and Wild-Places [18], attempt to address these challenges, but they lack key elements critical for wildfire scenarios, such as dynamic wildfire states, multi-modal observations, and controllable environmental conditions. Furthermore, these datasets typically rely on a single observation platform (either aerial or ground-based), limiting their adaptability to real-world applications that require multi-agent coordination.

To bridge these gaps, we present WildfireX-SLAM, a pioneering SLAM dataset specifically designed for wildfire and forest environments. Built on Unreal

Engine 5 and AirSim, this dataset provides synchronized aerial and ground observations, including high-quality RGB, thermal, and depth imagery, accompanied by ground-truth camera poses. WildfireX-SLAM enables realistic simulations with controllable environmental factors, such as wildfire types, weather, and lighting conditions, offering a flexible platform for diverse SLAM-related tasks, including wildfire emergency response, forest mapping, and swarm-based perception. With 5.5k aerial images collected over a 16 km^2 forest map, WildfireX-SLAM represents a comprehensive resource for SLAM research in these challenging environments.

In addition to introducing the dataset, we perform extensive benchmarking on WildfireX-SLAM, revealing critical challenges and unique behaviors exhibited by 3DGS-based SLAM methods in wildfire and forest scenarios. These experiments not only expose the limitations of current techniques but also highlight opportunities for developing robust methods tailored to these environments. We believe WildfireX-SLAM will serve as a valuable tool for the community, accelerating research on emergency response, environmental monitoring, and beyond.

In summary, our contributions are as follows:

1. **Development of the WildfireX-SLAM Dataset:** We present WildfireX-SLAM, a large-scale and high-quality SLAM dataset specifically designed for wildfire environments. This dataset highlights key attributes of wildfire scenarios, including complex dynamic forest environments, diverse wildfire types, and variable weather conditions. WildfireX-SLAM consists of aerial imagery captured by unmanned aerial vehicles (UAVs) flying at low altitudes below 50 meters above ground, supplemented with additional data modalities such as depth, supporting a wide range of SLAM-related tasks.
2. **Creation of a Flexible Data Collection Framework:** We developed a robust framework leveraging Unreal Engine 5 and AirSim to automate the collection of high-quality forest data. This framework allows researchers to flexibly control wildfire states, weather conditions, and sensor configurations, simplifying data acquisition for various task setups. It is a valuable community resource, enabling users to generate advanced datasets tailored to their specific research needs.
3. **Extensive Benchmarking and Insights:** We conducted comprehensive experiments on the WildfireX-SLAM dataset to uncover key challenges in performing SLAM in wildfire and forest environments. These insights aim to drive future research efforts and advancements in this domain.

2 Related Works

Recently, benefiting from the explicit map representation and real-time high-fidelity rendering capabilities provided by 3D Gaussian Splatting (3DGS) [17], SLAM technology has made significant advancements in both real-time performance and mapping quality, bringing it closer to practical deployment. 3DGS-based methods, such as SplaTAM [16], MonoGS [23], and GS-SLAM [32], have achieved state-of-the-art (SOTA) performance in terms of map reconstruction quality and speed within static environments. DG-SLAM [32] attempts to address tracking challenges in dynamic environments by filtering out dynamic objects, achieving SOTA performance in dynamic indoor scenes. However, most of these approaches remain confined to controlled indoor scenarios, with limited exploration of their application in outdoor emergency response contexts, where SLAM could play a crucial role. The primary underlying reason for this limitation lies in the absence of relevant datasets.

In the field of SLAM, numerous representative datasets, such as MIT-DARPA [14], Rawseeds [8], KITTI [10], NCLT [4], Oxford RobotCar [22], KAIST Day/Night [31], 4-Seasons [6], ComplexUrban [15], UrbanLoco [30], TUM-RGBD [28], EuRoC [2], TUM-VI [26], OpenLORIS [27], M2DGR [33], Newer College [24], and Hilti SLAM [12], have been developed. These datasets cover a wide range of real-world challenges in structured environments, providing invaluable opportunities for algorithm validation and improvement. However, emergency SLAM applications often operate in unstructured environments that lack clear, regular, or well-defined features. Forest environments exemplify such conditions, presenting monotonic textures and limited structural cues, which pose a series of challenges to SLAM systems. Currently, the following datasets focus on forest-specific scenarios:

SLOAM [5] dataset utilizes LiDAR scans in dense forest environments to enable semantic-aware SLAM, providing ground-truth labels for terrain and tree instances. It addresses vegetation occlusion and sparse geometric features, supporting applications in forest inventory and autonomous navigation under unstructured conditions.

FinnForest [1] and FinnWoodlands [19] datasets can be jointly used to address forest environment perception. FinnForest captures multi-sensor seasonal data (RGB, IMU, GNSS) for evaluating SLAM robustness under natural lighting and terrain dynamics, while FinnWoodlands provides static scenes with stereo RGB, LiDAR point clouds, and semantic instance labels (tree species, terrain types) to advance fine-grained environmental understanding and depth completion models.

Wild-Places [18] dataset is a longitudinal LiDAR-based collection designed to evaluate long-term autonomy in unstructured natural environments. It comprises 63k georeferenced submaps captured over 14 months across seasonal variations (e.g., vegetation growth/decay), addressing challenges in place recognition and semantic segmentation under GNSS-denied conditions. With multi-temporal revisits and domain shifts induced by natural environmental dynamics, it supports robustness analysis of lidar odometry and retrieval systems in dense forestry scenarios.

BotanicGarden [21] dataset provides multi-sensor trajectories (spinning/MEMS LiDARs, stereo cameras, IMUs) spanning 17.1 km in a 48k m^2 botanical garden, focusing on SLAM performance in texture-poor, slope-varied terrains. It features synchronized ground-truth maps, annotated semantic obstacles, and traversability labels, specifically targeting navigation challenges in vegetation-dense, GNSS-deprived ecosystems with weak visual features and complex geometric structures.

Although the aforementioned datasets have made significant contributions to SLAM in forest environments, their lack of wildfire-specific elements renders them unsuitable for SLAM tasks in wildfire scenarios. Furthermore, these datasets rely on a single observation platform, either ground-based vision or aerial perspectives, and their lighting and weather conditions are entirely dependent on the time of data collection, making them uncontrollable. The statistics and properties of these related datasets are summarized in Table 1.

In contrast, the WildfireX-SLAM dataset proposed in this paper is the first specifically designed for wildfire observation. Beyond incorporating wildfire elements, our dataset provides more comprehensive observation perspectives, including both near-ground and aerial views. Moreover, lighting and weather conditions are fully controllable, which enables the realistic simulation of the challenges that SLAM and navigation tasks face in wildfire emergency response scenarios. Our dataset can also be extended to support multi-agent swarm-based emergency environment perception tasks. This scalability allows for the simulation and study of coordinated operations involving multiple UAVs and unmanned ground vehicles in dynamic wildfire scenarios, further enhancing its applicability to real-world emergency response and collaborative SLAM research.

Table 1. Comparison of statistics and properties between our *WildfireX-SLAM* dataset and related datasets.

Dataset	Data	Scene	Viewpoint	Source	Lighting	Weather	Forest/Tree	Flame/Smoke
S3E [7]	RGB-D & Heat & Lidar	Campus	Ground	Real	✗	✗	✓	✗
KITTI 2 [9]	RGB-D	City	Ground	Synthetic & Real	✓	✓	✗	✗
BASEPROD [11]	RGB-D & Heat	Desert	Ground	Real	✗	✗	✗	✗
SLOAM [5]	Lidar	Forest	Aerial	Real	✗	✗	✓	✗
FinnForest [1]	RGB	Forest	Ground	Real	✓	✓	✓	✗
FinnWoodlands [19]	RGB-D & Lidar	Forest	Ground	Real	✓	✓	✓	✗
Wild-places [18]	Lidar	Forest	Ground	Real	✓	✓	✓	✗
Botanicgarden [21]	RGB & Lidar	Forest	Ground	Real	✓	✓	✓	✗
Ours	RGB-D & Heat	Forest	Aerial+Near-ground	Synthetic	✓	✓	✓	✓

3 WildfireX-SLAM Dataset

The WildfireX-SLAM dataset aims to establish an unprecedented and challenging benchmark for SLAM and navigation in the context of wildfire emergency response by providing comprehensive wildfire scene maps observed from UAVs. In addition to RGB images, we provide depth images, normal maps, and thermal images observed by sensors on UAVs. Furthermore, we include their motion data, such as current position, velocity, acceleration, and camera poses, to support various other tasks. Furthermore, we offer flexible control over environmental factors, such as forest complexity, the types, colors, brightness, and spread rates of smoke and fire elements, as well as the direction and intensity of light and the density of fog, to simulate complex and dynamic real-world scenarios. Section 3.1 details the data construction process, while Sect. 3.2 present comprehensive statistics and characteristics of the dataset.

3.1 Dataset Construction

Construction of Wildfire Scenarios. Due to the limitations of human resources and the priority of emergency rescue, deploying swarms of UAVs and unmanned vehicles to collect sufficient data in real-world wildfire scenarios is impractical. Therefore, we constructed high-fidelity wildfire scenarios using Unreal Engine 5. The rationale behind using UE and Nigara to simulate wildfire as well as the effectiveness of sim-to-real transfer have been discussed and shown in the FireFly dataset [13]. Our synthetic wildfire scene seamlessly integrates two originally independent Unreal Engine 5 systems: the Electric Dreams Environment Sample Project, used for procedural biome generation with PCG-controlled terrain and Quixel Megascans assets, and the M5 VFX Niagara Fire System, which enables Niagara-based particle simulations with 4096 $\times$ 4096 fire and smoke textures. By embedding these systems into a unified UE5 project, we synchronize environmental parameters via Unreal Blueprints, dynamically linking PCG-generated variables such as vegetation density and slope gradients to the combustion properties of the M5 VFX system, including flame spread velocity and smoke particle density. This integration leverages Unreal Engine 5's native interoperability, allowing the creation of wildfire scenarios in which fire propagation adapts to procedurally generated fuel distributions. The constructed wildfire scenario, as illustrated in Fig. 1, encompasses a 16-square-kilometer forest environment featuring diverse forest landscapes. The wildfires can be observed from any perspectives and at different altitudes.

Fig. 2. Diversity of the *WildfireX-SLAM* dataset.

Data Collection Tool. SLAM and navigation in wildfire scenarios using UAVs require the dense collection of 2D images, depth maps, and thermal images to learn accurate scene representations and to avoid hazardous areas. In Unreal Engine 5, collecting such data is a complex process that involves setting up adjustable camera parameters to capture specific viewpoints. Although Unreal Engine 5 provides the Movie Render Queue plugin for high-quality image rendering, manually specifying keypoint positions, rotations, and frame numbers can be both time-consuming and inflexible. To address this challenge, we modified AirSim, an open-source simulator developed by Microsoft, and customized its code for our specific recording requirements. Our modified AirSim provides a high-fidelity simulation platform for aerial and ground vehicles, integrating realistic physics and environmental dynamics. It is equipped with APIs for automating data collection, supporting the generation of required sensor data. By incorporating AirSim, we significantly reduced the need for manual annotations and improved the efficiency of our data collection process.

Data Collection Using UAVs. During the data collection, UAVs are deployed to collect multi-modal data through predefined or manual flight paths, all executed within the same dynamic environment. This ensures that the dataset can simulate the concurrent operation of multiple UAVs, supporting research on swarm-based wildfire perception. Each UAV is equipped with a RGB-D camera, both mounted centrally underneath the UAV body, with their orientations can be changed in real-time according to the flight modes. Considering the complexity and diversity of flight patterns in wildfire environments, the data collection encompasses exploratory flights during periods when the wildfire locations are unknown, as well as target-search flights, manual flights, and scanning flights such as Z-scan, spiral-scan, and circular-scan after the wildfire areas have been identified, as illustrated in Fig. 3. The flight speed is maintained at 3 meters per second, with obstacle avoidance dynamically adjusting the path. All data, including RGB-D images and motion parameters (e.g., position, velocity, orientation), is recorded at 30 Hz to support dynamic wildfire mapping and SLAM research.

3.2 Dataset Characteristics

Large-Scale, Diversity, and Low-altitude. The WildfireX-SLAM dataset spans a 16 square kilometer forest environment, a significant scale compared to most existing datasets. It includes 5.5k aerial images captured by UAVs, ensuring comprehensive coverage of the dynamic wildfire scenario. Unlike other datasets [1,18,19,21], which focus on isolated and small-scale scenes, WildfireX-SLAM provides a densely captured, unified forest environment from various viewpoints, as shown in Fig. 2. Furthermore, unlike datasets such as S3E [7] and KITTI 2 [9], which limits coverage to urban region, or BASEPROD [11], which does not feature complete forest contexts, WildfireX-SLAM enables robust analysis of forest-specific SLAM and navigation. Additionally, unlike other wildfire datasets collected from UAVs flying at high altitudes [13,20], our dataset

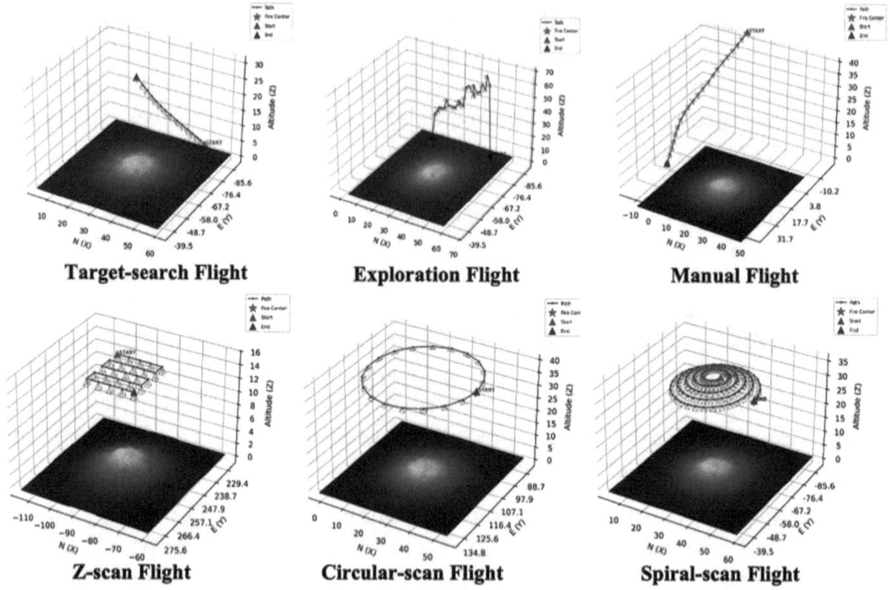

Fig. 3. Examples of flight path for data collection in *WildfireX-SLAM* dataset.

focuses on low-altitude flights below 50 m above ground. This relatively rare but feasible flying altitude provides richer environmental details, which are crucial for wildfire perception and emergency response.

Controllable Environments. One of WildfireX-SLAM's unique strengths lies in its ability to simulate dynamic, controllable environments. Researchers can manipulate key environmental factors, including flame and smoke dynamics, lighting angles and intensities, and fog density and height, as demonstrated in Fig. 1 (b) and (c). These controls allow for realistic simulation of complex wildfire conditions, supporting experiments with diverse scene configurations that would be challenging to capture in real-world data. Moreover, Unreal Engine facilitates the emulation of camera noise, such as motion blur and defocus blur, further enhancing realism. This flexibility ensures that WildfireX-SLAM bridges the gap between synthetic data and real-world applications.

Multiple Attributes. As demonstrated in Fig. 1 (d), Aprting from RGB-D data, the WildfireX-SLAM can also provide a wide range of intermediate data products, including thermal images, normals, and decomposed reflectance components. These attributes are critical for advanced tasks such as SLAM, depth estimation, and navigation research. Thanks to our integrated pipeline, these properties can be extracted at minimal additional computational cost, eliminating the prohibitive expenses associated with acquiring such data in real-world scenarios.

Applications. By exploring SLAM and navigation algorithms on WildfireX-SLAM, researchers can bridge the gap between synthetic and real-world applications. The dataset's high-quality imagery, precise camera poses, dynamic wildfire scenarios, and diverse sensor attributes make it invaluable for wildfire monitoring, collaborative navigation, and other research areas. Furthermore, its comprehensive design facilitates studies on forest-specific challenges, paving the way for effective real-world deployment.

4 Experiments

In this section, we primarily investigate the accuracy of camera pose estimation as well as the quality of map reconstruction and rendering. Furthermore, we identify and highlight the challenges faced by representative 3DGS-based methods when applied to this task.

4.1 Dataset and Metrics

WildfireX-SLAM Benchmark. WildfireX-SLAM dataset is divided into two benchmark subsets. Benchmark A consists of two sequences of UAV low-altitude flights (within 15 m above ground), where Sequence 2 contains wildfire presence and Sequence 1 does not. This benchmark is designed to compare the impact of wildfires on various SLAM algorithms. Benchmark B is a more comprehensive benchmark intended to assess algorithm performance across four challenging wildfire scenarios. Challenge Scenario 1 (CS 1) evaluates the effects of dynamic non-rigid objects in the environment, such as wildfires and falling leaves, on SLAM; it includes four sequences: the first two sequences are collected in the same environment without and with wildfire, respectively, while the latter two sequences are from another wildfire environment with one sequence lacking leaves and the other containing leaves. Challenge Scenario 2 (CS 2) examines the impact of lighting conditions on wildfire SLAM, comprising three sequences collected at the same scene during morning, noon, and late night. Challenge Scenario 3 (CS 3) assesses the influence of weather conditions, containing two sequences—one without fog and one with fog. Challenge Scenario 4 (CS 4) integrates all adverse environmental factors, including wildfire, falling leaves, fog, coupled with poor lighting conditions during late night.

Metrics. To evaluate the tracking and mapping performance of each baseline on our dataset, we adopt widely used metrics for camera tracking accuracy and map quality as in [25]. For camera tracking, we report the Root Mean Square Error (RMSE) of the Absolute Trajectory Error (ATE) in meters. For map quality, we report photometric rendering metrics, including Peak Signal-to-Noise Ratio (PSNR), Structural Similarity Index (SSIM), and Learned Perceptual Image Patch Similarity (LPIPS).

4.2 Baselines

We aim to evaluate the performance of current 3DGS-based SLAM methods on the WildfireX-SLAM dataset to explore the challenges posed by dynamic wildfire emergency scenarios. The decision not to select other SLAM methods designed for dynamic environments is based on the fact that 3DGS-based SLAM currently achieves SOTA performance in both mapping quality and speed, making it particularly well-suited for emergency wildfire scenarios that demand high real-time capabilities. Moreover, its potential in dynamic environments remains limited exploration. Although these methods currently rely on substantial memory resources, this limitation is acceptable and is expected to be addressed in the future. For our experiments, we selected two of the most representative 3DGS-based methods MonoGS and SplatAM. In all experiments, we utilized the official implementations of these baseline methods.

4.3 Results and Discussion

Table 2 presents the performance of different algorithms under wildfire and non-wildfire conditions. The presence of fire significantly degrades the localization and mapping performance of the tested algorithms. Specifically, PSNR nearly drops by half, while ATE RMSE increases by more than double. A comparative analysis across the algorithms reveals that the 3DGS-based methods demonstrate localization capabilities comparable to the traditional ORB-SLAM3. Concurrently, their mapping performance surpasses that of the NeRF-based Co-SLAM. Considering the image rendering speed, 3DGS-based SLAM remains the most promising research direction for wildfire scenarios.

Table 3 presents the mapping results of Splatam across different scenes, highlighting the significant challenges that wildfire environments pose to SLAM systems, which manifest in three main aspects. (a) Dynamic non-rigid objects and

Table 2. Evaluations of MonoGS, Splatam, Co-SLAM and ORB-SLAM3 on Benchmark A of the *WildfireX-SLAM* dataset.

Scene	MonoGS [23]					Splatam [16]					Co-SLAM [29]					ORB-SLAM3 [3]
	ATE	PSNR ↑	SSIM ↑	LPIPS ↓	Depth L1 ↓	ATE	PSNR ↑	SSIM ↑	LPIPS ↓	Depth L1 ↓	ATE ↓	PSNR ↑	SSIM ↑	LPIPS ↓	Depth L1 ↓	ATE ↓
Sequence 1	2.5	22.94	0.622	0.455	1.07	3.1	22.23	0.652	0.428	0.83	6.30	21.54	0.702	0.353	1.44	2.10
Sequence 2	12.5	13.47	0.562	0.649	6.13	13.4	13.34	0.551	0.549	4.70	52.8	6.77	0.412	0.635	7.60	9.24
Overall	7.50	18.21	0.592	0.552	3.60	8.25	17.79	0.602	0.488	2.77	29.55	14.16	0.557	0.494	4.52	5.67

Table 3. Evaluations of 3DGS-based SLAM on Benchmark B of the *WildfireX-SLAM* dataset.

Metrics	CS 1				CS 2		CS 3			CS 4		CS 5
	w/o fire	w fire	w/o leaf	w leaf	w/o	w	morn	noon	night	w/o	w	
ATE RMSE ↓	95.2213	98.2351	148.8109	154.2466	51.5609	59.618	98.4574	92.8674	92.1550	44.8714	59.7346	108.0769
PSNR ↑	15.60	14.77	17.96	18.02	17.50	16.70	14.04	15.12	17.81	42.06	51.94	44.74
Depth RMSE ↓	12.1128	14.8473	4.9928	6.7917	7.7263	9.9687	20.5924	12.0614	11.2009	28.6708	24.9421	17.4935
MS-SSIM ↑	0.614	0.609	0.789	0.705	0.700	0.670	0.575	0.650	0.771	0.763	0.870	0.832
LPIPS ↓	0.652	0.664	0.600	0.673	0.514	0.673	0.725	0.642	0.509	0.326	0.200	0.351

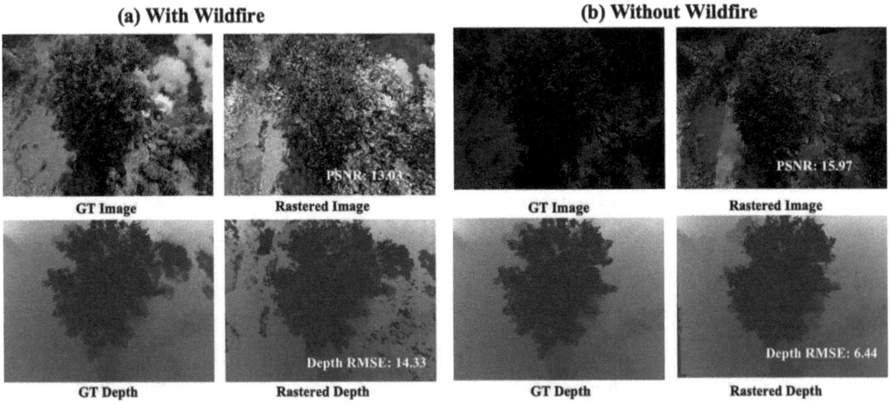

Fig. 4. Mapping results of 3DGS-based SLAM on Challenge Scenario 1 of the *WildfireX-SLAM* dataset.

Fig. 5. Mapping results of 3DGS-based SLAM on Challenge Scenario 2 of the *WildfireX-SLAM* dataset.

repetitive textures with limited features severely challenge SLAM localization and 3DGS fitting. This performance degradation due to such objects is evident in the results for CS 1 and CS 2 in the table. As shown in Fig. 4, dynamic rigid objects like wildfires affect both image and depth fitting, with a particularly severe impact on depth: the Depth RMSE increases from 6.44 without fire to 14.33 with fire. (b) Illumination plays a critical role in SLAM performance. From Fig. 5 and Table 3 results for CS 3, it is seen that as lighting improves from morning to noon, SLAM performance correspondingly improves. However, it is noteworthy that the PSNR metric appears to fail under very low light conditions. Specifically, Fig. 5 shows that at night, Depth RMSE worsens compared to noon, while PSNR paradoxically improves due to loss of scene details and simpler image rendering in dark conditions, which complicates tracking. This observation

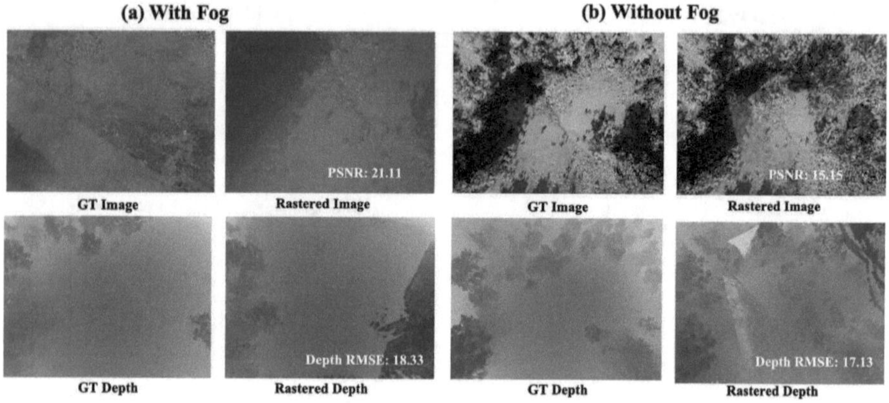

Fig. 6. Mapping results of 3DGS-based SLAM on Challenge Scenario 3 of the *WildfireX-SLAM* dataset.

suggests the need for more effective and innovative metrics tailored for nighttime scenarios. (c) Fog causes degradation in localization accuracy. Results from CS 4 in Table 3 and Fig. 6 indicate that fog leads to effects similar to nighttime conditions on SLAM, with improved mapping but declined localization performance, fundamentally due to light obstruction.

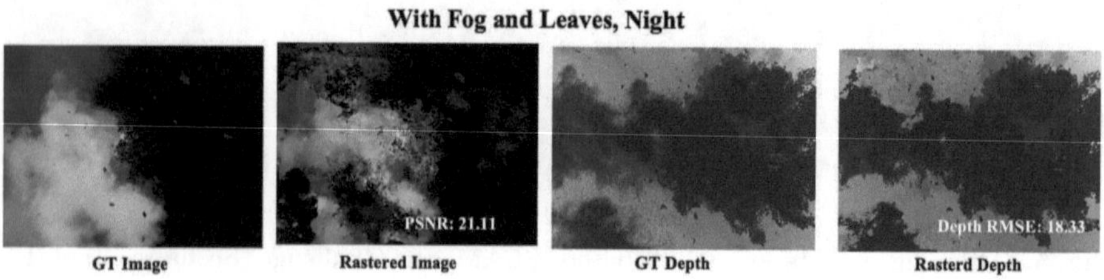

Fig. 7. Mapping results of 3DGS-based SLAM on Challenge Scenario 4 of the *WildfireX-SLAM* dataset.

The most challenging scenario, CS 5, combines all adverse factors: wildfire, falling leaves, fog, and nighttime conditions. As shown in Fig. 7, this compound environment severely impacts both visual quality and depth estimation, with Depth RMSE reaching 18.33. Notably, while PSNR appears improved (44.74), this paradoxically results from simplified rendering in extremely dark, obscured conditions rather than actual performance gains. The high ATE RMSE of 108.08 confirms severe localization degradation, underscoring the need for robust SLAM methods tailored to extreme wildfire scenarios.

5 Conclusion

In this paper, we introduced WildfireX-SLAM, a groundbreaking SLAM dataset designed to tackle the unique challenges of wildfire and forest environments.

With its diverse and realistic data captured through both aerial and near-ground perspectives, augmented depth images, WildfireX-SLAM provides an invaluable resource for advancing SLAM technologies. Beyond the dataset itself, we offer a flexible data collection framework using Unreal Engine 5 and our modified AirSim, enabling precise control over environmental factors like wildfire conditions and weather. This work goes beyond addressing technical gaps, it lays a critical foundation for robotics and computer vision research in safety-critical domains, empowering innovations in real-time emergency response and collaborative navigation in unstructured, dynamic wild environments.

References

1. Ali, I., et al.: FINNForest dataset: a forest landscape for visual slam. Robot. Auton. Syst. **132**, 103610 (2020)
2. Burri, M., et al.: The EUROC micro aerial vehicle datasets. Int. J. Robot. Res. **35**(10), 1157–1163 (2016)
3. Campos, C., Elvira, R., Rodríguez, J.J.G., Montiel, J.M., Tardós, J.D.: ORB-SLAM3: an accurate open-source library for visual, visual–inertial, and multimap slam. IEEE Trans. Robot. (2021)
4. Carlevaris-Bianco, N., Ushani, A.K., Eustice, R.M.: University of MichiGan north campus long-term vision and lidar dataset. Int. J. Robot. Res. **35**(9), 1023–1035 (2016)
5. Chen, S.W., et al.: SLOAM: semantic lidar odometry and mapping for forest inventory. In: IEEE Robotics and Automation Letters (RA-L) (2020)
6. Choi, Y., et al.: Kaist multi-spectral day/night data set for autonomous and assisted driving. IEEE Trans. Intell. Transp. Syst. **19**(3), 934–948 (2018)
7. Feng, D., et al.: S3E: a multi-robot multimodal dataset for collaborative slam. IEEE Robot. Autom. Lett., pp. 1–8 (2024). https://doi.org/10.1109/LRA.2024.3490402
8. Fontana, G., Matteucci, M., Sorrenti, D.G.: RAWSeeds: building a benchmarking toolkit for autonomous robotics. Methods Exp. Tech. Comput. Eng., 55–68 (2014)
9. Gaidon, A., Wang, Q., Cabon, Y., Vig, E.: Virtual worlds as proxy for multi-object tracking analysis. In: Proceedings of the IEEE Conference on Computer Vision and Pattern Recognition, pp. 4340–4349 (2016)
10. Geiger, A., Lenz, P., Urtasun, R.: Are we ready for autonomous driving? The KITTI vision benchmark suite. In: 2012 IEEE Conference on Computer Vision and Pattern Recognition, pp. 3354–3361. IEEE (2012)
11. Gerdes, L., et al.: BASEPROD: the BARDENAS semi-desert planetary rover dataset. Sci. Data **11**(1), 1054 (2024)
12. Helmberger, M., Morin, K., Berner, B., Kumar, N., Cioffi, G., Scaramuzza, D.: The HILTI slam challenge dataset. IEEE Robot. Autom. Lett. **7**(3), 7518–7525 (2022)
13. Hu, Y., et al.: Firefly: a synthetic dataset for ember detection in wildfire. In: Proceedings of the IEEE/CVF International Conference on Computer Vision, pp. 3765–3769 (2023)
14. Huang, A.S., Antone, M., Olson, E., Fletcher, L., Moore, D., Teller, S., Leonard, J.: A high-rate, heterogeneous data set from the DARPA urban challenge. Int. J. Robot. Res. **29**(13), 1595–1601 (2010)
15. Jeong, J., Cho, Y., Shin, Y.S., Roh, H., Kim, A.: Complex urban dataset with multi-level sensors from highly diverse urban environments. Int. J. Robot. Res. **38**(6), 642–657 (2019)

16. Keetha, N., et al.: SPLATAM: splat, track & map 3D gaussians for dense RGB-D slam. In: CVPR (2024)
17. Kerbl, B., Kopanas, G., Leimkühler, T., Drettakis, G.: 3D gaussian splatting for real-time radiance field rendering. ACM Trans. Graph. **42**(4), 139 (2023)
18. Knights, J., Vidanapathirana, K., Ramezani, M., Sridharan, S., Fookes, C., Moghadam, P.: Wild-places: a large-scale dataset for lidar place recognition in unstructured natural environments. In: 2023 IEEE International Conference on Robotics and Automation (ICRA), pp. 11322–11328. IEEE (2023)
19. Lagos, J., Lempiö, U., Rahtu, E.: Finnwoodlands dataset. In: Gade, R., Felsberg, M., Kämäräinen, J.K. (eds.) Image Analysis, pp. 95–110. Springer Nature Switzerland, Cham (2023). https://doi.org/10.1007/978-3-031-31435-3_7
20. Li, Y., et al.: Sim2Real-fire: a multi-modal simulation dataset for forecast and backtracking of real-world forest fire. Adv. Neural. Inf. Process. Syst. **37**, 1428–1442 (2024)
21. Liu, Y., et al.: BotanicGarden: a high-quality dataset for robot navigation in unstructured natural environments. IEEE Robot. Autom. Lett. **9**(3), 2798–2805 (2024)
22. Maddern, W., Pascoe, G., Linegar, C., Newman, P.: 1 year, 1000 km: the oxford Robotcar dataset. Int. J. Robot. Res. **36**(1), 3–15 (2017)
23. Matsuki, H., Murai, R., Kelly, P.H.J., Davison, A.J.: Gaussian splatting slam. In: CVPR (2024)
24. Ramezani, M., Wang, Y., Camurri, M., Wisth, D., Mattamala, M., Fallon, M.: The newer college dataset: Handheld lidar, inertial and vision with ground truth. In: 2020 IEEE/RSJ International Conference on Intelligent Robots and Systems (IROS), pp. 4353–4360. IEEE (2020)
25. Sandström, E., Li, Y., Van Gool, L., R. Oswald, M.: Point-slam: dense neural point cloud-based slam. In: International Conference Computing Vision (2023)
26. Schubert, D., Goll, T., Demmel, N., Usenko, V., Stückler, J., Cremers, D.: The tum vi benchmark for evaluating visual-inertial odometry. In: 2018 IEEE/RSJ International Conference on Intelligent Robots and Systems (IROS), pp. 1680–1687. IEEE (2018)
27. Shi, X., et al.: Are we ready for service robots? The OpenLoris-scene datasets for lifelong slam. In: 2020 IEEE International Conference on Robotics and Automation (ICRA), pp. 3139–3145. IEEE (2020)
28. Sturm, J., Engelhard, N., Endres, F., Burgard, W., Cremers, D.: A benchmark for the evaluation of RGB-D slam systems. In: 2012 IEEE/RSJ International Conference on Intelligent Robots and Systems, pp. 573–580. IEEE (2012)
29. Wang, H., Wang, J., Agapito, L.: Co-slam: joint coordinate and sparse parametric encodings for neural real-time slam. In: Proceedings of the IEEE/CVF Conference on Computer Vision and Pattern Recognition (CVPR), pp. 13293–13302 (2023)
30. Wen, W., et al.: URBANLOCO: a full sensor suite dataset for mapping and localization in urban scenes. In: 2020 IEEE International Conference on Robotics and Automation (ICRA), pp. 2310–2316. IEEE (2020)
31. Wenzel, P., et al.: 4Seasons: a cross-season dataset for multi-weather slam in autonomous driving. In: Pattern Recognition: 42nd DAGM German Conference, DAGM GCPR 2020, Tübingen, Germany, September 28–October 1, 2020, Proceedings 42, pp. 404–417. Springer (2021). https://doi.org/10.1007/978-3-030-71278-5_29
32. Yan, C., et al.: GS-SLAM: dense visual slam with 3D gaussian splatting. In: CVPR (2024)
33. Yin, J., Li, A., Li, T., Yu, W., Zou, D.: M2DGR: a multi-sensor and multi-scenario slam dataset for ground robots. IEEE Robot. Autom. Lett. **7**(2), 2266–2273 (2021)

LGF-Net: Integrating Local and Global Features in a Dual-Branch Architecture for Tooth Segmentation in CBCT Images

Yu Ao[1,2], Hongze Han[1], Yuqin Li[1], Yu Miao[1], and Weili Shi[1,2](✉)

[1] School of Computer Science and Technology, Changchun University of Science and Technology, Changchun, China

[2] Zhongshan Institute of Changchun University of Science and Technology, Changchun, China

shiweili@cust.edu.cn

Abstract. Medical image segmentation is a critical task for accurately delineating target and pathological regions. Despite significant progress in deep learning, existing methods still struggle to integrate local features with global contextual information, especially in tooth CBCT images with complex and irregular morphology. To address these challenges, we propose LGF-Net, a dual-branch encoder–decoder network that fuses local and global features to enhance segmentation accuracy and robustness. In the encoder, multiple DSsm modules combine Mamba mechanisms with dense connections, with each Local–Global Block (LG-Block) extracting local details and global context in parallel before feature fusion. An Adaptive Cross-scale Feature Fusion (ACFF) module at the bottleneck promotes multi-level feature interaction, while the decoder progressively restores spatial resolution through upsampling and skip connections. Experiments on two widely used CBCT datasets demonstrate that LGF-Net achieves state-of-the-art performance, significantly outperforming mainstream methods in DSC and IoU, and maintains high accuracy even in the presence of metallic artifacts, highlighting its robustness and clinical potential.

Keywords: Global context · Convolutional · neural network · Deep learning · Tooth segmentation · Vision Mamba

1 Introduction

Oral diseases, including dental caries, periodontitis, and tooth loss, are highly prevalent worldwide and severely impact patients' quality of life and health. Cone-Beam Computed Tomography (CBCT), offering high resolution and low radiation dose, is essential for clinical diagnosis and treatment planning. Accurate tooth segmentation from CBCT enables extraction of morphological and spatial information, supporting implant design, orthodontic analysis, and preoperative navigation, thus improving precision and personalization of interventions.

J. Lokoč et al. (Eds.): MMM 2026, LNCS 16413, pp. 547–560, 2026.
https://doi.org/10.1007/978-981-95-6957-1_39

Manual segmentation, however, is time-consuming, subjective, and dependent on clinician expertise, while early automated methods, such as thresholding, edge detection, and region growing, perform poorly under complex backgrounds, irregular arrangements, or metal artifacts. Deep learning has substantially advanced medical image segmentation. Convolutional neural networks (CNNs), exemplified by U-Net [1], hierarchically extract features from low-level edges and textures to high-level semantics. Yet, their limited receptive field hinders capture of long-range dependencies and global context. Strategies such as dilated convolutions [1], image pyramids, prior-guided methods, and multi-scale feature fusion aim to enhance cross-scale and global representation. In CBCT tooth images, however, irregular morphology and complex arrangement make local features alone insufficient to represent the overall structure, limiting segmentation accuracy and robustness.

To overcome the limitations of CNNs in capturing global features, Transformer architectures have been applied to medical image segmentation. Notably, TransUNet [2] combines CNN-based local feature extraction with Transformer-based global modeling, achieving superior performance across diverse medical imaging tasks. Swin-UNet [3] balances local and global information through hierarchical sliding windows, while UCTransNet [4] and nnFormer [5] leverage self-attention to effectively capture long-range dependencies. However, Transformers typically entail high computational cost, constraining their practical deployment. Recently, State Space Models (SSMs) [6] have shown efficiency in modeling long sequences. Mamba [7], incorporating input-dependent selection mechanisms, enhances global dependency modeling and offers a potential alternative to Transformers for visual tasks. Studies such as VM-UNet [8] and U-Mamba [9] embed Visual State Space (VSS) modules into U-shaped networks, achieving preliminary integration of global and local features, yet their representation of fine-grained local structures remains limited. This is particularly critical in CBCT tooth images, where global features capture overall dentition and jaw arrangement, while local features delineate fine structures, including crown-root boundaries, enamel–dentin transitions, and interproximal contacts. Reliance solely on global features risks losing fine details, whereas using only local features fails to capture spatial constraints. Therefore, integrating Mamba's global modeling capabilities with CNN-based local representations is essential for robust and high-precision tooth segmentation.

To address these challenges, we propose a dual-branch encoder–decoder network for tooth segmentation, designed to fuse local and global features and balance their extraction for improved segmentation performance. Considering the complementary characteristics and limitations of Mamba and convolutional neural networks, we introduce a Dense-connected Selective State-space Mamba (DSsm) module in the encoder. This module combines Mamba with dense connections to enhance feature representation while preserving structural consistency. Within DSsm, the Local–Global Block (LG-Block) performs dual-branch feature extraction and fusion, capturing comprehensive semantic information. At the deepest downsampling layer, an Adaptive Cross-Scale Feature

Fusion (ACFF) module is incorporated to further strengthen multi-scale feature representation. The decoder progressively restores spatial resolution through upsampling and skip connections, effectively preserving both local details and global consistency, thereby significantly improving segmentation accuracy and robustness.

The main contributions of this work are summarized as follows:

1. This paper propose a dual-branch encoder–decoder network for tooth segmentation, where the DSsm module integrates Mamba state-space modeling with dense connections to jointly capture fine-grained local details and global semantics.

2. To mitigate the loss of global structure and insufficient local detail representation, we design the LG-Block, which leverages dual-branch feature extraction to propagate global semantics into local representations, enabling deeper semantic geometric fusion.

3. Experiments on two benchmark datasets show that our method achieves superior DSC and IoU scores, confirming its accuracy and robustness in tooth segmentation.

2 Related Work

2.1 Local Feature Extraction Methods

Convolutional neural networks (CNNs) have achieved remarkable success in tooth segmentation, largely due to their efficiency in capturing local features. U-Net, proposed by Ronneberger et al. [10], employs an encoder–decoder architecture with skip connections to aggregate multi-scale local features, establishing itself as a foundational method for CBCT-based tooth segmentation. To enhance local feature representation, several U-Net variants have been developed. For example, Res-UNet [13] incorporates residual connections to improve gradient flow, while U-Net++ [11] introduces nested architectures with deep supervision to enable multi-level feature fusion. For dental-specific segmentation tasks, tailored approaches have been proposed: ToothNet [14] leverages a 3D region proposal network to localize teeth and extract structural details; Cui et al. [14] extend 2D Mask R-CNN into 3D; and Zhong et al. [15] utilize pose regression for precise mandibular alignment prior to regional segmentation, thereby improving boundary delineation. Despite these advances, CNN-based methods remain limited by their restricted receptive fields, which hinder the modeling of long-range dependencies and global contextual information.

2.2 Local-Global Feature Extraction Methods

Compared with CNNs, which excel at local feature extraction but struggle with long-range dependencies, Transformers [16] employ self-attention to effectively capture global context. For example, U-Transformer [17] enhances spatial resolution recovery via multi-head self-attention, while Valanarasu et al. [19] introduced a local–global framework that fuses shallow geometric and textural details with deep positional representations. Despite their advantages, the quadratic complexity of self-attention makes Transformers computationally prohibitive for 3D CBCT tooth segmentation. To address this, the recently proposed Mamba architecture, grounded in state space models (SSMs), achieves linear complexity while maintaining strong global modeling capability. LightM-UNet [20] incorporates residual visual Mamba layers with residual scaling to reinforce long-range dependency learning. SegMamba [21] introduces a tri-directional Mamba combined with gated spatial convolution and feature uncertainty estimation, thereby improving feature reuse and preserving structural integrity. T-Mamba [23] further strengthens both spatial and frequency-domain representations through shared positional encoding and frequency fusion. Nevertheless, these Mamba-based frameworks remain limited by insufficient integration of fine-grained local context during the encoding process, highlighting the need for architectures that jointly exploit local detail and global dependency.

3 Methods

3.1 Overview of the LGF-Net Method

This paper proposes a dental segmentation network, as illustrated in Fig. 1, which adopts a typical U-shaped architecture composed of an encoder, skip connections, and a decoder. During the encoder stage, the input image first undergoes initial downsampling, reducing the feature size to $2C \times \frac{H}{2} \times \frac{W}{2} \times \frac{D}{2}$. The features then sequentially pass through three encoder layers, each containing a DSsm module. The DSsm modules are responsible for feature extraction and dense connectivity, with feature sizes after successive downsampling becoming $4C \times \frac{H}{4} \times \frac{W}{4} \times \frac{D}{4}$, $8C \times \frac{H}{8} \times \frac{W}{8} \times \frac{D}{8}$, $16C \times \frac{H}{16} \times \frac{W}{16} \times \frac{D}{16}$ Within the DSsm modules, the Local–Global Block (LG-Block) play a central role, efficiently extracting and fusing local and global features through a dual-branch design. This allows the module to simultaneously preserve fine-grained boundary information and capture global contextual dependencies, achieving an effective integration of local and global representations. At the bottom of the encoder, an Adaptive Cross-scale Feature Fusion (ACFF) module is introduced to jointly model and integrate features at different resolutions, further enhancing the interaction between local and global information. During the decoder stage, spatial resolution is progressively restored through upsampling, while skip connections are employed to merge corresponding encoder features, ensuring that both low-level local details and high-level global semantics are preserved. The network ultimately outputs the segmented dental structures.

3.2 Dual Selective State-Space Mamba Module(DSsm)

To address the limitations of convolutional neural networks in modeling long-range dependencies while preserving local boundary details, we design the Dual Selective State-space Mamba (DSsm) module. This module combines the Mamba architecture with dense connections to enable efficient feature extraction and

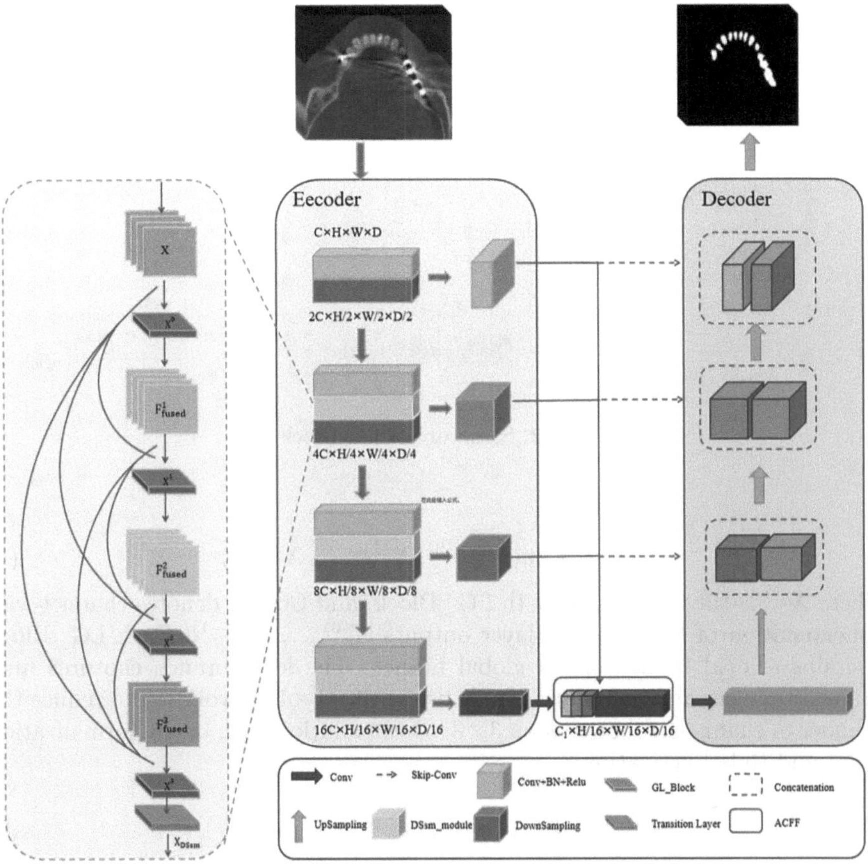

Fig. 1. Overall Architecture of the LGF-Net.

enhancement. The Local–Global Block (LG-Block), as the core component of DSsm in Fig. 2, employs a dual-branch structure. In the local branch, convolutional operations enhance fine-grained textures and boundary features to accurately delineate tooth contours and adjacent structures. In the global branch, the Mamba Selective State-space Modeling mechanism captures long-range dependencies over a large receptive field and extracts global contextual information. Subsequently, local and global features are fused along the channel dimension, achieving an integrated modeling of long-range dependencies and preservation of fine local details within a single module.

Specifically, the DSsm module processes three-dimensional feature maps $X \in R^{C \times H \times W \times D}$, where C denotes the number of channels, and H, W, D represent the spatial dimensions. To fully exploit hierarchical information and improve gradient propagation, outputs from all previous LG_Blocks are densely concatenated along the channel dimension:

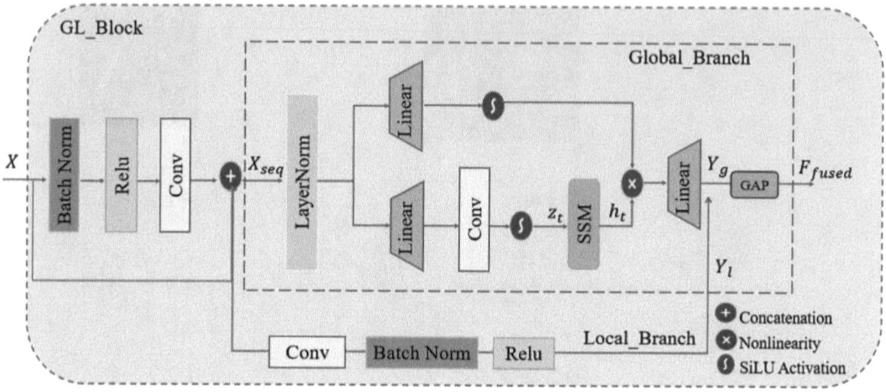

Fig. 2. Structure of LG-Block.

$$X^{(l)} = \mathrm{Concat}(X^{(0)}, X^{(1)}, \ldots, X^{(l-1)}) \tag{1}$$

where $X^{(l)}$ is the input to the l-th LG_Block, and Concat denotes channel-wise concatenation of the previous layer outputs $X^{(0)}, \ldots, X^{(l-1)}$. Each LG_Block contains a local branch and a global branch. The local branch captures fine-grained textures and boundary details using a $1 \times 1 \times 1$ convolution to reduce the number of channels, followed by a $3 \times 3 \times 3$ convolution with batch normalization (BN) and ReLU activation:

$$F_l = \mathrm{ReLU}\left(\mathrm{BN}\left(\mathrm{Conv}_{3\times3\times3}\left(\mathrm{Conv}_{1\times1\times1}(X)\right)\right)\right) \tag{2}$$

where the $1 \times 1 \times 1$ convolution reduces the number of channels, the $3 \times 3 \times 3$ convolution extracts local spatial and boundary features, BN normalizes the feature maps to stabilize training, and ReLU introduces non-linearity. The global branch models long-range dependencies. The 3D input is first flattened into a sequence $X_{\mathrm{seq}} \in R^{(HWD)\times C}$ in raster-scan order, then projected into a latent space:

$$Z = X_{\mathrm{seq}} W_p \tag{3}$$

where $W_p \in R^{C\times d}$ is a learnable projection matrix and d is the latent state dimension. The sequence is recursively processed by a Selective State-space Model (SSM)[16]:

$$h_t = \mathrm{SSM}(h_{t-1}, z_t) \tag{4}$$

where $h_t \in R^d$ is the latent state at position t, and $z_t \in R^d$ is the projected feature vector at the same position. The sequence output is reshaped back to the 3D spatial domain:

$$F_g = \mathrm{Reshape}(H) \in R^{C\times H\times W\times D} \tag{5}$$

3.3 Adaptive Cross-Scale Feature Fusion (ACFF)

Due to the significant differences in spatial resolution and channel numbers among feature maps at different downsampling stages, directly performing feature enhancement and fusion may lead to scale and channel mismatches, thereby reducing the effectiveness of feature interaction. To address this, an Adaptive Cross-scale Feature Fusion (ACFF) module is designed, as illustrated in Fig. 3, to achieve unified representation and efficient fusion across multiple scales. Specifically, the output features from the downsampling stages are given by:

$$X_l \in R^{C_l \times H_l \times W_l \times D_l}, \quad l \in \{1, 2, 3, 4\} \tag{6}$$

where X_1, X_2, X_3 denote the output features from the first three downsampling stages, and X_4 denotes the output features from the fourth downsampling stage (note: original text may have a typo, adjusted for logical consistency). Here, C_l and H_l, W_l, D_l indicate the number of channels and spatial dimensions, respectively. Since the spatial resolution of the deepest feature X_4 is selected as the unified target, the shallow features X_1, X_2, X_3 are first downsampled to the same spatial resolution through adaptive max-pooling:

$$\tilde{X}_l = \text{MaxPool}\left(X_l;\ (H_4, W_4, D_4)\right), \quad \tilde{X}_l \in R^{C_l \times H_4 \times W_4 \times D_4} \tag{7}$$

Subsequently, to further unify the channel dimensions, a $1 \times 1 \times 1$ convolution is applied to each branch for channel mapping, followed by normalization and a nonlinear activation function:

$$M_l = \sigma\left(\text{BN}\left(\text{Conv}_{1\times1\times1}\left(\tilde{X}_l\right)\right)\right), \quad M_l \in R^{C_m \times H_4 \times W_4 \times D_4}, \quad l \in \{1, 2, 3\} \tag{8}$$

where C_m denotes the unified number of channels, which is typically set to match the channel number of the deepest feature X_4. The deepest feature X_4 is also subjected to the same channel mapping operation, resulting in $M_4 \in R^{C_m \times H_4 \times W_4 \times D_4}$. After achieving spatial and channel alignment, all branch features are concatenated along the channel dimension.

$$F = [M_1,\ M_2,\ M_3,\ M_4] \in R^{4C_m \times H_4 \times W_4 \times D_4} \tag{9}$$

Finally, a $3 \times 3 \times 3$ convolutional layer is applied to the concatenated features for fusion and channel compression, yielding the final multi-branch-fused feature:

$$A = \sigma\left(\text{BN}\left(\text{Conv}_{3\times3\times3}\left(F\right)\right)\right) \in R^{C_{\text{out}} \times H_4 \times W_4 \times D_4} \tag{10}$$

where C_{out} denotes the number of output channels. This design not only ensures consistency among features from different downsampling stages but also fully exploits the complementarity between shallow fine-grained information and deep semantic features, thereby enhancing the representational capacity of the fused features.

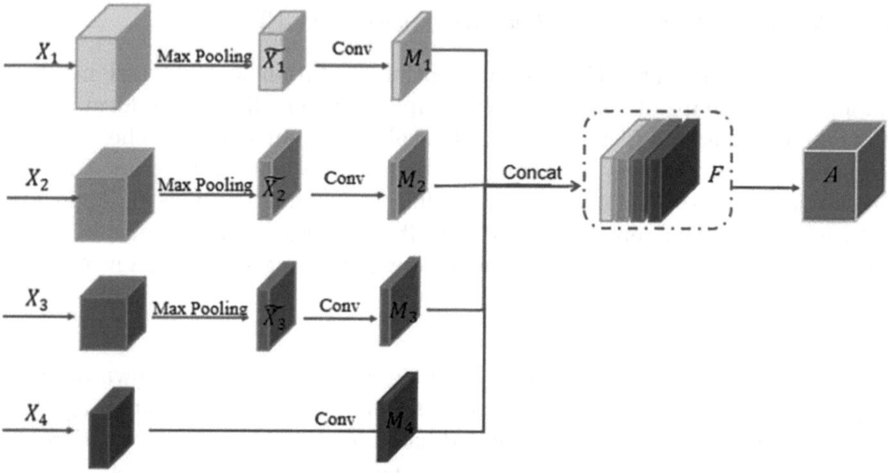

Fig. 3. Structure of ACFF Module.

3.4 Loss Function

We adopt a hybrid loss combining Dice and Binary Cross-Entropy (BCE):

$$\mathcal{L} = \alpha \mathcal{L}_{\text{Dice}} + \beta \mathcal{L}_{\text{BCE}}, \tag{11}$$

where

$$\mathcal{L}_{\text{Dice}} = 1 - \frac{2 \sum_{i=1}^{N} p_i g_i + \epsilon}{\sum_{i=1}^{N} p_i + \sum_{i=1}^{N} g_i + \epsilon}, \tag{12}$$

$$\mathcal{L}_{\text{BCE}} = -\frac{1}{N} \sum_{i=1}^{N} \left[g_i \log(p_i) + (1 - g_i) \log(1 - p_i) \right], \tag{13}$$

with p_i and g_i denoting the prediction and ground truth, N the number of pixels/voxels, and ϵ a smoothing term.

4 Experiments

4.1 Dataset and Metric

The effectiveness of the proposed method was validated on two commonly used dental segmentation datasets. The first dataset, provided by Cui et al. [35], consists of 4,938 CBCT scans collected from 15 medical centers, covering complex clinical cases such as metal artifacts, crowded dentition, missing teeth, and malocclusion. In this study, a total of 129 scans were utilized. The second dataset, ToothFairy2024 [36], comprises 417 CBCT scans with resolutions ranging from 0.2 to 1.0 mm. To comprehensively evaluate the model performance, five-fold cross-validation was adopted, where four folds were used for training and one fold for testing in each iteration. The final results were averaged across the five folds to ensure evaluation stability and reliability.

To further assess segmentation performance on CBCT images, two widely used evaluation metrics were employed: Dice Similarity Coefficient (DSC) and Intersection over Union (IoU). Their formulations are given as follows:

$$\mathrm{DSC} = \frac{2\,|A \cap B|}{|A| + |B|}, \quad \mathrm{IoU} = \frac{|A \cap B|}{|A \cup B|} \tag{14}$$

where A and B denote the predicted and ground-truth regions, respectively.

4.2 Implementation Details

All CBCT images were resampled to a uniform voxel size of $0.5 \times 0.5 \times 0.5\,\mathrm{mm}^3$ and randomly cropped into sub-volumes of $160\times$ 160 $\times$ 96. To enhance data diversity, random elastic deformations, Gaussian noise, and spatial displacements were applied, and all input volumes were normalized to the range [0,1]. Experiments were conducted using PyTorch 2.0.1 on a single NVIDIA V100 32 GB GPU. The Adam optimizer was employed with an initial learning rate of 0.0006, weight decay of 0.00005, and momentum of 0.8. Gradient accumulation and a ReduceLROnPlateau scheduling strategy were used to stabilize training. The model was trained for 15 epochs with a batch size of 1.

4.3 Experimental Results

To validate the effectiveness of the proposed dual-branch network for fusing local and global features, LGF-Net was qualitatively and quantitatively compared with three state-of-the-art models. The comparison includes local feature extractors (DenseNet, MultiResUNet3D), which capture fine textures and boundaries but lack long-range dependency modeling, and a global feature extractor (SegMamba), which incorporates contextual information through attention mechanisms. All models were trained and evaluated under identical settings to ensure fairness. As shown in Table 1, DenseNet achieved 92.4% DSC and 85.91% IoU on the Cui dataset, while MultiResUNet3D performed poorly (77.52% DSC, 77.19% IoU). SegMamba improved performance to 92.75% DSC and 86.60% IoU, but still showed limitations. LGF-Net outperformed all baselines, reaching 93.25% DSC and 86.75% IoU on Cui and 94.74% DSC and 87.24% IoU on ToothFairy, consistently demonstrating superior segmentation performance by effectively integrating local and global feature representations.

Figure 5 and Fig. 6 present qualitative comparisons of different networks on the two dental segmentation datasets, with red boxes marking mis-segmented regions. On the Cui dataset, DenseNet and MultiResUNet3D suffer from under- or over-segmentation caused by metallic artifacts and missing teeth, as local convolutions capture fine textures but fail to model global dependencies. SegMamba alleviates these issues by incorporating global context, yet still produces local errors in crowded or impacted teeth. In contrast, LGF-Net preserves local

Table 1. Comparison quantitative results of different segmentation methods on Cui dataset and Toothfairy dataset.

Dataset	Evaluation metrics							
	DSC				IoU			
	DenseNet [26]	MultiResUNet3D [27]	SegMamba [22]	**LGF-Net**	DenseNet [26]	MultiResUNet3D [27]	SegMamba [22]	**LGF-Net**
Cui dataset [24]	92.47%	77.52%	92.75%	**93.25%**	85.91%	77.19%	86.60%	**86.75%**
ToothFairy dataset [25]	92.55%	85.12%	93.29%	**94.74%**	82.90%	75.23%	85.92%	**87.24%**

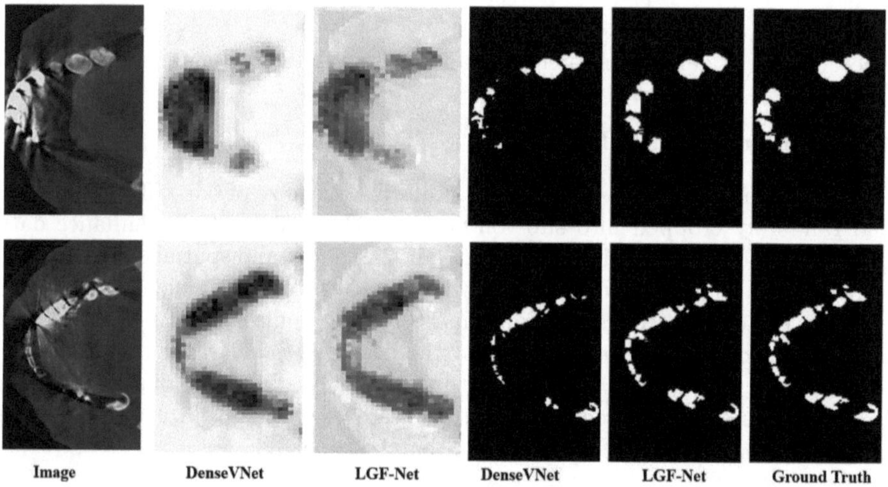

Fig. 4. Visualization of feature responses and segmentation results of DenseNet and LGF-Net.

boundaries while accurately modeling global arrangements, yielding smooth and complete contours. Similar patterns are observed on the ToothFairy dataset: DenseNet and MultiResUNet3D show boundary inaccuracies, while SegMamba improves global consistency but struggles with tightly packed teeth. LGF-Net consistently achieves the best results, effectively segmenting complex dentitions without local errors or boundary oversmoothing.

4.4 Ablation Experiment

To validate the effectiveness of the DSsm and ACFF modules, ablation studies were conducted on the Cui and ToothFairy datasets in Table 2. Beginning with the DenseNet baseline, DSsm (DS-Net) and ACFF (DF-Net) modules were progressively introduced, yielding four comparative settings. The baseline, constrained by local convolutional features, achieved DSC and IoU of 92.47% and 85.91% on Cui and 92.55% and 82.90% on ToothFairy, reflecting limited global context modeling. Incorporating DSsm improved performance to 93.11% and 86.26% on Cui and 94.03% and 85.77% on ToothFairy, indicating better global and boundary representations. Adding ACFF further increased results to 93.25% and 86.75% and 94.74% and 87.24%, respectively, underscoring the benefit of

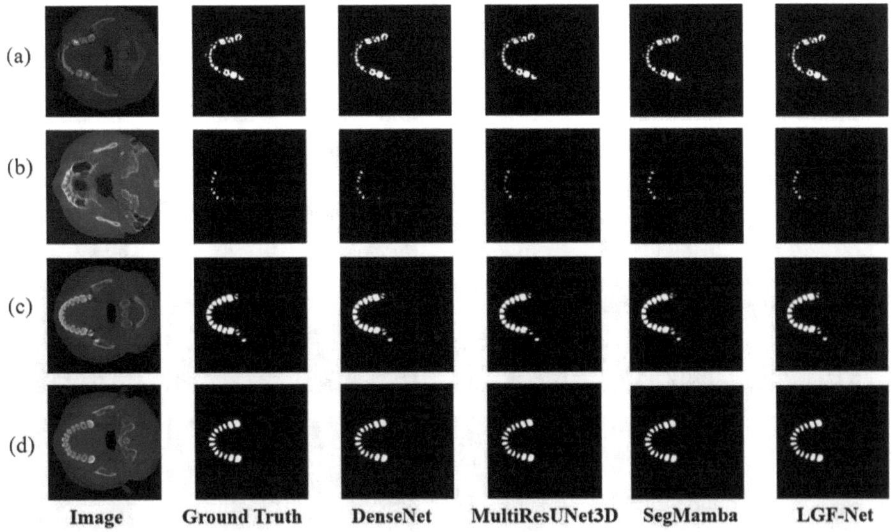

Fig. 5. Visual comparison of different segmentation methods on the Cui dataset [24]: (a) metallic artifacts, (b) missing teeth, (c) impacted teeth, and (d) crowded teeth.

multi-scale fusion. The final LGF-Net achieved the best performance across both datasets, confirming the effectiveness and generalizability of integrating global and local features.

To investigate the impact of local–global feature fusion on segmentation, we visualized feature responses of DenseNet and LGF-Net at the final downsampling stage in Fig. 4. The first column shows the input images, while the second and third columns display feature heatmaps from DenseNet and LGF-Net, respectively. DenseNet exhibits dispersed responses, reflecting reliance on local convolutions and limited representation of overall tooth arrangement. In contrast, LGF-Net responses are concentrated on tooth regions, preserving global structural consistency and emphasizing boundary details. The fourth and fifth columns present the final segmentation outputs, with ground truth in the sixth column. LGF-Net demonstrates more accurate boundary delineation by effectively integrating global and local information. These qualitative observations align with quantitative results, confirming the efficacy of dual-branch LG-Blocks in enhancing tooth segmentation.

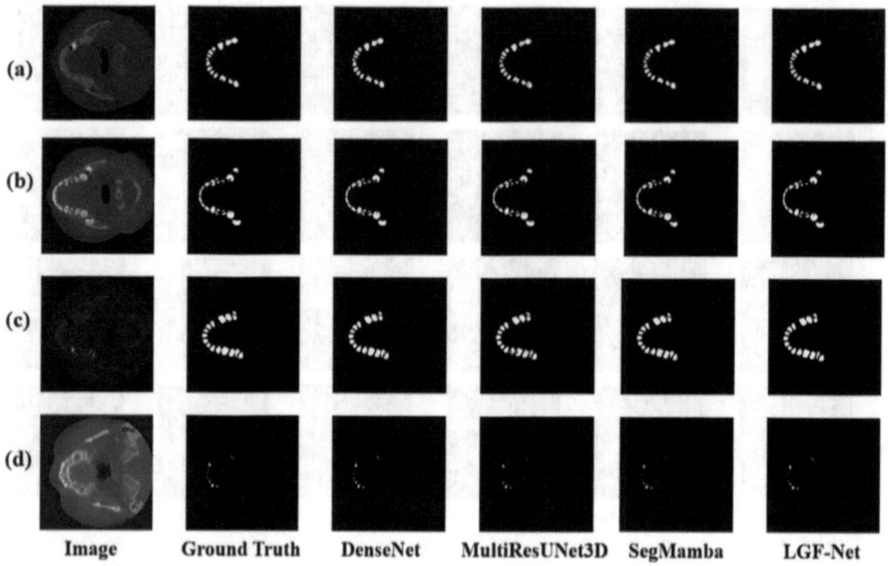

Fig. 6. Visual comparison on the ToothFairy2024 dataset [25]: (a) metallic artifacts, (b) missing teeth, (c) impacted teeth, and (d) crowded teeth.

Table 2. The ablation experiments of LGF-Net.

Dataset	Evaluation metrics							
	DSC				IoU			
	DenseNet [26]	DS-Net	DF-Net	LGF-Net	DenseNet [26]	DS-Net	DF-Net	LGF-Net
Cui dataset [24]	92.47%	93.01%	93.11%	**93.25%**	85.91%	85.93%	86.26%	**86.75%**
ToothFairy dataset [25]	92.55%	93.72%	94.03%	**94.74%**	83.90%	85.02%	85.77%	**87.24%**

5 Conclusion

In this work, we propose LGF-Net, a dental segmentation network that employs multi-layer DSsm modules, combining the Mamba architecture with dense connections for efficient feature extraction. Within each DSsm module, Local–Global Blocks (LG-Blocks) adopt a dual-branch design: one branch captures fine-grained local boundary details, while the other models global contextual semantics, enabling precise feature fusion. Additionally, an Adaptive Cross-scale Feature Fusion (ACFF) module is introduced to enhance multi-level feature interaction. Experimental results show that LGF-Net outperforms existing methods in DSC and IoU on two widely used dental datasets, demonstrating superior capability in simultaneously preserving local details and global structure.

References

1. He, F. et al. Brain tumor image segmentation network based on dual attention mechanism. In International Conference on Intelligent Computing, 125-136 (Springer, 2023)
2. Chen, J., Lu, Y., Yu, Q., et al. TransUNet: Transformers Make Strong Encoders for Medical Image Segmentation. arXiv:2102.04306 (2021)
3. Cao, H., Wang, Y., Chen, J., Jiang, D., Zhang, X., Tian, Q., et al. Swin-Unet: Unet-Like Pure Transformer for Medical Image Segmentation. In: Karlinsky, L., Michaeli, T. and Nishino, K., Eds., Computer Vision—ECCV 2022 Workshops, Springer, 205–218 (2023)
4. Wang, H., Cao, P., Wang, J., Zaiane, O.R.: Uctransnet: Rethinking the Skip Connections in U-Net from a Channel-Wise Perspective with Transformer. Proceedings of the AAAI Conference on Artificial Intelligence **36**, 2441–2449 (2022)
5. Zhou, H.Y., Guo, J., Zhang, Y., et al. nnFormer: Interleaved Transformer for Volumetric Segmentation. arXiv:2109.03201 (2021)
6. Gu, A., Goel, K., Ré, C.: Efficiently modeling long sequences with structured state spaces. arXiv:2111.00396 (2021)
7. Gu, A., Dao, T.: MAMBA: linear-time sequence modeling with selective state spaces. arXiv:2312.00752 (2023)
8. Ruan J., Li J., Xiang, S.: VM-UNet: vision mamba UNet for medical image segmentation. arXiv preprint arXiv:2402.02491 (2024)
9. Liu, Y., Tian, Y., Zhao, Y., et al.: VMamba: Visual state space model. arXiv:2401.10166 (2024)
10. Ronneberger, O., Fischer, P., Brox, T.: U-Net: convolutional networks for biomedical image segmentation. In: Navab, N., Hornegger, J., Wells, W.M., Frangi, A.F. (eds.) MICCAI 2015. LNCS, vol. 9351, pp. 234–241. Springer, Cham (2015). https://doi.org/10.1007/978-3-319-24574-4_28
11. Zhou, Z., Rahman Siddiquee, M.M., Tajbakhsh, N., Liang, J.: UNet++: a nested U-Net architecture for medical image segmentation. In: Stoyanov, D., et al. (eds.) DLMIA/ML-CDS -2018. LNCS, vol. 11045, pp. 3–11. Springer, Cham (2018). https://doi.org/10.1007/978-3-030-00889-5_1
12. Oktay, O., Schlemper, J., Folgoc, L.L., et al.: Attention U-Net: learning where to look for the pancreas. arXiv:1804.03999 (2018)
13. Xiao, X., Lian, S., Luo, Z., Li, S.: Weighted Res-UNet for high-quality retina vessel segmentation. In: 2018 9th International Conference on Information Technology in Medicine and Education (ITME), Hangzhou, 19–21 October 2018, pp. 327–331 (2018)
14. Cui Z., Li C., Wang W.: ToothNet: automatic tooth instance segmentation and identification from cone beam CT images. In: Proceedings of the IEEE/CVF Conference on Computer Vision and Pattern Recognition, pp. 6368–6377 (2019)
15. Chung, M., Lee, M., Hong, J., et al.: Pose-aware instance segmentation framework from cone beam CT images for tooth segmentation. Comput. Biol. Med. **120**, 103720 (2020)
16. Vaswani, A., Shazeer, N., Parmar, N., Uszkoreit, J., Jones, L., Gomez, A.N., et al.: Attention is all you need. In: Proceedings of the 2017 Advances in Neural Information Processing Systems, pp. 5998–6008 (2017)
17. Petit, O., Thome, N., Rambour, C., Themyr, L., Collins, T., Soler, L.: U-Net transformer: self and cross attention for medical image segmentation. Mach. Learn. Med. Imag. 12, 966 (2021). https://doi.org/10.1007/978-3-030-87589-3_28

18. Zhang, Y., Liu, H., Hu, Q.: TransFuse: fusing transformers and CNNs for medical image segmentation. In: de Bruijne, M., Cattin, P.C., Cotin, S., Padoy, N., Speidel, S., Zheng, Y., Essert, C. (eds.) MICCAI 2021. LNCS, vol. 12901, pp. 14–24. Springer, Cham (2021). https://doi.org/10.1007/978-3-030-87193-2_2
19. Valanarasu, J., Oza, P., Hacihaliloglu, I., Patel, V.: Medical transformer: gated axial-attention for medical image segmentation. MICCAI, vol. 12901, pp. 36–46 (2021). https://doi.org/10.1007/978-3-030-87193-2_4
20. Liao, W., Zhu, Y., Wang, X., et al.: LightM-UNet: Mamba assists in lightweight UNet for medical image segmentation. arXiv preprint arXiv:2403.05246 (2024)
21. Xing, Z., Ye, T., Yang, Y., et al.: SegMamba: long-range sequential modeling Mamba for 3D medical image segmentation. In: Linguraru, M.G., et al. (eds.) International Conference on Medical Image Computing and Computer-Assisted Intervention (MICCAI), pp. 578–588. Springer Nature Switzerland, Cham (2024). https://doi.org/10.1007/978-3-031-72111-3_54
22. Liu, J., Yang, H., Zhou, H.Y., et al.: SWIN-UMamba: Mamba-based UNet with ImageNet-based pretraining. In: Linguraru, M.G., et al. (eds.) International Conference on Medical Image Computing and Computer-Assisted Intervention (MICCAI), pp. 615–625. Springer Nature Switzerland, Cham (2024). https://doi.org/10.1007/978-3-031-72114-4_59
23. Hao, J., He, L., Hung, K.F.: T-Mamba: frequency-enhanced gated long-range dependency for tooth 3D CBCT segmentation. arXiv e-prints arXiv:2404.01065 (2024)
24. Cui, Z., et al.: A fully automatic AI system for tooth and alveolar bone segmentation from cone-beam CT images. Nat. Commun. **13**(1), 2096 (2022)
25. Bolelli, F., et al.: Segmenting maxillofacial structures in CBCT volume. In: IEEE/CVF Conference on Computer Vision and Pattern Recognition (CVPR). IEEE (2025)
26. Gibson, E., et al.: Automatic multi-organ segmentation on abdominal CT with dense V-networks. IEEE Trans. Med. Imaging **37**, 1822–1834 (2018)
27. Ibtehaz, N., Rahman, M.S.: MultiResUNet: rethinking the U-Net architecture for multimodal biomedical image segmentation. Neural Netw. **121**, 74–87 (2020)

MCN-CL: Multimodal Cross-Attention Network and Contrastive Learning for Multimodal Emotion Recognition

Feng Li[1], Ke Wu[1], and Yongwei Li[2(✉)]

[1] School of Computer and Information Engineering, Anhui University of Finance and Economics, Anhui 233030, China
[2] State Key Laboratory of Cognitive Science and Mental Health, Institute of Psychology, Chinese Academy of Sciences, Beijing 100045, China
liyw@psych.ac.cn

Abstract. Multimodal emotion recognition plays a key role in many domains, including mental health monitoring, educational interaction, and human-computer interaction. However, existing methods often face three major challenges: unbalanced category distribution, the complexity of dynamic facial action unit time modeling, and the difficulty of feature fusion due to modal heterogeneity. With the explosive growth of multimodal data in social media scenarios, the need for building an efficient cross-modal fusion framework for emotion recognition is becoming increasingly urgent. To this end, this paper proposes Multimodal Cross-Attention Network and Contrastive Learning (MCN-CL) for multimodal emotion recognition. It uses a triple query mechanism and hard negative mining strategy to remove feature redundancy while preserving important emotional cues, effectively addressing the issues of modal heterogeneity and category imbalance. Experiment results on the IEMOCAP and MELD datasets show that our proposed method outperforms state-of-the-art approaches, with Weighted F1 scores improving by 3.42% and 5.73%, respectively.

Keywords: Multimodal cross-attention network · Contrastive learning · Pyramid squeeze attention · Multimodal emotion recognition

1 Introduction

Emotion recognition technology, as a cutting-edge field in human-computer interaction research, is evolving from unimodal analysis to multimodal fusion [1–3]. With the widespread adoption of mobile internet technology, video calls and voice messages have become the mainstream forms of interaction on social media [4], dominating modern communication methods. Additionally, psychological research findings indicate that during emotional expression, some users primarily rely on paralinguistic cues, such as the melody, tone, and rhythm of speech, to convey emotions, while others tend to express inner feelings more through facial

J. Lokoč et al. (Eds.): MMM 2026, LNCS 16413, pp. 561–574, 2026.
https://doi.org/10.1007/978-981-95-6957-1_40

micro-expressions [5]. These converging trends underscore the growing urgency to develop sophisticated multimodal emotion recognition systems. By integrating complementary data streams, such systems aim to overcome the inherent limitations of unimodal analysis, thereby capturing the complex, interwoven nature of human emotions with greater accuracy and depth for a more holistic and reliable understanding of affective states.

Although significant progress has been made in multimodal emotion recognition, current cross-modal fusion methods still face three fundamental limitations that hinder their effective promotion in practical applications. First, the redundancy and heterogeneity problems in cross-modal fusion need urgent resolution. Traditional multimodal fusion methods, such as the MVN-based feature connection [6] or the EmoCaps capsule network [7], fail to model the differences between modalities. This results in a masking effect where large amounts of redundant cross-modal information cover up key affective patterns [8,9]. For example, Hu et al. [10] developed MM-DFN, which employs gating mechanisms to dynamically adjust modal weights. However, its feature connection strategy still cannot eliminate the spectral semantic coupling noise between text and speech [11]. Second, there is a clear deficiency in modeling dynamic micro-expressions. Current visual architectures, such as the ResNet-101 model [12], primarily rely on global pooling or static attention mechanisms. This makes it difficult for them to accurately capture the spatiotemporal dynamic features of facial action units (FAUs), thus leading to inaccurate recognition and analysis of micro-expressions. Finally, the discriminative bias problem is caused by extreme category imbalance. Cross-entropy loss-based methods, such as MultiEMO [13] and TelME [14], yield low F1 scores for sparse classes.

Therefore, to tackle multimodal emotion recognition challenges, this paper proposes the Multimodal Cross-Attention Network and Contrastive Learning (MCN-CL). Our model utilizes cross-attention to effectively integrate features from acoustic, visual, and linguistic modalities, while contrastive learning refines the latent representations to better distinguish between emotional states. The main contributions are summarized as follows

- We propose a PSA-enhanced Visual Feature Extraction Module. This module precisely captures facial dynamic patterns and enhances the resolution of emotional features through multi-scale feature extraction and channel attention interaction.
- We present a multimodal cross-attention network and contrastive learning enhancement module, which progressively integrates information from text, audio, and visual modalities via multi-layer bidirectional multi-head cross-attention mechanisms to generate comprehensive feature representations.
- We introduce a contrastive learning enhancement module. By integrating a hard negative sample mining strategy, it strengthens the model's capacity to distinguish difficult samples and refines the alignment of cross-modal representations.

- We conduct extensive experiments on the IEMOCAP and MELD datasets, respectively. The results demonstrate that our approach outperforms previous state-of-the-art methods.

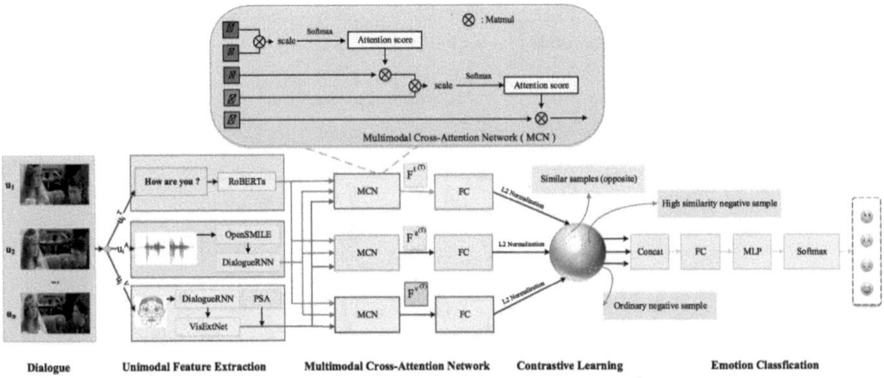

Fig. 1. The framework of our proposed method.

2 Proposed Method

This section details the method proposed in this paper, whose core architecture is shown in Fig. 1. It mainly consists of three steps: unimodal feature extraction, multimodal cross-attention network, and contrastive learning enhancement.

2.1 Task Definition

Given a dialogue sequence containing N speakers, denoted as $U = \{u_1, u_2, \ldots, u_n\}$, each utterance u_i consists of three modalities: visual(u_i^V), speech(u_i^A), and text (u_i^T). The objective is to predict the corresponding emotional category for each utterance $y_i \in \{1, 2, \ldots, K\}$, and K is the total number of emotion categories.

2.2 PSA-Enhanced Visual Feature Extraction Module

In this study, we design a one-dimensional PSA module for time-series visual feature modeling tasks in Fig. 2. By leveraging multi-scale spatiotemporal feature decomposition and dynamic channel attention recalibration, this module effectively alleviates the problems of limited receptive field and local detail loss commonly found in traditional models.

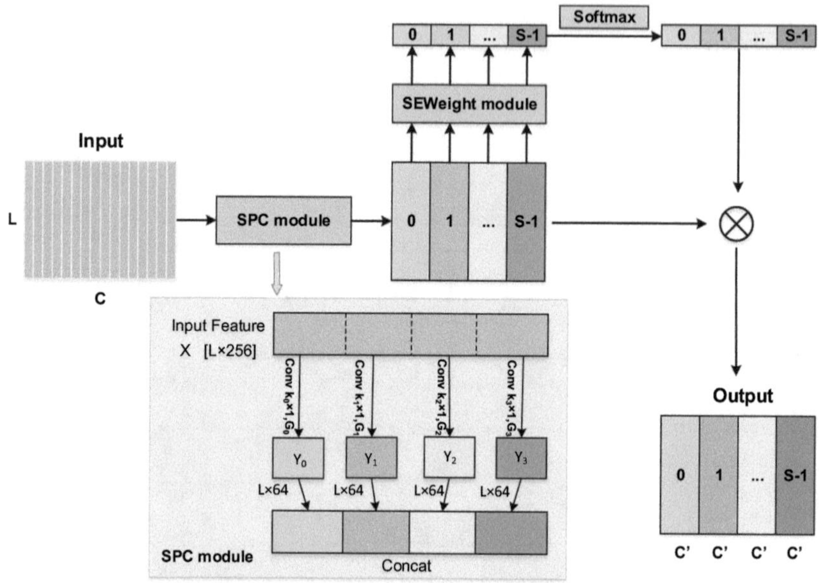

Fig. 2. The overview of PSA-enhanced Visual Feature Extraction Module.

Multi-scale One-Dimensional Temporal Convolution Operation To extract the dynamic patterns of FAUs in the temporal dimension, this section extends the original PSA's two-dimensional spatial convolution to a one-dimensional temporal convolution. The input feature tensor is denoted as $X \in R^{L \times C}$, where L represents the number of video frames and C = 256 indicates the number of feature channels.

We design S=4 parallel sets of standard convolution kernels with different kernel sizes, specifically $3 \times 1, 5 \times 1, 7 \times 1$, and 9×1,to extract sequence features ranging from short-term to long-term. Following the grouped convolution strategy, each set processes C/S channels. The feature output of the s-th branch is expressed as follows

$$Y_s = \text{Conv}_s\left(X_s\right) \tag{1}$$

where X_s is the input of the s-th branch. $\text{Conv}_s(\cdot)$ represents a standard convolution operation with a kernel size of $2 * (s+1) + 1$, and Y_s is the feature output for each scale.

Finally, all scale outputs are concatenated to obtain the fused feature Y as follows

$$Y = \text{Concat}\left(Y_0, Y_1, Y_2, Y_3\right) \tag{2}$$

Channel Attention Interaction. To dynamically adjust the importance of each channel, we introduce the SEWeight mechanism [2] to weight the fused feature Y. First, global average pooling (GAP) is applied to Y along the time dimension to obtain the channel-level representation as follows

$$\mathrm{Z} = \mathrm{GAP}(\mathrm{Y}) \tag{3}$$

In this context, it reduces the temporal dimension (L) of Y to 1, resulting in a channel-level representation $Z \in R^{1\times c}$, where each element Z_c represents the average value of channel c across all time steps. This allows the model to focus on the most relevant channels for the given task.

Subsequently, it Z is passed through two fully connected layers to generate an attention weight vector as follows

$$\mathrm{W} = \mathrm{Sigmoid}\left(\mathrm{FC}_2\left(\mathrm{ReLU}\left(\mathrm{FC}_1(\mathrm{Z})\right)\right)\right) \tag{4}$$

where FC_1 and FC_2 denote the fully connected layer. FC_1 transforms the channel-level representation Z from dimension C to C/4. The ReLU activation function is applied after FC_1 to introduce non-linearity. FC_2 then transforms the output back to dimension C. The Sigmoid activation function is applied after FC_2 to produce the channel attention coefficient W, where each element W_c indicates the importance weight of the channel c.

Finally, the attention coefficient W is applied to the original fusion feature Y through element-wise multiplication in broadcast mode as follows

$$C^v = \mathrm{Y} \odot \ \mathrm{W} \tag{5}$$

where $\odot$ represents per-channel multiplication. This operation adjusts the importance of each channel Y based on the learned attention weights W. And C^v is aligned with the dimensions of the text and speech modalities to facilitate subsequent cross-modal interaction operations.

2.3 Multimodal Cross-Attention Network

The three modalities of text, audio, and visual, bidirectional multi-head cross-attention are first used to achieve cross-modal information interaction between any two modalities. Taking the text modality as an example, the features obtained from its interaction with the audio modality are then input into another bidirectional multi-head cross-attention as queries (Q). Here, the visual modality serves as keys (K) and values (V) for further feature fusion and refinement. Each interaction involves different modality combinations, ensuring that each modality can obtain unique context information from the others. This process enables more comprehensive and in-depth cross-modal information fusion.

First, multi-head projection transformations are applied separately to the query, key, and value as follows

$$Q_h^{ta^{(j)}} = F^{t^{(j-1)}} W^{Q_h^{ta^{(j)}}} \tag{6}$$

$$K_h^{ta^{(j)}} = C^a W^{K_h^{ta^{(j)}}} \tag{7}$$

$$V_h^{ta^{(j)}} = C^a W^{V_h^{ta^{(j)}}} \tag{8}$$

where $W^{Q_h^{ta^{(j)}}}$, $W^{K_h^{ta^{(j)}}}$ and $W^{V_h^{ta^{(j)}}}$ are the parameters of the multi-head projection matrix, H is the number of attention heads, and $h \in \{1, \ldots H\}$.

Then, for each attention head, the weighted attention output is computed as follows

$$A_h^{ta^{(j)}} = \text{Softmax}\left(\frac{Q_h^{ta^{(j)}} K_h^{ta^{(j)}}}{\sqrt{d_{K_h^{ta^{(j)}}}}}\right) V_h^{ta^{(j)}}, h \in \{1, \ldots H\} \tag{9}$$

All attention heads are concatenated as follows

$$MH^{ta^{(j)}} = \text{Concat}\left(A_1^{ta^{(j)}}, A_2^{ta^{(j)}}, \ldots, A_H^{ta^{(j)}}\right) W^{O^{ta^{(j)}}} \tag{10}$$

where $W^{O^{ta^{(j)}}}$ is the output projection matrix.

The updated feature representation is as follows

$$F^{tav^{(j)}} = \text{Add \& Norm}\left(F^{ta^{(j)}} + MH^{tav^{(j)}}\right) \tag{11}$$

Next, perform multi-head attention computation on the updated text modality and visual modality as follows

$$Q_h^{\text{tav}^{(j)}} = F^{\text{ta}^{(j)}} W^{K_h^{\text{tav}^{(j)}}} \tag{12}$$

$$K_h^{\text{tav}(j)} = C^v W^{Q_h^{\text{tav}(j)}} \tag{13}$$

$$V_h^{\text{tav}(j)} = C^v W^{v_h^{\text{tav}(j)}} \tag{14}$$

where $W^{Q_h^{\text{tav}(j)}}$, $W^{K_h^{\text{tav}(j)}}$ and $W^{v_h^{\text{tav}(j)}}$ are the parameters of the multi-head projection matrix.

By stacking T layers of this mechanism, the MCN can iteratively refine cross-modal representations, leading to more robust and context-aware feature fusion. The value of T is dataset-dependent; details are provided in the ablation study section. Each layer builds upon the outputs of the previous one, enabling the network to capture both local and global dependencies across modalities. This layered architecture not only enhances the model's expressive capacity but also improves its adaptability to complex multimodal data.

Similarly, the audio and visual modalities are also calculated with cross-attention with other modalities, obtaining updated representations $F^{a^{(j)}}$ and $F^{v^{(j)}}$. The multi-head bidirectional design can capture fine-grained inter-modal dependencies and ensures smooth information reconstruction through residual normalization and feedforward networks.

2.4 Contrastive Learning

For the text, audio, and visual features extracted by the MCN: $F^{a^{(T)}} F^{t^{(T)}} F^{v^{(T)}}$ we introduce a contrastive learning mechanism to further optimize the consistency of crossmodal representations. Specifically, we perform contrastive learning on the features of each modality separately to enhance the model's understanding and discrimination of the intrinsic structures across modalities. The mapped features are L2-normalized so that they are on the unit sphere, which facilitates the subsequent calculation of cosine similarity as follows

$$z_{\text{text}} = \text{ReLU}\left(W_{\text{text\_}fc1} F^{t^{(T)}} + b_{\text{text\_}fc1}\right) \tag{15}$$

$$z_{\text{text}} = W_{\text{text\_}fc2} z_{\text{text}} + b_{\text{text\_}fc2} \tag{16}$$

$$z_{\text{audio}} = \text{ReLU}\left(W_{\text{audio\_}fc1} F^{a^{(T)}} + b_{\text{audio\_}fc1}\right) \tag{17}$$

$$z_{\text{audio}} = W_{\text{audio\_}fc2} z_{\text{audio}} + b_{\text{audio\_}fc2} \tag{18}$$

$$z_{\text{visual}} = \text{ReLU}\left(W_{\text{visual\_}fc1} F^{v^{(T)}} + b_{\text{visual\_}fc1}\right) \tag{19}$$

$$z_{\text{visual}} = W_{\text{visual\_}fc2} z_{\text{visual}} + b_{\text{visual\_}fc2} \tag{20}$$

where $W_{()_{f}c1}$ and $W_{()_{f}c2}$ are the weight matrices of the fully connected layer, and $b_{()_{f}c1}$, $b_{()_{f}c2}$ are the bias terms.

Subsequently, the mapped features are L2-normalized so that they are on the unit sphere, which facilitates the subsequent calculation of cosine similarity as follows

$$z'_{\text{text}} = \frac{z_{\text{text}}}{\|z_{\text{text}}\|_2} \tag{21}$$

$$z'_{\text{audio}} = \frac{z_{\text{audio}}}{\|z_{\text{audio}}\|_2} \tag{22}$$

$$z'_{\text{visual}} = \frac{z_{\text{visual}}}{\|z_{\text{visual}}\|_2} \tag{23}$$

For each modality, we construct a supervised contrastive loss function, designed to bring closer the distances between samples of the same class while pushing farther apart the distances between samples of different classes. For the i-th sample, its corresponding modality's loss function is defined as follows

$$L_{\text{contrast}} = -\frac{1}{|\mathcal{P}_i|}\sum_{p\in\mathcal{P}_i}\log\frac{\exp\left(\text{sim}\left(z_i, z_p\right)/\tau\right)}{\sum_{k=1}^{N}1_{[k\neq i]}\exp\left(\text{sim}\left(z_i, z_k\right)/\tau\right)+\sum_{p\in\mathcal{P}_i}\exp\left(\text{sim}\left(z_i, z_p\right)/\tau\right)} \tag{24}$$

where $\mathcal{P}_i$ represents a set of positive samples of the same kind as sample $i \cdot \text{sim}\left(z_i, z_j\right)$ represents the cosine similarity of the features of samples i and j in the latent space. τ is a temperature parameter used to control the sharpness of the similarity distribution. $1_{[k\neq i]}$ is an indicator function, which takes $k \neq i$ when B is true; otherwise, it takes 0.N the total number of samples in the batch.

To enhance the model's ability, we introduce a hard negative mining strategy. We select the top 30% most similar negative sample subset to participate in loss computation, while the remaining negative samples are used solely for normalization. Specifically, for each sample i, we compute its similarity with all negative samples and then select the k most similar negative samples to participate in loss computation as follows

$$\mathcal{N}_i^{\text{hard}} = \{j \mid j \notin \mathcal{P}_i \cup \{i\}, \text{sim}\left(z_i, z_j\right) \geq \text{top}\,k\left(\text{sim}\left(z_i, z_{\mathcal{N}_i}\right)\right)\} \tag{25}$$

$$L_{\text{contrast}}^{\text{hard}} = -\frac{1}{|\mathcal{P}_i|}\sum_{p\in\mathcal{P}_i}\log\frac{\exp\left(\text{sim}\left(z_i, z_p\right)/\tau\right)}{\sum_{n\in\mathcal{N}_i^{\text{hard}}}\exp\left(\text{sim}\left(z_i, z_n\right)/\tau\right)+\sum_{p\in\mathcal{P}_i}\exp\left(\text{sim}\left(z_i, z_p\right)/\tau\right)} \tag{26}$$

where $\mathcal{N}_i^{\text{hard}}$ represents the set of difficult negative samples for sample i.

Finally, we fuse the contrastive loss of text, audio, and visual modalities to obtain the final contrastive loss as follows

$$L_{\text{total\_}contrast} = \alpha \cdot L_{\text{contrast\_}text} + \beta \cdot L_{\text{contrast\_}audio} + \gamma \cdot L_{\text{contrast\_}visual} \tag{27}$$

where α, β, and γ are weighting coefficients used to balance the contributions of the modal contrast losses.

2.5 Emotion Classification

After optimizing the features through the MCN-CL module, the multimodal feature representations f_{text} , f_{audio} , and f_{visual} are concatenated to form a comprehensive feature vector. This vector first passes through a fully connected layer, which integrates and transforms the features to map the high-dimensional concatenated features into a feature space suitable for classification. Then, the feature vector is input into a 2-layer Multilayer Perceptron (MLP) with ReLU activation, where the MLP further extracts and fuses features through nonlinear transformations, enhancing the model's expressive capability and ability to capture complex patterns. Finally, a Softmax layer is used for emotion recognition, selecting the emotion label with the highest probability as the prediction. The process is illustrated as follows

$$h_{\text{concat}} = [f_{\text{text}}, f_{\text{audio}}, f_{\text{visual}}] \tag{28}$$

$$h_{fc} = FC\left(h_{\text{concat}}\right) \tag{29}$$

$$h_{\text{mlp}} = MLP\left(h_{fc}\right) \tag{30}$$

$$\hat{y} = \text{Softmax}\left(W_{\text{softmax}} h_{\text{mlp}} + b_{\text{softmax}}\right) \tag{31}$$

where h_{concat} is the concatenated feature vector, h_{fc} is the output of the fully connected layer, h_{mlp} is the output of the MLP, and W_{softmax} and b_{softmax} are the weight matrix and bias term of the Softmax layer, respectively.

3 Experiment Evaluation

3.1 Datasets

In this study, we evaluated the proposed method on two datasets: IEMOCAP [20] and MELD [21], respectively.

IEMOCAP: This dataset is a database containing approximately 12 h of English audio-visual data, encompassing video, speech, facial motion capture, and text transcriptions. The database is structured around dyadic conversations, where actors interact through either improvisation or specifically designed scripted scenarios. These scenarios were carefully selected to elicit emotional expressions. The dataset provides six categorical emotion labels (such as anger, happiness, sadness, frustration, neutral, and excitement) as well as dimensional annotations (including valence, activation, and dominance).

MELD: This dataset is a multimodal, multi-speaker dialogue dataset generated from dialogue scenes of the TV show 'Friends'. The dataset contains a total of

13,708 utterances, 1,433 dialogues, and 304 different speakers. Specifically, unlike two-person dialogue datasets like IEMOCAP, MELD's dialogues include three or more speakers. Each utterance in the dialogue is annotated with one of seven emotional category labels, including anger, disgust, fear, joy, neutral, sadness, and surprise.

3.2 Setting and Metric

This study was conducted on PyTorch 2.0.0, Python 3.8.0 and CUDA Toolkit 12.1. A unified cross-modal feature processing flow is adopted in the experiment: text features are extracted by the RoBERTa-base model, and the dimensionality is reduced from 768 to 256 by a fully connected layer; speech features are extracted by OpenSMILE tool, and then linearly transformed to 256 dimensions; visual features are processed by ResNet-101 and Pyramid Squeeze Attention module, and 1000-dimensional features are extracted and projected to 256-dimensional space. On the IEMOCAP and MELD datasets, we adopt the Weighted F1 as evaluation metrics [23,24].

3.3 Results and Discussion

In this section, the experiment results are reported to evaluate the proposed method on the IEMOCAP and MELD datasets, as shown in Table 1 and Table 2, respectively.

On the IEMOCAP dataset in Table 1, the model achieved a Weighted F1 score of 74.22%. Meanwhile, on the MELD dataset, the Weighted F1 score reached as high as 73.1%, significantly outperforming the second-best method TelME (67.37%) by 5.73%. This consistently excellent performance across datasets validates the generalization capability of the framework design, particularly demonstrating advantages in multimodal feature interaction and dialogue context modeling.

In MELD's fear detection, the performance reaches 43.42%, an improvement of 16.45% compared to TelME (26.97%). Meanwhile, the cross-dataset anger recognition performance (IEMOCAP: 77.35% vs. MELD: 64.45%) ranked first, indicating that the model can effectively integrate the acoustic features of speech intonation with the implicit emotional cues of textual semantics.

3.4 Ablation Study

In this study, we conducted systematic ablation experiments, with the results shown in Table 3 and Table 4. The experiments demonstrate that after removing the PSA module, the model's Weighted F1 score on the IEMOCAP dataset decreased by 2.53%, and on the MELD dataset by 0.55%, with a significant decline in performance on fine-grained emotional categories such as happiness, fear, and disgust. Specifically, in the MELD dataset, the F1 score for the fear category dropped by 3.0%, and for the disgust category by 2.28%, highlighting the critical role of the PSA module in temporal feature modeling.

Table 1. Experiment results on the IEMOCAP dataset. The best results are highlighted in bold.

Methods	Hap.	Sad.	Neu.	Ang.	Exc.	Fru.	Weighted F1
BC-LSTM [15]	34.43	60.87	51.81	56.73	57.95	58.92	54.95
DialogueRNN [16]	33.18	78.8	59.21	65.28	71.86	58.91	62.75
DialogueGCN [17]	51.87	76.76	56.76	62.26	72.71	58.04	63.16
IterativeERC [22]	53.17	77.19	61.31	61.45	69.23	60.92	64.37
MMGCN [18]	42.34	78.67	61.73	69	74.33	62.32	66.22
MVN [6]	55.75	73.3	61.88	65.96	69.5	64.21	65.44
MM-DFN [10]	42.22	78.98	66.42	69.77	75.56	66.33	68.18
GA2MIF [19]	46.15	**84.5**	68.38	70.29	75.99	66.49	70
CRRGM [25]	52.4	76.8	70.1	65.4	77.3	63.6	70.4
TelME [14]	-	-	-	-	-	-	70.4
UniMSE [26]	-	-	-	-	-	-	70.66
MALN [27]	55.5	81.8	64.1	69.1	**78**	71.4	70.8
Ours	**68.22**	82.41	**76.6**	**77.35**	67.01	**71.9**	**74.22**

Furthermore, when both the MCN-CL framework is removed simultaneously, the model performance degrades significantly. The Weighted F1 score on the IEMOCAP dataset drops by 21.17% (from 74.22% to 53.05%), and on the MELD dataset, it decreases by 2.74% (from 73.10% to 70.36%). Particularly in extreme cases, such as the happy class on the IEMOCAP dataset, its F1 score drops sharply from 68.22% to 2.41%, while on the MELD dataset, the detection of fear and disgust categories completely fails (F1 score of 0). This result indicates that the multimodal cross-attention network can effectively capture cross-modal complementary information, and contrastive learning enhancement effectively alleviates the class imbalance problem.

4 Conclusions

In this paper, we propose a novel method for multimodal emotion recognition using visual pyramid squeeze attention with contrastive cross-attention. The designed Pyramid Squeeze Attention module significantly enhances the ability to capture temporal features of facial action units, such as eyelid micro-tremors and lip muscle tension fluctuations, through multi-scale one-dimensional temporal convolution operations and channel attention interaction. Additionally, the multimodal cross-attention network and contrastive learning enhancement network, with a triple-query architecture and hard negative sample mining strategy, effectively resolve the acoustic-semantic coupling problem in highly confused categories, achieving an F1 score of 43.42% for sparse categories (e.g., fear) on the MELD dataset. Experiment results on the IEMOCAP and MELD datasets demonstrate that our proposed method achieves Weighted F1 scores of 74.22%

Table 2. Experiment results on the MELD dataset. The best results are highlighted in bold.

Methods	Neu.	Sur.	Fea.	Sad.	Joy	Dis.	Ang.	Weighted F1
BC-LSTM [15]	73.8	47.7	5.4	25.1	51.3	5.2	38.4	55.9
DialogueRNN [16]	76.23	49.59	0	26.33	54.55	0.81	46.76	58.73
DialogueGCN [17]	76.02	46.37	0.98	24.32	53.62	1.22	43.03	57.52
IterativeERC [22]	77.52	53.65	3.31	23.62	56.63	19.38	48.88	60.72
MMGCN [18]	-	-	-	-	-	-	-	58.65
MVN [6]	76.65	53.18	11.7	21.82	53.62	21.86	42.55	59.03
MM-DFN [10]	77.76	50.69	-	22.93	54.78	-	47.82	58.65
EmoCaps [7]	77.12	63.19	3.03	42.52	57.5	7.69	57.54	64
GA2MIF [19]	76.92	49.08	-	27.18	51.87	-	48.52	58.94
TelME [14]	80.22	60.33	26.97	43.45	65.67	26.42	56.7	67.37
UniMSE [26]	-	-	-	-	-	-	-	65.51
MALN [27]	82	58.6	21.2	43	64.3	17.6	52.4	66.9
Ours	**85.16**	**74.9**	**43.42**	**46.83**	**67.94**	**28.37**	**64.45**	**73.1**

Table 3. Experiment results on the IEMOCAP dataset.

Methods	Hap.	Sad.	Neu.	Ang.	Exc.	Fru.	Weighted F1
w/o PSA	62.56	79.65	72.69	76.55	63.64	70.99	71.69
w/o MCN-CL	2.41	11.03	57.49	63.54	69.4	70.26	53.05
Ours	68.22	82.41	76.6	77.35	67.01	71.9	74.22

and 73.10%, respectively. In the future, we will advance multimodal learning to address high-frequency class bias, cross-modal noise decoupling, and modeling of extremely sparse samples.

Acknowledges. This work was supported in part by the National Natural Science Foundation of China (Grant No. 12574510), the Beijing Natural Science Foundation (No. L257021), the Key Project of Anhui Provincial University Scientific Research Plan (No. 2024AH050018), and the Key Project in Natural Sciences of Anhui University of Finance and Economics (No. ACKYB23016).

Table 4. Experiment results on the MELD dataset.

Methods	Neu.	Sur.	Fea.	Sad.	Joy	Dis.	Ang.	Weighted F1
w/o PSA	84.79	74.28	40.45	45.73	67.69	26.09	64.04	72.55
w/o MCN-CL	84.42	76.33	0	43.22	69.02	0	58.47	70.36
Ours	85.16	74.9	43.42	46.83	67.94	28.37	64.45	73.1

References

1. Li, F., Liu, R., Wang, B.: MIMCL: Multilayer interaction module with contrastive learning for speech emotion recognition, In: International Conference on Multimedia Retrieval, pp. 724–732 (2025)
2. Erkantarci, B., Bakal, G.: An empirical study of sentiment analysis utilizing machine learning and deep learning algorithms, J. Comput. Soc. Sci. 1–17 (2023)
3. Li, F., Luo, J., Xia, W.: Wavfusion: towards wav2vec 2.0 multimodal speech emotion recognition, In: International Conference on Multimedia Modeling, Springer, pp. 325–336 (2025)
4. Park, S., Shim, H.S., Chatterjee, M., Sagae, K., philippe Morency, L.: Multimodal analysis and prediction of persuasiveness in online social multimedia, ACM Trans. Interactive Intell. Syst. (TiiS) 6, 1 – 25 (2016)
5. Le, P.N., Epps, J., Choi, E.H.C., Ambikairajah, E.: A study of voice source and vocal tract filter based features in cognitive load classification, In: 2010 20th International Conference on Pattern Recognition, pp. 4516–4519 (2010)
6. Ma, H., Wang, J., Lin, H., Pan, X., Zhang, Y., Yang, Z.: A multi-view network for real-time emotion recognition in conversations . Knowl. Based Syst. **236**, 107751 (2021)
7. Li, Z., Tang, F., Zhao, M., Zhu, Y.: Emocaps: emotion capsule based model for conversational emotion recognition, In: Findings (2022)
8. Cambria, E., Hazarika, D., Poria, S., Hussain, A., Subramanyam, R.B.V.: Benchmarking multimodal sentiment analysis, ArXiv abs/1707.09538 (2017)
9. Wang, H., Li, X., Ren, Z., Wang, M., Ma, C.: Multimodal sentiment analysis representations learning via contrastive learning with condense attention fusion, Sensors (Basel, Switzerland) 23 (2023)
10. Hu, D., Hou, X., Wei, L., Jiang, L.-X., Mo, Y.: Mm-DFN: multimodal dynamic fusion network for emotion recognition in conversations, IEEE International Conference on Acoustics, Speech and Signal Processing (ICASSP), pp. 7037–7041 (2022)
11. Ghosh, S., Anwar, T.: Depression intensity estimation via social media: a deep learning approach . IEEE Trans. Comput. Soc. Syst. **8**, 1465–1474 (2021)
12. He, K., Zhang, X., Ren, S., Sun, J.: Deep residual learning for image recognition, In: 2016 IEEE Conference on Computer Vision and Pattern Recognition (CVPR), pp. 770–778 (2022)
13. Shi, T., Huang, S.-L.: Multiemo: an attention-based correlation-aware multimodal fusion framework for emotion recognition in conversations, In: Annual Meeting of the Association for Computational Linguistics (2023)
14. Yun, T., Lim, H., Lee, J., Song, M.: Telme: teacher-leading multimodal fusion network for emotion recognition in conversation, In: North American Chapter of the Association for Computational Linguistics (2024)
15. Poria, S., et al.: Context-dependent sentiment analysis in user-generated videos, In: Annual Meeting of the Association for Computational Linguistics (2017)
16. Majumder, N.., Poria, S., Hazarika, D., Mihalcea, R., Gelbukh, A., Cambria, E.: Dialoguernn: an attentive RNN for emotion detection in conversations, In: AAAI Conference on Artificial Intelligence (2018)
17. Ghosal, D., Majumder, N., Poria, S., Chhaya, N., Gelbukh, A.: Dialoguegcn: a graph convolutional neural network for emotion recognition in conversation, In: Conference on Empirical Methods in Natural Language Processing (2019)

18. Hu, J., Liu, Y., Zhao, J., Jin, Q.: Mmgcn: multimodal fusion via deep graph convolution network for emotion recognition in conversation, arXiv:2107.06779 (2021)
19. Li, J., Wang, X., Lv, G., Zeng, Z.: Ga2mif: Graph and attention based two-stage multi-source information fusion for conversational emotion detection . IEEE Trans. Affect. Comput. **15**, 130–143 (2022)
20. Poria, S., Hazarika, D., Majumder, N., Naik, G., Cambria, E., Mihalcea, R.: Meld: a multimodal multi-party dataset for emotion recognition in conversations, ArXiv abs/1810.02508 (2018)
21. Poria, S., Hazarika, D., Majumder, N., Naik, G., Cambria, E., Mihalcea, R.: Meld: a multimodal multi-party dataset for emotion recognition in conversations, ArXiv abs/1810.02508 (2018)
22. Lu, X., Zhao, Y., Wu, Y., Tian, Y., Chen, H., Qin, B.: An iterative emotion interaction network for emotion recognition in conversations, In: International Conference on Computational Linguistics (2020)
23. Liang, T., Lin, G., Feng, L., Zhang, Y., Lv, F.: Attention is not enough: mitigating the distribution discrepancy in asynchronous multimodal sequence fusion. In: Proceedings of the IEEE/CVF International Conference on Computer Vision, pp. 8148–8156 (2021)
24. Lv, F., Chen, X., Huang, Y., Duan, L., Lin, G.: Progressive modality reinforcement for human multimodal emotion recognition from unaligned multimodal sequences. In : Proceedings of the IEEE/CVF Conference on Computer Vision and Pattern Recognition, pp. 2554–2562 (2021)
25. Chen, G., Liu, S., Bi, X., Chen, Q., Meng, T.: Crgmr: a contextualized RGAT and graphtransformer method for multimodal emotion recognition, In: 2024 43rd Chinese Control Conference (CCC), pp. 8286–8291 (2024)
26. Hu, G., Lin, T.-E., Zhao, Y., Lu, G., Wu, Y., Li, Y.: Unimse: towards unified multimodal sentiment analysis and emotion recognition, In: Conference on Empirical Methods in Natural Language Processing (2022)
27. Ren, M., Huang, X., Liu, J., Liu, M., Li, X., Liu, A.: Maln: multimodal adversarial learning network for conversational emotion recognition, IEEE Trans. Circuits Syst. Video Technol. **33**, 6965–6980 (2023)

CTDiff : A Lightweight Hybrid Diffusion Network for Low-Light Endoscopic Image Enhancement

Guodong Wei[1], Jiayu Yu[1], Yu Ao[1,4], YuQin Li[1,4], Yuanyuan Guan[1], Weili Shi[1,3](✉), Yu Miao[1,4], and Zhengang Jiang[1,2]

[1] Changchun University of Science and Technology, Changchun 130022, China
swl@cust.edu.cn

[2] Jilin Province Cross-regional Cooperation Science and Technology Innovation Center of Intelligent Technology and Instrument for Precise Diagnosis and Treatment, Changchun 130022, China

[3] Jilin Provincial Key Laboratory of Intelligent Computing in Medical Image, Changchun University of Science and Technology, Changchun 130022, China

[4] Zhongshan Institute of Changchun University of Science and Technology, Zhongshan 528437, China

Abstract. In minimally invasive surgery, endoscopic imaging is often compromised by insufficient illumination, which can degrade image quality and increase both the difficulty and risk of surgical procedures. In recent years, denoising diffusion models have shown considerable promise for enhancing low-light images, demonstrating strong capabilities in detail restoration and noise suppression. However, their high computational demands and limited ability to model low-dimensional features restrict their suitability for real-time clinical use.To overcome these limitations, we propose a lightweight hybrid diffusion framework, termed CTDiff, which combines a generator and a discriminator. Within the generator, local and global low-dimensional features are first extracted and fused through the convolutional neural network and the Swin Transformer, enabling initial enhancement of low-light images. This is followed by the iterative inverse diffusion process to refine residual noise and blurred regions, thereby further improving image quality. A dynamic resolution strategy is also introduced during generation to reduce computational overhead. In the discriminator, the global–local discriminative mechanism is employed to regulate brightness distribution and maintain structural consistency, enhancing both the realism and stability of the enhanced outputs. Experiments conducted on the simulated low-light endoscopy dataset demonstrate that the proposed method achieves notable improvements in brightness restoration and detail preservation. Moreover, the model enhances the accuracy of downstream surgical instrument segmentation, underscoring its potential for clinical application.

Keywords: minimally invasive surgery · low-light endoscopic images · image enhancement · diffusion model

J. Lokoč et al. (Eds.): MMM 2026, LNCS 16413, pp. 575–587, 2026.
https://doi.org/10.1007/978-981-95-6957-1_41

1 Introduction

Minimally invasive surgery has become an important component of modern clinical practice due to its advantages of reduced trauma, faster recovery, and a lower risk of postoperative complications. Endoscopes, as the primary visualisation tools in such procedures, rely on high-resolution imaging systems to provide surgeons with real-time and intuitive views of tissues and critical anatomical structures. This visual guidance is essential for improving surgical precision and safety. However, the restricted operating space in minimally invasive surgery often results in inadequate illumination, which produces endoscopic images with low brightness, diminished contrast, and loss of fine details. Such limitations make it more difficult to distinguish anatomical structures accurately, thereby compromising surgical precision and safety, while potentially prolonging operation time and increasing the likelihood of complications. Consequently, the development of efficient and robust low-light endoscopic image enhancement methods, capable of improving brightness and restoring fine structural details, is of significant importance for ensuring both the safety and accuracy of minimally invasive procedures.

Traditional low-light image enhancement (LLIE) methods typically rely on physical degradation models and hand-crafted priors, which characterise light attenuation and imaging degradation mechanisms to achieve brightness compensation and detail restoration. For instance, Guo et al. [1] estimated the illumination map and applied local adaptive enhancement, while using sparse constraints to model the brightness distribution, in order to improve overall brightness and local contrast. Zhang et al. [2] introduced a degradation modeling and decoupling strategy, in which brightness information and other degradation components were modeled and restored separately to improve image quality. Although these methods are effective under certain conditions, their reliance on prior assumptions limits their generalization ability in complex lighting environments, leading to insufficient model adaptability and difficulty in handling diverse situations in clinical surgical scenarios. In addition, for medical images with blurred tissue edges and high noise, such methods still have limitations in detail restoration and noise suppression.

Traditional LLIE methods often struggle with modelling complex structures and adapting across different scenes. To overcome these limitations, researchers have increasingly turned to deep learning (DL) approaches, which offer stronger feature representation and improved cross-domain generalisation. Convolutional Neural Networks (CNNs), in particular, support end-to-end feature learning without the need to explicitly model the underlying physics of image formation. This allows CNNs to extract multi-level features, thereby achieving notable progress in brightness restoration, detail reconstruction, and noise suppression. For instance, Guo et al. [3] designed a lightweight CNN that predicts pixel-level higher-order mapping curves and introduced a zero-reference loss function, enabling unsupervised training even without paired data. Similarly, Zamir et al.'s MIRNet [4] maintains spatial fidelity through a full-resolution branch, while incorporating parallel multi-scale branches to capture contextual information. It further employs cross-scale feature interaction and attention mechanisms for feature fusion. Its successor, MIRNet v2 [5], integrates a learnable feature correction module and a cross-stage residual recalibration strategy, improving restoration quality and detail preservation while keeping model complexity manageable. While CNN-based approaches are effective in extracting local features, they remain limited in capturing long-range dependencies and in handling more complex visual contexts. To address this issue, researchers have begun incorporating Transformer architectures, which are better suited for modelling global information. For example, Xu et al. [6] proposed a method that leverages a signal-to-noise ratio (SNR) map as prior knowledge, combining parallel Transformer and CNN branches to capture both global dependencies and local details. An SNR mask attention mechanism is then applied to suppress noise in low-SNR regions, enabling spatially adaptive enhancement. Although these approaches show promising results in enhancing natural low-light images, challenges remain in achieving effective global–local feature integration and accurate structural restoration, particularly in detail-rich and complex domains such as endoscopic imaging.

With the advancement of deep feature modelling, researchers have increasingly explored the use of Generative Adversarial Networks (GANs) [7] for image enhancement. The adversarial interplay between the generator and discriminator allows the generator to learn feature distributions that more closely resemble those of real images, thereby improving perceptual quality and detail fidelity. For instance, Jian et al. [8] developed a framework with both global and local discriminators within an unsupervised adversarial training setting, integrating self-regularised perceptual loss and attention mechanisms to enhance low-light images without requiring paired training data. Similarly, Ma et al. [9] proposed a non-paired bidirectional GAN architecture, incorporating lighting regularisation and structural preservation losses to improve both illumination balance and structural consistency in medical images. Although GAN-based approaches can achieve notable improvements in visual quality, they remain challenged by issues of training stability, controllability, and artefact suppression. These limitations are particularly pronounced in medical image enhancement, where accurate structural preservation is critical, thereby constraining their practical applicability.

In recent years, Denoising Diffusion Probabilistic Models (DDPMs) [10] have demonstrated notable progress in image restoration and generation. These models reconstruct high-quality images from noise through a gradual reverse denoising process, offering strong structural fidelity and detail preservation. For instance, Jiang et al. [11] proposed a wavelet-domain conditional diffusion mechanism, executing the diffusion process layer by layer in the frequency domain and incorporating a high-frequency detail restoration module to further enhance low-light image quality. Despite their effectiveness, DDPMs are computationally demanding, which limits their applicability in real-time tasks such as endoscopic imaging. To address this, Zhou et al. [12] introduced a multi-level resolution progressive sampling strategy based on a pyramid architecture, improving sampling efficiency and integrating a global corrector to mitigate global degradation. Chen et al. [13] further reduced model complexity by applying adaptive resolution and pruning downsampling components, combined with a chrominance balancer to correct spectral shifts, thereby lowering computational overhead while preserving enhancement performance. Although these approaches alleviate some computational burdens, low-dimensional feature extraction capabilities remain limited, and the sampling process continues to be resource-intensive. Recent studies have explored combining DDPMs with other deep learning techniques to accelerate sampling [14–16]. For example, Bai et al. [15] employed a CNN branch for initial feature extraction and shallow reconstruction, followed by an inverse diffusion process for fine-grained detail restoration, enhancing overall image quality. Nevertheless, the CNN branch remains limited in capturing long-range dependencies, which reduces its representational capacity for endoscopic images with complex tissue structures, and computational demands continue to pose a significant challenge.

To address these issues, this paper proposes a lightweight hybrid diffusion framework for low-light endoscopic image enhancement, named CTDiff. The goal is to improve image quality while controlling model complexity and computational cost. The main contributions of this paper are as follows:

- We design a low-light image enhancement framework for endoscopy (CTDiff). After the CNN–Transformer branch performs feature learning and initial image reconstruction, the generator applies an inverse diffusion process to further optimise reconstruction and preserve details, bringing enhanced results closer to the target. The discriminator constrains lighting distribution and structural consistency through a global-local joint discrimination mechanism, improving the authenticity and stability of the enhanced images.
- We propose a Hybrid CNN–Transformer Block (HCTB) that fuses local and global low-dimensional features. Combined with CBAM, it forms a joint attention mechanism across channel and spatial dimensions to emphasise anatomical structures while suppressing background signals.
- Experiments on two simulated low-light endoscopy datasets show that the proposed model and its modules perform effectively. In downstream surgical instrument segmentation tasks, CTDiff also demonstrates potential clinical value.

2 Method

As illustrated in Fig. 1, the CTDiff model comprises two primary components: a generator and a discriminator. The generator is responsible for restoring low-light endoscopic images to normal-light conditions. While the discriminator evaluates both the illumination distribution and structural consistency of the generated images through a global–local joint discrimination mechanism. This mechanism serves to regulate the authenticity and stability of the generator's outputs. The following sections provide a detailed description of the generator architecture, the discriminator design, and the associated loss functions employed in the CTDiff framework.

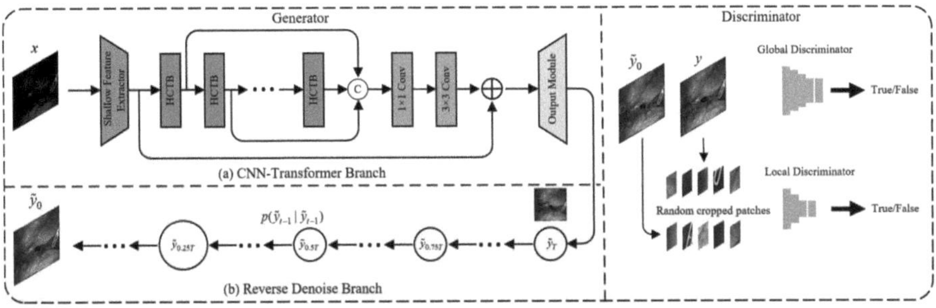

Fig. 1. An overview of the CTDiff network architecture we designed. See Sect. 2 for a detailed discussion.

2.1 Generator

As shown in the left-hand side of Fig. 1, the generator comprises two components: a CNN-Transformer branch and a denoising branch. Figure 1 (a) illustrates that the CNN-Transformer branch comprises a Shallow Feature Extractor (SFE), multiple Hybrid CNN–Transformer Blocks (HCTB), a Dense Feature Fusion (DFF) module and an Output Module (OPM).

SFE uses a two-dimensional convolution operation to map the input image $x \in \mathbb{R}^{3\times W\times H}$(where W and H represent the width and height of the image, respectively) to a high-dimensional feature space to obtain basic structural and texture representations. These shallow features are then fed into a series of cascaded HCTB modules as input for subsequent deep modelling to capture structural details and global dependencies in complex low-light scenes layer by layer.

HCTB is the core component of the CNN–Transformer branch, which structure is shown in Fig. 2. It combines the convolutional attention mechanism with the cross-region modelling capability of the Swin Transformer [17]. The module first constructs a joint attention mechanism in the channel and spatial dimensions through the Convolutional Block Attention Module (CBAM) [18] to highlight key anatomical structures and suppress redundant background signals. The attention-optimised features are then input into a Swin Transformer block (Swin Transformer Block, STB) composed of two stacked Swin Transformer layers (Swin Transformer Layer, STL), alternately executing window multi-head self-attention (W-MSA) and shifted window multi-head self-attention (SW-MSA) to effectively fuse local and global information. To address the limitations of Transformers in translation-invariant processing, HCTB introduces standard convolutional layers after global modelling to compensate for local structural information and restore positional sensitivity, and reintroduces CBAM at the tail to further enhance the response in salient regions. Its processing process can be formally represented as: Its processing process can be formally represented as:

$$F_{out} = CBAM(Conv(STB(CBAM(F_{in})))) \quad (1)$$

F_{in} and F_{out} represent the input and output feature maps of the module, respectively.

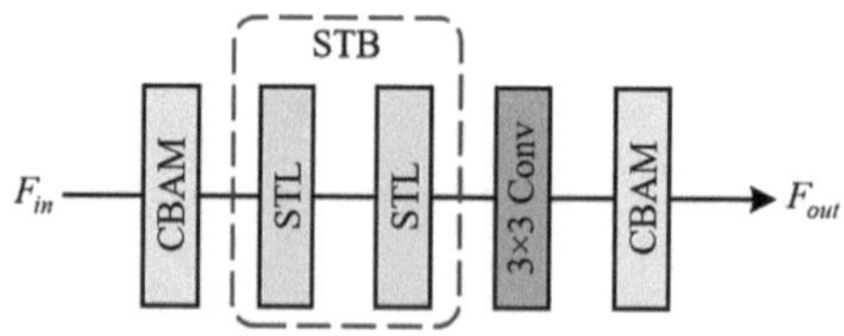

Fig. 2. Overview of the HCTB module. Detailed information is provided in Sect. 2.1.

The output features from all HCTB modules are concatenated and fed into the DFF module for feature fusion. The DFF comprises two cascaded convolutional layers and incorporates a global residual connection to mitigate gradient vanishing and ease the training process. Finally, the OPM projects the fused features back into pixel space via two-dimensional convolution, producing a preliminary enhanced image $\tilde{y}_T \in \mathbb{R}^{3\times W\times H}$. Accordingly, the computational process of the CNN-Transformer hybrid branch can be formally defined as:

$$\begin{aligned} F_{\mathrm{SFE}} &= H_{\mathrm{SFE}}(x) \\ F_{\mathrm{HBCTs}} &= H_{\mathrm{HBCTs}}(F_{\mathrm{SFE}}) \\ F_{\mathrm{DFF}} &= H_{\mathrm{DFF}}(F_{\mathrm{HBCTs}}) + F_{\mathrm{SFE}} \\ \tilde{y}_T &= H_{\mathrm{OPM}}(F_{\mathrm{DFF}}) \end{aligned} \quad (2)$$

Building on this, to further improve detail recovery, we introduce the reverse denoising process of DDPM as a reverse denoising branch in the generator, as illustrated in Fig. 1 (b). This design is inspired by prior studies [14–16], which integrate diffusion models with deep networks to lower training overhead and enhance downstream task performance. Consequently, the reverse denoising branch is employed to refine the output of the CNN–Transformer hybrid branch, yielding the final generated result $\tilde{y}_0$.

Furthermore, relevant experiments [12,13] also indicate that in real-world scenarios with noise interference, moderate downsampling has a limited impact on the final restoration quality. Therefore, in the reverse denoising process, we adopt an adaptive resolution strategy, which adjusts the resolution adaptively according to the noise level. When the noise is removed to a certain extent, the resolution is restored to the original resolution. This reverse denoising process can be modelled as a Markov chain over time steps t :

$$p_\theta(\tilde{y}_{0:T}) = p_\theta(\tilde{y}_T) \prod_{t=1}^{T} p_\theta(\tilde{y}_{t-1} \mid \tilde{y}_t),$$

$$p_\theta(\tilde{y}_{t-1} \mid \tilde{y}_t) = \begin{cases} N\left(\tilde{y}_{t-1}; \sqrt{\hat{\gamma}_{t-1}}(\tilde{y}_t \uparrow \frac{r_{t-1}}{r_t}), (1-\hat{\gamma}_t)\zeta\right), & \text{when } r_t < r_{t-1}, \\ N\left(\tilde{y}_{t-1}; \frac{\sqrt{1-\hat{\gamma}_t}\,\beta_t}{\hat{\gamma}_{t-1}} \epsilon_\theta(\tilde{y}_t) + \frac{\sqrt{\gamma_t}\,(1-\hat{\gamma}_{t-1})}{1-\hat{\gamma}_t} \tilde{y}_t, \frac{1-\hat{\gamma}_{t-1}}{1-\hat{\gamma}_t} \beta_t \zeta\right), & \text{else.} \end{cases} \tag{3}$$

r_t represents the resolution of the image at time step$t$; $\uparrow \frac{r_{t-1}}{r_t}$ represents the upsampling operation based on the resolution change.

2.2 Discriminator

As shown in the right half of Fig. 1, this paper adopts a global-local discriminator [8] architecture in the discriminator design to enhance the discriminative ability during the image enhancement process. Specifically, both the global discriminator and the local discriminator are based on the PatchGAN structure to judge the authenticity of the image. The global discriminator evaluates the lighting of the image from a holistic perspective, aiming to improve the global lighting consistency of the enhanced image. Meanwhile, the local discriminator randomly crops multiple local regions from both enhanced and real images and trains the discriminator to determine their source (real or generated), thereby enhancing the network's ability to model and distinguish local details. This discriminator structure helps the network generate more realistic and natural lighting effects in local regions, effectively avoiding issues such as overexposure or underexposure in the images.

2.3 Loss Function

It is worth emphasising that although the network structure proposed by this research includes both a generator and a discriminator, its working mechanism differs fundamentally from that of GAN. Unlike GAN, in which the generator and discriminator engage in mutual competition through adversarial training, our generator and discriminator do not directly interfere with each other's parameter updates. Instead, they are optimised independently, and the results calculated by their respective loss functions are weighted and summed in the overall objective, thereby synergistically promoting the overall performance of the network. Specifically, the generator is responsible for generating enhanced images from low-light inputs and uses the Smooth L1 loss function as the sole optimisation objective, with its hyperparameter ε empirically set to 1. The Smooth L1 loss is defined as follows:

$$L_{\text{generator}} = \begin{cases} \frac{1}{2}(y-\tilde{y})^2/\varepsilon, & \text{if } |y-\tilde{y}| < \varepsilon \\ |y-\tilde{y}| - 0.5\varepsilon, & \text{otherwise} \end{cases} \tag{4}$$

y Indicates images under normal lighting conditions. This loss function provides strong gradient feedback when the error is small, and controls the gradient amplitude when the error is large, which helps to enhance the stability and robustness of the model.

The discriminator is used to distinguish between normal light images and enhanced images generated by the generator, and is optimised using the Charbonnier loss function, whose hyperparameter empirical value is set to ϕ . The loss function is defined as follows:

$$L_{\text{discriminator}} = \frac{1}{2}\left(\sqrt{\|D(y)\|^2 + \phi^2} + \sqrt{\|1 - D(\tilde{y})\|^2 + \phi^2}\right) \tag{5}$$

Finally, we combine the losses of the generator and discriminator in a weighted manner to form a joint optimisation objective function for image enhancement tasks:

$$L_{\text{total}} = L_{\text{generator}} + \lambda L_{\text{discriminator}}. \tag{6}$$

the default value for the hyperparameter λ balancing the loss weights of the generator and discriminator is 0.5. This design can achieve comprehensive improvements in both the big picture and details while maintaining independent optimisation of the two, thereby obtaining higher-quality enhancement results.

3 Experiment

3.1 Dataset

We conduct our experiments on two publicly available laparoscopic datasets, EndoVis17 [19] and EndoVis18 [20] , which contain 2,235 images from 14 videos and 2,400 images from 10 videos, respectively. Because obtaining true low-light and normal-light paired images in real endoscopic procedures is challenging, we synthesize low-light images following the approaches in [13,15,21,22], applying random Gamma correction and brightness reduction to normal-light images, with the gamma factor γ sampled from [1.5, 3.0] and the brightness scaling coefficient β from [0.1, 0.5] to simulate various levels of underexposure and light attenuation. According to the data splitting strategies in [13,23], EndoVis17 is divided into 1,800 training images and 1,200 test images, and EndoVis18 into 1,639 training images and 596 test images. During training, all images are resized to 256×256 and augmented with random flipping.

3.2 Implementation Details

We compare the performance of our CTDiff against state-of-the-art (SOTA) low-light image enhancement (LLIE) methods, including traditional algorithms (LIME [1], DUAL [2]), CNN-based methods (Zero-DCE [3], SNR-Aware [6], MIRNet [4], MIRNet v2 [5]), GAN-based methods (EnlightenGAN [8], Still-GAN [9]), and recent diffusion-based methods (LLCaps [15], DiffLL [11], PyDiff [12], LighTDiff [13]). All comparative models are trained under the same conditions and data augmentation strategies as CTDiff. For evaluation, we adopt three commonly used image quality metrics: Peak Signal-to-Noise Ratio (PSNR) [24], Structural Similarity Index (SSIM) [25], and Learned Perceptual Image Patch Similarity (LPIPS) [26]. In addition, we report processing efficiency using Frames Per Second (FPS). To assess practical utility in medical tasks, we evaluate enhanced images on downstream segmentation. Specifically, the Transformer-based MATIS [27] model is trained on normal-light images and tested on enhanced low-light images without further fine-tuning. Segmentation performance is quantified using Intersection over Union (Iou), mean IoU (mIoU) and mean class-wise IoU (mcIoU), reflecting the ability to preserve organ and lesion boundaries.

3.3 Results

We evaluated the performance of the proposed CTDiff method against existing approaches, as shown in Table 1 and Fig. 3, using both quantitative and qualitative assessments. Compared with other methods, CTDiff achieves the best performance across all metrics. Specifically, it improves PSNR by 1.93 dB on the EndoVis17 dataset and by 1.62 dB on the EndoVis18 dataset compared with LighTDiff. The SSIM reaches 96.27% and 95.83% on EndoVis17 and EndoVis18, respectively. CTDiff also demonstrates superior performance on the reference-free perceptual metric LPIPS. Although its processing speed (FPS) is 16.67, slightly lower than ElightenGAN's 21.61, CTDiff outperforms ElightenGAN in

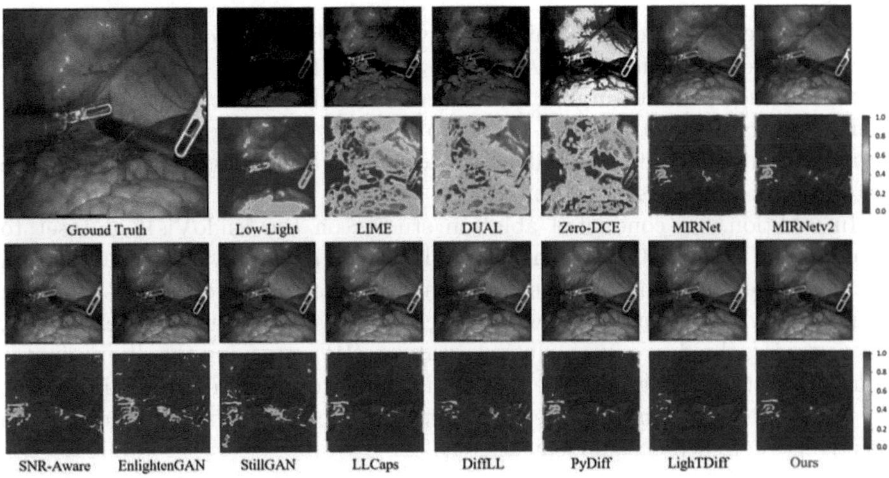

Fig. 3. The quantitative results for CTDiff compared with SOTA approaches on EndoVis17. The first row shows the enhanced images for different LLIE baselines, and the second row shows the reconstruction error heat maps. Blue to red indicates the error from small to large. Zoom to see the details. (Color figure online)

Table 1. Comparison of low-light enhancement models on EndoVis17 and EndoVis18 datasets.

Model	Efficiency	EndoVis17			EndoVis18			EndoVis17 Seg			EndoVis18 Seg		
	FPS$\uparrow$	PSNR$\uparrow$	SSIM$\uparrow$	LPIPS$\downarrow$	PSNR$\uparrow$	SSIM$\uparrow$	LPIPS$\downarrow$	mIoU$\uparrow$	IoU$\uparrow$	mcIoU$\uparrow$	mIoU$\uparrow$	IoU$\uparrow$	mcIoU$\uparrow$
LIME	1.48	12.13	34.99	0.3216	12.38	39.44	0.3241	35.18	11.42	9.31	39.12	14.06	10.78
DUAL	1.30	12.09	35.94	0.3221	12.20	44.71	0.3290	35.73	12.19	9.68	40.01	14.81	11.32
Zero-DCE	14.26	11.23	31.97	0.4667	10.64	38.91	0.5060	32.14	10.23	7.54	36.05	12.49	9.14
MIRNet	3.25	27.94	93.26	0.0796	28.25	92.96	0.0920	57.06	54.43	31.72	59.82	56.37	33.88
MIRNetv2	5.23	31.88	95.84	0.0380	30.84	95.16	0.0515	70.34	65.92	39.41	75.28	68.03	42.12
SNR-Aware	11.82	31.01	94.24	0.0574	29.81	93.67	0.0675	66.57	61.13	34.82	73.91	66.17	40.13
EnlightenGAN	**21.61**	27.21	90.58	0.0821	26.54	90.46	0.0888	53.84	49.61	26.88	56.17	50.33	28.54
StillGAN	4.86	29.99	91.57	0.0600	27.27	90.48	0.0810	54.79	51.03	28.21	57.39	52.10	29.72
LLCaps	1.90	31.44	95.87	0.0372	30.98	95.17	0.0511	69.21	63.98	38.11	76.14	68.29	43.21
DiffLL	4.84	30.21	92.62	0.0820	28.28	91.45	0.1096	70.82	65.87	39.76	77.36	69.01	44.07
PyDiff	11.84	31.02	94.99	0.0468	31.62	95.31	0.0469	70.41	64.36	39.44	76.91	68.55	43.67
LighTDiff	13.89	32.92	95.46	0.0386	31.74	94.73	0.0555	70.95	65.22	40.34	77.08	68.81	44.12
CTDiff (Ours)	16.67	**34.85**	**96.27**	**0.0289**	**33.24**	**95.83**	**0.0371**	**71.35**	**67.54**	**42.25**	**79.45**	**69.65**	**46.36**

terms of image enhancement quality, particularly in brightness restoration, structural preservation, and detail recovery.

Figure 3 presents qualitative results and corresponding error heatmaps for all compared methods on the EndoVis17 and EndoVis18 datasets. CTDiff not only enhances image brightness but also preserves key structures and edge information. Compared with EnlightenGAN and StillGAN , the error heatmaps show lower errors in critical regions, indicating higher fidelity in maintaining structural details.

We further assessed the clinical utility of CTDiff in a downstream surgical instrument segmentation task. As shown in Table 1, CTDiff outperforms other methods in segmenting surgical instruments. On the synthetic EndoVis17 and EndoVis18 datasets, it achieves higher IoU, mIoU, and mcIoU scores, with the best results highlighted in bold, demonstrating its effectiveness in low-light image restoration and edge preservation.

In addition, we conducted ablation studies on the EndoVis17 dataset to evaluate the contribution of the network design and individual components, as summarized in Table 2. The experiments considered the following strategies: (I) replacing CBAM in HCTB with the enhanced spatial attention (ESA) [28], (II) removing the dynamic resolution strategy, (III) removing the discriminator branch, and (IV) using only the global discriminator. The results indicate that the removal of any component leads to performance degradation, confirming the effectiveness of each design element.

Table 2. Ablation study results on the proposed model.

CBAM	Unconstant Resolution	Discriminator	Local Discriminator	Param (M)	RSNR↑	SSIM↑	LPIPS↓
✗	✗	✗	–	99	32.90	95.63	0.0327
✓	✗	✗	–	98.91	33.20	96.00	0.0317
✗	✓	✗	–	19.43	32.50	95.70	0.0340
✗	✗	✓	✓	117.11	33.10	96.10	0.0308
✓	✓	✗	–	**19.34**	32.80	95.85	0.0328
✓	✗	✓	✓	117.02	**35.10**	**96.35**	**0.0284**
✗	✓	✓	✓	37.54	34.05	96.11	0.0308
✓	✓	✓	✗	30.49	34.20	96.19	0.0300
✓	✓	✓	✓	37.45	34.85	96.27	0.0289

4 Conclusion

The proposed CTDiff framework consists of a generator and a discriminator. The generator integrates convolutional neural networks with Swin Transformers to effectively extract and fuse both local and global features, restoring low-light endoscopic images while enhancing brightness, contrast, and structural details. The discriminator employs a global–local joint discrimination mechanism to ensure illumination and structural consistency. Although CTDiff demonstrates strong performance in image enhancement and downstream tasks, challenges remain due to the limited availability of real low-light endoscopic data and the need for further optimisation of the network architecture. Future work will focus on improving model efficiency and robustness, as well as refining the architecture to enhance practical clinical applicability.

Acknowledgements. This work was supported by a grant from the National Natural Science Foundation of China (No. 82472118) and Jilin Provincial Department of Science and Technology (No. YDZJ202401620ZYTS).

References

1. Guo, X., Li, Y., Ling, H.: LIME: Low-light image enhancement via illumination map estimation. IEEE Trans. Image Process. **26**(2), 982–993 (2016)
2. Zhang, Q., Nie, Y., Zheng, W.S.: Dual illumination estimation for robust exposure correction. Comput. Graph. Forum, **38**(7), 243-252 (2019)
3. Guo, C., Li, C., Guo, J., et al.: Zero-reference deep curve estimation for low-light image enhancement. In: Proceedings of the IEEE/CVF Conference on Computer Vision and Pattern Recognition, pp. 1780–1789 (2020)
4. Zamir, S.W., Arora, A., Khan, S., et al.: Learning enriched features for real image restoration and enhancement. In: European Conference on Computer Vision. Cham: Springer International Publishing, pp. 492–511 (2020)
5. Zamir, S.W., Arora, A., Khan, S., et al.: Learning enriched features for fast image restoration and enhancement. IEEE Trans. Pattern Anal. Mach. Intell. **45**(2), 1934–1948 (2022)

6. Xu, X., Wang, R., Fu, C.W., et al.: Snr-aware low-light image enhancement. In: Proceedings of the IEEE/CVF Conference on Computer Vision and Pattern Recognition, pp. 17714–17724 (2022)
7. Goodfellow, I., Pouget-Abadie, J., Mirza, M., et al.: Generative adversarial networks. Commun. ACM **63**(11), 139–144 (2020)
8. Jiang, Y., Gong, X., Liu, D., et al.: Enlightengan: deep light enhancement without paired supervision. IEEE Trans. Image Process. **30**, 2340–2349 (2021)
9. Ma, Y., Liu, J., Liu, Y., et al.: Structure and illumination constrained GAN for medical image enhancement. IEEE Trans. Med. Imaging **40**(12), 3955–3967 (2021)
10. Ho, J., Jain, A., Abbeel, P.: Denoising diffusion probabilistic models. Adv. Neural. Inf. Process. Syst. **33**, 6840–6851 (2020)
11. Jiang, H., Luo, A., Fan, H., et al.: Low-light image enhancement with wavelet-based diffusion models. ACM Trans. Graph. (TOG) **42**(6), 1–14 (2023)
12. Zhou, D., Yang, Z., Yang, Y.: Pyramid diffusion models for low-light image enhancement. arXiv preprint arXiv:2305.10028 (2023)
13. Chen, T., Lyu, Q., Bai, L., et al.: LighTDiff: surgical endoscopic image low-light enhancement with T-diffusion. In: International Conference on Medical Image Computing and Computer-Assisted Intervention. Cham: Springer Nature Switzerland, pp. 369–379 (2024)
14. Pandey, K., Mukherjee, A., Rai, P., et al.: Diffusevae: efficient, controllable and high-fidelity generation from low-dimensional latents. arXiv preprint arXiv:2201.00308 (2022)
15. Bai, L., Chen, T., Wu, Y., et al.: Llcaps: learning to illuminate low-light capsule endoscopy with curved wavelet attention and reverse diffusion. In: International Conference on Medical Image Computing and Computer-Assisted Intervention. Cham: Springer Nature Switzerland, pp. 34–44 (2023)
16. Chen, T., Li, Y., Zhang, X.: MWRD (mamba wavelet reverse diffusion)—an efficient fundus image enhancement network based on an improved state-space model. Electronics **13**(20), 4025 (2024)
17. Woo, S., Park, J., Lee, J.Y., et al.: Cbam: convolutional block attention module. In: Proceedings of the European Conference on Computer Vision (ECCV), pp. 3–19 (2018)
18. Liu, Z., Lin, Y., Cao, Y., et al.: Swin transformer: hierarchical vision transformer using shifted windows. In: Proceedings of the IEEE/CVF International Conference on Computer Vision, pp. 10012–10022 (2021)
19. Allan, M., Shvets, A., Kurmann, T., et al.: 2017 robotic instrument segmentation challenge. arXiv preprint arXiv:1902.06426 (2019)
20. Allan, M., Kondo, S., Bodenstedt, S., et al.: 2018 robotic scene segmentation challenge. arXiv preprint arXiv:2001.11190 (2020)
21. Girshick, R.: Fast r-CNN. In: Proceedings of the IEEE International Conference on Computer Vision, pp. 1440–1448 (2015)
22. Lore, K.G., Akintayo, A., Sarkar, S.: LLNet: a deep autoencoder approach to natural low-light image enhancement. Pattern Recogn. **61**, 650–662 (2017)
23. González, C., Bravo-Sánchez, L., Arbelaez, P.: Isinet: an instance-based approach for surgical instrument segmentation. In: International Conference on Medical Image Computing and Computer-assisted Intervention. Cham: Springer International Publishing, pp. 595–605 (2020)
24. Huynh-Thu, Q., Ghanbari, M.: Scope of validity of PSNR in image/video quality assessment. Electron. Lett. **44**(13), 800–801 (2008)

25. Wang, Z., Bovik, A.C., Sheikh, H.R., et al.: Image quality assessment: from error visibility to structural similarity. IEEE Trans. Image Process. **13**(4), 600–612 (2004)
26. Zhang, R., Isola, P., Efros, A.A., et al.: The unreasonable effectiveness of deep features as a perceptual metric. In: Proceedings of the IEEE Conference on Computer Vision and Pattern Recognition, pp. 586–595 (2018)
27. Ayobi, N., Pérez-Rondón, A., Rodríguez, S., Matis: masked-attention transformers for surgical instrument segmentation. In: IEEE 20th International Symposium on Biomedical Imaging (ISBI). IEEE, vol. 2023, pp. 1–5 (2023)
28. Liu, J., Tang, J., Wu, G.: Residual feature distillation network for lightweight image super-resolution. In: European Conference on Computer Vision. Cham: Springer International Publishing, pp. 41–55 (2020)

Hierarchical Cross-Modality Interaction for Unified Video-Text Retrieval Modeling

Tianshi Xu[1], Zhengzheng Sun[1,2(✉)], Yizheng Hu[1,2(✉)], Junyuan Shang[1,2], and Si Wu[1]

[1] South China University of Technology, Guangzhou, China
[2] GRGBanking Equipment Co., Ltd., Guangzhou, China
sunzz1996@foxmail.com, huyizheng@hotmail.com

Abstract. Existing video-text retrieval (VTR) methods suffer from two primary limitations: they typically extract features from each modality in isolation and struggle to effectively align information across different granularity levels. These issues neglect potential contextual cues between the two modalities and can lead to a trade-off between capturing global context and fine-grained details. To address these challenges, we introduce the Hierarchical Cross-Modality Interaction (HCMI) framework, a novel approach that performs bidirectional, multi-level feature interaction during training. Our framework systematically aligns features at patch-word, frame-sentence, and global video-sentence levels, leveraging a dual attentive module for deep feature enhancement and a cross-modality adaptor for efficient feature distillation. Crucially, HCMI requires only a single modality's input during inference, making it highly practical for large-scale retrieval tasks. Our method achieves new state-of-the-art results on several challenging benchmarks, including MSR-VTT, DiDeMo, VATEX, MSVD and Activity-Net, demonstrating its significant advantages.

Keywords: Bidirectional VTR · Cross Attention · Feature Alignment

1 Introduction

Video-Text Retrieval (VTR), a fundamental task in cross-modal understanding, aims to bridge the semantic gap between visual and textual data [1,12,29]. Early approaches focused on learning a joint embedding space where corresponding video and text pairs are projected closely [20,28]. More recently, the advent of powerful vision-language pre-trained models, particularly CLIP [19], has significantly advanced the state-of-the-art. Many leading methods now adapt CLIP's image and text encoders for the video domain [15–17], typically by adding temporal modeling modules to aggregate frame-level features into a global video representation before computing cross-modal similarity [6,16]. To capture more nuanced alignments, subsequent works have explored finer-grained matching, such as aligning individual frames with sentences [7,15] or even video patches with words [32].

J. Lokoč et al. (Eds.): MMM 2026, LNCS 16413, pp. 588–601, 2026.
https://doi.org/10.1007/978-981-95-6957-1_42

Despite these advancements, existing methods suffer from two primary limitations. First, they typically extract features from a single source modality in isolation. The video encoder processes visual content without any knowledge of the text query it will be compared against, and vice versa. This approach neglects the potential of using the target modality's information as a contextual cue to guide the extraction of more discriminative and relevant features. For instance, given the query "a person slicing a cucumber", an ideal video encoder would amplify features related to the specific action of "slicing" and the object "cucumber", while suppressing irrelevant background details. Second, current models often struggle to effectively align features across different levels of granularity. Methods that operate at a global level (video-to-sentence) may capture overall context but miss crucial fine-grained details, while patch-to-word alignment excels at details but can lose the broader semantic context. This creates a need for a unified framework that can coherently integrate multi-level correspondences.

To address these challenges, we introduce the Hierarchical Cross-Modality Interaction (HCMI) framework, a novel approach for unified video-text retrieval. Our key idea is to enable multi-level, bidirectional interaction between modalities during the feature encoding process itself, while still allowing for efficient single-modality inference. HCMI systematically addresses the density and granularity mismatch between video and text by performing hierarchical alignment at the fine-grained (patch-word), coarse-grained (frame-sentence), and global (video-sentence) levels. This is achieved through a two-stage training paradigm where cross-modal interaction enriches the representations. Crucially, during inference, the model only requires a single modality's input to generate these enhanced features. To summarize, our main contributions are as follows:

- We propose the Hierarchical Cross-Modality Interaction (HCMI) framework, a novel paradigm that performs bidirectional, multi-level feature interaction during training yet requires only a single modality for efficient inference, resolving the common trade-off between performance and retrieval speed.
- We design a Dual Attentive Module (DAM) for deep bidirectional feature enhancement, along with a Cross-Modality Adaptor (CMA) for single-modality feature distillation, effectively integrating hierarchical information to balance fine-grained details with global context.
- We achieve new state-of-the-art results on several challenging benchmarks, including MSR-VTT, DiDeMo, VATEX, MSVD and Activity-Net, demonstrating the significant advantages of our proposed method.

The remainder of this paper is organized as follows. Sect. 2 reviews related work; Sect. 3 details the HCMI architecture, DAM and CMA modules, and the training protocol; Sect. 4 presents experimental results and analyses; and the last section concludes with directions for future work.

2 Related Work

This section reviews three major research directions in video-text retrieval: architectural paradigms balancing efficiency and accuracy, mechanisms for cross-modal alignment, and recent advances in hierarchical modeling with pre-trained foundation models. These directions highlight the limitations of existing methods and motivate our hierarchical framework.

Architectural Paradigms in Video-Text Retrieval. Video-text retrieval models are typically built on two paradigms: dual-encoders and cross-encoders.

Dual-encoders encode video and text independently into a shared embedding space, enabling efficient large-scale retrieval through pre-computation and nearest-neighbor search [11,16]. While scalable, their lack of cross-modal interaction restricts fine-grained alignment and reduces accuracy on nuanced tasks.

Cross-encoders jointly process both modalities with deep interaction layers, often Transformers, to capture complex inter-modal dependencies [17]. This improves retrieval accuracy but prohibits pre-computation, making them computationally expensive and unsuitable for large-scale deployment. Hybrid strategies that combine dual-encoder and cross-encoder strengths [11] partly alleviate this trade-off, but scalability versus accuracy remains a core challenge.

Cross-Modal Alignment and Interaction Mechanisms. Retrieval performance critically depends on effective alignment across modalities.

Contrastive learning has become the dominant paradigm, aligning video-text pairs by maximizing similarity for positives and minimizing it for negatives [4,26]. Despite its success, standard formulations can suffer from biased normalization of similarity distributions. This issue has motivated refinements such as normalized contrastive learning [18].

Attention-based fusion, particularly cross-attention in Transformer architectures, has become the standard mechanism for deep interaction [17,21]. Cross-attention allows queries to selectively attend to relevant visual or temporal segments, enabling context-aware alignment. However, flat attention often fails to capture multi-level semantics (e.g., matching a sentence to a video while also aligning verb phrases to actions) [5,8], underscoring the need for structured, hierarchical integration.

Hierarchical Modeling and Pre-Trained Foundation Models. To better capture video complexity, recent works explore hierarchical modeling and the use of large-scale vision-language models (VLMs).

Hierarchical approaches explicitly align features at multiple granularities, such as object-word, action-phrase, and event-sentence levels [13,24,30]. These methods improve interpretability and enable multi-level semantic reasoning.

Meanwhile, VLMs like CLIP, which are pre-trained on massive image-text datasets, provide powerful general-purpose embeddings transferable to video-text retrieval [14,25]. Despite their success, image-based pre-training limits temporal modeling, and such models often excel at global alignment but underperform at fine-grained object-level understanding. Augmenting them with fine-grained modules reintroduces scalability issues, leaving unresolved the central gap: how to achieve robust multi-level alignment without sacrificing efficiency.

3 Method

We propose a hierarchical cross-modality interaction framework to address the granularity mismatch and information imbalance in video-text retrieval. The framework is organized into two stages. In stage I, a cross-attention-based hierarchical interaction module learns robust alignment patterns at multiple levels (patch-word, frame-sentence, and video-text). The resulting cross-modal representations are treated as latent soft labels, which encode enriched semantic correspondences. In stage II, we introduce a label pattern distillation strategy that leverages a lightweight adaptor network to project original CLIP features into the cross-attentive feature space. This process enables each video-text pair to inherit its alignment pattern from stage I and refine its modality-specific features accordingly. By coupling hierarchical interaction with knowledge distillation using a single source, our method produces more discriminative and generalizable retrieval features and achieving superior video-text matching performance.

3.1 Problem Definition and Motivation

The Video Text Retrieval task aims to retrieve the most relevant videos from a video database given a text query, or conversely, to retrieve the most relevant texts from a text database given a video query.

Formally, let $V = \{v_1, v_2, ..., v_N\}$ be a set of videos and $T = \{t_1, t_2, ..., t_M\}$ be a collection of texts. For *Video-to-Text Retrieval*: Given a query video $v_i \in V$, the objective is to find a ranked list $t'_1, t'_2, ..., t'_k$ (where $t'_j \in T$), such that t'_j are ordered by their relevance to v_i from highest to lowest. This is particularly useful for generating captions or summaries for unannotated videos. For *Text-to-Video Retrieval*: Given a query text $t_j \in T$, the objective is to find a ranked list $v'_1, v'_2, ..., v'_k$ (where $v'_j \in V$), such that v'_j are ordered by their relevance to t_j from highest to lowest. This scenario is common in video search engines where users input textual descriptions.

3.2 Stage I: Hierarchical Cross-Modality Interaction for Pattern Label Producing

As illustrated in Fig. 1, Stage I builds a *hierarchical cross-modality interaction* (HCMI) mechanism to mitigate the information-density gap between video and text. We progressively align the two modalities at fine (patch–word), coarse (frame–sentence), and global (video–sentence) granularities via cross-attention, and then fuse them into modality-specific *pattern features*. These fused features act as implicit *soft labels* to supervise Stage II. Since HCMI consumes *both* modalities during inference (which is impractical in retrieval), Stage II adopts an adaptor to distill these interaction patterns so that robust retrieval features can be produced from a *single* modality.

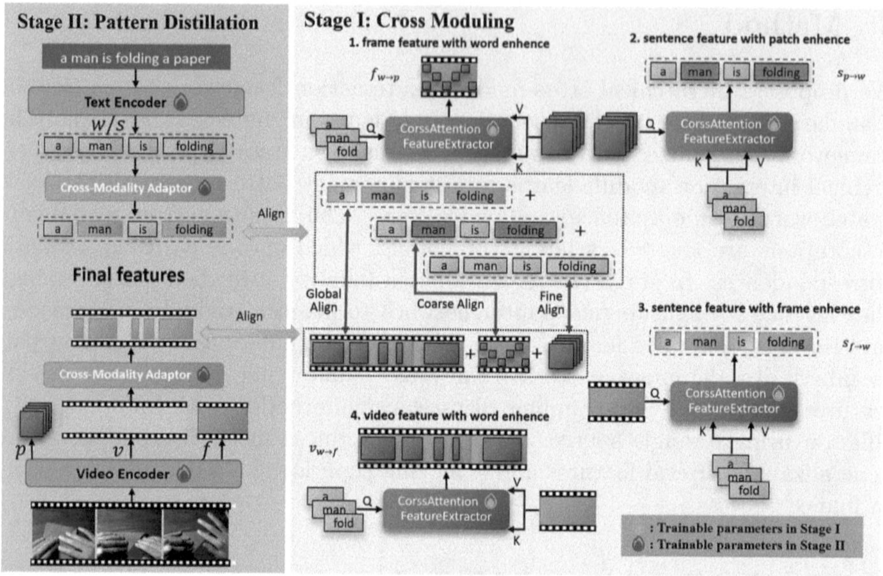

Fig. 1. The main pipeline of the proposed method for multi-modal feature generation.

Original Feature Extraction. We use CLIP encoders for initialization and keep notation compact: the text encoder yields word features w and a sentence feature s; the video encoder yields patch features p, frame features f, and a global video feature v. These serve as the basis for the subsequent cross-modal interactions.

Fine Alignment Between Original Patch and Word Features. At the most granular level, we align local visual patches with lexical tokens to ground words in concrete regions. We directly use the original features:

$$v_{\text{fine}} = p, \quad t_{\text{fine}} = w \tag{1}$$

Here, cross-attention provides the alignment signal: words query patches (and vice versa) to expose patch–word correspondences that guide higher-level interactions.

Coarse Alignment Between Enhanced Frame and Sentence Features. Building on the fine level, we form frame–sentence interactions enhanced by lower-level cues:

$$v_{\text{coarse}} = \lambda_2 \cdot f_{w \to p} + f, \quad t_{\text{coarse}} = \alpha_2 \cdot s_{p \to w} + s, \tag{2}$$

where $f_{w \to p}$ is obtained by cross-attention with *words as queries* and *patches as keys/values* (word→patch), and $s_{p \to w}$ is analogously obtained with *patches as queries* and *words as keys/values* (patch→word). This level captures event-centric semantics beyond isolated regions or tokens.

Global Alignment Between Enhanced Video and Sentence Features. At the highest level, we align the overall video with the whole sentence while injecting coarse-level context:

$$v_{\text{global}} = \lambda_3 \cdot v_{w\to f} + v, \quad t_{\text{global}} = \alpha_3 \cdot s_{f\to w} + s, \tag{3}$$

where $v_{w\to f}$ uses *words as queries* over *frames as keys/values* (word→frame) to emphasize salient temporal segments, and $s_{f\to w}$ uses *frames as queries* over *words as keys/values* (frame→word) to strengthen sentence representations with global visual context.

Final Alignment Between Enhanced Visual and Textual Features. We fuse multi-level cues into compact modality-specific pattern features that supervise stage II:

$$v_{\text{final}} = \lambda_1 \cdot p + \lambda_2 \cdot f_{w\to p} + \lambda_3 \cdot v_{w\to f}, \tag{4}$$

$$t_{\text{final}} = \alpha_1 \cdot w + \alpha_2 \cdot s_{p\to w} + \alpha_3 \cdot s_{f\to w}, \tag{5}$$

where $\lambda_1, \lambda_2, \lambda_3$ and $\alpha_1, \alpha_2, \alpha_3$ are learnable weights. Intuitively, v_{final} aggregates patch-, frame-, and video-guided interactions, and t_{final} aggregates word- and sentence-level interactions informed by visual evidence. We treat the well pre-trained $\{v_{\text{final}}, t_{\text{final}}\}$ as the **pattern labels** for distillation in stage II, transferring cross-modal interaction patterns to single-modality encoders.

3.3 Stage II: Label Pattern Distillation for Cross Feature Modeling

As shown in stage II of Fig. 1, we introduce a Cross-Modality Adaptor (CMA) after the CLIP encoders to refine unimodal representations. Formally, given the original sentence embedding s and video embedding v extracted by CLIP encoders, CMA performs a nonlinear mapping:

$$s_{re} = s + ws', \quad v_{re} = v + wv', \tag{6}$$

where $s' = \text{CMA}_t(s)$ and $v' = \text{CMA}_v(v)$ denote the enhanced features for retrieval in a semantically enriched space. This transformation is supervised by the cross-modal interaction patterns pre-trained in stage I, which serve as soft labels to guide the adaptor. In this way, the raw features are encouraged to implicitly attend to modality-specific details revealed by the other modality, rather than relying solely on unimodal evidence.

The key motivation of CMA is that, in retrieval scenarios, paired text–video inputs cannot be used for cross-attention fusion without causing data leakage. To address this, CMA aligns unimodal features with the stronger cross-attention features from stage I, enabling them to inherit complementary semantic cues. This not only reduces the modality gap but also produces more expressive and semantically aligned representations for retrieval.

3.4 Custom Modules for Feature Enhancement

To overcome the limitations of shallow fusion mechanisms that struggle with complex inter-dependencies, we introduce the Dual Attentive Module (DAM), illustrated in Fig. 2. The DAM enables effective cross-modal interaction beyond simple feature aggregation by: (1) *enhancing* the target modality through selective incorporation of salient source context, and (2) *distilling* source features into compact, relevant representations, thus reducing redundancy. This cascaded attention strategy supports semantic enhancement across multiple levels of granularity.

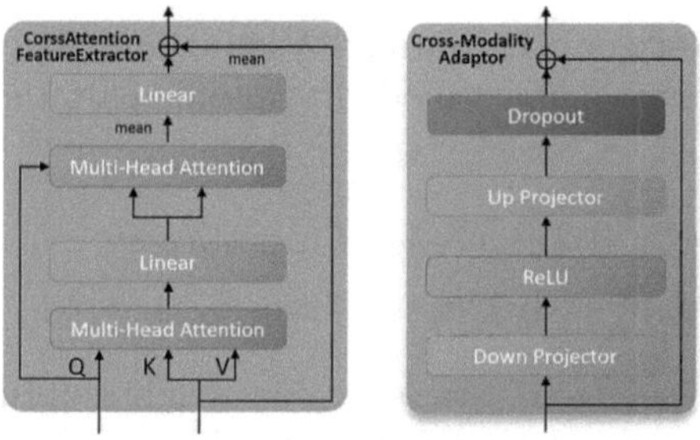

Fig. 2. Custom Modules for Feature Enhancement: DAM and CMA.

Dual Attentive Module. DAM employs a dual-layer residual attention mechanism. Given a query $\mathbf{Q}$ from the target modality, and key $\mathbf{K}$ and value $\mathbf{V}$ from the source, the first attention extracts broad context, which is then refined by a second pass using the original query. Formally:

$$\mathbf{F}_{int} = \text{Linear}_1(\text{MHA}_1(\mathbf{Q}, \mathbf{K}, \mathbf{V})) \tag{7}$$

$$\text{DAM}(\mathbf{Q}, \mathbf{K}, \mathbf{V}) = \mathbf{V} + \text{Linear}_2(\text{mean}(\text{MHA}_2(\mathbf{Q}, \mathbf{F}_{int}, \mathbf{F}_{int}))) \tag{8}$$

where $\mathbf{F}_{int}$ denotes the intermediate features. The output is an enhanced version of $\mathbf{V}$ via a residual connection. Applying DAM symmetrically (text-to-video and video-to-text) yields a robust bidirectional fusion framework.

Cross Modality Adaptor. In stage II, CMA maps original features into interaction-enhanced ones, guided by the patterns pre-trained in stage I. It follows a bottleneck design: input features are projected to a lower-dimensional space, activated non-linearly, then restored, with dropout for regularization. A residual connection fuses the transformed and original features, enriching them with fine-grained cross-modal details. Formally, for input $\mathbf{x}$:

$$\text{CMA}(\mathbf{x}) = \mathbf{x} + \text{Dropout}\Big(W_u\, \sigma\big(W_d \mathbf{x}\big)\Big), \tag{9}$$

where W_d and W_u are the down- and up-projectors, $\sigma(\cdot)$ is ReLU, and Dropout denotes regularization.

3.5 Learning Objective

Cross-Modal Fusion Loss for Stage I. To train our model and align the final feature representations (v_{final} and t_{final}), we employ a symmetric contrastive loss. This objective function, commonly known as InfoNCE, aims to minimize the distance between corresponding (positive) video-text pairs while maximizing the distance to non-corresponding (negative) pairs within a batch.

Given a batch of N video-text pairs, the loss is computed bidirectionally: for video-to-text retrieval ($\mathcal{L}_{v\to t}$) and text-to-video retrieval ($\mathcal{L}_{t\to v}$).

The video-to-text contrastive loss for a single video v_i is defined as:

$$\mathcal{L}_{v\to t}^{(i)} = -\log \frac{\exp(\text{sim}(v_i, t_i)/\tau)}{\sum_{j=1}^{N} \exp(\text{sim}(v_i, t_j)/\tau)} \tag{10}$$

Symmetrically, the text-to-video contrastive loss for a single text t_i is:

$$\mathcal{L}_{t\to v}^{(i)} = -\log \frac{\exp(\text{sim}(t_i, v_i)/\tau)}{\sum_{j=1}^{N} \exp(\text{sim}(t_i, v_j)/\tau)} \tag{11}$$

Here, $\text{sim}(a, b)$ denotes the cosine similarity between feature vectors a and b. The term t_i represents the positive text sample corresponding to video v_i, while t_j (for $j \neq i$) are the negative text samples. Temperature τ is a hyperparameter that scales the similarity scores.

To enforce semantic alignment across multiple levels of feature granularity, we extend the contrastive loss to operate on several hierarchical feature pairs. For each level s in the set $S = \{\text{global, coarse, fine, final}\}$, a bidirectional contrastive loss $\mathcal{L}_s$ is computed between the corresponding visual (v_s) and textual (t_s) features:

$$\mathcal{L}_s = \frac{1}{N} \sum_{i=1}^{N} \left(\mathcal{L}_{v_s\to t_s}^{(i)} + \mathcal{L}_{t_s\to v_s}^{(i)} \right) \tag{12}$$

The total loss $\mathcal{L}_{\text{total}}$ for the batch is the weighted sum of these individual losses, controlled by hyperparameters λ_s:

$$\mathcal{L}_{\text{total}} = \sum_{s\in S} \lambda_s \mathcal{L}_s + \mathcal{L}_{re}, \tag{13}$$

where $\mathcal{L}_{re}$ is the retrieval loss proposed in UCoFiA [24].

By minimizing this comprehensive loss, the model learns to project visual and textual features from multiple hierarchical levels into a shared embedding space. This multi-level supervision ensures that semantically related content from different modalities is located closely together, regardless of the feature granularity.

Pattern Distillation Loss for Stage II. We adopt a cosine similarity loss to align the CMA-transformed features with their corresponding soft labels pre-trained and saved in stage I. Formally, the loss is defined as:

$$\mathcal{L}_{\text{distill}} = 2 - \cos(v', v_{\text{final}}) - \cos(s', t_{\text{final}}) \tag{14}$$

$$\mathcal{L}_{\text{total}} = \mathcal{L}_{re} + \lambda_{\text{distill}} \mathcal{L}_{\text{distill}}, \tag{15}$$

where v' and s' denote the distilled features produced by CMAs, and v_{final} and t_{final} represent cross-modal attentive features pre-trained in stage I.

4 Experimental Setup

4.1 Datasets and Evaluation Metrics

We evaluate our HCMI method on five popular video-text retrieval datasets:

- MSR-VTT [27] contains 10,000 videos, each annotated with 20 text captions. 9,000 videos are used for training, and each video is trained with all 20 captions. Results are reported on the remaining 1,000 videos using the selected caption set (the JSFusion [28] test set).
- DiDeMo [2] contains 10,000 videos and 40,000 captions. All captions of a given video are concatenated into a single query for both training and testing. The training, validation and test set contain 8,395, 1,065 and 1,004 videos, respectively.
- VATEX [23] contains 34,991 videos. We use HGR 's [5] split, which uses 25,991 videos for training, 1,500 and 1,500 for validation and testing. This dataset is evaluated with the multiple caption evaluation setting following [24].
- MSVD [3] contains 1,970 videos. The training, validation and test sets contain 1,200, 100 and 670 videos, respectively. This dataset is evaluated with the multiple caption evaluation setting.
- Activity-Net [10] contains 20,000 videos and 100,000 captions. All captions of a given video are concatenated into a single query for both training and testing. The training, validation and test sets contain 10,000, 5,000 and 5,000 videos, respectively.

All videos are compressed to 3 fps and resized to 224 pixels in width or height to speed up the training and evaluation process, according to [16].

We use standard retrieval metrics to evaluate retrieval performance, including recall at rank K (R@K) with K=1,5,10, median rank (MdR) and mean rank (MnR).

4.2 Implementation Details

Our baseline method is UCoFiA [24]. Following that, we use the pre-trained CLIP ViT-B/32 [19] weights to initialize the visual encoder of our HCMI model. The model architecture and dataset settings are the same as in UCoFiA.

During the first stage, the number of attention heads is set to 8. All λ_s for $s \in S$ in $\mathcal{L}_{\text{total}}$ are set to 0.1, and the temperature τ is also set to 0.1. During the second stage, the down projector projects the feature dimension from 512 to 32, and the up projector projects the hidden state from 32 to 512. Both λ_{distill} and w are set to 1.0 for all datasets except for MSR-VTT, where w is set to 2.0. During both stages, the learning rate is set to $1e-7$ for the visual and text encoder and $1e-4$ for other models. For the first stage, all datasets are trained for 5 epochs, and for the second stage, the number of epochs are 5, 5, 10, 15, 20 for MSVD, VATEX, DiDeMo, MSR-VTT, and Activity-Net datasets, respectively.

5 Experimental Results

In this section, we compare our HCMI method with several prior methods on five video-text retrieval datasets.

5.1 Comparison to State-of-the-Art Methods

MSR-VTT Dataset. As shown in Table 1, HCMI achieves a 1.1% improvement on text-to-video R@1 metric compared to the baseline UCoFiA method and a 0.5% improvement compared to the recent competitor TC-MGC. For text-to-video R@5 and R@10 and video-to-text metrics, HCMI also shows significant improvement over UCoFiA and comparable results to TC-MGC. These results verify that our cross-modality interaction is useful for video-text retrieval.

Table 1. Retrieval results on MSR-VTT. Results of other models came from TC-MGC paper [9]. **Bold** denotes the best and underline denotes the second best.

Method	Text → Video					Video → Text				
	R@1↑	R@5↑	R@10↑	MdR↓	MnR↓	R@1↑	R@5↑	R@10↑	MdR↓	MnR↓
CLIP4Clip [16]	44.5	71.4	81.6	**2.0**	15.3	42.7	70.9	80.6	**2.0**	11.6
X-Pool [7]	46.9	72.8	82.2	**2.0**	14.3	44.4	73.3	**84.0**	**2.0**	9.0
CenterCLIP [31]	44.2	71.6	82.1	**2.0**	15.1	42.8	71.7	82.2	**2.0**	10.9
DRL [22]	47.5	73.8	83.6	**2.0**	13.3	46.3	72.7	82.5	**2.0**	9.5
UCoFiA [24]	46.8	72.6	82.4	**2.0**	13.9	45.3	72.7	82.2	**2.0**	9.8
X-CLIP [17]	46.1	73.0	83.1	**2.0**	13.2	46.8	73.3	**84.0**	**2.0**	9.1
TC-MGC [9]	47.4	**74.8**	**84.2**	**2.0**	**12.8**	45.9	**74.5**	83.3	**2.0**	**8.6**
HCMI (ours)	**48.0**	72.1	82.5	**2.0**	13.0	**46.9**	73.0	82.5	**2.0**	9.6

DiDeMo and VATEX Datasets. Table 2 shows the results on the DiDeMo and VATEX datasets. In the text-to-video task, HCMI outperforms all compared models, achieving 2.5% and 3.2% improvements in R@1 compared to the second-best TC-MGC results. In the video-to-text task, HCMI also achieves 0.9% - 1.0% improvement in R@1 and delivers comparable result in R@5 and R@10.

MSVD and Activity-Net Datasets. Table 3 shows the text-to-video results on the MSVD and the Activity-Net datasets. The results of the UCoFiA method are obtained by re-training using their source code. HCMI achieves a significant improvement on the Activity-Net dataset and a modest improvement on the MSVD dataset, compared to the baseline UCoFiA method.

Table 2. Retrieval results on DiDeMo and VATEX. Results of other models came from TC-MGC paper [9]. **Bold** denotes the best and underline denotes the second best.

	Method	Text → Video					Video → Text				
		R@1↑	R@5↑	R@10↑	MdR↓	MnR↓	R@1↑	R@5↑	R@10↑	MdR↓	MnR↓
DiDeMo	CLIP4Clip [16]	40.8	67.8	77.4	**2.0**	20.8	40.7	67.3	77.5	**2.0**	15.4
	TS2-Net [15]	42.5	69.3	77.8	**2.0**	18.8	42.2	68.2	78.6	**2.0**	13.6
	UCoFiA [24]	45.2	72.4	81.3	**2.0**	13.9	44.8	72.0	81.5	**2.0**	10.7
	X-CLIP [17]	44.7	72.7	80.6	**2.0**	15.9	44.5	72.3	82.0	**2.0**	12.6
	TC-MGC [9]	45.7	72.7	81.9	**2.0**	15.0	45.4	73.3	**82.9**	**2.0**	10.5
	HCMI (ours)	**48.2**	**75.4**	**83.6**	**2.0**	**13.7**	**46.3**	**73.4**	82.7	**2.0**	**10.1**
VATEX	CLIP4Clip [16]	58.7	89.3	94.7	**1.0**	3.7	74.8	96.7	98.6	**1.0**	1.8
	TS2-Net [15]	59.5	89.8	95.0	**1.0**	3.6	74.6	96.3	98.9	**1.0**	1.8
	UCoFiA [24]	59.3	88.8	94.3	**1.0**	4.1	73.0	96.1	98.9	**1.0**	1.9
	X-CLIP [17]	59.1	88.9	94.2	**1.0**	3.9	74.8	**97.3**	**99.0**	**1.0**	1.8
	TC-MGC [9]	60.0	89.5	94.9	**1.0**	3.8	75.3	97.0	98.6	**1.0**	1.9
	HCMI (ours)	**63.2**	**91.2**	**95.7**	**1.0**	**3.2**	**76.3**	97.2	98.7	**1.0**	**1.7**

Table 3. Retrieval results on MSVD and Activity-Net. † denotes re-training. **Bold** denotes the best.

Method	MSVD					Activity-Net				
	R@1↑	R@5↑	R@10↑	MdR↓	MnR↓	R@1↑	R@5↑	R@10↑	MdR↓	MnR↓
CLIP4Clip [16]	45.2	75.5	84.3	**2.0**	10.3	40.5	72.4	**98.2**	**2.0**	7.5
TS2-Net [15]	-	-	-	-	-	41.0	73.6	84.5	**2.0**	8.4
UCoFiA† [24]	45.7	75.6	84.6	**2.0**	10.4	42.3	73.2	84.0	**2.0**	7.4
HCMI (ours)	**45.8**	**76.7**	**85.6**	**2.0**	**9.9**	**45.3**	**75.7**	85.8	**2.0**	**6.9**

5.2 Hyperparameter Tuning and Ablation Study

We conduct an ablation study on DiDeMo to analyze the effect of our method.

Hyperparameter Tuning of the Enhanced Feature Factor w**.** As shown in Eq. (6), w controls the balance between original CLIP features and CMA-enhanced features. Table 4 shows that text-to-video retrieval performance first improves with larger w and peaks at $w = 1.0$, where the best R@1 and MnR are obtained. This suggests that moderate enhancement is most beneficial, while excessive weighting may suppress useful original features.

Ablation of Using Different Enhanced Features. We conduct an ablation study to analyze the effectiveness of the proposed enhanced features, as shown in Table 5. The Baseline represents our full proposed method with all features

Table 4. Comparison of different hyper parameter w. **Bold** denotes the best.

w	Text → Video					Video → Text				
	R@1↑	R@5↑	R@10↑	MdR↓	MnR↓	R@1↑	R@5↑	R@10↑	MdR↓	MnR↓
0.1	46.5	74.5	84.0	**2.0**	14.6	43.9	72.3	81.4	**2.0**	10.3
0.3	47.2	75.2	83.7	**2.0**	14.0	46.1	73.4	83.0	**2.0**	10.2
0.5	47.3	75.1	83.4	**2.0**	13.9	45.7	73.2	**83.3**	**2.0**	**10.1**
1.0	**48.2**	**75.4**	83.6	**2.0**	**13.7**	46.3	73.4	82.7	**2.0**	**10.1**
2.0	47.4	75.2	**84.1**	**2.0**	13.9	**47.1**	73.4	82.7	**2.0**	10.3
5.0	47.9	74.8	84.0	**2.0**	14.1	46.6	**73.7**	82.9	**2.0**	10.5

enabled. We observed a performance degradation across all ablated components. The w/o $\mathcal{L}_{\text{distill}}$ variant, which removes the distillation loss and its associated pre-trained feature alignment from stage I, shows the most significant drop in performance. This highlights the crucial role of stage I pre-trained features in improving the overall performance of our method. When analyzing the individual contributions of the enhanced features, the w/o v' and w/o s' configurations also show clear declines in performance, confirming that both the enhanced visual and enhanced textual features are essential to the success of our framework.

Table 5. Comparison of different experiment settings

Method	Text → Video					Video → Text				
	R@1↑	R@5↑	R@10↑	MdR*downarrow*	MnR*downarrow*	R@1↑	R@5↑	R@10↑	MdR*downarrow*	MnR*downarrow*
Baseline	48.2	75.4	83.6	2.0	13.7	46.3	73.4	82.7	2.0	10.1
w/o $\mathcal{L}_{\text{distill}}$	43.2	69.5	79.7	2.0	16.2	41.7	69.3	79.9	2.0	11.8
w/o v'	46.0	72.3	81.4	2.0	15.9	45.7	72.3	81.6	2.0	12.2
w/o s'	46.5	74.1	82.8	2.0	14.0	44.7	72.1	81.3	2.0	10.4

6 Conclusion

In this paper, we introduce the HCMI framework to overcome the limitations of isolated feature extraction and suboptimal multi-granularity alignment in existing video-text retrieval approaches. HCMI enhances feature representations through multi-level, bidirectional cross-modal interactions during training, while requiring only single-modality input at inference. Extensive experiments show that our framework achieves state-of-the-art performance across multiple benchmarks, striking an effective balance between retrieval accuracy and efficiency. In the future, we plan to extend HCMI to more diverse modalities and investigate its potential in large-scale, real-world retrieval scenarios.

Acknowledgment. This work was supported by Guangdong S&T Programme (2024B0101040008).

References

1. Alayrac, J.B., et al.: Flamingo: a visual language model for few-shot learning. arXiv preprint arXiv:2204.14198 (2022)
2. Anne Hendricks, L., Wang, O., Shechtman, E., Sivic, J., Darrell, T., Russell, B.: Localizing moments in video with natural language. In: Proceedings of the IEEE International Conference on Computer Vision, pp. 5803–5812 (2017)
3. Chen, D., Dolan, W.B.: Collecting highly parallel data for paraphrase evaluation. In: Proceedings of the 49th Annual Meeting of the Association for Computational Linguistics: Human Language Technologies, pp. 190–200 (2011)
4. Chen, L., Deng, Z., Liu, L., Yin, S.: Multilevel semantic interaction alignment for video-text cross-modal retrieval. IEEE Trans. Circuits Syst. Video Technol. **34**(7), 6559–6575 (2024)
5. Chen, S., Zhao, Y., Jin, Q., Wu, Q.: Fine-grained video-text retrieval with hierarchical graph reasoning. In: Proceedings of the IEEE/CVF Conference on Computer Vision and Pattern Recognition, pp. 10638–10647 (2020)
6. Fang, H., Xiong, P., Xu, L., Chen, Y.: Clip2video: Mastering video-text retrieval via image clip. arXiv preprint arXiv:2106.11097 (2021)
7. Gorti, S.K., Vouitsis, N., Ma, J., Golestan, K., Volkovs, M., Garg, A., Yu, G.: X-pool: cross-modal language-video attention for text-video retrieval. In: Proceedings of the IEEE/CVF Conference on Computer Vision and Pattern Recognition, pp. 5006–5015 (2022)
8. Jiang, H., Zhang, J., Huang, R., Ge, C., Ni, Z., Song, S., Huang, G.: Cross-modal adapter for vision-language retrieval. Patt. Recogn. **159**, 111144 (2025)
9. Jing, X., Yang, G., Chu, J.: Tc-MGC: text-conditioned multi-grained contrastive learning for text-video retrieval. Inf. Fusion **121**, 103151 (2025)
10. Krishna, R., Hata, K., Ren, F., Fei-Fei, L., Carlos Niebles, J.: Dense-captioning events in videos. In: Proceedings of the IEEE International Conference on Computer Vision, pp. 706–715 (2017)
11. Lei, J., Chen, X., Zhang, N., Wang, M., Bansal, M., Berg, T.L., Yu, L.: Loopitr: combining dual and cross encoder architectures for image-text retrieval. arXiv preprint arXiv:2203.05465 (2022)
12. Lei, J., Li, L., Zhou, L., Gan, Z., Berg, T.L., Bansal, M., Liu, J.: Less is more: clipbert for video-and-language learning via sparse sampling. In: Proceedings of the IEEE/CVF Conference on Computer Vision and Pattern Recognition, pp. 7331–7341 (2021)
13. Li, W., Wang, S., Zhao, D., Xu, S., Pan, Z., Zhang, Z.: Multi-granularity and multi-modal feature interaction approach for text video retrieval. arXiv preprint arXiv:2407.12798 (2024)
14. Liu, S., Fan, H., Qian, S., Chen, Y., Ding, W., Wang, Z.: Hit: hierarchical transformer with momentum contrast for video-text retrieval. In: Proceedings of the IEEE/CVF International Conference on Computer Vision, pp. 11915–11925 (2021)
15. Liu, Y., Xiong, P., Xu, L., Cao, S., Jin, Q.: Ts2-net: Token shift and selection transformer for text-video retrieval. In: European Conference on Computer Vision, pp. 319–335. Springer (2022)
16. Luo, H., et al.: Clip4clip: an empirical study of clip for end to end video clip retrieval and captioning. Neurocomputing **508**, 293–304 (2022)
17. Ma, Y., Xu, G., Sun, X., Yan, M., Zhang, J., Ji, R.: X-clip: end-to-end multi-grained contrastive learning for video-text retrieval. In: Proceedings of the 30th ACM International Conference on Multimedia, pp. 638–647 (2022)

18. Park, Y., et al.: Normalized contrastive learning for text-video retrieval. arXiv preprint arXiv:2212.11790 (2022)
19. Radford, A., et al.: Learning transferable visual models from natural language supervision. In: International Conference on Machine Learning, pp. 8748–8763. PMLR (2021)
20. Torabi, A., Tandon, N., Sigal, L.: Learning language-visual embedding for movie understanding with natural-language. arXiv preprint arXiv:1609.08124 (2016)
21. Vaswani, A., et al.: Attention is all you need. Adv. Neural Inf. Process. Syst. **30** (2017)
22. Wang, Q., Zhang, Y., Zheng, Y., Pan, P., Hua, X.S.: Disentangled representation learning for text-video retrieval. arXiv preprint arXiv:2203.07111 (2022)
23. Wang, X., Wu, J., Chen, J., Li, L., Wang, Y.F., Wang, W.Y.: Vatex: a large-scale, high-quality multilingual dataset for video-and-language research. In: Proceedings of the IEEE/CVF International Conference on Computer Vision, pp. 4581–4591 (2019)
24. Wang, Z., Sung, Y.L., Cheng, F., Bertasius, G., Bansal, M.: Unified coarse-to-fine alignment for video-text retrieval. In: Proceedings of the IEEE/CVF International Conference on Computer Vision, pp. 2816–2827 (2023)
25. Wu, P., He, X., Tang, M., Lv, Y., Liu, J.: Hanet: hierarchical alignment networks for video-text retrieval. In: Proceedings of the 29th ACM International Conference on Multimedia, pp. 3518–3527 (2021)
26. Xu, H., et al.: Videoclip: contrastive pre-training for zero-shot video-text understanding. arXiv preprint arXiv:2109.14084 (2021)
27. Xu, J., Mei, T., Yao, T., Rui, Y.: MSR-VTT: a large video description dataset for bridging video and language. In: Proceedings of the IEEE Conference on Computer Vision and Pattern Recognition, pp. 5288–5296 (2016)
28. Yu, Y., Kim, J., Kim, G.: A joint sequence fusion model for video question answering and retrieval. In: Proceedings of the European Conference on Computer Vision (ECCV), pp. 471–487 (2018)
29. Zellers, R., et al.: Merlot reserve: neural script knowledge through vision and language and sound. In: Proceedings of the IEEE/CVF Conference on Computer Vision and Pattern Recognition, pp. 16375–16387 (2022)
30. Zhang, D., Wang, Z., Hu, Z., Wu, X.J.: HTVR: hierarchical text-to-video retrieval based on relative similarity. Patt. Recogn., 112145 (2025)
31. Zhao, S., Zhu, L., Wang, X., Yang, Y.: Centerclip: token clustering for efficient text-video retrieval. In: Proceedings of the 45th International ACM SIGIR Conference on Research and Development in Information Retrieval, pp. 970–981 (2022)
32. Zou, X., Wu, C., Cheng, L., Wang, Z.: Tokenflow: rethinking fine-grained cross-modal alignment in vision-language retrieval. arXiv preprint arXiv:2209.13822 (2022)

SE-EEND: A Structurally Enhanced End-to-End Neural Diarization System

Penghao Ma, Guangcun Wei(✉), Chuike Kong, Shuo Li, and Jianfeng Fang

College of Intelligent Equipment, Shandong University of Science and Technology, Taian, Shandong 271019, China
weigc@sdust.edu.cn

Abstract. In recent years, End-to-End Neural Diarization (EEND) has emerged as a prominent research direction in the field of speech processing, owing to its ability to jointly model speaker activity detection and speaker classification within a unified framework. However, most existing EEND models rely on Transformer-based encoders, which often encounter performance bottlenecks when dealing with complex conditions such as speech overlaps, rapid speaker transitions, and fine-grained boundary modeling. To address these challenges, we proposes a novel end-to-end neural diarization system named Structurally Enhanced EEND (SE-EEND), based on a MULTI-CONVFORMER encoder architecture. This architecture incorporates multi-scale convolutional kernels and gating mechanisms, enhancing the model's capacity to capture both local and global dependencies at varying temporal resolutions. Furthermore, a Bi-Path Residual Block (BPRB) is introduced at the encoder front-end, which integrates standard and dilated convolutions in parallel to structurally enrich the input features. Extensive evaluations across multiple publicly available datasets demonstrate that SE-EEND consistently outperforms mainstream EEND systems in terms of Diarization Error Rate (DER). Notably, on real-world meeting corpora such as AISHELL-4, AliMeeting, AMI, DIHARD-III and RAMC, SE-EEND achieves state-of-the-art performance.

Keywords: Speaker Diarization · End-to-End · Neural Diarization

1 Introduction

Speaker Diarization (SD) aims to assign audio segments to the correct speakers, effectively aligning speech content with the corresponding speaker identities to answer the question "who spoke when" in multi-speaker environments [1]. Recent technological advances have further intensified the demand for accurate speaker classification in downstream tasks such as automatic speech recognition [2].

Traditionally, SD has been formulated as a speaker clustering problem, where each speech segment is assigned a speaker label. Most clustering-based diarization systems follow a modular pipeline comprising Voice Activity Detection

J. Lokoč et al. (Eds.): MMM 2026, LNCS 16413, pp. 602–615, 2026.
https://doi.org/10.1007/978-981-95-6957-1_43

(VAD), speaker embedding extraction, and clustering [3,4]. For a given utterance, each segment is typically represented using speaker embedding vectors such as i-vectors, d-vectors, or x-vectors. After assigning speaker labels to segments, those with the same label are grouped into clusters. Recent advances in more discriminative embeddings [5] and improved feature extractors [6] have further boosted the performance of clustering-based SD systems.Despite their overall effectiveness, clustering-based methods face two major limitations. First, they struggle with overlapping speech involving more than two speakers, as each segment can only be associated with a single embedding. Second, these modular systems cannot benefit from joint optimization, which limits their overall diarization accuracy.

To address these challenges, End-to-End Neural Diarization (EEND) methods have been introduced. Unlike traditional clustering-based pipelines, EEND formulates speaker diarization as a multi-label classification task, thereby eliminating the need for separate modules such as voice activity detection (VAD) and overlap detection. Empirical results on both simulated and real-world datasets have shown that EEND consistently outperforms clustering-based approaches [7,8]. Building on this foundation, Self-Attentive EEND (SA-EEND) incorporates self-attention mechanisms to simultaneously model global frame-to-frame dependencies, further enhancing diarization performance. Recent advances in EEND have primarily focused on the integration of additional embedding representations [9–12] or attractor modules [13–17], which often require customized model architectures, loss functions, and data preparation strategies.

However, EEND still shows limitations in modeling rapid speaker transitions, short-term overlaps, and local dynamic variations. Although transformer architectures excel at capturing global context, they often lack the granularity needed to model fine-grained local changes. Furthermore, deep stacking may introduce information dilution and channel redundancy, which affect speaker discriminability. To mitigate these issues, prior work has explored Time-Dilated Convolutions Neural(TDCN) [18] and heterogeneous networks like Conformer-based EEND(CB-EEND) [19], yet challenges remain regarding shallow feature degradation and limited receptive field diversity.

To further enhance the model's representation capability for complex multi-speaker speech, we propose a structurally enhanced end-to-end speaker diarization system, termed Structurally Enhanced EEND (SE-EEND). The key contributions of this work are summarized as follows:

- In SE-EEND, we substitute the Transformer encoder with a MULTI-CONVFORMER encoder. This encoder integrates multi-scale depthwise separable convolutions and gated mechanisms, enabling more effective modeling of both global dependencies and fine-grained local variations.
- A Bi-Path Residual Block (BPRB) module, is introduced before the encoder, combining standard and dilated convolutions to reinforce structural awareness while balancing local continuity and broader receptive field coverage.
- The proposed SE-EEND system is validated on multiple real-world multi-speaker meeting corpora. Experimental results show that it outperforms exist-

ing EEND systems in terms of Diarization Error Rate (DER), particularly under challenging conditions with high overlap ratios and frequent speaker transitions.

2 Review of SA-EEND

In this section, we briefly review the SA-EEND model, which is used as the baseline in our study.

To extract input feature vectors, the SA-EEND model applies a 25-ms Hamming window with a 10-ms shift to each utterance. For each windowed frame, a 23-dimensional log-scaled Mel filterbank analysis is performed. Subsequently, every group of 15 consecutive frames is concatenated into a single feature vector $\mathbf{x}_t \in \mathbb{R}^F$, where $F = 345$. These vectors are extracted with a stride of 10 frames, resulting in a sequence $X = [\mathbf{x}_1, \cdots, \mathbf{x}_T]$, where T denotes the total number of characteristic vectors.

Next, the sequence X is processed through a linear transformation followed by layer normalization to obtain the initial input embeddings $E^0 = [\mathbf{e}_1^0, \cdots, \mathbf{e}_T^0]$ for a series of P encoder blocks in the EEND model. Typically, $P = 4$ as described in [8]. The p-th encoder block derives self-attentive features based on the embeddings from the $(p-1)$-th layer, as formulated in the following:

$$\mathbf{e}_t^0 = \text{Norm}(\text{Linear}^F(\mathbf{x}_t)) \in \mathbb{R}^D \tag{1}$$

$$E^p = \text{Encoder}_p^D(E^{p-1}), \quad (1 \leq p \leq P) \tag{2}$$

After passing through all P encoder blocks, the final output embeddings E^P are passed through a linear transformation followed by a sigmoid function to obtain the posterior probabilities of the S speakers at time t, denoted as $\hat{\mathbf{y}}_t = [\hat{y}_{t,1}, \cdots, \hat{y}_{t,S}]$, as shown below:

$$\hat{\mathbf{y}}_\mathbf{t} = \text{sigmoid}(\text{Linear}^D(\mathbf{e}_t^P)) \tag{3}$$

A permutation invariant scheme is used to optimize SA-EEND during training [20]. That is, the loss is computed between the predicted labels $\hat{\mathbf{y}}_t$ and the ground truth labels $\mathbf{y}_t = [y_{t,1}, \cdots, y_{t,S}] \in \{0,1\}^S$. The total loss function $\mathcal{L}_d$ is therefore defined as:

$$\mathcal{L}_\text{d} = \frac{1}{TS} \min_{\phi \in \text{perm}(S)} \sum_{t=1}^{T} H\left(\mathbf{y}_t^\phi, \hat{\mathbf{y}}_t\right) \tag{4}$$

where $perm(S)$ denotes the set of all possible permutations of the speaker indices, and $\mathbf{y}_t^\phi \in \{0,1\}^S$ represents the ground truth labels rearranged according to the permutation ϕ. In Equation (4), $H(\mathbf{y}_t, \hat{\mathbf{y}}_t)$ denotes the binary cross-entropy, which is defined as:

$$H(\mathbf{y}_t, \hat{\mathbf{y}}_t) = \sum_s -y_{t,s} \log \hat{y}_{t,s} - (1 - y_{t,s}) \log(1 - \hat{y}_{t,s}) \tag{5}$$

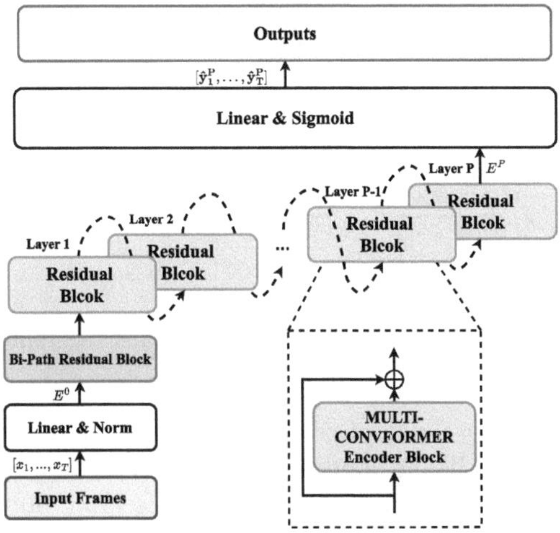

Fig. 1. Overall architecture of the proposed SE-EEND model.

3 Proposed Method

In recent years, the MULTI-CONVFORMER architecture [21] has demonstrated remarkable performance as a novel hybrid encoding structure in Automatic Speech Recognition (ASR) tasks. By combining multi-scale depthwise separable convolutions with multi-head self-attention mechanisms, it enables joint modeling of local acoustic patterns and global contextual dependencies. This synergy enhances spatiotemporal representation and modeling flexibility in sequence modeling tasks.Inspired by this architectural design, we propose a structurally enhanced end-to-end neural diarization system, named SE-EEND, as illustrated in Fig. 1.

SE-EEND adopts the MULTI-CONVFORMER encoder as its backbone and explicitly introduces residual connections into each encoder block to facilitate deeper feature propagation and improve gradient stability. While retaining the cooperative strengths of local convolutions and global attention mechanisms inherent in the original architecture, the added residual structure helps mitigate the degradation of deep stacking, enhancing the model's robustness in modeling dynamic speaker transitions. In addition, SE-EEND integrates a front-end Bi-Path Residual Block, which fuses standard and dilated convolutions in parallel to enrich the structural representation of input features, particularly in terms of local continuity and contextual coverage.

3.1 Bi-Path Residual Block

In speaker diarization tasks, especially in scenarios with frequent speaker switches within short utterances and extensive overlapping speech, the model

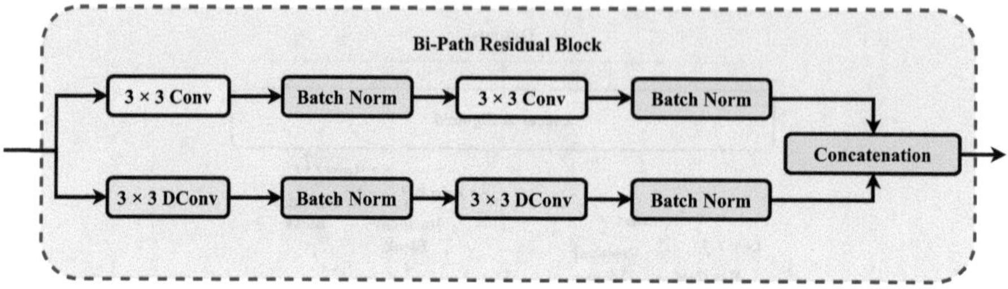

Fig. 2. Architecture of the proposed Bi-Path Residual Block.

must strike a balance between capturing global contextual information and preserving local acoustic details. Expanding the receptive field is an important strategy for improving modeling capacity. Although stacking global modeling modules such as multi-layer attention can enhance long-term dependency modeling, excessive reliance on global pathways often leads to the degradation of local feature representations, reducing the model's sensitivity to short-term speaker changes and boundary cues. To address this issue, we propose a Bi-Path Residual Block, as illustrated in Fig. 2. This module integrates local and global modeling pathways in parallel, effectively combining short-term dynamic modeling with long-range contextual awareness. The BPRB consists of two functionally complementary branches: The local pathway employs standard one-dimensional convolution to capture speaker variations within short temporal windows, maintaining temporal continuity and preserving detailed speech dynamics. The global pathway adopts dilated convolution with an enlarged dilation rate to expand the temporal receptive field, thereby enhancing the model's ability to capture speaker patterns across wider time spans. However, the sparse sampling nature of dilated convolution may weaken local continuity.

To integrate the complementary strengths of both branches, the outputs are concatenated along the channel dimension and passed through convolution layer for dimensionality compression and fusion. The final output is a structurally enhanced feature representation, formally defined as follows:

$$x_{s+1}^{\mathrm{loc}} = \mathrm{BatchNorm}\left(f_s\left(\mathrm{BatchNorm}\left(f_s(x_s)\right)\right)\right) \tag{6}$$

$$x_{s+1}^{\mathrm{glob}} = \mathrm{BatchNorm}\left(f_s\left(\mathrm{BatchNorm}\left(f_s^k(x_s)\right)\right)\right) \tag{7}$$

$$x_{s+1} = \mathrm{Conv}\left(\mathrm{Concat}\left[x_{s+1}^{\mathrm{loc}}, x_{s+1}^{\mathrm{glob}}\right]\right) \tag{8}$$

Here, x_s denotes the input feature of the current layer, while x_{s+1}^{loc}, x_{s+1}^{glob} and x_{s+1} represent the outputs of the local path, global path, and the final fused feature, respectively. The function $f_s(\cdot)$ denotes a standard one-dimensional convolution operation, and $f_s^k(\cdot)$ refers to a dilated convolution with dilation rate k.BatchNorm$(\cdot)$ denotes batch normalization, and Conv$(\cdot)$ represents feature fusion along the channel dimension.

3.2 MULTI-CONVFORMER Encoder Block

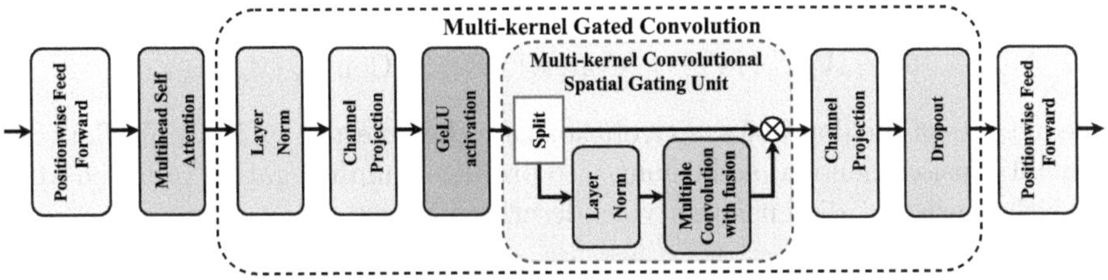

Fig. 3. Structure of the MULTI-CONVFORMER Encoder Block used in SE-EEND.

In the proposed SE-EEND framework, the core encoder adopts a structurally enhanced module based on the MULTI-CONVFORMER architecture. Each MULTI-CONVFORMER encoder layer consists of four sequentially stacked functional modules, an initial Feed-Forward Network (FFN) module, a Multi-Head Self-Attention (MHSA) module, a Multi-kernel Gated Convolution, and a second FFN module.

The first FFN module performs point-wise linear transformations followed by non-linear activation to enhance the abstraction of frame-level features. The MHSA module applies multi-head attention to globally model temporal dependencies across different time steps. The intermediate Multi-kernel Gated Convolution further captures local pattern variations from multi-scale temporal contexts, effectively improving the model's ability to represent speaker boundaries and overlapping speech regions. The final FFN module acts as a semantic fusion layer, further enhancing contextual coherence and the overall modeling depth.

As illustrated in Fig. 3, in a single encoder layer, the output sequence from the MHSA module is denoted as $A = \{a_1, a_2, \ldots, a_T | \ a_j \in \mathbb{R}^d\}$. First, this output is normalized using layer normalization to stabilize temporal feature distributions. Then, a channel projection is applied to expand the dimensionality from d to an intermediate dimension d_{inter}, followed by a GELU activation function to enhance non-linear representation. This results in an intermediate representation $\hat{A} = \{\hat{a}_1, \hat{a}_2, \ldots, \hat{a}_T, |\hat{a}_j \in \mathbb{R}^{d_{\text{inter}}}\}$. The intermediate features $\hat{A}$ are then passed into the Multi-kernel Convolutional Spatial Gating Unit(M-CSGU) module for multi-scale local context modeling and gated feature fusion. Finally, a second channel projection is applied to map the features back from $\mathbb{R}^{d_{\text{inter}}}$ to the original dimension $\mathbb{R}^d$, followed by dropout for regularization.

The M-CSGU module adopts a dual-branch interaction structure to model multi-scale spatial dependencies across the channel dimension. The input feature $\hat{A}$ is evenly split along the channel axis into two sub-representations. The first half of the channels, referred to as Z_l, functions as the main information branch, and the second half, referred to as Z_r, serves as the input to the gating pathway. Formally, this operations are as follows:

$$[Z_l, Z_r] = \left[\hat{A}[:, : d'], \text{LayerNorm}\left(\hat{A}[:, d' :]\right)\right] \tag{9}$$

Subsequently, Z_r is fed into P parallel depthwise convolutional branches, where the kernel sizes are defined by the set $K = \{k_1, k_2, \cdots, k_p\}$. Each branch captures local acoustic patterns under a different receptive field:

$$[V_1, V_2, \ldots, V_P] = [\text{Conv}_{k_1}(Z_r), \ldots, \text{Conv}_{k_P}(Z_r)] \tag{10}$$

The outputs of the depthwise convolutions are then merged by Multiple Convolution with fusion, denoted as Fusion$(\cdot)$, to produce a unified gating representation $\tilde{Z}_r$ with consistent dimensionality, as described below:

$$\tilde{Z}_r = \text{Fusion}([V_1, V_2, \ldots, V_P]) \tag{11}$$

Finally, the main path feature Z_l and the gated path output $\tilde{Z}_r$ are realized as feature modulation via Hadamard product to form a spatially gated interaction mechanism, as described below:

$$\hat{C} = Z_l \odot \tilde{Z}_r \tag{12}$$

The output $\hat{C}$ is then projected back to the original dimensionality d, followed by a dropout operation for regularization. This output serves as the final representation of the encoder layer and is forwarded to subsequent modules.

4 Experimental Setup

4.1 Datasets

This study conducts a comprehensive evaluation of the proposed model on several publicly available speaker diarization datasets, including AISHELL-4 [22], AliMeeting [23], AMI [24], DIHARD-III [25], RAMC [26]. These datasets span a wide range of languages, microphone configurations, and overlapping speaker conditions, offering both diversity and challenges for evaluation. More information about each dataset can be found in Table 1.

Table 1. Dataset statistics: number of files, speakers per session, and total hours for train/dev/test splits.

Dataset	Train			Development			Test		
	Files	Spk	Hrs	Files	Spk	Hrs	Files	Spk	Hrs
AISHELL-4	191	3–7	107.53	–	–	–	20	5–7	12.72
AliMeeting	209	2–4	111.36	8	2–4	4.20	60	2–4	10.78
AMI	136	3–5	80.67	18	4	9.67	16	3–4	9.06
DIHARD-III	–	–	–	254	1–10	34.15	259	1–9	33.01
RAMC	289	2	149.65	19	2	9.89	43	2	20.64

For datasets with officially provided training/validation/test splits, we strictly follow the standard partitioning protocols. For datasets that only provide training and test sets, we reserve 20% of the training data as a validation set, which is used for hyperparameter tuning and early stopping, while the test set is used for final evaluation. In multi-channel recording scenarios, we simplify the feature extraction process by performing weighted mixing of all audio channels to produce a single-channel input. For the AliMeeting dataset, we utilize the far-field 8-channel microphone array recordings adopted in the M2MeT Challenge to further investigate the impact of different microphone configurations on model performance.

4.2 Implementation Details

All experiments in this study were conducted in a computing environment equip with an NVIDIA GeForce RTX 4090D GPU, ensuring both efficiency and stability during training and inference. The input features to the model are 23-dimensional log-Mel filterbank coefficients, extracted from raw audio using a sliding window of 25 ms frame length and 10 ms frame shift. To enhance contextual modeling capability, each frame is concatenated with 7 frames before and after, forming a 15-frame contextual window. As a result, the dimensionality of each input frame becomes 23×15. To meet the modeling requirements of deep neural networks on long sequences and to improve computational efficiency, the concatenated feature sequence is downsampled by a factor of 10, reducing the temporal resolution to one frame per 100 ms. The final input to the model is therefore a sequence with a 100 ms sampling interval and a feature dimension of 345. During training, to manage GPU memory usage and maintain training stability, the length of each audio segment is fixed at 500 frames. All training samples are split into non-overlapping 50-second segments, which are uniformly fed into the model. The proposed SE-EEND model is constructed using a stacked MULTI-CONVFORMER encoder architecture comprising 5 encoder layers, each configured with 4 attention heads and an attention dimension of 1024. The convolution module within each encoder layer adopts a multi-scale parallel kernel design, with kernel sizes set to $K = \{7, 15, 23, 31\}$, enabling local feature modeling across different receptive field scales. The internal dimension of each position-wise feed-forward module is set to $d_{ff} = 1024$.

For model optimization, we adopt the Adam learning rate scheduler with warm-up proposed in [23], with 30,000 warm-up steps to alleviate early-stage training instability and accelerate convergence. The model is trained for a total of 150 epochs with a batch size of 64 per epoch. To further improve robustness and generalization, the final model used for evaluation is obtained by applying sliding average over the last 10 epochs' parameters.

4.3 Evaluation Metric

To evaluate the classification performance of the model in multi-speaker speech scenarios, we adopt the standard metric Diarization Error Rate (DER) [27] as the primary evaluation criterion. DER measures the overall discrepancy between the system-generated speaker labels and the reference annotations, and is one of the most widely used performance metrics in speaker diarization tasks.DER consists of the following three components:

False Alarm (FA): The amount of time that non-speech segments are incorrectly detected as speech.

Missed Speech (Miss): The amount of time that actual speech is not detected by the system.

Speaker Confusion (SC): The amount of time that speech is attributed to the wrong speaker.

The overall DER is computed by normalizing the sum of these three error components by the total duration of speech, as follows:

$$\text{DER} = \frac{\text{FA} + \text{Miss} + \text{SC}}{\text{Total Speech Duration}} \tag{13}$$

In all experiments, we follow the standard protocols used in NIST and DIHARD evaluations by setting the collar toleranceto 0.25 s. This means that prediction errors occurring within 250 milliseconds before or after a reference speaker change point are ignored, thus reducing the impact of annotation uncertainty and minor alignment shifts on evaluation results.

5 Results

5.1 Performance Comparison

We compare SE-EEND with several representative baselines. Our main results, can be seen in Table 2. Across all benchmarks, SE-EEND consistently outperforms. Notably, the model achieves significant DER reductions on AliMeeting and DIHARD-III, both of which feature frequent speaker overlaps, spontaneous speech, and adverse acoustic environments. This demonstrates the model's robustness in handling complex interaction patterns and overlapping speech segments. On AISHELL-4 and RAMC, which mainly contain Mandarin and telephony-style conversations, SE-EEND also achieves clear improvements, indicating strong cross-lingual and cross-domain generalization capabilities. Even on the well-studied AMI dataset, where many systems have saturated in performance, SE-EEND delivers competitive results, validating its effectiveness in mature benchmarks. The representative model is described as follows:

- VAD + VBx + OSD [28]: A classic modular pipeline that separates diarization into voice activity detection, Bayesian clustering, and overlap detection stages.
- SA-EEND [8]: A Transformer-based end-to-end model that directly predicts frame-level speaker labels.
- EEND-EDA [29]: An extension of SA-EEND that introduces an encoder-decoder attractor mechanism to handle variable speaker numbers and improve overlap modeling.
- PyAnnote 3.1 [30]: A widely used open-source toolkit with pretrained models and a flexible inference pipeline, often used as a practical baseline in real-world applications.

- DiaPer [28]: A large-scale pretrained system with strong generalization across datasets, although it requires substantial training resources and may lack fine-tuned adaptation to specific conditions.

Table 2. Speaker diarization error rate (DER%) comparison on multiple datasets. Bold values indicate best results among compared models.

Method	AISHELL-4	AliMeeting	AMI	DIHARD-III	RAMC
VAD+VBx+OSD [28]	15.84	28.84	22.42	20.28	18.19
SA-EEND [8]	–	28.02	27.70	22.64	–
EEND-EDA [29]	–	–	15.80	20.69	–
PyAnnote 3.1 [30]	13.20	23.30	18.00	21.30	22.20
DiaPer [28]	31.30	26.27	30.49	22.77	18.69
SE-EEND (ours)	**12.46**	**17.03**	**13.62**	**19.41**	**16.95**

5.2 Ablation Experiment and Analysis

To further verify the independent contributions and performance impact of the two structural innovations proposed in SE-EEND, we conducted a series of ablation experiments. We removed or replaced the MULTI-CONVFORMER encoder, the Bi-Path Residual Block, and evaluated model performance on the AliMeeting test set. All experiments were conducted under identical training configurations and evaluation protocols, with only the structural components varied. Table 3 summarizes the diarization error rate for each ablation setting.

First, we replaced the MULTI-CONVFORMER encoder in SE-EEND with the standard Transformer encoder used in SA-EEND, keeping all other components unchanged. The DER increased significantly from 17.03% to 23.87%. This result highlights the advantage of MULTI-CONVFORMER in modeling complex speech boundaries, overlapping segments, and speaker transitions. The improvement is mainly attributed to its integration of multi-scale depthwise separable convolutions and gating mechanisms, which enhance local receptive field modeling while preserving global attention capabilities.

Next, we removed the Bi-Path Residual Block placed before the encoder, keeping the rest of the architecture intact. The DER rose to 19.62%, showing a slight performance drop compared to the full model. This confirms that BPRB provides effective structural enhancement in the early feature encoding stage. By introducing parallel paths of standard and dilated convolutions, BPRB successfully balances local continuity with broader contextual awareness.

To further verify the applicability and transferability of the proposed Bi-Path Residual Block across different model architectures, we integrated this module into another representative end-to-end speaker diarization system CB-EEND to construct an enhanced variant named RC-CB-EEND. To ensure consistency and fairness in comparison with SA-EEND and SE-EEND, we removed the original SpecAugment and convolutional subsampling modules from the CB-EEND architecture, thereby eliminating structural interference. As shown in Table 4, RC-CB-EEND achieves a significantly lower DER on the AliMeeting test set compared to both the original CB-EEND and SA-EEND. This demonstrates that the proposed structural enhancement design is not only effective within the SE-EEND framework but can also be transferred to and reliably improve the modeling capabilities of other types of EEND systems. The BPRB module enhances local continuity modeling and boundary detection performance at the feature front-end.

Table 3. Ablation study of model architecture components (DER% on AliMeeting dataset). Lower values indicate better performance.

Architecture Configuration	MULTI-CONVFORMER	BPRB	DER (%)
Replaced with Transformer Encoder		✓	23.87
BPRB Removed	✓		19.62
Full Model (SE-EEND)	✓	✓	**17.03**

Table 4. Effect of the Bi-Path Residual Block on diarization performance of EEND variants on the AliMeeting dataset. The DER is composed of three components: Miss (Missed Speech), FA (False Alarm) and SC (Speaker Confusion).

Method	Miss (%)	FA (%)	SC (%)	DER (%)
SA-EEND [8]	7.71	5.21	15.10	28.02
CB-EEND [19]	4.77	3.00	13.90	21.67
RC-CB-EEND	**4.35**	**2.98**	10.93	18.26
SE-EEND (Ours)	6.08	3.88	**7.07**	**17.03**

6 Conclusions

To address the limitations of existing end-to-end diarization systems in handling speaker transitions, overlapping speech, and complex acoustic patterns, we propose SE-EEND, a structurally enhanced end-to-end speaker diarization model. While retaining the modeling strengths of the EEND framework, SE-EEND introduces key architectural improvements to boost representational and discriminative power in challenging multi-speaker scenarios. SE-EEND replaces the conventional Transformer encoder with a locally sensitive MULTI-CONVFORMER encoder, which combines multi-scale gated convolutions and multi-head self-attention to jointly capture local variations and global context. Additionally, we introduce a Bi-Path Residual Block that fuses standard and dilated convolutions in parallel, enhancing the modeling of boundary transitions and temporal continuity.

Evaluations on real-world datasets, AliMeeting, AMI, AISHELL-4, DIHARD-III, and RAMC, show that SE-EEND consistently outperforms baselines such as SA-EEND, EEND-EDA, and PyAnnote in DER. The model demonstrates strong robustness and generalization, particularly in highly overlapped or spontaneous speech. Ablation and transfer studies further validate the effectiveness and portability of each proposed component.

References

1. Park, T.J., Kanda, N., Dimitriadis, D., Han, K.J., Watanabe, S., Narayanan, S.: A review of speaker diarization: recent advances with deep learning. Comput. Speech Lang. **72**, 101317 (2022)
2. Kanda, N., et al.: Joint speaker counting, speech recognition, and speaker identification for overlapped speech of any number of speakers. In: Interspeech 2020, pp. 36–40 (2020). https://doi.org/10.21437/Interspeech.2020-1085
3. Lin, Q., Yin, R., Li, M., Bredin, H., Barras, C.: LSTM based similarity measurement with spectral clustering for speaker diarization. In: Interspeech 2019, pp. 366–370 (2019). https://doi.org/10.21437/Interspeech.2019-1388
4. Landini, F., Profant, J., Diez, M., Burget, L.: Bayesian hmm clustering of x-vector sequences (vbx) in speaker diarization: theory, implementation and analysis on standard tasks. Comput. Speech Lang. **71**, 101254 (2022)
5. Jung, J.w., et al.: In search of strong embedding extractors for speaker diarisation. In: ICASSP 2023-2023 IEEE International Conference on Acoustics, Speech and Signal Processing (ICASSP), pp. 1–5. IEEE (2023)
6. Heo, H.S., Kwon, Y., Lee, B.J., Kim, Y.J., Jung, J.w.: High-resolution embedding extractor for speaker diarisation. In: ICASSP 2023-2023 IEEE International Conference on Acoustics, Speech and Signal Processing (ICASSP), pp. 1–5. IEEE (2023)
7. Fujita, Y., Kanda, N., Horiguchi, S., Nagamatsu, K., Watanabe, S.: End-to-end neural speaker diarization with permutation-free objectives. In: Interspeech 2019, pp. 4300–4304 (2019). https://doi.org/10.21437/Interspeech.2019-2899
8. Fujita, Y., Kanda, N., Horiguchi, S., Xue, Y., Nagamatsu, K., Watanabe, S.: End-to-end neural speaker diarization with self-attention. In: 2019 IEEE Automatic Speech Recognition and Understanding Workshop (ASRU), pp. 296–303. IEEE (2019)
9. Wang, C., Li, J., Fang, X., Kang, J., Li, Y.: End-to-end neural speaker diarization with absolute speaker loss. In: Interspeech 2023, pp. 3577–3581 (2023). https://doi.org/10.21437/Interspeech.2023-656
10. Fung, I., Samarakoon, L., Broughton, S.J.: Robust end-to-end diarization with domain adaptive training and multi-task learning. In: 2023 IEEE Automatic Speech Recognition and Understanding Workshop (ASRU), pp. 1–7. IEEE (2023)

11. Yu, Y., Park, D., Kim, H.K.: Auxiliary loss of transformer with residual connection for end-to-end speaker diarization. In: ICASSP 2022-2022 IEEE International Conference on Acoustics, Speech and Signal Processing (ICASSP), pp. 8377–8381. IEEE (2022)
12. Zeghidour, N., Teboul, O., Grangier, D.: Dive: end-to-end speech diarization via iterative speaker embedding. In: 2021 IEEE Automatic Speech Recognition and Understanding Workshop (ASRU), pp. 702–709. IEEE (2021)
13. Samarakoon, L., Broughton, S.J., Härkönen, M., Fung, I.: Transformer attractors for robust and efficient end-to-end neural diarization. In: 2023 IEEE Automatic Speech Recognition and Understanding Workshop (ASRU), pp. 1–8. IEEE (2023)
14. Chen, Z., Han, B., Wang, S., Qian, Y.: Attention-based encoder-decoder network for end-to-end neural speaker diarization with target speaker attractor. In: Interspeech 2023, pp. 3552–3556 (2023). https://doi.org/10.21437/Interspeech.2023-1228
15. Hao, F., Li, X., Zheng, C.: End-to-end neural speaker diarization with an iterative adaptive attractor estimation. Neural Netw. **166**, 566–578 (2023)
16. Rybicka, M., Villalba, J., Dehak, N., Kowalczyk, K.: End-to-end neural speaker diarization with an iterative refinement of non-autoregressive attention-based attractors. In: INTERSPEECH, pp. 5090–5094 (2022)
17. Horiguchi, S., Watanabe, S., García, P., Xue, Y., Takashima, Y., Kawaguchi, Y.: Towards neural diarization for unlimited numbers of speakers using global and local attractors. In: 2021 IEEE Automatic Speech Recognition and Understanding Workshop (ASRU), pp. 98–105. IEEE (2021)
18. Maiti, S., Erdogan, H., Wilson, K., Wisdom, S., Watanabe, S., Hershey, J.R.: End-to-end diarization for variable number of speakers with local-global networks and discriminative speaker embeddings. In: ICASSP 2021-2021 IEEE International Conference on Acoustics, Speech and Signal Processing (ICASSP), pp. 7183–7187. IEEE (2021)
19. Liu, Y.C., Han, E., Lee, C., Stolcke, A.: End-to-end neural diarization: From transformer to conformer. In: Interspeech 2021, pp. 3081–3085 (2021). https://doi.org/10.21437/Interspeech.2021-1909
20. Yu, D., Kolbæk, M., Tan, Z.H., Jensen, J.: Permutation invariant training of deep models for speaker-independent multi-talker speech separation. In: 2017 IEEE International Conference on Acoustics, Speech and Signal Processing (ICASSP), pp. 241–245. IEEE (2017)
21. Prabhu, D., Peng, Y., Jyothi, P., Watanabe, S.: Multi-convformer: extending conformer with multiple convolution kernels. In: Interspeech 2024, pp. 232–236 (2024). https://doi.org/10.21437/Interspeech.2024-2384
22. Fu, Y., et al.: Aishell-4: An open source dataset for speech enhancement, separation, recognition and speaker diarization in conference scenario. In: Interspeech 2021, pp. 3665–3669 (2021). https://doi.org/10.21437/Interspeech.2021-1397
23. Yu, F., et al.: M2met: The icassp 2022 multi-channel multi-party meeting transcription challenge. In: ICASSP 2022-2022 IEEE International Conference on Acoustics, Speech and Signal Processing (ICASSP), pp. 6167–6171. IEEE (2022)
24. Kraaij, W., Hain, T., Lincoln, M., Post, W.: The AMI meeting corpus. In: Proc. International Conference on Methods and Techniques in Behavioral Research, pp. 1–4 (2005)
25. Ryant, N., et al.: The third dihard diarization challenge. In: Interspeech 2021, pp. 3570–3574 (2021). https://doi.org/10.21437/Interspeech.2021-1208

26. Yang, Z., et al.: Open source magicdata-ramc: A rich annotated mandarin conversational (RAMC) speech dataset. In: Interspeech 2022, pp. 1736–1740 (2022). https://doi.org/10.21437/Interspeech.2022-729
27. Vaswani, A., et al.: Attention is all you need. Adv. Neural Inf. Process. Syst. **30** (2017)
28. Landini, F., Stafylakis, T., Burget, L., et al.: Diaper: end-to-end neural diarization with perceiver-based attractors. IEEE/ACM Trans. Audio Speech Language Process. (2024)
29. Horiguchi, S., Fujita, Y., Watanabe, S., Xue, Y., Garcia, P.: Encoder-decoder based attractors for end-to-end neural diarization. IEEE/ACM Trans. Audio Speech Language Process. **30**, 1493–1507 (2022)
30. Plaquet, A., Bredin, H.: Powerset multi-class cross entropy loss for neural speaker diarization. In: Interspeech 2023, pp. 3222–3226 (2023). https://doi.org/10.21437/Interspeech.2023-205

SPADE: Attention-Guided Split Diffusion for Precise Spatial Control in Interior Layout Image Generation

Wenzheng Liu[1], Ming Yuan[1,3], Yizhou Wang[1], Lianghao Shen[2,3(✉)], Xiaofeng Wang[2], Qianqian Xing[2], Ronghui Cao[1], Xiaoyong Tang[1], Tan Deng[1], and Cheng Fu[1]

[1] School of Computer Science and Technology, Changsha University of Science and Technology, Changsha, China
{liuwenzheng,23408030291,202208170220,caoronghui,tangxy, dengtan0510,202308170227}@csust.edu.cn

[2] College of Computer, National University of Defense Technology, Changsha, China
{shenlianghao,xf_wang,xingqianqian12}@nudt.edu.cn

[3] KylinSoft, Changsha, China

Abstract. Diffusion models generate highly realistic images from text, enabling wide applications. However, they often struggle with ambiguous object positioning for multiple objects in a prompt. While recent methods like Complementary Regional Diffusion define positions via segmentation, constraining diffusion regions may reduce object recognizability and fail to solve the positioning problem, primarily due to insufficient pixel-level control over object regions during noise prediction. To address these issues, we propose Spatial Positioning via Attention-guided Diffusion without Extra Training (SPADE). First, we introduce Split Noise Prediction Method, which partitions prediction regions for individual objects prior to noise prediction, significantly enhancing positional accuracy and object recognizability. Second, we propose Split Attention Prediction Method, integrating Region-Specific Weighted and Split Parameters attention mechanisms. This enhances positional accuracy while countering overall image integrity degradation introduced by the Split Noise Prediction Method. Finally, quantitative evaluation on T2I-CompBench shows SPADE surpasses Complementary Regional Diffusion by 27.6% in spatial relationship. Qualitative comparisons further confirm superior prompt alignment in generated images. Consequently, SPADE significantly enhances diffusion models' practicality in interior layout design. Our code is available at https://github.com/yuanming0415/SPADE.

Keywords: Regional Diffusion · Split Noise Prediction · Region-Specific Weighted Attention · Split Parameters Attention

W. Liu and M. Yuan—Equal contribution.

J. Lokoč et al. (Eds.): MMM 2026, LNCS 16413, pp. 616–630, 2026.
https://doi.org/10.1007/978-981-95-6957-1_44

1 Introduction

Recent research on diffusion models [1,24] has led to remarkable advancements. Numerous application studies have emerged, including DreamBooth [25] and Collaborative Diffusion [12]. However, existing models often struggle with precise spatial positioning when generating multiple objects. Several strategies have been proposed to address this issue, such as complementary regional diffusion [33]. These methods offer innovative solutions to improve positional accuracy. However, a major limitation persists: restricting diffusion regions for precise spatial positioning may degrade object recognizability. Conversely, enhancing recognizability often reduces positional accuracy. Furthermore, improvements in spatial accuracy remain limited in magnitude. This constitutes a significant challenge.

To overcome this limitation, we propose Spatial Positioning via Attention-guided Diffusion without Extra Training (SPADE), which is developed based on the following strategies: **(1) Split Noise Prediction Method** adjusts noise regions to enable region-specific predictions, where each distinct region corresponds to a different object. It utilizes integrated noise prediction in subsequent denoising steps. After several denoising iterations, the initial positions of objects within the image can be preliminarily determined. By dividing the image into object-specific regions, this approach reduces gradient conflicts and improves both object positional accuracy and recognizability. **(2) Split Attention Prediction Method** enhances positional accuracy through attention weighting and region control, and it also addresses the image integrity degradation introduced by the Split Noise Prediction Method. In this mechanism, the Query comes from the hidden states of the noisy image, while the Key and Value are derived from the hidden states of the base and region prompts. Firstly, the **Region-Specific Weighted Attention Mechanism** enhances object positioning by adaptively increasing the Query matrix weights of target regions, thereby strengthening their association with corresponding objects and improving positional accuracy. Secondly, the **Split Parameters Attention Mechanism** improves positional accuracy by modifying the region of the Query matrix of an object, directly influencing pixel values via controlled pixel-level adaptation. Experimental results show notable improvements in both qualitative and quantitative evaluations. The generated interior layout images are compared with those from baseline models RPG and SDXL in Fig. 1.

Our main contributions are summarized as follows:

- We propose the Split Noise Prediction Method, which partitions noisy images into distinct regions for per-region noise prediction, allowing different regions to specialize in generating different objects and thereby enhancing positional accuracy and recognizability of objects in generated images.
- We propose Split Attention Prediction Method, incorporating Region-Specific Weighted Attention Mechanism and Split Parameters Attention Mechanism. This method adjusts Query weights in specific regions while integrating their parameters regionally. It thereby enhances positional accuracy and reduces overall image integrity degradation from the Split Noise Prediction Method.

Prompt: This study room features a cozy, minimal design with wooden floors and soft lighting. On the left side stands a sleek wooden desk with a lamp and organizer. To the right, a tall leafy green plant in a ceramic pot brings nature indoors, enhancing tranquility and balance.

Prompt: This living room blends modern style with comfort, featuring a clean layout and wooden accents. The upper side holds a wall-mounted TV and cabinet with drawers and shelves. Below, a spacious fabric sofa offers relaxed seating, creating a balanced and inviting environment.

Prompt: This modern office features a clean, organized design with neutral tones. On the left wall hangs a vibrant abstract mural, energizing the space. To the right, a sleek office desk with drawers and a monitor stand provides a functional, comfortable workstation.

Fig. 1. Compared to RPG and SDXL, our proposed SPADE demonstrates superior capability in conveying spatial relationships of objects described in complex prompts within generated interior layout images (colored text highlights critical information).

- On the specialized positional evaluation benchmark T2I-CompBench, our model demonstrates a 27.6% improvement over the state-of-the-art models. This work further establishes theoretically and experimentally derived optimal parameter configurations for generating two-object layout images.

2 Related Works

2.1 Text-Guided Diffusion Models for Image Generation

Diffusion models [24] are a highly promising category of generative models. Their applications span optics [21], molecular studies [36], biological research [23], medicine [30], security [9], and social domains [4], demonstrating broad utility. In recent studies, models such as ContextDiff [34], UNIMO-G [16], MT-Diffusion [3], UPainting [17], and Swinv2-Imagen [15] have significantly improved the multimodal processing capabilities of diffusion models. Meanwhile, growing attention has been given to understanding modal attributes within multimodal diffusion frameworks [14]. In recent advancements, the integration with Rectified Flow [18,19] has contributed to the development of Stable Diffusion 3 [6], further optimizing the original DDIM-based diffusion process. Although some work has addressed the spatial positioning of objects in generated images [13,33], achieving high positional accuracy while preserving recognizability of objects remains a key challenge. Moreover, advances in foundational architectures such as Transformers [28] and GAN optimization techniques [7] have provided essential building blocks for recent diffusion frameworks [20,22].

2.2 Interior Layout Design

Research on interior layout design has increased in recent years. Employing color image modeling and virtual technology improves the dissemination efficiency of 3D interior designs [27]. Vision Transformer (ViT) [29] detects interior images and assesses their authenticity. Image feature extraction algorithms [35] help reduce the high cost and inefficiency of traditional methods. Moreover, Generative Adversarial Networks (GANs) [26] capture the complexity of user preferences in interior layout design. Although significant research exists on interior layout design, effective solutions for addressing object positioning inaccuracies remain relatively underdeveloped. To this end, we incorporate attention mechanisms and diffusion models to the field of interior layout design to address this problem.

3 The Proposed Model

3.1 Overview of Proposed SPADE

In this section, we introduce our model SPADE. SPADE primarily comprises two modules, each specifically designed for distinct phases: the pre-noise prediction stage and the noise prediction process itself. The overall architecture is

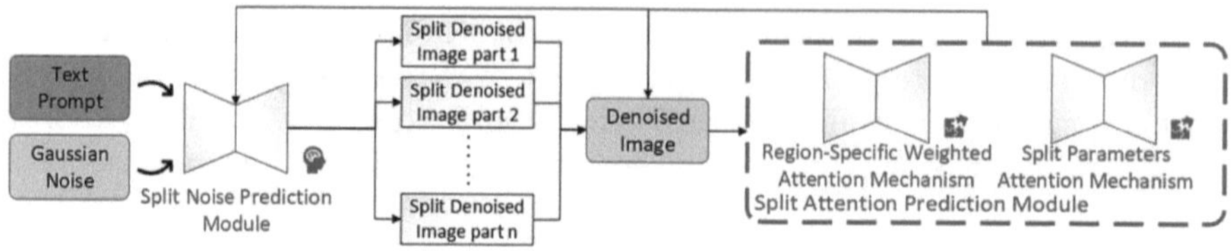

Fig. 2. The architecture diagram of our SPADE.

shown in Fig. 2. First, the diffusion process begins with complete Gaussian noise, which is denoised separately by the Split Noise Prediction Module to roughly determine object placement, thereby enhancing positional accuracy and object recognizability in the generated images. Second, after several steps of Split Noise Prediction, the remaining denoising steps transition to the Split Attention Prediction Module. This further refines positional accuracy while effectively mitigating the resulting decline in overall image integrity introduced by the Split Noise Prediction Module, thus ultimately generating high-fidelity images.

3.2 Split Noise Prediction Module

In this module, the noisy image is segmented, and each segment is individually processed by the U-Net, as shown in the flowchart in Fig. 3, to predict the noise corresponding to its associated object. The segmented noisy images are then reassembled into a complete noisy image, which is fed into the scheduler to predict the next-step image. The Noise Segmentation Coefficient (NSC) represents the proportion of each region relative to the entire image. Let s denote NSC in the Split Noise Prediction Module, and let w denote the width of the latent map. The overlap width is $sw + (1 - s)w - w = (2s - 1)w$. We discuss the following cases depending on the value of s:

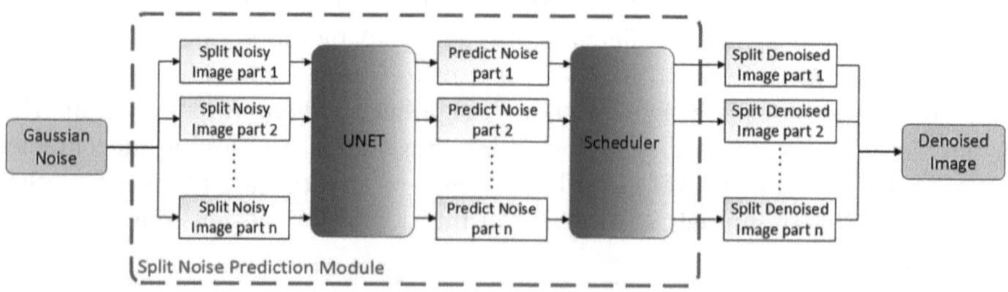

Fig. 3. The flowchart of Split Noise Diffusion.

No Overlap($s \leq 0.5$)*.* Gaps introduce discontinuity through zero-valued parameters, enhancing object positional accuracy at the expense of spatial compression, which may consequently reduce object recognizability.

Critical Overlap$(0.5 < s < 0.625)$. Appropriate overlap ensures smooth blending. When $(2s-1)w \in 0 \sim \frac{w}{4}$, it empirically balances object positional accuracy and recognizability. The predicted noise parameters in the overlapping regions are summed:

$$\epsilon_{merged}(x) = \epsilon_1(x) + \epsilon_2(x), x \in [(1-s)w, sw] \tag{1}$$

Excessive Overlap$(s \geq 0.625)$. When object regions are segmented overly large, these over-summed regions introduce conflicting gradients. This leads to pixel-level artifacts and visual incoherence.

3.3 Split Attention Prediction Module

Region-Specific Weighted Attention Mechanism. To enhance control over specific regions of the image, we introduce a Region-Specific Weighted Matrix $W_{region} \in R^{N*N}$, which adjusts the attention mechanism's Query matrix weights of specific regions during attention computation. The Region Importance Weights (RIW) represent the weighted values assigned to specific regions, with all other regions defaulting to a weight value of 1. This approach aims to fine-tune the focus degree of specific regions. As illustrated in Fig. 4.

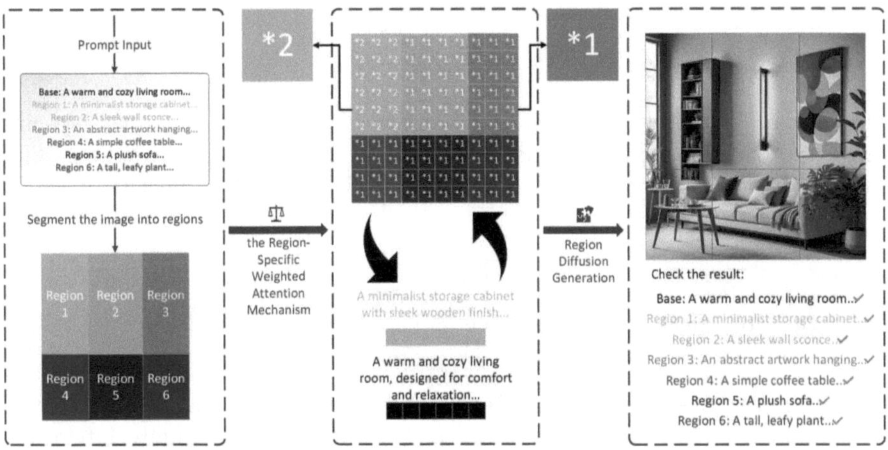

Fig. 4. The illustration of Region-Specific Weighted Attention Mechanism.

The weighted Query matrix is computed as follows:

$$Q_{weighted} = Q \odot W_{region} \tag{2}$$

where $\odot$ represents element-wise multiplication (Hadamard product). The proposed Region-Specific Weighted Matrix W_{region} enables the application of differentiated weights across regions. Specifically, if a region is of higher focus degree, its corresponding weights in W_{region} will be larger, thereby placing greater emphasis on that region during attention computation. The attention weights, now modified by the region-specific weights, are computed as follows:

$$A_{region} = Softmax(\frac{Q_{weighted}K^T}{\sqrt{d_k}}) \tag{3}$$

where A_{region} represents the modified attention weights, incorporating the region-specific weights. The term $\frac{Q_{weighted}K^T}{\sqrt{d_k}}$ computes the similarity between the weighted Query matrix and the Key matrix, where d_k denotes the dimensionality of the hidden states of the region prompt.

The modified attention matrix is then applied to the Value matrix to produce the attention output:

$$Y_{region} = A_{region}V \tag{4}$$

This attention output Y_{region} is characterized by attention concentrated on specific regions. When W_{region} increases in target regions, the corresponding regions in $Q_{weighted}$ are scaled up. This directly increases the dot-product similarity $Q_{weighted}K^T$ for those regions, sharpening the softmax distribution. Consequently, the attention matrix A_{region} becomes more concentrated on the specific regions, suppressing other regions. For two regions with the initial Query matrices s_1 and s_2, their attention weights after scaling by α become:

$$A_{region} = \frac{e^{\alpha s_1}}{e^{\alpha s_1} + e^{\alpha s_2}} \tag{5}$$

Once α reaches a critical value, attention weight parameters for the specific region approach 1, and further scaling yields negligible gradient updates. This critical value can be obtained by taking the derivative with respect to A_{region}:

$$\frac{\partial A_{region}}{\partial \alpha} \approx 0 \tag{6}$$

This saturation effect limits additional gains in positional accuracy.

Split Parameters Attention Mechanism. The Query matrix is cropped by positional matrix w to form A_{obj} components, then regionally integrated into a unified Query matrix. The illustration is shown in the Fig. 5. The Split-Attention Coefficient (SAC) measures the proportion of central truncation within object regions. When SAC is too large, it increases pixel spacing between objects, hindering proximity constraints like "near" or "next to". Since it does not modify the hidden states of the region prompt Value matrix, object features remain identifiable. Therefore, recognizability is largely unaffected by changes in SAC.

The base prompt attention weights A_{base} and region prompt attention weights A_{obj} are fused via a weighting parameter base ratio β:

$$A_{final} = \beta A_{base} + (1 - \beta)A_{obj} \tag{7}$$

Increasing the base ratio reduces object positional accuracy, as the influence of the region prompt weakens. Due to the lack of region-specific modulation in base prompt features, reducing β amplifies the contribution of A_{obj} but introduces conflicting gradients at region boundaries:

$$\frac{\partial \mathcal{L}}{\partial A_{obj}} \approx (1-\beta)\frac{\partial \mathcal{L}}{\partial A_{final}} \tag{8}$$

Large β weakens feature gradients from $Aobj$, reducing regional learning. Therefore, higher base ratios impair object positional accuracy. Split Noise Base Ratio (SNBR) and Split Attention Base Ratio (SABR) refer to base ratios in the denoising of Split Noise and Attention Prediction, respectively. Although $Abase$ and $Aobj$ differ in focus, both contain object features, with $Abase$ also aiding object generation.

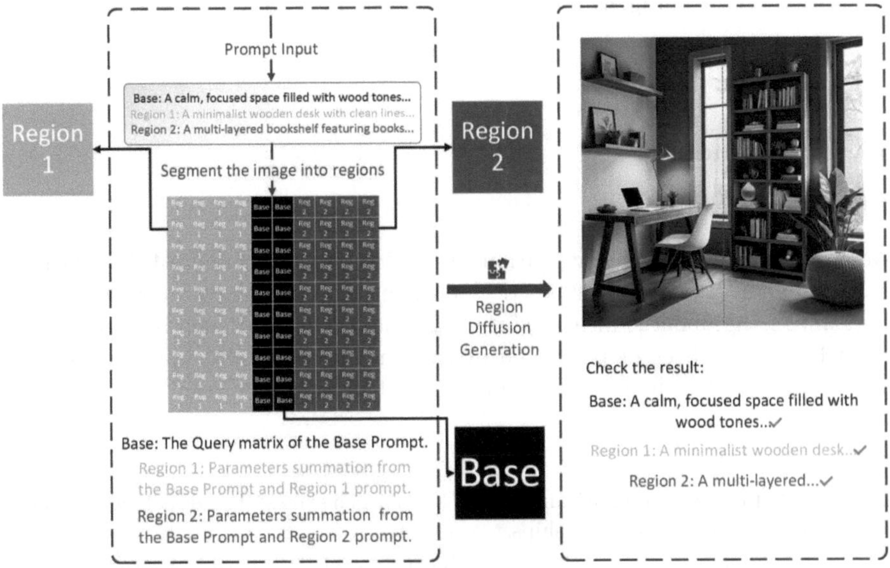

Fig. 5. The illustration of Split Parameters Attention Mechanism.

3.4 Integration Strategy

Split Noise Prediction and Split Attention Prediction are integrated as illustrated in Fig. 6. When determining the optimal step counts for Split Noise Prediction and Split Attention Prediction, we follow these principles: Firstly, the Split Noise Prediction step count should be sufficient to form regional constraints through iterative denoising. Secondly, a proportional number of Split Attention Prediction steps must be retained to balance positional accuracy, object recognizability and overall image integrity. The Split Noise Prediction Step Count (SNSC) refers to the steps assigned to Split Noise Prediction in a 20-step process, with the rest for Split Attention Prediction. We generated 3000 interior layout images, and the results indicate that the model performs best when the SNSC is set to 2 or 3. Employing higher SNSC values may compromise overall image integrity and fall short of achieving complete problem resolution.

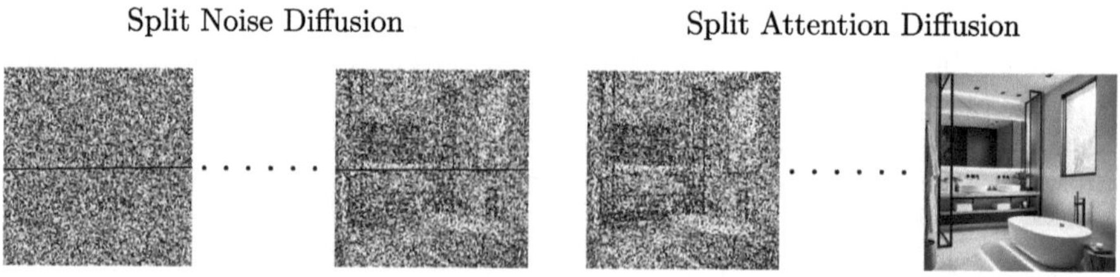

Fig. 6. Illustration of Module Integration.

4 Experiments

4.1 Main Results

We perform quantitative evaluation on a Xeon(R) Platinum 8481C platform featuring an NVIDIA RTX 4090 GPU, utilizing SDXL as the diffusion backbone. Text-image alignment is measured through the T2I-CompBench benchmark [10, 11], which specializes in positional evaluation and capability quantification of multi-object text-to-image generation. Table 1 presents the results.

Table 1. Performance comparison on T2I-CompBench shows SPADE achieves superior results in object spatial relationships.

Model	Object Spatial Relationship Score
Stable Diffusion v2	0.147
PixArt-α	0.191
SDXL	0.212
IterComp	0.250
Recaption, Plan & Generate	0.294
SPADE (ours)	**0.57**

A vibrant space features a star night light on the wall. To its right is an interactive wall. On the floor, a foam puzzle mat lies left, while a bed sits on the right.

Fig. 7. With an increased number of regions, our model generates objects at the target locations, demonstrating superior performance over RPG.

4.2 Increasing Diffusion Region

By subdividing broader regions into smaller, more precise sections, SPADE unlocks a higher level of structural flexibility in interior layout generation. As demonstrated in Fig. 7, this finer granularity captures subtle spatial relationships while preserving object coherence, even under complex prompts. Consequently, SPADE produces layouts that are not only accurate in positioning but also natural, functional, and visually harmonious.

4.3 Extending Diffusion Backbones

Our SPADE exhibits strong adaptability and can seamlessly integrate with a wide range of diffusion backbones, a characteristic that echoes the flexible integration trends in modern diffusion pipelines [2]. The generation results of different diffusion backbones are shown in Fig. 8.

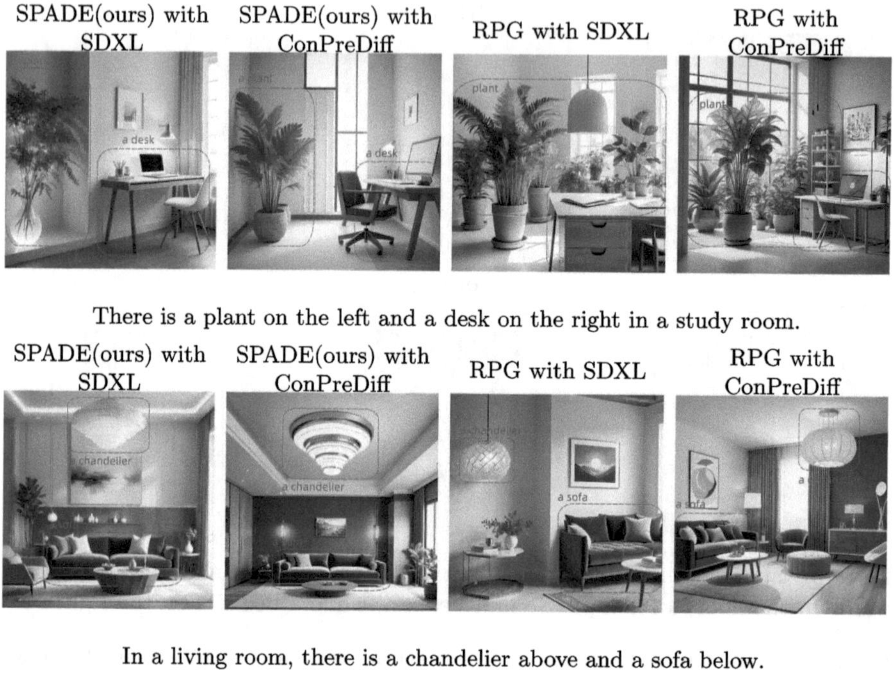

Fig. 8. SPADE can be extended to various diffusion backbones such as SDXL and the recent ConPreDiff, while demonstrating enhanced positioning capability over RPG.

4.4 Model Parameter Optimization

We generated 3000 interior layout images per parameter configuration, all featuring equally distributed two-object layouts. These were evaluated using T2I-CompBench [10,11] to measure object recognizability scores and positional accuracy scores across all generated images. Subsequently, average values were calculated for both metrics, followed by manual inspection of overall image integrity.

For the Noise Segmentation Coefficient (NSC), a value of 0.5 achieved the best balance between positional precision and visual coherence. Lower values constrained spatial realism, while higher values caused gradient conflicts and blurred edges. The 0.5 setting allowed precise placement with natural edges and strong overall fidelity. The trend of the change is illustrated in Fig. 9.

The Region Importance Weights (RIW) performed optimally at approximately 2, providing stronger attention to key regions and improving positional accuracy. Weights below this diluted spatial cues, while higher weights offered little gain and risked attention instability. RIW around 2 effectively balanced precision and stability, as shown in Fig. 10.

For the Split Attention Coefficient (SAC), 0.1 proved ideal. It maintained clear spatial relationships without unnaturally separating objects. Values too low blurred positional distinctions, while higher values disrupted adjacency cues.

SAC at 0.1 offered consistent relational accuracy and visual clarity. The curve of variation is depicted in Fig. 11.

The Split Noise Step Count (SNSC) worked best at 2 or 3 steps in the 20-step process. This provided enough regional anchoring for accurate positioning without over-segmentation. Lower counts reduced accuracy, while higher counts harmed global image coherence. The trend is displayed in Fig. 12.

Balancing the Split Noise Base Ratio (SNBR) and Split Attention Base Ratio (SABR) was key. Setting SNBR at 0.2 and SABR at 0.8 ensured high positional fidelity and visual realism. These values stabilized gradient behavior while harmonizing global attention effects. As illustrated in Fig. 13 and 14.

We identified six optimal parameter configurations for two-object layout images, which consistently deliver superior accuracy, stability, and visual coherence. These configurations remain highly effective for the same scenario, even after switching to different datasets, demonstrating strong generalization capability and robustness across diverse generation environments.

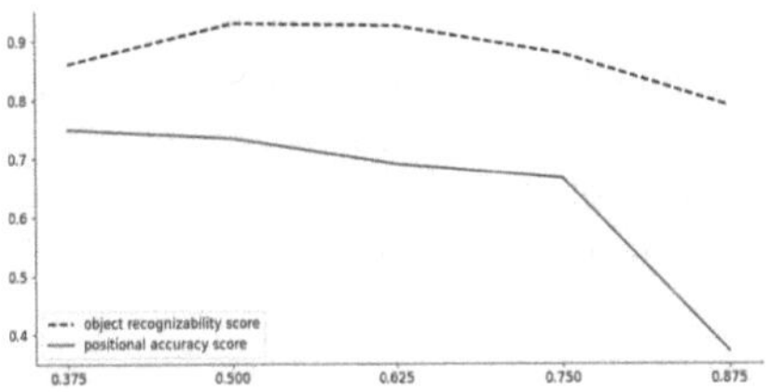

Fig. 9. SPADE performance curve with varying NSC.

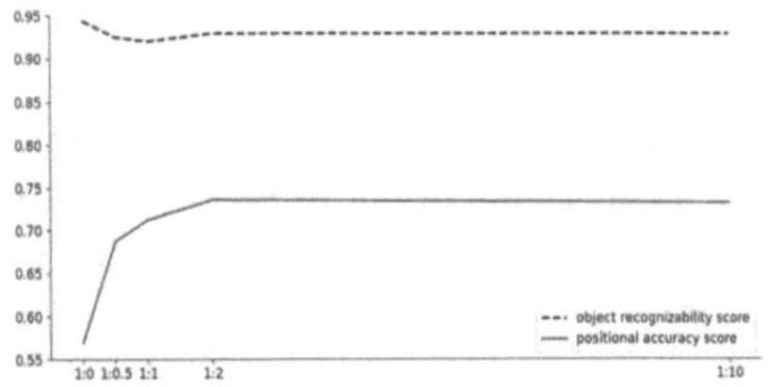

Fig. 10. SPADE performance curve with varying RIW.

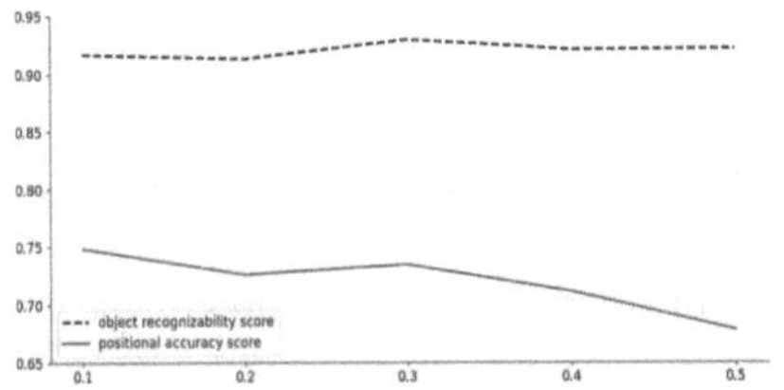

Fig. 11. SPADE performance curve with varying SAC.

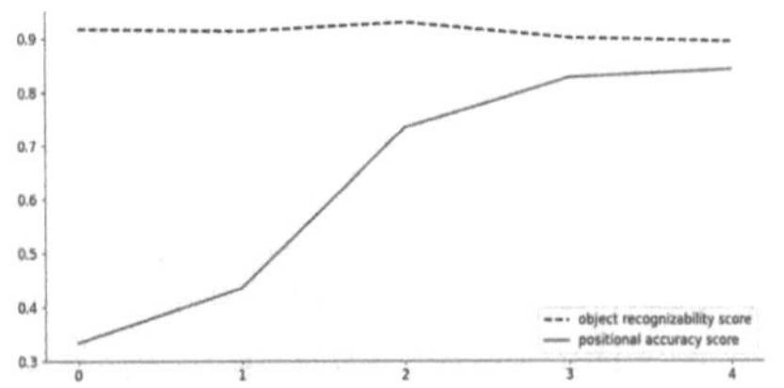

Fig. 12. SPADE performance curve with varying SNSC.

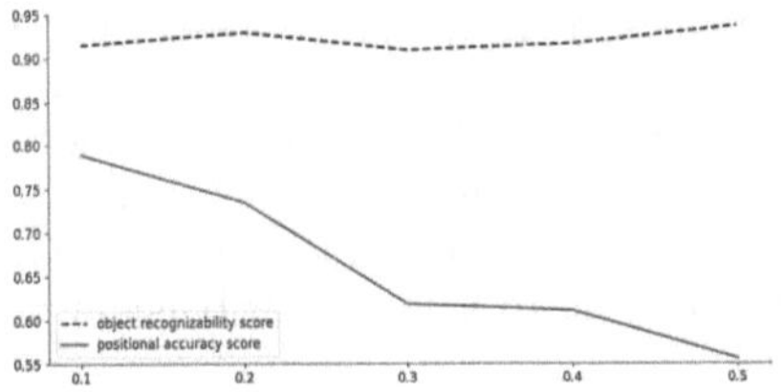

Fig. 13. SPADE performance curve with varying SNBR.

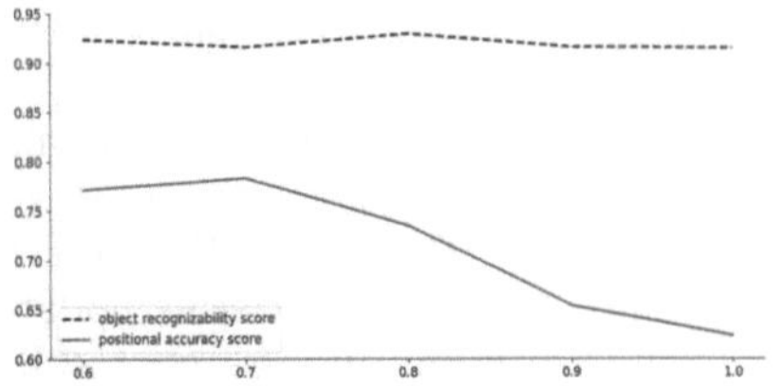

Fig. 14. SPADE performance curve with varying SABR.

5 Conclusion

In this paper, we have proposed a novel diffusion model called SPADE for interior layout design. We have divided the diffusion procedure into Split Noise Prediction and Split Attention Prediction stages. Furthermore, we have introduced two improved attention mechanisms: Region-Specific Weighted Attention Mechanism and Split Parameters Attention Mechanism, both in the Split Attention Prediction stage. Our model has enhanced positional accuracy of generated objects while preserving object recognizability and overall image integrity. We have evaluated qualitative and quantitative performance using a large dataset of generated interior layout images, obtaining competitive results. For future research, we posit that integrating reasoning and adaptation techniques from large language models, such as chain-of-thought prompting [31], step-by-step reasoning [32], and efficient fine-tuning approaches like LoRA [8], alongside the capabilities of emerging foundation models [5], may further enhance spatial reasoning and controllability in multi-modal generation tasks.

Acknowledgments. The project was supported by Scientific Research Fund of Hunan Provincial Education Department, 24B0308, and Scientific research project of Hunan Provincial Department of Science and Technology, 2024RC3165, 2025JJ50409, and the National Natural Science Foundation of China (Grant No. 62372064).

References

1. Betker, J., et al.: Improving image generation with better captions. Comput. Sci. **2**(3), 8 (2023). https://cdn.openai.com/papers/dall-e-3.pdf
2. Cao, J., Liu, J., Kitani, K., Zhou, Y.: Multi-modal diffusion for hand-object grasp generation. arXiv preprint arXiv:2409.04560 (2024)
3. Chen, C., et al.: Diffusion models for multi-modal generative modeling. arXiv preprint arXiv:2407.17571 (2024)
4. Cheng, L., Li, S.: Diffuspoll: Conditional text diffusion model for poll generation. In: Findings of the Association for Computational Linguistics ACL 2024, pp. 925–935 (2024)
5. Dubey, A., et al.: The llama 3 herd of models. arXiv preprint arXiv:2407.21783 (2024)
6. Esser, P., et al.: Scaling rectified flow transformers for high-resolution image synthesis. In: Forty-first International Conference on Machine Learning (2024)
7. Heusel, M., Ramsauer, H., Unterthiner, T., Nessler, B., Hochreiter, S.: GANs trained by a two time-scale update rule converge to a local NASH equilibrium. Adv. Neural Inf. Process. Syst. **30** (2017)
8. Hu, E.J., et al.: Lora: low-rank adaptation of large language models. ICLR **1**(2), 3 (2022)
9. Hu, X., Li, S., Ying, Q., Peng, W., Zhang, X., Qian, Z.: Establishing robust generative image steganography via popular stable diffusion. IEEE Trans. Inf. Forensics Secur. (2024)

10. Huang, K., Duan, C., Sun, K., Xie, E., Li, Z., Liu, X.: T2i-compbench++: an enhanced and comprehensive benchmark for compositional text-to-image generation. IEEE Trans. Pattern Anal. Mach. Intell. (2025)
11. Huang, K., Sun, K., Xie, E., Li, Z., Liu, X.: T2i-compbench: a comprehensive benchmark for open-world compositional text-to-image generation. Adv. Neural. Inf. Process. Syst. **36**, 78723–78747 (2023)
12. Huang, Z., Chan, K.C., Jiang, Y., Liu, Z.: Collaborative diffusion for multi-modal face generation and editing. In: Proceedings of the IEEE/CVF Conference on Computer Vision and Pattern Recognition, pp. 6080–6090 (2023)
13. Jia, C., Luo, M., Dang, Z., Dai, G., Chang, X., Wang, M., Wang, J.: Ssmg: spatial-semantic map guided diffusion model for free-form layout-to-image generation. In: Proceedings of the AAAI Conference on Artificial Intelligence. vol. 38, pp. 2480–2488 (2024)
14. Kong, C., Kwak, N.: Analyzing multimodal objectives through the lens of generative diffusion guidance. arXiv preprint arXiv:2302.10305 (2023)
15. Li, R., Li, W., Yang, Y., Wei, H., Jiang, J., Bai, Q.: Swinv2-imagen: hierarchical vision transformer diffusion models for text-to-image generation. Neural Comput. Appl. **36**(28), 17245–17260 (2024)
16. Li, W., Xu, X., Liu, J., Xiao, X.: Unimo-g: unified image generation through multimodal conditional diffusion. arXiv preprint arXiv:2401.13388 (2024)
17. Li, W., et al.: Upainting: unified text-to-image diffusion generation with cross-modal guidance. arXiv preprint arXiv:2210.16031 (2022)
18. Lipman, Y., Chen, R.T., Ben-Hamu, H., Nickel, M., Le, M.: Flow matching for generative modeling. arXiv preprint arXiv:2210.02747 (2022)
19. Liu, X., Gong, C., Liu, Q.: Flow straight and fast: learning to generate and transfer data with rectified flow. arXiv preprint arXiv:2209.03003 (2022)
20. Ma, Y., Yang, H., Wang, W., Fu, J., Liu, J.: Unified multi-modal latent diffusion for joint subject and text conditional image generation. arXiv preprint arXiv:2303.09319 (2023)
21. Minello, G., Bicciato, A., Rossi, L., Torsello, A., Cosmo, L.: Graph generation via spectral diffusion. arXiv preprint arXiv:2402.18974 (2024)
22. Nair, N.G., Valanarasu, J.M.J., Patel, V.M.: Maxfusion: Plug&play multi-modal generation in text-to-image diffusion models. In: European Conference on Computer Vision, pp. 93–110. Springer (2024)
23. Ngasa, E.E., Jang, M.A., Tarimo, S.A., Woo, J., Shin, H.B.: Diffusion-based Wasserstein generative adversarial network for blood cell image augmentation. Eng. Appl. Artif. Intell. **133**, 108221 (2024)
24. Rombach, R., Blattmann, A., Lorenz, D., Esser, P., Ommer, B.: High-resolution image synthesis with latent diffusion models. In: Proceedings of the IEEE/CVF Conference on Computer Vision and Pattern Recognition, pp. 10684–10695 (2022)
25. Ruiz, N., Li, Y., Jampani, V., Pritch, Y., Rubinstein, M., Aberman, K.: Dreambooth: fine tuning text-to-image diffusion models for subject-driven generation. In: Proceedings of the IEEE/CVF Conference on Computer Vision and Pattern Recognition, pp. 22500–22510 (2023)
26. Sakabe, K., Ono, K., Adamidis, P., Masuda, N.: Generating interior images with latent user preferences through GANs. In: 2024 IEEE Congress on Evolutionary Computation (CEC), pp. 1–7. IEEE (2024)
27. Sun, L., Tan, D.: Distributed 3d interior environment design system based on color image model. Informatica **49**(10) (2025)
28. Vaswani, A., et al.: Attention is all you need. Adv. Neural Inf. Process. Syst. **30** (2017)

29. Wang, H.: Vision transformer-based framework for AI-generated image detection in interior design. Informatica **49**(16) (2025)
30. Wang, W., et al.: A two-stage generative model with cycle GAN and joint diffusion for MRI-based brain tumor detection. IEEE J. Biomed. Health Inform. (2024)
31. Wei, J., et al.: Chain-of-thought prompting elicits reasoning in large language models. Adv. Neural. Inf. Process. Syst. **35**, 24824–24837 (2022)
32. Xu, G., Jin, P., Hao, L., Song, Y., Sun, L., Yuan, L.: Llava-cot: let vision language models reason step-by-step, (2024). https://arxiv.org/abs/2411.10440
33. Yang, L., Yu, Z., Meng, C., Xu, M., Ermon, S., Bin, C.: Mastering text-to-image diffusion: recaptioning, planning, and generating with multimodal LLMs. In: Forty-first International Conference on Machine Learning (2024)
34. Yang, L., et al.: Cross-modal contextualized diffusion models for text-guided visual generation and editing. In: The Twelfth International Conference on Learning Representations (2024)
35. Zhao, Y.: Interior space design method considering image feature extraction algorithms. IEEE Access (2024)
36. Zhu, H., Xiao, T., Honavar, V.G.: 3m-diffusion: latent multi-modal diffusion for text-guided generation of molecular graphs. arXiv e-prints pp. arXiv–2403 (2024)

Question-Aware Spatial-Temporal Reasoning in Patch for Audio-Visual Question Answering

Feifei Xu, Wenjing Zhu, Dongyang Li(✉), and Puzhe Li

Shanghai University of Electric Power, Shanghai, China
{xufeifei,dongyangli.ldy}@shiep.edu.cn,
{y24208145,lipuzhe}@mail.shiep.edu.cn

Abstract. Audio-Visual Question Answering (AVQA) aims to answer textual questions by perceiving, understanding and integrating audio-visual information, which requires not only question-aware multimodal reasoning, but also precise spatial-temporal modeling to capture subtle dynamics. However, existing methods often rely on discrete, coarse-grained Top-K frame selection, ignoring temporal continuity. Some of them introduce question semantics in the early stage to enhance detail focus, but their guiding effect diminishes during reasoning. Although recent graph networks improve fine-grained modeling, the static connections still constrain dynamic adaptation to specific questions. To address these limitations, we propose QSTP, a framework that constructs patch-level spatial-temporal graphs, dynamically adjusts neighbor patch weights via question-aware aggregation, and continuously attends to information highly relevant to the question. We design an audio-driven visual gating module to highlight audio-related regions for better audio source localization, and a question-guided fusion strategy is put forward to further refine cross-modal alignment before answer prediction. Extensive experiments on the MUSIC-AVQA and MUSIC-AVQA-R datasets demonstrate that QSTP achieves state-of-the-art performance across various complex reasoning tasks.

Keywords: Audio-Visual Question Answering · Spatio-Temporal Graph Reasoning · Multimodal Fusion · Question-Aware

1 Introduction

Audio-Visual Question Answering (AVQA) is an attractive task in multimodal learning, aiming to comprehensively understand information from audio, visual, and textual modalities. As shown in Fig. 1, given a long audible video and a related question, AVQA requires accurately capturing key audio-visual cues relevant to the question in order to predict the correct answer for the given question.

AVQA methods can be broadly divided into two categories: (1) **Mainstream efficient expert models.** For example, Li et al. [11] employ a Top-K mechanism to select the time segments most relevant to the question, but the selection

J. Lokoč et al. (Eds.): MMM 2026, LNCS 16413, pp. 631–645, 2026.
https://doi.org/10.1007/978-981-95-6957-1_45

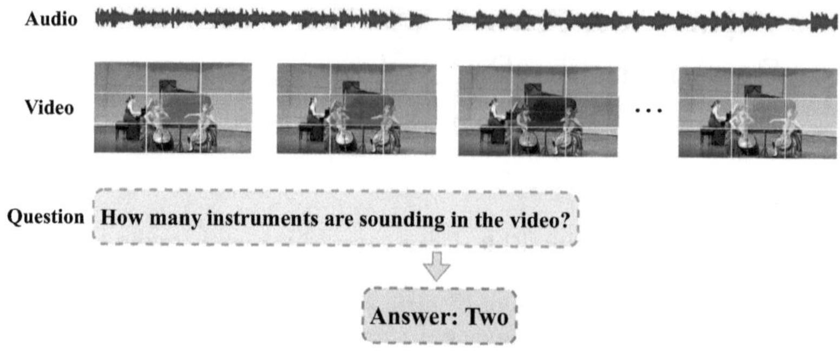

Fig. 1. An Example of Audio-Visual Question Answering task.

process relies solely on visual cues and uses discrete sampling, ignoring temporal continuity. Kim et al. [9] propose a MoE model that adaptively assigns weights to both continuous and non-continuous time spans, enabling better modeling of complex temporal dependencies. However, it still operates mainly at the frame level or on coarse-grained regions, lacking sustained attention to local details, and the guidance of question semantics gradually weakens during reasoning. Li et al. [15] build multiple graph network structures to model cross-modal relationships, but their graph connections are mostly predefined and lack dynamic adaptation to question semantics, making it difficult to precisely align key information relevant to the question. (2) **Large-scale pretraining models.** For example, VALOR [2] introduces a unified high-level audio-visual-language representation and demonstrates strong transferability to AVQA tasks. However, such methods typically require large amounts of multimodal data for fine-tuning, along with massive parameter sizes and high computational costs.

To address the above challenges in AVQA, we propose a patch-level question-aware dynamic spatial-temporal reasoning framework (QSTP), which follows the mainstream branch, aiming to build an expert AVQA model with efficient and powerful modules while using only limited training data. QSTP suggests a fine-grained, question-aware dynamic reasoning framework that jointly models the spatial structure and temporal relations of the graph to precisely locate key multimodal information. Specifically, instead of treating each video frame as a whole, we directly construct a spatial-temporal graph over fine-grained visual patches. More importantly, we design a novel question-aware aggregation mechanism, which dynamically adjusts the weights of neighboring patch nodes during each layer of graph propagation according to question semantics. This allows each node to prioritize aggregating spatial-temporal information that is highly relevant to the question, thereby gradually forming a spatial-temporal context closely aligned with question semantics and enabling more precise cross-modal reasoning.

Furthermore, we regard audio features as dynamic probes and employ a cross-modal attention mechanism to compute frame-wise audio-visual alignment scores for all visual patches. This highlights visual regions closely related to sound events while suppressing irrelevant background noise. Finally, all processed multimodal features are fused under question guidance to obtain the answer.

In summary, our main contributions are as follows:

- We propose QSTP, a patch-level question-aware dynamic spatial-temporal reasoning framework for fine-grained multimodal reasoning and precise modeling of subtle spatial-temporal dynamics.
- we construct spatial-temporal patch graphs and dynamically adjust neighboring node weights via a question-aware aggregation mechanism, enabling continuous focus on information highly relevant to the question while jointly modeling spatial structure and temporal dependencies.
- We design an audio-driven visual gating module to emphasize audio-related regions for better audio source localization, and develop a question-guided fusion strategy to further refine cross-modal alignment before answer prediction.
- QSTP achieves state-of-the-art performance on two datasets and, with only limited training data and model parameters, delivers comparable performance to large-scale pretrained models.

2 Related Work

2.1 Audio-Visual Question Answering.

The Audio-Visual Question Answering (AVQA) task [2,16] aims to exploit the multimodal characteristics of audio and video to acquire accurate answers to textual questions and achieve comprehensive cross-modal understanding. With the introduction of benchmark datasets such as MUSIC-AVQA [13], AVQA [29], and Pano-AVQA [31], researches on fine-grained reasoning [12,24] for long-duration and panoramic videos have developed rapidly. Early approaches mainly relied on global audio-visual representations and adopted a two-stage pipeline to perform spatial and temporal semantic alignment separately [11,13,29], which ignored maintaining the temporal continuity, making it difficult to capture dynamic changes across frames. TSPM [11] and QA-TIGER [9] introduced question semantics to locate key temporal segments, but question guidance tended to diminish during subsequent reasoning, which was not conducive to fine-grained temporal reasoning and audio-visual alignment. PSOT [15] built multiple graph structures for cross-modal relationship modeling, however, its graph connectivity was often predefined, constraining adaptability to diverse questions. There were also works exploring AVQA from the perspectives of dataset bias [10,17,18], missing modalities [20,21], large-scale pretraining [4,25,26], and adapter-based architectures [3,16]. Although significant progress has been made, existing methods still face challenges in sustaining question guidance, and modeling temporal continuity. To address these limitations, we target fine-grained patch-level spatial-temporal modeling and embed question semantics throughout multi-layer graph reasoning, enabling precise cross-modal alignment and progressive reasoning.

2.2 Graph Neural Networks in QA Tasks.

Graph Neural Networks (GNNs), with their strong capability in modeling relational structures and enabling information propagation among nodes, have been widely applied in traditional QA tasks, such as visual QA [19,28], video QA [4,27], and text-based QA [7,22]. In AVQA, mainstream approaches predominantly adopt Transformer architectures, with some exploring Mamba [8,30]. However, existing graph-based methods often exhibit limited adaptability to question semantics, restricting their effectiveness in complex reasoning. Our approach integrates a GNN module with a question-aware mechanism, deeply embedding question semantics into the graph reasoning to enhance temporal event inference, achieving competitive performance in complex AVQA scenarios.

3 Method

QSTP achieves reliable audio-visual reasoning by explicitly modeling patch-level spatial-temporal relations with a question-aware graph network, which is then dynamically aligned and filtered by the audio modality. Figure 2 illustrates the overall architecture.

3.1 Input Representation.

The input video sequence is uniformly split into T non-overlapping $1s$ segments, each containing corresponding audio and visual information.

Visual Representation. Each visual segment is processed using a pretrained CLIP. Specifically, the visual input of each segment is divided into M patches, with a special [CLS] token added at the beginning of the sequence to capture global semantic information. Then, features are extracted in two forms: Frame-level features $v = \{v^t\}_{t=1}^{T} \in R^{T\times D}$ are obtained from the [CLS] token output for each segment t, representing the global semantics of the frame. Patch-level features $p = \{p^t\}_{t=1}^{T} \in R^{T\times M'\times D}$ are derived by compressing the original M patch tokens in each segment into M' tokens using Token Merging (ToMe) [1], effectively retaining more fine-grained spatial detail.

Audio Representation. For each segment t , we use VGGish [6] pretrained on AudioSet [5] to extract audio features $a^t \in R^D$. The complete sequence of audio features is represented as $a = \{a^t\}_{t=1}^{T} \in R^{T\times D}$.

Question Representation. The tokenized input question is processed using the CLIP text encoder to extract the sentence-level features $q_s \in R^D$ and the word-level features $q_w \in R^{N\times D}$, where N is the number of tokens including padding.

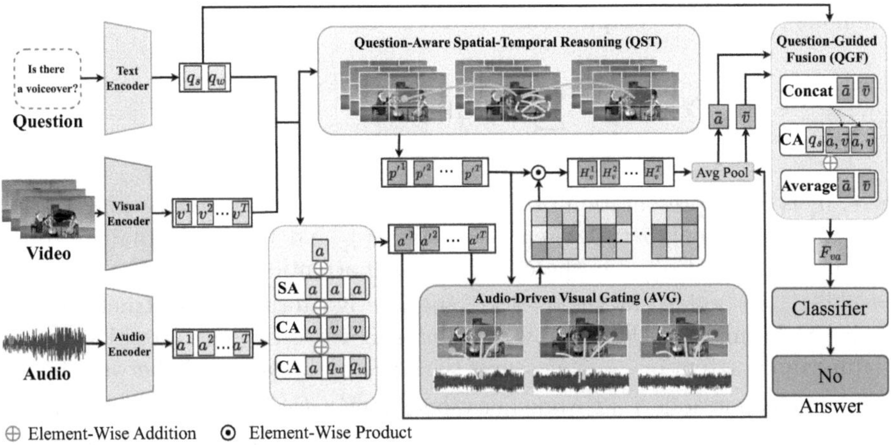

Fig. 2. The overall architecture of QSTP. The QST module integrates question semantics into the visual modality to progressively enhance spatial-temporal modeling. The AVG module leverages audio signals to highlight relevant visual regions. Finally, the QGF module fuses multimodal representations under question guidance to produce the final answer prediction.

3.2 Question-Aware Spatial-Temporal Reasoning

To effectively capture fine-grained spatial structures and dynamic temporal relations in videos, we propose a question-aware spatial-temporal reasoning module. Guided by question semantics, this module explicitly constructs a spatial-temporal graph for the visual modality to model intra-frame spatial dependencies and inter-frame temporal dynamics, while progressively enhancing the semantic representation of the audio modality. This design facilitates the integration of question information across modalities, enabling more effective multimodal reasoning.

Specifically, we first construct a visual spatial-temporal graph, denoted as $\mathcal{G}_v = (\mathcal{V}_v, \mathcal{E}_v)$, where the nodes $\mathcal{V}_v$ consist of patch-level features p extracted from each frame of the video sequence. The edges $\mathcal{E}_v$ encode both spatial and temporal relations between patches, including two types of connections: Spatial edges connect all patches within the same frame at time step t to model spatial layout and semantic relations across different visual regions. Temporal edges link patches at the same spatial position across adjacent frames t and $t+1$ to capture temporal continuity and dynamics.

To incorporate question semantics into visual reasoning, we design a question-aware graph attention network with L stacked layers. This network leverages a multi-head attention mechanism to dynamically adjust neighbor weights based on question relevance, enabling each visual patch node to aggregate information that is most relevant to the question during feature updating. At each layer l, the node i update is computed as:

$$z_i^{l+1} = W_o \cdot \text{Concat}\left(\left[\sum_{j\in\mathcal{N}(i)} \alpha_{ij,\mathcal{H}}^l W_{\mathcal{H}}^l h_j^l\right]_{\mathcal{H}=1}^{H}\right), \quad h_i^{l+1} = z_i^{l+1} + h_i^l, \tag{1}$$

here, h_i^l is the representation of node i at layer l, $\mathcal{N}(i)$ denotes its neighboring nodes, H is the number of attention heads, $W_{\mathcal{H}}^l$ and W_o are learnable projection matrices, and $\alpha_{ij,\mathcal{H}}^l$ is the question-aware attention weight. The attention weight $\alpha_{ij,\mathcal{H}}^l$ is guided by the question and jointly captures both structural similarity and semantic relevance by incorporating three types of attention scores: (1)Inter-node attention $(\hat{n}_{ij}^l)$ measures the structural relationship between node i and its neighbor node j. (2)Node-question attention$(\hat{h}_{iq}^l)$ measures the semantic relevance between node i and question q_s. (3)Question-neighbor attention$(\hat{t}_{qj}^l)$ measures the semantic relevance between question q_s and neighbor node j. These scores are computed as follows:

$$\hat{n}_{ij}^l = \frac{Q_i^l \cdot K_j^l}{\sqrt{d_k}}, \; \hat{h}_{iq}^l = \frac{Q_i^l \cdot K_q^l}{\sqrt{d_k}}, \; \hat{t}_{qj}^l = \frac{Q_q^l \cdot K_j^l}{\sqrt{d_k}}, \tag{2}$$

where d_k is the dimensionality of the key vectors. The corresponding query and key vectors are generated through linear projections as follows: For each graph node i, the projections are defined as $Q_i^l = W_Q^l h_i^l, K_i^l = W_K^l h_i^l$, where W_Q^l and W_K^l are learnable parameter matrices. For the input sentence-level embedding q_s, the projections are defined as $Q_q^l = W_Q'^l q_s, K_q^l = W_K'^l q_s$, where $W_Q'^l$ and $W_K'^l$ are also learnable matrices.

Next, we fuse the three attention scores via a tunable hyperparameter $\gamma \in (0, 0.5)$ weighted sum and softmax-normalize over neighbors to obtain the final attention weight $\alpha_{ij,\mathcal{H}}^l$:

$$A_{ij}^{\mathcal{H}} = (1-2\gamma)\hat{n}_{ij}^l + \gamma\hat{h}_{iq}^l + \gamma\hat{t}_{qj}^l, \; \alpha_{ij,\mathcal{H}}^l = \frac{\exp(A_{ij}^{\mathcal{H}})}{\sum_{v\in\mathcal{N}(i)} \exp(A_{iv}^{\mathcal{H}})}. \tag{3}$$

By stacking L such modules, each visual patch node can progressively integrate spatial-temporal context highly relevant to question semantics. The node representations from the final layer $h^{(L)}$ constitute the set of question-aware spatial-temporal patch features p' for the following steps.

Concurrently, the audio features a are enhanced through a multi-head attention fusion module that progressively refines the representation. First, a multi-head Self-Attention (SA) layer captures internal temporal dependencies by using a as the query, key, and value. Next, the audio representation is refined through two rounds of multi-head Cross-Attention (CA): In the first round, it uses visual features v as the key and value, with a as the query; In the second round, word-level question feature q_w is applied as the key and value to align with question information. Then, the outputs from the SA and two CA layers are combined with residual connections to generate the question-aware audio features

a', enhancing temporal coherence and retainning alignment with the question, as follows:

$$a' = a + SA(a, a, a) + CA(a, v, v) + CA(a, q_w, q_w). \tag{4}$$

This mechanism effectively improves the temporal coherence of audio features and the alignment with the question, collaborating with the graph reasoning module to provide a richer foundation for multimodal fusion.

3.3 Audio-Driven Visual Gating.

We propose an audio-driven visual gating module to dynamically highlight the most relevant visual patches per frame, enabling fine-grained alignment between audio and vision. Specifically, at each time step t, the question-aware audio features a'^t and the corresponding enhanced visual patch features p'^t are projected into a shared space to obtain query and key vectors: $Q_a^t = W_Q^t a'^t$, $K_p^t = W_K^t p'^t$, where W_Q^t and W_K^t are learnable projection matrices. We then employ the scaled dot-product attention to compute the audio-visual alignment score S_{align}^t:

$$S_{align}^t = \text{Softmax}\left(\frac{Q_a^t \cdot {K_p^t}^\top}{\sqrt{d_k}}\right), \tag{5}$$

where d_k is the key dimension, and S_{align}^t denotes the attention weights over visual patches in frame t guided by the corresponding audio. The complete alignment matrix is formed by concatenating attention weights across all time steps: $S_{align} = \{S_{align,t}\}_{t=1}^T$. To further emphasize audio-relevant visual regions, we apply a gating mechanism:

$$H_v = p' \odot S_{align}^T, \tag{6}$$

where $\odot$ denotes element-wise product, p' represents the visual patch features across all frames, and $S_{align}^\top$ is the transposed alignment matrix after dimension expansion. Finally, the resulting visual representation H_v exhibits improved semantic focus and cross-modal consistency, offering more accurate and reliable features for subsequent multimodal fusion and answer prediction.

3.4 Question-Guided Fusion and Answer Prediction

Through the above modules, we obtain enhanced audio a' and patch-level visual features H_v. For visual modality, the segment-level feature $\bar{v}$ can be obtained from H_v via average pooling operation on the spatial dimension. Concurrently, the enhanced audio sequence a' is also pooled over the temporal dimension to yield a global audio feature $\bar{a}$. Next, we fuse them to derive the final audio-visual representation F_{va}:

$$F_{va} = \text{CA}(q_s, [\bar{a}, \bar{v}], [\bar{a}, \bar{v}]) + \text{Avg}(\bar{a}, \bar{v}), \tag{7}$$

where $[\cdot, \cdot]$ denotes concatenation and Avg represents element-wise averaging. This step ensures that both audio and visual features contribute to the final representation in a question-guided manner. Finally, the fused representation F_{va} is fed into a linear classifier followed by a Softmax layer to perform answer prediction, yielding a probability distribution over C candidate answer classes. The model is trained using cross-entropy loss: $\mathcal{L}_{qa} = -\sum_{c=1}^{C} y_c \log \hat{y}_c$, with the answer selected as the class with the highest probability.

4 Experiments

4.1 Datasets

Our experiments are conducted on two most commonly used datasets: MUSIC-AVQA [13] and its extended version MUSIC-AVQA-R [18]. MUSIC-AVQA is a comprehensive benchmark specifically designed for audio-visual question answering (AVQA), consisting of 9,288 music performance videos (over 150 h), 22 musical instruments, and more than 45K question-answer pairs. The questions are categorized into three types based on different modalities: audio-visual, visual-only, and audio-only, with audio-visual questions accounting for the majority and spanning five reasoning types: existential, counting, location, comparative, and temporal. The dataset is split into training, validation, and test sets.The extended MUSIC-AVQA-R dataset further introduces 211,572 restructured questions, with a particular emphasis on rare cases and out-of-distribution samples, significantly enhancing the evaluation of models in terms of spatial-temporal reasoning and complex multimodal interaction robustness.

4.2 Evaluation Metric

We evaluate the model performance on each question type and use answer prediction accuracy (%) as the metric [9,15], calculated as the ratio of correct answers to the total number of questions for each type.

4.3 Implementation Details

The video is sampled at 1 fps, with audio features extracted using VGGish [6], while visual features and questions are processed using the CLIP-ViT-L/14 encoder. The initial patch-level visual tokens are reduced using the ToMe [1] technique to improve computational efficiency. To ensure cross-modal consistency, all features are linearly projected into a common 512-dimensional space. For model configuration, the number of question-aware spatial-temporal graph layers L and the hyperparameter γ are empirically set to 3 and 0.2, respectively. Each multi-head attention module is configured with 8 heads and 0.1 dropout probability. The model is trained using the Adam optimizer with an initial learning rate of 1e-4, which decays by a factor of 0.1 every 8 epochs. The batch size and the number of training epochs are set to 32 and 15, respectively. Results are averaged over three test runs. All experiments are conducted on a single NVIDIA RTX 4090 GPU.

Table 1. Experimental results on the MUSIC-AVQA test set. The top-2 results are highlighted.

Method	Audio QA			Visual QA			Audio-Visual QA						Avg.
	Count	Comp	Avg.	Count	Local	Avg.	Exist	Count	Local	Comp	Temp	Avg.	
AVSD [23]	72.47	62.46	68.78	66.00	74.53	70.31	80.77	57.93	64.03	62.85	61.07	65.44	67.32
Pano-AVQA [31]	75.71	65.99	72.13	70.51	75.76	73.16	82.09	61.30	65.38	63.67	62.04	66.97	69.53
AVST [13]	77.78	67.17	73.87	73.52	75.27	74.40	82.49	69.88	64.24	64.67	65.82	69.53	71.59
COCA [10]	79.84	67.68	75.42	75.10	75.43	75.27	**83.50**	66.64	69.67	64.12	65.57	69.74	72.21
LAVISH [16]	82.10	65.49	75.98	78.95	81.47	80.22	81.68	75.49	66.09	63.76	68.00	71.25	74.47
AV-Mamba [8]	81.61	62.79	74.67	78.86	79.67	79.27	82.49	74.94	65.33	61.31	68.13	70.62	73.63
SH-Mamba [30]	82.30	63.64	75.42	78.53	81.31	79.93	82.89	72.65	67.93	61.31	68.37	70.64	74.12
PSTP [12]	73.94	65.66	70.89	77.11	77.39	77.25	76.21	72.25	**71.85**	**71.75**	68.98	72.57	73.52
APL [14]	82.40	**70.71**	78.09	76.52	82.78	79.69	83.00	73.28	66.63	64.76	65.94	70.96	74.53
TSPM [11]	83.58	63.97	76.35	81.70	84.65	83.20	81.28	77.15	70.87	66.12	70.44	73.35	76.49
QA-TIGER [9]	84.07	67.00	77.78	83.54	85.31	84.43	82.29	77.94	71.63	63.22	68.61	72.96	76.85
PSOT [15]	84.37	65.15	77.28	83.63	85.63	84.64	82.49	78.26	71.52	66.58	71.29	73.59	77.17
QSTP	**85.05**	68.18	**78.83**	**85.05**	**87.02**	**86.04**	82.19	**78.50**	71.74	66.49	**71.41**	**74.25**	**78.19**

4.4 Quantitative Results and Analysis

To evaluate our model's effectiveness, we evaluated QSTP against existing AVQA methods across Audio (A-QA), Visual (V-QA), and Audio-Visual (AV-QA) tasks, considering both question types and overall averages. QSTP was trained on the MUSIC-AVQA training set and evaluated using its validation set on both the MUSIC-AVQA and MUSIC-AVQA-R datasets.

MUSIC-AVQA. As shown in Table 1, compared with works from the mainstream AVQA branch, QSTP achieved an overall accuracy of 78.19%, surpassing the previous best method PSOT (77.17%), and obtained 78.50% and 71.41% in the complex reasoning tasks AV-Counting and AV-Temporal, respectively, demonstrating significant performance advantages. These results verify the effectiveness of our QSTP model.

In addition, compared with models adopting the emerging Mamba architecture, such as AV-Mamba (73.63%) and SH-Mamba (74.12%), QSTP achieved notable performance gains of 4.5% and 4.07%, respectively, indicating its superior capabilities in modeling audio-visual spatial-temporal dependencies and question-semantic associations.

We further compared QSTP with large-scale pretrained LLM-based methods. Although models such as VALOR [2], pretrained on 33.5M tri-modal samples, achieve an accuracy of 78.50%, they have a parameter size of 593M and heavily rely on high-quality audio-visual-text data. In contrast, our method achieves comparable performance with only 9k samples and 142M parameters, fully demonstrating the efficiency and practicality of the QSTP model for specific downstream tasks.

Table 2. Experimental results on the MUSIC-AVQA-R test set for (a) Head (frequent) and (b) Tail (rare) answer categories. † indicates results reproduced using official codes.

Method	Audio QA		Visual QA		Audio-Visual QA					Avg.
	Count	Comp	Count	Local	Exist	Count	Local	Comp	Temp	
	(a) Head									
AVSD[23]	54.12	60.65	60.03	74.51	66.26	46.17	32.65	51.77	**40.29**	54.14
Pano-AVQA[31]	50.96	50.54	46.89	67.36	52.29	52.01	44.15	45.74	34.97	49.93
AVST[13]	56.42	62.34	59.84	63.31	73.38	48.29	35.34	**53.25**	40.18	54.62
TSPM† [11]	81.65	67.66	78.29	80.58	69.15	77.09	42.24	52.07	39.23	67.76
QA-TIGER† [9]	82.67	71.75	81.30	**84.76**	72.84	76.70	**48.22**	37.55	36.85	68.81
QSTP	**83.02**	**72.50**	**82.00**	84.44	**73.76**	**77.20**	46.50	39.00	38.88	**69.20**

Method	Audio QA		Visual QA		Audio-Visual QA					Avg.
	Count	Comp	Count	Local	Exist	Count	Local	Comp	Temp	
	(b) Tail									
AVSD[23]	46.39	48.09	12.78	60.79	61.34	8.01	40.18	65.55	41.90	48.45
Pano-AVQA[31]	41.73	43.43	20.77	63.77	21.30	22.70	**64.67**	40.34	49.62	42.40
AVST[13]	41.73	**57.96**	19.43	54.23	77.10	9.37	40.67	62.16	37.98	49.55
TSPM† [11]	71.80	49.56	47.56	73.18	**82.79**	38.64	57.37	68.86	49.36	63.55
QA-TIGER† [9]	75.82	43.11	**54.59**	75.59	78.56	33.55	64.65	**80.47**	62.96	66.45
QSTP	**76.19**	55.64	53.60	**76.04**	80.90	**40.03**	60.51	78.88	**65.25**	**68.45**

MUSIC-AVQA-R. Without explicit bias-handling techniques, our model achieves 69.20% on Head (frequent) and 68.45% on Tail (rare) questions, as shown in Table 2, highlighting its strong temporal modeling and question-aware capabilities.

4.5 Ablation Study

We conduct ablation experiments on the three main modules proposed in QSTP, namely the question-aware spatial-temporal reasoning, audio-driven visual gating, and question-guided fusion, with the results shown in Table 3. Compared with the complete model, removing these modules results in average performance drops of 1.60%, 1.15%, and 0.65%, respectively, indicating that these modules can complement each other when used simultaneously, with the question-aware spatial-temporal graph contributing most significantly to performance improvement. When all three modules are removed, the performance drops to the lowest level, where audio and visual features are directly fused for answer prediction without any reasoning mechanism. These results not only validate the effectiveness of each module but also highlight the core role of the proposed graph neural network structure in enhancing multimodal spatial-temporal reasoning capability for the AVQA task.

Similarly, we further analyze the key hyperparameter γ and the number of GNN layers L. As shown in Fig. 3 (a), when γ is small, the attention on spatial-temporal relations between patches dominates, resulting in insufficient perception of question semantics. Conversely, when γ is large, the question relevance is overemphasized, which in turn weakens the role of patch information in spatial-temporal modeling. When $\gamma = 0.20$, the model reaches best performance, indicating an optimal balance between preserving spatial-temporal relations and integrating question semantics, thereby significantly enhancing the effectiveness of multimodal spatial-temporal reasoning. As shown in Fig. 3 (b), increasing L initially improves the ability to capture more saptail-temporal dependencies, but an excessive number of layers lead to over-smoothing and degraded performance. The best performance is achieved at $L = 3$, striking a balance between expressive capacity and training stability.

Table 3. Ablation study of the proposed framework.

Question-Aware Spatial-Temporal Reasoning	Audio-Driven Visual Gating	Question-Guided Fusion	A-QA	V-QA	AV-QA	Avg.
✓	✓	✓	**78.83**	**86.04**	**74.25**	**78.19**
–	✓	✓	77.47	84.02	72.78	76.59
✓	–	✓	76.85	84.85	73.39	77.04
✓	✓	–	77.90	85.22	73.78	77.54
–	–	✓	76.60	83.77	72.47	76.20
–	–	–	76.35	83.03	72.25	75.84

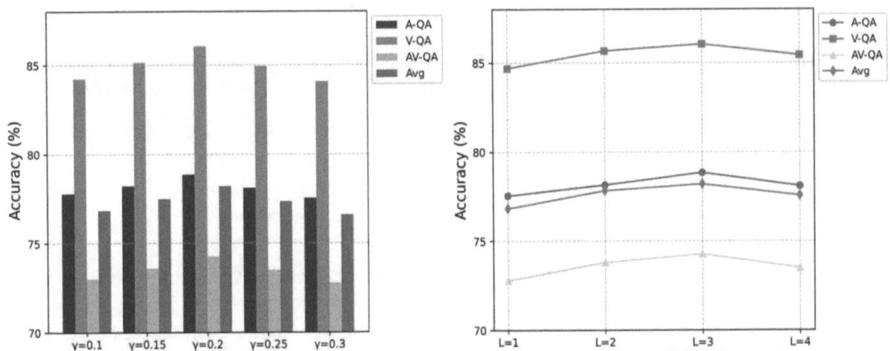

(a) The influence of hyperparameter γ. (b) The influence of GNN Layers L.

Fig. 3. Comparison of model performance under different settings: (a) The influence of hyperparameter γ; (b) The influence of GNN Layers L.

4.6 Qualitative Results and Analysis

We conduct a quantitative analysis on six representative AVQA examples, covering five fixed question types: existential, counting, location, comparative, and temporal. As shown in Fig. 4, QSTP accurately predicts the answers for all five types and outperforms the latest AVQA methods, QA-TIGER [9] and PSOT [15]. Specifically, in (a) counting questions, the model can precisely count the number of instruments sounding; in (b) comparative questions, it can effectively distinguish rhythm differences between different instruments; in (c, e) temporal questions, it can accurately determine the order of events or correctly identify the sequence in which specific instruments appear; in (d) location questions, it can accurately pinpoint the position of the target instrument; and in (f) existential questions, it can correctly judge the presence or absence of the target object or sound. These results clearly demonstrate the significant advantages of QSTP's question-aware spatial-temporal reasoning in multimodal reasoning tasks.

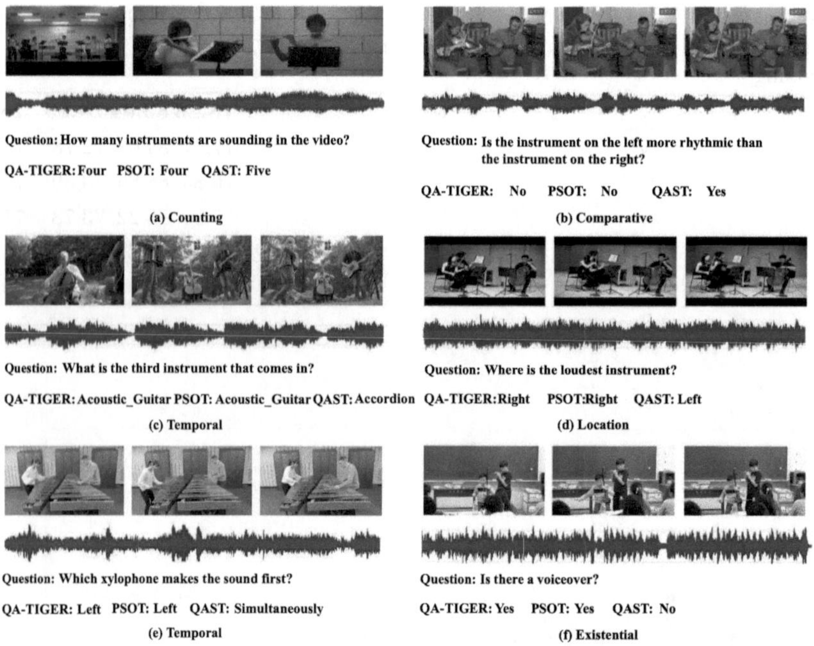

Fig. 4. Quantitative analysis on six representative AVQA examples.

5 Conclusion

In this paper, we propose QSTP, a patch-level question-aware dynamic spatial-temporal reasoning framework to address limitations in temporal continuity, question semantic guidance, and graph adaptability for Audio-Visual Question Answering (AVQA). Specifically, we construct spatial-temporal graphs at a fine-grained visual patch level to jointly model intra-frame spatial dependencies and inter-frame temporal relationships, and propose a question-aware aggregation mechanism to dynamically adjust neighbor weights, enabling continuous focus on question-relevant spatial-temporal information. In addition, we design an audio-driven visual gating module to emphasize audio-related regions for better audio source localization, and put in a question-guided multimodal fusion strategy to refine answer prediction. Extensive experiments on the MUSIC-AVQA and MUSIC-AVQA-R datasets demonstrate that QSTP achieves state-of-the-art performance across various complex reasoning tasks and matches the performance of large-scale pretrained methods while using significantly fewer data and parameters. In summary, QSTP not only provides a novel and efficient solution for AVQA but also showcases the broad application prospects of fine-grained question-aware spatial-temporal modeling in multimodal reasoning.

Acknowledgments. This work was supported by the Shanghai Municipal Education Commission Artificial Intelligence Plan (Z2024-119).

References

1. Bolya, D., Fu, C.Y., Dai, X., Zhang, P., Feichtenhofer, C., Hoffman, J.: Token merging: your vit but faster. arXiv preprint arXiv:2210.09461 (2022)
2. Chen, S., Het al.: Valor: Vision-audio-language omni-perception pretraining model and dataset. arXiv preprint arXiv:2304.08345 (2023)
3. Duan, H., Xia, Y., Mingze, Z., Tang, L., Zhu, J., Zhao, Z.: Cross-modal prompts: adapting large pre-trained models for audio-visual downstream tasks. Adv. Neural. Inf. Process. Syst. **36**, 56075–56094 (2023)
4. Fei, H., et al.: Video-of-thought: Step-by-step video reasoning from perception to cognition. arXiv preprint arXiv:2501.03230 (2024)
5. Gemmeke, J.F., et al.: Audio set: an ontology and human-labeled dataset for audio events. In: 2017 IEEE International Conference on Acoustics, Speech and Signal Processing (ICASSP), pp. 776–780. IEEE (2017)
6. Hershey, S., et al.: CNN architectures for large-scale audio classification. In: 2017 IEEE International Conference on Acoustics, Speech and Signal Processing (ICASSP), pp. 131–135. IEEE (2017)
7. Hu, R., Singh, A., Darrell, T., Rohrbach, M.: Iterative answer prediction with pointer-augmented multimodal transformers for textvqa. In: Proceedings of the IEEE/CVF Conference on Computer Vision and Pattern Recognition, pp. 9992–10002 (2020)
8. Huang, Z., Li, J., Zhao, W., Guo, Y., Tian, Y.: Av-mamba: cross-modality selective state space models for audio-visual question answering. In: Proceedings of the IEEE/CVF Conference on Computer Vision and Pattern Recognition Workshop (CVPRW), pp. 1–4 (2024)
9. Kim, H., Jung, I., Suh, D., Zhang, Y., Lee, S., Hong, S.: Question-aware gaussian experts for audio-visual question answering. In: Proceedings of the Computer Vision and Pattern Recognition Conference, pp. 13681–13690 (2025)

10. Lao, M., Pu, N., Liu, Y., He, K., Bakker, E.M., Lew, M.S.: Coca: collaborative causal regularization for audio-visual question answering. In: Proceedings of the AAAI Conference on Artificial Intelligence. vol. 37, pp. 12995–13003 (2023)
11. Li, G., Du, H., Hu, D.: Boosting audio visual question answering via key semantic-aware cues. In: Proceedings of the 32nd ACM International Conference on Multimedia, pp. 5997–6005 (2024)
12. Li, G., Hou, W., Hu, D.: Progressive spatio-temporal perception for audio-visual question answering. In: Proceedings of the 31st ACM International Conference on Multimedia, pp. 7808–7816 (2023)
13. Li, G., Wei, Y., Tian, Y., Xu, C., Wen, J.R., Hu, D.: Learning to answer questions in dynamic audio-visual scenarios. In: Proceedings of the IEEE/CVF Conference on Computer Vision and Pattern Recognition, pp. 19108–19118 (2022)
14. Li, Z., Guo, D., Zhou, J., Zhang, J., Wang, M.: Object-aware adaptive-positivity learning for audio-visual question answering. In: Proceedings of the AAAI Conference on Artificial Intelligence. vol. 38, pp. 3306–3314 (2024)
15. Li, Z., Zhou, J., Zhang, J., Tang, S., Li, K., Guo, D.: Patch-level sounding object tracking for audio-visual question answering. In: Proceedings of the AAAI Conference on Artificial Intelligence. vol. 39, pp. 5075–5083 (2025)
16. Lin, Y.B., Sung, Y.L., Lei, J., Bansal, M., Bertasius, G.: Vision transformers are parameter-efficient audio-visual learners. In: Proceedings of the IEEE/CVF Conference on Computer Vision and Pattern Recognition. pp. 2299–2309 (2023)
17. Liu, X., Dong, Z., Zhang, P.: Tackling data bias in music-AVQA: crafting a balanced dataset for unbiased question-answering. In: Proceedings of the IEEE/CVF Winter Conference on Applications of Computer Vision, pp. 4478–4487 (2024)
18. Ma, J., et al.: Look, listen, and answer: overcoming biases for audio-visual question answering. Adv. Neural. Inf. Process. Syst. **37**, 9507–9531 (2024)
19. de Oliveira Souza, B.C., Aasan, M., Pedrini, H., Rivera, A.R.: Selfgraphvqa: a self-supervised graph neural network for scene-based question answering. In: Proceedings of the IEEE/CVF International Conference on Computer Vision, pp. 4640–4645 (2023)
20. Park, K.R., Lee, H.J., Kim, J.U.: Learning trimodal relation for audio-visual question answering with missing modality. In: European Conference on Computer Vision, pp. 42–59. Springer (2024)
21. Park, K.R., Oh, Y., Kim, J.U.: Enhancing audio-visual question answering with missing modality via trans-modal associative learning. In: ICASSP 2024-2024 IEEE International Conference on Acoustics, Speech and Signal Processing (ICASSP), pp. 5755–5759. IEEE (2024)
22. Reddy, K., Pal, M.: Contextual graph transformer: a small language model for enhanced engineering document information extraction. arXiv preprint arXiv:2508.02532 (2025)
23. Schwartz, I., Schwing, A.G., Hazan, T.: A simple baseline for audio-visual scene-aware dialog. In: Proceedings of the IEEE/CVF Conference on Computer Vision and Pattern Recognition, pp. 12548–12558 (2019)
24. Shen, X., et al.: Fine-grained audible video description. In: Proceedings of the IEEE/CVF Conference on Computer Vision and Pattern Recognition, pp. 10585–10596 (2023)
25. Shu, F., Zhang, L., Jiang, H., Xie, C.: Audio-visual LLM for video understanding. arXiv preprint arXiv:2312.06720 (2023)
26. Tang, Y., Shimada, D., Bi, J., Xu, C.: Avicuna: Audio-visual llm with interleaver and context-boundary alignment for temporal referential dialogue. arXiv preprint arXiv:2403.16276 2 (2024)

27. Wang, Y., Haruta, S., Zeng, D., Vizcarra, J., Kurokawa, M.: Multi-object event graph representation learning for video question answering. arXiv preprint arXiv:2409.07747 (2024)
28. Wang, Y., Yasunaga, M., Ren, H., Wada, S., Leskovec, J.: VQA-GNN: Reasoning with multimodal knowledge via graph neural networks for visual question answering. In: Proceedings of the IEEE/CVF International Conference on Computer Vision, pp. 21582–21592 (2023)
29. Yang, P., et al.: Avqa: A dataset for audio-visual question answering on videos. In: Proceedings of the 30th ACM International Conference on Multimedia, pp. 3480–3491 (2022)
30. Yang, Z., Li, W., Cheng, G.: Shmamba: structured hyperbolic state space model for audio-visual question answering. arXiv preprint arXiv:2406.09833 (2024)
31. Yun, H., Yu, Y., Yang, W., Lee, K., Kim, G.: Pano-avqa: Grounded audio-visual question answering on 360deg videos. In: Proceedings of the IEEE/CVF International Conference on Computer Vision, pp. 2031–2041 (2021)

R^2-Mesh: Reinforcement Learning Powered Mesh Reconstruction via Geometry and Appearance Refinement

Haoyang Wang, Liming Liu, and Xinggong Zhang(✉)

Peking University, Beijing, China
{haoyang.wang,llm}@stu.pku.edu.cn, zhangxg@pku.edu.cn

Abstract. Mesh reconstruction from Neural Radiance Fields (NeRF) is widely used in 3D reconstruction and has been applied across numerous domains. However, existing methods typically rely solely on the given training set images, which restricts supervision to limited observations and makes it difficult to fully constrain geometry and appearance. Moreover, the contribution of each viewpoint for training is not uniform and changes dynamically during the optimization process, which can result in suboptimal guidance for both geometric refinement and rendering quality. To address these limitations, we propose R^2-Mesh, a reinforcement learning framework that combines NeRF-rendered pseudo-supervision with online viewpoint selection. Our key insight is to exploit NeRF's rendering ability to synthesize additional high-quality images, enriching training with diverse viewpoint information. To ensure that supervision focuses on the most beneficial perspectives, we introduce a UCB-based strategy with a geometry-aware reward, which dynamically balances exploration and exploitation to identify informative viewpoints throughout training. Within this framework, we jointly optimize SDF geometry and view-dependent appearance under differentiable rendering, while periodically refining meshes to capture fine geometric details. Experiments demonstrate that our method achieves competitive results in both geometric accuracy and rendering quality.

Keywords: Mesh Reconstruction · Reinforcement Learning · Viewpoint Selection

1 Introduction

Mesh reconstruction of 3D scenes is profoundly essential in domains such as virtual reality [23], medical imaging [11], and robotics [16]. High-quality 3D meshes enable immersive interaction, precise analysis, and reliable perception. However, reconstructing meshes directly from RGB images remains a challenging problem. Real-world scenes often contain occlusions, non-uniform lighting, and complex textures. These factors, together with the need to capture fine-grained

J. Lokoč et al. (Eds.): MMM 2026, LNCS 16413, pp. 646–660, 2026.
https://doi.org/10.1007/978-981-95-6957-1_46

geometric structures, make it difficult to obtain faithful reconstructions from image observations.

The recent advent of Neural Radiance Fields (NeRF) [5,18,20] has brought transformative progress to 3D reconstruction. NeRF excels at modeling volumetric density and view-dependent appearance, achieving photorealistic novel view synthesis. Motivated by this capability, numerous works [13,26,31,32] employ NeRF as an intermediate representation for mesh extraction. Typical strategies either transform radiance fields into Signed Distance Fields (SDFs) followed by Marching Cubes [15], or refine explicit meshes under differentiable rendering supervision [4,30].

While these approaches have made notable progress, they typically rely on the given training set images. In this setting, supervision is restricted to a limited number of captured images, which is often insufficient to fully constrain geometry and appearance. Moreover, the contribution of each viewpoint for training is not only unequal but also changes dynamically as training progresses, so fixed viewpoints cannot provide the most effective guidance throughout the optimization process.

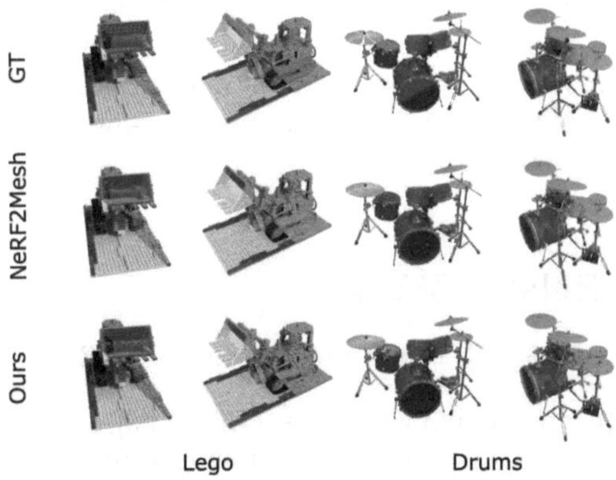

Fig. 1. Visualization of our overall performance on the NeRF-synthetic dataset. Our method achieves highly competitive and robust performance in both mesh rendering quality and geometric quality.

Our key insight is to go beyond the fixed captured images by exploiting the generative capacity of NeRF itself. NeRF can synthesize high-quality images from arbitrary camera poses, effectively serving as a source of additional pseudo-supervision. These NeRF-rendered views introduce complementary perspective information that is otherwise unavailable in the original dataset, thereby enriching the training signal and strengthening geometric constraints.

However, not all views are equally informative, and selecting unhelpful or redundant viewpoints may introduce noise or inefficiency. To address this, we propose an online viewpoint selection strategy based on the Upper Confidence Bound (UCB) algorithm [1]. This reinforcement learning strategy dynamically balances exploration and exploitation, automatically identifying the most beneficial viewpoints at each stage of training. Combined with a geometry-aware reward, our method ensures that both real and NeRF-rendered images provide supervision from perspectives that maximally improve reconstruction.

Based on these insights, we propose R^2-**Mesh**, a reinforcement learning framework for high-quality mesh reconstruction. The framework jointly optimizes SDF geometry and view-dependent appearance under differentiable rendering supervision. Periodically, meshes are extracted from the evolving SDF and refined, allowing both geometry and connectivity to adapt dynamically over time. The overall performance is illustrated in Fig. 1.

In summary, the key contributions of our work are as follows:

- We leverage NeRF-rendered images as additional pseudo-supervision, enriching the training signal with diverse and high-quality viewpoints that go beyond the original captures.
- We propose a UCB-based online viewpoint selection strategy with a geometry-aware reward, dynamically identifying the most informative views as training evolves.
- We present R^2-Mesh, a joint optimization framework of SDF and appearance, enabling progressive and topology-aware mesh refinement for high-fidelity reconstruction.

2 Related Work

2.1 Mesh Reconstruction from NeRF

NeRF [18] and its variants [3,6,20,29] have significantly advanced 3D scene reconstruction by learning volumetric functions from multi-view images. Some works attempt to reconstruct meshes simultaneously during NeRF training. A representative example is MobileNeRF [4], which directly generates polygonal meshes for real-time rendering. However, the produced meshes often lack smoothness since the topology is fixed at the initial stage and cannot adapt to complex geometry.

Recent studies [13,14,21,24,31–33] instead adopt Signed Distance Functions (SDFs) to represent geometry. Several of them leverage the ray-marching formulation of NeRF to optimize the SDF, followed by iso-surfacing algorithms such as Marching Cubes [15] or Dual Contouring [9] for mesh extraction. While effective, this post-processing often causes detail loss and surface artifacts, degrading mesh quality.

To mitigate these issues, some methods refine the extracted meshes. NVdiffrec [21] employs differentiable mesh rendering to optimize the SDF, but random initialization may cause instabilities and floating artifacts. NeRF2Mesh [30] refines vertex positions and face density based on rendering errors, but vertex connectivity remains fixed. Moreover, all these methods rely solely on the given training images, restricting supervision and overlooking that the training value of each viewpoint varies dynamically during optimization.

In contrast, our approach ensures a well-initialized representation, supports flexible adjustment of vertices and connectivity, and incorporates adaptive viewpoint selection to provide more effective supervision.

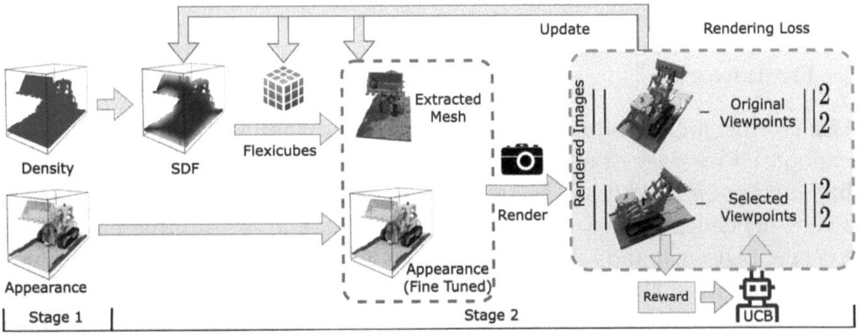

Fig. 2. Our Framework, R^2-Mesh. In stage 1, we initialize the geometry and view-dependent appearance representation based on NeRF. This initial phase results in a coarse SDF grid and a set of candidate viewpoints rendered by NeRF model for enhancement in the subsequent stage. Then in stage 2, for each training iteration, we take two steps. We first select the optimal NeRF-rendered viewpoints based on UCB strategy to incorporate into the training dataset. We then simultaneously refine both the geometry and the appearance representation. After training is complete, we obtain the final mesh.

2.2 Best Views Selection in 3D Scenes

Selecting best views in 3D scenes is a critical issue in computer vision and 3D reconstruction, directly impacting the quality of final rendering and model reconstructions. Traditional approaches often rely on heuristic or predetermined viewpoint strategies, such as uniform sampling on a sphere or predefined camera trajectories. While straightforward, these strategies lack adaptability and often fail to capture the most informative views in scenes with complex geometry or strong occlusions.

Some recent works [8,10,25] in radiance fields incorporate uncertainty estimation to guide the selection process. By allocating more supervision to regions of high uncertainty, they improve rendering quality. However, these approaches are designed for volumetric ray sampling in radiance field training, and are less applicable to our focus on surface-based mesh reconstruction, where the supervision requirements differ significantly.

Additionally, several studies investigate next-best-view (NBV) planning [12,17,22], primarily in robotic exploration of unknown environments. These methods aim to minimize the number of views required for complete coverage, whereas our objective is different: we aim to enhance mesh reconstruction quality by online selecting the most beneficial viewpoints from available candidates, based on their actual performance gains during training.

3 Method

3.1 Framework

In this section, we introduce our framework, R^2-**Mesh**, as illustrated in Fig. 2. We adopt a two-stage training process for mesh reconstruction. In stage 1, we utilize Instant-NGP [20] as the primary architecture to initialize the geometry and view-dependent appearance of the mesh (Sect. 3.2). This initial phase results in a coarse SDF grid and a set of candidate viewpoints rendered by NeRF model for enhancement in the subsequent stage. Then in stage 2, for each training iteration, we take two steps. We first select the optimal combination of viewpoints based on UCB values to incorporate into the training dataset (Sect. 3.3). We then simultaneously refine both the geometry and the appearance representation (Sect. 3.4). Upon completion of the training, we utilize xatlas [34] for UV unwrapping and export the resulting object mesh.

3.2 Efficient 3D Scene Initialization (Stage 1)

In the first stage, we adopt the approach from [30], which builds on the Instant-NGP architecture to efficiently train a NeRF model for initializing 3D scene representation. The geometry is learned through a multi-resolutional density grid combined with a shallow Multilayer Perceptron (MLP). For appearance, the representation is decomposed into diffuse color and view-dependent specular components. The pixel color can be calculated by the following equations:

$$C(r) = \sum_{i=1}^{N} T_i(1 - \exp(-\sigma_i \Delta t_i))\mathbf{c}_i, \tag{1}$$

where $C(r)$ is the accumulated color along ray r, T_i is the transmittance from the camera to the i-th point, calculated as $T_i = \exp\left(-\sum_{j=1}^{i-1} \sigma_j \Delta t_j\right)$. σ_i represents the density at the i-th point, Δt_i is the distance between the i-th and $(i+1)$-th sampled points, and $\mathbf{c}_i$ is the combined diffuse and specular color at the i-th point.

To optimize the NeRF model, an mean squared error (MSE) loss is used to calculate the squared differences between the predicted and ground truth colors of the pixels:

$$L_{\text{MSE}} = \frac{1}{N} \sum_{i=1}^{N} (\hat{C}_i - C_i)^2, \tag{2}$$

where $\hat{C}_i$ is the predicted color and C_i is the ground truth color of the i-th pixel, with N representing the total number of pixels.

After training is complete, we extract a density grid at the specified resolution and convert it into an SDF grid. Density represents the amount of material at a point and is inherently positive, while SDF values indicate the shortest distance to the nearest surface, allowing them to be both positive and negative to distinguish between exterior and interior regions. Therefore, we set a threshold θ to represent the iso-surface and convert density values into a coarse SDF grid using the following formula:

$$\mathrm{SDF} = \begin{cases} \frac{\sigma-\theta}{\max(\sigma)-\theta}, & \text{if } \sigma > \theta \\ \frac{\sigma-\theta}{\theta}, & \text{otherwise} \end{cases}, \tag{3}$$

where σ represents the density value at a point, and θ represents the iso-surface.

This transformation establishes our initial SDF grid, which we further refine in stage 2.

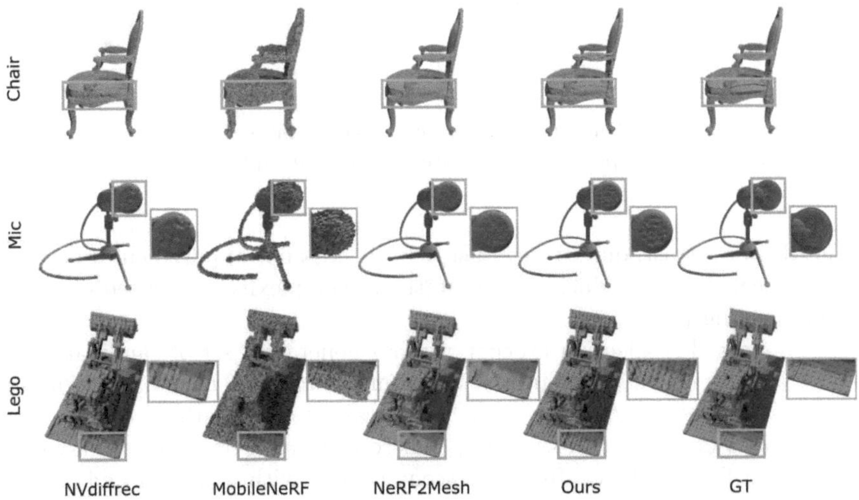

Fig. 3. Mesh reconstruction quality on the NeRF-synthetic dataset. Our method produces significantly finer and more detailed mesh geometry compared to previous approaches.

3.3 UCB-Based Adaptive Viewpoint Enhancement (Stage 2)

After stage 1 converges, our goal in this stage is to refine the mesh by leveraging additional supervision beyond the initial training views. However, a key challenge lies in selecting the most informative viewpoints for continued refinement. As training progresses, both the model parameters and the rendering losses evolve, causing the effectiveness of each viewpoint to change over time. Relying on a fixed set of views may lead to redundant or suboptimal supervision. To address this, we propose an online reinforcement learning strategy based on the Upper Confidence Bound (UCB) algorithm, which adaptively selects viewpoints at each iteration by considering the current model state.

Many state-of-the-art reinforcement RL methods, such as DQN [19] and PPO [27], introduce additional networks for inference, which results in significant time

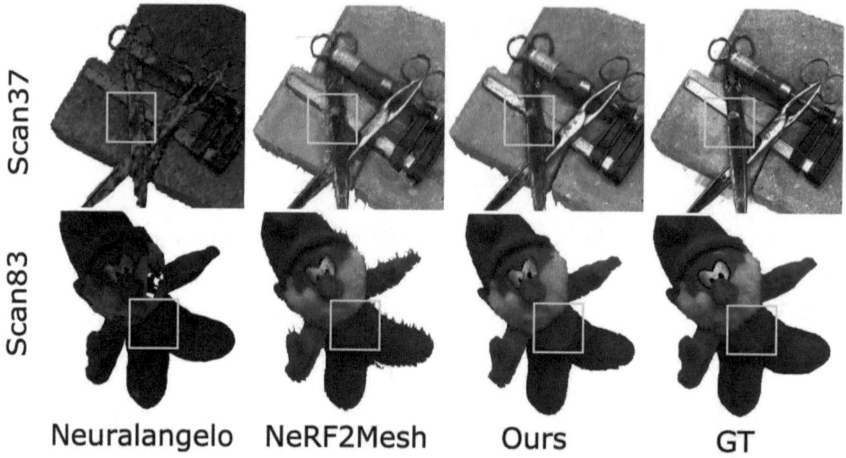

Fig. 4. Visualization of rendering quality on the DTU dataset. Our method produces more accurate surface appearances compared to previous approaches.

overhead. In contrast, our strategy using UCB does not require neural networks for inference, thereby reducing computational complexity and speeding up the decision-making process.

To support this selection process, we first generate a set of candidate viewpoints by rendering the scene from n uniformly distributed camera poses around a virtual sphere centered at the scene. Specifically, we define the candidate viewpoint set as $V_{\text{NeRF}} = \{v_1, v_2, \ldots, v_n\}$, where each v_i corresponds to a camera located at a fixed radius r from the scene center, with elevation and azimuth angles sampled uniformly. These candidate views serve as the action space $\mathcal{A}$ for the UCB algorithm.

At each iteration, we compute the UCB value for every candidate viewpoint (action) as:

$$\text{UCB}_a(t) = \hat{r}_a(t) + c\sqrt{\frac{2\ln t}{N_a(t)}}, \tag{4}$$

where $\hat{r}_a(t)$ represents the empirical mean reward for action a up to time t, c is a constant controlling the degree of exploration, t represents the total number of steps t, and $N_a(t)$ denotes the number of times action a has been selected. Here, time t refers to the t-th iteration in the stage 2 training process. At initialization, the UCB value for each action is set to a large value, ensuring that every action is explored at least once.

For each training iteration, we randomly sample m viewpoints from the original training set V_{train} as ground truth supervision. In parallel, we evaluate the UCB values for all candidate viewpoints in the action space $\mathcal{A} = V_{\text{NeRF}}$, and select the top-$k$ actions with the highest UCB values. The corresponding viewpoints are treated as pseudo-ground-truth for this iteration. The mesh model is then trained using both the m ground truth images and the k NeRF-rendered pseudo ground truth images.

To guide the mesh refinement toward improved quality at novel views, we define the reward for each action a as the sum of two components:

$$r_a = \alpha\, r_{\text{color}} + (1-\alpha)\, r_{\text{geo}}, \quad \alpha \in [0,1].$$

The **color reward** r_{color} evaluates how well the mesh captures the appearance at a given viewpoint. We use the MSE to enforce accurate pixel-level color alignment and the Learned Perceptual Image Patch Similarity (LPIPS) to capture perceptual differences that reflect high-level structural consistency. Formally,

$$r_{\text{color}} = \text{MSE}(\text{Mesh}_a, \text{NeRF}_a) + \text{LPIPS}(\text{Mesh}_a, \text{NeRF}_a).$$

To further leverage NeRF's rich geometric information, we define a **geometry reward** r_{geo}, which encourages alignment in the visible object regions between the mesh and the NeRF rendering. We compare their binary foreground indicators obtained by thresholding the rendered depth map. The reward is defined as:

$$r_{\text{geo}} = \text{MSE}(\mathbb{1}_{\text{Mesh}_a}, \mathbb{1}_{\text{NeRF}_a}).$$

Here, $\mathbb{1}_{\text{NeRF}_a}$ denotes the binary visibility map derived from NeRF's depth rendering:

$$\mathbb{1}_{\text{NeRF}_a}(u,v) = \begin{cases} 1, & \text{if } D_{\text{NeRF}}(u,v) < \tau \\ 0, & \text{otherwise} \end{cases},$$

where $D_{\text{NeRF}}(u,v)$ is the NeRF-predicted depth at pixel (u,v), and τ is a fixed depth threshold to distinguish foreground regions.

By combining both appearance and geometric cues from NeRF, this reward formulation promotes improved rendering fidelity and reconstruction accuracy under both training and unseen viewpoints.

3.4 Geometry and Appearance Refinement (Stage 2)

After stage 1, we obtain a coarse representation of the scene in the form of an SDF grid and a view-dependent appearance field. While this initialization captures the overall geometry and appearance, it often lacks fine surface details and exhibit artifacts in poorly supervised regions. To achieve high-fidelity reconstruction, it is essential to further refine both geometry and appearance based on more precise surface supervision.

However, directly optimizing mesh vertex positions on a fixed topology limits the model's ability to capture complex geometry. To address these issues, we refine both the SDF grid and the appearance representation based on surface rendering of the mesh. By directly optimizing the SDF, we enable continuous updates to both the mesh vertex positions and their connectivity, allowing the mesh geometry to flexibly adapt to complex surfaces as training progresses.

To ensure smooth gradient flow and enable topology-aware refinement, we adopt FlexiCubes [28], which augments each SDF grid vertex with learnable deformation and weight parameters. These attributes allow the extracted mesh to deform continuously and adjust its connectivity during optimization.

At each training iteration, we take two steps. We first select the optimal viewpoints to enhance the training dataset based on our UCB strategy. We then extract the mesh from the SDF using FlexiCubes and render it with nvdiffrast [21], enabling end-to-end optimization of both geometry and appearance through differentiable rendering.

To optimize the parameters, the overall loss functions are:

$$L = \lambda_{\text{color}} L_{\text{color}}(x) + \lambda_{\text{TV}} L_{\text{TV}} + \lambda_{\text{dev}} L_{\text{dev}}, \tag{5}$$

where $L_{\text{color}}(x) = \sqrt{(x - x^*)^2 + \epsilon^2}$ is the Charbonnier loss [2], with x as the predicted value, x^* as the ground truth, and ϵ as a small constant to improve robustness to outliers. L_{TV} is the Total Variation (TV) regularization term [3] applied to the SDF grid to reduce floaters and enhance smoothness, while L_{dev} is the FlexiCubes regularizer used to suppress mesh artifacts. Upon completing the training process, we employ xatlas to perform UV unwrapping on the generated mesh and subsequently export the final object mesh.

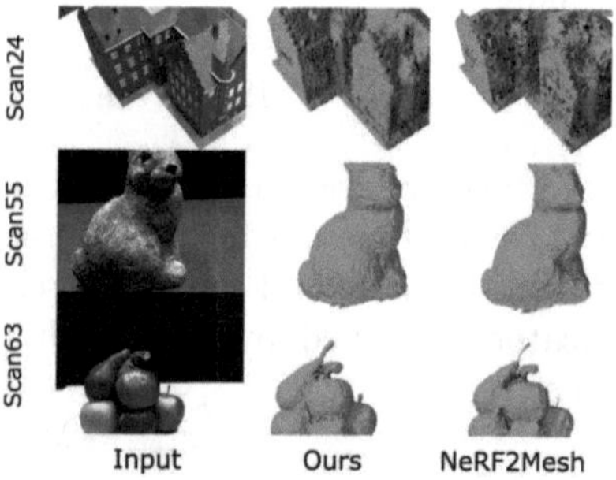

Fig. 5. Mesh reconstruction quality on the DTU dataset. Our method delivers better performance than previous methods.

4 Experiments

4.1 Implementation Details

Datasets. To evaluate our approach and baseline approaches, we use 15 scenes from the real-world DTU dataset [7] and all scenes from the NeRF-synthetic [18] dataset to assess both the reconstructed mesh quality and the rendering quality. The DTU dataset provides 49 or 64 images per scene captured from fixed viewpoints in a controlled indoor setup, while the NeRF-synthetic dataset contains 100 training images per scene rendered from synthetic objects along a predefined spherical trajectory. Experimental results demonstrate that our approach achieves excellent rendering and mesh reconstruction outcomes across most real-world and synthetic scenes.

Training Details. In the first stage, we train the NeRF model for 30,000 steps. At the end of the training, we extract an SDF grid at a resolution of 128^3. For stage 2, we train the model for 90,000 steps to ensure that both the UCB and the model converge. The optimizer used is Adam, with the learning rates for both the FlexiCubes parameters and the SDF set to 1×10^{-4}. The experiments are all conducted on a high-performance computing platform equipped with a GeForce RTX 3090 GPU.

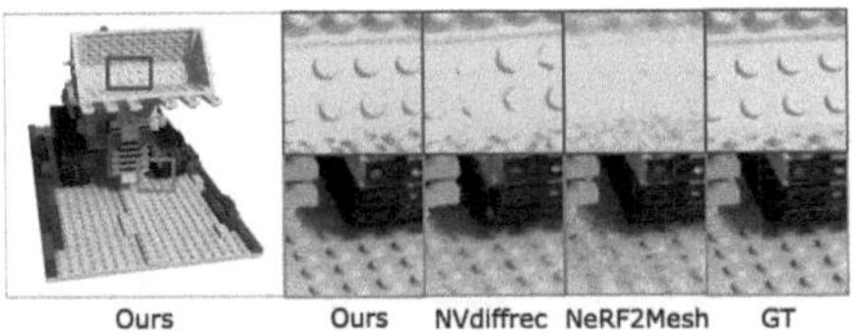

Fig. 6. Visualization of rendering quality on the NeRF-synthetic dataset. Our method can render more detailed geometry.

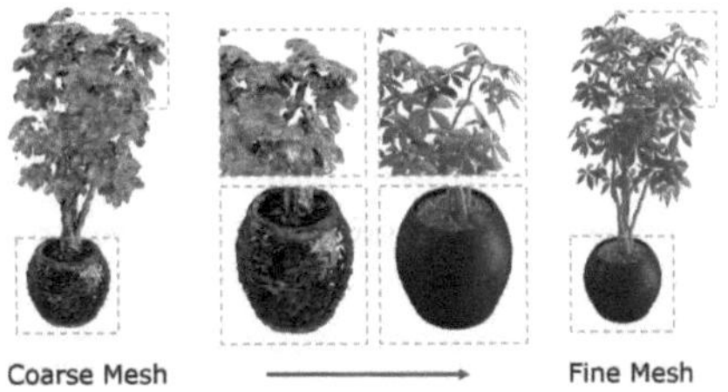

Fig. 7. Comparison between coarse and fine mesh.

4.2 Evaluation

Mesh Quality. To evaluate the geometric accuracy of our mesh reconstructions, we utilize the Chamfer Distance (CD) metric which quantifies the average closest point distance between the vertices of the reconstructed mesh and the ground truth. We establish MobileNeRF [4], NVdiffrec [21], NeuS2 [32], Neuralangelo [13], NeRFMeshing [26] and NeRF2Mesh [30] as baselines for our experiments and find that our results significantly outperform the previous works. The detailed outcomes of our mesh reconstruction are presented in Table 1, Table 2, Fig. 3 and Fig. 5.

Table 1. Chamfer Distance (↓) on the NeRF-synthetic dataset compared to the ground truth geometry. Our method achieves competitive results compared to other methods. Red and orange indicate the first and second best performing algorithms for each scene.

Method	Lego	Hotdog	Chair	Drums	Ficus	Mic	Ship	Materials	Mean
NVdiffrec	1.9	**3.1**	9.4	3.4	0.6	1.0	38.4	1.2	7.38
NeuS2	2.0	3.4	4.1	1.8	0.6	0.9	19.6	1.3	4.22
NeRF2Mesh	2.1	3.7	**1.0**	2.5	**0.4**	1.2	35.9	1.2	6.00
NeRFMeshing	**1.5**	4.2	**1.0**	1.8	0.8	1.3	10.4	1.4	2.80
Ours	1.6	**3.1**	3.3	**1.6**	**0.4**	**0.8**	**10.4**	**0.5**	**2.71**

Rendering Quality. We employ PSNR, SSIM (Structural Similarity Index Measure), and LPIPS metrics to evaluate the rendering quality of our mesh

Table 2. Chamfer Distance (↓) on the DTU dataset compared to the ground truth geometry. Our method achieves competitive results compared to other methods. Red and orange indicate the first and second best performing algorithms for each scene.

Method	Scan 24	37	40	55	63	65	69	83	97	105	106	110	114	118	122	Mean
NeuS2	0.56	0.76	0.49	0.37	**0.92**	0.71	**0.76**	1.22	1.08	0.63	0.59	**0.89**	**0.40**	0.48	0.55	0.69
NeRF2Mesh	0.59	0.89	0.51	0.45	1.03	**0.65**	0.83	1.27	1.03	**0.62**	0.72	1.05	0.66	0.72	0.49	0.77
Ours	**0.51**	**0.65**	**0.38**	**0.32**	0.96	0.68	0.79	**1.12**	**0.82**	0.87	**0.57**	1.08	0.44	**0.41**	**0.44**	**0.67**

reconstruction. Compared to baselines, our methodology demonstrated superior performance in the majority of the tested scenes. Note that NeRFMeshing only reports PSNR on its benchmarks, NVdiffrec results are not reported on the DTU dataset, as the official release did not include the dataset reader and the MLP-based geometry representation required for evaluation. The detailed results of this comparative analysis are presented in Table 3, Fig. 4 and Fig. 6. These metrics highlight our approach's advancements in delivering high-fidelity reconstructions across various complex scenarios.

Table 3. Rendering quality comparisons on the NeRF-synthetic (SYN) and the DTU dataset. Compared to Neuralangelo, NVdiffrec, NeRFMeshing and NeRF2Mesh, our method achieves better rendering quality. Red and orange indicate the first and second best performing algorithms for each scene. (VE: Viewpoint Enhancement)

Method	PSNR ↑ SYN	PSNR ↑ DTU	SSIM ↑ SYN	SSIM ↑ DTU	LPIPS ↓ SYN	LPIPS ↓ DTU
Neuralangelo	27.12	21.98	0.91	0.88	0.091	0.25
NeRFMeshing	27.31	-	-	-	-	-
NVdiffrec	28.76	-	0.93	-	0.049	-
NeRF2Mesh	29.11	22.46	0.93	**0.90**	0.080	0.21
Ours (w/o VE)	29.26	22.84	**0.94**	**0.90**	0.048	**0.13**
Ours	**29.55**	**23.20**	**0.94**	**0.90**	**0.046**	**0.13**

4.3 Ablation Studies

We conduct an ablation study to evaluate the effectiveness of our mesh refinement method, the online UCB-based adaptive viewpoint enhancement strategy, and the geometric reward r_{geo}. We compare the performance of the full model against variants that exclude mesh refinement (RF), viewpoint enhancement (VE), and geometric reward. We also assess the effectiveness of the UCB strategy by comparing it to random and greedy viewpoint selection baselines. In the greedy strategy, we select viewpoints nearest to the one with the highest loss at each training iteration.

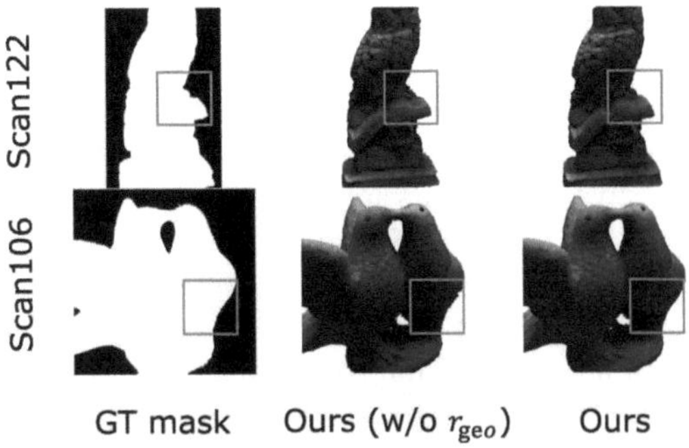

Fig. 8. Ablation study on geometric reward. With the geo reward, the mesh shows cleaner boundaries and fewer artifacts.

Table 4. Ablation study of our design choices. The results show that both the mesh refinement (RF) and viewpoint enhancement (VE) improve the overall performances.

VE	RF	PSNR ↑		SSIM ↑		LPIPS ↓	
		SYN	DTU	SYN	DTU	SYN	DTU
✓	✓	**29.55**	**23.20**	**0.94**	**0.90**	**0.046**	**0.13**
✗	✓	29.26	22.84	0.94	0.90	0.048	0.13
✗	✗	15.43	16.10	0.72	0.65	0.27	0.42

Table 5. Ablation study of our viewpoint enhancement strategies. Our UCB strategy outperforms both random and greedy strategies.

Strategy	PSNR ↑		SSIM ↑		LPIPS ↓	
	SYN	DTU	SYN	DTU	SYN	DTU
Ours	**29.55**	**23.20**	**0.94**	**0.90**	**0.046**	**0.13**
Ours (w/o r_{geo})	29.47	23.07	0.94	0.90	0.046	0.13
Greedy	29.18	22.87	0.93	0.90	0.052	0.15
Random	29.38	23.08	0.94	0.90	0.047	0.13

As shown in Table 4, when viewpoint enhancement is removed, the rendering quality decreases, primarily because the model lacks the ability to incorporate diverse perspectives that contribute to more detailed and accurate reconstructions. When mesh refinement is further removed, there is a significant decline in the rendering quality of the mesh. The comparison between fine mesh and coarse mesh is shown in Fig. 7. Additionally, removing the geometric reward r_{geo} results in increased artifacts along object boundaries, as shown in Fig. 8.

Table 5 shows the UCB strategy outperforms both random and greedy selection methods. This is because the greedy approach tends to overfit to the currently known best viewpoints, potentially missing out on more informative perspectives. In contrast, the UCB method effectively balances exploration and exploitation, allowing for the online selection of the optimal viewpoints.

5 Conclusion

In summary, we present R^2-Mesh, a framework that leverages an online reinforcement learning strategy based on the UCB algorithm to adaptively select the most informative viewpoints for mesh reconstruction. Starting from a NeRF-trained coarse SDF and appearance field, our method refines geometry and appearance through differentiable mesh extraction and rendering. By combining NeRF-rendered pseudo-supervision with UCB-guided viewpoint selection, our method achieves substantial improvements in geometric fidelity and rendering quality over existing approaches.

Acknowledgments. This work was sponsored by the NSFC grant(62431017). We gratefully acknowledge the support of State Key Laboratory of Media Convergence Production Technology and Systems, Key Laboratory of Intelligent Press Media Technology. Xinggong Zhang is the corresponding author.

References

1. Auer, P., Cesa-Bianchi, N., Fischer, P.: Finite-time analysis of the multiarmed bandit problem. Mach. Learn. **47**, 235–256 (2002)
2. Barron, J.T., Mildenhall, B., Verbin, D., Srinivasan, P.P., Hedman, P.: MIP-NERF 360: unbounded anti-aliased neural radiance fields. In: Proceedings of the IEEE/CVF Conference on Computer Vision and Pattern Recognition, pp. 5470–5479 (2022)
3. Chen, A., Xu, Z., Geiger, A., Yu, J., Su, H.: TenSorf: tensorial radiance fields. In: European Conference on Computer Vision, pp. 333–350. Springer (2022). https://doi.org/10.1007/978-3-031-19824-3_20
4. Chen, Z., Funkhouser, T., Hedman, P., Tagliasacchi, A.: MobileNERF: exploiting the polygon rasterization pipeline for efficient neural field rendering on mobile architectures. In: Proceedings of the IEEE/CVF Conference on Computer Vision and Pattern Recognition, pp. 16569–16578 (2023)
5. Fridovich-Keil, S., Meanti, G., Warburg, F.R., Recht, B., Kanazawa, A.: K-planes: explicit radiance fields in space, time, and appearance. In: Proceedings of the IEEE/CVF Conference on Computer Vision and Pattern Recognition, pp. 12479–12488 (2023)
6. Fridovich-Keil, S., Yu, A., Tancik, M., Chen, Q., Recht, B., Kanazawa, A.: PLENOXELS: radiance fields without neural networks. In: Proceedings of the IEEE/CVF Conference on Computer Vision and Pattern Recognition, pp. 5501–5510 (2022)
7. Jensen, R., Dahl, A., Vogiatzis, G., Tola, E., Aanæs, H.: Large scale multi-view stereopsis evaluation. In: Proceedings of the IEEE Conference on Computer Vision and Pattern Recognition, pp. 406–413 (2014)

8. Jin, L., Chen, X., Rückin, J., Popović, M.: Neu-NBV: next best view planning using uncertainty estimation in image-based neural rendering. In: 2023 IEEE/RSJ International Conference on Intelligent Robots and Systems (IROS), pp. 11305–11312. IEEE (2023)
9. Ju, T., Losasso, F., Schaefer, S., Warren, J.: Dual contouring of Hermite data. In: Proceedings of the 29th Annual Conference on Computer Graphics and Interactive Techniques, pp. 339–346 (2002)
10. Kim, H., Yang, H., Kim, T., Kim, Y., Kim, J.H., Zhang, B.T.: ActiveNEUS: active 3D reconstruction using neural implicit surface uncertainty. arXiv preprint arXiv:2405.02568 (2024)
11. Kong, F., Wilson, N., Shadden, S.: A deep-learning approach for direct whole-heart mesh reconstruction. Med. Image Anal. **74**, 102222 (2021)
12. Krainin, M., Curless, B., Fox, D.: Autonomous generation of complete 3D object models using next best view manipulation planning. In: 2011 IEEE International Conference on Robotics and Automation, pp. 5031–5037. IEEE (2011)
13. Li, Z., et al.: NeuralAngelo: high-fidelity neural surface reconstruction. In: Proceedings of the IEEE/CVF Conference on Computer Vision and Pattern Recognition, pp. 8456–8465 (2023)
14. Long, X., Lin, C., Wang, P., Komura, T., Wang, W.: SparSeneus: fast generalizable neural surface reconstruction from sparse views. In: European Conference on Computer Vision, pp. 210–227. Springer (2022). https://doi.org/10.1007/978-3-031-19824-3_13
15. Lorensen, W.E., Cline, H.E.: Marching cubes: a high resolution 3D surface construction algorithm. In: Seminal Graphics: Pioneering Efforts that Shaped the Field, pp. 347–353 (1998)
16. Lu, J., Liu, F., Girerd, C., Yip, M.C.: Image-based pose estimation and shape reconstruction for robot manipulators and soft, continuum robots via differentiable rendering. In: 2023 IEEE International Conference on Robotics and Automation (ICRA), pp. 560–567. IEEE (2023)
17. McGreavy, C., Kunze, L., Hawes, N.: Next best view planning for object recognition in mobile robotics. In: CEUR Workshop Proceedings, vol. 1782. CEUR Workshop Proceedings (2017)
18. Mildenhall, B., Srinivasan, P.P., Tancik, M., Barron, J.T., Ramamoorthi, R., Ng, R.: NERF: representing scenes as neural radiance fields for view synthesis. Commun. ACM **65**(1), 99–106 (2021)
19. Mnih, V., et al.: Human-level control through deep reinforcement learning. Nature **518**(7540), 529–533 (2015)
20. Müller, T., Evans, A., Schied, C., Keller, A.: Instant neural graphics primitives with a multiresolution hash encoding. ACM Trans. Graph. (TOG) **41**(4), 1–15 (2022)
21. Munkberg, J., et al.: Extracting triangular 3D models, materials, and lighting from images. In: Proceedings of the IEEE/CVF Conference on Computer Vision and Pattern Recognition, pp. 8280–8290 (2022)
22. Naazare, M., Rosas, F.G., Schulz, D.: Online next-best-view planner for 3D-exploration and inspection with a mobile manipulator robot. IEEE Robot. Autom. Lett. **7**(2), 3779–3786 (2022)
23. Navarro, F., Fdez, J., Garzón, M., Roldán, J.J., Barrientos, A.: Integrating 3D reconstruction and virtual reality: a new approach for immersive teleoperation. In: Ollero, A., Sanfeliu, A., Montano, L., Lau, N., Cardeira, C. (eds.) ROBOT 2017. AISC, vol. 694, pp. 606–616. Springer, Cham (2018). https://doi.org/10.1007/978-3-319-70836-2_50

24. Oechsle, M., Peng, S., Geiger, A.: UniSurf: unifying neural implicit surfaces and radiance fields for multi-view reconstruction. In: Proceedings of the IEEE/CVF International Conference on Computer Vision, pp. 5589–5599 (2021)
25. Pan, X., Lai, Z., Song, S., Huang, G.: ActiveNERF: learning where to see with uncertainty estimation. In: European Conference on Computer Vision, pp. 230–246. Springer (2022). https://doi.org/10.1007/978-3-031-19827-4_14
26. Rakotosaona, M.J., Manhardt, F., Arroyo, D.M., Niemeyer, M., Kundu, A., Tombari, F.: NERFMeshing: distilling neural radiance fields into geometrically-accurate 3D meshes. In: 2024 international conference on 3D Vision (3DV), pp. 1156–1165. IEEE (2024)
27. Schulman, J., Wolski, F., Dhariwal, P., Radford, A., Klimov, O.: Proximal policy optimization algorithms. arXiv preprint arXiv:1707.06347 (2017)
28. Shen, T., Munkberg, J., Hasselgren, J., Yin, K., Wang, Z., Chen, W., Gojcic, Z., Fidler, S., Sharp, N., Gao, J.: Flexible ISOSurface extraction for gradient-based mesh optimization. ACM Trans. Graph. **42**(4), 37 (2023)
29. Sun, C., Sun, M., Chen, H.T.: Direct voxel grid optimization: super-fast convergence for radiance fields reconstruction. In: Proceedings of the IEEE/CVF Conference on Computer Vision and Pattern Recognition, pp. 5459–5469 (2022)
30. Tang, J., et al.: Delicate textured mesh recovery from nerf via adaptive surface refinement. In: Proceedings of the IEEE/CVF International Conference on Computer Vision, pp. 17739–17749 (2023)
31. Wang, P., Liu, L., Liu, Y., Theobalt, C., Komura, T., Wang, W.: NEUS: learning neural implicit surfaces by volume rendering for multi-view reconstruction. arXiv preprint arXiv:2106.10689 (2021)
32. Wang, Y., Han, Q., Habermann, M., Daniilidis, K., Theobalt, C., Liu, L.: Neus2: fast learning of neural implicit surfaces for multi-view reconstruction. In: Proceedings of the IEEE/CVF International Conference on Computer Vision, pp. 3295–3306 (2023)
33. Yariv, L., Gu, J., Kasten, Y., Lipman, Y.: Volume rendering of neural implicit surfaces. Adv. Neural. Inf. Process. Syst. **34**, 4805–4815 (2021)
34. Young, J.: Xatlas (2021)

DPERC: Direct Parameter Estimation for Mixed Data with Random Missingness

Tuan L. Vo[1], Uyen Dang[3], Thu Nguyen[2(✉)], Pål Halvorsen[2], Michael A. Riegler[2], and Binh T. Nguyen[3,4,5]

[1] Télécom Paris, Institut Polytechnique de Paris, Paris, France
[2] Simula Metropolitan, Oslo, Norway
thu@simula.no
[3] University of Science, Ho Chi Minh City, Vietnam
[4] Vietnam National University, Ho Chi Minh City, Vietnam
[5] AISIA Research Lab, Ho Chi Minh City, Vietnam

Abstract. The covariance matrix is a foundation in numerous statistical and machine learning applications, such as Principal Component Analysis, Correlation Heatmap, etc. However, missing values within datasets present a formidable obstacle to accurately estimating this matrix. While imputation methods offer one avenue for addressing this challenge, they often entail a trade-off between computational efficiency and estimation accuracy. Consequently, attention has shifted towards direct parameter estimation, given its precision and reduced computational burden. In this paper, we propose **Direct Parameter Estimation for Randomly Missing Data with Categorical Features (DPERC)**, an efficient approach for direct parameter estimation tailored to mixed data that contains missing values within continuous features. Our method exploits information from categorical features to improve covariance estimation for continuous variables, effectively leveraging the structure of mixed data. Through comprehensive evaluations, we show that DPERC is not only a valuable tool for visualizing correlation heatmaps but also performs competitively against the state-of-the-art techniques by improving up to 23.47% of the sum of squared errors.

Keywords: Missing data · Parameter estimation

1 Introduction

In practice, datasets are often incomplete, and missing values arise for many reasons, such as human error, misunderstanding, and equipment malfunctions. Missing data can hinder analysis, modeling, and the explainability of the built model [22]. A critical task in case of missing data is estimating the covariance matrix, which provides essential information about the relationships between different variables in a dataset [16]. Also, various applications, such as hypothesis testing, modeling uncertainty with confidence intervals, parameter estimation, and principal component analysis, also rely on the estimation of the covariance matrix. In finance, slight improvements in covariance matrix estimation

J. Lokoč et al. (Eds.): MMM 2026, LNCS 16413, pp. 661–675, 2026.
https://doi.org/10.1007/978-981-95-6957-1_47

result in better precision matrix estimation [21] and more accurate predictions of asset behaviors and correlations, thereby enhancing the performance of financial models. This improvement translates into more effective risk assessment [6] and better-optimized investment strategies [5,23]. In fields like signal processing, improved covariance matrix estimation can enhance techniques for noise reduction, facilitating clearer extraction of signals [18].

For handling missing data, Nguyen *et al.*. illustrated that the two-step procedures (imputation then parameter estimation) can be computationally expensive and less efficient compared to direct parameter estimation, such as PMF [12] and DPER [13]. However, these parameter estimation algorithms are designed for scenarios where the data only comes from continuous features. In other words, these algorithms only rely on continuous features for estimation and do not exploit the categorical features under mixed data. We hypothesize that categorical variables can provide important information that aids the quality of the estimation for continuous features. Therefore, we propose the DPERC algorithm (Direct Parameter Estimation for Randomly Missing Data with Categorical Features) as an endeavor to enhance DPER in estimating the covariance matrix. This is done by selecting a categorical variable to maximize the log-likelihood, treating it as an artificial label, and applying the DPER algorithm [13]. Our contribution can be summarized as follows: (i) We propose DPERC, an improvement over DPER for mixed-type data for two situations: single-class data and multiple-class data; (ii) We provide the theoretical foundation to select a suitable categorical feature for the proposed algorithm; (iii) We demonstrate that DPERC can outperform DPER and imputation techniques in estimating the covariance matrix and correlation heatmaps.

2 Related Works

Typically, there are two major approaches for parameter estimation: (1) utilizing imputation methods and then estimating the covariance matrix from imputed data; (2) directly estimating parameters from missing data.

The first approach imputes missing values, i.e., filling in the missing values with plausible values. Basic options can be mean, mode, and median imputation, where the missing values are imputed based on the mean, mode, or median, respectively. Many machine learning-based techniques have been developed for imputation as well. For example, k-nearest neighbor imputation (KNNI) [17] utilizes KNN to estimate missing values based on the characteristics of similar neighboring data points. As another example, multiple Imputation by Chained Equations (MICE [4]) iteratively imputes missing values using regression models based on other variables in the dataset. Next, Bayesian network imputation [8] models probabilistic dependencies among variables, while multiple imputation with Deep Denoising Autoencoders [1] learn complex patterns to reconstruct missing entries. Another group of techniques is based on matrix completion, where matrix factorization is used to exploit the inherent low-rank structure and thereby infer the missing entries. Typical techniques in this category include Polynomial Matrix Completion [7] and Soft-Impute [10].

To handle data that contains both categorical and continuous features, we can mention the technique based on Random Forest, like missForest [15,19], which operates by averaging predictions from numerous regression trees. Another approach is based on a principal component method [2], the prediction of the missing values is based on the similarity between individuals and on the relationships between variables. With large datasets that contain many missing values, the imputation method based on multilevel singular value decomposition (SVD [9]) can be considerable.

The second approach can overcome the drawback of the first one, which can be computationally inefficient. The group of methods relies on Maximum Likelihood such as the Expectation-Maximization (EM [11]) and its improved name Equalization-Maximization (EqM [20]); Under various assumptions for the data that contains missing values, a group of techniques estimating the maximum likelihood estimators for a multivariate normal distribution, consisting of EPEM [12], which can be applied to monotone missing data, PMF [14] when the missing data only occurs in some features, and DPER can be used for randomly missing data. However, when dealing with mixed-type data, these methods only use the information from continuous features and ignore the categorical features.

In summary, although many approaches have been developed to address the problem of missing data, important limitations remain. Imputation-based methods can handle mixed-type datasets but often suffer from error propagation, which may affect parameter estimation. On the other hand, the direct estimation methods mentioned avoid imputation and yield efficient covariance estimates, but completely ignore the categorical part of the mixed data. To the best of our knowledge, no existing direct estimation method effectively integrates categorical information into the estimation process. This motivates our work on DPERC, a method that extends direct estimation to mixed-type data by leveraging categorical features to enhance covariance matrix estimation.

3 Preliminaries

We briefly summarize DPER algorithms [13] for directly estimating parameters for randomly missing data, starting with their theoretical ground.

Theorem 1. *Assume that we have a dataset with G classes and the assumption of equal covariance matrices, where each sample from the $g^{th}(1 \leq g \leq G)$ class follows a bivariate normal distribution with mean $\boldsymbol{\mu}^{(g)} = (\mu_1^{(g)}, \mu_2^{(g)})^T$,and arrange the data into the following pattern:*

$$\boldsymbol{x}^{(g)} = \begin{pmatrix} x_{11}^{(g)} \dots x_{1m_g}^{(g)} & x_{1m_g+1}^{(g)} \dots x_{1n_g}^{(g)} & * & \dots & * \\ x_{21}^{(g)} \dots x_{2m_g}^{(g)} & * \quad \dots \quad * & x_{2n_g+1}^{(g)} & \dots & x_{2l_g}^{(g)} \end{pmatrix}.$$

Algorithm 1. DPER for multiple-class data

Input: Dataset $\mathbf{X}$ has p continuous features $\mathbf{f}_1, ..., \mathbf{f}_p$ and $\mathbf{y}$ as labels with G classes.
Output: the covariance matrix estimation $\widehat{\boldsymbol{\Sigma}} = (\widehat{\sigma}_{ij})_{i,j=1}^{p}$.

Estimate $\widehat{\boldsymbol{\mu}}^{(g)}$: $\widehat{\mu}_i^{(g)} \leftarrow$ mean of all observations of class g in $\mathbf{f}_i$.
$(\widehat{\sigma}_{ii})_{i=1}^{p} \leftarrow$ uncorrected sample variance of all observations in $\mathbf{f}_i$
for $1 \leq i \leq p$ **do**
  for $1 \leq j \leq i$ **do**
    $\widehat{\sigma}_{ij} = \widehat{\sigma}_{ji} \leftarrow$ root of the polynomial (1), closest to case deletion estimation
  end for
end for

So, each column represents an observation, and $x_{ij} \in \mathbb{R}$ is an entry. Denoting

$$A = \sum_{g=1}^{G} m_g, s_{11} = \sum_{g=1}^{G}\sum_{j=1}^{m_g}(x_{1j}^{(g)} - \widehat{\mu}_1^{(g)})^2, s_{22} = \sum_{g=1}^{G}\sum_{j=1}^{m_g}(x_{2j}^{(g)} - \widehat{\mu}_2^{(g)})^2,$$

$$s_{12} = \sum_{g=1}^{G}\sum_{j=1}^{m_g}(x_{2j}^{(g)} - \widehat{\mu}_2^{(g)})(x_{1j}^{(g)} - \widehat{\mu}_1^{(g)}),$$

Then, the resulting estimators obtained by maximizing the likelihood of the data L w.r.t $\mu_1^{(g)}, \sigma_{11}, \mu_2^{(g)}, \sigma_{22}$, and σ_{12} are:

$$\widehat{\mu}_1^{(g)} = \frac{1}{n_g}\sum_{j=1}^{n_g} x_{1j}^{(g)}, \widehat{\mu}_2^{(g)} = \frac{\sum_{j=1}^{m_g} x_{2j}^{(g)} + \sum_{j=n_g+1}^{l_g} x_{2j}^{(g)}}{m_g + l_g - n_g}, \widehat{\sigma}_{11} = \frac{\sum_{g=1}^{G}\sum_{j=1}^{n_g}(x_{1j}^{(g)} - \widehat{\mu}_1^{(g)})^2}{\sum_{g=1}^{G} n_g},$$

$$\widehat{\sigma}_{22} = \frac{\sum_{g=1}^{G}\left[\sum_{j=1}^{m_g}(x_{2j}^{(g)} - \widehat{\mu}_2^{(g)})^2 + \sum_{j=n+1}^{l_g}(x_{2j}^{(g)} - \widehat{\mu}_2^{(g)})^2\right]}{\sum_{g=1}^{G}(m_g + l_g - n_g)},$$

and $\widehat{\sigma}_{12}$ which is the root of the following polynomial:

$$P(\sigma_{12}) = s_{12}\sigma_{11}\sigma_{22} + (\sigma_{11}\sigma_{22}A - s_{22}\sigma_{11} - s_{11}\sigma_{22})\sigma_{12} + s_{12}\sigma_{12}^2 - A\sigma_{12}^3. \quad (1)$$

Based on this theoretical ground, the authors developed the DPER algorithm [13] for multiple-class data as in Algorithm 1.

4 DPERC Algorithm for Single-Class and Multiple-Class Mixed Data

We begin this section by formalizing the fundamental idea and theoretical grounds that justify treating categorical variables as class surrogates under a common-covariance rationale. We then propose two concrete procedures: a DPERC algorithm for single-class mixed data and an extension to multi-class settings.

4.1 Fundamental Idea and Theoretical Grounds

We assume $\mathbf{X} = (\mathbf{X_{cont}}, \mathbf{X_{cate}})$ is a single class mixed-type data, where $\mathbf{X_{cont}}$ has p continuous features and $\mathbf{X_{cate}}$ has q categorical features $\mathbf{c}_1, \mathbf{c}_2, \ldots, \mathbf{c}_q$. The DPER algorithm only uses the information from continuous features in $\mathbf{X_{cont}}$ and ignores the categorical features in $\mathbf{X_{cate}}$. This means information from the categorical features, which could be useful, is not utilized. Therefore, we propose to select the most useful categorical feature to be used as artificial class features. After that, we treat the covariance matrix estimation problem under the artificial class as if we are estimating the common covariance matrix under the assumption of equal covariance matrices for these artificial classes.

To be more specific, to estimate the covariance matrix $\boldsymbol{\Sigma}$ of $\mathbf{X_{cont}}$, we treat each feature $\mathbf{c}_i$ of $\mathbf{X_{cate}}$ as a new artificial class. Moreover, if $\mathbf{c}_i$ has G_i categories, we assume that $\boldsymbol{\Sigma}^{(1)} = \boldsymbol{\Sigma}^{(2)} = \cdots = \boldsymbol{\Sigma}^{(G_i)}$, where $\boldsymbol{\Sigma}^{(g)}$ is the covariance matrix of the gth class which has $\mathbf{c}_i = g$. Then, we use the DPER multiple-class algorithm for the input data $(\mathbf{X_{cont}}, \mathbf{c}_i)$, which consists of the continuous features ($\mathbf{X_{cont}}$ and the artificial labels $\mathbf{c}_i$). Suppose that we have q categorical features. Then, we begin with seeking a selection criterion for choosing the categorical feature $\mathbf{c}_i$ that yields an estimator of $\boldsymbol{\Sigma}$ that outperforms DPER, i.e.,

$$\|\widehat{\boldsymbol{\Sigma}}_{DPERC} - \boldsymbol{\Sigma}\|_F \leq \|\widehat{\boldsymbol{\Sigma}}_{DPER} - \boldsymbol{\Sigma}\|_F,$$

where $\|.\|_F$ is the Frobenius norm.

Following the strategy of analyzing pairs of features as in DPER, we consider a dataset a data $\boldsymbol{x}$ with two continuous features and one categorical feature $\mathbf{c}$ with G categories:

$$\boldsymbol{x} = \begin{pmatrix} x_{11} \; x_{12} \ldots x_{1N} \\ x_{21} \; x_{22} \ldots x_{2N} \end{pmatrix},$$

where $\mathbf{u}_j^{(g)} = (x_{1j}^{(g)}, x_{2j}^{(g)})$ belong to g^{th} category and suppose that the number of observations of the gth category is n_g, for $1 \leq g \leq G$. We denote the *weighted average pairwise distance* of the gth category of data $\boldsymbol{x}$ follow $\mathbf{c}$ as:

$$d_{\mathbf{c}}^{(g)} = \sum_{1 \leq i < j \leq n_g} \frac{1}{n_g} (\mathbf{u}_i^{(g)} - \mathbf{u}_j^{(g)})^T \boldsymbol{\Sigma}^{-1} (\mathbf{u}_i^{(g)} - \mathbf{u}_j^{(g)}),$$

which presents the average of the distances between any two elements of data $\boldsymbol{x}$.

Theorem 2. *Suppose that we have a single-class data $\boldsymbol{x}$ from a bivariate normal distribution with mean $\boldsymbol{\mu}$ and covariance matrix $\boldsymbol{\Sigma}$, and a categorical feature $\mathbf{c}$ which has G categories $1, 2, \ldots, G$. Moreover, the observations of the gth category be $\mathbf{u}_1^{(g)}, \mathbf{u}_2^{(g)}, \ldots, \mathbf{u}_{n_g}^{(g)}$ with mean $\boldsymbol{\mu}^{(g)}$, where $1 \leq g \leq G$. We denote:*

$$\delta^{(g)} = \sum_{i=1}^{n_g} (\mathbf{u}_i^{(g)} - \boldsymbol{\mu})^T \boldsymbol{\Sigma}^{-1} (\mathbf{u}_i^{(g)} - \boldsymbol{\mu}) \text{ and } \Delta^{(g)} = (\boldsymbol{\mu}^{(g)} - \boldsymbol{\mu})^T \boldsymbol{\Sigma}^{-1} (\boldsymbol{\mu}^{(g)} - \boldsymbol{\mu}).$$

Then, for every $g = 1, 2, \ldots, G$, $\delta^{(g)} = d_{\mathbf{c}}^{(g)} + n_g \Delta^{(g)}$.

Proof. Consider class g^{th} of $\mathbf{c}$, expanding $\delta^{(g)}$ we get

$$\begin{aligned}\delta^{(g)} &= \sum_{i=1}^{n_g}(\mathbf{u}_i^{(g)} - \boldsymbol{\mu})^T\boldsymbol{\Sigma}^{-1}(\mathbf{u}_i^{(g)} - \boldsymbol{\mu})\\ &= \sum_{i=1}^{n_g}[(\mathbf{u}_i^{(g)})^T\boldsymbol{\Sigma}^{-1}\mathbf{u}_i^{(g)} - (\mathbf{u}_i^{(g)})^T\boldsymbol{\Sigma}^{-1}\boldsymbol{\mu} - \boldsymbol{\mu}^T\boldsymbol{\Sigma}^{-1}\mathbf{u}_i^{(g)} + \boldsymbol{\mu}^T\boldsymbol{\Sigma}^{-1}\boldsymbol{\mu}]\\ &= \sum_{i=1}^{n_g}(\mathbf{u}_i^{(g)})^T\boldsymbol{\Sigma}^{-1}\mathbf{u}_i^{(g)} - \sum_{i=1}^{n_g}(\mathbf{u}_i^{(g)})^T\boldsymbol{\Sigma}^{-1}\boldsymbol{\mu} - \sum_{i=1}^{n_g}\boldsymbol{\mu}^T\boldsymbol{\Sigma}^{-1}\mathbf{u}_i^{(g)} + n_g\boldsymbol{\mu}^T\boldsymbol{\Sigma}^{-1}\boldsymbol{\mu}.\end{aligned}$$

Expanding $n_g\Delta^{(g)}$ then:

$$\begin{aligned}n_g\Delta^{(g)} &= n_g(\boldsymbol{\mu}^{(g)} - \boldsymbol{\mu})^T\boldsymbol{\Sigma}^{-1}(\boldsymbol{\mu}^{(g)} - \boldsymbol{\mu}) = n_g\left(\frac{1}{n_g}\sum_{i=1}^{n_g}\mathbf{u}_i^{(g)} - \boldsymbol{\mu}\right)^T\boldsymbol{\Sigma}^{-1}\left(\frac{1}{n_g}\sum_{i=1}^{n_g}\mathbf{u}_i^{(g)} - \boldsymbol{\mu}\right)\\ &= \frac{1}{n_g}\left(\sum_{i=1}^{n_g}\mathbf{u}_i^{(g)}\right)^T\boldsymbol{\Sigma}^{-1}\left(\sum_{i=1}^{n_g}\mathbf{u}_i^{(g)}\right) - \sum_{i=1}^{n_g}(\mathbf{u}_i^{(g)})^T\boldsymbol{\Sigma}^{-1}\boldsymbol{\mu} - \sum_{i=1}^{n_g}\boldsymbol{\mu}^T\boldsymbol{\Sigma}^{-1}\mathbf{u}_i^{(g)} + n_g\boldsymbol{\mu}^T\boldsymbol{\Sigma}^{-1}\boldsymbol{\mu}.\end{aligned}$$

$$\begin{aligned}\Rightarrow \delta^{(g)} - n_g\Delta^{(g)} &= \sum_{i=1}^{n_g}(\mathbf{u}_i^{(g)})^T\boldsymbol{\Sigma}^{-1}\mathbf{u}_i^{(g)} - \frac{1}{n_g}\left(\sum_{i=1}^{n_g}\mathbf{u}_i^{(g)}\right)^T\boldsymbol{\Sigma}^{-1}\left(\sum_{i=1}^{n_g}\mathbf{u}_i^{(g)}\right)\\ &= \frac{1}{n_g}\sum_{i=1}^{n_g}\mathbf{u}_i^T\boldsymbol{\Sigma}^{-1}(n_g\mathbf{u}_i - \sum_{\substack{j=1\\ j\neq i}}^{n_g}\mathbf{u}_j) = \frac{1}{n_g}\sum_{i=1}^{n_g}\mathbf{u}_i^T\boldsymbol{\Sigma}^{-1}\sum_{\substack{j=1\\ j\neq i}}^{n_g}(\mathbf{u}_i - \mathbf{u}_j)\\ &= \frac{1}{n_g}\sum_{1\le i<j\le n_g}^{n_g}(\mathbf{u}_i^{(g)} - \mathbf{u}_j^{(g)})^T\boldsymbol{\Sigma}^{-1}(\mathbf{u}_i^{(g)} - \mathbf{u}_j^{(g)}) = d_{\mathbf{c}}^{(g)}.\end{aligned}$$

Therefore, $\delta^{(g)} = d_{\mathbf{c}}^{(g)} + n_g\Delta^{(g)}, \quad g = 1, 2, \ldots, G$, and the proof follows.

Now, from [13], we know that the log-likelihood of data $\boldsymbol{x}$ is

$$L = C - \frac{N}{2}\log|\boldsymbol{\Sigma}| - \frac{1}{2}\sum_{i=1}^{N}(\mathbf{u}_i - \boldsymbol{\mu})^T\boldsymbol{\Sigma}^{-1}(\mathbf{u}_i - \boldsymbol{\mu}).$$

The last term can be rewritten as

$$\sum_{i=1}^{N}(\mathbf{u}_i - \boldsymbol{\mu})^T\boldsymbol{\Sigma}^{-1}(\mathbf{u}_i - \boldsymbol{\mu}) = \sum_{g=1}^{G}\sum_{j=1}^{n_g}(\mathbf{u}_j^{(g)} - \boldsymbol{\mu})^T\boldsymbol{\Sigma}^{-1}(\mathbf{u}_j^{(g)} - \boldsymbol{\mu}) = \sum_{g=1}^{G}\delta^{(g)}.$$

By Theorem 2, we can rewrite L to

$$L = C - \frac{N}{2}\log|\boldsymbol{\Sigma}| - \frac{1}{2}\sum_{g=1}^{G}\delta^{(g)} = C - \frac{N}{2}\log|\boldsymbol{\Sigma}| - \frac{1}{2}\sum_{g=1}^{G}(d_{\mathbf{c}}^{(g)} + n_g\Delta^{(g)}).$$

Algorithm 2. DPERC for single-class mixed data

Input: A data $\mathbf{X} = (\mathbf{X_{cont}}, \mathbf{X_{cate}})$ where $\mathbf{X_{cont}}$ has p continuous features $\mathbf{f}_1, \mathbf{f}_2, ..., \mathbf{f}_p$ and $\mathbf{X_{cate}}$ has q categorical features $\mathbf{c}_1, \mathbf{c}_2, \ldots, \mathbf{c}_q$.
Output: the covariance matrix estimate $\widehat{\boldsymbol{\Sigma}} = (\widehat{\sigma}_{ij})_{i,j=1}^p$.
Procedure:

Estimate $\widehat{\boldsymbol{\mu}}$: $\widehat{\mu}_i \leftarrow$ mean of all observations in $\mathbf{f}_i$
$\widehat{\boldsymbol{\Sigma}}_D \leftarrow$ using DPER for single-class $\mathbf{X_{cont}}$ data
$(\widehat{\sigma}_{ii})_{i=1}^p \leftarrow$ uncorrected sample variance of all observations in $\mathbf{f}_i$
for $1 \leq i \leq p$ **do**
    for $1 \leq j < i$ **do**
        for $1 \leq k \leq q$ **do**
            $\mathcal{D}_{\mathbf{c}_k} \leftarrow \sum_{g=1}^{\#\mathbf{c_k}} n_g (\widehat{\boldsymbol{\mu}}_{ij}^{(g)} - \widehat{\boldsymbol{\mu}}_{ij})^T \widehat{\boldsymbol{\Sigma}}_{D(ij)}^{-1} (\widehat{\boldsymbol{\mu}}_{ij}^{(g)} - \widehat{\boldsymbol{\mu}}_{ij})$
        end for
        $\mathcal{D}_c \leftarrow \{\mathcal{D}_{\mathbf{c_1}}, \mathcal{D}_{\mathbf{c_2}}, \ldots, \mathcal{D}_{\mathbf{c}_q}\}$
        $\mathbf{c} \leftarrow \arg\min_c \mathcal{D}_c$
        $\widehat{\sigma}_{ij} = \widehat{\sigma}_{ji} \leftarrow$ using DPER for multiple-class on $(\mathbf{f}_i, \mathbf{f}_j)$ and $\mathbf{c}$ as a class
    end for
end for

We want to maximize L, and therefore minimize $\sum_{g=1}^{G}(d_{\mathbf{c}}^{(g)} + n_g \Delta^{(g)})$. Here, we could control the value of $n_g \Delta^{(g)}$ by $\mathbf{c}$. It follows that we can choose the categorical feature $\mathbf{c}$ so that $\sum_{g=1}^{G} n_g \Delta^{(g)}$ is minimal. Intuitively, if the sum of average pairwise distances between all samples in a group is small, for each class, the observations are clustered together in one cluster. Therefore, in such a case, one can expect the sum of the distances from all samples to the central mean to be small if the sum of distances of all class means to the central mean is small. In other words, we choose the categorical feature $\mathbf{c}$ such that the elements in the group of category g^{th} do not have much difference. That is the standard to choose a good categorical feature.

4.2 DPERC for Single-Class Mixed Data

In this section, we will detail the DPERC algorithm for estimating the covariance matrix for single-class mixed data. To begin with notations, $\widehat{\boldsymbol{\mu}}$ is the estimated mean of the data. Recall that for each categorical feature $\mathbf{c}$ which has G different labels, we treat it as a class feature, so this separates the data $\mathbf{X_{cont}}$ into G class data. Therefore, we denote $\widehat{\boldsymbol{\mu}}^{(g)}$ is the estimated mean of gth class data where $1 \leq g \leq G$. Next, denote $\widehat{\boldsymbol{\mu}}_{ij}, \widehat{\boldsymbol{\mu}}_{ij}^{(g)}$ for a vector consists i^{th} and j^{th} elements of $\widehat{\boldsymbol{\mu}}$ and $\widehat{\boldsymbol{\mu}}^{(g)}$, respectively. In addition, $\widehat{\boldsymbol{\Sigma}}_{D(ij)}^{-1} = \begin{pmatrix} \lambda_{ii} & \lambda_{ij} \\ \lambda_{ji} & \lambda_{jj} \end{pmatrix}$ is the submatrix of $\widehat{\boldsymbol{\Sigma}}_D^{-1}$ with the corresponding elements.

The DPERC algorithm for single-class mixed data is presented in Algorithm 2. The algorithm input requires mixed data and returns its covariance matrix estimation, and the procedure is as follows: First, we estimate the mean vector by computing, for each feature, the sample mean after case deletion. Next, we obtain an initial covariance estimate $\widehat{\Sigma}_D$, used solely to guide the choice of a categorical feature; any reasonable estimator may be employed at this stage. The diagonal entries of the target covariance are then set to the uncorrected sample variances. We subsequently focus on the off–diagonal terms σ_{ij} $(i \neq j)$. For each candidate categorical feature $\mathbf{c}$, we compute the corresponding *weighted average pairwise distance* using the 2×2 principal submatrix $\widehat{\boldsymbol{\Sigma}}^{-1}_{D(ij)}$, collect these values into the set $\mathcal{D}_{\mathbf{c}}$, and select the feature that minimizes this criterion. Finally, conditional on the selected $\mathbf{c}$, we estimate σ_{ij} by applying the multi-class DPER procedure to the pair $(\mathbf{f}_i, \mathbf{f}_j)$ with $\mathbf{c}$ serving as the class indicator. To provide more intuition, Fig. 1 illustrates the visual workflow of Algorithm 2.

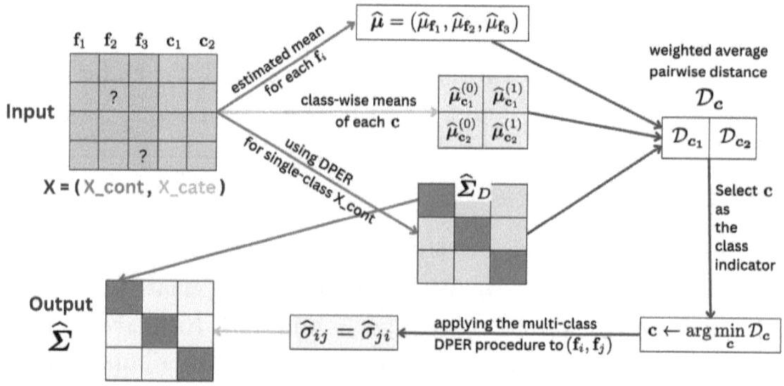

Fig. 1. Conceptual workflow of the DPERC algorithm for single-class mixed data. Here, we assume that $\mathbf{X_{cont}} = (\mathbf{f}_1, \mathbf{f}_2, \mathbf{f}_3)$ and $\mathbf{X_{cate}} = (\mathbf{c}_1, \mathbf{c}_2)$ where each categorical feature consists of two groups.

There are two remarks about the algorithm implementation:

1. Similar to DPER, we also need a mild assumption of bivariate normality on each pair of continuous features. Moreover, we need to go through all the categorical features for each pair $(\mathbf{f}_i, \mathbf{f}_j)$ to choose the good one for estimating each σ_{ij}. However, one more point to note, we do not estimate the first covariance matrix for each pair of features. Instead, we use the corresponding principal sub-matrix of $\widehat{\Sigma}_D$ in step 2.
2. Note in step 7, when the missing rate is high, if the element i^{th} of $\widehat{\boldsymbol{\mu}}^{(g)}$ is missing, we set this element to zero.

Example: Consider a toy dataset:

$$\boldsymbol{X} = \begin{pmatrix} 1 & 7 & 0\,1 \\ 3 & 4 & 0\,0 \\ \text{NaN} & 2 & 0\,1 \\ 2 & \text{NaN} & 1\,0 \\ 4 & 5 & 1\,1 \end{pmatrix},$$

where the first two columns are continuous features $(\mathbf{f}_1, \mathbf{f}_2)$ and the last two columns are categorical variables $(\mathbf{c}_1, \mathbf{c}_2)$. First, we estimate the mean vector of continuous features: $\widehat{\boldsymbol{\mu}} = (\widehat{\mu}_{\mathbf{f}_1}, \widehat{\mu}_{\mathbf{f}_2}) = (2.5, 4.5)$. We then apply the single-class DPER algorithm to the continuous subset $\mathbf{X}_{\mathbf{cont}} = (\mathbf{f}_1, \mathbf{f}_2)$, yielding the preliminary covariance estimate $\widehat{\boldsymbol{\Sigma}}_D = \begin{pmatrix} 1.25 & -1.118 \\ -1.118 & 3.25 \end{pmatrix}$. Next, we estimate the mean of each class of each categorical feature: $\widehat{\boldsymbol{\mu}}^{(0)}_{\mathbf{c}_1} = (2.0, 4.33)$, $\widehat{\boldsymbol{\mu}}^{(1)}_{\mathbf{c}_1} = (3.0, 5.0), \widehat{\boldsymbol{\mu}}^{(0)}_{\mathbf{c}_2} = (2.5, 4.0), \widehat{\boldsymbol{\mu}}^{(1)}_{\mathbf{c}_2} = (2.5, 7.0)$. We subsequently evaluate class distances: $\mathcal{D}_{\mathbf{c_1}} = 2.3, \mathcal{D}_{\mathbf{c_2}} = 0.26$. After that, the best categorical feature is selected based on the class distances: $\mathbf{c} = \arg\min_{\mathbf{c}} \mathcal{D}_{\mathbf{c}} = \mathbf{c}_2$. Finally, the covariance between $(\mathbf{f}_1, \mathbf{f}_2)$ is refined using the DPER algorithm for multiple-class data with $\mathbf{c}_2$ as the class variable: $\widehat{\sigma}_{12} = \widehat{\sigma}_{21} = -1.127$. Thus, the final covariance matrix estimated by DPERC is $\widehat{\boldsymbol{\Sigma}} = \begin{pmatrix} 1.25 & -1.127 \\ -1.127 & 3.25 \end{pmatrix}$.

Table 1. Datasets used in the experiment.

Dataset	# Features	# Continuous features	# Categorical features	# Classes	# Samples
Bank	21	10	10	2	4168
Heart	15	6	8	2	32561
Statlog	21	7	13	2	1000
Student	33	16	17	1	395

4.3 DPERC Algorithm for Multiple-Class Data

In case data comes from multiple classes without the assumption of equal covariance matrices, one can apply Algorithm 2 multiple times, as the multiple-class problem boils down to many single-class problems.

Specifically, given a mixed data $(\mathbf{X}_{\mathbf{cont}}, \mathbf{X}_{\mathbf{cate}}, \mathbf{y})$ where $\mathbf{y}$ has M classes. Let $\mathbf{X}^{(m)}$ be the data from the mth class and $\mathbf{X}^{(m)}_{\mathbf{cont}}$ be its corresponding continuous part. For each class m where $1 \leq m \leq M$, we want to estimate the covariance matrix of this class without the assumption of equal covariance matrices. Now, we denote $\mathbf{X}_{\mathbf{c}}$ as a feature of G categories from $\mathbf{X}_{\mathbf{cate}}$. Then, we assume $\boldsymbol{\Sigma}^{(m)}_1 = \boldsymbol{\Sigma}^{(m)}_2 = \cdots = \boldsymbol{\Sigma}^{(m)}_G$ where $\boldsymbol{\Sigma}^{(m)}_i$ is the covariance matrix of the mth class, i.e., the part that has$\mathbf{X}^{(m)}_{\mathbf{c}} = i$. Therefore, we can use Algorithm 2 for $(\mathbf{X}^{(m)}_{\mathbf{cont}}, \mathbf{X}^{(m)}_{\mathbf{cate}}, \mathbf{y}^{(m)})$ to estimate the covariance matrix of class m^{th}.

5 Experiments

5.1 Experimental Settings and Datasets

To illustrate the efficiency of our algorithm in covariance estimation, we conduct two comparisons. First, we benchmark DPERC against state-of-the-art baselines, including DPER [13], missForest [19], KNNI [3], Soft-Impute [10], and MICE [4], using the sum of squared errors (**SSE**) between the ground truth $\boldsymbol{\Sigma}$ and its estimate $\widehat{\boldsymbol{\Sigma}}$: SSE $= \|\widehat{\boldsymbol{\Sigma}} - \boldsymbol{\Sigma}\|_F^2$. Second, to isolate the incremental benefit of DPERC over DPER, we compare *only* these two methods on the *off-diagonal* entries, since both produce identical diagonal estimates. Let $A^\star := A - \text{diag}(A)$ denote the off-diagonal part of a matrix A. We report the SSE: $SSE^\star = \|\widehat{\boldsymbol{\Sigma}}^\star - \boldsymbol{\Sigma}^\star\|_F^2$, together with the percentage improvement of DPERC over DPER, Imp(%) $= 100\left(\frac{SSE^\star_{\text{DPER}}}{SSE^\star_{\text{DPERC}}} - 1\right)$, and also the best-case improvement (MaxImp%) as the maximum of Imp(%) across repetitions.

To investigate the effects of missing data on the correlation heatmap, we follows [16] and use: Correlation Heatmap $\boldsymbol{R} = \boldsymbol{D}^{-1}\,\boldsymbol{\Sigma}\,\boldsymbol{D}^{-1}$, where $\boldsymbol{D}$ is diagonal with $D_{ii} = \sqrt{\Sigma_{ii}}$. Local MSE Difference Heatmaps for Correlation $\Delta\boldsymbol{R}_{\text{squared}} = \Delta\boldsymbol{R} \circ \Delta\boldsymbol{R}$, where $\Delta\boldsymbol{R} = \boldsymbol{R}_{\text{true}} - \boldsymbol{R}_{\text{est}}$ and $\circ$ denotes the Hadamard product.

All experiments were run on Google Colaboratory[1] and repeated 20 times; we report mean ± standard deviation. The datasets for all experiments are from the Machine Learning Database Repository at the University of California[2]: Bank Marketing, Statlog, Heart Disease, and Student Performance. More detail of these data is shown in Table 1. Note that we only consider missing values in continuous features; the categorical features are fully observed. We simulate missing data for each dataset with missing rates ranging from 20%-80%. Here, the missing rate is defined as the proportion of missing entries to the total number of entries.

5.2 Results and Analysis

Table 2 reports SSEs for covariance–matrix estimation across various datasets and missing rates. Note that for each dataset, rows correspond (top to bottom) to missing rates of 20%, 35%, 50%, 65%, and 80%. One can see that as missingness increases, SSE generally rises for most methods, while missForest is the exception with a decreasing trend. DPERC achieves the lowest SSE in every setting and markedly outperforms imputation-based approaches; for instance, at 80% missingness on Bank dataset, DPERC yields 29.22 versus 150.75 for the next best imputation method (MICE), i.e., over a fivefold reduction.
Table 3 summarizes the gains of DPERC over DPER. On average (Imp metric), DPERC improves by roughly 0.4%–6.15%. For the multi-class datasets (Bank, Heart, Statlog), the gains lie between 2.33% and 6.15%, whereas for the single-class dataset (Student) they are about 0.5%. This is because DPERC refines

1 https://colab.google/.
2 https://archive.ics.uci.edu/ml.

Table 2. SSE for covariance-matrix estimation. For each dataset, rows correspond (top to bottom) to missing rates of 20%, 35%, 50%, 65%, and 80%.

Datasets	DPERC	DPER	MissForest	KNNI	Soft-Impute	MICE
Bank	**3.72 ± 1.16**	3.86 ± 1.19	1147.51 ± 149.90	12.02 ± 4.53	22.24 ± 5.12	10.25 ± 2.07
	7.28 ± 2.80	7.57 ± 2.88	990.13 ± 116.76	56.19 ± 13.24	63.02 ± 10.66	25.92 ± 6.24
	12.32 ± 4.81	12.81 ± 4.90	887.30 ± 144.36	129.14 ± 19.33	122.77 ± 13.94	55.77 ± 19.59
	20.20 ± 5.32	21.22 ± 5.72	873.49 ± 86.32	196.24 ± 23.16	196.72 ± 22.14	93.63 ± 19.69
	29.22 ± 7.46	30.48 ± 8.08	849.63 ± 80.32	256.11 ± 37.83	288.56 ± 33.94	150.75 ± 27.63
Heart	**1.01 ± 0.84**	1.02 ± 0.84	161.04 ± 30.63	8.12 ± 3.83	8.87 ± 3.89	8.85 ± 3.90
	1.81 ± 1.48	1.82 ± 1.48	149.42 ± 24.90	20.34 ± 5.94	22.40 ± 6.11	22.61 ± 6.09
	2.36 ± 2.24	2.39 ± 2.25	150.87 ± 27.63	33.28 ± 6.74	37.09 ± 7.04	46.14 ± 12.16
	5.03 ± 5.23	5.07 ± 5.23	133.51 ± 46.54	45.31 ± 9.39	51.37 ± 9.44	66.72 ± 14.79
	7.97 ± 8.44	8.02 ± 8.45	141.84 ± 40.87	56.48 ± 8.70	64.62 ± 9.13	91.90 ± 24.47
Statlog	**1.52 ± 0.34**	1.56 ± 0.35	56.18 ± 4.45	6.10 ± 0.86	6.34 ± 0.90	4.41 ± 0.51
	2.88 ± 0.65	2.96 ± 0.66	56.08 ± 5.04	16.90 ± 2.01	16.03 ± 1.65	9.39 ± 0.85
	4.74 ± 0.88	4.85 ± 0.88	58.61 ± 5.38	28.87 ± 2.89	28.51 ± 2.45	13.84 ± 2.27
	7.21 ± 1.78	7.35 ± 1.79	63.69 ± 3.10	40.86 ± 3.18	42.77 ± 3.09	23.86 ± 4.44
	10.18 ± 1.92	10.46 ± 2.02	72.36 ± 4.04	51.30 ± 3.66	56.74 ± 3.49	32.06 ± 5.41
Student	**2.23 ± 0.32**	**2.23 ± 0.32**	184.15 ± 10.24	6.65 ± 0.83	6.98 ± 0.87	4.32 ± 0.60
	4.40 ± 0.50	4.42 ± 0.51	162.35 ± 8.57	16.00 ± 1.62	17.95 ± 1.23	11.78 ± 5.46
	6.95 ± 0.84	6.98 ± 0.84	156.82 ± 7.80	33.73 ± 3.24	33.25 ± 1.64	14.90 ± 1.12
	11.13 ± 1.58	11.18 ± 1.58	156.08 ± 4.91	61.00 ± 2.36	53.80 ± 2.99	23.74 ± 2.98
	16.40 ± 2.21	16.47 ± 2.21	163.42 ± 6.18	88.15 ± 4.22	76.77 ± 5.24	37.24 ± 5.80

the SSE within each class, so datasets with more classes benefit more. For the best-case improvement (MaxImp metric), the ranking is Bank (23.47%), Statlog (11.74%), Heart (9.60%), and Student (1.58%). This ordering aligns with the richness of categorical information in each dataset, more categorical features give DPERC more opportunities to select an informative feature and thus achieve larger gains.

5.3 Correlation Visualization

On the Statlog dataset, we visualize correlation heatmaps derived from the covariance estimates in Sect. 5 and compare them against the ground-truth correlation matrix using the Local MSE Difference for Correlation. Following the standard color scheme (Fig. 2), positive correlations appear in cooler blues, negative correlations in warmer magentas, and zeros in white; color intensity increases as coefficients approach ±1. At the level of raw correlation heatmaps, the visual differences among methods are not obvious.

The Local MSE difference maps (Fig. 3) sharpen these contrasts. Here, red intensity encodes elementwise squared deviations from the ground truth (white). As missingness rises, discrepancies become more pronounced overall. Notably, at 65% missingness, DPERC and DPER remain close to the ground truth, while missForest, KNNI, and Soft-Impute exhibit marked errors concentrated around

Table 3. Comparison of DPER and DPERC in covariance matrix estimation.

Missing rate	Bank dataset				Heart dataset			
	DPERC	DPER	Imp(%)	MaxImp(%)	DPERC	DPER	Imp(%)	MaxImp(%)
20%	1.38 ± 0.43	1.44 ± 0.45	5.17 ± 3.20	13.52	0.12 ± 0.04	0.13 ± 0.04	3.10 ± 1.71	5.72
35%	2.68 ± 0.98	2.82 ± 1.01	5.75 ± 3.74	14.48	0.24 ± 0.08	0.25 ± 0.08	2.38 ± 2.34	6.45
50%	4.72 ± 1.56	4.96 ± 1.62	5.55 ± 5.51	23.47	0.40 ± 0.17	0.42 ± 0.18	3.97 ± 2.54	9.60
65%	8.13 ± 2.32	8.64 ± 2.54	6.15 ± 3.70	11.01	0.68 ± 0.39	0.70 ± 0.39	3.61 ± 2.17	6.73
80%	12.05 ± 3.72	12.68 ± 3.99	5.21 ± 4.03	17.75	1.08 ± 0.47	1.10 ± 0.47	2.92 ± 1.76	6.87
Missing rate	Statlog dataset				Student dataset			
	DPERC	DPER	Imp(%)	MaxImp(%)	DPERC	DPER	Imp(%)	MaxImp(%)
20%	0.62 ± 0.14	0.63 ± 0.15	2.70 ± 1.67	6.06	1.98 ± 0.26	1.98 ± 0.26	0.39 ± 0.31	1.08
35%	1.18 ± 0.28	1.22 ± 0.29	3.28 ± 1.50	5.27	4.02 ± 0.51	4.04 ± 0.51	0.50 ± 0.30	1.09
50%	2.03 ± 0.43	2.09 ± 0.43	2.80 ± 2.24	7.32	6.46 ± 0.78	6.49 ± 0.78	0.49 ± 0.42	1.49
65%	3.13 ± 0.90	3.20 ± 0.91	2.33 ± 0.92	3.77	10.45 ± 1.62	10.50 ± 1.62	0.46 ± 0.45	1.58
80%	4.40 ± 0.96	4.54 ± 1.00	3.22 ± 3.28	11.74	15.34 ± 2.37	15.42 ± 2.37	0.48 ± 0.32	1.07

Fig. 2. Correlation Heatmaps for Statlog dataset across missing rates from 20% to 80%.

strongly correlated pairs. MICE shows the largest departures, characterized by broad, diffuse deviations rather than localized peaks.

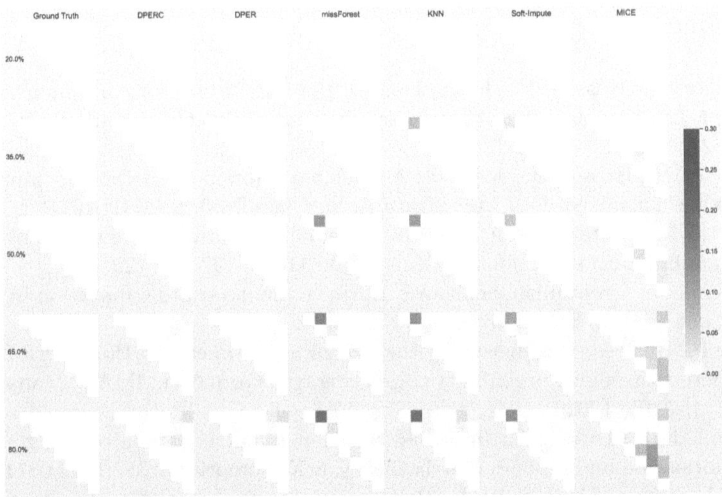

Fig. 3. Local MSE Difference Heatmaps for Correlation for the Statlog dataset across missing rates from 20% to 80%.

6 Conclusion and Future Works

In this article, we propose the algorithm DPERC to estimate the covariance matrix directly from mixed data. Our proposed algorithm not only keeps the strength of DPER but also efficiently exploits the categorical features by treating them as a class under the assumption of equal covariance matrices. In particular, the technique presents choosing a categorical feature that satisfies Theorem 2 to improve the estimate for each $\sigma_{ij}(i \neq j)$. The experiments show that compared to DPER then, DPERC helps significantly enhance the covariance matrix estimation up to 23.47%, especially when data come from multiple classes and have many categorical features.

For future works, first, note that our methodology provides a condition for choosing a good, not the best, categorical feature. This is because we assume $n_g \Delta^{(g)}$ is considerably smaller than $d_{\mathbf{c}}^{(g)}$. However, we explore that $n_g \Delta^{(g)}$ can be higher compared to $d_{\mathbf{c}}^{(g)}$ in some cases. It can lead to a future direction for a better strategy when selecting categorical features. Next, our work only assumes that the data has randomly missing values in continuous features. When the labels of the class contain missing values, even if categorical features have missing values, the categorical features may be utilized to handle this issue.

References

1. Audigier, V., Husson, F., Josse, J.: Multiple imputation for continuous variables using a Bayesian principal component analysis. J. Stat. Comput. Simul. **86**(11), 2140–2156 (2016)
2. Audigier, V., Husson, F., Josse, J.: A principal component method to impute missing values for mixed data. Adv. Data Anal. Classif. **10**, 5–26 (2016)
3. Beretta, L., Santaniello, A.: Nearest neighbor imputation algorithms: a critical evaluation. BMC Med. Inform. Decis. Mak. **16**(3), 197–208 (2016)
4. Buuren, S.v., Groothuis-Oudshoorn, K.: mice: multivariate imputation by chained equations in R. J. Stat. Softw., 1–68 (2010)
5. Date, P., Bustreo, R.: Measuring the risk of a non-linear portfolio with fat-tailed risk factors through a probability conserving transformation. IMA J. Manag. Math. **27**(2), 157–180 (2016)
6. Deshmukh, S., Dubey, A.: Improved covariance matrix estimator using shrinkage transformation and random matrix theory. arXiv preprint arXiv:1912.03718 (2019)
7. Fan, J., Zhang, Y., Udell, M.: Polynomial matrix completion for missing data imputation and transductive learning. In: Proceedings of the AAAI Conference on Artificial Intelligence, vol. 34, pp. 3842–3849 (2020)
8. Hruschka, E.R., Hruschka, E.R., Ebecken, N.F.: Bayesian networks for imputation in classification problems. J. Intell. Inf. Syst. **29**(3), 231–252 (2007)
9. Husson, F., Josse, J., Narasimhan, B., Robin, G.: Imputation of mixed data with multilevel singular value decomposition. J. Comput. Graph. Stat. **28**(3), 552–566 (2019)
10. Mazumder, R., Hastie, T., Tibshirani, R.: Spectral regularization algorithms for learning large incomplete matrices. J. Mach. Learn. Res. **11**, 2287–2322 (2010)
11. Nelwamondo, F.V., Mohamed, S., Marwala, T.: Missing data: a comparison of neural network and expectation maximization techniques. Curr. Sci., 1514–1521 (2007)
12. Nguyen, T., Nguyen, D.H., Nguyen, H., Nguyen, B.T., Wade, B.A.: EPEM: efficient parameter estimation for multiple class monotone missing data. Inf. Sci. **567**, 1–22 (2021)
13. Nguyen, T., Nguyen-Duy, K.M., Nguyen, D.H.M., Nguyen, B.T., Wade, B.A.: DPER: direct parameter estimation for randomly missing data. Knowl.-Based Syst. **240**, 108082 (2022)
14. Nguyen, T., Phan, N.T., Hoang, H.V., Halvorsen, P., Riegler, M.A., Nguyen, B.T.: PMF: efficient parameter estimation for data sets with missing data in some features. Available at SSRN 4260235 (2022)
15. Nguyen, T., Vo, T.L., Halvorsen, P., Riegler, M.A.: Imputation using training labels and classification via label imputation. arXiv preprint arXiv:2311.16877 (2023)
16. Pham, N.H., et al.: Correlation visualization under missing values: a comparison between imputation and direct parameter estimation methods. In: Rudinac, S., et al. (eds.) International Conference on Multimedia Modeling, pp. 103–116. Springer Nature Switzerland Cham (2024). https://doi.org/10.1007/978-3-031-53302-0_8
17. Sanjar, K., Bekhzod, O., Kim, J., Paul, A., Kim, J.: Missing data imputation for geolocation-based price prediction using KNN-MCF method. ISPRS Int. J. Geo Inf. **9**(4), 227 (2020)
18. de Santi, N.S., Abramo, L.R.: Improving cosmological covariance matrices with machine learning. J. Cosmol. Astropart. Phys. **2022**(09), 013 (2022)

19. Stekhoven, D.J., Bühlmann, P.: MissForest-non-parametric missing value imputation for mixed-type data. Bioinformatics **28**(1), 112–118 (2012)
20. Stoica, P., Xu, L., Li, J.: Parameter estimation with missing data via equalization-maximization. In: Proceedings.(ICASSP 2005). IEEE International Conference on Acoustics, Speech, and Signal Processing, 2005, vol. 4, pp. iv–57. IEEE (2005)
21. Vo, T.L., Do, Q.H., Nguyen, T., Halvorsen, P., Riegler, M.A., Nguyen, B.T.: The effects of data imputation on covariance and inverse covariance matrix estimation. IEEE Access (2024)
22. Vo, T.L., Nguyen, T., Lopez-Ramos, L.M., Hammer, H.L., Riegler, M.A., Halvorsen, P.: Explainability of machine learning models under missing data. arXiv preprint arXiv:2407.00411 (2024)
23. Zhang, Y., Tao, J., Yin, Z., Wang, G.: Improved large covariance matrix estimation based on efficient convex combination and its application in portfolio optimization. Mathematics **10**(22) (2022). https://www.mdpi.com/2227-7390/10/22/4282

Author Index

J. Lokoč et al. (Eds.): MMM 2026, LNCS 16413, pp. 677–679, 2026.
https://doi.org/10.1007/978-981-95-6957-1

MIX
Papier aus verantwortungsvollen Quellen
Paper from responsible sources
FSC® C105338

If you have any concerns about our products,
you can contact us on
ProductSafety@springernature.com

In case Publisher is established outside the EU,
the EU authorized representative is:
Springer Nature Customer Service Center GmbH
Europaplatz 3, 69115 Heidelberg, Germany

Printed by Libri Plureos GmbH
in Hamburg, Germany